A Dictionary of

Mechanical Engineering

Tony Atkins is Emeritus Professor of Mechanical Engineering at the University of Reading and Visiting Professor at Imperial College. He has authored several books and over 150 journal articles.

Marcel Escudier is Emeritus Harrison Professor of Mechanical Engineering at the University of Liverpool. He has written many journal articles and a book, *The Essence of Engineering Fluid Mechanics* (Prentice Hall, 1998).

SEE WEB LINKS

Some entries in this dictionary have recommended web links. When you see the above symbol at the end of an entry go to the dictionary's web page at www.oup.com/uk/reference/resources/mechanicalengineering, click on **Web links** in the Resources section and locate the entry in the alphabetical list, then click straight through to the relevant websites.

A Dictionary of
Mechanical Engineering

TONY ATKINS

AND

MARCEL ESCUDIER

OXFORD
UNIVERSITY PRESS

OXFORD
UNIVERSITY PRESS

Great Clarendon Street, Oxford, OX2 6DP,
United Kingdom

Oxford University Press is a department of the University of Oxford.
It furthers the University's objective of excellence in research, scholarship,
and education by publishing worldwide. Oxford is a registered trade mark of
Oxford University Press in the UK and in certain other countries

First Edition published in 2013

Impression: 1

British Library Cataloguing in Publication Data
Data available

Library of Congress Cataloging in Publication Data
Data available

ISBN 978-0-19-958743-8

Printed in Great Britain by
Clays Ltd, St Ives plc

Preface

Mechanical engineering is a broad subject, encompassing as it does dynamics (including vibrations), fluid mechanics, stress analysis, thermodynamics (including heat transfer) and design together with such topics as combustion, control, instrumentation and measurement, lubrication and robotics. The first five topics may be regarded as the core of mechanical engineering. None of these topics is exclusive to mechanical engineering and one or more form part of aeronautical, chemical and civil engineering. There is further overlap with acoustics, bioengineering, electrical engineering, electronics, environmental engineering and materials science. Indeed, much of modern mechanical engineering takes place at the interfaces between disciplines. There are very few man-made artefacts in which mechanical engineering does not play a major role in the design, development and manufacturing processes. For example, in the most advanced computers, the design and manufacture of high-precision moving parts and their temperature control would not be possible without the application of mechanical-engineering principles.

The selection of terms to be defined in this *Dictionary of Mechanical Engineering* has been to some degree subjective and it is inevitable that we have included some that others would have omitted and vice versa. For example, a practising mechanical engineer is expected to be aware of legal obligations, particularly where health and safety are concerned, cost and business implications, project management, etc. We have taken the view that important as these considerations are, they are not part of mechanical engineering as such.

This dictionary is intended to be suitable for a wide range of readers: mechanical-engineering specialists, from students to researchers and professional engineers, as well as journalists, lawyers, politicians and others with a need for definitions of mechanical-engineering terms. As well as native-English speakers we expect that the dictionary will be of value to those whose mother tongue is not English. Engineering terms tend to have the same meanings throughout the English-speaking world, but variants from British English, mainly American English, are included parenthetically where there is an essential difference other than a minor difference in spelling. Every definition is based upon information from a number of sources: textbooks, reference books (including existing dictionaries), manufacturers' websites and publications, and research papers. We have also made extensive use of the internet and become aware that many definitions there are traceable to a single source and not always correct.

Although we have tried to provide succinct definitions in physical rather than patent or mathematical language, some mechanical-engineering terms rely on a relatively advanced knowledge of mathematics, physics or chemistry. This is especially true in the areas of stress analysis and fluid mechanics. We have assumed that the interested reader will either have the required knowledge or be able to access it from appropriate textbooks and specialist sources.

Mechanical engineering is a mature but ever-developing subject. We have included any terms we regard as relevant to the twenty-first century and avoided terms that must now be considered obsolete. To strengthen relevance to the present day, there are many entries concerning such 'modern' topics as micro- and nano-technology, renewable energy and composite materials. We note, however, that the prefix 'nano' has been used in recent currency to mean all things small but we find that the prefix 'micro' is just as frequently used to mean the same thing. Furthermore, many of these 'micro/nano' terms seem contrived to us. We also note that the actual scale implied by 'micro' is different in different applications.

Where appropriate, we have included the symbol (or symbols) most commonly used to represent a particular property or quantity, such as σ for stress and γ for the ratio of specific heats. However the meaning of many symbols depends upon their context from which the intended meaning is almost always apparent. For example while σ is generally used to represent stress in both solid and fluid mechanics, it is also one of the symbols for surface tension (another being γ). When double suffixes are employed for stress and strain, the first suffix denotes the normal to the plane and the second the direction so that σ_{xx} is the stress acting on the plane perpendicular to the x-axis and in the direction of the x-axis, and is a normal stress; σ_{xy} is the stress acting on the plane perpendicular to the x-axis and in the direction of the y-axis, and is a shear stress. However, following common practice, when different symbols are used for normal and shear stress (σ and τ) and normal and shear strains (ε and γ) we use single suffixes for σ and ε.

With very few exceptions, such as plane and solid angles which are ratios, physical quantities have units, such as kilogram (kg) for mass, metre (m) for length, and second (s) for time. In fact the value of any physical quantity in the absence of its units is practically meaningless. We have employed the International System of Units (Système international d'unités or SI) units throughout this dictionary as well as the self-consistent set of values for the fundamental physical constants provided by the Committee on Data for Science and Technology (CODATA) in 2006. A few non-SI units, such as the minute and hour, are still in general use and so are retained here. In a few cases, non-SI units are found in practice to be convenient to use, such as kilometers per hour (km/hr or kph) for wind speed rather than m/s. Many other non-SI units are still in widespread use, particularly in the United States, but generally regarded as obsolete. So far as the majority of these is concerned, such as the pound, we have defined the base units and their conversion to SI units but not the many combinations that occur, such as pounds per square inch or foot pounds. The units we have given for a particular quantity are either the base SI units or the base SI units with the SI prefix typically associated with it in practice. Unfortunately there is no space to provide historical context, for example regarding famous scientists and engineers whose names are now associated with units such as newton and non-dimensional quantities such as the Mach number.

We should like to thank Dr David Brookfield for providing the entries on the topic of control, for advice on a number of other topics including vibrations and for reviewing the majority of entries. We also are grateful to other colleagues who have clarified certain matters for us. However, we take full responsibility for any errors in the dictionary. The figures used to illustrate the definitions of turbofan engine, turbojet engine, turboprop engine, turboshaft engine, and two-speed turbojet engine are taken from the 1986 edition of "the jet engine" published by Rolls-Royce plc and appear by kind permission of Rolls-Royce plc.

Tony Atkins
University of Reading

Marcel Escudier
The University of Liverpool

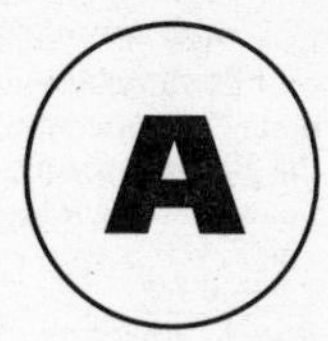

ablation cooling The cooling of a surface exposed to very high external gas temperature which causes the surface material to sublime, melt or decompose. The chemical process absorbs heat while the mass flow of material away from the surface blocks the heat flux from the hot gas. A **heat shield (ablation shield)** on a space vehicle consists of sacrificial cladding in the form of ceramic tiles that vaporize under the extreme frictional heat generated on re-entry into the earth's atmosphere, thus limiting the rise in temperature of the vehicle itself. *See also* HEAT OF ABLATION.

abradant The differently-sized grits of hard materials such as emery employed for grinding, polishing, etc.

abrasion The removal of surface material by the scratching action of hard particles, either deliberately (abrasive papers, abrasive cleaning, abrasive machining) or as a consequence of operation (wear). For intentional abrasion, grits may be fixed to a rigid or flexible backing or be airborne. *See also* SAND BLASTING; EROSION.

SEE WEB LINKS
- Abrasives terminology

ABS *See* ANTI-LOCK BRAKING SYSTEM.

absolute encoder A device which provides a continuous digital output representing angle or position. The output is normally obtained from a number of photocells mounted along a radial line and detecting circumferential patterns on a rotating glass disc. For example, for a resolution of 1 part in 2^{10} (1 part in 1024), 10 photocells would be mounted over the disk shown in the diagram and thus the output would be 10 parallel binary digits representing the angle of the disc.

absolute entropy The entropy value for a substance relative to absolute zero. For any pure crystalline substance at absolute zero the absolute entropy is zero. *See also* PRACTICAL ENTROPY.

absolute expansion The true volumetric expansion of a liquid with temperature, after account is taken of any expansion of the container in which it is held. *See also* APPARENT EXPANSION.

absolute humidity *See* SPECIFIC HUMIDITY.

absolute manometer A manometer that measures **absolute pressure**, i.e. pressure measured relative to a perfect vacuum. Absolute pressure cannot be negative.

absolute specific gravity The ratio of the weight of any volume of a substance to the weight of an equal volume of a reference substance at the same temperature, often water at 4°C, both measured in a vacuum to avoid any effect of buoyancy.

absolute temperature (Unit K) A temperature T measured relative to **absolute zero**, the lowest temperature achievable at which molecular motion vanishes so that a body would

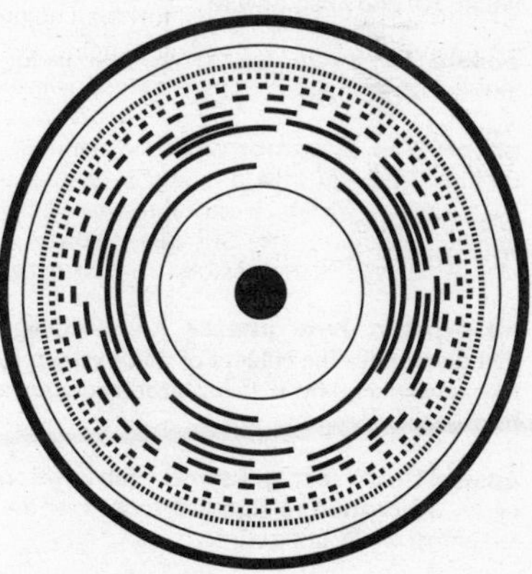

absolute encoder

a

have zero heat energy. Absolute zero is 0 K or −273.15°C, where the **Kelvin (K)** is the basic SI temperature unit equal to 1/273.16 of the thermodynamic temperature of the triple point of water, which is 273.16 K (i.e. 0.01°C). The Kelvin is equal in magnitude to the degree **Celsius (°C)**. The **Kelvin temperature scale (Kelvin absolute temperature scale)** is an absolute or thermodynamic temperature scale derived from the Celsius scale: T(K) = T(°C) + 273.15. The **Rankine absolute scale** is derived from the **Fahrenheit** scale such that T(R) = T (°F) + 459.67, i.e. a scale relative to 0 R or -459.67°F where R is the **Rankine** degree symbol and °F is the Fahrenheit symbol.

absolute viscosity *See* DYNAMIC VISCOSITY.

absorber 1. An auxiliary vibratory system that favourably modifies the vibration characteristics of a main system. **2.** Part of an absorption refrigeration system in which the refrigerant is absorbed by a transport medium. *See also* VAPOUR-ABSORPTION REFRIGERATION CYCLE. **3.** The **absorber plate** is the receiver of a concentrating solar collector where the radiation is absorbed.

absorptance (absorptivity, absorption coefficient, α) The fraction of radiant flux incident on a surface that is absorbed by the surface. The term also applies to absorption of radiation by a volume of fluid. For a semi-transparent surface, $\alpha + \rho + \tau = 1$, where ρ is the reflectance and τ is the transmittance.

absorption The process whereby a fluid permeates a porous solid, or a gas is dissolved by a liquid. *See also* ADSORPTION.

absorption cycle *See* VAPOUR-ABSORPTION REFRIGERATION CYCLE.

absorption dynamometer A dynamometer in which input work or power is dissipated by mechanical friction, electrical resistance, hydraulic resistance, etc. *See also* HYDRAULIC DYNAMOMETER; TRANSMISSION DYNAMOMETER.

absorption hygrometer An instrument that determines the content of water vapour in the atmosphere by it being absorbed into a hygroscopic medium.

absorption refrigeration (absorption cycle, absorption refrigerator) *See* VAPOUR-ABSORPTION REFRIGERATION CYCLE.

absorption tower *See* PACKED COLUMN.

ABS polymer A class of thermoplastic copolymer consisting of the three monomers **a**cryonitrile, **b**utadiene and **s**tyrene that has improved properties (particularly toughness) over the individual constituents.

accelerated testing A type of testing, for rates of wear, fatigue, corrosion etc., in which failure times are reduced by employing greater loads, more frequent power cycling, higher vibration levels, higher humidity, higher temperatures, greater potential differences etc. than would be encountered in normal operation.

accelerating frame of reference (moving frame of reference, non-inertial frame of reference) When the frame of reference of an observer is accelerating with respect to a fixed (inertial or Newtonian) frame of reference, it appears to the moving observer that additional terms are required to satisfy Newton's second law of motion. Where a reference frame S′ has, relative to a fixed frame S, motion of translation only, let $\boldsymbol{a_o}$ be the acceleration of S′ relative to S. If a particle has acceleration $\boldsymbol{a}$ relative to frame S and $\boldsymbol{a'}$ relative to S′, then $\boldsymbol{a} = \boldsymbol{a_o} + \boldsymbol{a'}$. In the fixed frame S, the law of motion is $\boldsymbol{F} = m\boldsymbol{a}$ where m is the mass of the particle and $\boldsymbol{F}$ is the force acting on it. In S′, the law is $m\boldsymbol{a'} = \boldsymbol{F} - m\boldsymbol{a_o}$ and it appears that it holds in the linearly accelerating providing that a **fictitious force** $-m\boldsymbol{a_o}$ is added to $\boldsymbol{F}$. This applies for every particle of a body. Fictitious forces that bring frame S′ to rest are variously called the **d'Alembert force, inertial force, effective force,** or **pseudo force.** Where a moving frame S′ rotates with variable angular velocity $\boldsymbol{\omega}$ about an axis fixed in a fixed frame S, the acceleration $\boldsymbol{a}$ of a particle with respect to the fixed frame S is given by $\boldsymbol{a} = \boldsymbol{a'} + \boldsymbol{a_t} + \boldsymbol{a_c}$ where $\boldsymbol{a'}$ is the acceleration with respect to S′, $\boldsymbol{a_t}$ is called the **acceleration of transport** and $\boldsymbol{a_c}$ is the **Coriolis acceleration**, both given with respect to S′. $\boldsymbol{a_t} = (d\boldsymbol{\omega}/dt) \times \boldsymbol{r} + \boldsymbol{\omega}(\boldsymbol{\omega} \cdot \boldsymbol{r}) - r\omega^2$ and $\boldsymbol{a_c} = 2\boldsymbol{\omega} \times \boldsymbol{v}$ where t is time, $\boldsymbol{r}$ is the position vector of the particle with respect to the fixed frame and $\boldsymbol{v}$ is the velocity of the particle with respect to the moving frame. In S the force $\boldsymbol{F}$ applied to a particle of mass m is given by $\boldsymbol{F} = m\boldsymbol{a}$, but to an observer moving with the particle in frame S′, $m\boldsymbol{a'} = \boldsymbol{F} - m\boldsymbol{a_t} - m\boldsymbol{a_c}$ giving rise to two fictitious forces $-m\boldsymbol{a_t}$ (**centrifugal force**) and $-m\boldsymbol{a_c}$ (**Coriolis force**). Even when $\boldsymbol{\omega}$ is constant, frame S′ is still accelerating owing to the vector $\boldsymbol{\omega}$ of constant magnitude continuously changing direction. In this case, $m\boldsymbol{a'} = \boldsymbol{F} + mr\omega^2 - 2m\boldsymbol{\omega} \times \boldsymbol{v}$ and the true forces X and Y along fixed Cartesian axes, the origin of which passes through the centre of rotation, are

$X = m(\ddot{x} - 2\omega\dot{y} - \omega^2 x)$ and $Y = m(\ddot{y} + 2\omega\dot{x} - \omega^2 y)$ in which $2m\omega\dot{y}$ *and* $-2m\omega\dot{x}$ are the components of the Coriolis force, and $m\omega^2 x$ and $m\omega^2 y$ are the components of the centrifugal force. When the point of rotation is itself accelerating at $\boldsymbol{a_o}$ in the Newtonian frame S, $ma' = \boldsymbol{F} - m\boldsymbol{a_o} - m\boldsymbol{a_t} - m\boldsymbol{a_c}$. The frame of reference that we employ in everyday life is the earth, which rotates with respect to the astronomical frame with an angular velocity 7.29 x 10^{-5} rad/s. For such slow-moving frames, the fictitious forces $-m\boldsymbol{a_o}$, $-m\boldsymbol{a_t}$, and $-m\boldsymbol{a_c}$ may not be noticeable, although they are important geographically, the centrifugal force being responsible for the equatorial bulge on the earth and the Coriolis force for the trade winds. Fictitious forces also become important for an aeroplane making a sharp turn or pulling out of a dive where the accelerations are of order *g* or greater. *See also* ANGULAR ACCELERATION; POINSOT METHOD; TANGENTIAL ACCELERATION.

acceleration (Unit m/s^2) The rate of change of linear velocity with respect to time. It is a vector quantity. *See also* ANGULAR ACCELERATION.

acceleration due to gravity (acceleration of free fall, gravitational acceleration, *g*) The acceleration of a freely-falling body in a vacuum, with a mean value at sea level of approximately 9.81 m/s^2.

acceleration-error constant When the reference (demand) input to a control system is parabolic, the output signal will also be parabolic in steady state. The signal that is constant in this situation is the acceleration and thus for a parabolic input the steady-state error, referred to as the acceleration-error constant, is the error in the acceleration.

acceleration of transport *See* ACCELERATING FRAME OF REFERENCE.

accelerator pump A pump which injects additional fuel when the throttle is opened rapidly to maintain constant the equivalence ratio of the mixture delivered to the cylinder of a piston engine.

accelerometer An electro-mechanical transducer used to measure acceleration. *See also* PIEZORESISTIVE ACCELEROMETER.

accessible resource (technical potential) The maximum annual energy that could be extracted from the accessible part of the available resource carried by a renewable energy resource using current mature technology.

accommodation The ability of a robot to respond to changes in the environment.

accumulator A fluid-energy storage device, analogous to a capacitor in an electric circuit. *See also* FLUID CAPACITOR.

accuracy *See* ERROR.

acetylene torch *See* OXY-ACETYLENE TORCH.

acfm *See* ACTUAL CUBIC FEET PER MINUTE.

ACHE *See* AIR-COOLED HEAT EXCHANGER.

acid rain Any form of precipitation, including rain, snow, sleet, fog, dew, and particulates, which contains higher than normal levels of sulphuric and nitric acids. These acids result from natural and man-made emissions of sulphur dioxide and oxides of nitrogen, respectively. **Acid soot** consists of unburned carbon particles, typically larger than 10 μm, in the atmosphere contaminated with sulphuric acid. The particles result from poor combustion of fossil fuels.

Ackerman linkage A steering linkage on a motor vehicle that approximately gives rolling without slipping of both wheels about the turning point. This is achieved by having the inner stub axle (on the inside of the turning curve) move through a greater angle than the outer stub axle.

acme thread *See* SCREW.

acoustic droplet ejection The ejection of small droplets from the surface of a liquid that results from focusing high-intensity ultrasound near the surface.

acoustic emission (stress-wave emission) Sound emitted by some materials when deformed under load. It arises from stress

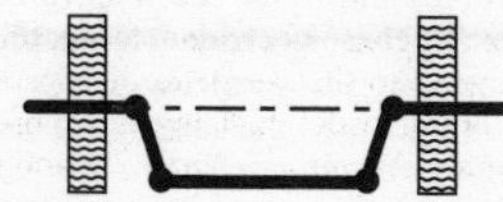
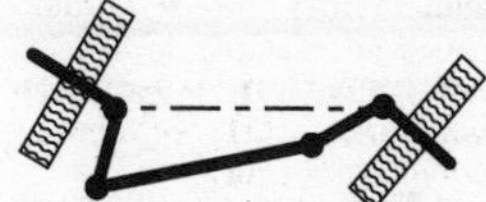

Ackerman linkage

waves emitted by sudden dislocation motion in crystals, slip, crack growth, etc.

acoustic energy density (sound energy density, w) (Unit J/m^3) The sum of the kinetic-energy density $w_{kin} = ½\,\rho\langle u^2\rangle$ and the compressional-energy density $w_{comp} = ½\,\langle p^2\rangle/c^2$ at a point in an acoustic field, where ρ is the fluid density, u is the magnitude of the particle velocity, p is acoustic pressure, c is the sound speed of the fluid and $\langle\dots\rangle$ denotes a time average.

acoustic fatigue *See* FATIGUE.

acoustic levitation of droplets The use of acoustic-radiation forces to suspend liquid droplets.

acoustic power (sound power) (Unit W) The rate of flow of acoustic energy across a specified surface.

acoustic pyrometer A non-intrusive pyrometer based on the principle that the sound speed in a gas is proportional to the square root of its absolute temperature.

acoustics The science and engineering of sound; its production, propagation, control, interaction with materials, etc.

acoustic separation The separation of particles in a fluid using standing acoustic waves, typically ultrasound, to drive them to nodal points (**acoustic particle concentration**).

acoustic streaming A steady fluid flow that results from a sound source or oscillating boundary.

acoustic velocity *See* SOUND SPEED.

actinometer An instrument that measures the intensity of radiant energy, such as a pyroheliometer or solarimeter.

action The direct application of a force. *See also* NEWTON'S LAWS OF MOTION; REACTION.

activation energy (E_a, U) (Unit kJ/mol) The minimum energy for a chemical reaction to occur or for processes such as diffusion to take place in crystals.

active accommodation The use of information from sensors, for example, in a vision system, that allows a robot to show accommodation to the environment.

active-cavity radiometer An instrument that gives an absolute reading of solar radiation. The solar beam falls on an absorbing surface whose temperature rise is compared with that of an identical absorber heated electrically.

active solar system (active solar-heating system, active space heating) A solar-powered heating or cooling system. It requires mechanical components, such as motors, pumps, thermosyphons and valves, for operation.

active vibration suppression The reduction of undesirable vibration in components by feedback control.

actual cubic feet per minute (acfm) An obsolete (i.e. non-SI) measure of volumetric flow rate; the volume of a gas flowing per minute at actual operating pressure and temperature, as opposed to the corresponding volume flow rate at STP.

actual power (actual horsepower) (Unit W or hp) The power delivered at the output shaft of an engine, before subsequent transmission through a gearbox etc. *See also* BRAKE POWER.

actual value The output of a plant that is being controlled, i.e. the controlled variable. Not directly accessible by the control system, as it can only be measured by a sensor which may distort the measurement.

actuating system A system in which an electrical, pneumatic or hydraulic input supplied to an **actuator** produces force, torque, or displacement, usually in a controlled way.

adaptive branch A robot programming instruction which can cause a branch in programme execution in response to signals from one or more sensors, and is thus used to provide accommodation to the environment.

adaptive control function That part of a control system which, by responding to measurement of changes in the controlled plant, adapts the parameters of the controller so as to improve the performance of the system (**adaptive control**).

adaptive robot A robot which responds to changes in the environment, i.e. a robot which shows accommodation.

adaptive system The ability of any engineering system to respond to changes in the environment, thus improving the performance in that environment.

ADC *See* ANALOGUE-TO-DIGITAL CONVERTER.

added mass (induced mass, virtual inertia) The mass which has to be added to that of a body being accelerated through a fluid to account for the fact that some of the fluid surrounding the body is also accelerated. The calculation of added mass in general is complicated but, for a sphere in potential flow, the added mass is ½ $\rho\mathcal{V}$ where ρ is the fluid density and $\mathcal{V}$ is the volume of the sphere.

addendum In a spur gear, the radial distance between the pitch circle and the addendum circle. There are corresponding definitions for bevel gears, worm gears, and screw threads. *See also* DEDENDUM; SCREW; TOOTHED GEARING.

addendum angle *See* TOOTHED GEARING.

additive-layer manufacturing (additive manufacturing, ALM, AM) A development of versatile three-dimensional printing processes employed for rapid prototyping, to manufacture actual production components from metals and plastics. Layers are generated by software that has taken a series of digital slices through a computer-aided design of the object. Adhesion between powdered layers, each 20-30 μm thick, is achieved by application of a liquid binder or by sintering with a laser or an electron beam. Some machines deposit filaments of molten plastic. *See also* SUBTRACTIVE MANUFACTURING.

adherend A component part bonded with **adhesive (glue)**, a liquid or gel that hardens and bonds together surfaces to which it is applied.

adhesion The normal or tangential grip between surfaces in contact. **Adhesive strength**, with unit Pa, is the normal or shear stress required to separate components bonded together by an adhesive. **Adhesion work**, with unit J, is the work done by forces when surfaces are separated. *See also* BOND STRENGTH.

SEE WEB LINKS

- Adhesives terminology

adiabat A curve representing the relationship between any pair of state variables at the beginning and end of an adiabatic thermodynamic process. *See also* SHOCK ADIABAT.

adiabatic film-cooling effectiveness *See* FILM-COOLING EFFECTIVENESS.

adiabatic index *See* SPECIFIC HEAT.

adiabatic lapse rate (Γ) (Unit K/km) The rate of decrease of an atmospheric variable, usually air temperature T, with altitude z assuming changes are adiabatic. For dry air, the adiabatic lapse rate is given by $\Gamma_D = -dT/dz = g/C_P$ where g is the acceleration due to gravity and C_P is the specific heat of air at constant pressure. With $g = 9.81$ m/s^2 and $C_P = 1.005$ kJ/kg.K, $\Gamma_D = 9.8$ K/km but the normal or environmental lapse rate is usually taken to be 6.5 K/km.

adiabatic process A thermodynamic process in which there is no heat transfer to or from the surroundings.

adiabatic saturator A device for determining relative and specific humidity by adding water continuously to a stream of unknown humidity and measuring the temperature and pressure of the entering and exiting streams.

adiabatic wall temperature (adiabatic surface temperature, recovery temperature) (Unit K) In high-speed gas flow over an insulated surface, the temperature attained by the surface as a consequence of viscous dissipation within the boundary layer. *See also* RECOVERY FACTOR.

adjustable-pitch propeller *See* PROPELLER.

admission The instant when the inlet valve allows the working fluid to enter the cylinder of an internal-combustion engine or a steam engine. *See also* FULL ADMISSION; PARTIAL ADMISSION.

admission port The passage by which a working fluid enters the cylinder of an engine.

adsorption The adhesion of gas or liquid molecules or dissolved solids to a surface. Silica gel and activated charcoal (also called active carbon) are widely used adsorbers. *See also* ABSORPTION.

adsorption dehumidifier A dehumidifier that operates by bringing moist air into contact with the surface of a hygroscopic solid such as silica gel.

advance 1. *See* IGNITION ADVANCE. **2.** *See* PROPELLER.

advanced gas-cooled reactor *See* NUCLEAR FISSION.

advanced ignition A phenomenon in which the spark in a spark-ignition engine ignites the fuel-air mixture close to the end of the compression stroke before the piston reaches top-dead centre.

advance ratio (advance coefficient) *See* PROPELLER.

advection 1. The transport of a substance or conserved property, such as enthalpy, by a fluid flow such as a boundary-layer flow. Advection differs from convection, which encompasses both advection and diffusion. In meteorology, oceanography, and river flows, advection is primarily horizontal.
2. In FEM, advection is the process of mapping solution variables from an old mesh to a new mesh. In the ALE approach, advective terms appear in the conservation equations to account for independent mesh motion as well as material motion.

adverse pressure gradient In a boundary-layer, a streamwise pressure gradient that opposes the flow of a fluid, reduces the wall shear stress and leads to earlier transition to turbulence. If the gradient is severe, flow separation may occur. In the case of an aerofoil, flow separation leads to stall. *See also* FAVOURABLE PRESSURE GRADIENT.

aeolian anemometer (eolian anemometer) An instrument used to estimate flow speed V based on the measurement of the frequency of vortex shedding f from a cylinder of diameter d in cross flow. Over a limited range of Reynolds numbers, the Strouhal number is constant with a value of about 0.2 so that $V = 5fd$.

aerated flow A flow of a liquid in which air is either dissolved or dispersed throughout as bubbles.

aeration The process whereby air is mixed with a liquid. The air, usually in bubble form, may be passed through the liquid or the liquid passed through the air. Some of the air dissolves in the liquid. Venturi tubes, spray nozzles and compressed-air jets are all used as **aerators**.

aerobic digestion Digestion by micro-organisms, in the presence of oxygen, of matter dissolved or suspended in waste. *See also* ANAEROBIC DIGESTION.

aerodynamic balance An instrument that measures the forces and moments on a model in a wind tunnel.

aerodynamics The mechanics of motion of air and other gases, including the interaction of the air and solid surfaces moving relative to it.

aerodyne A heavier-than-air aircraft where lift is produced either by forward movement through the air or by direct engine thrust. *See also* AEROSTAT.

aeroelasticity The elastic deformation of bodies subjected to aerodynamic loads.

aeroengine The power unit of an aircraft, including a piston, gas turbine, turboprop, jet, or rocket engine.

aerofoil (airfoil) A structure, the shape of which gives rise to lower pressures on one side (**suction surface**) compared with the other (**pressure surface**) when moving through a fluid. The **leading edge** is the front part of an aerofoil about which an oncoming flow divides, while the **trailing edge** is the rear edge. The **camber line** is a curve constructed midway between the upper and lower surfaces of an aerofoil, while the **chord line** (**chord**) is a straight line between the leading and trailing edges, the length of which is termed the **chord length**. The **angle of attack** (α) is the angle between the chord line and the vector direction of the relative velocity between the body and the fluid through which it is moving. Aircraft wings, propeller, turbine, and compressor blades all have aerofoil cross sections. Vertical lift (**upthrust**) forces are produced on the wing of an

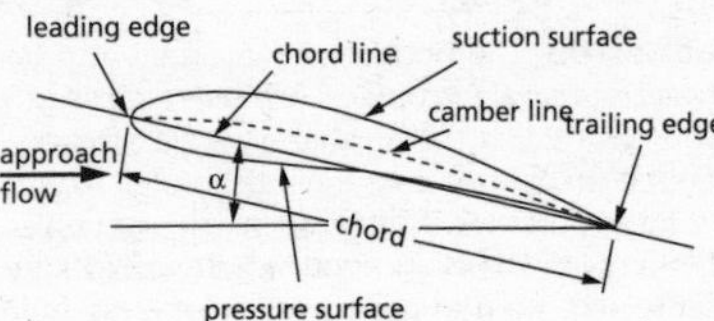

aerofoil

aircraft when moving through the air or vertical **downforce** on the spoiler (inverted aerofoil) of a road vehicle. *See also* SUBSONIC LEADING EDGE; SUPERSONIC LEADING EDGE.

aerogenerator Any device, such as a wind turbine, which uses wind energy to generate electricity.

aerometer *See* HYDROMETER.

aerosol A colloidal suspension of fine solid particles or liquid droplets in air or any other gas. In a typical **aerosol generator**, liquid is passed through a nozzle or sprayed onto a rapidly spinning disc.

aerospace engineering The design and construction of aircraft, rockets, earth satellites, space vehicles, and their power units. It includes control, launching and guidance.

aerostat A lighter-than-air system for which the aerodynamic lift is provided by buoyancy. The main structural component is a bag-like envelope filled with a low-density gas, such as helium or hydrogen. Examples include **airships**, gas **balloons** and hot-air balloons. *See also* AERODYNE; DIRIGIBLE.

aerostatics The study of gases in which there is no relative motion between gas particles. For a gas at rest, it concerns the variation of pressure, density, and temperature with altitude. *See also* FLUID MECHANICS.

A-frame *See* SHEAR LEGS.

afterburner An extension of the exhaust of a turbofan or turbojet engine into which additional fuel is injected and burned to provide increased thrust. Primarily used on supersonic military aircraft. *See also* REHEAT.

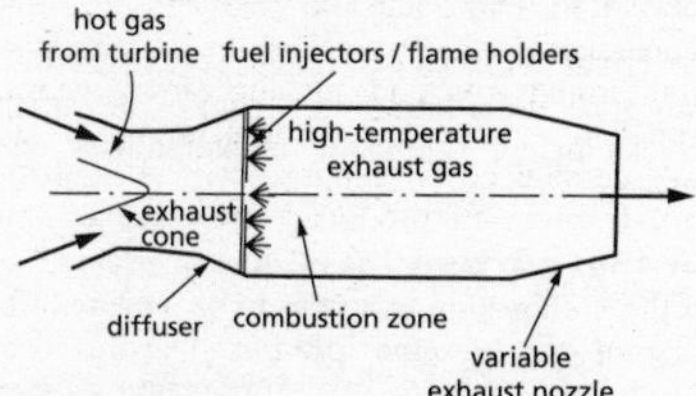

afterburner

aftercondenser A condenser used to condense the atmospheric stage of process fluid in a multi-stage ejector. Employed in air-conditioning and refrigeration systems and steam power plants.

afterfilter The final filter in a gas-flow system, such as in air-conditioning, ventilation, and machining processes, that removes fine (typically sub-micron) solid particles and liquid droplets from the flow.

after running (run on, dieseling) Continued running of an internal-combustion engine after the ignition is switched off due to continued supply of fuel.

after top-dead centre The position of the piston in a reciprocating engine during downward strokes. *See also* TOP-DEAD CENTRE.

A_g The symbol for the uniform strain in a tension test specimen before necking or fracture. *See also* RP, RM.

ageing (aging) Changes in microstructure, and hence mechanical properties, that occur as supersaturated solid solutions formed by quenching alter with time to the equilibrium phases. A well-known example relates to the precipitation-hardened aluminium-copper system that becomes stronger with time. The process may be accelerated by heating (artificial ageing), but too long a time at too high a temperature leads to over-ageing, when the benefits are lost.

aggregative fluidization (boiling fluidization) The type of fluidization that occurs when the fluid and solids have significantly different densities, as occurs in gas/solid systems and some liquid/solid systems. Particle-free voids rise through the fluidized bed, the surface of which takes on the appearance of a boiling liquid.

agitator A device to mix fluids and suspensions or keep them in motion by stirring or shaking.

AGR *See* NUCLEAR FISSION.

air aspirator valve *See* ASPIRATED AIR INJECTION.

air-atomizing oil burner A burner employing oil as a fuel that is dispersed into fine droplets by passing the fuel and compressed air through an atomizing nozzle.

air bearing *See* BEARING.

air bleeder A valve for removing air from a hydraulic system.

airborne waste Gases, particulates and vapours in the atmosphere resulting from evaporation, chemical, or combustion processes. *See also* SMOG.

air-bound Said of an air blockage in a pipe that restricts liquid flow. *See also* AIRLOCK (1).

air box A forward-facing inlet duct, connected to an internal-combustion engine, that directs air to the engine. Typically used on Formula 1 and other high-performance motor vehicles.

air brake 1. A brake on a vehicle that is operated by compressed air. **2.** An absorption dynamometer that dissipates power through a fan or propeller. **3.** A flap on an aeroplane or racing car that increases drag to reduce speed.

air-breathing An air-breathing engine is one for which air is the oxidant in the combustion process.

air chamber *See* PULSATION DAMPER.

air classifier (air elutriator) A device in which an airstream, which may be swirling, sorts particles by a combination of size, shape, and mass.

air cleaner A device, such as a filter, hydrocyclone, or electrostatic precipitator, that removes particles and aerosols from a flow of air.

air composition The sea-level composition (in per cent by volume) of air at a temperature of 15°C and a pressure of 1 atm is mainly 78.084% nitrogen, 20.947% oxygen and 0.934% argon. The remaining 0.035% consists of carbon dioxide, neon, helium, methane, krypton, hydrogen, oxides of nitrogen, xenon, ozone, iodine, carbon monoxide, and ammonia. Different sources give slightly different figures for the composition. Not included are water vapour (typically 0.4%) and pollutants such as sulphur dioxide.

air compressor A turbomachine that draws in air and delivers it at higher pressure, temperature, and density. It can be of axial, fan, reciprocating, or rotary design.

air conditioning The process of controlling the temperature and humidity in rooms, buildings, aircraft, passenger vehicles, etc. More generally it includes control of dust, levels of radiant heat, etc.

air-cooled condenser A heat exchanger, in which the cooling medium is air, used to condense the exhaust steam from a steam turbine, the condensate being returned to the boiler.

air-cooled engine An internal-combustion engine directly cooled by airflow, rather than by water flowing through the engine block being cooled by a radiator.

air-cooled heat exchanger (ACHE) A pressure vessel in which fluid flowing through finned tubes is cooled by ambient air forced over the exterior surfaces of the tubes. Examples include the radiator of a motor vehicle, lubrication-oil coolers, and steam condensers.

air curtain (air door) A stream of temperature-controlled air blown across an open entrance to limit heat transfer with the environment and allow the interior space to be air conditioned.

air-cushion vehicle A vehicle that rides over land or water on a cushion of air maintained by rotors or fans. *See also* HOVERCRAFT.

air cycle *See* AIR-STANDARD CYCLE.

air-cycle refrigeration *See* GAS REFRIGERATION CYCLE.

air ejector A Venturi nozzle in which low pressure is created at the throat by high-speed air flow. The low pressure can be used to suck another fluid into the Venturi for transport or mixing. *See also* JET PUMP.

air eliminator A device used in liquid-flow systems to remove air and other gases. An important application is to flow metering, where air or gas in the flow can lead to large errors.

air elutriator *See* AIR CLASSIFIER.

air engine (air motor) A heat engine, of either reciprocating or rotary design, for which the energy source is compressed air.

air filter A device that reduces the concentration of solid particles in a stream of air. Applications include ventilation, engine intakes, and air conditioning.

air-fuel mixture The mixture of air and fuel in the combustion chamber of an engine. The mixture may be either pre-mixed or mixed in the chamber. *See also* STOICHIOMETRIC.

air-fuel ratio (air/fuel ratio) The ratio of air to fuel in an air-fuel mixture. For a gaseous fuel the ratio is expressed by volume; for a solid or liquid fuel, by mass. *See also* STOICHIOMETRIC.

air gauge A non-contact device that uses a stream of compressed air to measure the clearance between a precision orifice or plug and a workpiece.

air injection *See* ASPIRATED AIR INJECTION.

air-line lubricator (line oiler) A device used to add lubricant via an air line to compressed-air-driven equipment.

airlock 1. (air pocket) A pocket of gas trapped at a high point in a liquid-filled piping system that restricts or even prevents flow. *See also* AIR-BOUND; AIR-RELEASE VALVE; BLEED VALVE. **2.** A chamber between two systems at different pressures, to enable movement of people between the two to take place.

air-mixing plenum The chamber in an air conditioning system in which fresh outdoor air is mixed with recirculating air.

air motor *See* AIR ENGINE.

air nozzle A convergent nozzle with one or more exit holes through which compressed air flows at high speed and entrains a large volume of ambient air. Applications include liquid blowoff, chip removal in cutting, cleaning, cooling, and drying.

air-pollution control The treatment of exhaust gases from piston engines, gas turbines, furnaces, boilers, etc., so as to limit the release of gaseous and particulate pollutants into the atmosphere to acceptable levels. Control devices include wet and dry scrubbers, filters, cyclones, baghouses, electrostatic precipitators, catalytic converters, adsorption systems, and exhaust-gas recycling.

air preheater *See* PREHEATER.

air pump A machine for providing a flow of air or for increasing or decreasing the mass and pressure of air in a closed container. The term pump is more usual when the working fluid is a liquid, while compressor is more usual for gases. *See also* VACUUM PUMP.

air purge Removal of gas or particulate contaminants in a piping system or pressure vessel by passing clean air through it.

air receiver (receiver) A pressure vessel in a compressed-air system which stores the air received from a compressor, condenses and traps moisture in the air, prevents short-cycle loading and unloading of the compressor, and equalizes pressure throughout the system.

air regulator A valve that supplies air at constant pressure regardless of flow variation or upstream pressure. *See also* PRESSURE REGULATOR.

air reheater A heat exchanger employed to reheat the working fluid from a turbine or from a compressor after drying. *See also* RANKINE CYCLE WITH REHEAT.

air-release valve (air valve) 1. A valve that releases air trapped at a high point in a pressurized pipeline filled with a liquid. *See also* AIR POCKET. **2.** A valve that prevents liquid in a water-supply, wastewater, or sewage system from coming in contact with the sealing mechanism by creating an air gap.

air resistance The drag force on a body moving through air, or the forces imposed on a stationary body or other structure by air flow, including wind.

air scoop 1. A cowl or duct projecting from an aircraft or motor vehicle and facing the airflow that allows ambient air direct access at increased pressure (ram effect) to the engine air intake or to create an interior flow, for example for ventilation. **2.** A device inserted into a water pipeline that reduces the water speed so that any air in the flow rises above the water surface and can be removed.

airscrew *See* PROPELLER.

air separator *See* AIR CLASSIFIER.

airship (dirigible) An aerostat with forward thrust provided by propellers or other engines and steered by rudders. The main types are blimps, which are large gas balloons with no internal support structure, and rigid airships, which have a full internal frame.

airspeed indicator An instrument that displays the speed of an aircraft through the air, usually in knots, based on the output of a pitot-static tube.

air spring (pneumatic spring) A spring whose stiffness is provided by compression of air in a cylinder or bellows.

air-standard assumptions In an air-standard cycle: 1. The working fluid is air that circulates in a closed loop and behaves as an ideal gas; 2. All the processes that make up the cycle are reversible; 3. The combustion process is replaced by a heat-addition process from an external source; 4. The exhaust process is replaced by a heat-rejection process to an external sink that restores the working fluid to its initial state. *See also* COLD-AIR-STANDARD ASSUMPTIONS.

air-standard cycles (heat-engine cycles) Thermodynamic cycles for which the air-standard assumptions are applicable. They provide simplified models of the gas processes that occur in an internal-combustion engine, and can be used to study qualitatively the influence of major parameters on the performance of an actual engine. Examples of air-standard cycles are the Otto cycle for petrol engines, the Diesel

a

cycle, the Joule or Brayton cycle for gas-turbine engines, the Sabathé dual-combustion cycle, the Stirling cycle, and the Ericsson cycle. *See also* GAS CYCLE.

air-standard efficiency The efficiency of any air-standard cycle.

air-standard engine Any heat engine operating on an air-standard cycle.

air starting valve A valve that admits compressed air to the cylinders of a diesel engine for starting purposes.

air-suspension system Suspension of a vehicle by air springs.

air thermometer *See* GAS THERMOMETER.

air valve *See* AIR-RELEASE VALVE.

air-vapour mixtures *See* GAS-VAPOUR MIXTURES.

air washer A combined humidifier and air purifier.

Airy points The locations along a nominally straight horizontal bar at which the bar should be supported to minimize deflexions under its own weight. For a bar of rectangular cross section and length L, if there are two symmetrically spaced supports, they should be separated by the distance $L/\sqrt{3}$.

Airy stress function One of a number of stress functions by means of which elasticity problems are solved. In two dimensions, a stress function ϕ satisfies $\nabla^4\phi = 0$ that is the same as the equations of equilibrium and compatibility when the normal and shear stress components are written as second partial derivatives of ϕ. *See also* BIHARMONIC EQUATION.

albedo A measure of how strongly an object reflects the sun's radiation. Defined as the ratio of total reflected to incident electromagnetic radiation.

alcohol thermometer (alcohol-in-glass thermometer) A liquid-in-glass thermometer using ethyl alcohol as the working fluid, usually dyed red or blue.

ALE *See* ARBITRARY LAGRANGIAN EULERIAN METHODOLOGY.

aliasing The generation of spurious low-frequency signals in sampled data due to sampling at intervals longer than required by the Nyquist–Shannon sampling theorem.

aligning drift *See* DRIFT.

alignment The adjustment of the relative positions and orientation of one or more objects. *See also* SHAFT ALIGNMENT.

Allen® screw A screw or bolt having a hexagonal recess in the head for tightening using an Allen® key.

allowable load (Unit N) The maximum load at a given location, and in a given direction, that may safely be applied to a component or structure.

allowable stress (σ_{ALLOW}) (Unit Pa) The maximum stress permitted at any point in a component or structure, taking into account a factor of safety, f. Possible choices are $\sigma_{ALLOW} = \sigma_Y/f$ and $\sigma_{ALLOW} = \sigma_{UTS}/f$ where σ_Y is the yield stress of the material and σ_{UTS} is its ultimate tensile strength. *See also* FATIGUE.

allowance An intentional difference in sizes of mating components, prescribed to give different grades of fit. *See also* TOLERANCES.

alloy A mixture of two or more metals in different proportions to obtain properties better than any one of them alone.

SEE WEB LINKS

- Ferrous alloy types
- Non-ferrous alloy types

all-translational system A robot with no rotational joints such that the movement of each joint is along a straight-line path.

ALM *See* ADDITIVE-LAYER MANUFACTURING.

alpha-type Stirling engine *See* STIRLING ENGINE.

alternating stress (Unit Pa) Originally, stresses of changing sign (tension-to-compression-to-tension, etc.) in a component produced by alternating forces acting in opposite directions, but now generally used to describe stresses that vary but may keep the same sign, as produced by periodic, out-of-balance, or vibrational loads. *See also* FATIGUE.

alternative energy (alternate energy) Energy sources that are renewable and do not have the undesired consequences of fossil fuels and nuclear energy. *See also* RENEWABLE ENERGY.

alternative-fuel engine An internal-combustion engine that can operate either as a diesel engine using diesel fuel or as a gas engine using gaseous fuel. *See also* DUAL-FUEL ENGINE.

alternator A relatively small electrical generator, usually driven by an internal-combustion engine to operate its ignition and charge the battery.

altitude Vertical height measured relative to a specified datum such as sea level.

altitude chamber (hypobaric chamber) A chamber in which conditions at different altitudes are simulated by a combination of appropriate pressures, temperatures and relative humidity. *See also* HYPERBARIC CHAMBER.

AM *See* ADDITIVE-LAYER MANUFACTURING.

Amagat's law (Amagat's law of additive volumes, Leduc's law) The volume of a mixture of ideal gases is equal to the sum of the volumes (**partial volumes**) each gas would occupy if it existed alone at the same temperature and pressure. *See also* DALTON'S LAW.

ambient conditions The thermodynamic properties, including pressure, temperature, and humidity, of the environment surrounding a body or system.

American valve *See* SCHRADER VALVE.

ammonia absorption refrigerator *See* VAPOUR-ABSORPTION REFRIGERATION CYCLE.

Amontons friction (Coulomb friction) Friction between surfaces where the ratio of the frictional force F to the normal force N is constant and independent of the area in contact. *See also* ANGLE OF FRICTION; COEFFICIENT OF FRICTION.

amplifier A device by which the output of a hydraulic, pneumatic, or electrical source is increased. *See also* ATTENUATION; GAIN.

amplitude For a sinusoidal signal, the amplitude is the peak value. More generally, the peak-to-peak amplitude may be defined as the difference between the highest and lowest values of a periodically-varying quantity. An alternative is the root-mean square (RMS) amplitude, which is the square root of the time average of the squared signal.

amplitude-decay coefficient *See* DAMPING CONSTANT.

anaerobic digestion A process whereby organic waste is broken down by naturally-occurring bacteria in a controlled, oxygen-free environment to produce biogas. *See also* AEROBIC DIGESTION.

analogue readout A display on a meter or other measuring device where the reading is taken from a pointer on a scale, rather than being shown in digital form.

analogue-to-digital converter (ADC) A device which converts a continuous quantity to a sampled digital representation of that quantity.

anechoic chamber A room having all surfaces covered with sound-absorbing material, often in the form of wedges pointing into the room. The aim is to simulate free-field acoustic conditions.

anelasticity Literally 'not elastic', but in practice used for materials that display time-dependent recovery on unloading. *See also* INELASTIC.

anemometer (velocimeter) Any instrument that measures speed, and in some cases also direction, usually of a fluid flow but also applicable to surface movement. *See also* LASER-DOPPLER ANEMOMETER; LASER 2-FOCUS ANEMOMETER; PARTICLE-IMAGE VELOCIMETER.

aneroid Liquid free.

aneroid barometer An instrument that measures and records changes in atmospheric pressure using the expansion and contraction of a sealed bellows vacuum unit that is kept open by an internal spring.

aneroid calorimeter A calorimeter that uses a metal of high thermal conductivity such as silver, rather than stirred water, to absorb the heat released by a chemical reaction within the calorimeter.

angle factor *See* VIEW FACTOR

angle of advance 1. For a propeller, the angle between the relative velocity of the airstream and the plane of the propeller. The relative velocity is the vector sum of the free-stream velocity (in flight, the aircraft speed) and the propeller rotational velocity. **2. (ignition advance, spark lead)** In a spark-ignition engine, the angle before top-dead centre at which the spark occurs.

a

angle of approach (arc of approach) *See* TOOTHED GEARING.

angle of attack (α) The angle between a reference line on a lifting body and the vector direction of the relative velocity between the body and the fluid through which it is moving. In the case of an aerofoil, turbine, or compressor blade, the usual reference line is the chord line.

angle of contact The angle subtended at the centre of a pulley or sprocket wheel by the circumferential contact of a belt or chain.

angle of friction (friction angle, β) For a body in contact with a plane surface, the angle between the normal to the surface and the resultant force between the body and the surface. If the friction force is F and normal force is N, β is given by $F/N = \mu = \tan \beta$ where μ is the coefficient of friction. *See also* AMONTONS FRICTION.

angle of inclination *See* SCREW.

angle of obliquity (angle of pressure) *See* TOOTHED GEARING.

angle of recess (arc of recess) *See* TOOTHED GEARING.

angle of torsion (angle of twist) The angle relative to a chosen section through which another part of a component rotates when subjected to a torque.

angle ply laminate A fibre-reinforced composite structure made up of anisotropic layers, usually with as many orientated one way with respect to a reference axis as to the other.

ångström (Å) An obsolete (i.e. non-SI) unit of length of magnitude 10^{-10} m or 0.1 nm, sometimes still employed for atomic and crystallographic measurements.

angular acceleration (Unit rad/s^2) A vector quantity equal to the rate of change of angular velocity of a rotating body with respect to time. *See also* ACCELERATION; INSTANTANEOUS CENTRE; TANGENTIAL ACCELERATION.

angular accelerometer An instrument that measures angular acceleration.

angular-contact bearing *See* BEARING.

angular displacement For a rotating body, the angle through which a point on the body has rotated about the axis of rotation in a given time.

angular distortion Longitudinal warping in the torsion of a non-circular section.

angular frequency (circular frequency, ω) (Unit rad/s) The rate of change of phase in a sinusoid, i.e. $\omega = 2\pi f$, where f is the frequency.

angular gear *See* TOOTHED GEARING.

angular impulse (Unit N.m.s) The integral with respect to time of a torque applied to a body. If the body is free to rotate, the angular impulse is equal to the change in angular momentum of the body.

angular momentum (moment of momentum) (Unit N.m.s) A vector quantity. For a rigid body rotating about an axis of symmetry with angular velocity ω, the angular momentum is equal to the product of ω and the moment of inertia about the same axis.

angular pitch *See* TOOTHED GEARING.

angular velocity (tangential velocity, ω, Ω) (Unit rad/s) For an object moving along a curved path, the rate of change with respect to time of angular displacement. In terms of tangential velocity V_θ, $\omega = V_\theta/r$ where r is the local radius of curvature of the path. *See also* RPM.

anisotropic etching (anisotropic micromachining) A process consisting of chemical attack along a particular crystallographic plane of a material, such as silicon, used in microfabrication to create well-defined microscopic features of high aspect ratio.

anisotropy The characteristic of being directionally dependent, which can apply to physical properties of fluids and solids, such as thermal conductivity, elastic moduli, and yield strengths, as well as to flow properties such as turbulence intensities.

annealing 1. In metal alloys, the process of restoring ductility and lowering the strength of cold-worked components, by heating above the recrystallization temperature and cooling. **2.** With glasses, heating to some 400°C in order to relieve residual stresses imparted by working. The corresponding metallurgical term is stress relieving. *See also* LEHR.

annular flow (annular-flow regime) A gas-liquid or vapour-liquid pipe flow in which the dispersed gas or vapour phase occupies the central part of the cross section while the liquid forms a layer on the pipe wall. *See also* BOILING; BUBBLY FLOW; SLUG FLOW.

annular gear *See* TOOTHED GEARING.

annular nozzle A nozzle consisting of a circular centrebody and a concentric outer tube. If the annular cross section first decreases, then increases, the nozzle is said to be convergent-divergent.

annulus The space between two circles or circular cylinders, which may be concentric or eccentric.

anodize (anodizing) To give a metal, particularly an aluminium alloy, a protective oxide coating by electrolysis with the metal as the anode.

anomalous expansion A decrease in the volume of a fixed mass of material resulting from an increase in temperature, such as the behaviour of water between 0 and 4°C.

ANSI The acronym for **A**merican **N**ational **S**tandards **I**nstitute.

anthropomorphic configuration *See* ARTICULATED ROBOT.

anti-backlash gear A gear arrangement in which backlash is reduced or eliminated. The diagram shows a double (split) gear, one part of which is spring-loaded against the other that is fixed to the same shaft, which eliminates backlash when engaged to a gear wheel or rack. Two nuts, keyed together but spring loaded apart (a split-and-sprung nut) perform the same function for the lead screw on a lathe.

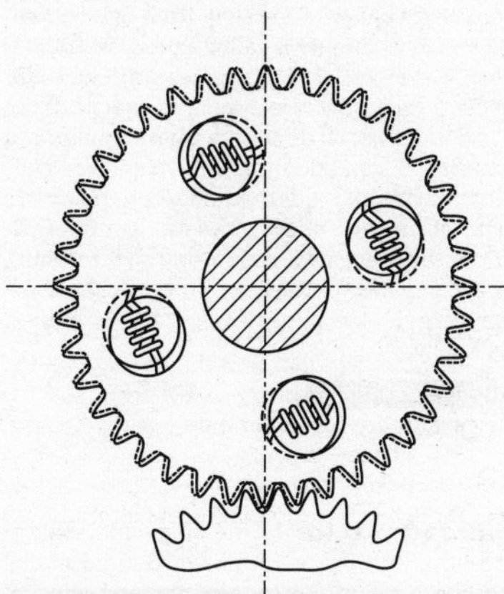

anti-backlash gear

anti-freeze A substance such as ethylene glycol added to the cooling system of a water-cooled engine to lower the freezing point of the cooling water and also inhibit the formation of rust and other deposits.

antifriction bearing (antifriction material) *See* BEARING.

anti-knock index (anti-knock rating) A measure of a fuel's ability to resist spontaneous ignition in an internal-combustion engine. *See also* OCTANE RATING; KNOCKING.

anti-lock braking system (ABS) The Sensor-controlled brakes on a motor vehicle that prevent wheel lockup when braking. *See also* TRACTION CONTROL.

antinode A point of maximum displacement in a one-dimensional stationary wave field; a line of maximum amplitude in a 2D field; and a surface of maximum amplitude in a 3D field. *See also* NODE.

antiphase Where the phase difference between two sinusoidal signals at the same frequency is π (i.e 180°).

anti-roll bar A device that reduces the outwards roll of the body of a motor vehicle when turning a corner.

anti-stiction technique In surface micromachining, a special drying process to avoid the sticking effect caused by capillary forces which occurs after rinsing.

anti-torque rotor The tail rotor of a helicopter which provides sideways thrust to counteract the torque where there is a single main rotor. It also gives directional control.

anti-vibration mounting A device to reduce or isolate the transmission of vibrations. *See also* DAMPER.

aperiodic An oscillation which has non-periodic or non-repeating cycles in either time or space.

apparent expansion The expansion of a liquid when the expansion of the container in which it is held is not accounted for. *See also* ABSOLUTE EXPANSION.

apparent force *See* ACCELERATING FRAME OF REFERENCE.

apparent motion *See* RELATIVE MOTION.

a

apparent slip In near-wall fluid flow, apparent departure from the no-slip boundary condition. Attributed in slurry flow to near-wall depletion of solid particles and in microchannel flow to a near-wall hydrophobic layer.

apparent viscosity (μ) (Unit Pa.s) For a non-Newtonian fluid, the ratio of the shear stress τ to the corresponding shear rate $\dot{\gamma}$, i.e. $\mu = \tau/\dot{\gamma}$. For a Newtonian fluid, μ is independent of $\dot{\gamma}$, but for many non-Newtonian liquids it is shear-rate dependent. *See also* SHEAR-THICKENING LIQUID; SHEAR-THINNING LIQUID.

apparent weight (apparent immersed weight) (Unit N) The difference between the actual weight of a body submerged in a fluid and the upthrust (buoyancy force) which it experiences. *See also* ARCHIMEDES PRINCIPLE.

applied mechanics (engineering mechanics, mechanics) The application of physical principles to practical problems, including the response of components, materials and structures to applied forces. The field is traditionally divided into statics and dynamics but may also include fluid mechanics. **Statics** is concerned with the analysis of the forces and moments on a body, structure or physical system in static equilibrium. **Kinematics** is the aspect of **dynamics** concerned with the displacements, velocities (**velocity analysis**) and accelerations of a mechanism or system without regard to the forces required for the motion. **Kinetics** is the aspect of dynamics that relates the unbalanced forces applied to particles, bodies and systems to the resulting velocities and accelerations through Newton's second law of motion, **rigid-body dynamics** being the study of the forces and couples required to produce motions of a rigid body. The assumption that a body is rigid (i.e. undeformable) simplifies dynamic analyses, but consideration of elastic and plastic deformations is often necessary to obtain solutions.

applied thermodynamics (engineering thermodynamics, thermodynamics) The science of the relationship between heat, work and the properties of systems and the ways in which heat energy from fuels can be converted into mechanical work. It involves the study of all aspects of energy use and energy transformation, including power generation, refrigeration, the relevant properties of the substances involved and the relationships between them. The principle of conservation of energy is a fundamental law of nature. In thermodynamics it is expressed as the first law of thermodynamics: in any process or interaction, energy can change from one form to another, such as thermal energy into mechanical energy, but the total amount of energy remains constant. The second law of thermodynamics is more concerned with the quality of energy, with all physically realistic processes occurring in the direction of decreasing energy quality.

Although the properties of any substance depend on the behaviour of its molecules, for the purposes of engineering a macroscopic approach is adequate. It is usual, therefore, to consider the behaviour of matter contained within a region of space termed a system, which may be closed, with energy but no material crossing its boundary, or open (a control volume), through which there is flow of material and energy. Any characteristic of a system is called a property. Intensive properties, such as pressure, density and temperature, are independent of the size of the system concerned in contrast to extensive properties, such as mass, volume and total energy.

Thermodynamics is concerned with equilibrium states. A system is in thermal equilibrium if the temperature is the same throughout. A system is in mechanical equilibrium if there is no change in pressure with time, although the pressure may vary from point to point. If a system involves two or more phases, it is in phase equilibrium when the mass of each phase is at an equilibrium level. A system in which no chemical reactions occur is in chemical equilibrium. Any change of a system from one equilibrium state to another is called a process, and it is usually the case that processes proceed sufficiently slowly that the system is practically in equilibrium at all times (a quasi-equilibrium process). Engineering thermodynamics is concerned with the quasi-equilibrium processes that occur in such devices as internal-combustion engines, steam and gas turbines, rocket engines, refrigerators, pumps, heat exchangers, and cooling towers. *See also* HEAT TRANSFER.

SEE WEB LINKS

- CODATA key values for thermodynamics

approach vector In the kinematic analysis of robots, a 4×4 matrix is obtained representing the position and direction of the end effector. The third column of this is the approach vector $\bar{A}$, a unit vector running in the direction that the end effector approaches the task. For example, where the end effector is a gripper, as shown in the diagram, the approach vector is parallel to

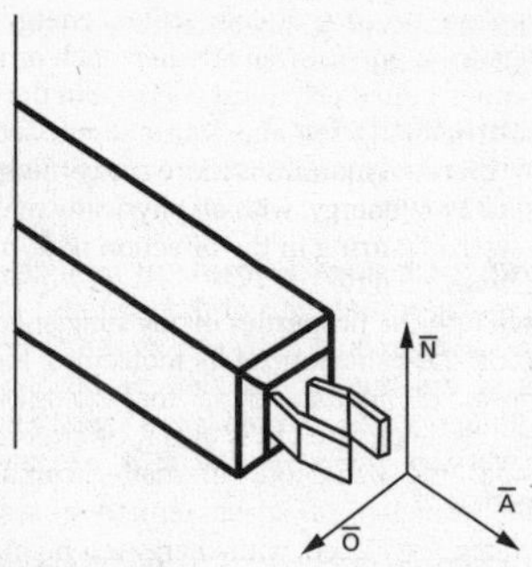

approach vector

the gripper fingers. *See also* NORMAL VECTOR; ORIENTATION VECTOR.

APU *See* AUXILIARY POWER UNIT.

Araldite® A synthetic epoxy resin adhesive suitable for bonding metals, glass, porcelain, plastics etc.

Arbitrary Lagrangian Eulerian methodology (ALE) An FEM modelling procedure that reduces to Lagrangian form on free boundaries, while maintaining Eulerian form at locations where significant deformation occurs, thus avoiding the need for re-meshing.

arbor A spindle of a milling machine used to hold a revolving cutting tool or, sometimes, the workpiece.

Archimedes number (*Ar*, Best number, Galileo number) A non-dimensional parameter that arises in the study of the motion of a solid object and a fluid where the density difference between the two is important, as in a fluidized bed. It is defined by $Ar = \rho_F g L^3(\rho_S - \rho_F)/\mu^2$ where ρ_F is the fluid density and μ is its viscosity, ρ_S is the solid density, L is a characteristic dimension of the solid object and g is the acceleration due to gravity. The factor 4/3 sometimes appears on the right-hand side of the definition.

Archimedes principle A body, wholly or partly immersed in a fluid, experiences an upthrust (the buoyancy force or apparent loss of weight) equal to the weight of the fluid displaced. The buoyancy force acts vertically upwards through the centroid of the submerged volume.

Archimedes screw A machine which comprises a rotating helical blade inside a close-fitting tube, which may be used to pump liquids, slurries (such as sewage), granular materials, etc. If water is poured into the top of an inclined or vertical Archimedes screw, the screw will rotate and can be used to drive an electrical generator.

Archimedes Wave Swing A machine for tidal-power energy generation consisting of two concentric, air-filled submerged cylinders. The inner lower cylinder is tethered to the ocean floor while the upper floater unit, which is closed at the top, moves up and down due to the variations in hydrostatic pressure caused by the wave motion. The relative movement of the two cylinders is used to generate electricity using a linear generator.

arc of approach *See* TOOTHED GEARING.

arc of contact (arc of action) *See* TOOTHED GEARING.

arc spraying A high-productivity thermal-spraying process in which a material is melted by a DC electric arc struck between two consumable wire electrodes and propelled by a gas jet on to a substrate to form a coating. *See also* FLAME SPRAYING; PLASMA SPRAYING.

arc welding *See* WELDING.

arm The part of a robot that carries the wrist and end effector and is used to make large movements of the robot.

Armco iron Iron that is almost pure ferrite (less than 0.03% carbon).

arm solution *See* INVERSE KINEMATICS.

Arrhenius equation An empirical relation that represents the dependence of the rate constant k of a chemical reaction on the absolute temperature T according to $k = A\exp(-E_a/\mathcal{R}T)$ where A is a constant, E_a is the activation energy and $\mathcal{R}$ is the universal gas constant.

articulated blade A helicopter rotor blade that can move in three ways as the rotor turns: vertically up and down (flapping), back and forth in the horizontal plane and tilt or feather (change pitch angle).

articulated robot (revolute configuration robot) A robot having a rotational waist joint, joint angle θ_1, with a vertical axis; a rotational shoulder joint, joint angle θ_2, with a horizontal axis; and a rotational elbow joint, joint angle θ_3, also with a horizontal axis. This configuration is termed anthropomorphic, as it emulates that of

a

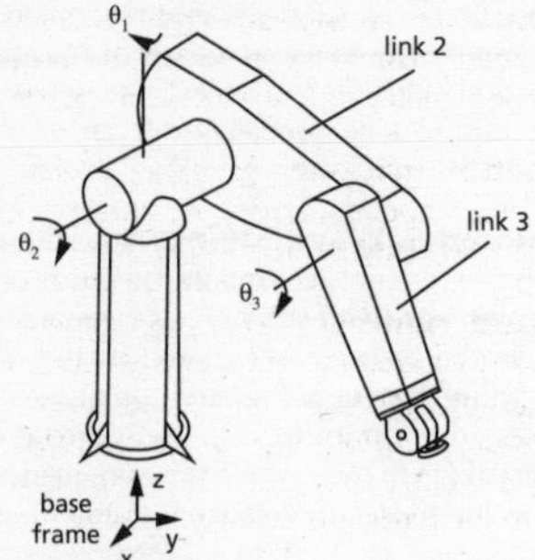

articulated robot

the human body. The diagram shows an idealized articulated robot.

articulated vehicle A motor vehicle, such as a lorry, bus, tram, or train, which has one or more vertical pivots to allow one part to turn relative to another, such as the cab of a lorry relative to its trailer.

articulation The joints and links of a robot, hence a description of the articulation of a robot gives a description of its configuration.

aspect ratio The ratio of the longest to the shortest dimension of a shape or object e.g. the chord-to-thickness ratio of an aerofoil or the span-to-chord ratio of a wing.

aspirated air injection (air injection, secondary air injection) An emissions-control technique involving the injection of fresh air into the exhaust gases from an internal-combustion engine using the sub-atmospheric pressure between exhaust-pressure pulses to draw the air through a one-way aspirator valve. The air oxidizes unburned hydrocarbons and carbon monoxide in the exhaust gas.

aspirator (eductor jet pump, jet pump) Any device that produces a partial vacuum using the Venturi effect. *See also* FILTER PUMP; JET PUMP.

assembly drawing An engineering drawing that shows how parts are assembled to produce a component or a complete machine. It may include sections to show internal features, dimensions that are critical for assembly, manufacturing information, and part numbers.

assembly line A system of mass production in which work is moved progressively from one operation to another, ultimately to give the final complete product.

asymmetric rotor A rotor for which the axis of rotation is not coincident with the centroid of its cross section.

asymptotic approximation In Bode plotting, the approximation of the gain and phase plots as straight lines derived from the poles and zeros of the transfer function by making the substitution $s = i\omega$, where s is the Laplace transform variable, $i = \sqrt{-1}$ and ω is the angular frequency.

asynchronous control In a control system, a method of control where the time allocated to a particular activity is not fixed in advance but depends on the actual time taken. For example, a CNC machine tool operates asynchronously because each step in the control programme is taken after the previous step is completed rather than after a fixed time.

at *See* TECHNICAL ATMOSPHERE.

Atan2 The two-argument inverse tangent function used in robot inverse kinematic analysis. By using two arguments corresponding to x and y coordinates, the function determines the angle θ within the range $-\pi \leq \theta \leq \pi$ from the tangent. This is unlike the standard inverse tangent function, which can only determine angles in the range $-\pi/2 \leq \theta \leq \pi/2$.

athodyd *See* PULSE-JET ENGINE.

Atkins number (*At*) The non-dimensional number required to satisfy the same scaling law for inertia forces and cracking forces when modelling ice forces on structures or on vessels in ice-covered towing tanks. $At = V^2\rho\sqrt{a}/K_C$ where V is velocity, ρ is ice density, a is a characteristic length of crack, and K_C is the critical stress intensity for fracture of the ice. It is K_C in model and prototype ice that should be scaled, not ice 'strength'. At is a generalization of the Cauchy number Cn to include cracks in a continuum since $At = Cn^2\sqrt{Ea/R}$ where E is Young's modulus and R is the fracture toughness of ice.

Atkinson cycle An air-standard thermodynamic cycle for an internal combustion engine, theoretically more efficient than the Otto cycle, in which the expansion ratio v_6/v_3 exceeds the compression ratio v_1/v_3. As shown on the diagram of pressure (p) *vs* specific volume (v), it consists of six internally reversible processes: isentropic compression (1-2), isopycnic heat addition (2-3), isobaric heating (3-4), isentropic expansion (4-5), isopycnic heat rejection and isobaric cooling (6-1). This is an overexpanded cycle in which the gas in the cylinder is expanded to the exhaust pressure p_5. A number of hybrid vehicles use an Atkinson-cycle engine combined with an electric motor.

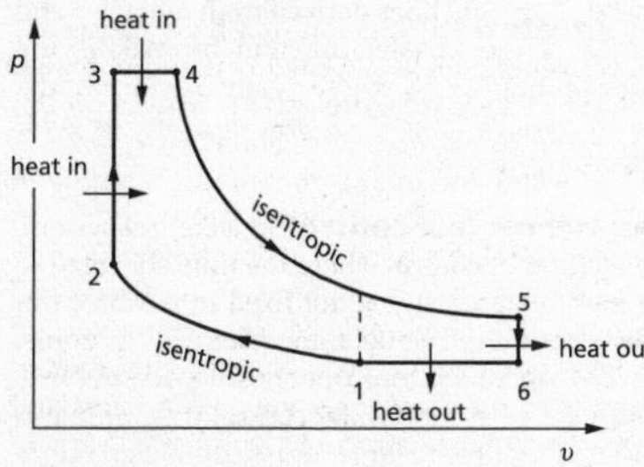

Atkinson cycle

SEE WEB LINKS

- Animation and explanation of the principles of the Atkinson engine

atm *See* ATMOSPHERE (2).

atmometer (atmidometer, evaporimeter) An instrument that measures the rate of evaporation of water from a surface into the atmosphere.

atmosphere 1. (atm) An obsolete (non-SI) unit of pressure equal to 101,325 Pa or 1.01325 bar and approximately equal to the **atmospheric pressure (barometric pressure)** measured at mean sea level. *See also* BAR. **2.** *See* STANDARD ATMOSPHERE.

atmospheric pollution *See* POLLUTANTS.

atomization The production of a spray of fine droplets from a liquid, such as diesel oil or petrol, by injection through a nozzle (an **atomizer**).

atomizing humidifier A humidifier that functions by spraying fine droplets of water into an airstream.

attemperation of steam *See* STEAM DESUPERHEATING.

attenuation The reduction of the amplitude and power of a signal or vibration. *See also* AMPLIFIER.

attenuator wave-energy device A wave-energy converter in which the principal axis is perpendicular to the incident wave fronts.

attraction gripper A robot end effector that picks up components using an attractive force such as magnetism or suction.

Atwood number (*A*) A non-dimensional number that arises in situations where a layer of fluid of high density ρ_H lies above one of lower density ρ_L. It is defined by $A = (\rho_H - \rho_L)/(\rho_H + \rho_L)$ The configuration leads to the Rayleigh–Taylor instability.

auger *See* SCREW CONVEYOR.

ausforming A thermomechanical treatment that produces cold-worked tempered-martensite microstructures in certain alloy steels that have extremely high strengths (of order 2.5 GPa UTS).

autoadaptivity The ability of a robot to respond to signals from one or more sensors and thus accommodate to changes in the environment.

autoclave 1. An oven containing gas (nitrogen, air, or carbon dioxide) at high temperature and pressure used in the bonding and curing of composite materials. **2.** An oven containing water vapour at high temperature (> 120°C) and pressure used for sterilization of medical equipment, glassware, waste, etc.

autoclave curing A means of producing fully-cured adhesives, polymers, composites, etc., in an autoclave.

autofrettage The shrinking of one thick-walled cylinder on to another so as to make more efficient use of material by inducing residual stresses that, when the component (a pressure vessel, large gun barrel, etc.) is under pressure, working stresses are distributed more evenly throughout.

autogyro (gyrocopter, gyroplane, rotaplane) A type of rotorcraft with an unpowered rotor in autorotation to develop lift, and an engine-powered propeller, usually in the pusher configuration, to provide thrust.

autoignition Premature ignition of the fuel in the combustion chamber of an internal-combustion engine in the absence of an external ignition source. It produces an audible noise called knock.

a

autoignition temperature The lowest temperature at which a substance will spontaneously ignite in air in the absence of a spark or flame. *See also* FLASH POINT.

automatic control (automatic regulation) The control of a plant using a control system.

automatic-control block diagram *See* BLOCK DIAGRAM (2).

automatic-control error coefficient *See* ERROR COEFFICIENT.

automatic-control frequency response *See* FREQUENCY RESPONSE.

automatic controller *See* CONTROL SYSTEM.

automatic-control servo valve *See* SERVO VALVE.

automatic-control stability *See* STABILITY.

automatic-control system *See* CONTROL SYSTEM.

automatic-control transient analysis *See* TRANSIENT ANALYSIS.

automatic indexing In robotics, the process of automatically determining the position and orientation of a component with respect to the robot base frame, so as to allow the component to be picked up or otherwise processed.

automatic regulation *See* AUTOMATIC CONTROL.

automation Mechanisms and systems that reduce or eliminate human labour; often applied specifically to manufacture and inspection on production lines.

automechanism A mechanical system or component that operates automatically, usually when a predetermined condition occurs: for example, a pump and a float switch that operates when a liquid rises above a pre-determined level.

autonomous energy system (stand-alone energy system) A sole source of electricity, usually small-scale, for applications remote from a grid, especially with energy storage in the system. Hydroelectric, photovoltaic, wind-power and other renewable systems are well suited to stand-alone applications.

autonomous robot A robot which can, by sensing information about its environment, work for a prolonged period of time, learn and adapt without human or other external intervention. For example, the NASA Mars Rover robot senses the surrounding terrain and can thus avoid such hazards as collision with boulders or falling over cliffs.

autorotation Rotation of a body caused by aerodynamic forces. In the operation of a helicopter, if there is no drive to the main rotor but it is free to rotate, autorotation is the state of flight where the rotor is turned by the action of air moving up through it and creating lift. An autogyro relies on autorotation to generate lift as the main rotor is unpowered. For fixed-wing aircraft, if the angle of attack exceeds the stalling angle, the situation is unstable and autorotation results in the aircraft spinning. Windmills and ram-air turbines also function on the basis of autorotation.

auxetic materials Those man-made materials for which Poisson's ratio is negative, so that the cross section expands when subjected to a longitudinal tensile stress and contracts when subjected to a longitudinal compressive stress.

auxiliary power unit (APU) In aircraft-gas-turbine applications, a small gas turbine used to provide start-up power, electrical and hydraulic power, and compressed air for cabin ventilation. In other applications, an APU may be a gas turbine or internal-combustion engine used to provide emergency power.

auxiliary rotor *See* ANTI-TORQUE ROTOR.

availability 1. (available energy) *See* EXERGY. **2.** The fraction of time for which a wind turbine is available to generate electrical power when the wind is blowing within its operating range.

available draught The reduced pressure of combustion gases in a furnace or boiler, either forced or due to the buoyancy of hot gases, which is used to draw in combustion air and remove products of combustion. *See also* INDUCED DRAUGHT; FORCED DRAUGHT; BALANCED DRAUGHT.

available head In a hydroelectric power system, the difference between the vertical height of the water level in the supply reservoir above the turbine inlet less the head loss due to friction and fittings in the duct leading to the turbine.

available resource (total resource) The total annual energy theoretically available from a renewable-energy source, such as ocean waves, the wind or the total incident solar energy. *See also* ACCESSIBLE RESOURCE.

average velocity *See* BULK VELOCITY.

averaging In instrumentation and control, improving the signal-to-noise ratio of the output of a sensor by taking a time average of a number of readings.

averaging pitot tube A cylindrical flow sensor which is installed spanwise across a pipe or duct in which there is flow. A number (typically 6) of upstream-facing holes in the cylinder communicate with an internal chamber to provide an average stagnation pressure which is sensed by one side of a differential pressure transducer. A single downstream hole is connected to the other side of the transducer to give an approximation to the static pressure. When calibrated, the difference between the pressures can be used to provide a measure of the flow rate through the duct. *See also* PITOT TUBE.

aviation fuel For gas-turbine engines, either unleaded kerosene (Jet A-1) or a naphtha-kerosene blend (Jet B) with higher calorific value (HCV), about 43,200 kJ/kg. For piston engines, a high-octane fuel called Avgas with HCV about 44,700 kJ/kg. *See also* CALORIFIC VALUE.

avionics Aviation electronics.

Avogadro constant (Avogadro number, N_A, L) The number of entities (photons, atoms or molecules) in one mol of substance. The 2006 CODATA value is $6.022114179 \times 10^{23}$.

SEE WEB LINKS

- Values of frequently used constants

Avogadro's law Equal volumes of two gases, at the same temperature and pressure, contain the same number of molecules.

avoirdupois weight Obsolete (non-SI) imperial units of weight based on the pound (lb) including the imperial long ton (2240 lb), the US short ton (2000 lb) and the ounce ($1/16^{th}$ lb).

axial engine *See* SWASHPLATE ENGINE.

axial fan (axial-flow fan) A fan with blades that cause a gas, usually air, to flow primarily parallel to the shaft around which the blades rotate.

axial-flow compressor A compressor with aerofoil-shaped rotor and stator blades which cause the working fluid to flow primarily parallel to the axis of rotation. *See also* COMPRESSOR BLADES.

axial-flow meter *See* TURBINE FLOW METER.

axial-flow pump A pump in which an impeller causes a liquid to flow primarily parallel to the pump shaft.

axial hydraulic thrust An unbalanced axial force caused mainly by the differences in pressure between the suction side and the discharge side of a pump impeller.

axial interference factor The ratio of the undisturbed relative airspeed ahead of a propeller or wind-turbine rotor to the relative airspeed behind the propeller or rotor.

axial load In general, a tensile or compressive load directed along the axis of a component. Strictly the load should pass through the centroid of the cross section to avoid inducing bending moments and be perpendicular to the plane of the section.

axial modulus *See* YOUNG'S MODULUS.

axial moment of inertia *See* POLAR SECOND MOMENT OF AREA.

axial pitch The pitch of a screw or gear measured along its axis. *See also* SCREW; TOOTHED GEARING.

axial plane A plane containing an axis of symmetry.

axial section A plane containing the axis of a gear, screw or shaft.

axial thrust Loading induced along the axis of a shaft by a propeller, fan blade, helical gearing, etc.

axial turbine (axial-flow turbine) A turbine in which the principal flow direction of the working fluid (gas, steam or water) is parallel to the axis of rotation of the rotor.

axial winding In the manufacture of filament-reinforced composite vessels, where the windings are mainly in an axial direction.

axis 1. *See* AXIS OF SYMMETRY. **2.** A line about which a body rotates or twists.

axis of freedom In a gyroscope, an axis about which a gimbal provides a degree of freedom.

axis of rotation The straight line about which a body, gear, screw, etc. rotates.

axis of symmetry (axis) The imaginary line about which an axisymmetric body is generated. In a body of uniform density, the centre of gravity lies on this axis.

axis of torsion (axis of twist) The line about which a body twists when subjected to a torque.

axle The cross-shaft that carries the wheels of a vehicle. In a live axle, the wheel is rigidly fixed thereto and power is transmitted; in a dead axle, the wheel turns on a stationary axle.

axlebox The assembly, including an oil reservoir, in which axles having plain bearings fit.

axle ratio The ratio of the rotation speed of a drive shaft to that of the driven shaft.

axle windup The torsional deflection of a shaft due to applied torque.

azeotrope A liquid mixture whose composition is the same as that of the vapour in equilibrium with it.

azimuthal velocity *See* TANGENTIAL VELOCITY.

Babbitt metal *See* BEARING.

backfire An explosion of fuel in a piston engine prior to closure of the inlet valve or of unburned fuel or hydrocarbon gases in the exhaust system.

background noise The unwanted electrical noise on the output of a sensor or instrument. Usually measured in terms of the signal-to-noise ratio in dB.

backing pump In vacuum systems, a pump that maintains a specified pressure at the outlet of a high-vacuum pump and exhausts to the atmosphere. *See also* ROUGHING PUMP.

backlash The amount of linear or rotary motion of a component of a mechanism before its motion is communicated to a second mating component (e.g. looseness or slop in gears). *See also* ANTI-BACKLASH GEAR.

back pressure **1.** In a piping system, the pressure that opposes fluid flow as a consequence of baffles, bends, valves, etc. **2.** The exhaust pressure for an internal-combustion engine.

back-pressure steam turbine A type of steam turbine used in industrial processes, such as district heating or drying, where there is a need for low- or medium-pressure (but usually above atmospheric) steam.

back-pressure valve A valve that maintains a set pressure at the outlet of a pump, for example when the system pressure is too low.

back solution *See* INVERSE KINEMATICS.

backup system Where high reliability is required from a system, such as in life-support and aerospace applications, redundancy can be provided by incorporating a second system in the design which takes over should the main system fail.

backward bent duct buoy (BBDB) A floating wave-energy conversion machine using an oscillating water column to drive an air turbine.

backward extrusion *See* EXTRUSION.

back work ratio In a gas turbine, the fraction of the turbine power needed to drive the compressor.

baffle (baffle plate) **1.** In a shell-and-tube heat exchanger, a plate used to support the tubes and control the direction of fluid flow. **2.** In a container used for transporting liquid (e.g. a fuel tanker), a plate, often perforated, installed to prevent sloshing when a partially filled container is accelerated or decelerated. **3.** In any flow system, a plate installed to condition fluid flow, including noise reduction. **4.** In acoustics, a partition that reduces interference between sound waves produced simultaneously by different parts of a loudspeaker.

baffle-type separator (vane-type separator) A steam separator containing a series of plates or baffles which removes water droplets from a steam–water mixture.

bagasse The crushed residue of sugar cane, which can be used as a fuel for boilers supplying steam to generate electricity and also provide process heat. Bagasse is rich in cellulose, which can be used to produce ethanol.

bag filter *See* DUST FILTER.

Bairstow number A non-dimensional group defined in the same way as, but now replaced by, Mach number.

balance **1.** A lever-arm device for measuring weight. **2.** The state in rotating components when all forces and moments are in equilibrium so that no vibration is produced. *See also* DYNAMIC BALANCING; STATIC BALANCING.

balanced draught *See* DRAUGHT (1).

balanced laminate In filament-reinforced composites made up of anisotropic layers, where the properties of the laminate are essentially the same in all directions.

b

balance piston (balance drum, dummy piston) A disc attached to the shaft of a turbine or compressor, to one side of which high or low pressure is applied to counteract the axial thrust produced by the pressure change across the machine. A form of thrust bearing.

balance shaft An eccentrically weighted rotating shaft that offsets vibrations in piston engines.

balance weight A corrective mass used in the static or dynamic balancing of a rotating object.

balancing *See* BALANCE (2).

balancing machine A device that assesses the state of static and dynamic balance of a rotating part, and indicates the magnitude and location of weights to be added to give balance.

Baldwin–Lomax model An eddy-viscosity model for the outer ('wake') region of a turbulent boundary layer.

ball-and-seat check valve (ball check valve) A check valve in which a ball is pressed against a circular seat by either a spring or its own weight.

ball-and-socket joint (ball joint) A joint in which one member having a ball on its end, fits into a socket on the end of another member, resulting in free motion within a cone.

ball bearing *See* BEARING.

ball bushing See BEARING.

ballistic limit The velocity of a given projectile that penetrates a given thickness of armour plate in more than 50% of firings.

ballistic pendulum A pendulum with a bob of large mass, swinging in one plane, into which a projectile such as a bullet is fired, the height of swing giving the velocity of the projectile through conservation of momentum.

ballistics The mechanics of projectiles, including the effects of air resistance, compressibility, shock waves, the Magnus effect, the Coriolis effect, and aerodynamic lift.

ball mill A mill for grinding and pulverizing materials, consisting of a horizontal rotating drum containing loose steel or ceramic balls.

ballonet An airbag used in a hybrid airship. Helium gas expels air from the ballonet during ascent. Fans draw air in during descent.

balloon A flexible bag filled with a gas, usually with a density less than that of air, e.g. helium or hydrogen. *See also* AEROSTAT.

ball race *See* BEARING.

ball screw and nut A nut and bolt assembly having semi-circular helical grooves, as opposed to threads, in which run ball bearings. On rotation of the nut, the balls move along the helix and carry the axial load. Balls reaching the end of the groove are recirculated back to the beginning. Such devices have low friction and very little backlash, and are used in some steering mechanisms.

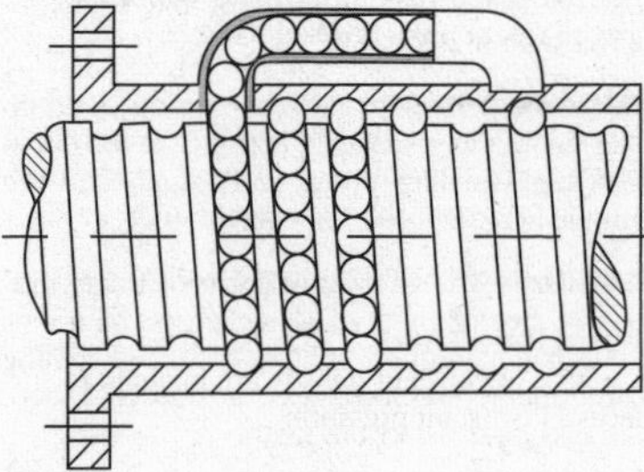

ball screw and nut

ball valve 1. A valve consisting of a ball with a hole through the centre, located within a spherical seat. There is full flow when the hole is aligned with the main flow direction, with progressively reduced flow as the ball is turned through 90° to the fully-shut position. **2.** A non-return valve (often spring-loaded), having a solid ball with limited movement located within a spherical seat. *See also* BALL CHECK VALVE.

band brake A brake consisting of a flexible band wrapped around the circumference of a wheel or drum. The band is anchored at one end and pulled against the wheel at the other. *See also* DYNAMOMETER.

band clutch A friction clutch in which drive is achieved by a band contracting on to the rim of the clutch.

band-pass filter A filter which blocks frequencies outside a specified range. A **high pass filter** passes signals at frequencies greater than a specified cut-off frequency. A **low-pass**

filter passes signals below a specified cut-off frequency.

bang-bang control (on-off control) A feedback-control method where the input to the plant switches between maximum and minimum without passing through intermediate states. For example, a domestic heating thermostat is a bang-bang controller because it switches the heater on at full power when the temperature is below the set point and switches it off when the temperature exceeds the set point.

Banki turbine (Michell turbine, Ossberger turbine) A cross-flow, impulse-type water turbine for very low heads in which a jet of water in the form of a flat sheet passes transversely through the turbine, so going through the runner twice. The thin runner blades which run horizontally across the turbine parallel to the axis of rotation are profiled in cross section.

bar A widely used, but non-SI, unit of pressure equal to 10^5 Pa (100 kPa or 0.1 MPa) and approximately equal to normal atmospheric pressure at sea level. *See also* ATMOSPHERE (2).

bar linkage *See* FOUR-BAR LINKAGE.

baroclinic fluid A stratified fluid in which surfaces of constant density are inclined to surfaces of constant pressure.

barometer An instrument used to measure atmospheric pressure.

barometric altimeter *See* PRESSURE ALTIMETER.

barometric pressure *See* ATMOSPHERE (1).

barostat A device which maintains constant pressure in a container, such as a balloon or a pressure vessel.

barotropic fluid A stratified fluid in which the density depends only on the pressure. Isobaric and isopycnic surfaces coincide, as there is a one-to-one correspondence between pressure and density.

barrage *See* TIDAL BARRAGE.

barrel 1. Any bored part of a component or machine, such as a cylinder. **2. (bbl)** A non-SI unit of volume used in the petroleum industry and equal to 42 US gallons or 159 litres.

barrel engine *See* SWASHPLATE ENGINE.

base circle *See* TOOTHED GEARING.

base cylinder *See* TOOTHED GEARING.

base diameter *See* TOOTHED GEARING.

base frame A Cartesian coordinate frame located at the base of a robot and fixed with respect to the task being performed by the robot.

base helix *See* TOOTHED GEARING.

base helix angle *See* TOOTHED GEARING.

base lead angle *See* TOOTHED GEARING.

base-load power The minimum level of demand from a power plant which must be available 24 hours a day, but can vary throughout any given day. Most base-load power is supplied by coal-fired, gas-fired, or nuclear power plants, but increasingly wind power, hydroelectric power, and other renewable-energy sources are being integrated into national grids. Peaks in demand (**peak-load power**) are satisfied by smaller, more responsive, but less efficient, plants typically powered by gas turbines.

base pitch *See* TOOTHED GEARING.

base pressure 1. In a vacuum system, the lowest pressure attainable at the vacuum pump inlet: typically 20 mTorr (*c.*3 Pa) for a good single-stage pump and 1 mTorr (*c.*133 mPa) for a two-stage pump. **2.** The static pressure on the downstream surface of a blunt body. **3.** Any pressure employed as a reference.

basic form of thread *See* SCREW.

basic size The nominal size of a part to which tolerances, limits, and fits are referenced.

Basquin's equation *See* FATIGUE.

bastard thread *See* SCREW.

BA thread *See* SCREW.

battery Two or more connected closed cells in which electrical energy is stored.

baulk ring *See* GEARBOX.

bayonet connector (bayonet fitting) A cylindrical male/female connector. The male part has two diametrically-opposed pins close to the connecting end, which engage with J-shaped slots close to the connecting end of the female part. Accidental disengagement is prevented by a compression spring within the female part that exerts an axial force on the male part. Applications include light-bulb fittings, camera lens mounts, and hose connections. *See also* GUILLEMIN COUPLING.

BBDB *See* BACKWARD BENT DUCT BUOY.

BDC *See* BOTTOM DEAD CENTRE.

bead height *See* WELDING.

beam A structural member the length of which is long compared with its cross-sectional dimensions. It is used to carry transverse and bending loads. *See also* H-BEAM; I-BEAM.

beam supports *See* FIXITY.

beam theory (beam relations) Small-deflexion elasticity relations that apply to the bending of beams. At any location along the beam $\sigma/y = M/I = E/r$ where σ is the tensile or compressive stress, y is the distance from the beam's neutral axis, M is the bending moment, I is the second moment of area of the cross section of the beam, E is Young's modulus, and r is the radius of curvature of the beam. *See also* FIBRE STRESS (2).

bearing A device that supports a component which rotates (a shaft), slides, or oscillates in or on it. The principal types are sliding bearings and rolling bearings. Sliding bearings can be designed to support either radial loading (**radial bearing**) or thrust loading (**thrust bearing**), while rolling bearings can support a combination of the two, as can an **angular-contact bearing.** A **bush** is a cylindrical sleeve forming a bearing surface for a shaft, a **ball bushing** being a ball bearing that permits axial movement of the shaft. A **collar bearing** resists the thrust of a shaft through collars. A **plain bearing (journal bearing)** consists of a hard-wearing, low-friction surface such as **Babbitt metal**, which is any one of a variety of 'white metal' **bearing alloys**. Such alloys may be tin-based (up to 92%) for heavily loaded bearings or lead-based (up to 85%) for less heavily loaded applications, and also contain copper (up to 11%) and antimony (up to 20%). The **journal** is that part of a shaft that turns in a bearing and is supported by it, and **journal friction** is the friction within a journal bearing resulting from resistance to shear of grease, oil or other lubricant. A **cone bearing** is a conical journal bearing running in a conical bush which acts as a combined journal and thrust bearing. In a **spherical plain bearing** the inner component has a spherical outer surface and the outer component a spherical inner surface of slightly larger diameter to allow for lubricant between the two surfaces. In a **hydrostatic bearing,** high-pressure oil, supplied by a pump, forms a lubricating film of uniform thickness between shaft and journal. A term implying very low friction applied to plain bearings and the materials from which they are manufactured is **antifriction bearing (antifriction material).** The bearing metal may be held in a **plummer block,** a split casting. A cylindrical bearing consisting of two semi-circular plain bearings is a **shell bearing**, a common application being to connecting rods, crankshafts and gudgeon pins for piston engines. A journal bearing with an arc length less than 360° is termed a **partial bearing.** A **sleeve bearing** is a plain bearing in which a shaft, or journal, rotates or oscillates within a sleeve. The relative motion is sliding. Lubricant may be introduced into the narrow gap between the sleeve and the shaft, or the sleeve may consist of a low-friction material such as Teflon and be self-lubricating. **Side leakage** is the flow of lubricant out of the ends of a journal bearing. A **porous bearing** is a self-lubricated bearing in which the bearing surface is made from sintered metal powder and impregnated with lubricant. The projected area of a shaft in its bearing is termed the **bearing area (bearing surface)**; the **bearing stress (bearing pressure)** is then the radial load carried by a bearing divided by the bearing area. The **load rating** is the radial load a bearing can support at a given rotation speed and length of time. For a journal bearing, an approximate equation (**Petroff's law**) for the coefficient of friction f, defined by $f = T/Wr$, T being the torque acting on the rotating shaft, W a small radial load acting on the bearing and r the shaft radius, is $f = 2\pi^2\mu Nr/Pc$ where μ is the dynamic viscosity of the lubricating fluid, N is the shaft rotation speed (rev/s), c is the clearance and P is the radial load per unit projected area given by $P = W/2rL$, where L is the bearing length. Analysis of the load-carrying capacity of bearings (**bearing theory**) ranges from simple **dry-friction models** to elastohydrodynamic theories of **full fluid-film lubrication**. The **Sommerfeld number (bearing characteristic number, *S*)** is a non-dimensional parameter which characterizes the load capacity of a journal bearing. It is defined by $S = (r/c)^2\mu N/P$ where r is the journal radius, c is the radial clearance, μ is the dynamic viscosity of the lubricating oil, N is the rotation speed of the journal and P is the load per unit of projected bearing area. There is a minimum in the plot of the bearing friction factor *vs* S. At high values of S, which corresponds to high rotation speeds and light loading, the plot follows Petroff's Couette-flow prediction, whereas at low values of S, as the average lubricant film thickness decreases, the friction factor rises progressively. There is thus a critical value of S, for any given bearing, which distinguishes between full-film and boundary lubrication.

A **gas bearing** is a journal or thrust bearing in which the lubricant is a pressurized gas, typically air (**air bearing**), fed into the bearing through an orifice or pores.

A **ball bearing** consists of hardened-steel balls, separated by a metal cage, rolling between inner and outer rings (called **ball races**). The inner race is forced onto a shaft while the outer race is carried in a housing. In a **split bearing** the outer race, the ball cage or bushing, and the bearing housing are split diagonally and bolted together during installation into a machine. A ball bearing that permits axial movement of the shaft is termed a **ball bushing**. A **shielded (ball) bearing** has a close-fitting shield that allows grease to be drawn in as the bearing cage assembly rotates. The grease is then spread into the outer race by centrifugal action. **Self-aligning ball bearings** have the inner ring and ball assembly contained within an outer ring with a spherical raceway to allow for angular misalignment resulting from deflection or improper installation. A **double-row bearing** has two parallel races, each containing a full complement of balls. A **roller bearing (rolling-contact bearing)** is similar to a ball bearing but with cylindrical rollers running between inner and outer races. A **needle-roller bearing** has rollers of very small diameter. In a **linear roller bearing,** used between flat surfaces in relative motion, the rollers recirculate in a caterpillar action. A **spherical roller thrust bearing** is a thrust bearing in which load is transmitted from one raceway to the other through barrel-shaped rollers, angled with respect to the bearing axis, which roll in a spherical outer race. In a **taper roller bearing,** which is designed to take axial thrust as well as normal journal loads, the rollers are tapered and the races conical. **Rating life** is the number of revolutions, or hours at a given constant speed, that 90% of a group of apparently identical ball or roller bearings will complete or exceed before failure.

For an individual bearing, the **bearing life (*L*)** is the total number of revolutions, or the number of hours at a constant rotational speed, for **bearing (fatigue) failure** to develop as revealed by spalling of the load-carrying surfaces. Increased fatigue life and decreased shaft slope at the bearing can be achieved by removal of the internal clearance in a ball bearing (**bearing preloading**) either by expanding the inner ring or shrinking the outer ring. For a rolling contact bearing subjected to a load F, $L = (C/F)^a$ where C is the bearing load rating; $a = 3$ for ball bearings and 10/3 for roller bearings. The **bearing load rating (basic dynamic capacity, specific dynamic capacity, *C*)** for a rolling-contact bearing is the constant radial load which a group of apparently identical bearings can endure for a rating life of one million revolutions of the inner ring. Surface damage due to lubrication failure in bearings is termed **scoring**. Another common cause of failure in roller bearings is damage to a surface in which minute permanent indentations are formed due to contact stresses (brinelling).

SEE WEB LINKS

- Bearing types and associated terms

bearing area (bearing surface) The projected area of a hole, such as a rivet hole, that carries a transverse load. *See also* BEARING.

bearing failure 1. A failure that occurs in a riveted or bolted joint when the transverse load divided by the bearing area results in a stress that leads to permanent plastic deformation. **2.** *See* BEARING.

beat The slow amplitude modulation caused by the interference of two single-frequency signals when the frequency difference is small. For signals with frequencies f_1 and f_2, the beat frequency is $(f_1 - f_2)$.

Beattie-Bridgeman equation An empirical equation of state for gases relating pressure p, absolute temperature T and $\bar{v}$ where $\bar{v} = Mv$, M being the molecular weight and v the specific volume. It can be written as

$$p = \frac{\mathcal{R}T}{\bar{v}^2}\left(1 - \frac{c}{\bar{v}T^3}\right)(\bar{v} + B) - \frac{A}{\bar{v}^2}$$

where $\mathcal{R}$ is the universal gas constant, $A = A_0(1 - a\bar{v})$ and $B = B_0(1 - b\bar{v})$. The constants a, b, c, A_0 and B_0 are listed for a range of gases, including air, argon, carbon dioxide, helium, hydrogen, nitrogen and oxygen. *See also* KEYES EQUATION.

Beaufort wind-force scale (Beaufort scale) An empirical scale of wind speed linked to observations of sea conditions. The scale ranges from 0 (wind speed less than 0.3 m/s, calm sea) to 12 (wind speed above 32.7 m/s, hurricane conditions).

Beer's law *See* MONOCHROMATIC ABSORPTION COEFFICIENT.

b

Bell–Coleman cycle *See* GAS REFRIGERATION CYCLE.

bell crank An L-shaped lever that rotates about a pivot situated where the two arms of the L meet. The angle between the arms can take any value from 0 to 360°, but 90° and 180° are common. Motion imposed at the end of one arm produces motion at the end of the other arm. Applications include aircraft control systems.

Belleville washer *See* DISC SPRING.

bell mouth A contracting inlet to a pipe or duct having the shape of a bell.

bellows 1. A device for forcing air through a nozzle. **2.** Hollow flexible structures having accordion-like walls employed as (a) expansion joints (**bellows expansion joints**) in a pipeline that accommodate expansion and contraction, thus limiting overall movement of the pipeline; (b) energy absorbing structures; (c) a sealed aneroid device that expands or contracts with changes in outside pressure.

belt drive Transmission of motion from one shaft to another by means of a continuous plain, or toothed, flexible band (**belt**) passing over pulleys. In contrast to chain drives, belt drives tend to be employed in low-torque applications. Reduction in transmissible power may occur due to stretch of a plain transmission belt which results in slack in the drive (**belt creep**) or slip of a belt on a driving or driven pulley (**belt slip**). *See also* V-BELT.

Beltrami criterion *See* FAILURE.

Bénard convection (Rayleigh–Bénard convection) A form of natural convection arising as a consequence of thermocapillary instability, in which horizontal counter-rotating vortices occur in a thin layer of liquid heated from below. *See also* MARANGONI CONVECTION.

Bénard–Marangoni convection *See* MARANGONI CONVECTION.

bendability The ease with which material may be bent without fracture. For ductile sheet metals of thickness t, $R/t \approx (1/2A_r - 1)$ where R is the radius of the inside of the sharpest achievable bend and A_r is the fractional value of the reduction in area at fracture in a tensile test.

bending The act of producing curvature in a component: elastic (recoverable) bending occurs in springs; plastic (permanent) bending occurs in sheet-metal forming.

bending moment (BM) The algebraic sum of the moments of all forces acting at a point in a body.

bending-moment diagram (moment diagram) The variation, plotted as the ordinate, of bending moment along the length of a loaded beam. The diagram shows a simply-supported beam of length l carrying three loads P_1, P_2, and P_3 spaced at distances a_1, a_2, and a_3 along its length measured from the left-hand end. For $x < a_1$, the shear force $V = R_1$ and the bending moment $M = R_1x$. For $a_1 < x < a_2$, $V = R_1 - P_1$ and $M = R_1x - P_1(x - a_1)$. Similarly for the other portions of the beam, from which the shear-force diagram and bending-moment diagrams may be drawn.

bending stress The tensile and compressive stresses developed in a component such as a beam as a result of the strains set up by the curvature induced by external loading.

Bendix drive A drive widely used to transmit starter-motor torque to a piston engine. It comprises a pinion carried by a **Bendix screw**, a coarse helical thread on a shaft, that slides axially to engage a ring gear when the screw rotates. The pinion slides out of engagement when the engine starts.

bent-axis pump A pump with a rotating cylinder block containing parallel pistons arranged radially around the cylinder centre line. The pistons are attached mechanically to a circular plate rotated by a shaft at an angle to the cylinder axis. The rotation causes the pistons to reciprocate within the cylinder. *See also* SWASHPLATE PUMP.

Berkovich indenter A hardness indenter employed particularly in nanoindentation, since its three faces should intersect in a point and thus avoid problems of bluntness, although in practice there will be a tip radius of some 50 nm.

Bernoulli obstruction flow meter (head meter, variable-head flow meter) A standardized in-line flow meter used to measure flow rates for fluid flow through a circular pipe by introducing a local reduction in cross sectional area that accelerates the fluid

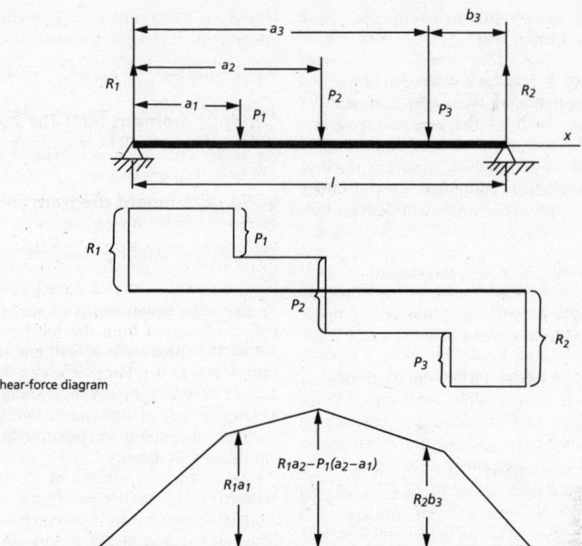

bending-moment diagram

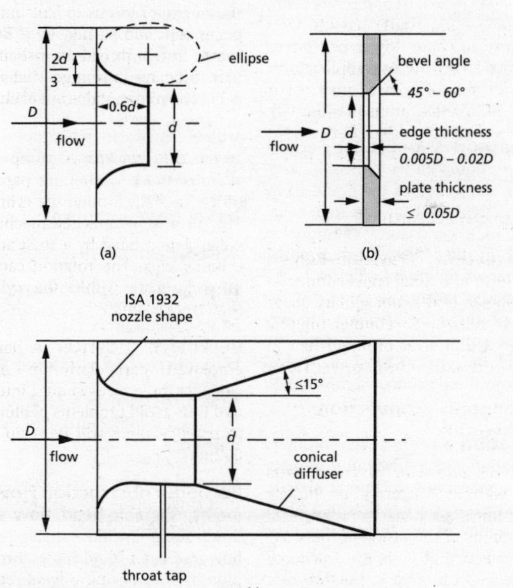

Bernoulli obstruction flow meter

b

to produce a static-pressure drop Δp. There are three principal types: (a) the flow nozzle, (b) the thin-plate orifice and (c) the Venturi nozzle. In each case Δp is measured between a location upstream of the meter and another downstream, such as the location of lowest pressure. Critical dimensions, tolerances and the location of pressure tappings are given in ISO 5167-1:2003. Although an estimate for the flow rate can be calculated using Bernoulli's equation, an accurate value requires calibration.

Bernoulli's equation In its most general form, an equation derived from the momentum and mass-conservation equations for the unsteady flow of an inviscid fluid along a streamline. It may be written in the following form:

$$\int_0 \frac{\partial V}{\partial t} ds + \int_0 \frac{dp}{\rho} + \frac{1}{2}V^2 + gz = constant$$

(**Bernoulli's constant**), where V is the fluid velocity, t is time, s is the distance measured along a streamline, p is the static pressure, ρ is the fluid density, g is the acceleration due to gravity, and z is the vertical height of the streamline above a horizontal datum plane. For the steady flow of a constant-density fluid along a streamline, the equation simplifies to

$$p + \frac{1}{2}\rho V^2 + \rho g z = constant$$

See also EULER'S EQUATION.

Berthelot method A method for studying the behaviour of a liquid under hydrostatic tension. A cylinder almost full of the liquid to be tested is sealed at both ends and heated, causing the liquid to expand until the tube is completely full at the 'filling' temperature. When the tube cools the liquid adheres to the cylinder walls and a tension is set up in the liquid until it ruptures at the 'breaking' temperature. The pressure immediately before the break is termed the critical tension. Typical values for water in the presence of glass are 30 to 50 bar.

Bessel's equation A second-order ordinary differential equation which arises in a number of engineering applications, including axisymmetric heat conduction, starting flow in a circular pipe and vibration of thin membranes. The equation may be written as

$$x^2\frac{d^2y}{dx^2} + x\frac{dy}{dx} + (x^2 - \upsilon^2)y = 0$$

where υ is a real constant. The solution is given in terms of Bessel functions of order υ.

Best number *See* ARCHIMEDES NUMBER.

beta ratio (β_D) A measure of the efficiency of a fluid filter to remove particles greater than a particular diameter D and defined as the number of such particles per unit volume upstream of the filter to the number downstream. The filter efficiency is $100\,(1 - 1/\beta_D)$.

beta-type Stirling engine *See* STIRLING ENGINE.

Betti's theorem *See* RECIPROCAL THEOREM.

Betz actuator-disc model (Betz momentum theory) A theory of wind-turbine performance based on the deceleration of air passing through the disc swept out by the blades of the turbine. According to **Betz criterion (Betz limit)**, the fraction of power extracted (the power coefficient) is given by $C_P = 4a(1 - a)^2$ where a is the interference factor, i.e. the fractional decrease in wind speed across the blades. The maximum value of $C_P = 0.59$, for $a = 1/3$, is known as Betz criterion.

Betz manometer (micromanometer) An instrument (resolution 0.1 mm of water) for accurately measuring small pressure differences. The basis of the manometer is a vertical glass tube, the lower end of which is submerged in a reservoir containing distilled water.

bevel The angle (when not a right angle) between adjacent lines, surfaces, shafts, etc. *See also* CHAMFER.

bevel gear *See* TOOTHED GEARING.

BHN *See* BRINELL HARDNESS NUMBER.

Bhp *See* BRAKE HORSEPOWER.

bias *See* ERROR.

bias-corrected variance *See* VARIANCE.

bias pressure In a fluidic device controlled by pressure difference, the magnitude of that difference.

biaxial stress Generalized loading of a body in a single plane with no loading normal to it. *See also* PLANE STRESS.

biaxial winding The helical filament winding of a composite cylinder with two sets of filaments. The helix angles are equal and opposite so as to give the same torsional stiffness in both directions. In cylinders used as pressure vessels, the helix angle is about ±55° with respect to the axis of the cylinder, which gives the same 2:1 hoop:axial stress ratio that membrane theory predicts for isotropic thin cylinders, thus giving balanced strength.

Bibby coupling A device for transmitting torque between shafts via flat springs mounted between curved surfaces. Owing to the non-linearity of the springs, the natural frequency of one element relative to the other varies with the amplitude, so that all the shock-absorbing properties of a spring coupling are maintained without the danger of setting up resonant torsional oscillations.

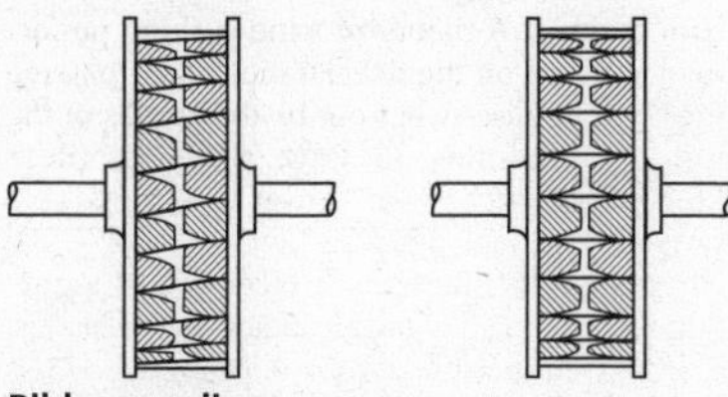

Bibby coupling

BIBO stability *See* BOUNDED INPUT BOUNDED OUTPUT STABILITY.

bidirectional laminate A reinforced composite made up of a series of unidirectional laminae orientated in two directions (often 90°).

bifilar micrometer *See* FILAR MICROMETER.

bifilar suspension The suspension of a lamina or rod-like body by two fine vertical wires of equal length. The polar moment of inertia of the body about a vertical axis passing through its centre of mass can be determined by measuring the frequency of small rotational oscillations about the axis. *See also* TRIFILAR SUSPENSION.

big end *See* CONNECTING ROD.

biharmonic equation For two-dimensional Stokes (or creeping) flow, the vorticity equation may be transformed to a biharmonic equation for the stream function ψ: $\nabla^4\psi = 0$. *See also* AIRY STRESS FUNCTION.

bilateral tolerance *See* TOLERANCES.

billet A piece of metal ready for forming, for example by forging or extruding.

bimetallic strip A strip formed by welding, riveting or brazing together two metals having different coefficients of expansion, which causes the strip to curl when its temperature changes. Typical combinations are steel and copper or steel and brass.

binary-cycle power plant *See* ORGANIC RANKINE CYCLE PLANT.

binary diffusion coefficient (*D*) For a mixture of two species A and B (**binary mixture**), such as two gases, the mass diffusion coefficients of A and B, D_{AB} and D_{BA}, are equal and defined by Fick's law of diffusion.

binary vapour cycle A power cycle which combines a steam cycle at relatively low temperature (the bottoming cycle) with a higher temperature cycle (the topping cycle) in which a working fluid such as mercury, sodium or potassium is used. The condenser of the topping cycle serves as the boiler for the bottoming cycle and the boiler for the topping cycle as the superheater for the steam cycle. The two thermodynamic cycles are shown on a temperature (T) *vs* specific entropy (s) plot. *See also* ORGANIC RANKINE CYCLE PLANT.

binder The metal matrix (e.g. cobalt) of sintered tungsten, titanium carbides etc., used for cutting tools and dies.

Bingham number *See* YIELD-STRESS FLUID.

Bingham plastic *See* YIELD-STRESS FLUID.

biocontrol system A control system where the input is derived from a biological system. For example, an arm or leg prosthesis controlled by muscle or nerve voltage signals.

biodiesel A substitute for diesel fuel derived from the oily seeds of sunflowers, oilseed rape, soya beans, etc. *See also* BIOFUEL; DIESOHOL.

bioenergy **1.** Energy derived from materials such as purpose-grown energy crops, including sugar cane, maize, wheat, and rice, as well as wood, straw, and animal waste, including sewage, manure, and animal litter. *See also* BAGASSE. **2.** A term sometimes used to cover biomass and biofuels together.

bioengineering (biological engineering) The application of engineering principles to biology, medicine, agriculture, etc. *See also* BIOMEDICAL ENGINEERING.

biomass The material of plants and animals, including their wastes and residues. It is organ-

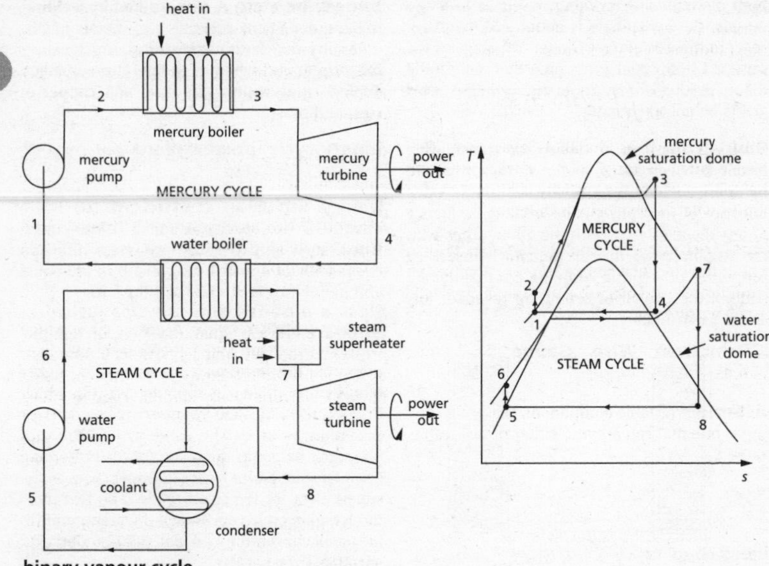

binary vapour cycle

ic, carbon-based material that reacts with oxygen in combustion and natural processes to release heat. **Biofuel** is biomass transformed by chemical and biological processes into fuels such as methane gas, liquid ethyl alcohol, methyl esters, oils (**bio oil**) and charcoal. **Biogas** is a mixture of methane and carbon dioxide produced from biomass, manure, sewage, municipal waste, plant material, and energy crops by anaerobic digestion or gasification of wood and other biomass. A **biomass gasifier** produces a gaseous mixture of hydrogen, carbon monoxide, carbon dioxide, and some methane by applying heat under pressure to biomass in the presence of steam and air.

biomass integrated gasification combined cycle A combined-cycle gas-turbine system in which the fuel for the gas turbine is the gaseous output from a biomass gasifier.

biomechanics The application of mechanical-engineering principles, including thermodynamics, fluid mechanics, and solid mechanics together with materials engineering, to biological systems.

biomedical engineering The application of engineering principles and methodology to the medical field. Examples involving mechanical engineering include the design and construction of artificial limbs and hearts, heart-lung machines, prosthetic eyes, and orthopaedic implants such as hip joints and pins to stabilize fractured bones. *See also* BIOENGINEERING.

biomimetics The practice of drawing inspiration from nature in order to make novel materials, instruments, sensors, etc.

biotechnical robot (teleoperator) A robot that requires continuous input from a human operator. Such robots are used in nuclear plant decommissioning and similar hazardous tasks.

biotechnology The United Nations Convention on Biological Diversity (1992) defines biotechnology as 'Any technological application that uses biological systems, living organisms, or derivatives thereof, to make or modify products or processes for specific use.' The term includes genetic engineering, cell- and tissue-culture technologies.

Biot–Fourier equation *See* HEAT TRANSFER.

Biot modulus *See* HEAT TRANSFER.

Biot–Savart law In fluid mechanics, if the vorticity distribution $\Omega(r',t)$ is known throughout a volume of fluid $\mathcal{V}$, the velocity $\boldsymbol{u}_{\mathcal{V}}$ induced by the vorticity can be calculated by integration using the equation

$$\boldsymbol{u}_v(\boldsymbol{r},t) = \frac{1}{4\pi}\iiint_v \xi^{-3}\boldsymbol{\Omega}(\boldsymbol{r}',t)\times\boldsymbol{\xi}d\mathcal{V}(\boldsymbol{r}')$$

which is known as the Biot–Savart law. The displacement vector $\boldsymbol{\xi} = \boldsymbol{r} - \boldsymbol{r}'$ extends from the point $\boldsymbol{r}'$ where the vorticity is located to the point $\boldsymbol{r}$ where the velocity is induced.

bipolar drive A method of switching the current in the coils of a stepper motor such that at some times two coils are simultaneously energised. *See also* UNIPOLAR DRIVE.

Birmingham Wire Gauge *See* WIRE GAUGES.

bistable A bistable component, flow, or structure is one that can assume either of two stable states.

bite In the rolling of materials, the maximum plate or sheet thickness Δh that can enter the rolls unaided (i.e. be gripped and taken in by the rolls) is given by $\Delta h = \mu^2 R$ where μ is the coefficient of friction and R is the roll radius.

blackbody (black body, ideal radiator) An ideal body that is a perfect emitter and absorber of radiation. It is a diffuse emitter that absorbs all incident radiation regardless of wavelength and direction. The rate of radiant energy emitted from a blackbody at some fixed temperature (**blackbody radiation**) is expressed by the Stefan–Boltzmann law, with the spectral-energy distribution given by Planck's function. **Blackbody temperature** is the temperature at which a blackbody emits the identical radiation heat flux as a given body.

black carbon *See* SOOT.

black start The process of restoring a shut-down power station to operation without relying on external energy sources. The required power may be derived from batteries or small diesel generators.

blade A thin component as in (a) a knife or cutting tool, (b) the vane of a fan or propeller, or the aerofoils of a turbine or compressor, or (c) the curved spreader of a bulldozer.

blade angle (blade pitch angle, stagger angle, ξ) In a row of turbine or compressor blades, the angle between the chord line of a blade and either the axial direction or a meridional reference plane. Some definitions incorporate the angle of attack.

blank A piece of material (often in sheet form) ready for deep drawing or other metal-forming operations.

blank flange A disc between two mating flanges that isolates part of a piping system.

blanking The punching of holes in metal sheet or plate where, in contrast to piercing, it is the part punched out (the blank) that is required.

Blasius pipe-friction law (Blasius equation) An empirical equation for calculating the pressure drop per unit length $\Delta p/L$ for fully-developed turbulent flow of a Newtonian fluid, through a hydraulically smooth circular pipe. The equation may be written as $f = 0.3164\, Re^{-0.25}$ where f is the Darcy friction factor, and Re is the Reynolds number, defined here in terms of the pipe diameter and the bulk flow velocity. The equation is valid for Mach numbers up to about 0.3 and Re in the range 4 x 10^3 to 1 x 10^5. *See also* COLEBROOK EQUATION; MOODY CHART; POISEUILLE FLOW.

Blasius problem The analysis of the longitudinal development of a laminar boundary layer for flow over a flat surface with constant free-stream velocity of a fluid of constant density and viscosity. Blasius recognized that the partial differential equations governing the problem admitted a similarity solution which then reduced them to a single third-order ordinary differential equation, which can be written as $f''' + \frac{1}{2}ff'' = 0$ where $f(\eta) \equiv \psi/\sqrt{\nu x U_\infty}$ and the similarity variable $\eta \equiv y/\sqrt{\nu x/U_\infty}$.

$$\psi = \int_0^y u\,dy$$

is the stream function, u being the velocity parallel to the surface a distance y from the surface. U_∞ is the free-stream velocity (i.e. the value of u once the effect of viscosity is negligible), ν is the kinematic viscosity of the fluid, and x is the streamwise distance from the start of the boundary layer. The boundary conditions are the no-slip condition $f'(0) = 0$, $f(0) = 0$ and $f'(\eta\rightarrow\infty) = 1$. The differential equation has no exact analytical solution (**Blasius solution**) and has to be solved numerically. The most important part of the solution is $f''(0) = 0.332$ from which the surface shear stress τ_S can be calculated since $\tau_S = \rho\nu(\partial u/\partial y|_S) = \rho\sqrt{\nu U_\infty^3/x}f''(0)$. *See also* FALKNER-SKAN SOLUTIONS.

b

blast freezer A freezer in which items are frozen rapidly by passage of very cold air at high speed.

blast furnace A large steel stack lined with refractory brick in which iron ore, coke, and limestone are dumped into the top while pre-heated air is blown in at the bottom. Iron oxides in the ore are reduced chemically to produce molten iron and liquid slag.

blast heater A heater in which high-speed air is forced to pass through a heat exchanger into a volume which is to be heated quickly.

blast wave (blast) The approximately spherical shock wave which results from the sudden release into a gas, such as the atmosphere, of high-density energy (i.e. the energy flow per unit volume is very large, as in an explosion).

bleeding 1. The process of releasing gas trapped in a hydraulic system or venting excess liquid or gas from a pipeline to the atmosphere through a **bleed valve**. **2.** The process of extracting steam between the stages of a multi-stage steam turbine to be used in feedwater heating or other processes. *See also* COMBINED HEAT AND POWER PLANT; REGENERATIVE RANKINE CYCLE.

blind hole A drilled hole that does not pass completely through a body. *See also* PILOT HOLE.

blind rivet A rivet that enables a connexion to be made from one side only of an assembly. *See also* POP RIVET.

blisk A turbomachine rotor comprising both the rotor blades and a disc as a single component. A **bling** is a development in which the bore is replaced by a fibre-reinforced ring.

block *See* CYLINDER BLOCK.

block and tackle A lifting device consisting of two or more pulley blocks (sheaves), often arranged vertically, and a length of rope or cable passing over each pulley in turn. The mechanical advantage is proportional to the number of pulleys.

block brake A vehicle brake in which a curved block of friction material or metal is forced against the rim of a rotating wheel.

block diagram 1. (flow chart, flow diagram) A graphical representation using standard symbols, such as rectangles, circles, triangles, and diamonds, to represent important steps in a process or components in a complex machine or system, such as a power plant. The technique may be applied to a wide range of processes including major engineering projects, manufacturing processes, chemical-production processes, and computer programs. The relations between the components are shown by connecting lines. **2.** A diagram that allows the overall behaviour of a control system to be determined from the behaviour of each of its elements or sub-systems, and a knowledge of the connections along which information flows between them. In such a diagram, each element is represented by its transfer function, that is, the Laplace transform of the ratio of the element output to the element input. For example, a simple closed-loop system with negative feedback having an input *X(s)*, an output *Y(s)*, a forward transfer function *G(s)* and a sensor transfer function *H(s)* can be represented as shown in the block diagram. By constructing this block diagram it can be seen that the overall system transfer function *Y(s)/X(s)* is given from the element transfer functions by

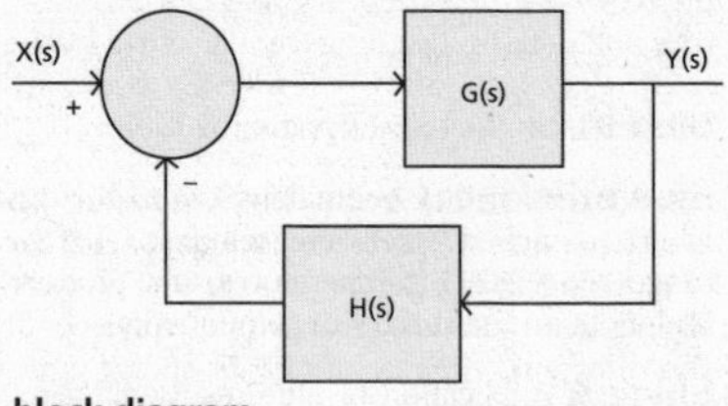

block diagram

$$\frac{Y(s)}{X(s)} = \frac{G(s)}{1 + G(s)H(s)}$$

blow back The sudden reversal of air flow through a carburettor, typically due to incorrect timing.

blow-by Leakage past the pistons of a piston engine during the compression stroke.

blow-down wind tunnel A type of transient-flow wind tunnel, used for high subsonic and supersonic flow conditions, in which a large pressurized vessel, containing air or an inert gas such as nitrogen, is connected to a low-pressure vessel, typically evacuated. The test section and other wind-tunnel components separate the two. A fast-opening valve or diaphragm initiates the flow, which is regarded as quasi-steady.

blower 1. *See* FAN. **2.** *See* EJECTOR.

blowing The process by which a gas is transferred through a porous surface over which there is a gas flow. The effect is used to decrease skin friction and increase heat transfer, for example in turbine-blade cooling. *See also* TRANSPIRATION.

blow moulding The manufacture of hollow polymer objects (e.g. bottles) by expanding, with internal air pressure, a hollow tube (**parison**) against the walls of a cavity mould.

blown flap A system in which either engine air or air from the pressure surface is blown tangentially over the suction surface of a trailing-edge flap on an aircraft wing to enhance lift.

blow out The potentially catastrophic result of the uncontrolled release of high-pressure gas when drilling for oil. The gas may be toxic (such as hydrogen sulphide) or explosive (such as methane). A **blow-out preventer (BOP)** is a system of valves, installed at the point where an oil well reaches the earth's surface or the sea bed, designed to control the gas release and so prevent a blow out.

blowtorch *See* OXYACETYLENE TORCH.

blue brittleness Anomalous loss of ductility when quenched steels are tempered in the range 250–350°C (temperatures that produce blue tints on the surface of components).

blueing Application of blue dye ('engineers' blue') to identify high spots on surfaces in contact or for marking out.

bluff body (blunt body) An unstreamlined body, often with sharp corners at the front and rear, having a high drag coefficient (typically > 1).

BM *See* BENDING MOMENT.

bmep *See* BRAKE MEAN-EFFECTIVE PRESSURE.

Bode plot (Bode diagram) The representation of the behaviour of a dynamic system by two graphs, one showing the gain of the system in decibel and the other showing the phase shift in degrees, both functions of frequency on a logarithmic scale. The plots can be measured, calculated precisely or approximated from the poles and zeros of the transfer function using straight lines, known as **Bode asymptotic approximations**, by substituting for the Laplace transform variable $s = i\omega$, where $i = \sqrt{-1}$. For example, a low pass filter with a cut-off frequency of 1000 rad/s will have the gain and phase plots shown in the figure. Here the asymptotic (straight line) approximations to the precise curves are shown as chain lines (- - -). The frequencies at which there is a change of slope of the gain or phase asymptotic approximations are known as **break frequencies** or **break points.**

The Bode plot can be used to investigate the stability of a system. The **Bode gain margin** is the amount by which the open-loop gain of the system expressed in dB, at the frequency at which the phase shift is 180°, (the **phase crossover frequency**), would have to be increased to reach zero. A high gain margin indicates that a system is further from instability than if the gain margin is lower. A gain margin of zero indicates instability or marginal stability. For example, consider a system with an open-loop transfer function *G(s)* of:

$$G(s) = \frac{1}{(s+1)(s-1)}$$

The Bode gain and phase plots for this system are shown in the diagram below.

From the gain plot, the gain is 0 dB up to a frequency of approximately 10^{-1} rad/s. As the phase shift is 180° over all frequencies, the gain margin is zero. Hence the system will be unstable or marginally stable.

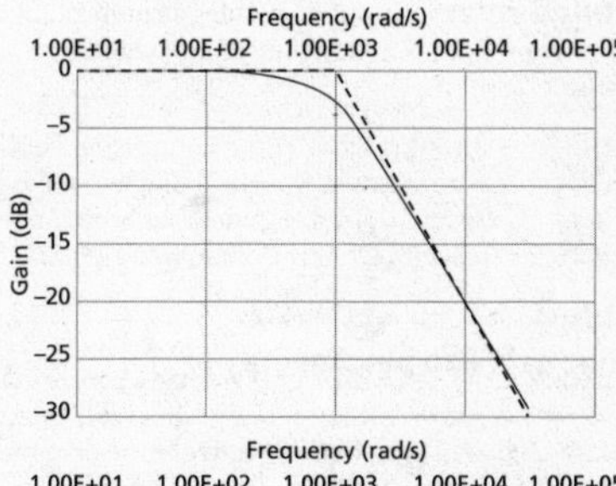

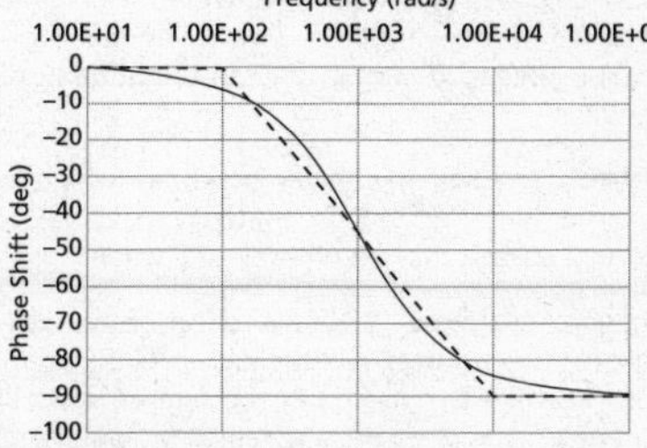

Bode plot for low-pass filter

The stability can also be investigated through the **Bode phase margin.** This is the amount by which the phase shift of the system, at the frequency at which the open-loop gain is 0 dB (the **gain crossover frequency**), would have to decrease to reach −180°. Hence, where θ is the phase shift at the gain crossover frequency, the phase margin is $\theta + 180°$. A high phase margin indicates that a system is further from instability than if the phase margin is lower. A phase margin of zero indicates either instability or marginal stability. For example, consider a system with an open loop transfer function *G(s)* of:

$$G(s) = \frac{1}{s(s+1)}$$

The Bode gain and phase plots for this system are shown in the Bode phase margin diagram.

From the gain plot the gain crossover frequency is 0.786 rad/s and plotting this frequency on the phase plot shows that the phase shift at this frequency is $\theta = -128°$. The phase margin is thus $180° + \theta = 52°$ and the Bode phase margin thus shows this system to be stable.

body centrode The locus of the instantaneous centre of a rotating body with respect to the motion of the body.

body force A force that acts throughout the volume of a body due to gravity (i.e. its weight), magnetism, Coriolis acceleration, centripetal acceleration, etc.

body rotation The waist-axis rotation in an articulated robot.

boiler *See* STEAM BOILER.

boiler capacity (boiler size) (Unit kg/h) The rate at which a boiler can generate steam. *See also* EQUIVALENT EVAPORATION.

boiler economizer *See* ECONOMIZER.

boiler efficiency The ratio of the enthalpy flow rate of the steam leaving a boiler to the product of the fuel mass flow rate and the calorific value of the fuel.

boiler feedtank A reservoir, typically rectangular in shape, in a steam-generation system

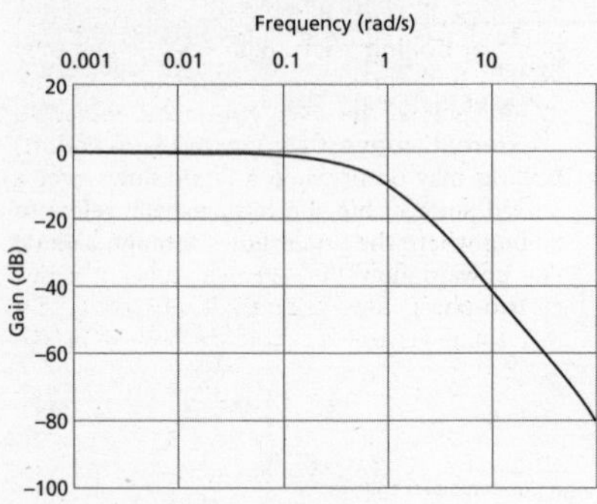

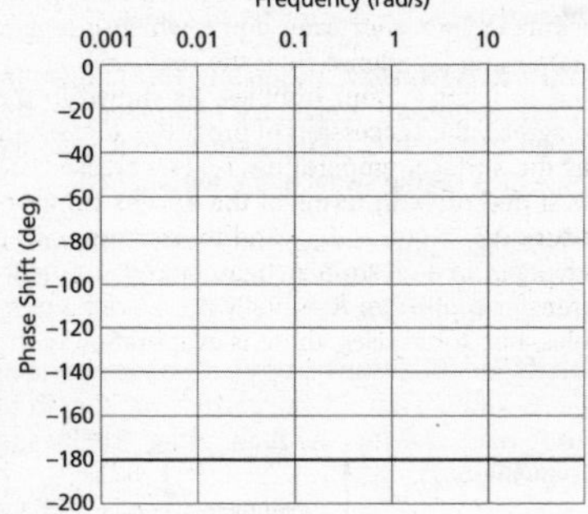

Bode gain and phase plots

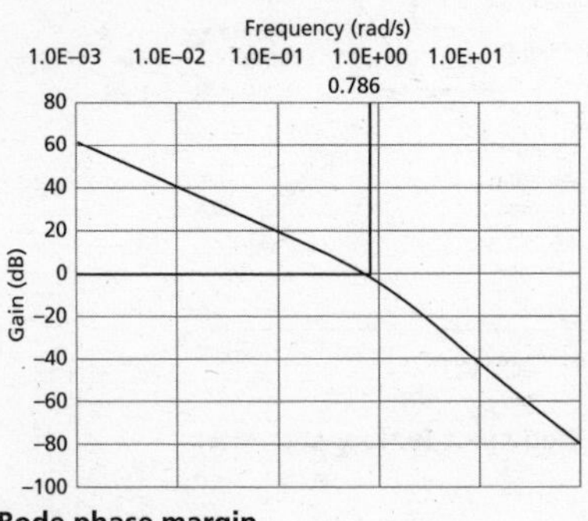

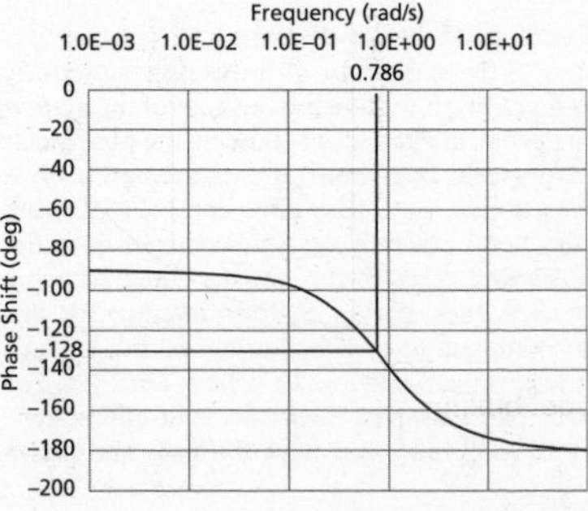

Bode phase margin

for returned condensate and treated make-up water used to supply the boiler.

boiler furnace (steam-generating furnace) A through-flow chamber accommodating the flames from the burners in a water-tube boiler.

boiler hydraulic test A structural integrity test of a new or repaired boiler by means of water under a pressure that exceeds the normal working pressure by at least 50%. Since water is practically incompressible, a fault leads to steady leakage rather than an explosion as would happen if a compressible fluid such as air were used. *See also* PRESSURE-VESSEL HYDRAULIC TEST.

boiling (boiling heat transfer) The process that occurs at a liquid–solid interface when the liquid is in contact with a surface maintained at a temperature sufficiently above the saturation temperature T_{SAT} of the liquid for vapour bubbles to be produced (**wall superheat**). The two principal categories are pool boiling and convective boiling.

In **pool boiling** any motion within the liquid is due to buoyancy from liquid which is locally hotter and less dense than the surrounding liquid or rising vapour bubbles. As shown in the diagram, the succession of processes that occur as the surface temperature T_W is increased are best described in terms of the **excess temperature** $\Delta T_E = T_W - T_{SAT}$ and the corresponding changes in the surface heat flux $\dot{q}''$ or heat-transfer coefficient h. Initially there are no bubbles, hot liquid rises, there is evaporation at the liquid surface and $\dot{q}''$ progressively increases with ΔT_E (natural convection). Above a certain temperature (point A), depending upon the fluid and the pressure, small bubbles begin to form at discrete nucleation sites on the heated surface, break away and rise towards the surface but dissipate within the liquid. As ΔT_E increases further, the rate of bubble formation increases and the bubbles are dissipated at the liquid surface, a process termed **nucleate boiling**. Eventually $\dot{q}''$ reaches a maximum (point B) and then decreases as the rate of bubble formation is so great that the bubbles blanket the heated surface. They coalesce to form a vapour film that covers the surface, a regime known as **film boiling**. The film-boiling regime is initially unstable (transition boiling) but if the surface temperature is sufficiently high, such that a significant fraction of the heat transfer may be through thermal radiation, stable conditions are established. The heat flux is a minimum during stable film boiling at the **Leidenfrost point** (C). The foregoing has assumed that boiling is temperature controlled (i.e. ΔT_E is the controlling quantity), but boiling can also be heat-flux controlled. Under heat-flux control ΔT_E increases abruptly at the peak (the **burnout point** or **boiling crisis**, path B to D) and may lead to burnout of the heating surface if the melting point of the heating surface is exceeded.

External convective (forced-convection) boiling may occur when a liquid flows over a heated surface, but the term usually refers to boiling where the liquid flows through a duct. For upward flow in a vertical tube, a series of two-phase flow regimes is observed with increasing vertical location, as shown in the

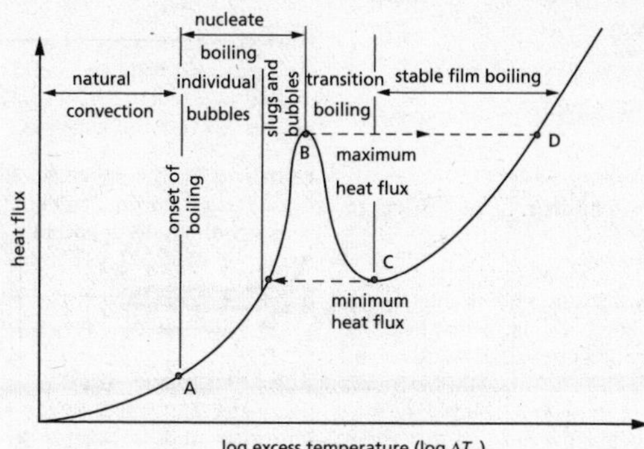

pool boiling

diagram of heat-transfer coefficient *vs* quality. Initially heat transfer is through forced convection. Bubbles first occur at the tube wall and are swept along by the flow. As the bubbles coalesce they form larger bubbles, then long vapour-filled slugs that occupy most of the tube cross section. Beyond the **bubbly-** and **slug-flow regimes** is the **annular-flow regime** consisting of a liquid film on the tube wall with a central core of vapour. The film breaks up exposing dry spots on the tube wall, droplets appear in the vapour core (**mist-flow regime**) and eventually evaporate. The working fluid finally changes from saturated liquid to saturated vapour which becomes superheated upon further heating.

Calculations of both pool and forced-convection boiling are heavily reliant on the numerous empirical correlations that have been developed for each flow regime.

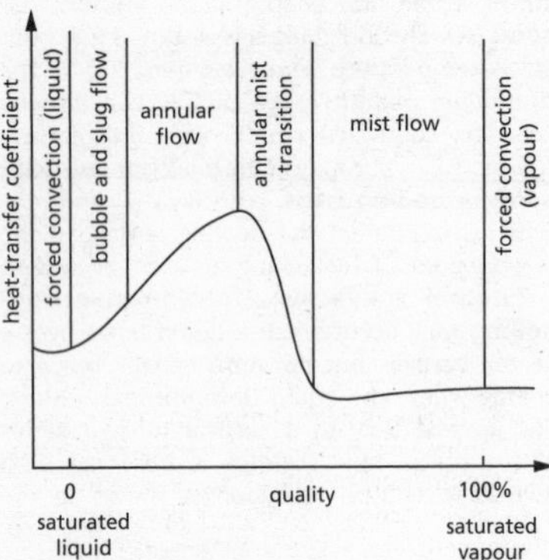

convective boiling

boiling fluidization *See* AGGREGATIVE FLUIDIZATION.

boiling-water reactor *See* NUCLEAR FISSION.

bolt (machine bolt) A fastener in the form of an externally-threaded cylinder with a head on one end that is inserted through holes in assembled parts that are then held together by a nut tightened on to the thread. A bolt has an unthreaded region below the head so that when located in a hole, transverse loads are borne by two plain cylindrical surfaces in contact, rather than (when using a machine screw) threads bearing against the plain surface of a hole. *See also* MACHINE SCREW; SHOULDER BOLT.

SEE WEB LINKS

- Nut, bolt and screw terminology
- Types of nut, bolt and screw and associated terms

bolt blank A rod on which a head has been formed, but on which no thread has been cut, from which bolts or screws may be made as required.

bolt stress The axial tensile stress induced in a bolt by tightening.

Boltzmann constant (*k*) A fundamental physical constant with the value 1.380654 x 10^{-23} J/K which arises in a number of areas of thermodynamics, including the statistical definition of entropy and Planck's function for the spectral blackbody emissive power. It can be shown that k is the ratio of the universal gas constant $\mathcal{R}$ (8.314472 J/K mol) to the Avogadro constant N_A (6.02214179 x 10^{23}/mol). Different sources give slightly different values for k, $\mathcal{R}$ and N_A; those given here are the 2006 CODATA values.

SEE WEB LINKS

- Values of frequently used constants

bomb calorimeter (oxygen bomb calorimeter) A device used to determine the calorific value (CV) of a fuel. A small mass of the fuel is burned in oxygen at constant volume in a small stainless steel pressure vessel (the bomb) immersed in a bath of water (the calorimeter), the temperature of which is monitored to determine CV.

bonded abrasive An abrasive material in grit form held within a matrix or sintered and shaped into blocks, sticks or wheels for grinding. *See also* COATED ABRASIVE.

bonding The joining together of components by adhesion, brazing, soldering, welding, etc., with or without the application of external pressure.

SEE WEB LINKS

- Adhesive types and associated terms

Bond number (*Bo*) A non-dimensional parameter that arises in the study of bubble formation and is defined by $Bo = \rho g d^2/\sigma$ where ρ is the fluid density and σ is the surface tension, g is the acceleration due to gravity and d is a characteristic length (usually the bubble

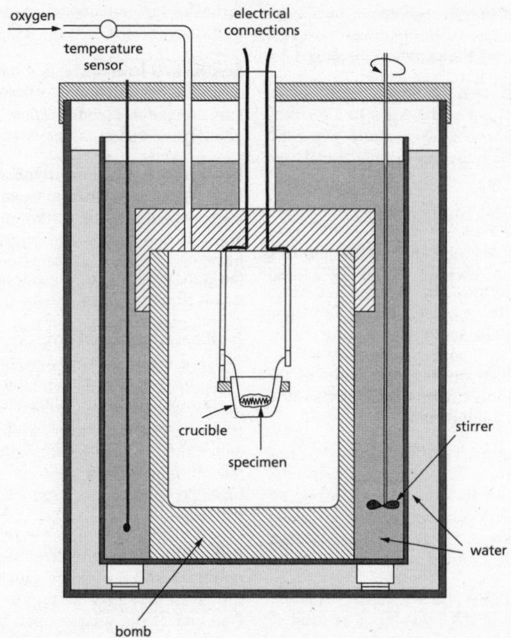

bomb calorimeter

diameter). *See also* EÖTVÖS NUMBER; MORTON NUMBER.

bond strength The tensile, compressive or shear stress at which joints fail, e.g. by fracture or excessive deformation.

booster The compressor stages, used to compress the core air, on the low-pressure shaft in a two-shaft turbojet engine.

booster pump A pump that increases the pressure within a pipeline.

boost pressure The increased air pressure provided to the inlet manifold of a piston engine by a supercharger or turbocharger. It is measured with a **boost gauge.** *See also* MANIFOLD PRESSURE.

bootstrap system A system or device, such as a turbocharger, that, although not self-starting, is self-sustaining once in operation.

BOP *See* BLOW OUT.

bore 1. The inside diameter of a gun barrel, pipe, tube, or cylinder, such as in a reciprocating compressor, pump, or piston engine. **2.** To drill or otherwise machine a long hole in a component.

borescope (boroscope) An inspection device consisting of a fibre-optic endoscope that enables inaccessible regions of, for example, an aeroengine to be viewed.

boring bar A stiff tool holder employed to machine internal surfaces.

boss (hub) 1. The centre of a wheel from which, originally, spokes radiated, though the term now also applies to disc wheels. A hub may rotate on an axle or be fixed to it. **2.** The component by which the blades of a propeller are attached to the engine shaft. **3.** A projecting knob on a casting or forging for fastenings to aid assembly, etc.

b

bottom blowdown valve A manually-operated valve attached to the lowest part of a boiler; used to eject sludge and sediment.

bottom dead centre (BDC, inner dead centre) The position of the crank in a reciprocating compressor, engine or pump when the piston is furthest from the cylinder head. *See also* TOP DEAD CENTRE.

bottom end *See* CONNECTING ROD.

bottoming 1. A piston striking the end of a cylinder. **2.** A tap reaching the end of a blind hole.

bottoming cycle *See* BINARY VAPOUR CYCLE.

bottoming drill A flat-ended drill that removes the cone formed by a conventional drill at the bottom of a blind hole.

boundary (boundary of a system) The real or imaginary surface that separates a thermodynamic system from its surroundings. A boundary has zero thickness so contains no mass and occupies no volume. *See also* CLOSED SYSTEM; ISOLATED SYSTEM; OPEN SYSTEM.

boundary conditions The values of parameters prescribed at the boundaries of solid or thermofluid systems being analysed; for example, the distribution of stress, displacement and slope in stress analysis; the heat flux or temperature in heat transfer.

boundary-element method A computational method for solving linear partial differential equations that have been expressed in integral form. In contrast to FEM, there are nodes only on the boundaries. Having established the correct boundary values, interior values are determined from decomposing the integral equation.

boundary friction The friction that occurs when bearing surfaces are neither completely dry, nor completely separated by a lubricant film.

boundary layer (hydrodynamic boundary layer, velocity boundary layer) The developing region within a fluid flowing over a surface which is affected by the viscosity of the fluid. The fluid velocity changes from zero at the surface to its free-stream value at the 'edge' of the boundary layer. Except at very low Reynolds numbers, a boundary layer is thin compared with its streamwise development length. The flow within a boundary layer may be laminar, turbulent or transitional. *See also* THERMAL BOUNDARY LAYER.

boundary lubrication A state of lubrication where the oil film is only a few molecules thick and comparable with surface roughness. *See also* HYDRODYNAMIC LUBRICATION.

bounded input bounded output stability (BIBO stability) A linear system is described as BIBO stable (the concept does not apply to non-linear systems) when a bounded input will always produce a bounded (i.e. non-infinite) output. Hence a continuous-time system is BIBO stable where the poles of the transfer function lie in the left half of the pole-zero plot on the complex plane and a discrete-time system is BIBO stable if the poles of the transfer function lie within the unit circle, i.e. have magnitude less than one. Stable systems show a response to a finite impulse input that decays with time, marginally stable systems a response that continues at a constant amplitude, and unstable systems a response that increases with time.

Bourdon gauge (spiral pressure gauge, spiral gauge) A pressure gauge consisting of a tube with oval cross section bent into a curve. One end of the tube is open to the unknown pressure and the other is sealed. A change of pressure alters the radius of the bend, causing the sealed end, which is linked to a dial, to move.

Boussinesq approximation In fluid flows, the assumption that density differences are important only where they are multiplied by the acceleration due to gravity. The approximation is valid if density differences are small.

Boussinesq problem In contact mechanics, the determination of the stresses and strains in a half space on the surface of which is applied a point load.

Bowden cable A wire or cable encased in a hollow concentric flexible tube fixed at both ends, by which axial motion of the wire or cable can be transmitted around corners.

bow shock *See* DETACHED SHOCK.

bow thruster A fluidic device installed under water in the bow of a marine vessel, in which a jet of fluid can be switched from one side to the other in order to steer the vessel.

boxer engine *See* VEE ENGINE.

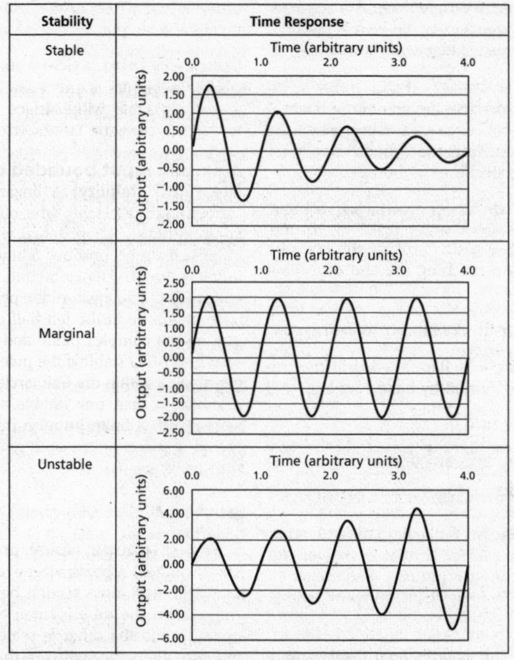

bounded input bounded output stability (BIBO stability)

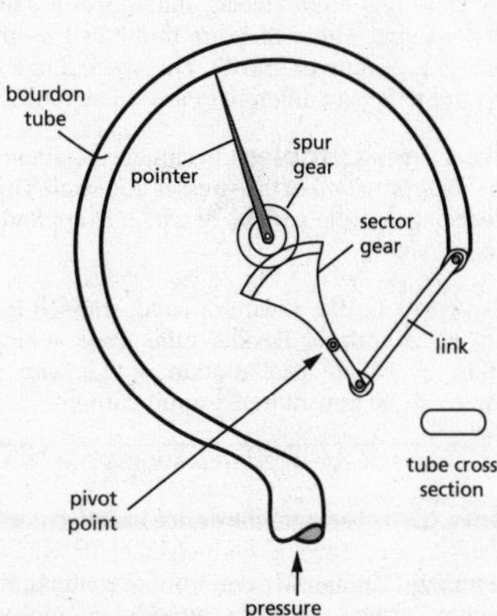

Bourdon gauge

Boyle's law The product of the pressure of a gas p and its volume $\mathcal{V}$ is constant at a given absolute temperature T. Together with Charles law, it yields the ideal gas equation $p\mathcal{V} = mRT$ where m is the mass of gas and R is the specific gas constant.

Boyle temperature (metacritical temperature) In the virial expansion for a real-gas equation of state in terms of pressure p, specific volume v and absolute temperature T, $pv = RT + B/v + (C/v)^2 + \ldots$, the Boyle temperature corresponds to $B = 0$. R is the specific gas constant and A, B, $C\ldots$ are the virial coefficients which depend on T. It is the temperature for which

$$\lim_{p\to 0} \frac{\partial (pv)}{\partial p} = 0$$

Boys gas calorimeter (gas calorimeter) An apparatus used to determine the calorific values of gaseous fuels based upon a counterflow heat exchanger in which water flowing in one direction is heated by burned gas flowing in the other.

brachiating motion Movement by swinging from point to point, as seen in primates such as gibbons. Some research robots have been developed using this form of motion.

brackish water Water with a salt content between that of fresh water (*c*.0.5 gm/*l*) and sea water (*c*.35 gm/*l*).

brake A mechanism for applying frictional resistance to the wheels of a moving vehicle, or to the driving shaft of machinery, to reduce the speed. *See also* DYNAMOMETER.

brake calliper *See* DISC BRAKE.

brake disc *See* DISC BRAKE.

brake duct *See* DISC BRAKE.

brake dynamometer *See* DYNAMOMETER.

brake fluid The hydraulic fluid used to transmit force to the pistons in disc brakes or the wheel cylinders in drum brakes. Requirements include a high boiling point and low hygroscopy.

brake lining The replaceable friction material that covers a brake shoe in an internally-expanding brake.

brake mean-effective pressure (bmep, mean effective pressure, mep) (Unit Pa) The average pressure on the face of a piston during the power stroke of a reciprocating engine, determined from the brake power *BP* (in W), and given by $bmep = BP.C/ALNn$ where A is the cross-sectional area of a piston in m^2, L is the stroke in m, N is the rotational speed in rps, n is the number of cylinders. The factor C is 2 for a four-stroke engine and 1 for a two-stroke engine. *See also* INDICATED POWER.

brake pad The replaceable friction material held in the calliper of a disc brake and applied to the surface of the brake disc.

brake power (brake horsepower, output power, *BP*) (Unit kW or hp) The power of an engine as measured by a brake dynamometer.

brake shoe The internally-expanding members within a brake drum on which the brake linings are attached, or the externally-contracting metal shoes that act directly on to the wheels of some railway vehicles.

brake specific fuel consumption *See* SPECIFIC FUEL CONSUMPTION.

brake thermal efficiency (η_{BT}) The ratio of the brake power of an engine *BP* to the rate at which chemical energy in the form of fuel is supplied to it, i.e. $\eta_{BT} = BP/\dot{m}_F Q_{net,v}$ where $\dot{m}_F$ is the fuel mass flow rate and $Q_{net,v}$ is the net calorific value of the fuel.

braking propeller A propeller in which the pitch of the blades may be altered to give reverse thrust.

Bramah's press *See* HYDROSTATIC PRESS.

branch transmittance The gain of a path in a block diagram or electrical network, usually a function of frequency.

Brayton cycle (Joule cycle) An air standard cycle that is the ideal cycle for a gas-turbine engine. As shown on the diagram of pressure (p) *vs* specific volume (v), it consists of four internally reversible processes: isentropic compression in a compressor (1–2), isobaric heat addition in a combustor (2–3), isentropic expansion in a turbine (3–4) and isobaric heat rejection. The cycle can be extended to include regeneration, reheating, and intercooling. *See also* BELL-COLEMAN CYCLE.

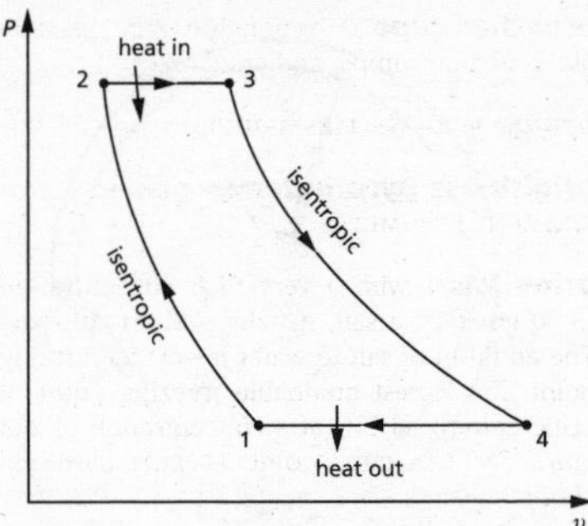

Brayton cycle

brazing Bonding of parts without melting the mating surfaces using a thin film of molten brazing alloy (usually 50/50 copper/zinc melting at some 900°C) that adheres to the contacting surfaces. As with soldering, fluxes are employed to ensure the brazing alloy fills the gap and wets the parent metals. **Silver solder (hard solder)**, an alloy of silver, copper, and zinc with a melting point of about 700°C, is suitable to bond brass, copper, and nickel silver. Brazing permits different metals and components of dissimilar size and mass to be joined, and is also capable of joining tungsten carbide, ceramics, and similar non-metallic materials.

breakaway point A point in the root-locus analysis of a control system at which two or more loci meet and then break away along separate paths.

breakaway wrist A robot wrist which is designed to allow free movement of the end effector without damage when excessive force is applied. It thus prevents damage to the arm or wrist due to such force.

break frequency (break point) The frequency, in a Bode plot, at which there is a step change in the gradient of the magnitude (gain) graph.

breaking load The load applied at some point to a component or structure which leads to fracture. The **breaking stress (breaking strength)** is the average stress at which a member breaks, given by the breaking load divided by the area over which it acts. *See also* FRACTURE MECHANICS.

breather pipe A ventilation pipe for fuel tanks, engine sumps, and crankcases.

bridge wall *See* FIRE BRIDGE.

brightness pyrometer *See* DISAPPEARING-FILAMENT PYROMETER.

brine Water with a very high concentration (> 50 gm/*l*) of a salt, usually sodium chloride. The addition of salt to water lowers its freezing point, the lowest attainable freezing point of brine being −21.1°C at a concentration of 233 gm/*l*. *See also* CRYOSCOPIC EFFECT; EBULLIOSCOPIC EFFECT.

Brinell hardness number (BHN, Bhn, HB) (Unit originally kgf/mm^2 but sometimes now Pa) The load required to indent a test surface permanently with a hard spherical indenter, divided by the spherical surface area of the indentation, and so a pressure. For many steels, the ultimate tensile strength (UTS in MPa) = 3.45 BHN (in kgf/mm^2). *See also* ROCKWELL HARDNESS; VICKERS HARDNESS.

brinelling *See* BEARING.

Brinkman number (*Br*) A non-dimensional parameter that characterizes viscous dissipation in fluid flow and is defined as $Br = \mu V^2/k\Delta T$ where μ is the dynamic viscosity of the fluid, k is its thermal conductivity, V is a characteristic velocity for the flow and ΔT is a characteristic temperature difference produced by the dissipation. *See also* NAHME–GRIFFITH NUMBER.

British Association screw thread *See* SCREW.

British Standard fine thread *See* SCREW.

British Standard pipe thread *See* SCREW.

British Standards Institution (BSI) The Institution in the UK responsible for the preparation and publication of standard specifications (**British standards**) for manufactured goods, their design, manufacture and testing.

British Standard Whitworth coarse thread *See* SCREW.

British Standard Whitworth fine thread *See* SCREW.

British Standard Wire Gauge *See* WIRE GAUGES.

British thermal unit (BTU, Btu) An obsolete (non-SI) British unit for energy. The energy required to raise the temperature of one pound of pure water at 68°F by 1°F so that 1 Btu ≈ 1.055 kJ. *See also* CALORIE; THERM.

brittle A brittle material is one that breaks, often suddenly, with no permanent deformation. Examples of brittle materials are some cast irons, glass, concrete, and some plastics.

brittle fracture The fracture of a component or structure in the globally-elastic range of loading, so that the broken pieces may be refitted to regain the original article. *See also* DUCTILE FRACTURE; EMBRITTLEMENT; FRACTURE MECHANICS.

brittle lacquer coating A lacquer that, when painted onto an unloaded body, reveals the direction of maximum tension stresses from the pattern of cracking produced in the lacquer when the body is loaded.

broach A bar-like cutting tool having multiple transverse teeth along its length, successive teeth cutting more deeply as the tool passes through the workpiece. Used to cut slots, finish holes. *See also* REAMER.

b

Brownian motion The continuous random movement of microscopic solid particles, *c.*1 μm in diameter, suspended in a fluid.

Brunt-Väisälä frequency (buoyancy frequency, *N*) (Unit Hz) The frequency at which a displaced parcel of fluid oscillates in a stratified fluid. Defined for meteorological applications by

$$N = \sqrt{\frac{g}{\theta}\frac{d\theta}{dz}}$$

where θ is the potential temperature, g is the acceleration due to gravity, and z is the vertical height. In oceanographic applications, if ρ is the potential density,

$$N = \sqrt{-\frac{g}{\rho}\frac{d\rho}{dz}}$$

BSF *See* SCREW.

BSI *See* BRITISH STANDARDS INSTITUTION.

BSP *See* SCREW.

BSW *See* SCREW.

BTU (Btu) *See* BRITISH THERMAL UNIT.

bubble A volume of gas or vapour submerged in a liquid or completely surrounded by a soap film. The bubble shape is determined by surface tension, the acceleration due to gravity, the gas volume, relative movement between the gas and the liquid, the extent of the liquid, and the densities of the gas and liquid. Very small bubbles tend to be spherical. *See also* DROP; DROPLET; SEPARATION BUBBLE.

bubble curtain A curtain-like barrier of bubbles produced by a continuous release of air through numerous small holes in a long pipe submerged beneath a liquid surface. The bubbles rise to the surface, dragging liquid as they do so, and form the barrier, which can suppress wave motion and the spread of contaminants such as an oil film on the surface.

bubble train flow A series of long gas bubbles separated by liquid slugs flowing through a capillary tube.

bubbly flow (bubble flow) A gas–liquid or vapour–liquid flow in which the dispersed gas or vapour phase is distributed throughout the continuous liquid phase in the form of discrete bubbles, which may or may not be of the same size and shape. *See also* ANNULAR FLOW; BOILING; SLUG FLOW.

bucket 1. A cup-shaped vane with a central dividing ridge attached to the periphery of the runner of an impulse water turbine such as a Pelton turbine. **2.** A rotor blade in a compressor or turbine.

Buckingham's similarity rules *See* PHYSICAL SIMILARITY.

Buckingham's theorem (Buckingham's pi theorem) A theorem stating that, if a physical problem is determined by n physical variables with j independent dimensions, the number k of independent non-dimensional groups (termed Π's) which can be formed is $k = n - j$. The procedure for determining the non-dimensional groups is termed dimensional analysis. In the majority of mechanical-engineering problems the dimensions considered are mass, length, time, and temperature.

buckling The unstable transverse displacement of a structural component, caused by compressive or shear loads. *See also* EULER FORMULAE.

buckling load The load needed to cause buckling, that is load-, not stress-, controlled.

buckytube *See* CARBON NANOTUBE.

buffer layer *See* LAW OF THE WALL.

built-in *See* FIXITY.

built-up edge An effect produced when, in machining, workpiece metal adheres to a cutting tool to form an unstable false tool geometry that results in poor surface finish.

bulb thermometer A simple thermometer consisting of a bulb attached to an evacuated stem in the form of a capillary tube closed at the other end. The bulb and the lower part of the stem are filled with a liquid, such as alcohol or mercury, which expands when heated and rises up the tube by an amount proportional to the temperature increase. *See also* LIQUID-IN-GLASS THERMOMETER; MERCURY THERMOMETER.

bulb turbine A tidal-power turbine in which the electrical generator is enclosed in a bulb-shaped enclosure held within a duct. Water flows through the duct, around the bulb and through a turbine runner at the downstream end of the bulb. *See also* TUBULAR TURBINE.

bulging 1. In forming, the production of convex shapes on the sides of hollow workpieces by hydraulic pressure or internal expansion of a rubber plug against the walls of a surrounding die or mould (**bulge forming**). **2.** In forming and ballistics, the production of a dome shape by bending action.

bulk density The average density of a mass of granular or powdered material at ambient conditions.

bulk modulus (Unit Pa)) **1.** (*K*) In linear elasticity, the ratio of hydrostatic stress p to volumetric strain $\Delta\mathcal{V}/\mathcal{V}$. **2.** (*B*) For a fluid, a measure of compressibility given by

$$B = \rho\left(\frac{\partial p}{\partial \rho}\right)_{S \text{ or } T}$$

where ρ is the fluid density, p is static pressure, and the subscripts denote whether the compression process is isentropic (S) or isothermal (T).

bulk strain *See* VOLUMETRIC STRAIN.

bulk temperature (bulk mean temperature, T_B) (Unit K or °C) The average temperature of fluid flowing through a duct, defined as the enthalpy flow rate of the flow divided by the product of the mass flow rate $\dot{m}$ and the specific heat of the fluid C_P, i.e.

$$T_B = \int_A \rho u h dA / \dot{m} C_P$$

where ρ is the fluid density, h is specific enthalpy, u is the velocity and A is the cross-sectional area of the duct.

bulk velocity (average velocity, V) (Unit m/s) For fluid flow in a pipe or duct, the average flow velocity, given by $V = \dot{m}/\rho A$ where $\dot{m}$ is the mass flow rate, ρ is the fluid density, and A is the duct cross-sectional area.

bulk viscosity (volume viscosity, second coefficient of viscosity, expansion viscosity, μ_B) A coefficient related to bulk fluid expansion or compression, rarely encountered in practice. If a fluid undergoes a finite dilatation rate Δ then the difference between the instantaneous pressure p_{inst} and the equilibrium pressure p_{eq} at the same temperature is given by $p_{eq} - p_{inst} = \mu_B \Delta$.

bump stop A cushioning device, often made of hard rubber, to limit the motion of the components of a vehicle suspension system.

bund A continuous barrier or low wall to contain a liquid spill or leak.

buoyancy force *See* ARCHIMEDES PRINCIPLE.

buoyancy frequency *See* BRUNT–VÄISÄLÄ FREQUENCY.

Burgers equation A model equation for one-dimensional unsteady flow of a viscous fluid which can be written

$$\frac{\partial u}{\partial t} + u\frac{\partial u}{\partial x} = \nu\frac{\partial^2 u}{\partial x^2}$$

where u is the fluid velocity at a location x after time t and ν is the kinematic viscosity of the fluid. The inviscid form ($\nu = 0$) can be solved by the method of characteristics and is a prototype for flows which develop shock-like discontinuities.

Burgers vortex A vortex flow having azimuthal velocity variation with radius $\upsilon(r)$ described by the equation

$$\upsilon = \frac{\Gamma}{2\pi r}\left[1 - exp(-\frac{r^2}{4\delta^2})\right]$$

where Γ is the circulation and δ is the core radius.

Burke–Plummer equation *See* ERGUN EQUATION.

burner A nozzle or nozzle array in which air and gaseous or liquid fuel are mixed and burned at the exit. Liquid fuel may be atomized by the nozzle. *See also* DUAL-FUEL BURNER; GAS BURNER; PRE-MIX GAS BURNER; PRESSURE-JET BURNER.

burner turndown *See* TURNDOWN RATIO.

burnout *See* BOILING.

burr A protrusion left on a guillotined edge, drilled hole, etc.

bursting disc (rupture disc) An elastomeric membrane or thin metal diaphragm designed to burst at a set pressure to prevent overpressure. A bursting disc is often installed on either the inlet or outlet of a safety valve.

burst pressure The pressure at which a pressure vessel will explode or otherwise fail.

bush *See* BEARING.

butterfly valve A valve in which a disc rotates on a shaft at right angles to the axis of a pipe to regulate flow. When open, the disc is edge-on to the flow and offers limited resistance. When closed, the disc is pressed against a seat in the valve body. *See also* THROTTLE VALVE.

butt joint The end-to-end joining of two plates either by welding or by overlapping plates that are bolted or riveted.

buttress thread *See* SCREW.

butt ring *See* SPLIT RING.

butt weld *See* WELDING.

bypass A duct or pipe through which fluid is made to flow instead of, or in addition to, its normal path.

bypass engine *See* TURBOFAN ENGINE.

bypass flow meter (shunt flow meter) A flow meter installed in a pipework bypass which may itself be part of the flow meter. An orifice plate is used to ensure a fraction of the main flow passes through the bypass.

bypass ratio In a turbofan engine, the ratio of the mass flow rate of the bypass stream to the mass flow rate through the core of the engine.

bypass valve A valve that directs flow through a bypass.

CAD *See* COMPUTER-AIDED DESIGN.

CAD/CAM *See* COMPUTER-AIDED DESIGN; COMPUTER-AIDED MANUFACTURING.

CAE *See* COMPUTER-AIDED ENGINEERING.

CAES *See* COMPRESSED-AIR ENERGY STORAGE.

cage The frame that holds and separates the balls in a ball bearing or the rollers in a roller bearing.

cal *See* CALORIE.

calandria **1.** A term sometimes used for a tube or bundle of tubes in a shell-and-tube heat exchanger, particularly one used for distillation or evaporation. **2.** The core of the **CANDU** (**Can**ada **d**euterium **u**ranium) nuclear reactor, a pressurized heavy-water reactor.

calibration **1.** Establishing constants in empirical formulae from experimental data or exact numerical solutions. **2.** Choice of disposable parameters or constants in FEM or CFD simulations, to make the calculations agree as closely as possible with experimental data or analytical solutions. **3.** Comparing the output of an instrument against an instrument of known high accuracy or a standard when the same input is applied to both instruments. A **calibration curve** is a plot of calibration data indicating the correct value for every reading indicated by an instrument, often incorporating a fitted curve such as a polynomial. A **calibrated** instrument is one that has undergone calibration.

Callendar and Barnes continuous-flow calorimeter An apparatus for measuring the mechanical equivalent of heat $\mathcal{J}$ (unit J/cal). It consists of two long concentric glass tubes with an electrical heating coil along the length of the inner tube through which there is a constant flow of water with mass flow rate $\dot{m}$. The space between the two tubes is evacuated and so insulates the inner tube from the surroundings. If the temperature rise of the water is $\Delta\theta$, $\mathcal{J} = \dot{E}/\dot{m}\Delta\theta$ where $\dot{E}$ is the rate at which electrical energy is supplied to the coil.

Callendar's thermometer The first resistance thermometer to use platinum wire as the resistance element. *See also* PLATINUM-RESISTANCE THERMOMETER.

calliper (caliper) **1.** A device (often with electronic readout) like dividers but with curved feet for measuring inside and outside dimensions; when the feet point in the same direction, the device is used to measure the pitch of gearing. **2.** *See* DISC BRAKE.

calorically-imperfect gas *See* PERFECT GAS.

calorically-perfect gas *See* PERFECT GAS.

caloric equation of state An equation for the dependence of the specific internal energy of a substance on its temperature and specific volume.

calorie (cal, gram calorie, small calorie) An obsolete (i.e. non-SI) unit of energy equal to 4.1855 J. It is the amount of energy needed to raise the temperature of 1 gram of pure air-free water from 14.5°C to 15.5°C at standard atmospheric pressure. *See also* BRITISH THERMAL UNIT.

Calorie *See* KILOCALORIE.

calorific value of a fuel (heating value of a fuel, heat value) (Unit kJ/kg for solid and liquid fuels, but also used is kJ/kmol and, for gaseous fuels, kJ/m^3) The energy released per unit mass of fuel in complete combustion with oxygen, i.e. the heat of reaction. Values are determined by burning fuel in a bomb calorimeter under specified conditions. The gross or higher calorific value, GCV or HCV, corresponds to the water in the combustion products being in the liquid phase and the net or lower calorific value, NCV or LCV, to the vapour phase. The two are related by

$$HCV = LCV + m_{H_2O}h_{fg}/m_{FUEL}$$

where h_{fg} is the latent heat of vaporization of water, m_{H_2O} is the mass of water produced for a mass of fuel m_{FUEL}. The values given are normally determined at constant pressure, but constant-volume values are also found.

calorifier *See* NON-STORAGE CALORIFIER; STORAGE CALORIFIER.

calorimeter An apparatus used to measure such quantities as the calorific value of a fuel, the enthalpy of reaction of a chemical reaction, and the thermal capacities of materials. The apparatus typically consists of an insulated vessel, which may be open or closed, surrounded by stirred water. Temperature changes in the water are analysed to determine the quantity of interest. *See also* BOMB CALORIMETER; BOYS GAS CALORIMETER; CALLENDAR AND BARNES CONTINUOUS-FLOW CALORIMETER.

CAM *See* COMPUTER-AIDED MANUFACTURING.

cam A component of a mechanism that imparts a prescribed reciprocating motion to a **cam follower,** the output element of a cam mechanism which is in contact with the cam profile. It is typically a rod which slides in a guide with a roller or shaped end (translating follower) or a pivoted arm (oscillating or rotating follower). The motion is determined by the shape of the cam surface (**cam profile)** and the geometry of contact with the follower. Rotating cams (profiled discs) that open and close valves in engines are examples, but there are linear (wedge) cams, and cylindrical cams in which the end of the follower runs in a groove profiled to give the desired motion. The profiled part of a cam by which movement is imparted to the follower is the **cam nose. Cam dwell** is that part of a cam profile where no motion of the follower takes place. For a rotating cam it is an arc of a circle with its centre at the centre of cam rotation. An **eccentric cam** is a cam with circular profile rotating off centre. Rotating cams are mounted on a **camshaft**, the rotation being effected by the **camshaft drive,** a toothed belt (as in the diagram), gear train, or chain through which power is transmitted from the crankshaft of an engine. The drive may also power peripherals such as the coolant pump. The tensioner transmits no power.

camber line *See* AEROFOIL.

CANDU reactor *See* CALANDRIA (2).

cantilever A beam that is built in at one end and unsupported at the other. *See also* FIXITY.

cap A cover, often in the form of a short cylinder, one end of which is closed. Typically used to close an orifice or pipe end, onto which it can be pushed, welded, screwed, or attached with fasteners.

capacity factor (load factor, plant factor, energy utilization factor, utilization factor) The ratio, for a power plant, of its annual output divided by its maximum possible output.

capillarity correction A part of the correction which must be applied to the reading of a mercury-in-glass barometer to account for the convex shape of the meniscus.

capillary effect (capillary action, capillarity) 1. The tendency of liquids to move along a capillary tube as a consequence of the surface-tension force which arises at the tube wall-liquid interface. If the liquid wets the surface, as is the case for water, a concave meniscus is formed in a vertical tube of radius R with its lower end submerged below the liquid surface. The liquid rises by an amount $2\sigma cos\theta/(\rho gR)$ where σ is the liquid-air surface tension, θ is the contact angle, ρ is the liquid density, and g is the acceleration due to gravity. If the liquid is non-wetting, as is the case for mercury, a convex

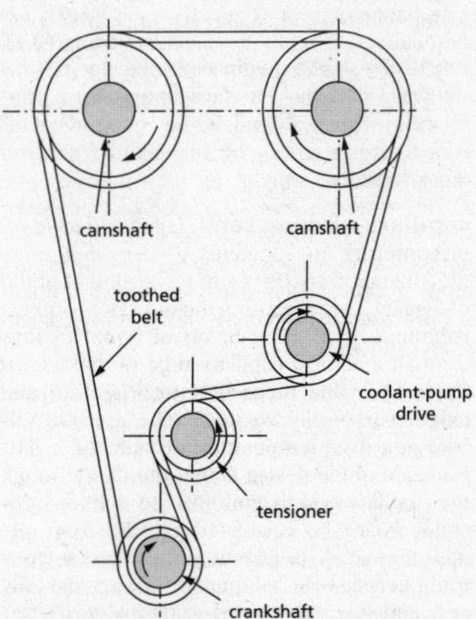

camshaft drive

meniscus is formed and the liquid level in the tube is depressed. **2.** Similar forces to those described in (1) arise in liquid flow through porous media and microchannels.

capillary filling The application of capillary action to introduce liquid into a microchannel **(capillary flow)** such as in a lab-on-a-chip.

capillary fitting (capillary joint) A pipe joint where a male end having soldering flux over its surface fits into a female end, and application of heat, typically using an LPG blowtorch, melts solder fed to the joint, which then is drawn into the annular gap by capillary action.

capillary-force valve (capillary valve) A valve that uses surface tension to control the flow of a liquid into a microchannel filled with a second immiscible fluid, usually a gas.

capillary number (*Ca*) A non-dimensional number that characterizes a liquid flow having an interface with an immiscible fluid (gas or liquid) affected by both viscous and surface-tension forces. It can be defined as $Ca = \mu V/\sigma$ where μ is the dynamic viscosity of the liquid, σ is the surface tension between the two fluids and V is a characteristic velocity of the flow. An important parameter for flow in a microchannel. *See also* BOND NUMBER; EÖTVÖS NUMBER.

capillary reactor A capillary tube, typically in the shape of a U, through which there is flow within which a chemical reaction occurs.

capillary tube A pipe of sufficiently small diameter that the interface between two immiscible fluids (liquid–liquid or gas–liquid) within the tube will be affected by surface-tension forces.

capillary viscometer 1. (suspended-level viscometer) An apparatus used to determine the kinematic viscosity ν of a Newtonian liquid of density ρ by measuring the time t for a known volume $\mathcal{V}$ (typically 5.6 ml) of liquid to flow through a vertical capillary tube of diameter d (typically in the range 0.64 to 6.76 mm) and length l (typically 90 mm). The apparatus is held at a fixed temperature, usually 20°C. The principle of the device is that the flow through the capillary tube is Poiseuille so that, in principle, ν can be calculated directly from the measurements. In practice, the viscometer is calibrated against a liquid of known viscosity ν_{CAL} and ν is determined from $\nu = \nu_{CAL}t/t_{CAL}$ where t_{CAL} is the time required for volume $\mathcal{V}$ of the calibration liquid to flow through the capillary tube. *See also* OSTWALD VISCOMETER; UBBLEHODE VISCOMETER. **2.** An apparatus used to determine the dynamic viscosity μ of a Newtonian fluid by measuring the pressure drop Δp and volumetric flow rate $\dot{Q}$ for flow through a capillary tube of length l and internal radius r. Assuming Poiseuille flow through the tube, μ is determined from the equation $\dot{Q} = \pi r^4 \Delta p/8\mu l$. *See also* SAYBOLT VISCOMETER.

capillary wave (ripple) A small-amplitude wave at the interface of two immiscible liquids or a liquid–gas interface in which the restoring force is due to surface tension. For capillary waves on a liquid–liquid interface, the dispersion relation is given by

$$\omega^2 = \frac{\sigma}{\rho_H + \rho_L}\left[\frac{2\pi}{\lambda}\right]^3$$

where ω is the angular frequency, σ is the surface tension, ρ_H and ρ_L are the densities of the heavier and lighter liquids, respectively, and λ is the wavelength. *See also* GRAVITY–CAPILLARY WAVE.

cap nut A nut with a blind threaded hole, for example a **dome nut** to cover the end of a bolt.

cap screw A bolt where the thread runs right up to the head and engages in a threaded hole or captive nut in an adjoining member. *See also* MACHINE SCREW.

capstan A winding drum having a vertical axis.

capstan lathe A lathe having a number of tools already set up in a moveable tool holder (the capstan), often hexagonal in shape, to perform a specified sequence of operations in mass production of components. The tools are brought into action by rotation of the capstan about a vertical axis. *See also* TURRET LATHE.

captive nut A nut attached loosely or rigidly to a sheet member that is too thin to thread and which engages with a cap screw.

capture efficiency The ratio of the net rate of heat transfer into a solar collector to the solar radiation incident on the collector (i.e. the product of the surface area and the irradiance).

carbide cutting tool A cutting tool made of sintered or hot-pressed 'hard-metal' carbides such as boron carbide, titanium carbide, tungsten carbide, etc., bonded with mixtures of cobalt, chromium and nickel.

SEE WEB LINKS

- Cutting-tool materials and associated terms

carbon capture and storage (CCS) To reduce its contribution to global warming, a system for removing CO_2 from combustion exhaust gases at source, concentrating it and removing to a remote storage location, such as an underground cavern.

carbon fibre A filament reinforcement used in composites. In addition to having a much greater fracture strength than glass fibre (~21 GPa *vs* ~1 GPa), it is also much stiffer (E = 750 GPa *vs* 70 GPa). In consequence working strains (deflexions) in a design using CFR composites can be kept at similar levels to those when using metals, unlike designs using GFRP, where strains and deflexions can be appreciable.

carbon nanotube (buckytube) A hollow carbon cylinder with a diameter of a few nanometres and a length-to-diameter ratio well in excess of a million, having exceptional mechanical properties, e.g. Young's modulus up to 1000 GPa and a fracture stress of some 60 GPa.

carburettor (carburetor) A device that supplies the air-fuel mixture to a spark-ignition engine. The air is drawn in by the downward movement of a piston and passes through a Venturi. The resulting low pressure sucks fuel through an atomizing jet and into the air flow, where it mixes. The intake air can flow either upwards through the main body into the engine intake manifold (**updraught carburettor**) or downwards (**downdraught carburettor**). Fuel-injection systems have replaced the carburettor in modern engines.

carburettor icing Ice formation on the surfaces of a carburettor, particularly the butterfly valve, owing to expansive cooling and evaporation of petrol which causes water vapour in the airflow to condense and freeze. Icing is particularly a problem with very humid air such as is encountered by an aircraft flying through clouds, but is preventable by carburettor heating.

carburizing *See* CASE HARDENING.

Cardan joint *See* HOOKE'S JOINT.

Cardan shaft *See* DRIVE SHAFT; PROPELLER SHAFT.

Cardan's suspension A system that uses three gimbals with orthogonal axes so as to support a component in a fixed orientation despite rotation of the mounting of the gimbals. It is typically used to support gyroscopes for navigational use and for experiments on bodies rotating about a fixed point (O in the diagram) as opposed to a fixed axis.

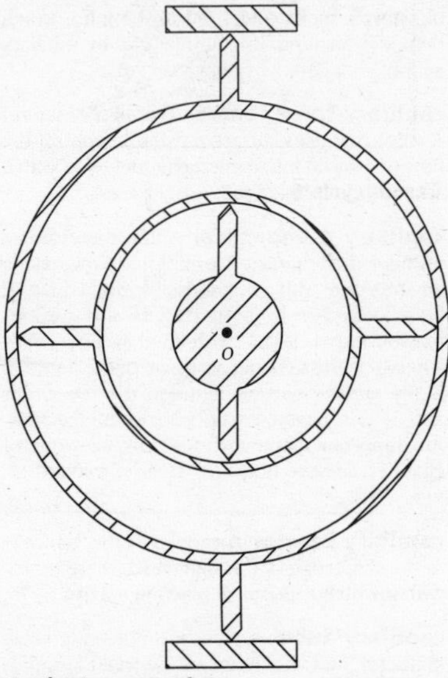

Cardan's suspension

cardice *See* DRY ICE.

Carnot cycle An ideal four-stage reversible cycle consisting of isothermal expansion (1–2) with the reception of heat from a heat source at high absolute temperature T_H; isentropic expansion (2–3); isothermal compression (3–4) with the rejection of heat to a heat sink at low absolute temperature T_L; and isentropic compression (4–1). The cycle efficiency **(Carnot efficiency)** η is given by $\eta = (T_H - T_L)/T_H$. The diagram shows the cycle as a plot of pressure (p) *vs* specific volume (v).

Carnot engine The diagram shows a possible arrangement for an engine operating on a Carnot steam cycle. Saturated water is evaporated in a boiler at constant pressure to

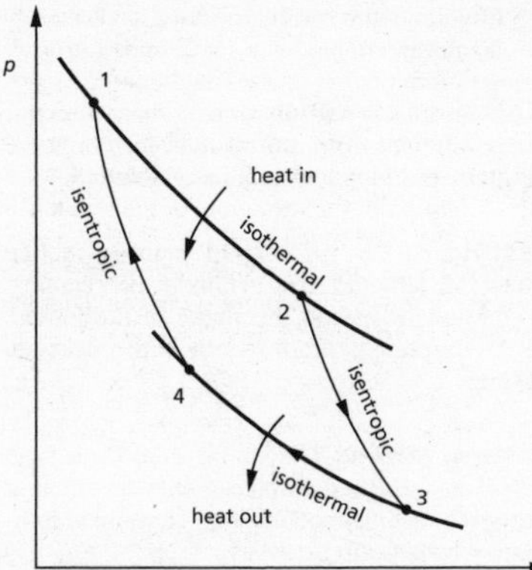

Carnot cycle

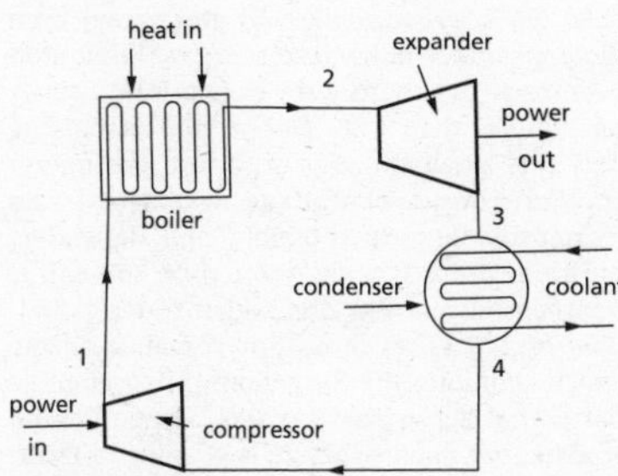

Carnot engine

produce saturated steam (steps 1–2); the steam is then expanded in a turbine or steam engine to produce power (2–3) and partially condensed at constant pressure (3–4). The cycle is completed by compression by a rotary or reciprocating compressor (4–1).

Carnot's law For an ideal gas, the difference between the specific heats at constant pressure C_P and constant volume C_V is equal to the specific gas constant R, i.e. $C_P - C_V = R$.

Carnot's theorem (Carnot's principles) (a) Any reversible heat engine operating between two constant-temperature reservoirs has the same thermal efficiency as any other. This is self-evident from the efficiency of the Carnot cycle. (b) The efficiency of an irreversible (i.e. actual) heat engine is always less than that of a reversible heat engine operating between the same temperatures. The two statements are proved by demonstrating that the violation of either one is a violation of the second law of thermodynamics.

carriage spring *See* SEMI-ELLIPTIC SPRING.

Cartesian-coordinate robot (rectangular-coordinate robot) A robot where the first three joints are translational, have axes normal to each other, and move in the three Cartesian coordinate directions of the base frame. The volume that the robot can reach is thus a hollow cube. The diagram shows an idealized Cartesian coordinate robot with the joint offsets d_1, d_2, and d_3 showing the movements of the first three joints. *See also* CYLINDRICAL-COORDINATE ROBOT.

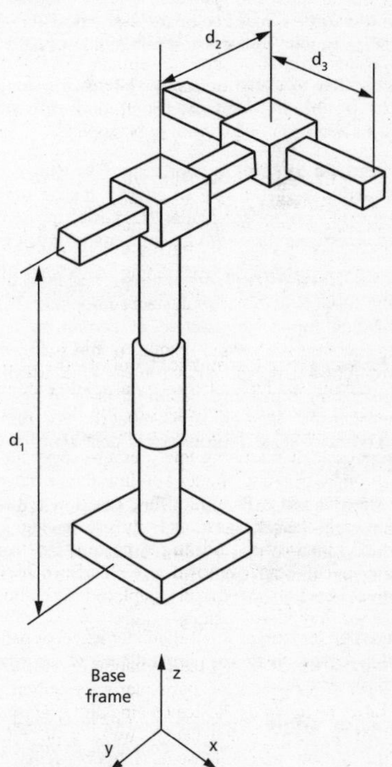

Cartesian-coordinate robot

cascade compensation Where a compensator is placed in series with a controller so as to improve the performance of the controller.

cascade control A control system in which the output of a primary controller determines the set point of a secondary controller that provides the input to the controlled plant. For such a system to operate correctly, the secondary controller must have a much shorter settling time than the primary controller so as to allow it to track changes in the set point output from the primary controller. For example, in a rolling mill the overall speed of the strip being rolled is controlled by a primary controller, but each drive motor will have a secondary controller to maintain the motor speed at that set by the primary controller.

cascade refrigeration cycle A cycle in which the refrigeration process involves two or more refrigeration cycles operating in series. A typical application is where a larger temperature range is required than is practical for a single vapour-compression refrigeration cycle.

cascade system In control systems, a system that has been divided into two or more parts so as to allow cascade control to be applied.

cascade tunnel A wind tunnel in which a straight, staggered row of equally spaced turbine or compressor blades (**cascade**) is used to investigate their aerodynamic characteristics.

case hardening (carburizing) Hardening of the surfaces of low-carbon steel components by heating in an atmosphere of carbon so as to increase the carbon content of the surfaces by diffusion up to, say, 0.6–0.8%, after which quenching transforms this outer layer to martensite that is then stress-relieved. *See also* ELECTRON BEAM HARDENING; LASER HARDENING; NITRIDING.

castellated nut A nut having slots across the hexagonal faces, used with a bolt having a drilled hole so that a split (cotter) pin may be inserted through both nut and bolt to prevent unscrewing.

caster (castor) A small-diameter wheel or ball that swivels on an axis perpendicular to the axis of the wheel.

caster angle The inclination of the kingpin (swivel pin) in a steering mechanism, as viewed along the axis of the steered wheels, that gives self-centring steering.

Castigliano's theorems The deflexion of an elastic body at the point of application of an external load is given by the partial derivative of the strain energy with respect to the component of the applied force in that direction. Strictly, there are two theorems, the one employing differentiation of the complementary strain energy for displacement under a load, and the other using differentiation of the strain energy to give the load causing a displacement, these energies being different from one another in non-linear elasticity but identical in linear elasticity.

casting **1.** The process of pouring molten metal into a mould so as to obtain, after cooling, a component having the shape of the mould. **2.** A component produced by the process of casting.

casting strains The strains that result from different shrinkage displacements in different parts of a casting upon cooling. They are accompanied by **casting stresses**.

cast iron A term for alloys of iron containing between 2% and 5% carbon. There are two principal types, called white and grey (gray) from the appearance of fracture surfaces. White iron is an extremely hard form of cast iron, usually having less than 4.3% carbon and containing only very small amounts of silicon. The microstructure consists of pearlite and cementite (but martensitic forms are possible) and, depending on the pearlite spacing, the tensile strength is between 250 and 500 MPa, with a Young's modulus of 170 GPa. Grey iron contains silicon, which promotes the formation of free graphite flakes that act as sharp cracks. Depending on whether the microstructure is pearlitic or ferritic, the tensile strength is between 140 and 420 MPa, with Young's modulus varying inversely with flake size between 80 and 140 GPa. Iron phosphide in the microstructure increases the fluidity and makes casting straightforward. Cast iron is brittle unless specially treated to give improved toughness and ductility. Among such modified types are malleable white iron (a traditional form of cast iron with average ductility, initially cast as white iron and then heat treated to produce graphite clusters called temper carbon; both pearlitic (whiteheart) and ferritic (blackheart) forms are made, the two names arising from shiny or dull fracture surfaces); ductile nodular grey iron and spheroidal grey cast iron, the ductility of which has been considerably improved by the addition of magnesium and/or cerium to molten grey cast iron; and meehanite iron, which would otherwise be a white iron of low silicon and carbon content, but to which is added in the melt calcium silicide, resulting in finely-dispersed carbon flakes. Alloying elements (nickel, etc.) may also be added to all types of cast iron to improve properties such as corrosion and wear resistance. *See also* CHILLED CASTINGS.

castor *See* CASTER.

cast steel Steel that is cast into shapes. It has superior properties to most cast irons, but is more expensive to produce.

catalytic converter A device installed in an exhaust system to reduce the toxicity of exhaust gases, typically using a catalyst, such as platinum, palladium, or rhodium, to oxidize carbon monoxide and unburned hydrocarbons and reduce oxides of nitrogen.

catastrophic failure A sudden and total failure of a large engineering structure such as an aeroengine, aircraft, space vehicle, bridge, or dam.

catenary The shape taken up by a uniform chain or cable having negligible bending stiffness when suspended between two points. For a cable with each end at the same height, and with the origin of axes at the lowest point, the equation is $y = (H/w)[cosh(wx/H) - 1]$ where y and x are vertical and horizontal distances, respectively, w is the weight per unit length of the chain, and H is the (constant) horizontal component of the tension T in the cable, given by $T = H + wy$. Problems are solved from known conditions, such as the values of the central dip and span, central dip and length of cable, or span and length of cable.

caterpillar (crawler vehicle) A vehicle that 'lays its own road' by running on endless belts, driven by toothed wheels, on each side. Used on soft ground where spreading of the load reduces contact stress.

Cauchy number (*Cn*) The non-dimensional number that has to be satisfied for equality of inertia and elastic forces, given by $Cn = V\sqrt{\rho/E}$ where V is velocity, ρ is density and E is Young's modulus. If Poisson's-ratio effects can be neglected, the compressional wave speed in a solid is given by $c = \sqrt{E/\rho}$ so that $Cn = V/c$ which is analogous to the Mach number in a gas. *See also* ATKINS NUMBER.

causal system A system where the output at a particular time T depends only on the inputs over the time t in the range $0 \leq t \leq T$. This is the case for normal physical systems which cannot anticipate future events. However, in the digital filtering of previously captured data, a non-causal filter is frequently used where the filter output at time T depends on inputs both before and after T.

cavitation The formation of vapour- or gas-filled cavities in a liquid due to reduction of the local pressure, often due to acceleration of the fluid, such as in flow through a convergent nozzle. If there is no dissolved gas in the liquid, vaporous cavitation occurs when the pressure falls below the saturated vapour pressure. If the bubbles are formed due to high temperature, the process is termed boiling. If there is dissolved gas, gaseous cavitation occurs due to pressure reduction, temperature increase, or diffusion (degassing). At the tips of marine propellers and in hydraulic machinery, the collapse of cavitation bubbles can cause noise and vibration and lead to surface damage in the form of pitting. Cavitation in a fluid flow is characterized by the non-dimensional **cavitation number (σ)** defined by $\sigma = 2(p_\infty - p_V)/\rho V^2$ where p_∞ is the ambient static pressure, p_V is the vapour pressure, ρ is the fluid density and V is the flow speed. *See also* EULER NUMBER.

cavitation-resistance inducer An axial-flow pump used upstream of a main pump in order to prevent cavitation in the latter by increasing the inlet head.

cavitation tunnel A closed-circuit recirculating water tunnel in which the static pressure can be reduced to sufficiently low levels for cavitation studies to be performed.

cavity radiator A heated chamber having a small hole through which radiation, approximating blackbody radiation, passes out.

cavity resonator *See* HELMHOLTZ RESONATOR.

CDD *See* DEGREE DAY.

Celsius temperature scale (centigrade temperature scale) A relative temperature scale now defined in terms of the Kelvin absolute temperature scale as °C = K − 273.15, where °C is the symbol for degrees Celsius. The scale was previously called the centigrade scale, with two fixed points: the melting point of ice (the ice point) as 0°C, and the boiling point of water (the steam point) as 100°C.

centi (c) An SI unit prefix indicating a multiplier of 0.01; thus centimetre is a unit of length equal to one one-hundredth of a metre or 10 mm.

centigrade heat unit (Celsius heat unit, CHU) An obsolete (i.e. non-SI) unit equal to the energy required to increase the temperature of one pound of pure, air-free water from 14.5°C to 15.5°C at a pressure of one standard atmosphere.

central gear *See* TOOTHED GEARING.

centreless grinding Grinding of a cylinder not supported along its axis, but by an underneath knife edge and a feeding wheel or pressure roller.

centre line 1. In an engineering drawing, a line of symmetry. **2.** An imaginary line along a pipe, duct, or shaft that defines an axis of symmetry. **3.** An imaginary straight line parallel to the intended direction of a surface located such that the areas above and below the line and the real wavy (rough) surface cancel out.

centre-line average (CLA, Ra) The arithmetic mean height of the peaks and valleys determined by a profilometer or optical interferometer in surface-roughness measurements.

$$\mathrm{Ra} = \frac{1}{L}\int_0^t |y(x)|dx$$

where y is the height of the surface above the mean line at distance x from the origin and L is the overall length of the profile under examination.

centre of buoyancy The point in a floating or submerged body, located at the centroid of the displaced volume, through which the upthrust acts. *See also* ARCHIMEDES PRINCIPLE; METACENTRE.

centre of flexure *See* SHEAR CENTRE.

centre of gyration The point in a rotating body about which its angular momentum is concentrated. *See also* RADIUS OF GYRATION.

centre of mass The point within a system or body at which all the mass can be imagined to be concentrated. If the system or body is of uniform density, its centre of mass coincides with its **centroid**. If the system is in a uniform gravitational field, it coincides with its **centre of gravity**, the fixed point through which the resultant gravitational body force (i.e. its weight) acts. When a rigid body is accelerated, the resultant force caused by its inertia acts through its centre of mass.

centre-of-mass coordinate system (centre-of-momentum coordinate system) A reference frame that moves at the velocity of the centre of mass of a body or system so that the latter is always at rest with respect to the reference frame and the total momentum of the system is zero. *See also* ACCELERATING FRAME OF REFERENCE.

centre of oscillation For a rigid body so suspended that it can rotate freely about a horizontal axis through a point O, the length L of a simple pendulum with the same period of oscillation is given by $L = c + k_c^2/c$ where c is the distance of the centre of gravity G of the body from O and k_c is the radius of gyration of the body about G. If the point of suspension is shifted to another point O′ on the line OG extended, and the periodic time remains unaltered, it follows that $\mathrm{OG} + k_c{}^2/\mathrm{OG} = \mathrm{GO'} + k_c{}^2/\mathrm{GO'}$, i.e. $\mathrm{OG.GO'} = k_c{}^2$. The point O′, related in this way to the point of suspension O, is called the centre of oscillation.

centre of percussion The location along the handle of a hammer, bat, etc. where, when struck at one end, there is zero impulsive reaction, i.e. no 'sting' to the hand of the person holding the shaft. For a uniform rod of length $2L$ struck at one end, the centre of percussion is at a distance $2L/3$ from the other end.

centre of pressure 1. The location, on an aerofoil or other body that develops lift, of the resultant lift force. **2.** The location, on a surface submerged in a liquid, of the resultant force due to the pressure acting on the surface. Because hydrostatic pressure increases with depth, the centre of pressure is generally below the centroid of the surface.

centre of twist *See* SHEAR CENTRE.

centre punch A conical-ended punch used to 'dot' the surface of a component where a hole is to be drilled in order to prevent chisel-ended drills from wandering. *See also* PILOT HOLE.

centrifugal Acting or moving in a direction away from the axis of rotation.

centrifugal acceleration *See* ACCELERATING FRAME OF REFERENCE.

centrifugal clutch A clutch that engages and disengages at a defined speed of rotation of the driving shaft, as when expanding friction shoes act against the inside of a drum.

centrifugal compressor A compressor in which kinetic energy is added to a fluid by radial acceleration in an impeller and then converted into a pressure increase by flow though a diffuser.

centrifugal fan A machine with a rotor consisting of a number of blades mounted around a hub and used for moving air or other gases. The gas enters the rotor axially and is discharged radially at increased pressure.

centrifugal force (Unit N) The inertial reaction force to the centripetal force. It is equal in magnitude but opposite in direction.

centrifugal pump A pump into which liquid enters axially through the eye of the casing and is then accelerated through an impeller, thereby increasing both its kinetic energy and pressure before being delivered to a ring diffuser (the volute) that further increases the liquid pressure and from which it leaves.

centrifuge A rotating device for separating liquids and suspended particles of different densities by centrifugal action.

centripetal Acting or moving in a direction towards the axis of rotation.

centripetal acceleration The radially-inward component of acceleration of a particle moving along a curved path. If the local radius of curvature is r and the instantaneous velocity of the particle is V, the centripetal acceleration is $V^2/r = r\omega^2$ where ω is the instantaneous angular velocity. It arises because the velocity vector, even if constant in magnitude, continually changes direction. For a particle of mass m moving along a curved path, the centripetal force is mV^2/r or $mr\omega^2$ directed towards the centre of rotation. *See also* ANGULAR ACCELERATION; RADIAL ACCELERATION; TANGENTIAL ACCELERATION.

centrode The locus of the instantaneous centre of rotation of a moving body.

centroid *See* CENTRE OF MASS.

cermet A metallic alloy strengthened by refractory particles such as alumina. *See also* METAL–MATRIX COMPOSITES.

cetane number (CN) A number that defines the ignition quality of a diesel fuel. The scale is defined by blends of two pure hydrocarbon reference fuels, n-cetane (or hexadecane) with CN = 100 and heptamethylnonane (HMN), a fuel with very low ignition quality (CN = 15). It is given by CN = percent n-cetane + 0.15 percent HMN. *See also* OCTANE NUMBER.

CFCC *See* CONTINUOUS FIBRE CERAMIC COMPOSITES.

CFD *See* COMPUTATIONAL FLUID DYNAMICS.

CFR engine *See* COOPERATIVE FUEL RESEARCH COMMITTEE ENGINE.

CFRP Carbon-fibre-reinforced plastic/polymer.

cgs system (cgs units) An obsolete (non-SI) system of units based on the centimetre, gram, and second.

chain A series of connected links, typically of steel. For lifting, pulling, securing, etc., each link is a closed loop, often in the form of a ring. For power transmission, the links are designed to mesh with the teeth of a sprocket wheel. In contrast to belt drives, chain drives tend to

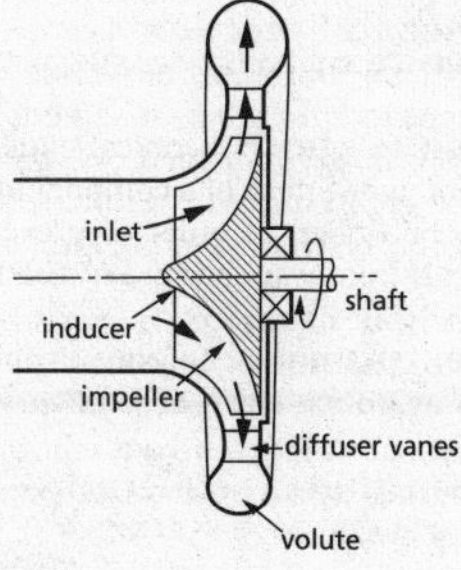

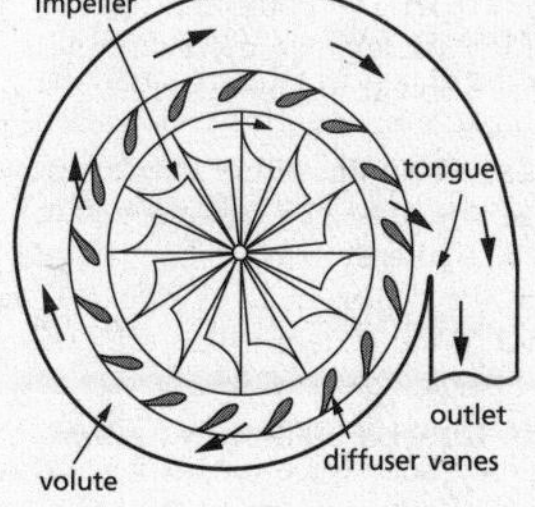

centrifugal pump

be employed in high-torque applications. *See also* KINEMATIC CHAIN; ROLLER CHAIN; SILENT CHAIN.

chain dimensioning On an engineering drawing, where the end point of one dimension is the starting point for the next. Parallel dimensioning is preferred as chain dimensioning can lead to the accumulation of tolerances. *See also* COMBINED DIMENSIONING.

chain hoist (differential chain hoist) A lifting device using a chain running over a sprocket to lift the load. The sprocket may be driven by a motor, or by force applied manually to another chain and sprocket with a gearbox joining the two sprockets to provide the mechanical advantage.

chain lines Thin lines on an engineering drawing with long sections separated by short dashes or dots. Principally used to denote centrelines and lines of symmetry.

chamfer An edge machined at an angle, typically 45°. *See also* BEVEL.

change of phase 1. A term used in thermodynamics for the change from one state (solid, liquid, vapour, or gas) to another. *See also* P-V-T DIAGRAM. **2.** A change within a given state, such as solid-state transformation in metals at different temperatures (e.g. austenite, ferrite, cementite, martensite in steels).

Chapman–Jouguet point A point on the detonation adiabat corresponding to sonic conditions downstream of a detonation wave. According to the **Chapman–Jouguet rule (Chapman–Jouguet condition),** the Mach number behind a detonation wave is unity, i.e. it propagates at the sonic velocity of the reacting gases. A **Chapman–Jouguet detonation** is a detonation wave with the Chapman–Jouguet point as the end state.

Chapman viscosity law A proportional approximation for the dependence of dynamic viscosity μ on absolute temperature T widely used in aerodynamics and given by $\mu = \mu_{REF}T/T_{REF}$ where μ_{REF} is the viscosity at a reference temperature T_{REF}.

characteristic equation *See* EQUATION OF STATE.

characteristic length *See* SCALING PARAMETER.

characteristics 1. In the propagation of acoustic waves through a gas with sound speed c, lines given by $t \pm x/c$ that trace the progress of the waves through the gas, where x represents distance and t time. *See also* COMPATIBILITY CONDITIONS. **2.** *See* SLIP-LINE FIELDS.

Charles law (Gay–Lussac law) The volume of a fixed mass of gas at constant pressure is proportional to its absolute temperature. *See also* BOYLE'S LAW; PERFECT GAS.

Charpy test (Charpy impact test, Charpy V-notch test) A notched-bar impact test in which the specimen, simply supported at both ends, is struck behind the notch by a pendulum. The decrease in height of the swing of the pendulum is a measure of the energy absorbed in fracturing the testpiece.

chart recorder An instrument that plots experimental data in the form of a dependent variable against an independent variable. It can be a Cartesian or a polar plot.

chatter 1. The noisy vibration of a cutting tool when the tool fixture is insufficiently stiff. **2.** The vibration that can occur in a control valve, sometimes referred to as flutter.

Chebyshev's mechanism *See* WATT'S MECHANISM.

check valve (clack valve, non-return valve) A mechanical device that allows fluid flow in one direction only. The numerous designs include ball, diaphragm, disc, lift, split disc, and swing check valves.

cheese head A cylindrical head on a screw or bolt. For driving, it may be slotted, or hexagonally recessed (Allen® screw).

chemical irreversibility *See* IRREVERSIBILITY.

chemical reaction A process in which two or more substances are mixed together and interact to form products. An **endothermic reaction** is accompanied by the absorption of energy. In an **exothermic reaction** chemical energy is released in the form of heat.

chilled castings Iron castings cooled at a rate that results in white iron (slow enough to avoid the formation of martensite, yet fast enough to prevent any silicon present from causing decomposition of cementite into iron and graphite).

chimney stack (chimney) *See* STACK.

choked flow (choking flow, critical flow) The flow of a compressible fluid through a contraction at the maximum possible flow rate $\dot{m}$. For a perfect gas,

$$\dot{m} = \left(\frac{2}{\gamma+1}\right)^{\frac{\gamma+1}{2(\gamma-1)}} \rho_0 c_0 A^*$$

where γ is the ratio of specific heats for the gas, ρ_0 is the stagnation density of the gas (i.e. $\rho_0 = p_0/RT_0$, R being the specific gas constant and p_0 is the stagnation pressure), c_0 is the sound speed corresponding to the absolute stagnation temperature T_O (i.e. $c_0 = \sqrt{\gamma RT_0}$) and A^* is the throat (minimum contraction) area (the superscript * denotes choked flow). The gas speed at the throat is equal to the speed of sound at that location, i.e. the Mach number equals unity. *See also* FANNO FLOW; RAYLEIGH FLOW.

choked nozzle A convergent or convergent-divergent nozzle for which choked flow occurs at the throat.

choke valve 1. A control valve for compressible fluids operating under choked-flow conditions. **2.** A valve, usually a butterfly valve, that limits air flow into the carburettor of a petrol engine to assist starting by providing a richer mixture. **3.** A heavy-duty valve, of plug or needle type, typically used in the oil industry to control liquid flow.

chopped strand mat (CSM) A thin mat formed of random discontinuous fibres (usually of glass) held together by resin, from which three-dimensional composite shapes may be manufactured which have quasi-isotropic in-plane properties. *See also* PREPEG.

chord *See* AEROFOIL.

chordal thickness *See* TOOTHED GEARING.

CHP plant *See* COMBINED HEAT AND POWER PLANT.

chuck A rotating cylindrical device for gripping drill bits and other rotating tools or work pieces. *See also* COLLET; FOUR-JAW CHUCK; JACOBS CHUCK; THREE-JAW CHUCK.

churn flow A regime of gas-liquid flow, with relatively high gas velocity, between the slug-flow and annular-flow regimes. *See also* BOILING.

CIM *See* COMPUTER-AIDED MANUFACTURING.

circlip (snap ring) An external or internal retaining ring that locates parts of circular cross section in an axial direction. It consists of an incomplete ring, with holes on either side of the gap, that may be expanded by a plier-like tool to pass into a groove in a shaft or contracted to pass into a groove in a bore.

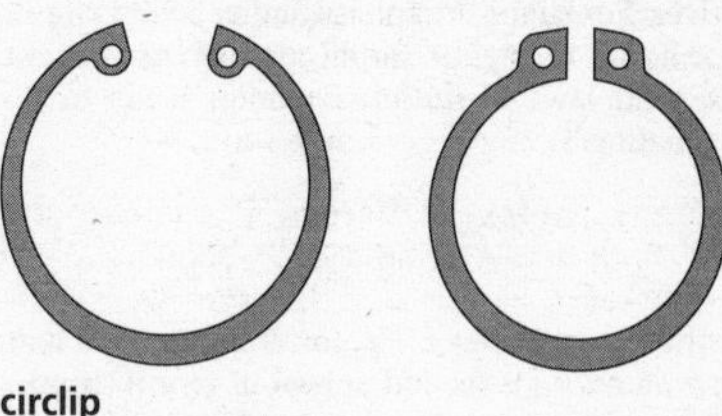

circlip

circular frequency *See* ANGULAR FREQUENCY.

circular motion The motion of a rigid body or fluid in which all particles move along circular paths about the same axis.

circular pitch (circumferential pitch) *See* TOOTHED GEARING.

circulation (Γ) (Unit m^2/s) Within a fluid flow, the anticlockwise line integral with respect to arc length around a closed curve C of the velocity component tangential to the curve. Thus $\Gamma = \oint_C V cos\alpha ds$ where V is the flow velocity, α is the angle between the velocity vector and the tangent to the curve and ds is an element of arc. *See also* KELVIN'S CIRCULATION THEOREM.

circumferential winding The manufacture of a continuous filament-reinforced composite tube by winding on a mandrel, often at angles to the axis of the pipe. *See also* AXIAL WINDING; HELICAL WINDING.

CLA *See* CENTRE-LINE AVERAGE.

clamped *See* BOUNDARY CONDITIONS; FIXITY.

Clapeyron-Clausius equation (Clausius–Clapeyron equation) An equation derived from the Clapeyron equation that can be used to determine the variation of saturation pressure p_{SAT} with saturation temperature T_{SAT} for a liquid-to-vapour phase change of a substance, according to

$$ln\left(\frac{p_2}{p_1}\right)_{SAT} = \frac{h_{fg}}{R}\left(\frac{1}{T_1} - \frac{1}{T_2}\right)_{SAT}$$

where h_{fg} is the latent heat of vaporization for the substance and R is the specific gas constant assuming the substance behaves as an ideal gas. For a solid-to-vapour phase change, h_{fg} can be replaced by the latent heat of sublimation.

Clapeyron equation The equation from which the Clapeyron–Clausius equation derives. For changes of phase, either solid to liquid or liquid to vapour, application of the first and second laws of thermodynamics leads to the equation

$$\frac{dp_{SAT}}{dT_{SAT}} = \frac{h_{fg}}{T_{SAT}(v_g - v_f)}$$

where p is pressure, T_{SAT} is the saturation temperature, h_{fg} is the latent heat of vaporization, v is specific volume and the subscripts g and f refer to the saturated vapour and liquid states, respectively. (Sometimes called the Clapeyron–Clausius equation, but not with the same meaning as the previous definition.)

classical control The design and analysis of control systems using frequency-domain methods that operate on the transfer function, i.e. on the Laplace transform of the ratio of the output to the input.

classical mechanics (Newtonian mechanics) A sub-field of mechanics based on Newton's laws of motion.

Clausius equation An equation of state for real gases that takes into account the fact that molecules cannot move to positions occupied by other molecules. It can be written $p(v - b) = RT$ where p is the gas pressure, v its specific volume, T its absolute temperature, R is the specific gas constant and b is the actual volume of 1 kg of the molecules. *See also* PERFECT GAS.

Clausius inequality (Clausius theorem) *See* SECOND LAW OF THERMODYNAMICS.

Clausius statement *See* SECOND LAW OF THERMODYNAMICS.

clean room A specially-prepared room in which a controlled atmosphere may be achieved, free of dust particles and other contaminants. Used for the assembly of delicate equipment, semi-conductor devices and the preparation of composite components.

clearance 1. The distance (if any) between mating components in an assembly. **2.** The distance between two moving parts, or a moving part and stationary part, in a machine (e.g. the gap between a piston and a cylinder head). **3.** With threads, the major clearance is the distance between the design form at the root of an internal thread and the crest of its mating external thread; the minor clearance is the corresponding dimension between the crest of an internal thread and the root of the external thread. *See also* SCREW.

clearance angle (relief angle) The angle between the underneath or flank of a cutting tool and the machined surface.

clearance fit A range of clearances ranging from close sliding to loose running, i.e. a fit in which the limits for the mating parts always permit assembly. *See* also LIMITS AND FITS.

clearance hole A hole of specified size such that a bolt, stud, etc. of the same nominal size will always pass through.

clearance volume The 'dead' volume above a piston, including the recess in the cylinder head, in a reciprocating compressor or engine when the piston is at top-dead centre.

clearness index The ratio of the solar radiation received by a horizontal surface in a prescribed period (usually one day) to the radiation that would have been received by a parallel extra-terrestrial surface in the same period.

cleavage fracture A fracture created by splitting (cleavage), as between layers in materials like slate or mica. In brittle metals, and brittle microconstituents in alloys, cleavage occurs along particular crystal planes. *See also* DUCTILE–BRITTLE TRANSITION.

clevis A U-shaped hook with holes at the ends through which a retaining bolt or pin (**clevis pin**) passes.

climb milling *See* DOWN MILLING.

clinometer *See* INCLINOMETER.

clip gauge A displacement gauge consisting of two thin strain-gauged cantilever arms attached through knife edges to a testpiece to give the load–line displacement in fracture mechanics test pieces, or used as an extensometer in tensile tests.

clock gauge *See* DIAL GAUGE.

closed cycle A thermodynamic cycle in which the working fluid is returned to the initial state at the end of the cycle and is recirculated. No fluid is exchanged with the surroundings, although there is energy transfer in the form of work and heat. Examples of idealized cycles include binary vapour cycle, Brayton or Joule cycle (**closed-cycle gas turbines**), Carnot cycle, combined gas-vapour cycle, Diesel cycle, Ericsson cycle, Otto cycle (petrol engines), Rankine cycle (**closed-cycle steam turbines**), Stirling cycle, and vapour-compression refrigeration cycle. *See also* OPEN CYCLE.

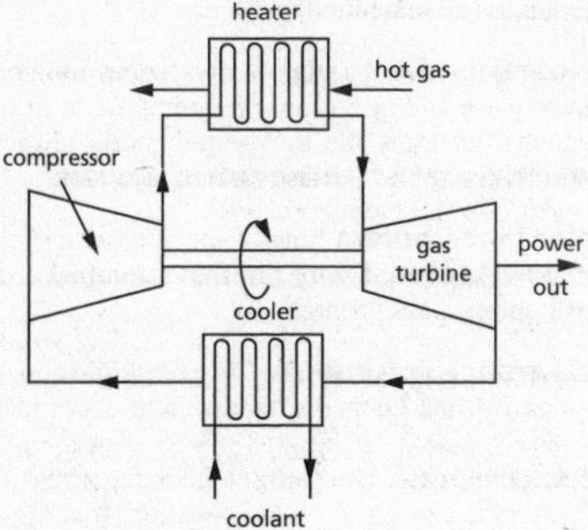

closed cycle turbine

closed-die forging The forming of a workpiece by compression within a pair of dies having the female form of the component to be manufactured (**closed dies**), superfluous metal being expelled as flash where the dies meet. *See also* OPEN-DIE FORGING.

closed feedwater heater *See* REGENERATIVE RANKINE CYCLE.

closed kinematic pair A kinematic pair where continuous contact between members is ensured by the constraints, as in pinned links.

closed loop A closed loop uses a signal from the output of a controlled plant to modify the input so as to attempt to make the output match a specified set point irrespective of disturbances. A **closed-loop control system** is one using one or more closed loops.

closed-loop gain The gain of a controller when the feedback loop is connected.

closed pair A pair of bodies both constrained so that no relative motion is possible.

closed system A closed thermodynamic system consists of a fixed amount of mass. No mass can cross its boundary although energy can, in the form of work or heat, and its volume can change. *See also* ISOLATED SYSTEM; OPEN SYSTEM.

close-off rating The maximum pressure difference to which a control valve may be subjected when fully closed. It is usually determined by the available actuator power.

close sliding *See* LIMITS AND FITS.

clutch A device for connecting and disconnecting rotating shafts, for example between an engine and a gearbox.

CNC *See* COMPUTER NUMERICALLY CONTROLLED.

coalescence-type separator A steam separator in which wet steam passes through an obstruction, such as a demister pad. Entrapped water droplets coalesce and fall to the bottom of the separator.

Coanda effect The tendency of a fluid jet flowing close to a flat or curved solid surface to attach to it due to entrainment of fluid into the jet.

coarse thread *See* SCREW.

coated abrasive An abrasive tool consisting of a flexible backing material, such as a woven cloth, paper or vulcanized fibre, a bond material, such as a glue or synthetic resin, and grit. *See also* BONDED ABRASIVE.

coaxial A term for components having a common axis such as concentric shafts.

COD *See* CRACK TIP OPENING DISPLACEMENT.

CODATA The Committee on Data for Science and Technology, which in 2006 provided a self-consistent set of values for the fundamental physical constants for international use. These values have been adopted within this dictionary.

SEE WEB LINKS

- Internationally recommended values of the fundamental physical constants

coefficient of area expansion *See* THERMAL EXPANSION.

coefficient of bulk viscosity *See* BULK VISCOSITY.

coefficient of cubical expansion *See* THERMAL EXPANSION.

coefficient of discharge (C_D) For flow through a nozzle or orifice plate, the ratio of the actual mass flow rate to the theoretical mass flow rate calculated assuming the flow to be isentropic. The **coefficient of velocity (velocity coefficient)** is the corresponding ratio of the actual average velocity to the theoretical value. For incompressible flow, the theoretical flow rate and velocity can be calculated using Bernoulli's equation. *See also* CONTRACTION COEFFICIENT.

coefficient of friction (coefficient of kinetic friction, coefficient of sliding friction, friction coefficient, *μ*) The ratio of the frictional force *F* to the normal force *N* between two surfaces in contact, i.e. $\mu = F/N$. Static friction is when there is no relative sliding; kinetic friction when there is. *See also* AMONTONS FRICTION.

coefficient of heat transfer *See* HEAT TRANSFER.

coefficient of linear expansion *See* THERMAL EXPANSION.

coefficient of mass transfer *See* MASS TRANSFER.

coefficient of performance (*COP*) A measure of the efficiency of a refrigerator or heat pump, defined as the ratio of the desired output to the required input. For a refrigerator, $COP_R = \dot{Q}_L/\dot{W}$ where $\dot{Q}_L$ is the rate of removal of heat from the refrigerated space and $\dot{W}$ is the power input into the compressor. For a refrigeration cycle, $\dot{W} = \dot{Q}_H - \dot{Q}_L$ where $\dot{Q}_H$ is the rate at which heat is transferred from the condenser to the surroundings. Thus $COP_R = 1/(\dot{Q}_H/\dot{Q}_L - 1)$ and it can be seen that COP_R can exceed unity. For a heat pump, $COP_H = \dot{Q}_H/\dot{W}$ where $\dot{Q}_H$ is the rate of heat transfer from the heat pump and $\dot{W}$ is again the power input into the compressor, and $COP_H = 1/(1 - \dot{Q}_L/\dot{Q}_H)$.

coefficient of permeability *See* PERMEABILITY.

coefficient of restitution (*e*) The ratio of the relative velocity of two colliding bodies after collision to that before. In perfectly elastic collisions $e = 1$; when all the impact energy is dissipated, $e = 0$.

coefficient of rigidity *See* SHEAR MODULUS.

coefficient of rolling friction The ratio of force parallel to a surface, on which an object rolls, to the normal force. Unlike sliding friction, rolling friction depends on the size of the contact patch and the radius of the rolling element, and the behaviour depends on whether the contact is elastic, viscoelastic, or plastic and on hysteresis losses.

coefficient of superficial expansion *See* THERMAL EXPANSION.

coefficient of thermal expansion *See* THERMAL EXPANSION.

coefficient of velocity *See* COEFFICIENT OF DISCHARGE.

coefficient of volumetric expansion *See* THERMAL EXPANSION.

coefficients of expansion *See* THERMAL EXPANSION.

coefficients of friction *See* COEFFICIENT OF FRICTION.

coextrusion The simultaneous extrusion through the same die of two or more materials in combination. *See also* PULTRUSION.

Coffin–Manson–Tavernelli relation *See* FATIGUE.

cog A tooth on the edge of a wheel (**cog wheel**), a series of which form a gear. Often applied to gears where the teeth are rudimentary and do not have involute or other precise form.

COGAS plant *See* COMBINED GAS AND STEAM PLANT.

cogeneration plant *See* COMBINED HEAT AND POWER PLANT.

coherent structure A term given to the larger eddies of turbulent shear flow, such as boundary layers, jets, and wakes, that show distinctive correlated patterns of motion.

cohesive strength A theoretical fracture strength for solids based on interatomic forces, approximately equal to $E/10$ where E is Young's modulus.

cohesive zone In fracture-mechanics modelling and simulation, the region at the crack tip over which an assumed traction (load-displacement) relation has to be overcome to permit initiation and propagation of a crack.

coil spring A spiral ('clockwork') or helical (cylindrical) spring.

coining A forging operation, employing a closely-fitting punch and die from which no metal is allowed to escape, in which the surface pattern on the punch and die is imprinted on the blank.

Colburn equation For fully-developed turbulent flow in a smooth pipe, an empirical formula for the Nusselt number given by $Nu = 0.023Re^{0.8}Pr^{1/3}$ for Prandtl numbers in the range $0.7 \leq Pr \leq 160$ and Reynolds numbers greater than 4000. *Nu* here is based on the pipe diameter *D* and *Re* on *D* and the bulk velocity of the flow.

Colburn *j*-factor (heat-transfer *j*-factor, j_H, j_M) A non-dimensional parameter that arises in convective-heat-transfer analysis and is defined as $j_H = St\ Pr^{\ 2/3}$ where *St* is the Stanton number and *Pr* is the Prandtl number. For mass transfer, the definition is $j_M = St_M\ Sc^{\ 2/3}$ where St_M is the Stanton number for mass transfer and *Sc* is the Schmidt number.

cold-air standard assumptions Air-standard assumptions for which the air has constant specific heats specified at 25°C.

cold joint *See* SOLDERING.

cold trap A device used to trap vapour from water and other solvents by passing it through stainless-steel or glass tubing cooled by liquid nitrogen, dry ice, a dry ice–alcohol slurry, or a dry ice–acetone slurry. The vapour condenses to liquid or ice crystals on the tube wall. A common application is to the inlet or outlet of a vacuum pump.

cold working The plastic deformation of a metal, by rolling (**cold rolling**), drawing, forging (**cold forging**), etc. at a temperature well below its recrystallization temperature, which results not only in permanent shape change but also increase in strength and loss of ductility owing to work hardening. *See also* HOT WORKING; STRAIN HARDENING; WARM WORKING.

Colebrook equation An empirical correlation for the Darcy friction factor *f* as a function of Reynolds number *Re* in fully-developed incompressible turbulent flow through smooth and rough pipes, and the basis for the Moody chart. It can be written as

$$\frac{1}{\sqrt{f}} = -2log_{10}\left(\frac{\varepsilon/D}{3.7} + \frac{2.51}{Re\sqrt{f}}\right)$$

where *D* is the pipe diameter, ϵ represents the equivalent roughness of the pipe-wall surface, and *Re* is based upon *D* and the bulk velocity of the flow. *See also* BLASIUS PIPE-FRICTION LAW.

collapse load The applied load at which a structure becomes a mechanism owing to the formation of sufficient plastic hinges for collapse to occur. *See also* PLASTIC DESIGN.

collapse mechanism The pattern and location of plastic hinges in a structure that would permit it to become a mechanism and thereby collapse. Various patterns may be envisaged for a given structure: that requiring least work (least collapse load over the same displacement) is the natural one.

collar A ring secured to, or integral with, a shaft to give axial location. *See also* CIRCLIP.

collar bearing *See* BEARING.

collector *See* SOLAR COLLECTOR.

collector efficiency *See* CAPTURE EFFICIENCY.

collet A type of chuck for holding small, circular workpieces or tools. It consists of a slotted sleeve with an external cone at the gripping end that holds the workpiece or tool when pulled against an internal cone on the lathe mandrel.

colloidal system A system in which microscopic particles are dispersed uniformly throughout another substance. Both the dispersed and the continuous phases may be solid, liquid or gas (although a gas-gas combination is not possible). *See also* AEROSOL; EMULSION; FOAM; GEL; HYDROSOL; SOL.

colour temperature The absolute temperature of an ideal blackbody that radiates light of the same colour as a given light source. The range is from about 1000 K (candlelight) to about 20 000 K (blue sky). For forging and hardening of steel, the range is from 873 K (brown-red) to 1473 K (white), and for tempering from 483 K (light yellow) to 603 K (grey-green).

column A vertical member of a structure that withstands axial compressive loads. *See also* PROP; STRUT.

column ends *See* FIXITY.

combined dimensioning The use of chain dimensioning and parallel dimensioning on the same engineering drawing.

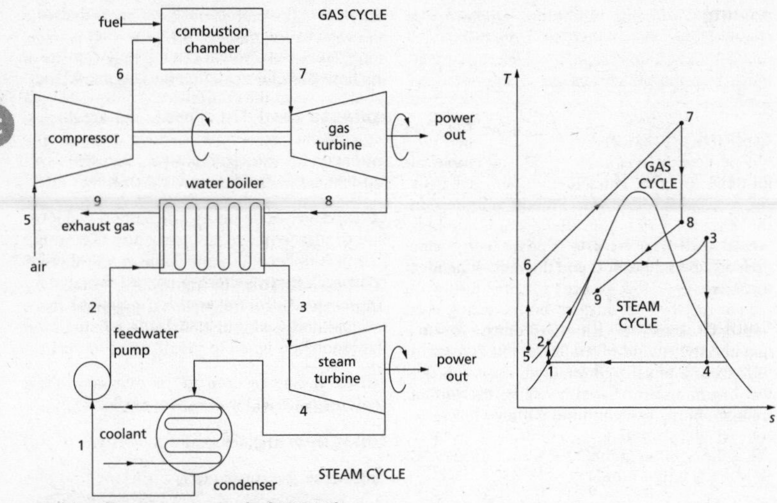

combined gas-vapour cycle (COGAS)

combined gas and steam plant (**COGAS plant, combined cycle power plant, CCPP, combined cycle gas turbine plant, CCGT plant**) A power plant in which the exhaust gases from a gas turbine are used in the production of steam for a steam turbine. COGAS plants are commonly used for marine propulsion.

combined gas–vapour cycle The thermodynamic cycle on which a combined gas and steam plant operates, such as that shown in the schematic diagram for an open-cycle gas turbine and a closed-cycle steam turbine. Exhaust gas from the gas turbine is the heat source for the water boiler. The gas and steam cycles are shown superimposed on a temperature (T) *vs* specific entropy (s) plot.

combined heat and power plant (**CHP plant, cogeneration plant, total-energy plant**) A plant for the simultaneous production of more than one useful form of energy from the same energy source, such as process heat and electric power. In the arrangement shown, partially expanded steam is extracted from the steam turbine and used to produce process heat.

combined stresses The stress state at a point in a component subjected to combination of axial, bending, torsional loadings etc., acting along all reference axes.

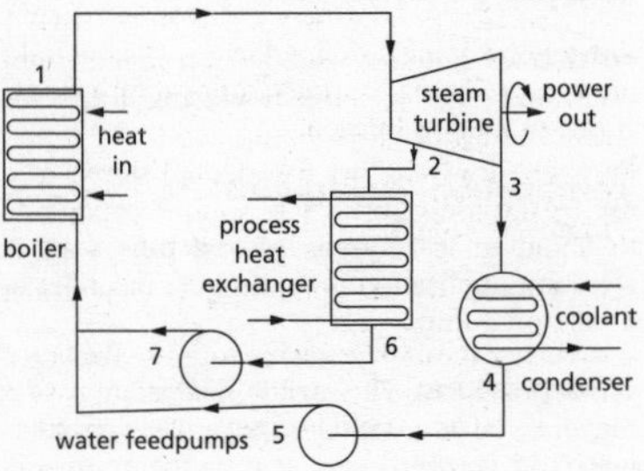

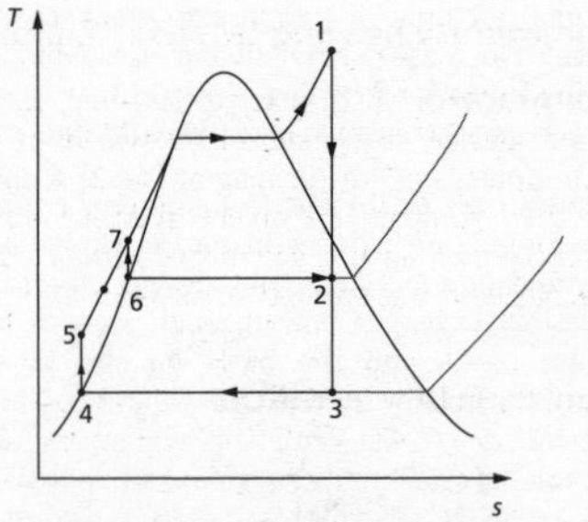

combined heat and power plant (COGEN)

combustible (inflammable) A term for substances that can be ignited and burned.

combustion An exothermic chemical reaction in which a fuel and an oxidant, typically air, react together to release a significant quantity of thermal energy in the presence of a flame. In the case of fluid fuels, ignition may be from an electric spark or compression of the fuel–oxidizer mixture. Solid fuels are usually ignited by a small flame. *See also* COMBUSTION EFFICIENCY.

combustion chamber 1. In a piston engine, the volume between the head of an individual cylinder and the crown of the piston in which the fuel–air mixture burns during each power stroke of the engine. **2. (combustor)** In a gas-turbine engine, a ramjet, an afterburner, or a rocket motor, the component, often cylindrical or annular in shape, in which the fuel–air mixture burns in a continuous-flow process.

combustion deposit Ash, carbon, and other incombustible solids, often due to impurities in the fuel, that build up on any surfaces exposed to products of combustion. They can lead to corrosion, reduced heat transfer and so higher flue-gas temperatures and reduced efficiency.

combustion efficiency (source efficiency, η_{COMB}) A performance measure for combustion equipment defined by η_{COMB} = amount of heat released during combustion/calorific value of the fuel burned.

combustion safeguard *See* THERMAL FLAME SAFEGUARD.

command The set point input to a controller.

comminution The fracture or pulverization of brittle materials into small fragments by passing the material through rollers, ball mills, or grinders.

common normal *See* TOOTHED GEARING.

community heating *See* DISTRICT HEATING.

comparator 1. In a control system, a two-input device that determines which of the two inputs is of greater magnitude. **2.** A device used to compare an object, typically during manufacture, against a standard master example. *See also* MECHANICAL COMPARATOR; OPTICAL COMPARATOR.

compatibility conditions (compatibility equations) 1. Mathematical relations between strains in solid mechanics that must arise because the six components of strain at a point can be expressed in terms of only three coordinate displacements. May be expressed in Cartesian, cylindrical, etc. coordinate systems. **2.** In the solution of the equations for two-dimensional supersonic flow, which is governed by a hyperbolic partial differential equation, the relationships between the Prandtl–Meyer function v and the local flow direction θ: $v \pm \theta$ are both constant along characteristics. Similarly in the theory of plasticity, slip-line fields are orthogonal curvilinear lines (characteristics) along which $p \pm 2k\varphi$ are constant, where p is hydrostatic stress, k is the shear yield stress and φ is the inclination of the characteristics.

compensation The improvement of the performance of a control system through the use of an additional control element, a compensator, frequently designed to modify the root locus of the system.

complementary strain energy The complementary strain energy is given by $\int_0^P \delta \, dP$ where δ is displacement and P is load. In contrast, the strain energy is given by $\int_0^\delta P \, d\delta$. In linear systems they have the same value, but in non-linear systems they do not. *See also* CASTIGLIANO'S THEOREMS; FRACTURE MECHANICS.

complex pendulum A system of two or more simple pendulums connected end-to-end. The resulting oscillations are chaotic.

complex plane The graphical representation of complex quantities, where the real parts are plotted on the abscissa and the imaginary parts on the ordinate. Used, for example, in control-system design to produce the pole-zero plot.

complex viscosity (η^*) (Unit Pa.s) The frequency-dependent viscosity function determined for a viscoelastic fluid by subjecting it to oscillatory shear stress. The dynamic viscosity (η') is the in-phase real part such that $\eta^* = \eta' - i\eta''$, η'' being the imaginary out-of-phase viscosity and $i = \sqrt{-1}$. *See also* LOSS MODULUS; STORAGE MODULUS.

compliance (Unit m/N) **1.** The reciprocal of stiffness, being the displacement resulting from the application of unit load. In the case of beams, flexibility is an alternative term used for compliance. **2.** For a spring, the inverse of spring rate.

composite material (composite structure) A general term used of two or more materials or structures acting in combination (e.g. concrete, reinforced concrete, filament-reinforced polymers, laminated materials, particulate-reinforced materials, flitched beams), resulting in values of strength, stiffness, or toughness greater than the base matrix material alone.

SEE WEB LINKS

- Composite materials

C

composite modulus (*E*) The Young's modulus of two or more materials in combination is given by the weighted average of the *E* values for the individual components. When loading parallel to the fibres in a filament-reinforced composite (assuming that the strains in every component are identical) $E = vf_1E_1 + vf_2E_2 + \ldots$. where vf_1 etc are the volume fractions and E_1 etc are the Young's moduli of the individual components (the Voigt upper bound). The composite strength (σ) is obtained in the same way i.e. $\sigma = vf_1\sigma_1 + vf_2\sigma_2 + \ldots$. where σ_1 etc. are the strengths of the individual components. When loading across the fibres in a filament-reinforced composite (assuming that the stresses in every component are identical) $1/E = vf_1/E_1 + vf_2/E_2 + \ldots$, i.e. the weighted harmonic mean (the Reuss lower bound). The strength in this case is that of the weakest component.

composite property In thermodynamics, a property defined in terms of the properties of a closed system and its surroundings, such as the non-flow exergy function.

composite wall In heat transfer, a wall consisting of two or more layers of material with different thermal conductivities. For a plane composite wall, with convective heat transfer on either side and an overall temperature difference ΔT, the heat flux $\dot{q}''$ is given by the equation $\Delta T/\dot{q}'' = 1/h_i + \Sigma_i\Delta_i/k_i + 1/h_o$ where h_i and h_o are the convective heat transfer coefficients on the outer surfaces, Δ_i is the thickness of layer i and k_i is its thermal conductivity. The term Δ_i/k_i is called the thermal or conduction resistance while $\dot{q}''/\Delta T$ is called the overall heat-transfer coefficient and given the symbol *U*, values of which are often quoted for double-glazed windows and cavity walls.

compounding In a steam engine (**compound steam engine**) or impulse turbine, the progressive reduction in pressure (expansion) across two or more stages in series. *See also* CURTIS STAGE; PRESSURE-COMPOUNDED STEAM TURBINE; VELOCITY-COMPOUNDED STEAM TURBINE.

compound pendulum A rigid body free to swing about an axis. *See also* SIMPLE PENDULUM.

compound steam-turbine system A steam-power plant where several turbines are connected together. *See also* CROSS-COMPOUND STEAM-TURBINE SYSTEM; TANDEM-COMPOUND STEAM-TURBINE SYSTEM.

compressed-air energy storage (CAES) A process of compressing air to 100 bar or more using off-peak electricity and storing it in an underground cavern, such as a disused mine. At times of peak demand the air is fed to a peaking gas turbine, resulting in much reduced fuel consumption.

compressed liquid A liquid subjected to a pressure greater than the saturation pressure corresponding to its temperature. *See also* SUBCOOLED LIQUID.

compressibility A measure of the reduction in volume or increase in density when a substance is subjected to an increase of pressure. Liquids and solids are normally considered incompressible, whereas gases are highly compressible. *See also* BULK MODULUS; INCOMPRESSIBLE; ISENTROPIC COMPRESSIBILITY; ISOTHERMAL COMPRESSIBILITY.

compressibility factor (*Z*-factor). A nondimensional parameter, dependent upon reduced temperature and reduced pressure, employed to indicate deviations from the ideal-gas equation ($Z \neq 1$), and given by $Z = pv/RT = p/\rho RT$, where p is the gas pressure, v is the specific volume of the gas and ρ $(= 1/v)$ is its density, R is the specific gas constant and T is the absolute temperature. *See also* PRINCIPLE OF CORRESPONDING STATES; PERFECT GAS.

compressible flow A gas flow in which the Mach number M is sufficiently high for the gas density to change significantly. For air, this is when $M > 0.3$.

compression 1. Loading, the principal effect of which is to squeeze and shorten a component or testpiece. **2.** The reduction in volume and increase in density of a substance as a consequence of increased pressure.

compressional wave speed (*c*) (Unit m/s) In a uniform, isotropic, elastic solid, plane compression (and tension) waves travel at speed $c = \sqrt{(K + 4G/3)/\rho}$ where K is the isentropic bulk modulus of the solid, G is its shear modulus and ρ is its density. In a long rod, thin enough that Poisson's-ratio effects do not produce lateral stresses to curve the wave front, $c \to \sqrt{E/\rho}$ from setting Poisson's ratio to zero in the previous expression for *c*.

compression crease A crease formed during the compression of composites having a high volume fraction of filaments. These composites fail in compression by forming a crease at an angle to the loading axis.

compression failure The reduction or removal of a component's load-bearing capacity in compression, caused by buckling, fracture, crease formation in fibre composites, etc.

compression fitting A screwed joint for pipework made resistant to leakage by permanent deformation of a closely-fitting ring, called a ferrule or olive, on tightening.

compression-ignition engine *See* DIESEL ENGINE.

compression member A structural component, the major loading on which is compressive. *See also* COLUMN; PROP; STRUT.

compression modulus *See* BULK MODULUS.

compression moulding The manufacture of components by moulding plastics (polymers) in granular or pellet form using heat and pressure. *See also* ROTATIONAL MOULDING; TRANSFER MOULDING.

compression pressure The pressure produced in a cylinder of a piston engine by compression of air in the absence of fuel.

compression ratio For a piston engine, if the swept volume is $\mathcal{V}_{SW}$ and the clearance volume is $\mathcal{V}_{CL}$, the compression ratio is given by $(\mathcal{V}_{SW} + \mathcal{V}_{CL})/\mathcal{V}_{CL}$, i.e. it is a volume ratio rather than a pressure ratio.

compression refrigeration *See* VAPOUR-COMPRESSION REFRIGERATION CYCLE.

compression ring *See* PISTON.

compression spring A spring that resists compression forces, usually in the form of a helix with separated coils (giving a linear axial stiffness) or a cone with separated coils (giving a non-linear axial stiffness). *See also* BELLVILLE WASHER.

compression stroke The stroke in a reciprocating compressor or engine during which the working fluid is compressed.

compression test The determination of the stress–strain curve of a material by axial loading of a specimen in compression. In brittle materials failure is in the elastic range; more ductile materials will yield before fracture; very ductile materials will plastically deform extensively before failure. The **compression strength (compressive strength)**, with unit Pa, is the compressive stress that causes failure in a component or structure. *See also* CRUSHING STRENGTH.

compression wave (dilatation wave) In a fluid or a solid, a progressive wave or wavefront that compresses the medium through which it propagates. *See also* RAREFRACTION WAVE; SHEAR WAVE.

compressive stress (Unit Pa) The compressive load per unit area at a point in a component. *See also* FORCE.

compressor A turbomachine, of either axial or radial type, that increases the pressure of a gas or vapour.

compressor blades The aerofoil-shaped vanes that form the rotor(s) and stator(s) of an axial-flow compressor. The aerodynamic design is more critical than is the case for turbine blades because there is an increase in pressure across each row of blades. The arrows in the diagram indicate the flow direction relative to the blades. *See also* VELOCITY TRIANGLE.

compressor bleed The removal of air before the final stage of a multistage compressor operating below design speed, to prevent the final stage from choking.

Comprex® A type of supercharger comprising a cylindrical rotor with longitudinal channels which is belt driven from the crankshaft of an internal-combustion engine. The inlet air is compressed by wave motion within the channels.

computational domain In CFD, FEM and other numerical methods, the area or volume within which calculations are performed and on the periphery of which the boundary conditions are specified.

computational fluid dynamics (CFD) The solution of fluid mechanics problems using a system of algebraic equations, based upon finite-difference, finite-volume or other discretization schemes to approximate the differential equations and the constitutive equations that govern fluid motion. All flows of Newtonian fluids are governed by the Navier-

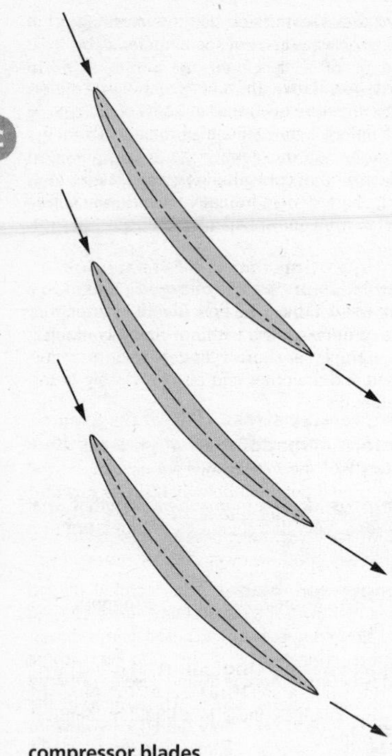

compressor blades

Stokes equations but in the case of turbulent flow, most calculations involve turbulence modelling, although direct numerical simulation is possible at relatively low Reynolds numbers.

computed path control In CNC or robotics, the use of a control program to determine the required path for the tool or end effector. This required path forms the set points for the motor controllers moving the machine tool or robot.

computer-aided design (CAD) Generally, design and calculations performed by computer; more specifically, the use of computer graphics and models to impart design concepts. **Computer-aided design and computer-aided manufacturing (CAD/CAM)** is where component dimensions resulting from CAD are passed by electronic means directly to machines for manufacture.

computer-aided engineering (CAE) The non-trivial use of computers in all branches of engineering, for example to solve equations, create engineering drawings and schematic diagrams, apply CAD/CAM, CFD, and other software packages.

computer-aided manufacturing (CAM, computer-integrated manufacturing, CIM) The use of computers in all branches of manufacturing, not only to control machines and robots for manufacturing and assembly, but also for process planning, and monitoring progress of materials and components during production, etc.

computer control The control of an engineering device or system by pre-programmed computer or by feedback control.

computer numerical control (CNC) A term relating to machine tools in which the movements of a tool and/or the workpiece are controlled by computer.

computer vision The digitization and processing of optical images/patterns by computer in order to recognize parts, orientation, etc. in manufacturing.

concentrated load A load on a component which is distributed over a very small area, idealized as the line load of a wedge or knife edge, and the point load of a cone.

concentration ratio For a concentrating solar collector, the ratio of the projected area of the concentrator facing the solar beam to the area of the receiver.

concentrator (concentrating solar collector) *See* SOLAR CONCENTRATOR.

concurrent engineering The integration of the procedures for product design, material selection and manufacturing method to include life-cycle analysis.

condensate strainer A filter in a steam plant used to remove particulate matter from condensate before it is added to feedwater.

condensation 1. The change of vapour into the liquid state when its temperature falls below the saturation temperature T_{SAT}. This usually occurs on a surface having a temperature (**condensation point, liquefaction point**) below T_{SAT} but can also occur spontaneously throughout the vapour. *See also* DROPWISE CONDENSATION; FILM CONDENSATION. **2. (condensate)** A liquid that has condensed from a vapour.

condensation shock For supersonic flow of a moist gas through a divergent nozzle, condensation occurs in the form of spontaneous nucleation at some point downstream of that at which the temperature falls to the saturation temperature. The condensation process proceeds rapidly, and results in a fairly thick discontinuity termed a condensation shock.

condensation trail *See* VAPOUR TRAIL.

condenser A heat exchanger in which a substance is changed from its vapour phase to its liquid phase by reducing its temperature to below the saturation temperature. *See also* SHELL-AND-TUBE CONDENSER; STEAM CONDENSER; SURFACE CONDENSER.

condenser vacuum The sub-atmospheric pressure imposed on the condenser of a steam-power plant which leads to an appreciable increase in overall efficiency.

condensing boiler A relatively small industrial or domestic boiler that burns sulphur-free natural gas so that the products of combustion do not contain sulphuric acid and can be allowed to condense on heat-transfer surfaces without danger of corrosion.

conductance *See* HEAT TRANSFER.

conduction resistance *See* HEAT TRANSFER.

conduction shape factor (*S*) (Unit m) For a solid body with two surfaces maintained at constant temperatures T and $T + \Delta T$, the rate of conductive heat transfer $\dot{Q}$ between the surfaces is given by $\dot{Q} = Sk\Delta T$ where k is the constant thermal conductivity of the solid medium and S is a factor that accounts for the geometry of the body, e.g. for a sphere of diameter D a distance z ($> D/2$) below the surface of a semi-infinite solid, $S = 2\pi D/(1 - D/4z)$.

conduction thickness *See* HEAT TRANSFER.

conduit A pipe, duct, or open channel through which there can be fluid flow.

cone bearing *See* BEARING.

cone clutch A friction clutch in which an internal cone moves axially in or out of engagement with an external cone. One or both surfaces is lined with high-friction material.

cone pulley A stepped pulley having several diameters which, when linked by a laterally-moveable transmission belt to a corresponding pulley, gives a series of speed ratios.

configuration In robotics, the description of the structure of a robot in terms of the type of each joint (i.e. translational or rotational) and the directions of the joint axes. There are five standard robot configurations: articulated (revolute), Cartesian-coordinate, cylindrical-coordinate, SCARA, and spherical-coordinate.

configuration factor *See* VIEW FACTOR.

confined flow The flow of a fluid through a pipe or duct, or flow within a closed container, for example a short cylinder with one end closed and the other end being a closely-fitting rotating disc.

conformal mapping (conformal transformation) The transformation, in two-dimensional potential-flow theory, of a function in one complex plane to another. This transformation preserves angles between lines in the two planes. *See also* JOUKOWSKI TRANSFORMATION.

confuser A converging duct. The opposite of a diffuser.

congruent melting point The temperature at which a solid substance at a specified pressure changes phase to a liquid of identical composition.

conical pendulum A pendulum in which the bob rotates in a horizontal circle with constant angular velocity about the vertical axis.

conical spring An open-coiled spring in the form of a cone which gives a non-linear axial stiffness. *See also* COILED SPRING.

conjugate action *See* TOOTHED GEARING.

conjugate teeth profile *See* TOOTHED GEARING.

connecting rod (con rod) A link that transmits power from one system to another, often changing linear to rotary motion, as in the rod connecting the piston to the crankshaft in a reciprocating compressor or pump or to the crankshaft in an internal-combustion engine, as in the diagram. The **big end (bottom end)** is the larger end that connects to the bearing on one of the crankpins of the crankshaft. The **little end (small end)** is joined by a gudgeon pin to the piston.

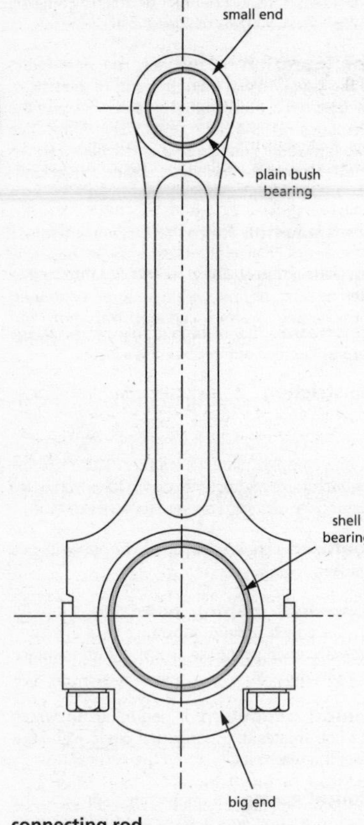

connecting rod

conservation equations The four equations or principles which form the basis of Newtonian mechanics are (a) matter cannot be created or destroyed (**conservation of mass**); (b) energy cannot be created or destroyed but may change its form (**conservation of energy**), a statement which derives directly from the first law of thermodynamics; (c) the total vector momentum remains constant in a system where bodies are subjected only to forces exerted by other bodies in the same system (**conservation of linear momentum**). This statement derives directly from Newton's second law of motion; (d) the total vector angular momentum remains constant in a system where bodies are subjected only to forces exerted by other bodies in the same system (**conservation of angular momentum**).

For fluid flow, each of the four equations is expressed in terms of the time rate of change of the conserved property contained in a control volume (CV), the net flow rate of that property across the surfaces of this volume (CS), and the effects of any external forces (linear-momentum conservation) or external moments (angular-momentum conservation) acting on CV or the net rate of heat transfer across CS less the power input into CV (energy conservation). If the vector velocity is $\boldsymbol{V}$ and the density of the fluid is ρ, then the mass-conservation equation (**continuity equation**) can be written as:

$$\frac{d}{dt}\int_{CV} \rho d\mathcal{V} + \int_{CS} \rho(\boldsymbol{V}\cdot\boldsymbol{n})dA = 0$$

where $\mathcal{V}$ represents the volume and A the surface area of CV and $\boldsymbol{n}$ is the unit normal vector directed out from the control surface.

The momentum-conservation equation can be written as

$$\frac{d}{dt}\int_{CV} \rho \boldsymbol{V} d\mathcal{V} + \int_{CS} \rho \boldsymbol{V}(\boldsymbol{V}\cdot\boldsymbol{n})dA = \Sigma \boldsymbol{F}$$

where $\Sigma \boldsymbol{F}$ represents the sum of all the external forces acting on CV. The angular-momentum-conservation equation can be written as

$$\frac{\partial}{\partial t}\int_{CV} (\boldsymbol{r}\times\boldsymbol{V})\rho\, d\mathcal{V} + \int_{CS} (\boldsymbol{r}\times\boldsymbol{V})\rho(\boldsymbol{V}\cdot\boldsymbol{n})dA = \Sigma \boldsymbol{M}$$

where $\boldsymbol{r}$ is the position vector from the axis of rotation to the elemental mass $\rho d\mathcal{V}$ and $\Sigma \boldsymbol{M}$ represents the sum of all the external moments acting on CV.

The energy-conservation equation can be written as

$$\frac{d}{dt}\int_{CV} e\rho\, d\mathcal{V} + \int_{CS} e\rho \boldsymbol{V}\cdot\boldsymbol{n}\, d\mathcal{V} = \dot{Q}_{in,CV} + \dot{W}_{in,CV}$$

where the heat-transfer rate $\dot{Q}_{in,CV}$ is the net rate of energy exchange between the control volume and its surroundings because of temperature differences, and $\dot{W}_{in,CV}$ is the work transfer rate into the control volume by the surroundings, i.e. the power input. $\dot{W}_{in,CV}$ includes power transferred across CS by a rotating shaft, by fluid pressure, and by shear stress. If u represents the specific internal energy of the fluid, g the acceleration due to gravity, and z the height above a datum level, then the stored energy $e = u + ½ V^2 + gz$, i.e. the sum of u, the kinetic energy per unit mass and the potential energy per unit mass.

conservative force field A system of forces where the work done on a body depends only upon the initial and final states, and is not dependent on the path taken (i.e. the process is reversible).

conservative property A property of a system whose value is invariant when other parameters vary.

consolute temperature The temperature at which two partially miscible liquids become fully miscible. Liquids which are fully miscible at high temperatures but separate into two liquid phases at lower temperatures have an **upper consolute temperature (upper critical solution temperature)**. Liquids which are fully miscible at low temperatures but separate at higher temperatures have a **lower consolute temperature**.

constant-force spring A spring that has the same restoring force regardless of displacement. The most common type takes the form of a coiled strip that, owing to tight coiling during manufacture, is pre-stressed (a steel measuring tape is an example). The uncoiling force is approximately constant as the change of curvature of the strip is approximately constant. Not to be confused with a clockwork spring from which power can be obtained.

constant-mesh gearbox A gearbox in which the pairs of gears giving different speed ratios are constantly in mesh, different ratios being obtained by connecting or disconnecting the relevant gear to the driving shaft.

constant-pressure gas thermometer An apparatus based on Charles law in which a rigid vessel is filled with a gas, usually hydrogen or helium, at low pressure and its volume measured as its temperature is increased while its pressure is maintained constant. The device must be calibrated at two fixed points, such as the ice and steam points. A **constant-volume gas thermometer (gas thermometer)** is similar but based on Boyle's law. The gas pressure is measured as its temperature is increased while its volume is maintained constant.

constant-speed propeller *See* PROPELLER.

constant-velocity universal joint (CV joint, homokinetic joint) A connexion that transmits constant angular velocity between two shafts that are neither necessarily in line nor whose axial position is necessarily fixed.

constitutive equation (constitutive relation) 1. In solid mechanics or fluid mechanics, an algebraic or numerical relation for the dependency of stress on deformation, strain, strain rate, temperature, etc. in a material. In solid mechanics such a relation is sometimes called an equation of state. Simple linear examples include Hooke's law and Newton's viscosity law. **2.** In heat transfer, a relation, such as Fourier's law of heat conduction, connecting heat flux with temperature gradient. *See also* FICK'S LAW.

constraint The restriction of one or more natural degrees of freedom of a system.

constriction A reduction in the cross-sectional area of a pipe or duct which, especially if sudden rather than gradual, can significantly reduce the flow rate for a given pressure difference.

constructive interference *See* DESTRUCTIVE INTERFERENCE.

contact angle (wetting angle, ϕ) At a gas-liquid-solid interface, the angle that the tangent to the liquid surface makes with the solid surface at the point of contact. If $\phi < 90°$, the liquid is said to wet the surface. For water-glass-air ϕ is practically zero, for kerosene-glass-air it is about 26° and for mercury-glass-air 130°. The **contact line** is the interface between a liquid and a gas on a solid surface. *See also* HYDROPHILIC; HYDROPHOBIC; SURFACE TENSION; YOUNG'S EQUATION.

contact area (contact patch) When two bodies having flat or curved surfaces are pressed together, the region where the surfaces touch is deformed ('locally flattened') into, for example, a rectangular contact area (two parallel cylinders in contact), a circular area (two spheres in contact) or generally an ellipse. *See also* CONTACT MECHANICS; ROLLING FRICTION.

contact fatigue strength *See* FATIGUE.

contact gear ratio (contact ratio) *See* TOOTHED GEARING.

contact mechanics The determination of **surface** and **sub-surface** stresses and strains when bodies in contact are loaded. For normal loading within the elastic range, the surface pressure distribution p over a general elliptical contact patch having semi-axes a and b is given by the ordinates of a semi-ellipsoid contained by the surface of contact, viz: $p = p_0\sqrt{[1 - (x/a)^2}$

$-(y/b)^2]$ where p_0 is the peak pressure. The total load F is given by the volume of the ellipsoid, i.e. $F = 2\pi abp_0/3$, so that the peak pressure is 1.5 times the average pressure $F/\pi ab$. The greatest shear stress, where plastic flow is likely to initiate, occurs beneath the surface at a depth of some $0.79b$ in the particular case of two cylinders in contact. In the presence of shear loading as well as normal loading (as in gearing), the location of the maximum shear stress moves nearer to the surface and, for coefficients of friction greater than about 0.1, is in the surface.

contact patch The contact area of a tyre. *See also* CONTACT AREA.

contact ratio *See* TOOTHED GEARING.

contact resistance *See* THERMAL CONTACT RESISTANCE.

contact strength The maximum allowable load between contacting bodies converted into a stress.

contact stresses *See* CONTACT MECHANICS.

contact surface 1. A surface that separates two fluids of different properties, such as water and air, hot air and cold air, or nitrogen and helium. It may be idealized as a surface of discontinuity, although for miscible fluids diffusion leads to thickening e.g. for gas–gas contact surfaces. **2.** The flattened surface at the contact patch between bodies pressed together.

continuity equation *See* CONSERVATION EQUATIONS.

continuous casting The casting of metal into a continuous stream rather than into ingots.

continuous cooling curve *See* TIME–TEMPERATURE TRANSFORMATION DIAGRAM.

continuous fibre ceramic composites (CFCC) Composite materials, such as silicon carbide (SiC) fibres in a matrix of polycrystalline alumina (Al_2O_3).

continuously-variable transmission (CVT, infinitely-variable transmission) A gearbox that can change steplessly through an infinite number of gear ratios between maximum and minimum values.

continuous path control Control of the position of a robot end effector where the required position is specified at all times, rather than as a series of discrete steps. For example, a robot performing arc welding would require continuous path control so that it could follow the seam being welded.

continuous time A term referring to a system in which information is continuously accessible, rather than being available only at discrete time intervals as occurs with a sampled data system.

continuum hypothesis (continuum assumption) The assumption that for any substance, a length scale exists which is much larger than the largest scale at which the molecular structure is important, but smaller than the scale at which there are significant spatial property variations, e.g. due to temperature or pressure variations or inhomogeneity of the substance itself or, in polycrystalline solids, anisotropy caused by different orientations of individual grains. In the majority of practical situations such a length scale is found to exist (typically order 1 μm) for most fluids and solids so that mathematical modelling of material behaviour is possible which takes no account of the molecular structure. The continuum hypothesis allows the definition of material properties, velocity, stress, etc. at any point within a substance. Exceptions in fluid mechanics include rarefied gas flow and the flow of fluids in microchannels where the mean-free path of the fluid molecules is of the same order as the flow dimensions. Exceptions in solid mechanics include the deformation of thin sheet material having a grain size comparable with the sheet thickness.

contraction 1. A reduction in the volume of a solid object, frequently due to a temperature reduction. **2.** A reduction in the cross-sectional area of a flow channel, usually gradual (for example, the confuser in a wind tunnel).

contraction coefficient (C_C) A factor that accounts for the area reduction of flow through a sharp-edged orifice. It is defined by $C_C = A_V/A_O$ where A_O is the orifice area and A_V is the minimum jet area. The section of the jet with area A_V is termed the *vena contracta*. *See also* COEFFICIENT OF DISCHARGE.

contraction crack A crack in a casting formed during cooling.

contraflexure *See* POINT OF CONTRAFLEXURE.

contrail *See* VAPOUR TRAIL.

contrarotating propellers Two propellers mounted one behind the other coaxially on the same shaft but rotating in opposite directions. The second propeller counteracts the swirling flow produced by the first.

contrate gear *See* TOOTHED GEARING.

control *See* CONTROL SYSTEM.

control accuracy *See* CONTROL SYSTEM.

control element *See* CONTROL SYSTEM.

control flow *See* FLUIDICS.

controllability *See* CONTROL SYSTEM.

controllable-pitch propeller *See* PROPELLER.

controlled rolled steel A high-strength, low-alloy (HSLA) steel with a microstructure of bainitic acicular ferrite.

controlled-strain mode *See* ROTATIONAL VISCOMETER.

controlled-stress mode *See* ROTATIONAL VISCOMETER.

controlled variable 1. (independent variable) In an experiment, a quantity that is kept constant or otherwise controlled. **2.** *See* CONTROL SYSTEM.

controller *See* CONTROL SYSTEM.

controller–structure interaction The interaction between the incomplete model in a controller and the full plant dynamics. Where a controller uses a mathematical model of the controlled plant, a reduction in the performance of the controller can be caused by feedback of signals from the sensor which represent unmodelled characteristics of the plant behaviour. As a result, the controller can excite these unmodelled characteristics in the plant.

control surface 1. The bounding surface of a control volume. **2.** In a control system, where the action of a controller is different for different set points or plant outputs, the control surface determines where switching between different control actions occurs. **3.** A moving surface exposed to the airflow over an aircraft, including the ailerons, elevators, flaps, and rudder, used by the pilot to control the aircraft's flight.

control system A collection of components used to ensure that the output of some system, referred to as the **plant**, behaves in a required way. The plant being controlled may have a single parameter as the output, such as the temperature of an oven, or numerous parameters, such as the angles of each joint of a robot. The control system takes information on the required plant output, known as the control system **reference input** or **set point**, and determines the input to be supplied to the plant to best achieve the required output.

Control systems have one of two different modes of operation: **open-loop** or **closed-loop**. In an open-loop control system, the input is determined from the reference input using a mathematical model of the plant behaviour. If the mathematical model is accurate and there are no external **disturbances** affecting the plant, the actual plant output, also known as the **control variable** or the **controlled variable**, will achieve the required value. In practice, models are imperfect and disturbances affect the plant so that the output will not accurately reach the required value. Closed-loop control attempts to overcome the problems caused by an imperfect model and disturbances by measuring the output using one or more **sensors** and evaluating the **error** in the output, which is the difference between the reference input and the actual output. This error is then used by the control system to determine the necessary plant input. The use of the error requires a signal to be fed back from the output to the control-system input, and thus closed-loop control is frequently referred to as **feedback control**, with the signal fed back being the **control-system feedback**. The **control accuracy** is the inverse of the error. Hence a low error represents a high control accuracy and vice versa.

The parts of a control system that apply the control equations to attempt to achieve the desired output are known as the **control elements**. For example, in a **PID controller** three elements are employed: one providing **proportional control**, the second **integral control**, and the third **derivative control**. The **controller** is the combination of all of the control elements. In designing the elements of a control system, one of two approaches may be taken: classical or modern control. In the classical method (**classical control**), the plant and control system are each analysed using the Laplace transform of the ratio of the output to the input. The resulting transform is the **transfer function** which describes the behaviour in the **frequency domain**, that is, as a function of frequency. Such an approach is only applicable where both the plant and the control system can be represented by linear differential equations. Where **modern control** design is applied, the plant is represented by its **states** and the analysis performed directly on the low-order differential equations relating these states. Non-linear behaviour can also be represented and analysed.

Because a knowledge of the behaviour of the plant is necessary when designing the controller, the performance of the controller will deteriorate if the plant behaviour subsequently changes. **Adaptive control** allows the controller to automatically track changes in the plant and change the controller to accommodate these.

However the controller is designed, there are three primary aims. Firstly, to ensure that the output is bounded, i.e. that the system is stable. Secondly, to ensure that the behaviour when the reference input changes is within specification, i.e. that the **transient behaviour** is satisfactory. Finally, to ensure that the error after transient behaviour has decayed is satisfactory, i.e. that the **steady state error** is small.

The formal design of a control system so as to show stability with the transient and error performance within specification despite external disturbances, measurement noise, and modelling errors is known as **robust-control design**. Many modern controllers are designed using **optimal control** methods in which one or more **performance measures** are selected and the parameters in the control system adjusted through to minimize the cost function or maximize the performance measure.

A formal measure of the ability of a system to respond to a change in the reference input (by changing from a specified initial state to specified final state within a finite time) is the **controllability.**

Although almost all control systems are now implemented using digital electronics and thus use **sampled data**, equations in **continuous time** are often used to design such systems. This is satisfactory provided that the interval between **samples** is short compared to the time taken for the plant to show a significant change in output. When the response of the plant is very rapid, **z-transforms** are used instead of Laplace transforms to represent sampled data.

SEE WEB LINKS

- Website of the International Federation of Automatic Control
- Website of the American Automatic Control Council
- Website of the IMech E Journal of Systems and Control Engineering
- Website of the Institution of Engineering and Technology Control and Automation Network
- Website of the American Institute of Electrical Engineering and Electronics Control Society
- Website of the IEEE Transactions on Automatic Control

control-system feedback *See* CONTROL SYSTEM.

control variable *See* CONTROL SYSTEM.

control volume (open system) In thermodynamics and fluid mechanics, a region in space chosen for study and which may move, deform, and allow mass flow across its boundaries, called the control surface.

convected energy *See* FLOW ENERGY.

convection *See* HEAT TRANSFER.

convection velocity The velocity of an identifiable flow structure, such as a turbulent eddy, which is normally lower than that of the bulk flow.

convective derivative In fluid flow, that part of the total change in a quantity associated with movement through regions of spatially different velocity. In Cartesian coordinates, for a quantity B, which could be a scalar such as pressure or temperature or a vector such as velocity (when it defines **convective acceleration**), it is given by

$$u\frac{\partial B}{\partial x}+v\frac{\partial B}{\partial y}+w\frac{\partial B}{\partial z}$$

and in vector notation as $(\boldsymbol{V}\cdot\nabla)B$ where $\boldsymbol{V}$ is the vector velocity with components u, v and w in the x, y, and z directions, respectively. *See also* LOCAL ACCELERATION; SUBSTANTIAL DERIVATIVE; TOTAL ACCELERATION.

convective heat-transfer coefficient *See* HEAT TRANSFER.

convective mass-transfer coefficient *See* MASS TRANSFER.

conventional milling *See* UPMILLING.

convergent–divergent nozzle (convergent–divergent duct) A flow nozzle with a convergent section (the confuser), in which (for an incompressible fluid) the fluid is accelerated, upstream of a divergent section (the diffuser) in which the fluid is decelerated. If the fluid is compressible, as in the case of a gas or vapour, and if the pressure difference is sufficiently great, the flow at the throat chokes (i.e. reaches sonic conditions) and supersonic velocities are achieved in the divergent section, possibly involving shock waves. *See also* VENTURI.

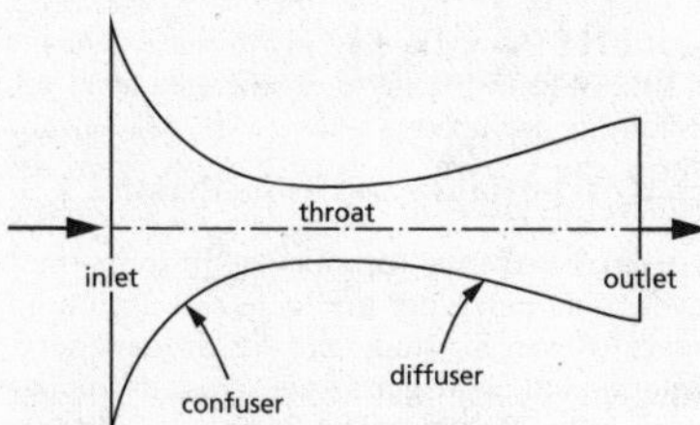

convergent-divergent nozzle

convergent duct (convergent nozzle) A duct or nozzle for which the cross-sectional area decreases with downstream distance. For subsonic flow, the velocity increases with distance, whereas for supersonic flow it decreases and the nozzle acts as a diffuser.

conversion efficiency *See* SOLAR-CELL EFFICIENCY.

convex involute helicoid *See* TOOTHED GEARING.

coolant A fluid used to limit the temperature of any device that is heated. Examples include water circulated through the radiator of a car engine where it is cooled prior to flowing through passages in the cylinder block; water flowing through the tubes of a surface condenser; a film of air that cools the surfaces of gas-turbine blades; a combustion chamber or a rocket engine exposed to high temperature gas flow; liquid sodium used to cool certain types of nuclear reactor; water-soluble oil used to cool a cutting tool during machining.

cooling coil A simple heat exchanger consisting of a coiled tube, typically of copper or stainless steel, through which is circulated a coolant such as a refrigerant, chilled water, or water mixed with ethylene glycol. Applications include air conditioning, process cooling, and refrigeration.

cooling correction In determining the calorific value of a fuel using a bomb calorimeter, the correction which must be applied to the measured temperature rise, which is slightly lower than the maximum that would be reached if the outer container was perfectly insulated.

cooling curves 1. Plots of temperature *vs* time as a substance cools. Plateaux of constant temperature (thermal arrest) in the smooth curve of falling temperature indicate freezing points in pure substances or eutectic (or eutectoid) points in mixtures and metal alloys; sharp changes of slope indicate the beginning and end of the temperature ranges over which non-eutectic mixtures transform to a different phase. Employed to construct phase or 'equilibrium' diagrams. **2.** *See* LUMPED-CAPACITY ANALYSIS.

C

cooling degree day *See* DEGREE DAY.

cooling fins (extended surfaces, finned surfaces, fins) Thin metal plates, studs, or pins which project into the cooling fluid from a surface to be cooled in order to increase the effective surface area for heat transfer. Applications include the cylinders of air-cooled piston engines and compressors, boiler superheater tubes, heat exchangers, electrical transformers, and electronic components. **Fin performance** is defined either as the ratio of the heat transfer rate from a surface with a fin to that without **(fin effectiveness, ε)** or as the ratio of the heat transfer rate from a fin to the heat transfer rate from a fin if its entire surface were at the base temperature **(fin efficiency, η)**.

cooling tower A large-scale cooler in which ambient air circulates and cools warm water from a power plant, large air-conditioning system, or industrial process. Depending upon how the airflow is driven, cooling towers are classified as natural draft, forced draft, or induced draft. The **cooling-tower range** is the difference between the temperature of the warm water entering T_I and that of the cool water leaving T_O. The **cooling-tower efficiency (η)** is the ratio of the cooling-tower range to the difference between T_I and the wet-bulb temperature of the ambient cooling air T_{WB}, i.e. $\eta = (T_I - T_O)/(T_I - T_{WB})$. The **cooling-tower approach** is $T_O - T_{WB}$. *See also* DRY COOLING TOWER; WET COOLING TOWER.

Cooperative Fuel Research Committee engine (CFR engine) A single-cylinder piston engine used for the testing of fuels, especially the determination of the Motor Octane and Research Numbers, as well as the testing of lubricants and research into fuels and lubricants generally.

coordinated-axis control The simultaneous control, in CNC and robot control, of all axis or joint drives, so as to achieve smooth movement of the tool or end effector along a profiled path.

coplanar forces A term for forces that act in a single plane.

C

copying machine Various types of cutting machine, including lathes and milling machines, in which the tool is guided by a template, thus producing identical components.

core 1. The inside material of a sandwich structural component (often foamed or a honeycomb). **2.** *See* GAS GENERATOR. **3.** *See* REACTOR CORE.

Coriolis acceleration (Coriolis force) *See* ACCELERATING FRAME OF REFERENCE.

Coriolis-type mass flow meter (Coriolis meter) A mass flow meter in which liquid flows through a vibrating U-tube. The Coriolis force exerted by the fluid distorts the tube, and the phase difference between the tube oscillation and the distortion is proportional to the mass flow rate.

correction time (settling time) The time taken for the output of a controlled plant to fall and remain within a specified percentage (typically 2%) of the final (steady-state) value.

corrective action The action of a control system in varying the plant input so as to minimize the error.

corresponding states *See* LAW OF CORRESPONDING STATES.

corrosion The deterioration of an exposed metal surface, usually due to electrochemical oxidation with its surroundings. A common example is the rusting of iron or steel in which iron oxides are formed in moist air or water. **Corrosion inhibitors** are chemicals applied to surfaces to decrease corrosion. *See also* PASSIVATION.

corrosion failure 1. The failure of a component or structure after corrosion has reduced the load-bearing area to an unsupportable level. **2.** A situation in which a mechanism cannot function owing to corrosion products preventing free movement at joints.

corrosion fatigue *See* FATIGUE.

cost function The variable that is minimized in system identification and other optimization methods to achieve the best parameter estimate.

cotter pin (cotter) 1. A tapered wedge or pin passing through a tapered slot or hole in one member and bearing against a second member to fix it in location. **2.** A split cotter is commonly called a split pin.

Couette flow The flow in the near-vicinity of a surface in which streamwise gradients of velocity are negligible and the variation of total shear stress τ with normal distance from the surface y is given by

$$\frac{\tau}{\tau_S} = 1 + \frac{\rho v_S \bar{u}}{\tau_S} + \frac{d\bar{p}}{dx}\frac{y}{\tau_S}$$

where τ_S is the wall shear stress, ρ is the fluid density, v_S is the normal component of velocity at the surface (zero unless there is blowing or suction through the surface), $\bar{u}$ is the mean flow velocity parallel to the surface at a distance y, and $d\bar{p}/dx$ is the streamwise gradient of the mean pressure $\bar{p}$. The equation for τ is valid for both laminar and turbulent flow, but for laminar flow can be further simplified since then $\tau = \mu du/dy$ where μ is the dynamic viscosity of the fluid. Laminar Couette flow is realized practically in the gap between two concentric cylinders of almost equal radii, one rotating relative to the other when $d\bar{p}/dx = 0$ and $u = \tau_S y/\mu$ if $v_S = 0$. *See also* TAYLOR–COUETTE FLOW.

Couette viscometer *See* ROTATIONAL VISCOMETER.

Coulomb damping *See* FRICTION DAMPING.

Coulomb friction *See* AMONTONS FRICTION.

Coulomb–Mohr fracture criterion *See* MOHR-COULOMB FRACTURE CRITERION.

Coulomb–Mohr yield criterion *See* MOHR-COULOMB YIELD CRITERION.

counterbalance *See* COUNTERWEIGHT.

counterbore A concentric enlargement of a hole to a limited depth.

counterflow heat exchanger (countercurrent-flow heat exchanger, contraflow heat exchanger) A heat exchanger, typically of shell-and-tube type, in which the overall directions of flow for the two working fluids are opposite.

countershaft (counter shaft, jackshaft) An intermediate shaft in gearing.

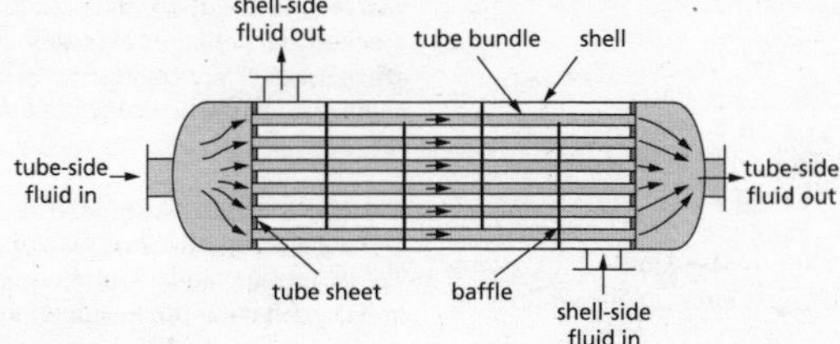

counterflow heat exchanger

countersinking The flaring out of the rim of a drilled hole to form a truncated conical depression to receive a screw having a conical head, thus giving a flush fitting.

counterweight **1.** A large, heavy weight close to the fulcrum of a beam, such as the jib of a crane, that balances a load at the other end. **2.** A weight, used to partially balance an object such as a lift cage, suspended over a pulley or sheave in order to reduce the torque needed to raise or lower the object.

couple If the resultant of two or more force vectors applied to an object is zero, the moment of those forces, tending to rotate the object about an axis, is termed a couple. Two parallel forces of equal magnitude F but with opposite sense, separated by a distance d, give a couple of magnitude Fd. *See also* TORQUE.

coupling **1.** Any mechanical fastening connecting two or more shafts, or parts of a mechanism, in order to transmit power. **2.** A device for connecting two vehicles.

coupling agent A compound that provides a chemical bond between two dissimilar materials, often otherwise non-bonding and incompatible. An example is the use of a silane to bond the reinforcement and the resin matrix of glass-fibre reinforced composite materials.

cowl (cowling) A shroud or rounded panel, usually of sheet metal or plastic, used on a motor vehicle or aircraft to reduce drag or to surround, or direct air into, an engine. *See also* AIR SCOOP; FAIRING.

crack A thin fissure-like defect in a component or structure across which material continuity is lost and which reduces the strength of the body. *See also* FRACTURE MECHANICS.

crack arrest Crack propagation that stops of its own accord when the energy release rate of the loaded component or structure falls below a critical value. If predictable, it can be incorporated into structural integrity assessments.

crack tip opening displacement (CTOD, crack opening displacement, COD) The amount of stretch at the tip of a crack in a loaded body prior to crack propagation. *See also* CRITICAL CRACK TIP OPENING DISPLACEMENT.

crank angle The angle between the crank of a slider-crank mechanism and a line from the crankshaft centreline to the piston axis.

crankcase The housing for the crankshaft of an engine. *See also* SUMP.

crankcase breather *See* BREATHER PIPE.

crank effort The force acting on an engine's crank pin.

crank press A press, the stroke of which is driven by a crank mechanism.

crankshaft The main shaft, of which the cranks are a part, of a reciprocating single-or multi-cylinder machine. Crankshafts may be built up in sections or forged as a single component. In an engine, the reciprocating motion of the pistons transmits power to the crankshaft and causes it to rotate, whereas in a pump the crankshaft is driven and its rotation causes the pistons to reciprocate. The **crank pin** is a short shaft parallel to the axis of the crankshaft but radially offset from it, to which is attached the big end of a connecting rod in a bearing. Sometimes the crank pin is supported at one end only (a **wrist pin**) but in built-up or one-piece forged crankshafts, the crank pin is supported by thick plates (**crank arms, crank webs**) at either end. The **crank throw** is the radial distance from the crank pin to the crankshaft axis and equal to half the stroke.

crawler vehicle *See* CATERPILLAR.

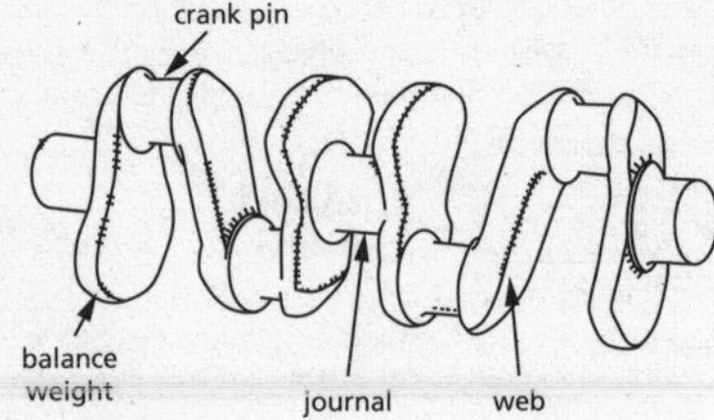

crankshaft

crazing A characteristic of some polymers below their glass transition temperature, whereby fine crack-like defects that produce bright reflections appear on loading. They are not true cracks, rather regions of highly plastically-deformed material interspersed with voids.

creep The deformation, which increases with time, of a material or structure under constant load (strictly constant stress). A typical creep curve of strain (ε) *vs* time (t) (**strain-time** or **creep diagram**) shows an immediate elastic and plastic strain when the load is applied, followed by a decreasing rate of strain with time (**primary** or **transient creep**) in which the increase of stress caused by workhardening exceeds the decrease in stress caused by thermal softening, resulting in a decrease in creep rate with time. That stabilizes into a steady increase of strain with time (**secondary or steady-state creep**), followed by an increasing rate of increase of strain leading to fracture (**tertiary creep**). For primary and secondary creep, ε *vs* t curves (**creep-time relations**) are often fitted by one of two empirical formulae, viz: $\varepsilon = \varepsilon_o + \alpha lnt + \kappa t$ or $\varepsilon = \varepsilon_o + \beta t^{1/3} + \kappa t$,

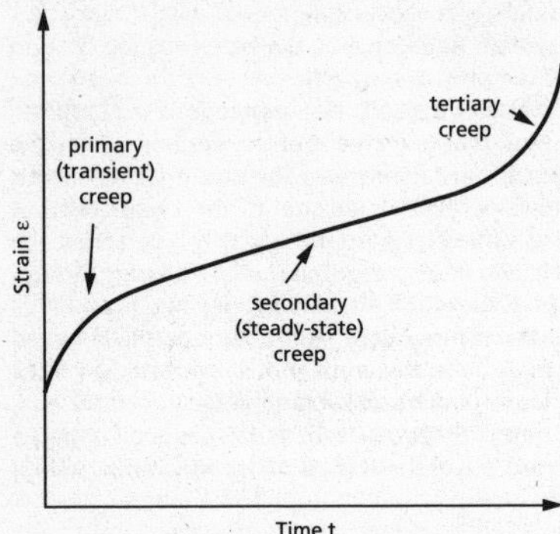

creep curve

where ε_o is the immediate strain; α, β and κ are creep coefficients; κt is steady-state creep; and αlnt and $\beta t^{1/3}$ are two forms of transient creep, called **logarithmic creep** and **Andrade creep** respectively. *See also* RECOVERY; RELAXATION.

creep fatigue The effects of combined creep and fatigue at high homologous temperatures T_M in metals, and in other materials such as polymers having time-dependent behaviour.

creeping flow *See* STOKES FLOW.

creep limit The maximum allowable stress under the action of which creep deformation of a material does not exceed a specified limit.

creep modulus The ratio of stress/strain at a chosen strain level obtained from creep tests, often plotted against time to show changes in stiffness.

creep rupture strength The fracture stress at the end of a creep test in metals and polymers, often plotted against time to give a stress-rupture curve.

creep–time relations Algebraic or numerical relations between creep strain and time at constant stress.

crest The highest point of a wave or of a screw thread. *See also* TROUGH.

crest clearance For screw threads and gearing, the radial clearance between the crest of a thread (or gear) and the root of the engaging thread (or gear).

criterion of performance 1. The coefficient of performance for a refrigerator or heat pump. **2.** A measure of efficiency for an internal-combustion engine or power plant, such as the overall efficiency, the brake thermal efficiency, the specific fuel consumption, the indicated thermal efficiency, or the mechanical efficiency.

critical angle of attack *See* STALL ANGLE.

critical compression ratio 1. The compression ratio for incipient knock of hydrocarbon fuels as determined in a variable-compression single-cylinder piston engine. **2.** The compression ratio at which an air–fuel mixture will spontaneously ignite due to the temperature increase produced.

critical crack length (Unit m) The crack length at which a given stress applied to a body will result in crack propagation and fracture. *See also* FRACTURE MECHANICS.

critical crack tip opening displacement (critical crack opening displacement) (Unit m) The crack tip opening displacement at which crack propagation occurs. *See also* FRACTURE MECHANICS.

critical damping The level of viscous damping in a system undergoing free vibrations that brings the system back to equilibrium in the shortest possible time without overshoot or oscillation. *See also* DAMPING RATIO.

critical density *See* CRITICAL STATE.

critical energy release rate (G_C) (Unit J/m^2) In linear elastic fracture mechanics, the critical rate of release of energy at which a crack will propagate. *See* FRACTURE TOUGHNESS.

critical exponent (α) A power-law exponent used to approximate the behaviour of a thermodynamic quantity in the immediate vicinity of the critical point. For example, the specific heat $C = k|T - T_C|^{\alpha}$ where T is the absolute temperature, T_C is the critical temperature, and k is a constant of proportionality.

critical fibre length In fibre-reinforced composites, the load in the matrix is transferred into the filaments (and vice versa) by the interfacial bond strength τ over a distance L_c given by $L_c = \sigma_{max}d/2\tau$ where σ_{max} is the stress in the fibre, and d its diameter. The critical fibre length is $2L_c$.

critical flow 1. *See* CHOKED FLOW. **2.** Open-channel flow for which the speed of propagation of small-amplitude surface waves is equal to the flow velocity, i.e. the Froude number equals unity (the **critical Froude number**).

critical frequency *See* RESONANCE.

critical heat flux *See* BOILING.

critical insulation thickness For certain geometries, including cylinders and spheres, there is a certain thickness of insulation that corresponds to a minimum heat transfer rate. Any further increase in the thickness leads to an increase in heat loss. For a circular cylinder, the critical radius of insulation is k/h where k is the thermal conductivity of the insulating material and h is the surface heat-transfer coefficient.

critical isobar The isobar on a T-v diagram, where T is the tempurature and v is the specific volume that passes through the critical point for a given substance.

critical isotherm The isotherm on a p-v diagram, where p is the pressure and v is the specific volume that passes through the critical point for a given substance.

critical load 1. The applied load that causes propagation of an existing crack of known length, and hence fracture of a component or structure. **2.** The applied load that results in buckling of a column of given end fixity.

critically-damped motion *See* CRITICAL DAMPING.

critical Mach number *See* CHOKED FLOW.

critical pressure (p^*) The static pressure at the throat of a choked nozzle. For isentropic flow of a perfect gas

$$\frac{p^*}{p_0} = \left(\frac{2}{\gamma + 1}\right)^{\gamma/(\gamma-1)}$$

where p_0 is the stagnation pressure and γ is the ratio of specific heats for the gas.

critical pressure ratio The ratio of the back pressure to the stagnation pressure that leads to choked flow in a convergent or convergent-divergent nozzle.

critical Rayleigh number The Rayleigh number for which laminar-to-turbulent transition occurs in natural convection.

critical Reynolds number The Reynolds number for which a flow undergoes transition from laminar to turbulent. The value depends upon the flow type and which characteristic length and velocity are chosen: for fully-developed pipe flow, the accepted value is 2300; for a flat-plate boundary layer it is about 5×10^5; for flow around a body it depends primarily upon the shape.

critical speed 1. In compressible flow through a convergent or convergent-divergent nozzle, the flow speed at the throat when the flow is choked. It is equal to the sound speed at the throat. **2.** *See* WHIRLING.

critical state (critical point) The thermodynamic state at which the saturated liquid and saturated vapour of a fluid have identical densities. The fluid is then at its **critical temperature**, **critical pressure**, and **critical density** (or **critical specific volume**). Above the critical temperature, the fluid cannot be liquefied by increasing the pressure.

C

critical stress-intensity factor *See* FRACTURE MECHANICS.

critical Taylor number The Taylor number for Couette flow in the annulus between rotating cylinders at which Taylor cells first appear.

cross-compound steam-turbine system A multi-cylinder steam-turbine arrangement in which several machines are connected in parallel to separate generators. The diagram shows a high- (HPT) and an intermediate-pressure turbine (IPT) cross-compounded with two low-pressure turbines (LPT). *See also* TANDEM-COMPOUND STEAM-TURBINE SYSTEM.

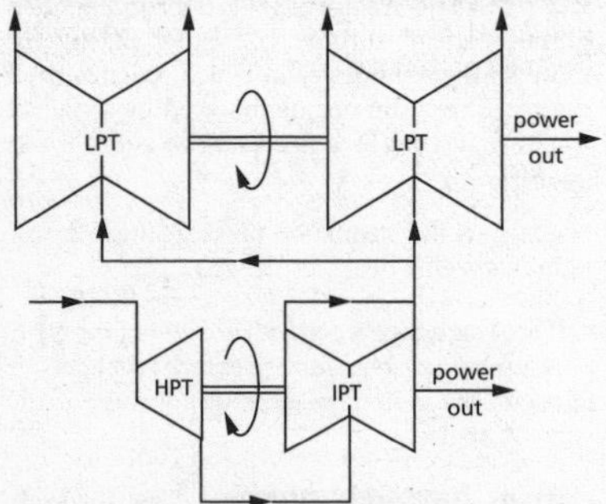

cross-compound steam-turbine system

cross-drum boiler A variant of the longitudinal-drum boiler in which the drum is transverse to the heat source.

crossed gears Gears that mesh together on non-parallel axes. *See also* TOOTHED GEARING.

crossed threads When the axis of a nut offered up to a bolt is not aligned with the axis of a bolt (or a screw to a threaded member), it may be possible for the threads to engage incorrectly and even for the nut or screw to advance a turn or more, but ultimately the misaligned threads become locked together. Forcing the nut in such circumstances may irretrievably damage the threads on both.

crossflow *See* SECONDARY FLOW.

cross-flow baffle In a shell-and-tube heat exchanger, a metal plate that forces the shell-side fluid to flow back and forth across the tubes, to enhance heat transfer, and also holds the tubes in position.

cross-flow heat exchanger A heat exchanger, typically of compact design, in which, on average, the two working fluids flow transverse to each other. *See also* PLATE-FIN HEAT EXCHANGER.

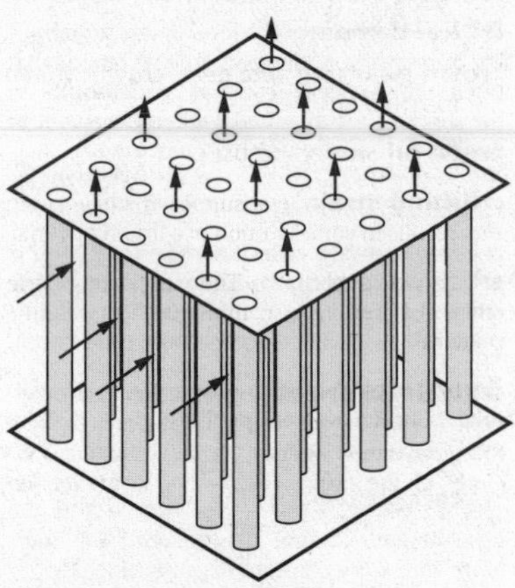

cross-flow heat exchanger

cross-flow turbine 1. *See* BANKI TURBINE. **2.** A vertical-axis wind turbine similar in design to the Banki water turbine that can generate power whatever the wind direction. *See also* SAVONIUS–DARRIEUS WIND TURBINE.

crosshead 1. A reciprocating member, sliding between guides, to which the piston rod is firmly attached on one side and to which the connecting rod is pinned on the other for the conversion of reciprocating into rotary motion. **2.** The moveable beam in a testing machine. **3.** A screwhead having slots in a + shape that takes a mating screwdriver.

Crossley meter (liquid-sealed meter, wet gas meter) A drum-type gas flow meter in which gas flows through a partially-submerged four-vane rotor, causing it to rotate. Flow rate is determined from the number of rotations in a given period of time.

cross-ply laminate A filament-reinforced composite formed of laminae that are laid up at fixed angles in successive layers.

cross section The shape corresponding to the intersection of an object with a cutting plane.

cross slide That part of a lathe attached to the saddle, over which it traverses at right angles to the lathe bed. The tool is fixed to the cross slide and may be moved to any position by the combined motion of the saddle along, and the cross slide across, the lathe bed.

crown 1. The sharp corner at the crest of some gearing. **2.** *See* PISTON.

crown gear (contrate gear, crown wheel) *See* TOOTHED GEARING.

crude oil *See* PETROLEUM.

crushing strain The supposed single strain that results in comminution of a brittle material.

crushing strength The supposed single stress that results in comminution of a brittle material.

cryogenics The study of processes and material behaviour at temperatures below about 123K (i.e. −150°C).

cryoscopic effect The reduction in the freezing point of solvents due to the addition of solutes e.g. salt added to water. *See also* EBULLIOSCOPIC EFFECT.

cryostat A vessel, similar to a vacuum flask, used to maintain low temperature levels, often using liquid helium.

crystalline fracture A fracture surface characterized by the shiny facets of transgranular cleavages as in steels below the ductile-brittle transition temperature.

CSM *See* CHOPPED STRAND MAT.

CTOD *See* CRACK TIP OPENING DISPLACEMENT.

cubical expansion coefficient *See* THERMAL EXPANSION.

cumulative damage Physical or microstructural damage in a component or structure accrued as the result of a number of separate events, leading to weakening.

Cunningham slip correction A correction factor applied to Stokes drag law for small particles in a gas flow. It becomes significant when the particle diameter d is comparable with the mean-free path of the gas λ. It can be written

$$C = 1 + \frac{2\lambda}{d}(a + be^{-cd/\lambda})$$

where a, b and c are empirical constants for the particular gas.

cup The outer cylinder of a rotational viscometer.

cup anemometer A device for measuring wind speed comprising a number of hemispherical cups, typically four, attached by horizontal radial arms to a vertical shaft. The rate of rotation of the shaft is a measure of the windspeed. *See also* ANEMOMETER.

cupping *See* DEEP DRAWING.

curing time The time required in such processes as thermoplastic moulding, the manufacture of fibre-reinforced composites, and the setting of epoxies, for the completion of the chemical reactions that bring properties to their design levels.

Curtis stage A stage of a velocity-compounded impulse steam turbine consisting of one or more stationary nozzles in which practically all the pressure drop occurs followed by a pair of rotors separated by a fixed ring of guide vanes (stator).

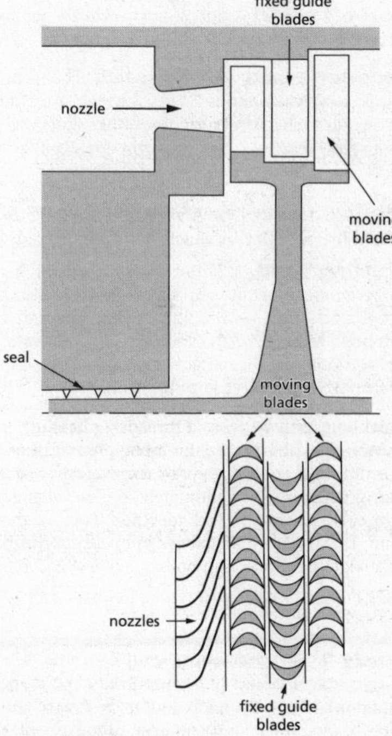

Curtis stage

curved-wall effect *See* COANDA EFFECT.

curvilinear motion Motion along a curved path.

cusped diffuser A diffuser, used in fluidic devices, with cusped recesses on either side close to the inlet in which stable vortices are created.

cut-in speed The wind speed at which a wind turbine begins to generate usable power; typically about 4 m/s (10 to 15 kph).

cut-off frequency The frequency at which the output of a device is reduced by 3dB with respect to the output within the passband. For example, in a low-pass filter, the frequency at which the output is reduced by 3dB with respect to that at lower frequencies. The cut-off frequency thus defines the passband, i.e. the range of frequencies passed by the filter with less than 3dB attenuation.

cut-off tool A parting tool to separate a finished part from barstock, etc.

cut-off wheel A thin abrasive wheel used to cut through, or cut slots in, a component.

cut-out speed (furling speed) The wind speed, typically about 20 to 35 m/s (70 to 130 kph) at which a wind turbine is either shut down for safety reasons or continues to operate at low efficiency, imposed by stall regulation.

cutter A blade or tool employed to cut or machine materials.

cutting fluid A liquid used in machining operations to cool and lubricate the cutting tool and wash away chips, swarf, and other debris. Mineral and synthetic oils, oil-water emulsions, and water are examples. *See also* MINIMUM-QUANTITY LUBRICATION.

cutting plane In an engineering drawing, a plane that passes through a component or assembly and is used to show the corresponding cross section.

CV joint *See* CONSTANT-VELOCITY UNIVERSAL JOINT.

CVT *See* CONTINUOUSLY-VARIABLE TRANSMISSION.

cycle 1. *See* THERMODYNAMIC CYCLE. **2.** The sequence of values of a periodically oscillating quantity over a complete period. **3.** A mechanical cycle, such as the four-stroke cycle of a piston engine.

cyclically-pivoting sail windmills A vertical-axis wind machine for which the pitch angle of each sail is cyclically adjusted to optimize its orientation with respect to the wind direction.

cyclic stress–strain curve The series of stress-strain curves obtained after repeated loading and unloading beyond the yield point. The curves may be zero-tension only, zero-compression, or zero-tension-zero-compression-zero-tension, etc. Such curves may stabilize or continue to expand. *See also* FATIGUE; SHAKEDOWN.

cyclic testing Determination of the mechanical properties of a material, component, or structure under variable loading conditions. *See also* FATIGUE.

cyclic train A set of gears in which one or more of the gear shafts rotates around a fixed axis. *See also* TOOTHED GEARING.

cycloidal gear teeth *See* TOOTHED GEARING.

cyclone (cyclone separator) A swirling-flow device with a conical body that flow enters tangentially and leaves axially. It is used to separate particles from a gas, vapour, or liquid, or liquid droplets from a gas or vapour. *See also* HYDROCYCLONE.

cyclonic-type separator A steam separator in which helical baffles cause the wet-steam flow to swirl. Water droplets are centrifuged to the outer wall and drain down to a steam trap.

cylinder 1. *See* ENGINE CYLINDER. **2.** The casing of a steam turbine, usually cast for high-and intermediate-pressure turbines, but low-pressure casings include some fabrication. There is often an inner and an outer casing.

cylinder block (block, engine block) The casting, typically of aluminium or magnesium alloy or cast iron, in which are machined the cylinders of a piston engine or a reciprocating pump.

cylinder bore The internal diameter of a cylinder in which a piston operates in a reciprocating engine or pump.

cylinder head The machined casting, typically of aluminium alloy or cast iron, that fits above the cylinder block of a piston engine and closes off the cylinders. It normally includes part of the combustion chambers and holes for the valves and spark plugs.

cylinder liner **(liner)** A replaceable thin metal cylinder or sleeve fitted within the cylinder of an engine or pump which may be replaced when worn, thus allowing continued use of the cylinder block. A **dry liner** in a piston engine has no contact with the coolant, whereas the outer surface of a **wet liner** is in direct contact with the coolant.

cylindrical-coordinate robot **(turret robot)** A robot having a rotational joint, joint angle θ_1, with a vertical axis above the base frame; a translational joint, joint offset d_2, with a vertical axis above the base frame; and a translational joint, joint offset d_3, with a horizontal axis attached to the second joint. The volume that the robot can reach is thus a hollow cylinder centred on the base frame. The diagram shows an idealized cylindrical-coordinate robot. *See also* CARTESIAN-COORDINATE ROBOT.

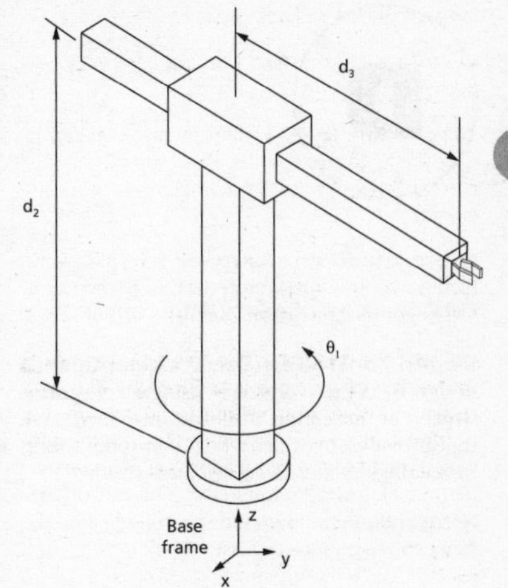

cylindrical-coordinate robot

cylindrical grinding Accurate finishing of cylindrical components using a high-speed abrasive grinding wheel.

DAC *See* DIGITAL-TO-ANALOGUE CONVERTER.

Dahlin controller (Dahlin's algorithm) A design of digital controller using z-transforms which, by accepting a slower response than that provided by a dead-beat controller, minimizes ringing after the set point is reached.

d'Alembert's force *See* ACCELERATING FRAME OF REFERENCE.

d'Alembert's paradox A paradox which states that an incompressible fluid offers no resistance to steady translational motion of any rigid body when the flow is everywhere irrotational, whereas such bodies do experience resistance (i.e. drag) to motion through a real, viscous fluid, and can also develop lift.

d'Alembert's principle The means by which a problem in dynamics is reduced to an equivalent problem in statics through the equations of equilibrium. Thus in a linkage, say, if kinematics gives the linear acceleration f and angular acceleration α of the centre of gravity of an element of mass m and moment of inertia I_G, the inertia force mf and couple $I_G\alpha$ are in equilibrium with the forces applied to it by the adjacent links, enabling those forces to be determined.

Dall tube A compact adaptation of the Venturi tube with an abrupt reduction in cross section at the inlet followed by a gradual area reduction and return to the full cross section after the throat. The permanent pressure loss exceeds that of a Venturi tube but is less than that for an orifice plate. Dall tubes are ideal for clean fluids but not suitable for fluids with suspended solids.

Dalton's law (Dalton's law of additive pressures) The pressure of a mixture of ideal gases is equal to the sum of the pressures (**partial pressures**) each gas would exert if it existed alone at the same temperature and volume. *See also* AMAGAT'S LAW; GIBBS–DALTON LAW.

damage 1. The deterioration of a component or structure in fault or accident conditions, lessening or preventing its ability to perform its intended function. **2.** The accumulation of defects or microcracks in the microstructure of a body loaded monotonically or in fatigue, which weakens the body and can lead to crack propagation and failure.

damage mechanics The theory of degradation in bodies, particularly fracture by accumulated microstructural damage. Analyses take two approaches: (a) the use of some critical integrated function of stress and strain at which cracking is initiated and propagated; (b) incorporation of damage in the stress–strain curves to reflect weakened material (**porous plasticity**). *See also* FRACTURE MECHANICS; JOHNSON–COOK FRACTURE EQUATION.

damage stress *See* FAILURE.

damage tolerance (defect tolerance) A design philosophy that takes into account initial imperfections, crack-growth rates and conditions at final fracture, and uses fracture mechanics to demonstrate that cracks should not grow to their critical length within the design life (or at least should be capable of ready detection). *See also* DAMAGE-TOLERANT DESIGN; FAIL SAFE; SAFE LIFE.

Damköhler number (*Da*) A non-dimensional number that measures the interaction, in turbulent combustion, between the turbulence and the chemical reactions. It is defined as $Da = \tau_T/\tau_L$ where τ_T is a characteristic eddy turnover time and τ_L is the laminar burning time. τ_T can be determined from $\tau_T = l_I/u'$ where l_I is the integral length scale of the turbulence and u' the rms velocity fluctuation. τ_L is given by $\tau_L = \delta_L/S_L$ where δ_L is the laminar flame thickness and S_L the laminar flame speed.

damp 1. To reduce the amplitude of an oscillating system. **2.** To extinguish a fire by cutting off the air supply. *See also* DAMPER; FIRE DAMP; FIRE DAMPER.

damped forced response The response of a damped mechanical system to harmonic excitation by an applied force or displacement.

damped harmonic oscillator An oscillator subject to damping which, if undamped, exhibits harmonic motion, i.e. displacement that is sinusoidal in time.

damped natural frequency (Unit Hz) The frequency of a vibrating system in the presence of damping. For a simple mass-spring-damper system with natural frequency $\omega_N = \sqrt{k/m}$ where k is the spring constant and m the mass, if the system is underdamped (i.e. damping less than critical damping), the damped natural frequency is given by $\omega_D = \omega_N\sqrt{(1 - \varsigma^2)}$ where ς is the damping ratio.

damper 1. A sliding or hinged plate for controlling the draught through a furnace. *See also* STACK DAMPER. **2.** A device for suppressing vibrations in a mechanical system by dissipating energy. *See also* FRICTION DAMPER; HARMONIC DAMPER; LANCHESTER DAMPER; TORSIONAL-VIBRATION DAMPER.

damping The loss of energy of an oscillating system by dissipation, internally or due to viscous damping in a fluid or surface friction, or by radiation. Damping material may be added to the vibrating parts of a structure. Damping is introduced intentionally in analogue measuring instruments to overcome the problem of a vibrating pointer. *See also* CRITICAL DAMPING; DEADBEAT; FRICTION DAMPING; LOGARITHMIC DECREMENT; OVERDAMPED; UNDERDAMPED.

damping capacity The ability of a material to absorb vibration and not transmit oscillations to attached components.

damping coefficient (*c*) (Unit N.s/m) For a linear dashpot, the ratio between the resistive force and the relative velocity between the piston and cylinder, determined by the viscosity of the damping fluid.

damping constant (amplitude decay coefficient, damping factor, *δ*) (Unit 1/s) The quantity in the exponential decay factor $e^{-\delta t}$ that describes the amplitude decay of damped free oscillations with time t. *See also* LOGARITHMIC DECREMENT.

damping function *See* VAN DRIEST DAMPING FUNCTION.

damping ratio (ζ) The ratio of the viscous-damping coefficient in a single-degree-of freedom system to its critical value.

Daniell hygrometer A dew-point hygrometer consisting of a bent glass tube terminating in two bulbs, one black and incorporating a thermometer, the other coated with muslin. The system contains ether, which is condensed in the black bulb. Ether poured over the muslin evaporates so cooling the bulb, condensing the vapour within and reducing the vapour pressure. The ether in the black bulb evaporates and cools causing dew to form on its surface.

darcy (*D*) *See* PERMEABILITY.

Darcy friction factor (Darcy–Weissbach friction factor, Moody friction factor, *f*) For pipe or duct flow, a non-dimensional measure of the surface shear stress defined by $f = 8\tau_S/\rho V^2$ where τ_S is the average surface shear stress at any axial location, ρ is the fluid density and V is the bulk flow velocity. The pressure loss Δp for fully-developed flow in a pipe of length L and diameter D is given by

$$\Delta p = f\frac{L}{D}\frac{1}{2}\rho V^2$$

(**Darcy–Weissbach equation**). *See also* FANNING FRICTION FACTOR; MOODY CHART.

Darcy's law A constitutive equation that describes fluid flow through porous media for Reynolds numbers up to about 1. The Reynolds number Re in this application is defined by $Re = \rho vd/\mu$ where d is a representative grain or void diameter, μ is the dynamic viscosity of the fluid, ρ is its density and v is the interstitial velocity given by $v = q/\phi$ where ϕ is the void fraction of the porous medium and q is the **Darcy flux (seepage velocity)** of the fluid, i.e. the volumetric flow rate per unit area. At the surface of the material it is termed the **face velocity**. Darcy's law can be written as $q = -\kappa\nabla p/\mu$ where κ is the permeability of the medium and ∇p is the applied pressure-gradient vector (< 0) in the flow direction. For the situation of a cylinder of cross-sectional area A and length L, Darcy's law reduces to $\dot{Q} = -\kappa A\Delta p/\mu L$ where $\dot{Q}$ is the volumetric flow rate along the cylinder due to a pressure difference Δp between its ends.

Differences that arise at Reynolds numbers above the applicability of Darcy's law (**For-**

chheimer flow) are accounted for by quadratic and higher-order terms, e.g.

$$-\frac{dp}{dx} = \frac{\mu q}{\kappa} + \beta \rho q^2 + cq^3$$

where dp/dx is the pressure gradient and β and c are dimensional constants. If the cq^3 term is omitted, the equation represents **Forchheimer's law**.

Darrieus wind turbine (Darrieus rotor) A type of wind turbine, usually with a vertical axis, typically with a two- or three-bladed rotor consisting of either straight (Δ type) or (as in the diagram) curved (ϕ type) blades having an aerofoil cross section. *See also* SAVONIUS–DARRIEUS WIND TURBINE.

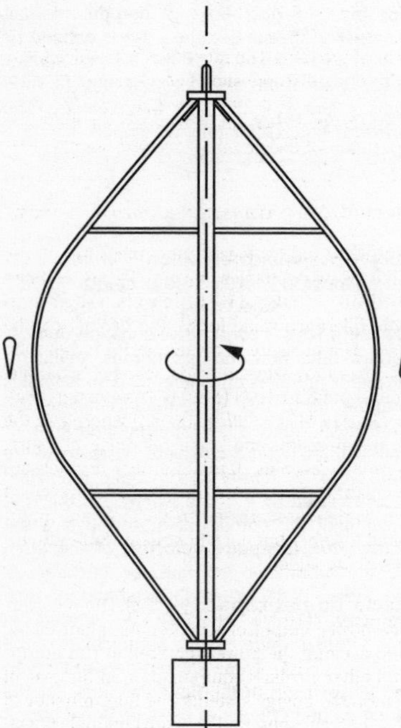

Darrieus wind turbine

dash (′) A symbol used, particularly in thermofluids, to denote per unit length, as in q'. Two dashes denote per unit area, as in q'', and three dashes per unit volume, as in q'''. For example, in heat transfer the heat flux is written $\dot{q}''$. *See also* DOT.

dashed line A broken line on an engineering drawing used to show a hidden outline or edge.

dashpot A device consisting of a piston (either loosely fitting or having holes) that slides in a cylinder of viscous oil and provides velocity- and time-dependent resistance to motion.

datum feature In an engineering drawing, any surface, edge, or axis of a component that has been chosen as the starting point for a dimension.

datum line (datum plane) A fixed reference line or plane from which distances and angles are measured.

datum point A fixed reference point from which distances are measured.

datum state (reference state) For any thermodynamic property, the temperature and pressure at which the internal energy and entropy are taken as zero. This is the triple point for the substance in its liquid state. For water, the values are 0.01°C (i.e. 273.16 K) and 6.112 mbar.

daylight The biggest gap available in a press or testing machine.

dB *See* DECIBEL.

dead axle A non-rotating axle.

dead band The range through which the output of a closed loop controlled plant can change without resulting in a change in the output of the controller.

dead-beat algorithm The mathematical method applied by a dead-beat controller.

dead-beat control A digital controller design method through z-transforms which aims for the most rapid response to a change in set point.

dead centre The point at which, in a crank mechanism, the piston connecting rod and crank are all in line so that there can be no driving moment. *See also* BOTTOM-DEAD CENTRE; TOP-DEAD CENTRE.

dead load A load on a component or structure that is steady with time, e.g. the self-weight of a bridge. *See also* LIVE LOAD.

dead space (dead volume) That volume of a gas-filled thermometer in which the gas is not at the same temperature as that being measured.

dead state In thermodynamics, a system that is in thermodynamic equilibrium with its environment, i.e. it is at the same temperature and pressure, has zero kinetic and potential energy, and is chemically inert. The dead-state temperature T_0 and pressure p_0 are usually taken as 25°C and 1 atm, respectively.

dead time The time delay between a change in the output of a controller (i.e. a change in the input to the plant) and the consequent change in the output of the plant. A long delay time will adversely affect the stability of the system.

dead-time compensation A compensation technique which uses a Smith predictor or similar method to predict the future change in output of a plant with dead time and thus improve the control of such a plant.

deadweight tester (deadweight pressure gauge, free-piston gauge, piston gauge) A device used for calibration of pressure gauges in which precisely-known pressures are generated by a weighted piston, usually spinning or oscillating to minimize friction, contained in a close-fitting cylinder and supported by a gas or liquid.

deaeration The removal of air or other dissolved gases from liquids, such as the feedwater for steam-generating boilers. *See also* DEGASIFICATION.

Dean number (*De*) A non-dimensional group used to characterize flow in curved pipes and ducts and defined, for a pipe of diameter D and radius of curvature of the centreline R, by $De = Re\sqrt{D/2R}$ where Re is the Reynolds number based on D and the bulk velocity.

debonding Said of an interface, glued joint, etc., when the adhesion breaks down and there is separation between surfaces.

Deborah number (*De*) A non-dimensional group used to characterize the flow of a visco-elastic fluid, in particular the degree to which elasticity manifests itself in response to deformation. It can be defined as the ratio of a characteristic relaxation time of the fluid λ to a time scale of the flow T, i.e. $De = \lambda/T$. T can be taken as L/V where L is a characteristic length for the flow geometry and V is a characteristic flow velocity. The choice for λ is far more difficult than that for T as in most situations there is a broad spectrum of relaxation times dependent on the constitutive model adopted for a given fluid. One possible choice is $\lambda = N_1/2\dot{\gamma}^2\mu$ where N_1 is the first normal stress difference and $\dot{\gamma}$ is a shear rate. If the fluid is modelled as a simple Maxwell material, G/μ is a logical choice for λ, G being an elastic modulus and μ a viscosity. Newtonian fluid behaviour is obtained in the limit $De \to 0$. In the limit $De \to \infty$ the fluid behaves much like a Hookean elastic solid. *See also* WEISSENBERG NUMBER.

deca (da) *See* INTERNATIONAL SYSTEM OF UNITS.

decay of oscillation *See* LOGARITHMIC DECREMENT.

deceleration The rate of decrease of velocity with time; the opposite of acceleration.

decelerometer An instrument for the measurement of deceleration. *See also* ACCELEROMETER.

deci (d) *See* INTERNATIONAL SYSTEM OF UNITS.

decibel (*G*) A logarithmic measure of the ratio of two variables, V_1 and V_2 say, defined by $G(\text{dB}) = 20log_{10}(V_1/V_2)$. By choosing a suitable reference value for V_2 the decibel can be used as a measure of a variable V_1, rather than of the ratio between variables. For example, sound pressure level (*SPL*) is expressed in decibel by $SPL(\text{dB}) = 20log_{10}(p/p_{ref})$ where p is the sound pressure and p_{ref} a reference pressure, normally taken as 20 μPa. Because decibels are a logarithmic measure, they are additive when two variables expressed in decibels are multiplied together but not when the variables themselves are added.

declared net capacity (dnc) The net average power output of a generating plant taking into account the power required to run pumps and other auxiliary equipment or, in the case of renewable-energy systems, the intermittency of the energy source. For conventional power plants, dnc is slightly less than rated capacity. For wind turbines, wave and tidal power, dnc is typically between 33% and 50% of rated capacity.

decompression chamber A chamber in which ambient-air pressure can be increased to levels found in deep-sea diving. It is used to gradually acclimatize divers back to normal conditions and avoid 'the bends'. *See also* HYPERBARIC CHAMBER.

decouple In analysis, where different processes occurring simultaneously may be considered to be unaffected by others. For example in convective heat transfer, the viscous (velocity) boundary layer is often analysed independently of the thermal boundary layer so long as the fluid properties can be regarded as constant. In elastoplastic mechanics, the works of elasticity, plasticity, friction, and fracture are sometimes evaluated independently and assumed not to affect each other. *See also* UNCOUPLE.

decrement 1. For damped free oscillations, the ratio of the amplitudes of successive cycles. **2.** In control-systems design, the ratio of the amplitude of the second overshoot to that of the first overshoot. *See also* ATTENUATION; INCREMENT; LOGARITHMIC DECREMENT.

dedendum In a spur gear, the radial distance between the pitch circle and the dedendum circle. There are corresponding definitions for bevel gears, worm gears, and screw threads. *See also* ADDENDUM; SCREW; TOOTHED GEARING.

dedendum circle *See* TOOTHED GEARING.

de Dion axle A form of independent rear suspension for motor vehicles in which the differential is fixed rigidly to the chassis while the driven wheels are kept parallel by a rigid (de Dion) tube.

deep drawing A forming process in which a peripherally-clamped blank of thin sheet is forced by a punch to flow into a die to make hollow shapes that eventually, by re-drawing and ironing, become items such as cartridge cases or beverage cans. The flange of the blank is subjected to compressive stresses while the drawn wall experiences tensile stresses that can lead to failure at the limiting draw ratio.

deep-well jet pump A combination of a centrifugal pump and a submerged nozzle-Venturi injector used for pumping water out of wells up to about 60m in depth.

defect tolerance *See* DAMAGE TOLERANCE.

deflagration The subsonic propagation of a zone of combustion through a combustible mixture. *See also* DETONATION.

deflector A flat or curved plate or baffle that causes a fluid stream to change direction.

deflexion (deflection) The linear or angular movement of a component, structure or assembly subjected to a force or torque.

deformation In solid mechanics, any change, reversible (elastic) or permanent (plastic), in the shape or size of parts of a body, or the whole body, caused by external or internal loading. It includes extension, compression, bending, and twisting. The same state of deformation in a body can appear as different combinations of elongation and shear. To state how much of the total strain is shear, it is split into dilatation and deviatoric components, the former changing volume but not shape, the latter shape but not volume. *See also* STRAIN.

deformation curve 1. *See* LOAD-DISPLACEMENT CURVE; STRESS–STRAIN CURVE. **2.** The shape taken up by beams bent in various ways. *See also* ELASTICA.

deformation law The deformation law for a Newtonian viscous fluid is

$$\tau_{ij} = -p\delta_{ij} + \mu\left(\frac{\partial u_i}{\partial x_j} + \frac{\partial u_j}{\partial x_i}\right) + \delta_{ij}\lambda div V$$

where τ_{ij} is the shear stress, p is the static pressure, δ_{ij} is the Kronecker delta, μ is the dynamic viscosity, V is the vector velocity with components u_i in the three coordinate directions x_i ($i = 1, 2, 3$), and λ is the coefficient of bulk viscosity. See also NAVIER–STOKES EQUATIONS; SHEAR RATE; STOKES HYPOTHESIS.

deformation rate Since by definition a fluid can deform continuously without limit, in fluid mechanics, deformation is defined in rate terms, i.e. changes with respect to time: translation in terms of the velocity components u, v and w in the three coordinate directions x, y and z; rotation ω_x, ω_y and ω_z in terms of velocity gradients such that

$$\omega_x = \frac{\partial w}{\partial y} - \frac{\partial v}{\partial z}, \quad \omega_y = \frac{\partial u}{\partial z} - \frac{\partial w}{\partial x}$$

and $\omega_z = \dfrac{\partial v}{\partial x} - \dfrac{\partial u}{\partial y}$;

shear strain rate ε_{xy}, ε_{yz} and ε_{zx} in terms of the velocity gradients such that

$$\varepsilon_{xy} = \frac{1}{2}\left(\frac{\partial v}{\partial x} + \frac{\partial u}{\partial y}\right), \quad \varepsilon_{yz} = \frac{1}{2}\left(\frac{\partial w}{\partial y} + \frac{\partial v}{\partial z}\right)$$

and $\varepsilon_{zx} = \dfrac{1}{2}\left(\dfrac{\partial u}{\partial z} + \dfrac{\partial w}{\partial x}\right)$;

and dilatation or extension, again in terms of velocity gradients

$$\varepsilon_{xx} = \frac{\partial u}{\partial x}, \; \varepsilon_{yy} = \frac{\partial v}{\partial y} \; \text{and} \; \varepsilon_{zz} = \frac{\partial w}{\partial z}.$$

These relationships are all derived using kinematic considerations. The quantities ω_x, ω_y and ω_z are the three components of the vorticity $\boldsymbol{\omega}$ which is the curl of the vector velocity $\boldsymbol{V}$, i.e. $\boldsymbol{\omega} = \nabla \times \boldsymbol{V}$. The sum of the three dilatation rates is equal to volumetric strain rate, which is governed by the mass-conservation equation. Similar expressions are employed in solid mechanics with displacements in place of velocities.

deformation thermometer Any instrument that determines temperature from a change in the shape or configuration of its components, such as a bimetallic strip or fluid-expansion thermometer.

deformation transitions Any changes in the mode of deformation, such as between continued elasticity and brittle fracture, continued elasticity and yielding, or elastoplastic flow and fracture.

degasification The removal of dissolved gases from liquids, typically by pressure reduction (**vacuum degasification**), heating or passing through a membrane. *See also* CAVITATION; DEAERATION.

degradation The reduction with time of the physical properties of a material.

degradation failure Failure of a system, component, or structure owing to material degradation.

degradation of energy Conversion of energy into forms of lower usefulness due to irreversibilities in energy transfer and conversion processes. The increase in entropy can be regarded as a measure of the degradation of energy.

degree (°) A measure of plane angle such that 1° is 1/360 of a complete revolution and equal to $\pi/180$ rad.

degree Celsius (degree centigrade, °C) *See* CELSIUS TEMPERATURE SCALE.

degree day (Unit °C) A crude and unscientific measure employed in the heating and ventilating industry of the influence of outside air temperature T_{EXT} on the heating and cooling (i.e. air-conditioning) energy requirements of a building. T_{EXT} is usually taken as the average (i.e. ½ [highest + lowest]) temperature on a given day. The **heating degree day** (HDD) measures the extent to which and for how long T_{EXT} falls below a base temperature T_{BASE}, taken as 15.5 or 18 °C. For the **cooling degree day** (CDD) T_{EXT} is measured above T_{BASE}.

degree Fahreneit (°F) *See* FAHRENHEIT TEMPERATURE SCALE.

degree Kelvin (K) *See* KELVIN TEMPERATURE SCALE.

degree of accuracy The accuracy of a measurement expressed as a proportion (percentage) of the quantity being measured. Thus a measurement of length to the nearest mm, say, has a degree of accuracy of 1% for length of 100 mm, but only 10% for a length of 10 mm. *See also* BIAS; ERROR; ERROR ANALYSIS; INACCURACY; INSTRUMENT ACCURACY; PRECISION.

degree of reaction 1. (reaction, Λ) In an axial-flow turbomachine, the fraction of the specific enthalpy change across a stage Δh_{STAGE} that occurs in the rotor blades Δh_{ROTOR}, i.e. $\Lambda = \Delta h_{ROTOR}/\Delta h_{STAGE}$. **2.** The extent to which a chemical reaction has reached completion.

degree of saturation (μ) An alternative to relative humidity used in the air-conditioning industry and defined by $\mu = m_S/m_{S,SAT}$ where m_S is the mass of water vapour in a given volume of a mixture of water vapour and dry air, and $m_{S,SAT}$ is the mass of water vapour that the mixture would contain if saturated at the same temperature. *See also* SPECIFIC HUMIDITY; PERCENTAGE SATURATION.

degree of supercooling Where a superheated vapour is expanded isentropically to a temperature T_R below the saturation temperature T_{SAT}, the difference $T_{SAT} - T_R$. The **degree of supersaturation** is the ratio of the actual pressure to the saturation pressure corresponding to T_R.

degree Rankine (R) *See* RANKINE TEMPERATURE SCALE.

degrees of freedom The number of variables required to specify the spatial position of a rigid body or the parts of a system. For example, a mass attached to a spring under the influence of gravity has one degree of freedom. A rigid body free to move in space has six degrees of freedom to specify its position along three Cartesian axes and rotation about them. The concept can be extended to a system of connected bodies.

degrees of superheat (superheat) The difference in temperature between a superheated vapour and saturated vapour at the same pressure.

d

dehumidification The removal of moisture from humid air by cooling it to below the dew point. It can also be accomplished by adsorption onto the surface of a material such as silica gel, or absorption by a substance such as calcium chloride solution.

de-icing The process of removing ice built up on a component or structure, such as aircraft wings and other exposed surfaces, by heating, chemical means, or mechanical means.

deionization The removal of mineral ions, such as cations from sodium, calcium, iron, and copper, and anions from chlorine and bromine, from water and other liquids.

delamination A mode of failure of composite materials, including radial-ply tyres, in which the layers separate due to repeated cyclic loading, impact, or weak bonding.

de Laval nozzle *See* CONVERGENT–DIVERGENT NOZZLE.

de Laval turbine A single-stage impulse steam turbine with steam admitted through one or more stationary nozzles onto an impulse wheel. *See also* RATEAU TURBINE.

delay period *See* IGNITION DELAY.

delay time In a component of a control system, the period between a change in the input and the consequent change in the output.

delivery *See* FAN CHARACTERISTIC.

delta wing A wing with a triangular planform, used for supersonic aircraft.

demand *See* SET POINT.

demister pad (demister blanket, mist eliminator, mist extractor) A porous pad of metal wire or plastic knitted mesh for the removal of liquid droplets from a gas or vapour stream.

demisting The removal of a condensate film from a glass or metal surface either by blowing warm air over it or by direct, often internal, heating.

Denavit and Hartenberg transform *See* HOMOGENEOUS TRANSFORM.

dense-air refrigeration cycle *See* REVERSED BRAYTON CYCLE.

densification Increasing the density of a material by compressing it or, in the case of a semi-porous solid material, by impregnating it with another substance.

densified refuse-derived fuel (d-RDF) Fuel derived from the combustible portion of refuse and other waste material which has been pulverized, compressed, and dried to produce solid-fuel pellets suitable for use in domestic boilers or co-combustion with coal in power plants.

densimeter An instrument used to determine the density or relative density of a solid or liquid.

densimetric Froude number *See* FROUDE NUMBER.

density (mass density, ρ) (Unit kg/m^3) The mass per unit volume of a substance that satisfies the continuum assumption. The reciprocal of specific volume.

SEE WEB LINKS

- Reference book companion website covering many material properties (thermal conductivity, density, thermal expansion coefficients, surface tension, viscosity, etc.)

density bottle (picnometer, pycnometer, relative-density bottle, specific-gravity bottle). A bottle or flask, typically of borosilicate glass, the volume of which $\mathcal{V}$ is known precisely at a specified temperature, used to determine the density ρ or relative density of a liquid at that temperature using the equation $\rho = m/\mathcal{V}$ where the mass of liquid m is obtained by weighing the container filled with the liquid and subtracting the weight of the container when empty. The bottle is usually fitted with a ground and polished stopper with a central capillary. Some bottles incorporate a thermometer.

density correction 1. An adjustment to the reading of a mercury barometer to account for the variation in mercury density with temperature. **2.** An adjustment to the reading of the airspeed indicator on an aircraft to account for the decreased density of the air at altitude or increased density at high speed. **3.** A correction applied to the static pressure increase created by a centrifugal fan to account for the difference between the actual air density and that at standard conditions. **4.** A correction applied to the power rating of a wind turbine to account for any difference in air density between the turbine location and that at standard conditions.

deposit gauge (deposition gauge) An instrument employed in air pollution studies for measuring the amount of pollutant deposited on a given area in a given time under given conditions.

deposition 1. The process of applying a thin film of material to a surface by chemical means, such as plating or chemical-vapour deposition, or by physical means such as sputtering or spraying. **2.** *See* SUBLIMATION. **3.** The process whereby small particles settle on a solid or liquid surface.

depth 1. The vertical distance below a datum surface, especially the sea surface. *See also* ALTITUDE; HEIGHT. **2.** The distance between the top and bottom of a hole, step in a surface, or a container.

depth gauge 1. A precision instrument, typically consisting of a machine-divided steel rule passing through a hardened-steel cross head, used to measure the depths of slots, holes, shoulders, projections, etc. **2.** A device used by divers to indicate the water depth.

depth of thread *See* SCREW.

Derailleur gear A gear system consisting of various size driving and driven sprockets on two parallel shafts connected by a continuous chain kept under tension, changing of gear being achieved by a lateral movement of the chain. Commonly used on bicycles.

derivative control (derivative action, derivative compensation) A control technique where the input to the plant is determined from the time derivative of the error. Normally employed in combination with proportional control. A **derivative element (derivative network)** produces an output proportional to the time derivative of the error, the constant of proportionality being the **derivative time constant** with unit s.

derrick crane A crane comprising a jib hinged to a stayed vertical mast. *See also* A-FRAME; SHEAR LEGS.

DES *See* DETACHED-EDDY SIMULATION.

desalination The removal of excess salt and other minerals from sea water or brackish water, typically using reverse osmosis.

désaxé engine A reciprocating engine in which the cylinders are offset to one side of the crankshaft centreline.

descaling The removal of chemical deposits from the surfaces of pipes, boilers, heat exchangers, etc.

describing function A technique for designing controllers for non-linear systems where the non-linear behaviour is approximated by a number of linear transfer functions applicable for different inputs.

desiccator A container with an airtight lid used for drying chemicals or maintaining a dry environment through the use of a hygroscopic material such as silica gel.

design (design engineering) *See* MECHANICAL-ENGINEERING DESIGN.

design code (design standard) A standard or specification for any aspect of engineering design, issued by national organizations such as ANSI, ASME, BSI, DIN, and ISO.

design factor *See* FACTOR OF SAFETY.

design for environment The incorporation into the engineering design process of considerations of any aspect that might have an impact on the environment, such as pollution prevention and conservation of resources.

design form of thread *See* SCREW.

design heating load The heating requirements based on a specified number of heating degree days, or required to maintain a building or other enclosed space at a specified temperature for a given outside temperature.

design load The greatest load that a component or structure is expected to experience under normal operating conditions. *See* FACTOR OF SAFETY.

design methodologies Design based on *s*–*N* fatigue curves in which a statistical mean life is determined for the expected service load spectrum is **safe-life design**. **Fail-safe design** relies on redundancies in a structure such that failure of one member does not lead to a cascade of fractures and hence complete failure. **Damage-tolerant design** employs fracture mechanics and takes into account initial imperfections (starter cracks), crack growth rates, and the conditions of final fracture.

design pressure The greatest pressure that a closed container is expected to experience under normal operating conditions. *See* FACTOR OF SAFETY.

d

design stress The greatest allowable stress in a component or structure that will not result in failure under normal operating conditions. *See* FACTOR OF SAFETY.

design thread form *See* SCREW.

desmodromic valve A valve for a piston engine that is mechanically closed by a camshaft and lever mechanism rather than by a spring operating on the tappet. Such valves are used primarily for high-performance (racing) engines.

desorption The process whereby a substance absorbed or adsorbed by a surface at low temperature returns to the gas phase as the temperature increases.

destructive distillation *See* PYROLISIS.

destructive interference The superposition of two harmonic wave fields in antiphase with the same frequency. The signal amplitudes subtract and the mean-square amplitude is less than that of either of the two components. The opposite occurs (**constructive interference**) when the two wave fields are in phase.

destructive testing Measurement of the mechanical properties of a material, component, or structure, by increased loading until the sample fails by fracture, collapse, or buckling.

desuperheater A device in which superheated steam is converted to saturated steam either by cooling it in a heat exchanger (indirect contact) or by spraying liquid water into it (direct contact).

detached-eddy simulation (DES) In CFD for turbulent flow near surfaces, a hybrid technique in which the near-wall regions are treated using a RANS (*see* REYNOLDS DECOMPOSITION) method and the outer flow with LES (*see* LARGE-EDDY SIMULATION).

detached shock (bow shock) A curved shock wave formed just upstream of a body in supersonic flow wherever the required turning angle exceeds that possible with an oblique shock, as in the case of a blunt body. The wave is approximately hyperboloid in shape but appears conical when viewed from a distance.

detail drawing An engineering drawing of an individual component that includes all the views, dimensions and tolerances together with material, surface-finish and other specifications needed to manufacture it.

detent A mechanism to resist or prevent the rotation of a wheel, shaft or spindle, usually in one direction only. Applications include rotary switches, clockwork motors, ratchets, and the scroll wheels on computer mice. *See also* PAWL.

determinate structure In a framework, when the loads in the members and the support reactions may be determined by statics alone. *See also* REDUNDANT STRUCTURE.

detonation adiabat A curve representing all possible end states for the combustion process in a detonation wave. It has the same form as the shock adiabat, taking into account the energy released in the combustion process, i.e. increased enthalpy. *See also* CHAPMAN–JOUGUET DETONATION.

detonation velocity The supersonic velocity of a detonation wave propagating into stationary reactants.

detonation wave (detonation, detonation front) The combination of a shock wave and a reaction zone propagating into reactants. The shock raises the reactants to sufficiently high temperature and pressure for an exothermic reaction to occur. *See also* DEFLAGRATION.

detuner *See* DYNAMIC DAMPER.

developing flow The changes in the distribution of velocity and other flow properties for flow of a viscous fluid through a pipe or duct of constant cross section which occur until fully-developed flow conditions are reached.

development The tasks required to perfect the design of a new product, the materials from which it should be made, and the most appropriate manufacturing process.

deviation The difference between the actual value and the desired value of a controlled variable.

deviatoric strain For a solid material, a principal strain system (ε_1, ε_2, ε_3) may be considered to consist of a volumetric strain $\Delta \mathcal{V}/\mathcal{V} = (\varepsilon_1 + \varepsilon_2 + \varepsilon_3)$ and deviatoric strain components $[\varepsilon_1 - (\Delta \mathcal{V}/3\mathcal{V})]$ etc.

deviatoric stress **1.** For a solid material, a principal stress system (σ_1, σ_2, σ_3) may be considered to consist of a uniform mean or hydrostatic stress $\sigma_H = (\sigma_1 + \sigma_2 + \sigma_3)/3$, which causes changes in volume, and deviatoric (shear) components ($\sigma_1 - \sigma_H$) etc., which cause changes in shape. **2.** For a moving fluid, the stress tensor σ_{ij} is the sum of an isotropic part $-p\delta_{ij}$ and a deviatoric part d_{ij}, where p is the pressure and δ_{ij} is the Kronecker delta. The non-isotropic part consists of the shear stresses and diagonal elements whose sum is zero and is due entirely to the fluid motion.

Dewar calorimeter (Dewar-flask calorimeter) A calorimeter in the form of a Dewar flask.

Dewar flask A glass or metal bottle with thin inner and outer walls separated by a narrow sealed evacuated space such that heat transfer between the inside and the surroundings is minimized. The glass surfaces exposed to vacuum are usually silvered to minimize thermal radiation.

dew cell A type of hygrometer used to determine the dew point. A wick treated with lithium chloride, which is highly hygroscopic, is wound over a heater. The wick absorbs moisture and its conductivity increases until an equilibrium temperature is reached at which evaporation begins and the conductivity starts to decrease. This temperature can be related to the dew point.

dewetting *See* SPREADING COEFFICIENT.

dew point (dew-point temperature) The temperature to which unsaturated moist air must be cooled at constant pressure for it to become saturated, i.e. for condensation to begin. In a **dew-point hygrometer** a metal mirror is cooled, typically by a Peltier device, until the dew point is reached, at which temperature moisture condenses on the mirror.

DHN *See* VICKERS HARDNESS NUMBER.

diabatic (non-adiabatic) A thermodynamic change of state of a system during which heat transfer takes place across the system boundary. *See also* ADIABATIC.

diagnostics Any method used to investigate a process, system, failure, software, etc. *See also* FAILURE ANALYSIS.

diagonal pitch The distance, when components in an assembly are staggered, between the position of a component in one row or column and the position of the corresponding component in the next row or column. The term is applied to rivets, turbine or compressor blades in a cascade, vortex generators on the surface of a wing, etc.

diagram efficiency (utilization factor, η_D) A measure of the effectiveness of energy extraction in one stage of a multistage axial turbine defined by

$$\eta_D = \frac{\textit{rate of doing work per unit mass of working fluid}}{\textit{energy supplied to unit mass of working fluid}}$$

dial gauge (clock gauge, dial indicator) A displacement gauge having a calibrated circular face and a pointer for the accurate measurement of small linear distances.

diametral pitch (p_d) *See* TOOTHED GEARING.

diamond hardness number (DHN) *See* VICKERS HARDNESS NUMBER.

diaphragm **1.** A thin flexible disc, typically of metal or plastic, that deflects in proportion to an applied pressure difference. **2.** A thin disc, of metal or plastic, designed to burst at a certain applied pressure difference. *See also* BURSTING DISC; SHOCK TUBE; SHOCK TUNNEL. **3.** In a stage of an impulse steam turbine, a disc to the periphery of which are attached the stator blades or nozzles. In some designs, the nozzles are machined into the diaphragm.

diaphragm check valve A non-return valve in which a flexible rubber diaphragm is placed in a mesh or perforated cone that points into the flow. The diaphragm material limits the application to fluids at temperatures and pressures below about 180°C and 16 bar.

diaphragm compressor A machine in which compression is achieved by the reciprocating motion of a flexible membrane of metal, plastic, or elastomeric material. This arrangement is ideally suited to pumping high-purity, toxic, or explosive gases. A similar machine incorporating check valves is a self-priming positive-displacement pump (**diaphragm pump, membrane pump**). *See also* DOUBLE-DIAPHRAGM PUMP.

diaphragm gauge An instrument for measuring a pressure difference in which the two pressures act on either side of a diaphragm that deflects in proportion to the difference. The movement may be linked to a mechanical indicator or converted into an electrical signal.

d

diaphragm meter A dry flow meter in which there are two or more interconnected chambers, each having a diaphragm in the wall. The chambers empty and fill alternately and the flow rate of gas is determined from the movement of diaphragms. Diaphragm meters are commonly used to monitor domestic and commercial gas supply. *See also* CROSSLEY METER.

diaphragm spring *See* DISC SPRING.

diaphragm valve (membrane valve) A heavy-duty valve that uses a flexible diaphragm to prevent flow if a certain pressure is exceeded. Diaphragm valves are well suited to control all gas and liquid flows, including liquid slurries and other particle-laden flows.

diathermal boundary (diathermous boundary) A boundary across which heat transfer is possible. The opposite of an adiabatic boundary.

diatomic gas A gas for which each molecule consists of two atoms. Examples are hydrogen, nitrogen, oxygen, the halogens, carbon monoxide, and nitric oxide as well as mixtures of these, including air. The specific-heat ratio γ for most diatomic gases at room temperature is 1.4. *See also* MONATOMIC GAS.

die 1. A tool having an appropriately-shaped hole through which material may be extruded or drawn. **2.** A tool employed in forging. *See also* CLOSED-DIE FORGING; OPEN-DIE FORGING. **3.** A block having the male or female shape employed in stamping operations. *See also* DIE SINKING. **4.** A thick circular disc, driven by a die wrench, with internally-threaded cutting edges for producing a screw thread. A **die wrench (die stock)** holds a screw-cutting die and has two projecting arms for applying the torque necessary to cut the thread. A die having a hexagonal or square shape (**die nut**), driven by a spanner, is used in confined spaces where a die wrench cannot be rotated through a full circle.

die casting A process in which molten metal, particularly alloys of aluminium, magnesium, copper, and zinc, is forced under pressure (10 to 200 MPa) into a reusable hardened-steel mould machined into a die.

diesel (diesel oil) The liquid hydrocarbon fuel used by diesel engines, principally dodecane, a paraffin.

diesel cycle The ideal air standard cycle, shown in the diagram as a plot of pressure (p) *vs* specific volume (v), for a diesel engine consisting of four parts: isentropic compression (1–2, from BDC to TDC), reversible isobaric heating (2–3, from TDC to the point where the heat input is cut off), isentropic expansion (3–4, from the cut-off point to BDC) and reversible isopycnic heat rejection (4–1, at BDC). The thermal efficiency is given by

$$\eta = \frac{\beta^{\gamma} - 1}{(\beta - 1)\gamma r_v^{\gamma - 1}}$$

where $\beta = v_3/v_2$ is the cut-off ratio and $r_v = v_1/v_2$ is the compression ratio. The ratio v_1/v_3 is termed the expansion ratio.

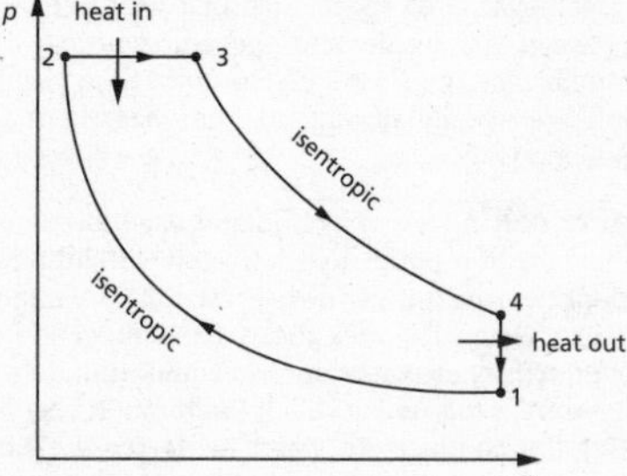

diesel cycle

diesel–electric locomotive A railway locomotive in which a diesel engine drives an electric generator that powers electric traction motors. In a **diesel–mechanical locomotive** the output of a diesel engine is connected directly to the axles via a gearbox.

diesel engine (compression–ignition engine) A piston engine operating on the diesel cycle in which the air is compressed to a temperature above the autoignition temperature of the fuel and combustion is initiated as the fuel is injected into the hot air. Diesel engines operate at higher compression ratios (typically in the range 12 to 24) than petrol engines.

SEE WEB LINKS

- Animation and explanation of the principles of the diesel engine

diesel generator An alternator driven by a diesel engine to produce electrical power.

dieseling (engine run on, run on) The tendency of a spark-ignition engine to continue running due to fuel ignition caused by a hot spot within the combustion chamber or cylinder, i.e. combustion without a spark.

diesel knock In a diesel engine, the noise associated with the rapid pressure rise caused by slow self-ignition of the fuel/air mixture prior to controlled combustion of the full charge.

die sinking Formation of a male or female pattern on a punch or die for use in coining.

diesohol An emulsion of ethanol (up to 15%) in diesel fuel, intended as an alternative fuel for diesel engines. *See also* BIODIESEL.

diesoline A high-quality diesel fuel.

differential (diff) A mechanism whose input or output velocity depends on the sum or difference of two other velocities, as in (a) a rack differential; (b) **(differential screw)** nuts on a screw thread having different pitches over different lengths of the screw; (c) the assembly of bevel or spur gears with two co-axial shafts and a third drive shaft employed on the back axle of motor vehicles.

differential chain hoist *See* CHAIN HOIST.

differential game A category of problems in game theory which are equivalent to optimal control and thus by analogy allow game-theory techniques to be used in the design of optimal controllers.

differential gap The width of a dead-band built into a control system such that no change in controller output occurs for a change of error within the dead band.

differential manometer *See* U-TUBE MANOMETER.

differential motion A mechanism in which the follower is driven from two independent sources, the net motion being the difference between the motions that would have occurred independently.

differential pressure gauge Any instrument used to measure pressure difference, for example a U-tube manometer. A **differential pressure transducer** is a pressure gauge for the accurate measurement of a pressure difference, typically using a metal diaphragm as the sensor and having an electrical output. If one side is connected to a known reference pressure such as vacuum, absolute pressure can be measured.

differential-producing primary device That part of a flow meter, such as a Bernoulli obstruction flow meter, that produces a pressure difference.

differential pulley block *See* BLOCK AND TACKLE.

differential-resistance windmill (cup-type windmill) A windmill with blades shaped to offer greater resistance to the wind on one side compared with the other.

differential scanning calorimetry An instrument used to investigate phase transitions and chemical reactions by measuring the difference between the thermal energy required to increase the temperature of a sample material and a reference material by the same amount in a controlled atmosphere. Properties which can be determined include phase-change temperatures, the glass-transition temperature, solid-solid transition temperatures and specific-heat capacity.

differential screw *See* DIFFERENTIAL (2).

differential steam calorimeter (Joly's steam calorimeter) An apparatus for measuring the specific heat of a gas at constant volume, C_V. It consists of two spherical containers immersed in steam, one evacuated, the other containing the sample gas. The spheres are suspended on opposite ends of a balance arm. The specific heat is calculated from the difference between the mass of steam that condenses on each sphere.

differential thermal analysis A technique similar to differential scanning calorimetry, but where the sample and reference material undergo the same degree of heating and the temperature response of each is monitored.

diffuse necking *See* NECK (1).

diffuser A duct or nozzle in which the kinetic energy of a flow is converted into pressure energy. For subsonic flow, the duct has an increasing cross-sectional area; for supersonic flow, the area decreases. *See also* CONFUSER; CONVERGENT–DIVERGENT NOZZLE.

diffuser efficiency (η_D) The ratio of the actual pressure rise across a diffuser Δp to the ideal pressure rise Δp_S. For an incompressible flow of density ρ with uniform velocity V at inlet and outlet, $\Delta p_S = ½\, \rho \Delta V^2$ where ΔV^2 is the change in the square of the flow velocity.

diffuse solar irradiance (Unit W/m^2) The rate of solar energy arriving at a horizontal plane on the Earth's surface resulting from scattering of the Sun's direct beam due to clouds and other atmospheric constituents. *See also* DIRECT SOLAR IRRADIANCE; GLOBAL SOLAR IRRADIANCE; IRRADIANCE.

d

diffuse surface (diffusely-emitting surface, Lambert surface) A surface for which the fraction of electromagnetic radiation reflected from it is independent of the angle of incidence. *See also* SPECULAR REFLECTION.

diffusion The transfer of an entity on a molecular level due to a gradient, for example momentum due to a velocity gradient, thermal energy due to a temperature gradient, or mass due to a concentration gradient. The **diffusion coefficient (diffusivity)** is the constant of proportionality for a diffusion process (for example, viscosity for momentum transfer, or thermal conductivity for conduction heat transfer). *See also* FICK'S LAW OF DIFFUSION.

diffusion bonding A process for joining two metals by interdiffusion of atoms across the interface brought about by local plastic deformation at elevated temperature, usually in a vacuum or inert atmosphere (e.g. nitrogen, argon, or helium).

diffusion flame A flame in which the fuel and oxidant mix together by laminar or turbulent diffusion, combustion occurring at the interface between the reactants. *See also* PREMIXED FLAME.

diffusion pump A high-vacuum (down to about 10^{-9} Pa) pump with no moving parts in which low gas pressure is created by momentum transfer to the gas molecules from high-speed vapour molecules. The vapour stream is produced by heating silicon oil or a synthetic organic liquid or, in certain applications, mercury. *See also* BACKING PUMP; ROUGHING PUMP.

digital control A form of control in which the control elements are implemented on a digital computer.

digital filter A filter which performs filtering by applying a mathematical process to sampled digital values representing the input variable.

digital-to-analogue converter (DAC) An electronic device that converts a digital signal to an analogue voltage or current. Such a converter is frequently required on the output of a control system using digital control.

dilatant liquid *See* SHEAR-THICKENING LIQUID.

dilatation (dilation) A change of volume caused by compression, temperature change, chemical action, etc. *See also* VOLUMETRIC STRAIN.

dilational wave *See* COMPRESSION WAVE.

dilatometer An instrument for measuring changes in volume of liquids or solids caused by a physical or chemical process.

dimension **1.** In mechanical-engineering applications, the basic (or primary or fundamental) dimensions are usually taken to be mass (M), length (L), time (T), and temperature (Θ). Almost any physical quantity has such dimensions which can be determined from its units or definition. Thus acceleration a has the unit m/s^2 and its dimensions are written $[a]$ = L/T^2. Since force F is mass × acceleration, its dimensions are written $[F]$ = ML/T^2. Writing $[q]$ indicates that it is only the dimensions of the quantity q that are concerned with no numerical value or magnitude. Angles and revolutions are dimensionless. *See also* DIMENSIONAL ANALYSIS. **2.** *See* DIMENSIONING.

dimensional analysis A systematic procedure for determining the k independent non-dimensional groups that are equivalent to the n dimensional variables with j independent dimensions that describe a particular physical problem. According to Buckingham's Π (pi) theorem, where Π indicates product, $k = n - j$. When the variables on which a phenomenon depends are known but not the functional relationship between them, dimensional analysis reveals the non-dimensional groups that are important. For example, at low speed, the drag force F on a sphere in a fluid flow depends upon its diameter D, the flow velocity V, the dynamic viscosity of the fluid μ, and its density ρ, i.e. $F = F(V, D, \mu, \rho)$. In this case $[F]$ = ML/T^2, $[V]$ = L/T, $[D]$ = L, $[\mu]$ = M/LT and $[\rho]$ = M/L^3. Thus we have $n = 5$ and $j = 3$ so that there will be two non-dimensional groups: $\Pi_1 = F/\rho V^2 D^2$ and $\Pi_2 = \rho VD/\mu$. If the governing equation for a problem is known, it is invariably advantageous to write it in non-dimensional form since the number of independent variables to consider is reduced by j. A simple example is $s = ut + \frac{1}{2}at^2$ which describes the distance s moved by a particle in time t if its initial velocity is u and its constant acceleration is *a*. In non-dimensional form the equation can be written $\Pi_1 = \Pi_2 + \frac{1}{2}\Pi_2^2$ where $\Pi_1 = as/u^2$ and $\Pi_2 = at/u$ so that the four dimensional variables, *s*, *t*, *u* and *a* have been reduced to two non-dimensional groups, Π_1 and Π_2. Knowing the non-dimensional groups that are important in a problem permits scaling experiments to be performed from which the behaviour of prototypes may be predicted by measurements on models. Dimensional analysis is routinely used in the study of fluid mechanics and heat transfer but is equally applicable in solid mechanics, fracture mechanics, geotechnics, and other areas of engineering and physics. *See also* IPSEN'S METHOD; PHYSICAL SIMILARITY; RAYLEIGH'S METHOD.

dimensional homogeneity Any equation that expresses the relationship between variables in a physical process must be dimensionally homogeneous; that is, each of its additive terms must have the same overall dimensions.

dimensioning The specification on an engineering drawing of the size (e.g. length, radius, angle, or spacing) and the relative location (e.g. angular position) of each feature of a component. The numerical values often include the tolerances. There should be no more dimensions than are necessary to manufacture the component. *See also* CHAIN DIMENSIONING; COMBINED DIMENSIONING; DIMENSION LINE; PARALLEL DIMENSIONING.

dimensionless number *See* NON-DIMENSIONAL NUMBER.

dimension line A line on an engineering drawing with a numeral above it that shows the length of a feature, usually in millimetres.

DIN Deutsches Institut für Normung, the German Institute for Standardization. *See also* DESIGN CODE.

dipole *See* DOUBLET.

direct-arc furnace An electric-arc furnace for high-melting-point metals such as steel in which the arc touches the metal that is contained in a refractory-lined hearth. *See also* INDIRECT-ARC FURNACE.

direct bonding In the production of microfluidic devices, the process of bonding two substrates of the same material, such as silicon wafers, glass slides, polymer components, ceramic tapes, and metal sheets.

direct-connected (directly-connected, direct-coupled, direct drive) A system in which an engine or motor drives another machine, such as a generator, pump, or compressor, with no intermediate gearing, belts, or chains.

direct-contact feedwater heater *See* REGENERATIVE RANKINE CYCLE.

direct digital control *See* DIGITAL CONTROL.

direct-drive arm A robot arm where the motor is directly coupled to the joint without a gearbox, harmonic drive, belt or chain.

direct-energy conversion A process in which an energy source is converted directly to electrical energy, such as using fuel cells or solar cells.

direct-expansion coil (direct-expansion evaporator coil) A refrigerant coil, usually finned, used to cool or dehumidify air in an air-conditioning system. The refrigerant temperature is lowered as it evaporates by passing through an expansion valve ahead of the coil.

direct expert control system The use of rules (i.e. an inference engine working with a knowledge base) within a control system so that the output is determined from the error by the rules. It is thus the application of an expert system as a controller.

direct extrusion (forward extrusion) *See* EXTRUSION.

direct injection (multi-point injection) The injection of fuel into the individual cylinders of a piston engine.

direct in-line variable-area flow meter (DIVA flow meter, force–balance meter) A variable-area inline flow meter in which the flow rate through a pipe is determined by measuring the force exerted on a spring-loaded cone situated in a converging section of the pipe. *See also* SPRING-LOADED VARIABLE-AREA FLOW METER.

direction The orientation of a line with respect to some datum.

directional gyroscope A horizontally mounted gyroscope gimballed to rotate about a vertical axis and remain pointing in the same direction.

directional valve (directional control valve) A component in a pneumatic or hydraulic system that directs or prevents fluid flow through selected passages connecting a power-input device (i.e. a pump) to a power-output device, such as a motor or cylinder.

direct metal laser sintering (DMLS) *See* LASER SINTERING.

d

direct numerical simulation (DNS) A numerical solution of the Navier–Stokes equations for a turbulent flow in which all time and length scales of the fluctuating motion are resolved. Except at Reynolds numbers too low to be of great practical relevance, the technique requires computer power far in excess of any computer available now or in the foreseeable future. DNS is a valuable research tool used to simulate idealized and simplified flow problems.

direct-return system A closed-loop heating or cooling system in which the working liquid is pumped through a series of loads, such as heat exchangers, in parallel and returned to the boiler or evaporator after the load closest to the pump. *See also* REVERSE-RETURN SYSTEM.

direct shear (simple shear) The shear stress arising in a component as a consequence of loading that results in a shear force only.

direct solar irradiance (Unit W/m^2) The rate of solar energy arriving at the Earth's surface from the Sun's direct beam. *See also* DIFFUSE SOLAR IRRADIANCE; GLOBAL SOLAR IRRADIANCE; IRRADIANCE.

dirigible *See* AIRSHIP.

disappearing-filament pyrometer (brightness pyrometer) An **optical pyrometer** in which temperature is determined by viewing a hot target through a telescope containing a calibrated lamp filament in the image plane. The lamp current is adjusted until the filament is indistinguishable from the target.

disc A flat circular sheet of metal, plastic, ceramic, or other material.

disc area (rotor-disc area) The area of the circle swept out by the tips of the blades of a helicopter rotor, propeller, or wind-turbine rotor.

disc brake A type of brake in which retardation is the result of **brake callipers**, activated mechanically, hydraulically, pneumatically, or electromagnetically, forcing pads of a friction material against the surface of a **brake disc**, a circular disc of metal, often cast iron, carbon fibre, or a ceramic matrix composite, rotating with the wheels of a motor vehicle. For applications to high-performance motor vehicles, the disc may be vented to enhance cooling and ambient air diverted onto it by a **brake duct**, typically of sheet metal or carbon fibre.

disc cam *See* CAM.

disc check valve A check valve similar to a sprung-ball check valve but with a disc rather than a ball.

disc clutch A clutch in which engine power is transmitted to the drive shaft by friction discs pressed against a flywheel driven directly by the engine.

disc-construction turbine stage An impulse-turbine stage in which the stator blades are supported by a disc-type diaphragm and the rotor blades are either attached directly to the shaft or held by a rotating disc. A **drum-construction turbine stage** is a reaction-turbine stage in which the stator blades are attached directly to the casing and the rotor blades are attached to a rotating drum.

disc coupling A flexible coupling consisting of one or more disc springs, typically stainless steel, held by a driving and a driven hub. This kind of coupling allows some axial, angular, and radial misalignment of the shafts which carry the hubs.

disc engine *See* NUTATING-DISC ENGINE.

Discflo® pump A laminar-flow rotary pump consisting of two or more ridged discs. It is practically impervious to clogging and widely used to pump wastewater.

discharge coefficient *See* COEFFICIENT OF DISCHARGE.

discharge head (Unit m) The static pressure difference Δp between the discharge port of a pump and the point of discharge, often at atmospheric pressure, converted to the height H of a vertical column of water using the equation $\Delta p = \rho g H$ where ρ is the density of water (usually taken as 10^3 kg/m^3) and g is the acceleration due to gravity.

disc loading The thrust of a propeller, or lift of a helicopter rotor, divided by the disc area.

disc meter *See* NUTATING-DISC METER.

discontinuity An idealization as infinitesimally thin of a narrow region, such as a contact surface, slip line, or shock wave, across which there are significant changes in properties, such as velocity or density.

disc pump *See* NUTATING-DISC PUMP.

discrete Fourier transform A sampled data implementation of the Fourier transform.

discrete-time system A control system where variables are known only at intervals of time and not continuously. The same as a sampled data system.

discretization 1. The process of converting a continuous signal into a series of discrete numerical values or samples, usually equally spaced in time. **2.** In numerical analysis, the representation of a continuous domain by a discrete (i.e. finite) number of grid points or adjacent grid elements or cells.

discretization error 1. The error introduced by the measurement of a variable as samples at discrete time intervals rather than continuously. This error occurs because any changes in the variable in the interval between samples will not be measured. **2.** In numerical analysis, the difference between the true value of a continuous quantity and the approximate value resulting from discretization of that quantity. For example, the derivative of a function $f(x)$ is defined as

$$f'(x) = \lim_{h\to 0}\frac{f(x+h)-f(x)}{h}$$

which can be approximated by

$$f'(x) = \frac{f(x+h)-f(x)}{h}$$

where h is small but non zero.

disc spring (Belleville washer, diaphragm spring) A disc manufactured from spring steel in the form of a frustum of a cone that carries a static or dynamic axial load and can be used singly or arranged in stacks.

disc valve 1. A non-return valve consisting of a fabric, ceramic, metal, or rubber disc held by a spring against a flat seat. The valve opens when the upstream pressure exerts a force on the disc that exceeds the spring force. **2.** A rotary valve sometimes used to time the admission of the fuel/air mixture into the cylinder of a two-stroke engine.

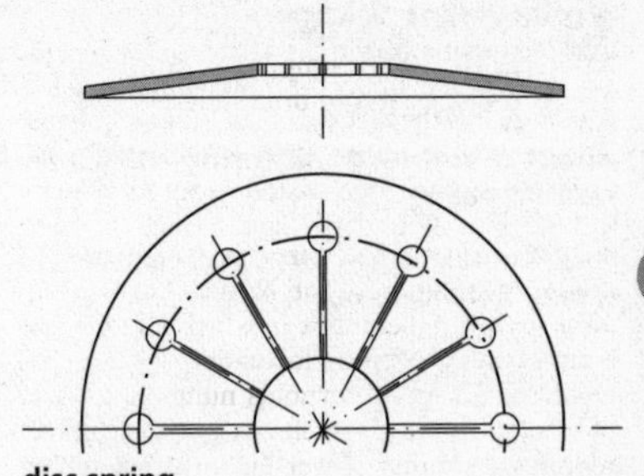

disc spring

disc wheel A wheel where the rim and hub are connected by a solid disc instead of spokes.

dishing The production, in ballistics and forming, of a domed shape as a result of membrane (stretching) action. *See also* BULGING.

dislocation A fault in the regularity of a crystal structure which may take the form of either a missing or an additional plane of atoms. Dislocations permit plastic shear to occur progressively over a slip plane, rather than the sliding of all atoms simultaneously (**block slip**), thus explaining why the practical shear strength of metals is about $G/1000$ (where G is the shear modulus) rather than $G/10$ as predicted by simple theory.

dispersion relation The relationship between the angular frequency of a wave ω and its angular wavenumber $k = 2\pi/\lambda$ where λ is the wavelength. For example, for transverse vibration of a uniform flexible string, $\omega = k\sqrt{T/\rho}$ where T is the string tension and ρ is the mass per unit length of the string.

displacement 1. The linear distance of a moving object from its initial to its final position, irrespective of the path taken. It is a vector quantity. **2.** For a body partially or totally immersed in a liquid, the volume of the immersed part of the body. Displacement may also refer to the mass or weight of this volume. *See also* ARCHIMEDES PRINCIPLE. **3.** The volume swept out per stroke by the piston of reciprocating engine or pump.

displacement compressor *See* POSITIVE-DISPLACEMENT COMPRESSOR.

displacement pump *See* POSITIVE-DISPLACEMENT PUMP.

displacement thickness (δ^*) (Unit m) A thickness that arises in the analysis of a boundary layer based upon the momentum-integral equation. It is the distance that a streamline just outside of a boundary layer is deflected away from the surface due to the reduced velocity within the boundary layer. For a two-dimensional boundary layer, from conservation of mass within the boundary layer,

$$\delta^* = \int_0^\infty \left(1 - \frac{u}{U_\infty}\right) dy$$

where $u(y)$ is the distribution of streamwise velocity within the boundary layer, y is the distance from the surface and U_∞ is the free-stream velocity. *See also* MOMENTUM THICKNESS.

displacer piston *See* STIRLING ENGINE.

dissipation **1.** The irreversible conversion of mechanical energy into thermal energy with an associated increase in entropy. **2.** (Unit m^2/s^2) In fluid flow, the dissipation of energy by viscosity due to velocity gradients. **3.** **(ε)** (Unit m^2/s^2) In turbulent flow, the dissipation of turbulent kinetic energy by the smallest eddies working against viscous stresses. It can be shown that

$$\varepsilon = 2\nu \overline{e'_{ij} e'_{ij}}$$

where ν is the kinematic viscosity of the fluid, e'_{ij} represents the gradient of the fluctuating velocity, and the overbar denotes a time average. In numerical analysis of turbulent-flow, ε is one of the quantities that has to be modeled. *See also* $k - \varepsilon$ MODEL.

dissipation integral A term that arises in the mechanical-energy integral equation for a boundary layer. For a two-dimensional flow it is given by

$$\int_0^\infty \tau \frac{\partial \bar{u}}{\partial y} dy$$

where τ is the time-averaged total shear stress a distance y from the surface and $\bar{u}$ is the time-averaged streamwise velocity at y. The integral can also be written as

$$\int_0^{U_\infty} \tau d\bar{u}$$

where U_∞ is the free-stream velocity.

dissociation **1. (thermal dissociation)** For a gas molecule, the breakdown at high temperatures (typically above 1500 K) into its constituent atoms or molecules. The process is endothermic and reversible. **2.** For an ionic compound, such as a salt, in solution the reversible breakdown into smaller particles, ions, or radicals.

dissociation constant (K) **1. (thermal equilibrium constant, equilibrium constant)** A constant that characterizes a chemical reaction at equilibrium. For a mixture of ideal gases A, B, C, and D undergoing the following reaction $\upsilon_A A + \upsilon_B B \leftrightarrow \upsilon_C C + \upsilon_D D$ it can be shown that

$$K = \frac{p_C^{\upsilon_C} p_D^{\upsilon_D}}{p_A^{\upsilon_A} p_B^{\upsilon_B}}$$

where p_i is the partial pressure of component i and υ_i is its stoichiometric coefficient. **2.** The product of the concentrations of dissolved ions in a solution in equilibrium divided by the concentration of the undissociated molecule.

distance lag *See* TRANSPORT DELAY.

distance lag error The position error in the output of a control system responding to a step input. *See also* VELOCITY LAG ERROR.

distance ratio The ratio of the movement at the input of a machine to the movement of its output in the same time interval. *See also* BACKLASH; MECHANICAL ADVANTAGE.

distortion energy The elastic strain energy associated with change of shape of a body.

distributed control system A control system where the control of different parts of a process is implemented in a number of coupled controllers.

distributed load (w, q) (Unit N/m) A load acting on a structure or component which is spread out rather than concentrated at a point. *See also* LOAD INTENSITY.

distributed numerical control The use of individual CNC controllers on a number of machine tools each receiving part programs from a single central computer.

distributor A device that supplies high voltage to the spark plugs of a spark-ignition engine at the appropriate time in the engine cycle. In modern engines the rotary-switch distributor has been replaced by an electronic system. *See also* DWELL ANGLE.

distributor gear The driving gear of a rotary distributor.

district heating (community heating) The distribution of thermal energy, generated in a centralized location, to residential and commercial consumers through city-wide heating networks.

disturbance An unintended signal applied to a control system that may affect the controlled output. For example, a controller for the temperature of an oven may suffer a disturbance due to changes in heat loss as a result of a varying ambient temperature around the oven.

dither 1. In control systems, a small continuously-applied force applied to a plant so as to maintain it in continuous small-amplitude motion and thus prevent stick–slip friction. **2.** In sampling (**discretization**), a small random electrical signal added to the analogue-to-digital converter input to reduce systematic errors.

dither mechanism An oscillating mechanism to reduce stiction between moving parts.

Dittus–Boelter correlation The empirical equation $Nu = 0.023\,Re^{0.8}\,Pr^{n}$ that can be used to describe convective heat transfer in turbulent pipe flow. Nu is the Nusselt number based upon the pipe diameter D, Re is the Reynolds number based upon D and the bulk velocity of the flow, and Pr is the Prandtl number. For cooling a fluid $n = 0.3$, and for heating $n = 0.4$. The range of applicability of the equation is $Re > 10000$, $0.7 < Pr < 160$ and $L/D > 10$ where L is the pipe length. *See also* SIEDER–TATE CORRELATION.

DIVA flow meter *See* DIRECT–INLINE VARIABLE-AREA FLOW METER.

divariant system A system that may be described by only two variables, e.g. a single-phase thermodynamic system where pressure and temperature are sufficient to define its state.

divergent nozzle (divergent duct) A nozzle (duct) the cross section of which increases from entry to exit. If the flow is subsonic, the nozzle acts as a diffuser.

diverter valve *See* BYPASS VALVE.

divided pitch *See* SCREW.

dividers A pair of hinged arms having pointed ends which are used like compasses for either measuring, transferring measurements, or scribing arcs by scratching.

dividing engine An instrument for making accurate subdivisions on linear, circular, or cylindrical scales of measurement.

dividing head An attachment used on machine tools for dividing the circumferences of workpieces for the accurate manufacture of splines, gears, etc. A **division plate**, consisting of a number of concentric rings of holes accurately dividing the circumference into various equal subdivisions, is employed to position a dividing head.

d

DMC *See* DOUGH-MOULDING COMPOUND.

DMLS *See* LASER SINTERING.

DNS *See* DIRECT NUMERICAL SIMULATION.

Doble vane The split, elliptical-shaped bucket design used for the runner of a Pelton turbine.

dog clutch (square jaw clutch) A clutch in which projections on one part fit into recesses on the other, one type being like an Oldham coupling where the shafts are in line. They are connected and disconnected by sliding along a splined shaft. Used, for example, in epicyclic gearboxes to lock particular gear wheels. *See also* POSITIVE CLUTCH.

dome nut *See* CAP NUT.

Donohue equation An empirical equation used to calculate the shell-side heat-transfer coefficient h for a shell-and-tube heat exchanger. It may be written as

$$\frac{hD}{k} = 0.2\left(\frac{GD}{\mu}\right)^{0.6} Pr^{0.33}\left(\frac{\mu}{\mu_W}\right)^{0.14}$$

where D is the outside diameter of a tube, k is the thermal conductivity of the shell-side fluid, μ is its dynamic viscosity at the bulk temperature, μ_W is the dynamic viscosity at the tube wall, Pr is the Prandtl number, and G is an average mass velocity for the shell-side fluid, taking into account the parallel and traverse nature of the flow.

Doppler current meter An acoustic instrument for measuring the flow speed of water based upon the Doppler effect, whereby a sound wave transmitted into the water is scattered by particles assumed to be moving at the water speed V. The frequency difference between the transmitted and scattered sound wave is given by $\Delta f = 2f_T V\cos\theta/c$ where f_T is the frequency of the transmitted wave, θ is the angle between the flow direction and the transmitted wave, and c is the speed of sound in the water.

Doppler effect The increase in frequency detected when an oscillating source moves towards a detector and the decrease when it moves away. It is the consequence of the change in the source–detector travel time.

d

Doppler ultrasonic flowmeter A Doppler current meter where the frequency of the sound waves is in the ultrasonic region, i.e. above about 20 kHz. *See also* ULTRASONIC FLOW METER.

dot (˙) A symbol used above a quantity to denote a time derivative, as in $\dot{x}$. Two dots denote a second derivative, as in $\ddot{x}$. *See also* DASH.

double-acting machine A reciprocating machine, such as a compressor, engine, or pump, in which the working fluid is compressed or moved alternately by both sides of the piston; that is, on both the upstroke and the downstroke.

double-block brake A clasp-brake system, typically used in locomotive applications, where braking force is applied separately on diametrically-opposed sides of a wheel. *See also* BLOCK BRAKE.

double-diaphragm pump A diaphragm pump in which there are two diaphragms connected by a sliding driving rod (the diaphragm rod). Check valves are used to ensure the diaphragms pump alternately.

double-entry compressor A centrifugal compressor in which a single impeller has blades on both sides or two impellers are mounted back to back. Gas enters axially from both sides and the compressed gas is discharged from the periphery.

double-flash power plant *See* FLASH-STEAM POWER PLANT.

double-flow turbine A turbine in which the working fluid enters radially in the middle of the casing and expands outwards in both axial directions. The arrangement has the advantage that the thrusts generated on the two sides largely cancel each other out. *See also* STEAM TURBINE.

double-helical gear (herringbone gear) *See* TOOTHED GEARING.

double Hooke's joint A universal joint, consisting of two Hooke's joints with an intermediate shaft, that eliminates variations in angular displacement and velocity between the driving and driven shafts.

double-integrating gyro A gyroscope with only one degree of freedom, having virtually no restraint of its spin about the output axis.

double pendulum Two pendulums, which may be simple or compound, suspended one below the other. Such pendulums are complex to analyse and can exhibit chaotic behaviour.

double-row bearing *See* BEARING.

double sleeve valve *See* SLEEVE VALVE.

double-slider coupling *See* OLDHAM COUPLING.

double-start thread *See* SCREW.

doublet (dipole) In potential-flow theory, the combination of a source and sink of equal strength q on the x-axis when the separation a between them is reduced to zero but qa is kept constant. The streamlines for a doublet are circles passing through the origin, tangent to the x-axis and with their centres on the y-axis. The combination of a uniform flow in the x-direction with a doublet represents irrotational flow around a circular cylinder.

double volute *See* VOLUTE.

dough-moulding compound (DMC) A fibre-reinforced polymer in an uncured or partially-cured condition, before manufacture into components.

dowel A headless cylindrical pin that fits into corresponding holes in mating components, thus ensuring relative location.

downcomer 1. A pipe through which fluid flows to a lower level. **2.** A tube in a steam boiler through which water flows down from the steam drum to the lower feedwater (mud) drum. *See also* WATER-TUBE BOILER.

downdraught carburettor *See* CARBURETTOR.

downforce *See* AEROFOIL.

down milling (climb milling) A cutting process in which the peripheral speed of the

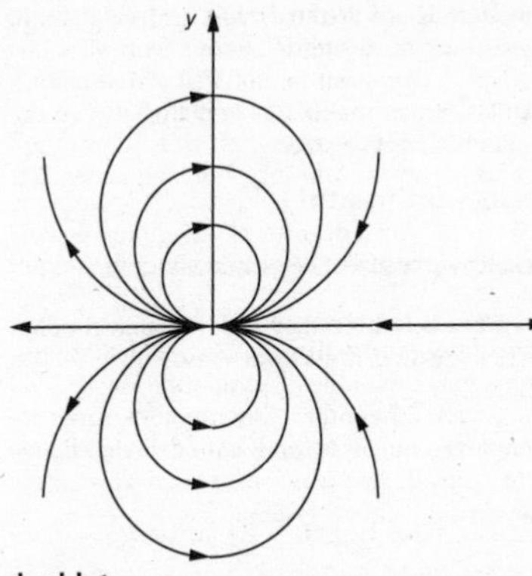

doublet

cutters of a rotating cylindrical tool and the linear feed of a workpiece have the same direction. It results in crescent-shaped chips that are initially thick, but become thin. Similar terminology applies in circular sawing. *See also* UPMILLING.

downstream Fluid in relative movement away from an object or location is said to be downstream of it. *See also* UPSTREAM.

draft *See* DRAUGHT (1).

drag (drag force, *D*) (Unit N) The aerodynamic or hydrodynamic force exerted on a body as it moves relative to a fluid acting in a direction opposite to that of the relative motion. The total **parasitic drag** is made up of two parts: the **friction drag (skin-friction drag)** due to the integrated effect of the shear stress acting on the body surface, and the **form drag (pressure drag)** due to the integrated effect of the static pressure acting on the surface. A sudden decrease in parasitic drag (**drag crisis**) can occur for flow over a body as the boundary-layer on the body's surface undergoes transition from laminar to turbulent flow. For a smooth sphere the drag crisis occurs at a Reynolds number of about 2×10^5. **Wave drag** is the resistance experienced by a body moving (a) at transonic or supersonic speeds as a result of the formation of a shock wave or (b) at or near a free liquid surface as a result of the formation of surface waves. **Induced drag** is that associated with the lift force produced by a wing.

The **drag coefficient** (C_D) is a non-dimensional quantity defined by $C_D = D/Aq$ where q is the dynamic pressure and A is an appropriate area, usually the projected area of the body in the flow direction.

drag-force flow meter (drag-body flow meter, target flow meter) A flow meter in which the drag force exerted on a body immersed in a duct flow is measured and used to estimate the flow rate. Such devices require calibration.

drag torque The additional torque required to rotate a shaft resulting from friction in the bearings and seals.

drape The ability of a fabric to hang closely over a body, as chain mail does in modern stab-resistant body armour.

draught 1. The act of pulling. **2. (draft)** The differential pressure between the air intake and the flue-gas exit in a furnace or boiler that causes air and flue gas to flow through. It can also mean the flow itself. **Balanced draught** is where the combustion air inflow matches the outflow of combustion gases. Draught can be brought about by the hydrostatic pressure difference which corresponds to the lower density of the hot gas flowing up a smoke stack and the higher density of the ambient air (**natural draught, stack effect**), by reducing the pressure on the exhaust side, for example with a fan or steam jets (**induced draught**) or by using fans at the intake (**forced draught, mechanical draught**). The draught can be regulated by a hinged plate installed at the stack outlet that operates like a butterfly valve (**stack damper**). Wall friction, and changes in cross section and flow direction, cause a reduction in the static pressure of air and flue gas as they flow through a furnace or boiler (**draught loss**). In a confined combustion system, **positive draught** is a draught pressure that exceeds atmospheric pressure. The pressure level (the **draught pressure**) in the stack can be measured using an inclined-tube manometer or other sensitive differential pressure gauge. The **overfire draught** is the gauge pressure, usually negative, measured directly above the combustion zone. *See also* COOLING TOWER.

draughtsman A person skilled in creating engineering drawings.

draught tube A flared tube, often in the form of an elbow with a vertical inlet and horizontal outlet, that guides the water flow from a hydraulic turbine into the tailrace. The draught tube is a form of diffuser that recovers some of the kinetic energy of the flow leaving the runner.

drawability The ease or otherwise by which metal sheet may be pressed or drawn into shapes without fracture, buckling, etc.

drawbar pull The force with which a locomotive or tractor pulls vehicles attached to it.

draw filing A finishing operation in which the file is placed across the workpiece, rather than along, to take fine cuts. Filing forces are much reduced owing to the 'slice-push' action of the highly-inclined teeth.

drawing 1. *See* ASSEMBLY DRAWING; DETAIL DRAWING; ENGINEERING DRAWING. **2.** The manufacture of metal wire, tubing, etc. by pulling the material through a die, generally in several steps using dies of decreasing size. *See also* DEEP DRAWING.

drawing types Engineering drawings may be of a single part (detail drawing) or an assembly of parts (assembly drawing).

d-RDF *See* DENSIFIED REFUSE-DERIVED FUEL.

Dremel® tool A multi-purpose, high-speed (up to about 40000 rpm) electric hand drill in which drill bits (as small as 0.3 mm diameter) and other accessories, such as routers, grinders, cutters, sanders, and engravers, are held in a collet inserted into a specially-designed chuck.

drift 1. A wedge or tapered rod used to free mating parts, such as morse tapers or tightly-fitting bolts, or to align holes in mating components. **2.** A progressive change with time in the controlled output of a system with a constant set point input.

drill 1. (drill bit) A rotating tool having cutting edges on one end leading to helical flutes along which chips escape. Used for boring cylindrical holes **2.** The machine that drives the tool in (1).

drill gauge A flat sheet having accurately drilled holes from which the diameters of drills can be determined.

drill press A bench-mounted drilling machine, or bench-mounted attachment for an electric hand drill.

drill sizes The diameters of standard metric drills that increase in 0.05 mm increments from 0.35 mm to 2.8 mm diameter, then in 0.1 mm increments up to 25 mm, noting that quarter and three-quarter mm diameters are also available starting at 3.25 mm. Imperial drills come in three ranges: **number sizes**, from No. 80 (0.0135 in diameter) to No. 1 (0.228 in diam.); **letter sizes**, from A (0.234 in diam.) to Z (0.413 in diam.); drills given as **inch sizes** start at $\frac{1}{64}$th in and go up to ½ in in $\frac{1}{64}$th in steps. *See also* TAPPING SIZE DRILL.

drinking water *See* POTABLE WATER.

drip-feed lubricator A device that supplies lubricating oil in drips from a reservoir to a plain bearing.

drive The means by which motion and power is (a) provided to a machine or device or (b) transferred within a system.

driven gear *See* TOOTHED GEARING.

drive shaft (Cardan shaft, tailshaft) The shaft through which power and torque are transmitted from a motor or engine to a machine or device.

drive-through teaching *See* PROGRAMMING BY DEMONSTRATION.

drive train In a motor vehicle, all the components that transmit torque and power from an engine to the driving wheels, including the clutch, gearbox, drive shaft, differential, and half shafts.

driving fit *See* LIMITS AND FITS.

driving gear *See* TOOTHED GEARING.

driving wheel 1. A wheel on a motor vehicle, train, etc., that applies tractive force to the road, rail, etc. thus propelling the vehicle. **2.** The first gear in a train of gears.

drop A small quantity of liquid, greater than about 10 μl in volume (2.6 mm diameter), that forms into an almost spherical shape in a gas due to surface tension. An even smaller quantity of liquid, below about 65 *nl* in volume (0.5 mm in diameter), forms a **droplet**. A **pendant drop** is a drop of liquid in a gas hanging from the end of a vertical tube under the influence of surface tension. Once a critical size is reached, the drop detaches itself.

drop forging A forging where the energy required for deformation is provided by a tup falling from a height. *See also* CLOSED DIE; OPEN DIE.

drop hammer *See* TUP.

droplet breakup Liquid droplets in a spray form smaller droplets due to two main mechanisms. According to the Taylor-analogy breakup model for low Weber numbers (*We*), breakup results from the droplet oscillating and distorting due to the effects of surface tension, drag, and viscosity. For high *We*, breakup is due to the relative velocity between the gas and droplet causing a Kelvin–Helmholtz instability. *We* here is defined by $We = \rho_G V^2 R/\sigma$ where ρ_G is the gas density, V is the relative velocity between the gas and the droplet, R is the droplet radius, and σ is the surface tension of the liquid.

droplet condensation (dropwise condensation) The formation of discrete droplets on a cold surface that is not wetted by the condensate. *See also* FILM CONDENSATION.

droplet ejection *See* ACOUSTIC DROPLET EJECTION.

drop test (drop tower) 1. Investigation of the mechanical properties of materials impacted at velocities up to 20 m/s in which a mass (the tup) falls from a known height onto a test-piece. Recorded data include impact velocity, energy and momentum, and rebound height. *See also* HOPKINSON BAR. **2.** Investigation of materials, structures, etc. in a microgravity environment produced by dropping from a great height (typically above 20 m). The drop-tower interior is often evacuated for such research.

drowned outflow (submerged outflow) The discharge from beneath an underflow gate when the liquid surface immediately downstream is submerged. *See also* FREE OUTFLOW.

drum 1. A cylinder, often with helical grooves, on which rope is wound in a hoist. **2.** Part of a friction-brake system where resistance is provided by shoes either contracting onto the outer surface of a drum, or expanding against the inner surface of a drum. **3.** Part of a boiler or pressure vessel. *See also* DRUM-TYPE BOILER; FEEDWATER DRUM; MUD DRUM; STEAM DRUM; WATER-TUBE BOILER.

drum brake *See* DRUM (2).

drum-construction turbine stage *See* DISC-CONSTRUCTION TURBINE STAGE.

drum-type boiler *See* CROSS-DRUM BOILER; LONGITUDINAL-DRUM BOILER.

dry air Air containing no water vapour.

dry-back boiler A shell boiler in which the hot-gas direction is reversed by a refractory-lined chamber on the outer plating of the boiler. *See also* WET-BACK BOILER.

dry-bulb temperature *See* WET-BULB TEMPERATURE.

dry cooling The cooling of an engine or other power plant by air rather than water, often in order to conserve water. Most motor vehicles use dry cooling with air passing through a radiator through which is circulated a coolant, usually water with ethylene glycol added to reduce the freezing point.

dry cooling tower A heat exchanger through which the water to be cooled is circulated and does not come in contact with the cooling air. *See also* WET COOLING TOWER.

dry friction The friction between two clean, dry solid surfaces, either at rest (static friction) or with relative sliding (kinetic or dynamic friction). *See also* AMONTONS FRICTION.

dry-friction model *See* BEARING.

dry ice (cardice) The solid form of carbon dioxide that exists below −78.5°C at atmospheric pressure.

dry joint *See* SOLDERING.

dry liner *See* CYLINDER LINER.

dry machining A machining operation in which no cooling or lubricant is used. *See also* MINIMUM QUANTITY LUBRICATION.

dryness fraction (quality, *x*) The mass of saturated vapour in unit mass of wet vapour. For a wet vapour, the value of a thermodynamic property y, such as specific energy, specific enthalpy, or specific volume, is given by $y = (1 - x)y_f + xy_g$ where y_f is the value of the property in its saturated-liquid state and y_g that of the saturated vapour, both at the temperature and pressure of the mixture. The **wetness fraction** is $1 - x$, i.e. the mass of saturated liquid in unit mass of wet vapour.

dry pipe A perforated pipe, to withdraw steam, installed above the liquid surface in the steam drum of a boiler. The perforated design minimizes moisture carry-over and provides for an even withdrawal of steam.

dry pipe sprinkler system A fire-protection system in which the pipes are filled with air or nitrogen, and water only enters the pipes when a valve is opened in the event of a fire.

dry saturated steam (dry steam, saturated steam) Steam that has changed completely to gas with no residual water content. *See also* SATURATED VAPOUR; SATURATED WATER.

dry saturated vapour line On a *p-v* or *T-v* diagram for a pure substance, the line that separates the wet-vapour (i.e. saturated liquid plus saturated vapour) states from the superheated-vapour states. If the pure substance is water, the vapour is termed steam.

dry scrubber A device for the removal of acid gases and other contaminants from flue gases and consisting of three main components: a gas cooling system, a reagent injection system, and a filter system. Typical reagents are hydrated lime and sodium bicarbonate, which are injected into the cooled gas as a fine dry powder. *See also* EXHAUST SCRUBBER; WET SCRUBBER.

dry-steam power plant (dry-steam energy system) A steam power plant supplied directly with superheated steam from a geothermal source. As shown in the diagram, after expansion through the turbine, the steam is condensed and the condensate reinjected into the geothermal reservoir. *See also* FLASH-STEAM POWER PLANT.

dry-sump engine An internal-combustion engine in which the lubricating oil is carried in an oil tank that is separate from the engine. *See also* WET-SUMP ENGINE.

dry vapour *See* SATURATED VAPOUR.

d_{sm} *See* SAUTER DIAMETER.

d_{32} *See* SAUTER DIAMETER.

D[3,2] *See* SAUTER DIAMETER.

dual combustion cycle (dual cycle) *See* SABATHÉ CYCLE.

dual control In optimal control, a method which aims to simultaneously control the errors in the estimation of the system parameters and the error in the controlled variable.

dual-fuel burner A burner designed to operate with gas as the main fuel but which includes a retractable oil burner for use when gas is unavailable.

dual-fuel engine An internal-combustion engine that can operate on one of two fuels, or a mixture of fuels; typically a diesel engine that can also run on liquid or compressed natural gas, although a pilot injection (about 10% of the total fuel energy) of diesel fuel is required to start combustion. *See also* MULTI-FUEL ENGINE.

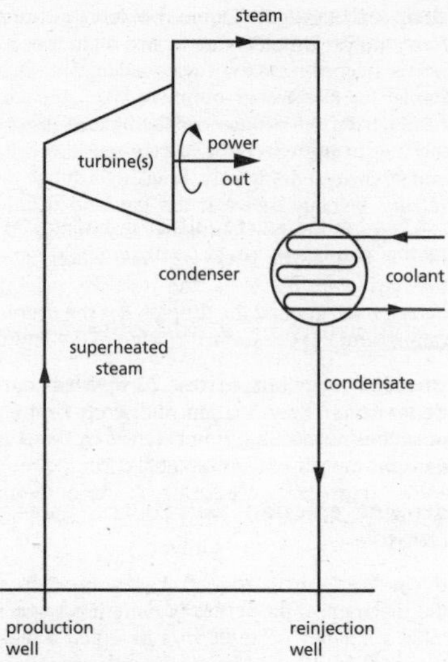

dry-steam power plant

dual ignition In a spark-ignition engine, the use of two spark plugs per cylinder, each supplied from a separate electrical source.

dual-mode control A control method where the controller operates in different ways under different input conditions. For example, to achieve a fast response in a control system responding to a step change in set point, a bang-bang method can be used. However, this method will not give a low steady-state error. Hence a dual-mode controller could initially employ bang-bang control and then switch to high-gain proportional control when the controlled variable is close to the set point.

dual phase steel A high-strength low-alloy (HSLA) steel with a microstructure of martensite precipitates in polygonal ferrite.

duckbill valve A one-piece elastomeric non-return valve which is conical in shape when closed and flared out like a diffuser when open.

duct A tube of any cross section through which there is fluid flow (**duct flow**). Examples are pipes, nozzles, diffusers, and confusers.

ducted fan A propulsion system in which a fan rotates inside a close-fitting co-axial duct or cowling. *See also* TURBOFAN ENGINE.

ducted propulsor A shrouded propeller employed to propel underwater vehicles or light aircraft.

ductile (ductile material) Describes a material that can be deformed permanently to large strains without fracture.

ductile-brittle transition temperature The narrow range of temperature below which materials, in laboratory tests, are brittle and above which they are ductile. Body-centred transition metals are brittle below about $0.1T_m$, and non-metals and intermetallic compounds below about $0.5T_m$ where T_m is the absolute melting point. Face-centred metals and alloys usually do not become brittle.

ductile fracture A fracture of a component or structure which is preceded by extensive ductile deformation so that the broken pieces cannot be re-fitted to regain the original size and shape of the component or structure. *See also* BRITTLE FRACTURE; FRACTURE MECHANICS.

ductile iron *See* CAST IRON.

Dufour effect The temperature gradient that occurs as a result of diffusion due to a concentration gradient. It is characterized by the non-dimensional **Dufour number (*Df*)**.

Dulong–Petit law To a good approximation, the specific heat of a solid is given by $C = 3kN_A = 24.94$ J/mol K where k is Boltzmann's constant and N_A is Avogadro's constant.

Dulong's formula A formula for estimating the higher calorific value (HCV) of a fossil fuel based upon the weighted heats of combustion of the constituent elements. It can be written as $\mathrm{HCV} = 338.9\mathrm{C} + 1433(\mathrm{H_2} - \mathrm{O_2}/8) + 94\mathrm{S}$ where C, H_2, O_2 and S represent the percentages by mass of carbon, hydrogen, oxygen, and sulphur, respectively.

dummy piston *See* BALANCE PISTON.

dump load A term for electrical space and water heaters incorporated into a diesel generator power-supply system to absorb power when other demands on the system are low, such as at night.

dump tank *See* MEASURING TANK.

dump valve A valve fitted to turbocharged engines to vent excess boost pressure when the throttle is closed.

duplex pump *See* SIMPLEX PUMP.

duplex tandem compressor A reciprocating compressor with two pistons driven by the same crank.

durability The ability of components and equipment to resist wear and to continue performing reliably over a long time.

duralumin® *See* PRECIPITATION HARDENING.

duration The period of time taken by an event or process.

durometer A testing machine used to determine the indentation, durometer, or Shore hardness of materials—particularly polymers, elastomers, and rubbers—by measuring the penetration depth of a cone-shaped indenter loaded with a calibrated spring.

dust Solid particles, typically less than 0.5 mm diameter, in the atmosphere, flue and exhaust gases, which may be either naturally occurring, the result of wear, or an industrial process. *See also* POLLUTANTS.

dust separator Any device for the removal of dust from a gas stream, including electrostatic precipitators, cyclone separators, and wet scrubbers. A **dust filter** is a fabric bag or screen that removes dust from a gas stream as it flows through.

duty cycle The fraction or percentage of time for which a component, machine, or system is operating.

dwell *See* CAM.

dwell angle For a rotary distributor, the degrees of rotation for which the points are closed.

dwell time The cure time required by the adhesive in e.g. the manufacture of laminates.

dynamic balancing The process, for a rotating body such as a crankshaft, of distributing the mass, including the addition of corrective masses (or balance weights), such that the second moment of inertia with respect to the axis of rotation and the radial direction is zero. The reactions at the bearings are then reduced to the static reactions and there is no vibration. A body or system that vibrates when component parts are rotating is **dynamically unbalanced**. *See also* BALANCE; STATIC BALANCING.

dynamic braking The use of an electric traction motor as a generator to assist in braking a vehicle. If the power generated is returned to the supply line it is termed regenerative; if dissipated as heat, it is rheostatic.

dynamic compressor A machine, such as a centrifugal or axial compressor, that compresses a gas by rotational rather than reciprocating motion.

dynamic coupling In vibrations, the existence of inertial terms depending on mass in the governing equations, so that there is only a force if there is a corresponding acceleration. *See also* STATIC COUPLING.

dynamic creep Creep that occurs under fluctuating load or temperature. *See also* FATIGUE.

dynamic damper (detuner) A device attached to a system to alter the latter's vibration characteristics, in particular to avoid frequencies such as those corresponding to critical speeds in rotating systems. An auxiliary vibrating mass driven from the main system by a non-linear spring is one form of dynamic damper. *See also* LANCHESTER DAMPER.

dynamic equilibrium *See* D'ALEMBERT'S PRINCIPLE.

dynamic flexibility The stiffness of a component, such as a beam in bending, when loaded at high rate or when vibrating.

dynamic friction *See* COEFFICIENT OF FRICTION; KINETIC FRICTION.

dynamic hardness (rebound hardness) The resistance of a material to local indentation by a rapidly-moving rigid indenter. In most practical methods the indenter is allowed to fall under gravity onto the surface of the material when the rebound height is a measure of the dynamic hardness. *See also* SHORE REBOUND SCLEROSCOPE.

dynamic load 1. Loading of a component or structure by a moving object whose point of application changes with time, e.g. the live load of a train passing over a bridge. *See also* DEAD LOAD. **2.** A load applied to a particular part of a component or structure in a short time interval. *See also* IMPACT.

dynamic load rating The allowable load on a component or structure when the loading is not static.

dynamic modulus 1. (Unit Pa) The elastic modulus of a material, defined as the ratio of stress to strain determined under oscillatory conditions. **2.** *See* UNRELAXED MODULUS.

dynamic pressure (*q*) (Unit Pa) In a fluid of density ρ flowing at velocity V, $q = ½\rho V^2$ which represents the kinetic energy per unit volume of fluid. In incompressible flow, it is the difference between the stagnation and static pressures. In compressible flow it depends upon the Mach number M and the static pressure p. For a perfect gas with specific-heat ratio γ, $q = ½\gamma p M^2$. *See also* BERNOULLI'S EQUATION.

dynamic range (Unit dB) The range of signals that can be accurately measured by a transducer, the lower limit being set by noise and the upper limit usually by non-linearity. It may also refer to the range of frequencies to which a transducer responds.

dynamics *See* APPLIED MECHANICS.

dynamic sensitivity The sensitivity of a sensor to a changing, rather than steady-state, input. It is frequently expressed in terms of the settling time or the frequency response.

dynamic similarity *See* PHYSICAL SIMILARITY.

dynamic stability *See* STABILITY.

dynamic stiffness (Unit N/m) The ratio between a dynamic excitation force and the resulting dynamic displacement. It can be both frequency- and amplitude-dependent, e.g. for elastomeric materials. *See also* STATIC STIFFNESS.

dynamic test Testing of components, materials or structures under high rates of loading.

dynamic viscosity 1. (absolute viscosity, viscosity, μ) (Unit Pa.s) The ratio, for a fluid, of the viscous shear stress to the rate of strain. For a Newtonian fluid, μ is a material property independent of the rate of strain but may vary with temperature and pressure. *See also* APPARENT VISCOSITY; KINEMATIC VISCOSITY. **2.** That part of the complex viscosity for a viscoelastic liquid associated with viscous dissipation.

SEE WEB LINKS

- Reference book companion website covering many material properties (thermal conductivity, density, thermal expansion coefficients, surface tension, viscosity, etc.)

dynamo A small electrical generator that supplies direct current.

dynamometer 1. (torque dynamometer) An instrument or machine for measuring the power output and torque of a piston engine, turbine, or any other prime mover. *See also* BRAKE DYNAMOMETER; HYDRAULIC DYNAMOMETER. **2.** An instrument for measuring forces by piezo-electric effect, strain gauges, etc.

dyne An obsolete (non-SI) unit of force in the cgs system, superseded by the newton, being the force acting on a mass of one gram that accelerates it by one cm/s^2.

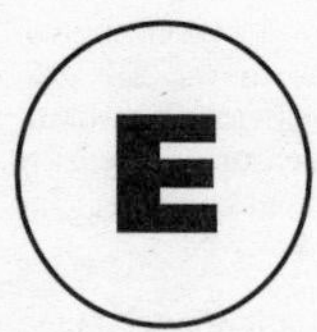

earing A term used in can manufacture, where the top edges of drawn cups undulate, with high points (ears) and valleys in between, rather than have a uniform height. Four ears are common, but up to eight may occur. Earing is caused by planar anisotropy in plastic flow and correlates well with the variation of the R-factor: ears occur at orientations of high R and vice versa.

ebb generation *See* TIDAL ENERGY.

ebullioscopic effect The increase in the boiling point of solvents due to the addition of solutes, e.g. salt added to water. *See also* CRYOSCOPIC EFFECT.

eccentric 1. A term for something not concentric, for example a circular component rotating about an axis that does not pass through its centre. **2.** A type of crank in which an eccentric disc is encircled by a narrow split bearing (the strap) that transmits reciprocating motion to valve gear.

eccentric annulus The space between two circular cylinders, one within the other with parallel but offset axes. The **eccentricity ratio** is the ratio of the eccentricity to the difference between the radii of the two cylinders.

eccentric cam *See* CAM.

eccentricity (*e*) (Unit m) **1.** The distance between the axis of rotation of a rotating body and its centre of gravity or, in an eccentric annulus, the distance between the parallel axes. **2.** The deviation from straightness in a strut.

eccentric load (eccentric loading) An applied load that does not pass through the centroid of a section.

eccentric-rotor engine A non-reciprocating type of internal-combustion engine, such as the Wankel engine, in which power is produced by one or more cylindrical but non-circular rotors revolving within a cylindrical but non-circular chamber.

echo sounder (sonic depth finder) A sonar device for measuring water depth or the depth of objects under water.

Eckert number (*Ec*) A non-dimensional number that arises in the study of high-speed convective heat transfer in which dissipation within the boundary layer is significant. It is defined by $Ec = U_\infty^2/(h_S - h_\infty)$ where U_∞ is the free-stream velocity, h_S is the specific enthalpy of the fluid at the surface and h_∞ is the free-stream value.

ECM 1. *See* ELECTROCHEMICAL MACHINING. **2.** *See* ENGINE CONTROL UNIT.

economic boiler (horizontal-return boiler) A three-pass, wet-back, fire-tube boiler where, as shown in the diagram, the water is first heated in the shell from below (first pass) and the flue gases then pass through horizontal fire tubes in one or two passes.

economizer (boiler economizer) A heat exchanger in the chimney of a steam plant in which the incoming feedwater is pre-heated by the flue gas, which also preheats the air to the boiler furnace.

ECU *See* ENGINE CONTROL UNIT.

eddy Within a turbulent flow, an energetic fluctuating body of fluid with a scale ranging from the smallest dissipative (Kolmogorov) scale to the largest, typically the shear-layer thickness.

eddy-current dynamometer *See* MAGNETIC DRAG DYNAMOMETER.

eddy viscosity (turbulent viscosity, μ_t) (Unit Pa.s) An effective viscosity used to model turbulent shear stress τ_t and defined, for a simple shear flow, through the equation $\tau_t = \mu_t \partial\bar{u}/\partial y$ where $\bar{u}$ is the time-averaged velocity a distance y from a surface. The term is also used for the eddy kinematic viscosity, $\nu_t = \mu_t/\rho$. The **eddy diffusivity (turbulent diffusivity, Γ_t)** is analogous to eddy viscosity for the

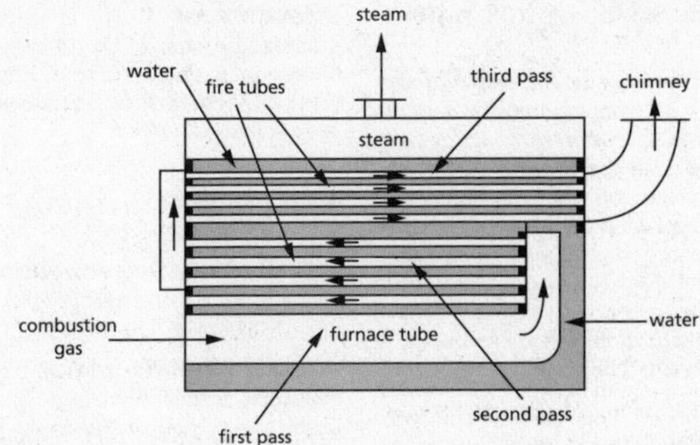

economic boiler

turbulent transport of any scalar quantity, e.g. **eddy conductivity (k_t),** with unit W/m.K, an effective thermal conductivity used to model turbulent heat flux according to $\dot{q}_t'' = -k_t \partial\bar{T}/\partial y$ where $\bar{T}$ is the time-averaged temperature a distance *y* from the surface.

edge tone A transverse periodic oscillation produced when a narrow fluid jet impinges on a sharp edge causing vortices to be shed alternately above and below the edge. The system can be tuned by adjusting the jet speed and/or the distance between the jet and the edge.

edge-tone amplifier A fluidic device in which edge tones excite oscillations in an acoustic resonator, such as a Helmholtz resonator, which in turn feed energy back into the edge tones.

Edison thread *See* SCREW.

EDM *See* ELECTRIC-DISCHARGE MACHINING.

eductor (eductor jet pump) *See* JET PUMP.

effective discharge area A nominal area for flow through a pressure relief valve used to determine the valve's flow capacity given the pressure difference across it, the fluid density and correction factors to allow for compressibility, the back pressure and the coefficient of discharge.

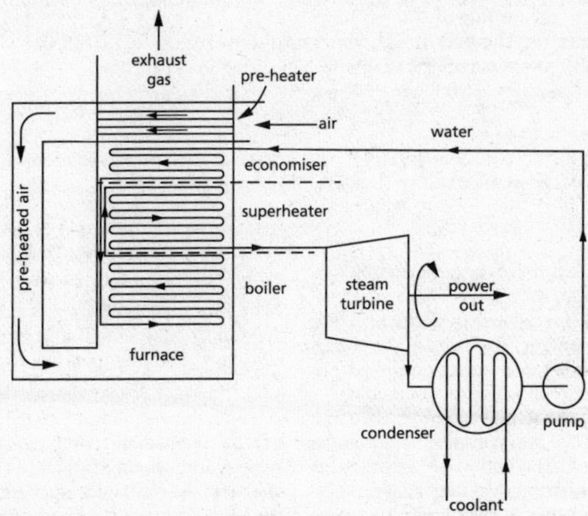

economizer

effective force *See* ACCELERATING FRAME OF REFERENCE.

effective inertia In forced torsional oscillation of a system of components, the lumped moment of inertia that represents an isolated part of the system, in the sense of giving the same angular acceleration when the same oscillating torque is applied at the point of isolation. For example, at some chosen point along a shaft having a series of discs (gears or rotors), each having different moments of inertia, the effect of all the discs to one side of the chosen point of isolation is represented by the effective inertia.

effective mass In forced linear oscillation of a system of components, the lumped mass that represents an isolated part of the system, in the sense of giving the same linear acceleration when the same oscillating force is applied at the point of isolation.

effectiveness 1. For a heating or cooling process, the ratio of the increase of exergy of the system to the loss of exergy to the surroundings. **2.** *See* HEAT-EXCHANGER EFFECTIVENESS. **3.** *See* COOLING FINS.

effective pitch *See* PROPELLER.

effective relative roughness *See* EQUIVALENT SAND ROUGHNESS.

effective rolling radius The radius of a hypothetical rigid, non-slipping wheel that performs over a road surface in the same way as a practical wheel having an inflated tyre.

effective sound pressure *See* ROOT–MEAN–SQUARE SOUND PRESSURE.

effective strain (equivalent strain, representative strain, ε_e^P) The single strain that represents the combined effects of multiaxial loading in total strain ('deformation') plastic flow:

$$\varepsilon_e^P = \frac{\sqrt{2}}{3}\sqrt{[(\varepsilon_x^P - \varepsilon_y^P)^2 + (\varepsilon_y^P - \varepsilon_z^P)^2 + (\varepsilon_z^P - \varepsilon_x^P)^2 + 6\{(\varepsilon_{xy}^P)^2 + (\varepsilon_{yz}^P)^2 + (\varepsilon_{zx}^P)^2\}]}$$

where ε_x etc. are the normal strains, and ε_{xy} etc. are the shear strains. The directions correspond to those of an *xyz*-Cartesian coordinate system. *See also* STRAIN.

effective strain increment (equivalent strain increment, representative strain increment, $d\varepsilon_e^P$) The single strain that represents the combined effects of multiaxial loading in incremental plasticity:

$$d\varepsilon_e^P = \frac{\sqrt{2}}{3}\sqrt{[(d\varepsilon_x^P - d\varepsilon_y{}^P)^2 + (d\varepsilon_y^P - d\varepsilon_z^P)^2 + (d\varepsilon_z^P - d\varepsilon_x^P)^2 + 6\{(d\varepsilon_{xy}^P)^2 + (d\varepsilon_{yz}^P)^2 + (d\varepsilon_{zx}^P)^2\}]}$$

where $d\varepsilon_x$ etc. are the increments of normal strain, and $d\varepsilon_{xy}$ etc. are the increments of shear strain. The directions correspond to those of an *xyz*-Cartesian coordinate system. *See also* STRAIN.

effective stress (equivalent stress, representative stress, σ_e) (Unit Pa) The single stress that represents the combined effects of multiaxial loading in plastic flow:

$$\sigma_e = \frac{\sqrt{2}}{3}\sqrt{[(\sigma_x - \sigma_y)^2 + (\sigma_y - \sigma_z)^2 + (\sigma_z - \sigma_x)^2 + 6\{(\tau_{xy})^2 + (\tau_{yz})^2 + (\tau_{zx})^2\}]}$$

where σ_x etc. are the normal stresses, and τ_{xy} etc. are the shear stresses. The directions correspond to those of an *xyz*-Cartesian coordinate system. *See also* STRESS.

effector *See* ACTUATING SYSTEM.

efficiency A measure of the performance of a machine or system, the ratio of the actual performance to the theoretical performance. For example, for an engine, the ratio of the actual power output to the rate of energy supplied. *See also* AIR-STANDARD EFFICIENCY; BOILER EFFICIENCY; BRAKE–THERMAL EFFICIENCY; CAPTURE EFFICIENCY; COOLING-TOWER EFFICIENCY; ENGINE EFFICIENCY; POLYTROPIC EFFICIENCY; POWER-PLANT EFFICIENCY; PROPELLER EFFICIENCY; PROPULSION EFFICIENCY; PUMP-HYDRAULIC EFFICIENCY; SMALL-STAGE EFFICIENCY; SOLAR-CELL EFFICIENCY; STAGE EFFICIENCY.

efficiency ratio For a thermodynamic cycle, the ratio of the actual efficiency to the ideal efficiency. For example, for any vapour cycle, the ideal efficiency is that of the Rankine cycle.

efflux velocity (Unit m/s) The average velocity with which a fluid flows from the outlet of a pipe, smokestack or opening in the wall of a container into the surroundings. It is generally taken as the volumetric flow rate divided by the area of the outlet.

effusion The flow of a gas through a hole that is small compared with the mean-free path of the gas. *See also* GRAHAM'S LAW.

EfW *See* ENERGY FROM WASTE.

EGS *See* ENHANCED GEOTHERMAL SYSTEM.

EHD lubrication *See* ELASTOHYDRODYNAMIC LUBRICATION.

ejector (ejector pump) *See* JET PUMP.

ejector condenser A type of jet pump in which steam is condensed in the divergent nozzle. Both the high-pressure driving fluid and the pumped fluid are usually steam, but cold air and water are also used as driving fluids.

Ekman layer A viscous boundary layer that develops on a stationary horizontal surface over which there is a geostrophic flow (i.e. a flow rotating about a vertical axis combined with a unidirectional flow parallel to the surface). An **Ekman spiral** is the curve in a polar diagram corresponding to the velocity variation with distance from the surface in an Ekman layer.

Ekman number (*Ek*) A non-dimensional parameter that arises in the study of swirling flows close to a surface. For a fluid with kinematic viscosity ν, $Ek = \nu/\Omega L^2$ where Ω is a characteristic angular velocity for the flow and L is a characteristic length scale. The Coriolis acceleration associated with the Earth's rotation is important in geophysical applications. The definition then used is $Ek = \nu/2\Omega L^2 \sin\phi$, ϕ being the latitude.

elastic (elastic deformation) Deformation of a solid that is reversible, path-independent, and fully recovered on unloading. *See also* LINEAR ELASTICITY.

elastica The collective name for the curved shapes taken up by thin beams at large elastic deflexions under different loadings.

elastic body A body whose deformation behaviour is assumed always to be elastic.

elastic buckling The buckling of a member when all stresses remain elastic.

elastic collision An impact in which the colliding bodies deform elastically so that all deformations are fully recovered. During the time of contact, before recovery, there is an apparent loss of the initial kinetic energy of the bodies which is temporarily stored in the elastically-deformed bodies as strain energy. *See also* COEFFICIENT OF RESTITUTION.

elastic constants *See* ELASTIC MODULUS.

elastic design The design of components and structures based on maintaining the working stresses everywhere below the yield point during normal operation. *See also* FACTOR OF SAFETY; PLASTIC DESIGN; THEORIES OF STRENGTH.

elastic fracture *See* FRACTURE MECHANICS.

elastic hysteresis (Unit J/m^3) The condition where a body fully recovers on unloading, but where the stress–strain paths followed during loading and unloading are different. The area between the two paths is called a hysteresis loop, and represents dissipation of energy per unit volume. Materials such as rubber exhibit high elastic hysteresis, hard metals far less so.

elasticity The time-independent deformation that disappears on unloading.

elastic limit (Unit Pa) The stress above which bodies, subjected to increasing load, no longer return to their original shape and size on unloading. *See also* PERMANENT SET; PROOF STRENGTH; YIELD POINT.

elastic modulus (Unit Pa) The ratio of stress to strain in linear elasticity. There are different moduli for different types of loading which are collectively known as elastic constants. *See also* BULK MODULUS; LAMÉ CONSTANTS; POISSON'S RATIO; SHEAR MODULUS; YOUNG'S MODULUS.

elastic recovery A body that is deformed into the plastic range to a stress σ will, upon unloading, recover the elastic component of the total strain, leaving the permanent plastic component. The recovery strain is given approximately by σ/E where E is Young's modulus. *See also* ELASTOPLASTICITY; SPRINGBACK.

elastic strain energy 1. (strain energy, internal work) (Unit J) The energy stored in an elastically-deformed body under load, given by $\int F d\delta$ that is $F\delta/2 = F^2/2k = k\delta^2/2$ for linear elasticity where F is the applied load, δ is displacement and k is the proportionality constant. **2.** (Unit J/m^3) The strain energy per unit volume for linear elasticity is $\sigma\varepsilon/2 = \sigma^2/2E = E\varepsilon^2/2$ where σ is stress, ε is strain and E is Young's modulus. *See also* COMPLEMENTARY STRAIN ENERGY.

e

elastic vibration The free or forced vibration of a component or structure in which all displacements remain within the elastic range.

elastohydrodynamic lubrication (EHD lubrication) The analysis of hydrodynamic lubrication taking account of the elastic deformation of the contacting members.

elastomer A material that at room temperature displays large reversible deformations. The behaviour is shown by polymers as well as rubbers.

elastomeric valve A valve in which the closing element is made from an elastomer. *See also* DUCKBILL VALVE.

elastoplastic fracture *See* FRACTURE MECHANICS.

elastoplasticity The deformation of a body into the plastic range, resulting in both irreversible plastic strains and reversible elastic strains. *See also* ELASTIC RECOVERY.

elbow 1. A fitting that connects the ends of two pipes at an angle, 45°, 90° and 180° being the most common. **2.** The third joint on an articulated robot corresponding to the human elbow.

elbow meter A flow meter in which mass flow rate $\dot{m}$ is determined by measuring the pressure difference Δp for flow through a 90° elbow between the outer and inner radii at the midpoint. If D is the inside diameter of the elbow, $\dot{m} = kD^2\sqrt{\rho\Delta p}$ where ρ is the fluid density and k is a calibration factor. Such meters are not of high accuracy.

electric-arc welding *See* WELDING.

electric-discharge machining (EDM, spark machining, spark erosion) A machining process in which material is removed from a workpiece immersed in a dielectric liquid by a series of electrical discharges (sparks) between an electrode (the tool) and the workpiece, as in the discharge of a capacitor. Like electrochemical machining (ECM), EDM is employed with difficult-to-machine alloys but, unlike ECM, material is lost from the electrode tool that wears away.

electrochemical machining (ECM) A manufacturing process working on the same principles as electrolytic polishing but capable of much greater rates of metal removal. The cathode has a reverse image of the shape or profile required in the workpiece (anode) towards which it is fed until profiling to the required depth is complete. Holes may also be made. Employed particularly with difficult-to-machine alloys.

electrochemical power generation The direct conversion of chemical energy into electrical energy in either a closed cell (battery) or an open system (fuel cell).

electrokinetic process A process in which an electric field induces fluid motion and vice versa. For example, a charged surface moves relative to a stationary liquid (electrophoresis), or a liquid moves relative to a stationary charged surface (electro-osmosis).

electrolysis The production of a chemical reaction by passing an electric current through a conducting liquid (the electrolyte), whereby positive ions migrate to the cathode and negative ions to the anode.

electrolytic machining The removal of one metal from another by means of electrolysis. It is the reverse of electroplating.

electrolytic polishing (electrochemical polishing, electropolishing) The preparation of metal surfaces for examination under the microscope in which the specimen is made the anode in an electrolytic cell. The reagents employed, voltage, temperature, and so on, depend on the particular alloy. It is also employed in manufacturing. *See also* ELECTROCHEMICAL MACHINING.

electromagnetic clutch *See* MAGNETIC CLUTCH.

electromagnetic flow meter (mag meter, magnetic flow meter) A non-intrusive flow meter used to measure the volumetric flow rate of an electrically-conducting fluid flowing through a pipe. A magnetic field is created in the pipe, and the voltage measured between two electrodes on opposite sides is proportional to the flow rate.

electron beam hardening A process very similar to laser hardening, but employing an electron beam.

electron beam welding *See* WELDING.

electronic control module *See* ENGINE CONTROL UNIT.

electron microscope An instrument that uses beams of electrons, rather than light, to obtain magnified images. *See also* SCANNING

ELECTRON MICROSCOPE; TRANSMISSION ELECTRON MICROSCOPE.

electroplating Deposition of one metal onto another using electrolysis. The metal to be plated forms the cathode in an electrolytic cell, and the metal to be deposited forms the anode.

electrostatic atomization A process in which a conducting liquid jet is broken into fine droplets by the application of a high-voltage electric field.

electrostatic precipitator A device for removing dust or other finely-divided particles from a gas by inducing charges on them using an electric field, so that they are then attracted to oppositely charged collector plates.

electrostatic separator A device for the selective sorting of charged solid particles in air by passing them through an electric field and subjecting them to gravitational or other body forces.

elliptic gear (elliptical gear) A gear wheel of elliptical cross section that rotates about a shaft passing through its focus and gives variable motion to the driven shaft.

elongation 1. The increase in length of a spring, testpiece, or component subjected to a tensile load. 2. The **extension** l_f of a tensile testpiece of a ductile material at fracture. Since the uniform extension prior to necking depends on the gauge length l_o of the specimen, but the subsequent extension due to necking is proportional to the diameter, the percentage-elongation given by $100(l_f - l_o)/l_o$ depends on the gauge length l_o of the testpiece.

elongational flow *See* EXTENSIONAL FLOW.

elongational viscosity *See* EXTENSIONAL FLOW.

elutriation A process for separating solid materials of different densities using a fluid flow. Typically upward flow through a tube carries lighter particles while the heavier particles fall downward. In some systems the flow enters the tube tangentially to create a cyclone effect.

embedded energy (GER, gross energy requirement) The total amount of energy, excluding incident solar energy, used in producing an energy crop: labour, fuel for machinery, fertilizer, etc.

embedded generators The relatively small-capacity electrical generators, such as gas turbines, diesel engines, and wind farms, that form part of a national or regional electricity grid.

embodied energy The commercial energy required to produce the materials for, construct, and later dispose of a machine, structure, or building. *See* LIFE-CYCLE ANALYSIS.

embrittlement The loss of ductility or fracture toughness of materials, either during processing or in service. *See also* BLUE BRITTLENESS; 475 °C EMBRITTLEMENT; HYDROGEN EMBRITTLEMENT; LIQUID–METAL EMBRITTLEMENT; TEMPER EMBRITTLEMENT.

emissions Gaseous and particulate pollutants released into the atmosphere and originating from combustion or other chemical processes, including landfill and natural sources such as volcanoes.

emission standards Legal limits placed by governments or other authorities on the maximum level of emissions permitted from a specified source, such as a motor vehicle or a power plant.

emissive power (radiant emissive power, radiant emittance, radiant exitance, *E*) (Unit W/m^2) The rate at which thermal-radiation energy is emitted per unit area of an emitting surface. For a blackbody at absolute temperature T the emissive power is σT^4 where σ is the Stefan–Boltzmann constant.

emissivity (emittance, thermal emissivity, ε) The ratio between the thermal radiation energy emitted by an actual surface to that radiated by a blackbody at the same temperature. It depends upon temperature, wavelength, and direction of emission. *See also* DIFFUSE SURFACE; EQUIVALENT BLACK-BODY TEMPERATURE; GREY BODY.

Emmons spot A randomly-occurring burst of turbulence following the instability and vortex stretching that develops in a laminar shear layer as part of the final transition to turbulence.

emulsified coolant *See* CUTTING FLUID.

emulsion A colloidal suspension of two or more immiscible liquids, such as an oil and water.

encastré *See* FIXITY.

encoder *See* ABSOLUTE ENCODER; INCREMENTAL ENCODER.

end conditions of columns Constraints at the points of compressive loading in buckling, idealized as free, pinned, or built-in. *See also* FIXITY.

end effector The device attached to the end of a robot arm to perform a given task. It may be a general-purpose gripper, an attraction gripper, or a specialized tool designed for the task.

end float (end play) A term for permissible axial movement in rotating shafts located in bearings for correct functioning of machinery.

end gas *See* ENGINE KNOCK.

end mill A cylindrical cutting tool having radial teeth on its end face.

end-of-arm speed The translational speed of a robot end effector.

endothermic reaction *See* CHEMICAL REACTION.

endplate A flat plate attached to the tip of a wing to minimize leakage of high-pressure air on the lifting surface to the low-pressure air on the suction surface, in order to reduce the strength of trailing vortices and induced drag.

end point The output of a plant controlled by a control system when steady state is reached, i.e. when transient behaviour has decayed completely.

end-point rigidity The stiffness of a robot arm and wrist against movement of the end effector caused by an applied force.

end-quench test *See* JOMINY END-QUENCH TEST.

endurance limit *See* FATIGUE.

endurance ratio (endurance-limit ratio) *See* FATIGUE.

energy (*E*) (Unit *J*) Energy, like heat and work, can only be defined in terms of changes with respect to an arbitrary reference or datum level. It is a scalar property implied by the First Law of Thermodynamics, according to which the increase in the total energy of a system during a process is equal to the heat transfer minus the work transfer across the system boundary during the process. The total energy stored in a system consists of three components: kinetic energy, potential energy, and internal energy.

For a system of mass m and velocity V, the kinetic energy KE is given by $KE = ½\,mV^2$. The potential energy $PE = mgz$, where g is the acceleration due to gravity and z is the elevation of the system above a datum plane. These are the macroscopic forms of energy. The microscopic form of energy is the internal energy U, which is the energy stored at a molecular or atomic level. For a single-phase substance, U is a thermodynamic property that depends upon temperature and pressure. More generally, U includes the latent energy associated with a phase change, the chemical energy associated with the chemical bonds in a molecule, and also nuclear energy.

energy-additive flow meter A flow meter, such as an electromagnetic or ultrasonic flow meter, that introduces energy into a flow to determine the flow rate. *See also* ENERGY-EXTRACTIVE FLOW METER.

energy balance (energy equation) According to the conservation of energy principle, the net change in the total energy of a system during a process is equal to the difference between the total energy entering and the total energy leaving the system during that process. *See also* ENERGY LOSSES.

energy content *See* CALORIFIC VALUE OF A FUEL.

energy crops Any plants grown specifically for use as fuel or conversion into other biofuels, including wood for burning, plants for fermentation into ethanol, and crops whose seeds are rich in oils. *See also* BIOENERGY.

energy density (Unit kJ/m^3) **1.** The energy per unit volume contained in a fuel. *See also* HIGHER CALORIFIC VALUE. **2.** A performance indicator for energy-storage devices, such as batteries and flywheels, expressed in terms of the amount of energy stored per unit volume. **3.** The area contained beneath a true stress–true strain diagram indicating the elastic energy stored or the energy dissipated in plasticity.

energy-extractive flow meter A flow meter, such as a turbine, pressure-drop or vortex-shedding flow meter, that extracts energy from a flow to determine the flow rate. *See also* ENERGY-ADDITIVE FLOW METER.

energy flow rate For a fluid stream with mass flow rate $\dot{m}$, the energy flow rate is $\dot{m}\,(h_0 + gz)$ where h_0 is the specific total enthalpy of the fluid, g is the acceleration due to gravity, and z is the altitude above a datum level.

energy from waste (EfW) The incineration of municipal solid waste with heat recovery which can be used for district heating or electrical-power generation in a steam-power plant.

energy integral equation (thermal-energy integral equation) An equation derived from the conservation of energy principle applied to the flow within a boundary layer. For two-dimensional flow over a body of revolution of radius R which varies with distance along the surface x, the equation can be written as

$$\dot{q}''_S = \frac{1}{R}\frac{d}{dx}\left(R\int_0^Y \rho u h_{0,REF}\,dy\right) - \rho_S v_S h_{0,REF,S}$$

where $\dot{q}''_S$ is the surface heat flux, ρ is the fluid density, u is the streamwise fluid velocity, and $h_{0,REF}$ is the stagnation enthalpy, all at distance y from the surface, v_S is the fluid velocity through the surface, and the subscript S refers to the surface. The distance Y is sufficiently large that all properties at Y have free-stream values. A restricted form for low-velocity, constant-property flow with $v_S = 0$ is

$$\dot{q}''_S = \frac{C}{R}\frac{d}{dx}[\Delta_2 R\rho U_\infty(T_S - T_\infty)]$$

where Δ_2 is the enthalpy thickness, C is the specific heat of the fluid, T is the temperature, and the subscript ∞ refers to free-stream conditions.

energy interactions For a closed system, those forms of energy that cross the system boundary: heat transfer and work.

energy losses A commonly-used misnomer in view of the conservation-of-energy principle, but used to mean energy converted into forms that are not used in a process, for example thermal energy from a heat engine dissipated to the surroundings or produced by friction in a machine or by surface drag.

energy payback ratio (energy ratio) The ratio of the total energy output of a power plant produced over its lifetime to the energy required to build, maintain, and fuel it. Hydropower has the highest ratio (170+), followed by large wind-turbine systems (*c.*20) and nuclear (15). Conventional coal-fired plants achieve 5 at best.

energy quality The proportion of the energy input to a device that can be converted to mechanical power. For example, electric motors have high quality as 95+% of the input energy can be converted into mechanical power. The quality of a nuclear, fossil, or biomass powered thermal power plant is about 33%, while a combined-cycle power station reaches about 50%.

energy storage Fossil fuels are a natural form of energy storage, but the term is usually taken to mean any of the many man-made devices or systems which enable energy to be stored for future use. Examples include: water pumped to high-altitude reservoirs for use in a hydro-electric power plant; a wound spring (potential energy); a rapidly-spinning flywheel (kinetic energy); refined petroleum and synthetic fuels, batteries, fuel cells (chemical energy); compressed air stored in underground caverns, salt domes, disused mines, etc. (internal energy); compressed hydrogen stored in underground reservoirs (chemical plus internal energy); ice, liquid nitrogen, liquid air, steam accumulator, molten salt (thermal energy).

energy utilization factor *See* CAPACITY FACTOR.

engine A machine that converts energy, including the chemical energy in a fuel and electrical energy, into mechanical energy, usually to produce power delivered through a rotating shaft or thrust. Examples include internal-combustion engines, gas and steam turbines, rocket engines, electric, hydraulic and pneumatic motors. *See also* HEAT ENGINE.

engine balance *See* BALANCING.

engine block *See* CYLINDER BLOCK.

engine capacity (engine displacement) (Unit cc or *l*) The total swept volume of all the cylinders in a piston engine. For non-reciprocating engines, such as the Wankel engine, an equivalent volume is defined.

engine configuration For a piston engine, the arrangement of the cylinders. *See also* IN-LINE ENGINE; VEE ENGINE.

engine control unit (ECU, ECM, electronic control module, engine management system) An electronic system that controls the air/fuel mixture, ignition and valve timing, idle speed, etc. on a modern petrol or diesel engine to ensure optimum performance. In some systems the transmission is also controlled.

engine cycle Either a series of ideal thermodynamic processes, such as compression, expansion, heat addition, or rejection, for a heat engine, or the actual processes that occur in a real engine. *See also* AIR-STANDARD CYCLES; FOUR-STROKE ENGINE; OTTO CYCLE.

engine cylinder A circular cylindrical chamber in which a piston operates machined in the block of a reciprocating engine, compressor, or pump. *See also* CYLINDER BORE.

e

engine displacement *See* ENGINE CAPACITY.

engine efficiency (η) The ratio of the power output of an engine P to the rate of energy input calculated from the product of the enthalpy of combustion Δh_0, essentially the latent energy, of the fuel and the mass flow rate $\dot{m}_F$, i.e. $\eta = P/\dot{m}_F \Delta h_0$. *See also* OVERALL THERMAL EFFICIENCY.

engine emissions (exhaust emissions) The pollutants contained in the exhaust gases from an internal-combustion engine.

engineering drawing (technical drawing) The graphical representation of components and assemblies, including projected views (projections), cuts, sections, dimensions and tolerances, material specifications, and surface finish specification. ISO 128 (in twelve parts) specifies the general principles of presentation. PP 7308 from BSI presents the basic principles.

engineering mechanics *See* APPLIED MECHANICS.

engineering notation For real decimal numbers, the representation in the form $a \times 10^n$, where a is a number from 1 to less than 1000 and n is a positive or negative multiple of 3. *See also* SCIENTIFIC NOTATION.

engineering plastics A loose description for those polymers that are accepted as engineering materials. The term is artificial, as suitability depends on circumstances.

engineering science (integrated engineering) The application of scientific knowledge to the solution of practical problems in engineering which are interdisciplinary and thus cross the traditional boundaries of civil/electrical/mechanical etc. engineering.

engineering strain (nominal strain, ε) For a component of initial length l_0 that becomes extended to length l when subjected to a tensile load, $\varepsilon = (l - l_0)/l_0$.

engineering stress (nominal stress, σ) Stress defined as load W divided by the initial area A_0 over which it acts, even if that area changes considerably under load, i.e. $\sigma = W/A_0$. *See also* TRUE STRESS.

engineering thermodynamics See APPLIED THERMODYNAMICS.

engineers' blue 1. A grease or other oily substance coloured with Prussian blue that, when smeared onto a reference surface, reveals high spots on a workpiece rubbed against it. **2. (layout blue)** A substance consisting of methylated spirits (denatured ethanol) coloured with Prussian blue, sprayed or brushed onto the surface of a workpiece to assist in marking up.

engine friction For a piston engine, the resistance to relative motion of all the moving parts, including the friction between the piston rings, piston skirt, and cylinder wall; friction in the gudgeon pins, big end, crankshaft, and camshaft bearings; friction in the valve mechanism, the gears or pulleys and belts which drive the camshaft, and other engine accessories. The frictional losses are generally measured using a motored hot engine. Sometimes included are the pumping losses and the losses associated with peripherals such as the cooling fan, water, oil, and fuel pumps. *See also* FRICTION POWER.

engine indicator diagram The graph of cylinder pressure *vs* cylinder volume recorded from a piston engine that shows the induction, compression, ignition, expansion, and exhaust phases of a cycle. Indicator diagrams can also be produced for steam engines. *See also* FOUR-STROKE ENGINE; INDICATED POWER.

engine knock (knock, pinging, pinking, spark knock) The noise emitted by a piston engine when essentially spontaneous ignition of the mixture of fuel, air and residual gas (the end gas) occurs ahead of the propagating flame. The result is an extremely rapid release of chemical energy accompanied by very high pressure and the propagation of sound waves (i.e. pressure fluctuations) across the combustion chamber. **Knock intensity** is the amplitude of the pressure fluctuations measured by a **knockmeter**. *See also* DIESEL KNOCK.

engine performance The performance of an engine represented by curves of power output (the power curve), torque, fuel and air consumption, exhaust emissions, cylinder pressure, and brake mean-effective pressure etc. *vs* output-shaft rotation speed.

engine run on *See* DIESELING.

engine torque (Unit N.m) The output power of an engine divided by the output-shaft rotation speed in rad/s.

Engler orifice viscometer An instrument used to quantify the viscous characteristics of fuel and lubricating oils, tars and other petroleum products in terms of Engler degrees, E°, where E° is the time for 200 ml of liquid to flow through an orifice divided by the time for 200 ml of distilled water to flow through the same orifice. The test is usually run at 20°C. For a liquid with relatively high kinematic viscosity ν (> 10 mm^2/s), E° is directly related to ν.

enhanced geothermal system (engineered geothermal system, EGS) A system or process in which cold water at high pressure is pumped down an injection well and through high-temperature rock (called hot dry rock), typically 5km below the surface, the hot water that emerges from a second well being used to generate electricity by a steam-power plant.

enthalpy (*H*) (Unit kJ) An extensive thermodynamic property of a substance equal to the sum of its internal energy U and the product of its pressure p and volume $\mathcal{V}$, i.e. $H = U + p\mathcal{V}$. The amount of energy absorbed or released in a phase-change process such as condensation, evaporation, solidification, melting, and sublimation, i.e. the latent heat associated with that phase change, is frequently termed the heat or enthalpy of the process, although in fact it is the specific enthalpy which is involved. In processes such as combustion, compression, cooling, formation of a compound, mixing, and chemical reaction, the same terminology is used, the unit being kJ/kmol or kJ/kg.

The **enthalpy** or **heat of combustion (h_C)** is the amount of heat released when 1 kmol or 1 kg of a fuel is burned completely at a specified temperature and pressure. For a chemical reaction, the **enthalpy of reaction** is the difference between the specific enthalpies of the products at a specified state and the enthalpies of the reactants at the same state for a complete reaction. The **enthalpy of formation (h_f)** of a substance is its specific enthalpy at a specified state due to its chemical composition. For all stable elements h_f is taken as zero at the standard reference state, i.e. $h_f^o = 0$.

The **enthalpy of compression (Δh)** is the increase in the specific enthalpy of a gas when compressed. For isentropic compression of an ideal gas with ratio of specific heats γ, $\Delta h = h_I[r^{(\gamma-1)/\gamma} - 1]$ where h_I is the initial enthalpy of the gas and r is the compression ratio. The **enthalpy of cooling** if a system of mass m cools by a temperature difference ΔT is the enthalpy reduction given by $m(C_P\Delta T + L)$ where C_P is the specific heat and L is the latent heat associated with any phase change, such as condensation, reverse sublimation, or freezing. The **enthalpy of mixing** is the amount of heat released or absorbed during a mixing process, which is zero for ideal solutions.

The enthalpies of condensation, crystallization, evaporation (vaporization), fusion, solidification, and sublimation are all referred to as latent heats.

enthalpy–entropy chart *See* MOLLIER DIAGRAM.

enthalpy thickness (Δ_2) (Unit m) A thickness that arises in the integral analysis of a thermal boundary layer. It can be defined by the equation

$$\Delta_2 = \int_0^\infty \frac{\rho u h_{0,REF}}{\rho_\infty U_\infty h_{0,REF,S}}, dy$$

where ρ is the fluid density, h_0 is its specific stagnation enthalpy, u is the fluid velocity a distance y from the surface, the subscript ∞ denotes free-stream conditions, S the surface and REF indicates that the enthalpy datum is the free-stream stagnation state $h_{0,\infty}$ such that $h_{0,REF} = h_0 - h_{0,\infty}$. For low-speed flow of a constant-property liquid or a perfect gas,

$$\Delta_2 = \int_0^\infty \frac{u}{U_\infty}\left(\frac{T - T_\infty}{T_S - T_\infty}\right) dy$$

where T is the temperature a distance y from the surface. *See also* STAGNATION CONDITIONS.

entrainment The process whereby a shear layer draws in fluid from its surroundings so that the volume flow rate within the shear layer continually increases with streamwise distance. For example, assuming laminar flow, a two-dimensional momentum source of strength K (unit kg.m^3/s) produces a volumetric flow rate $\dot{Q} = 3.302(K\nu x/\rho)^{1/3}$ a distance x from the origin, where ν is the kinematic viscosity of the fluid.

entrainment relation An integral boundary-layer equation based upon the principle of mass conservation. For a constant-property boundary layer, it can be written as

$$\frac{d\delta}{dx} - \frac{v_E}{U_\infty} = \frac{1}{U_\infty}\frac{d}{dx}[U_\infty(\delta - \delta^*)]$$

where δ is the boundary-layer thickness, v_E is the entrainment velocity, x is the streamwise distance along the surface, U_∞ is the free-stream velocity and δ^* is the displacement thickness.

entrainment velocity (v_E, v_∞) (Unit m/s) The component of velocity normal to the main flow direction at the outer edge of a shear layer that accounts for entrainment. If the volume flow rate within the shear layer is $\dot{Q}_{BL}$ then $v_E = d\dot{Q}_{BL}/dx$ where x is the streamwise distance along the surface.

entrance length For flow through a pipe or duct of uniform cross section, the distance from the inlet to the location where the flow can be regarded as fully developed.

entropy (S) (Unit kJ/K) An extensive thermodynamic property that arises as a consequence of the Second Law of Thermodynamics. The entropy change ΔS of a system during a process between states 1 and 2 is given by

$$\Delta S = S_2 - S_1 = \int_1^2 \left(\frac{dQ}{T}\right)_{INT.REV.}$$

where Q represents heat transfer and T the absolute temperature of the system. The subscript *INT. REV.* indicates that the heat-transfer process must be internally reversible, but since entropy is a property, the change is the same for any process between the specified end points. As with internal energy, for water in all its forms the zero level for entropy is taken as its triple point, 0.01°C and 6.112 mbar. For other fluids, −40°C and the corresponding saturation pressure are chosen. *See also* ISENTROPIC; TDS EQUATIONS.

entropy of vaporization For a liquid, the enthalpy of vaporization divided by its boiling point. *See also* TROUTON'S RULE.

entry flow *See* ENTRANCE LENGTH.

environment **1.** The immediate surroundings of a machine or device, such as a robot. **2.** The fluid in which a component or structure is immersed (e.g. humid air, corrosive gas or liquid, oil). **3.** In thermodynamics, the region beyond the immediate surroundings of a system the properties of which are unaffected by any process occurring within the system. **4.** The natural environment, including the earth's atmosphere, solar radiation, the rivers, lakes, and oceans.

environmental engineering The application of engineering principles to minimize damage to, or even to improve, the natural environment by pollution control, recycling, waste disposal, water treatment, etc.

environmental-impact analysis The quantitative estimation of the damage to the natural environment that will occur due to the construction and operation of a power plant, a processing operation, motor vehicles, aircraft, etc.

environmental stress cracking The fracture of materials as a consequence of chemical action. *See* STRESS CORROSION CRACKING.

environment simulator A system that artificially simulates features of the natural environment to which a machine might be exposed, such as very high or low temperatures in the case of a motor vehicle.

eolian anemometer *See* AEOLIAN ANEMOMETER.

Eötvös number (Eo) A non-dimensional parameter that arises in the study of bubbles and drops rising or falling in another fluid, including within a tube (e.g. slug flow), and defined by $Eo = \Delta\rho g d^2/\sigma$ where $\Delta\rho$ is the density difference and σ is the surface tension at the interface between the two fluids, g is the acceleration due to gravity, and d is a characteristic length (usually the drop diameter). *See also* BOND NUMBER; MORTON NUMBER.

epicyclic gear train (planetary gear train) Gearing in which one or more wheels (**planet gears, planetary gears, planet pinions**) mounted on a disc or spider (**planet carrier**) travel around the outside of another gear (**sun gear**) whose axis is fixed. It can also have gear wheels rotating around the inside of a fixed annular ring gear. A **solar gear** is where the sun wheel is fixed, there is a rotating spider and annulus, and the planet gears rotate about their own axes. For a **star gear** the planet carrier is fixed while the input sun wheel and output annulus rotate.

epicycloid The locus of a point on the circumference of a circle which rolls, without slipping, along the outside of a circular arc. *See also* TOOTHED GEARING.

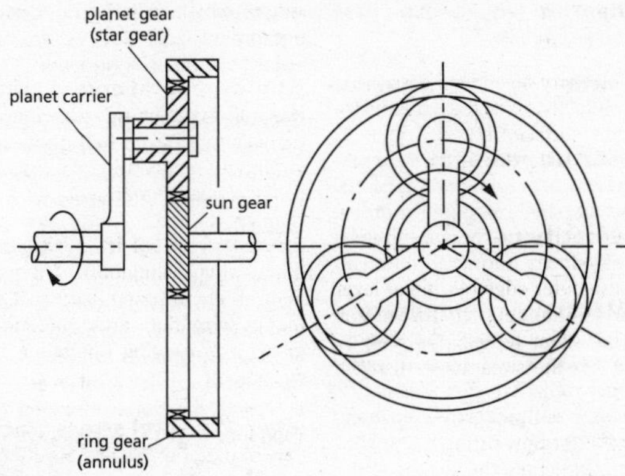

epicyclic gear train

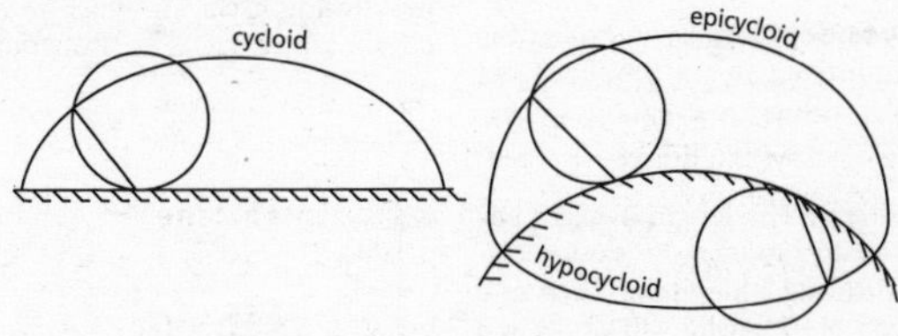

epicycloid

epitaxy The process of depositing μm-thick single crystal films on to single crystal substrates.

epitrochoidal engine A rotary internal-combustion engine, such as the Wankel engine, in which a rotor rotates within a cylinder of epitrochoidal cross section.

EP lubricants *See* EXTREME PRESSURE LUBRICANTS.

epoxy A range of thermosetting copolymers employed as structural adhesives and as matrices for filament-reinforced composite materials. Epoxies are formed by mixing epoxy resins (monomers having end expoxide groups) with polyamine hardeners, polymerization (curing) taking place over time at different temperatures. A wide range of physical and mechanical properties is possible.

equation of state 1. (characteristic equation, thermal equation of state, thermodynamic equation of state) In thermodynamics, any equation that relates the pressure, temperature, and specific volume of a substance in equilibrium. The term is sometimes used for thermodynamic properties other than specific volume. *See also* PERFECT GAS. **2.** *See* CONSTITUTIVE EQUATION (2).

equations of motion *See* NEWTON'S LAWS OF MOTION.

equatorial plane The plane, perpendicular to the axis of rotation of a body of symmetrical shape, passing through its centre of volume.

equilibrant *See* RESULTANT FORCE; RESULTANT MOMENT.

equilibrium The state in which the resultant force and resultant couple on a body are simultaneously zero, so that the body is either at rest or under uniform motion with respect to a fixed frame of reference. *See also* NEUTRAL EQUILIBRIUM; STABLE EQUILIBRIUM; UNSTABLE EQUILIBRIUM.

equilibrium constant *See* DISSOCIATION CONSTANT.

equilibrium diagram (equilibrium phase diagram) *See* PHASE DIAGRAM.

equivalence ratio *See* STOICHIOMETRIC MIXTURE.

equivalent bending moment In problems of combined elastic bending and torsional loading of shafts, the single bending moment that results in the same normal and shear stresses.

equivalent blackbody temperature (T_E) (Unit K) For a radiating body, the temperature that a blackbody would have in order to emit the same radiation flux density *E*, $T_E = (E/\sigma)^{1/4}$ where σ is the Stefan–Boltzmann constant.

equivalent diameter *See* HYDRAULIC DIAMETER.

equivalent evaporation For a boiler, the quantity of dry saturated steam produced per unit of fuel burned when the evaporation process occurs at 100°C. *See also* BOILER CAPACITY.

equivalent length The length of plain circular pipe that would produce the same pressure drop as a particular pipe fitting, such as a sudden contraction or expansion, an elbow or a fully-open valve.

equivalent orifice For flow through a non-circular opening, the circular sharp-edged orifice that would allow the same flow rate for the same pressure difference.

equivalent sand roughness (effective relative roughness, equivalent roughness, ε, $k_{s\ eq}$) (Unit m) For turbulent flow in a naturally-rough pipe of diameter *D*, the ε/D value of sand-grain roughness that matches the friction factor *f* for flow in the fully-rough region, where *f* is independent of the Reynolds number, ε being the sand-grain diameter. *See also* COLEBROOK EQUATION.

equivalent strain *See* EFFECTIVE STRAIN.

equivalent strain increment *See* EFFECTIVE STRAIN INCREMENT.

equivalent stress *See* EFFECTIVE STRESS.

equivalent viscous damping The level of viscous damping in an idealized system that produces the same energy dissipation per cycle at resonance as the non-viscous damping in the actual system.

ergonomics The part of design that ensures equipment and devices are optimized with regard to human well-being.

Ergun equation An empirical equation for the frictional pressure drop per unit length $\Delta p/L$ in flow through unconsolidated porous media. In terms of a friction factor defined by $f = (\Delta p/L)d/\rho U_0^2$ the equation is $f\phi^3/(1-\phi) = 150(1-\phi)/Re + 1.75$ where *d* is the particle diameter, ρ is the fluid density, U_0 is the superficial velocity, ϕ is the porosity and *Re* is the Reynolds number based upon *d* and U_0. The equation is valid for $Re < 10^3(1-\phi)$. If the right-hand side is taken as 1.75, the equation is termed the Burke–Plummer equation and is valid for $Re > 10^3(1-\phi)$. *See also* DARCY'S LAW.

Ericsson cycle A cycle, shown on a plot of pressure (*p*) *vs* specific volume (*v*), similar to the Stirling cycle, in which the isopycnic processes are replaced by isobaric processes (1–2 and 3–4).

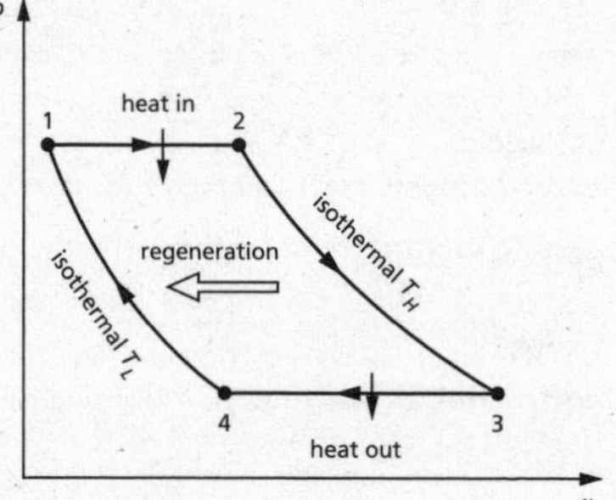

Ericsson cycle

Ericsson engine A heat engine operating on an approximation to the Ericsson cycle.

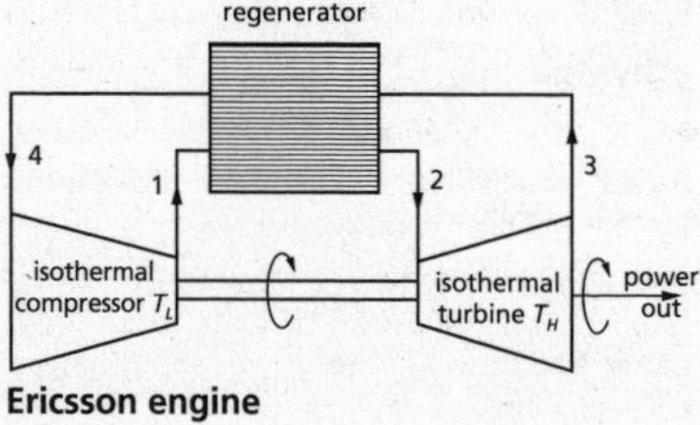

Ericsson engine

erosion The intentional or accidental removal of surface material by the action of abrasive particles in a fluid stream.

error 1. The difference between the measured and the true value of a measured quantity. **Accuracy** is the degree of correctness of a measurement, i.e. its closeness to the true value. Uncertainty (margin of error) is expressed as a range of values estimated to enclose the true value. **Instrument accuracy** is normally taken to mean the maximum error an instrument leads to when making a measurement, but more correctly it is the **inaccuracy. Systematic** or fixed **errors (bias)** are typically caused by imperfect calibration of an instrument including zero offset and interference of the environment with the measurement. A **random error (stochastic error)** is a contribution to uncertainty caused by non-systematic, non-repeatable and unpredictable variations in the measurement system. Such errors can be largely eliminated by calculating the mean of the measurements since positive and negative errors are equally probable. It is a form of random signal. **Precision** is the reproducibility of a measurement under unchanged conditions. *See also* UNCERTAINTY ANALYSIS. **2.** In a control system, the difference between the set point and the plant output. The **error signal** is the input to the control elements in the controller.

error analysis *See* UNCERTAINTY ANALYSIS.

error constant (**error coefficient**) The value of the open-loop transfer function of a system as s→0. It is determined by the system type and the set point input. *See also* POSITIONAL-ERROR CONSTANT; VELOCITY-ERROR CONSTANT.

errors of form *See* TOLERANCES.

etching 1. In metallography, the process of chemical attack of the polished surfaces of specimens in order to reveal the microstructure of the metal under an optical microscope.
2. The use of chemical agents to remove surface material from metals, semiconductor materials, glass, etc. Widely used in the semiconductor industry.

ethylene glycol A polyether commonly used as the basis for anti-freeze.

Euler buckling formulae Expressions for the greatest compressive load F_{crit} that a long, slender column can carry without failing by elastic instability (buckling). In general, $F_{crit} = n\pi^2 EI/L^2$ where E is Young's modulus, L is the column length, I is the second moment of area of the cross-section with respect to the axis about which sideways bending occurs, and n is a coefficient that accounts for different end fixity conditions of the column. For a strut that is pinned at both ends, $n = 1$; when built-in at both ends, $n = 4$. In the plastic range, the **Euler–Engesser formula** $F_{crit} = n\pi^2 E^* I/L^2$ applies where E^* is either (a) $E_t = (d\sigma/d\varepsilon)$ the tangent modulus or local slope of the workhardening stress (σ) *vs* strain (ε) curve; or (b) the reduced modulus $E_r = 4EE_t/[\sqrt{E} + \sqrt{E_t}]^2$. The use of E_t or E_r depends on whether the problem is viewed as being non-linear elastic or elastoplastic.

Eulerian description The analysis of the flow of material, fluid, or solid, relative to a frame of reference fixed in space. *See also* LAGRANGIAN DESCRIPTION.

Euler number (***Eu***) A non-dimensional pressure coefficient defined by $Eu = p/\rho V^2$ or $\Delta p/\rho V^2$ where p is the static pressure, Δp is a pressure difference, usually the difference between the static pressure at a point and a reference pressure, ρ is the fluid density, and V is the flow speed. The denominator is sometimes taken as $\frac{1}{2}\rho V^2$. *See also* CAVITATION NUMBER.

Euler's equation (**Euler's turbomachine equation**) For incompressible flow through a centrifugal or axial turbomachine, the power input or output P is given by $P = \dot{m}\omega\,\Delta(RV_\theta)$ where $\dot{m}$ is the mass flow rate through the machine, ω is the angular velocity of the impeller (or rotor), and $\Delta(RV_\theta)$ is the change in the product of the impeller radius R and the absolute circumferential velocity V_θ of the fluid between impeller inlet and outlet.

Euler's equation of motion A form of the momentum equation for the flow of an inviscid fluid which may be written $\rho d\mathbf{V}/dt = -\nabla p + \rho\mathbf{g}$ where ρ is the fluid density, $\mathbf{V}$ is the vector velocity, t is time, p is pressure (i.e. ∇p is the pressure gradient) and $\mathbf{g}$ is the acceleration due to gravity. Bernoulli's equation is an integral of Euler's equation.

eutectic In a binary alloy, the composition at which a single-phase liquid transforms at constant temperature on cooling to a solid consisting of two phases. For example, the single-phase liquid of 28.1% copper and 71.9% silver transforms at 780°C to a solid consisting of the two phases α (consisting of 7.9%Ag/92.1%Cu) and β (8.8%Cu/91.2%Ag).

eutectoid In a binary alloy, the composition at which a single-phase solid transforms at constant temperature on cooling to another solid consisting of two phases. For example, steel containing 0.77% carbon transforms at 727°C from single-phase austenite to pearlite that is made up of ferrite (almost pure iron) and cementite (Fe_3C).

e

evacuated solar collector (evacuated tube collector) A solar collector in the form of a tube through which flows a fluid, typically water. This tube is surrounded by a glass tube, the annulus between the two being evacuated to minimize convection losses.

Evans rotor A wind-turbine rotor with vertical axis and blades that change pitch for control and safe shutdown.

evaporation (vaporization, volatilization) The change of phase from liquid to vapour that occurs at the surface of a liquid when its temperature is higher than its saturation temperature or its pressure is lower than its vapour pressure. In contrast to boiling, there is no bubble formation. *See also* LATENT HEAT OF EVAPORATION.

evaporative condenser A heat exchanger in which the vapour to be condensed flows through a tube, typically in the form of a coil, over which flow air and water spray. Evaporation of the water lowers both the air and water temperatures, thus cooling the coil. An evaporative condenser can be regarded as a type of induced-draught cooling tower.

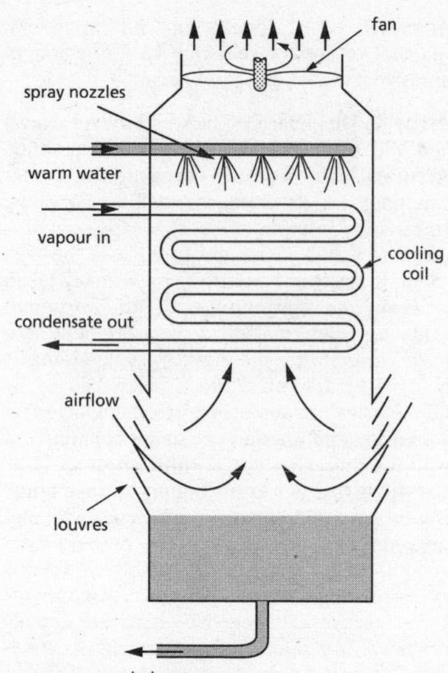

evaporative condenser

evaporative cooler A device in which hot dry air is forced through a wet cloth or pad. The air is cooled as some of the water evaporates by extracting from the air the latent heat required for evaporation. Applications include air conditioning and cooling the inlet air for a gas turbine.

evaporator 1. A type of heat exchanger in which a process liquid is evaporated by steam heating. **2.** The component in a refrigeration system in which low-temperature refrigerant evaporates by absorbing heat from the refrigerated space. *See also* VAPOUR-COMPRESSION REFRIGERATION CYCLE.

evaporimeter *See* ATMOMETER.

even pitch The pitch of a thread cut on a lathe, when it is a multiple of the pitch of the lathe's lead screw. *See also* FRACTIONAL PITCH.

eversion The process of pushing a tube on to a die that turns the walls outward, thus forming a shorter, larger-diameter tube.

exception handling The method used by a control system to respond to unexpected behaviour of the plant, for example, when the feedback signal is lost or movement occurs outside the normal range.

excess air *See* STOICHIOMETRIC MIXTURE.

excess temperature *See* BOILING.

exchange factor In radiant interchange among specular reflecting surfaces, the fraction of radiation from one surface that arrives at another. *See also* VIEW FACTOR.

excitation 1. The power supplied to an actuator. **2.** The process of causing vibration of a structure by the application of an oscillatory force or displacement.

exducer In an inward-flow radial gas turbine, the exit part of the rotor vanes that are curved to remove most of the swirl.

exergy (availability, available energy) (Unit kJ) The theoretical maximum amount of energy that can be obtained from a system interacting with an environment at fixed temperature and pressure. It is a composite thermodynamic property of a system and its environment. *See also* REVERSIBLE WORK; SPECIFIC PROPERTY.

exhaust cone In a turbojet engine, a fixed axisymmetric engine component, of conical, domed, or ogival longitudinal cross section, installed at the turbine exit within the exhaust-duct assembly. Together with the exhaust-gas properties, the annular space between the cone and the exhaust duct determines the exit velocity.

exhaust diffuser A diffuser at the outlet of a turbomachine, such as a turbine or compressor, that recovers some of the kinetic energy in the flow to reduce the pressure at the final stage and increase the overall efficiency.

exhaust emissions *See* ENGINE EMISSIONS.

exhaust fan A ventilation fan to remove air contaminated by fumes, smoke, steam, etc. from a building.

exhaust gas The mixture of gases produced during any combustion process, typically including carbon monoxide and dioxide, nitrogen and oxides of nitrogen, sulphur oxides, unburned hydrocarbons, water vapour, and soot particles. *See also* ENGINE EMISSIONS.

exhaust-gas analyser An instrument used to determine the composition of the exhaust gas from an engine both to check whether regulated emissions (unburned hydrocarbons (UHC), carbon monoxide (CO), and oxides of nitrogen (NO_x)) are within allowable limits and also to analyse engine performance. The analysis methods used include: a flame ionization detector (UHC); a non-dispersive infra-red detector (carbon dioxide and CO); a chemiluminescent analyser (NO_x). *See also* ORSAT APPARATUS.

exhaust-gas boiler *See* WASTE-HEAT BOILER.

exhaust-gas purifier *See* EXHAUST SCRUBBER.

exhaust manifold In a multi-cylinder internal-combustion engine, steel, or ceramic pipes through which exhaust gases flow from a number of engine cylinders to an exhaust pipe.

exhaust nozzle *See* PROPELLING NOZZLE.

exhaust pipe (exhaust) The final section of the exhaust system of an internal-combustion engine through which exhaust gases are discharged into the environment.

exhaust port (outlet port) A circular opening in the cylinder head of an internal-combustion engine through which the exhaust gases enter the exhaust system when the exhaust valve is open.

exhaust scrubber (exhaust-gas purifier) A device, such as a dry or wet scrubber or catalytic converter, that reduces the levels of gaseous pollutants in an exhaust stream.

exhaust silencer *See* SILENCER.

exhaust stator blades A stator located downstream of the last stage of an axial compressor or turbine designed to remove residual swirl from the exhaust gases.

exhaust stroke 1. (scavenging stroke) For an internal-combustion engine, that part of the cycle in which the piston or rotor forces the exhaust gases from the engine. *See also* FOUR-STROKE ENGINE; TWO-STROKE ENGINE. **2.** For a compressor or pump, that part of the cycle in which pressurized fluid is forced from the machine.

exhaust system The assembly of components, including the exhaust manifold, silencer, and exhaust pipe, through which exhaust gases flow from any combustion system, especially the cylinders of a piston engine. The latter often includes a catalytic converter.

exhaust turbine In a turbocharger, the radial turbine that is driven by the exhaust gas from an internal-combustion engine.

exhaust valve A valve in the cylinder head of a piston engine, opened by a cam and closed either by a spring or mechanically, that allows exhaust gas to enter the exhaust system during the exhaust stroke. *See also* DESMODROMIC VALVE.

exhaust vanes 1. A ring at the end of a turbojet tailpipe divided into hinged segments that can be used to adjust the outlet diameter. **2.** Steering vanes located in the exhaust of a rocket engine.

exosphere *See* STANDARD ATMOSPHERE.

exothermic reaction *See* CHEMICAL REACTION.

expander A tool that can locally increase the diameter of tubing.

expanding brake *See* INTERNALLY-EXPANDING BRAKE.

expanding reamer A reamer that is adjustable by means of a coned internal plug working within a longitudinally-slit section.

expansion 1. The increase in the volume of a solid body or mass of fluid when the temperature increases or the external pressure decreases. An exception is water for which the solid phase (i.e. ice) is less dense than the liquid phase. *See also* RAREFACTION. **2.** Any increase in the cross-sectional area of a flow channel, usually gradual. For example, the diffuser in a wind tunnel.

expansion coefficient *See* THERMAL EXPANSION.

expansion cooling The reduction in temperature that occurs when a gas expands. For example, an ideal gas with specific-heat ratio γ expands isentropically according to the equation $Tv^{\gamma-1}$ = constant where v is the specific volume and T is the absolute temperature.

expansion curve A plot of pressure *vs* specific volume for the expansion of a gas or vapour.

expansion engine A reciprocating engine in which high-pressure gas is drawn into a cylinder by downward movement of the piston. Once the inlet valve closes, the gas expands and cools, typically by 60°C. Upward movement of the piston then expels the cold air through the outlet valve. Applications include the liquefaction of gases.

expansion fan *See* PRANDTL–MEYER EXPANSION.

expansion fit *See* SHRINK FIT.

expansion joint A joint between components which permits movement during a change in temperature without the creation of stresses through distortion or buckling.

expansion ratio 1. For a duct with a sudden increase in cross-sectional area, the ratio of the downstream area to the upstream area. **2.** For a fixed mass of liquefied gas, the ratio of the volumes in the gaseous and liquefied states. **3.** For a rocket engine, the ratio of the nozzle-exit area to the area of the combustion chamber outlet. **4.** *See* DIESEL CYCLE.

expansion valve *See* THROTTLING VALVE.

expansion viscosity *See* BULK VISCOSITY.

expansion wave *See* PRANDTL–MEYER EXPANSION.

expansion work (*W*) (Unit kJ) For a mass of fluid that expands from a volume $\mathcal{V}_1$ to a volume $\mathcal{V}_2$,

$$W = \int_{\mathcal{V}_1}^{\mathcal{V}_2} p\, d\mathcal{V}$$

where p is the fluid pressure during the expansion process. If p is plotted *vs* the volume for the process, W is the area under the curve.

expansivity *See* THERMAL EXPANSION.

expert system A computer program that attempts to replicate the expertise of one or more human experts. It will normally employ a knowledge base containing information on the effect of different decisions, and an inference engine that applies rules to make decisions using input data and the knowledge base. An **expert control system** is one that includes an expert system.

explicit programming The process of programming a robot by entering into the program precise information about the required end effect or position and orientation, the time derivatives of these (for example, the end effector velocity), and other functions (for example, opening or closing the gripper).

explosion A sudden release of significant amounts of energy, either stored in the form of pressure or generated by chemical or nuclear reaction. The energy release is accompanied by rapid expansion of the gas released. *See also* DEFLAGRATION; DETONATION; INTENSE EXPLOSION.

explosive (explosive material) A substance that contains a high proportion of chemical energy which can be released rapidly, typically by heat or pressure. *See also* HIGH EXPLOSIVE; LOW EXPLOSIVE.

explosive energy For a pressure vessel, the work that a pressurized fluid would do if allowed to expand adiabatically to the state of the surroundings. *See also* EXPANSION WORK.

explosive forming The forming of metal parts in which an explosive charge is employed to send a uniform pressure wave through water to deform the workpiece against a shaped cavity or die.

explosive welding *See* WELDING.

extended surfaces *See* COOLING FINS.

extension *See* ELONGATION.

extensional flow (elongational flow) A stretching flow in which the distance between fluid particles on any streamline is continually increasing or decreasing. An example is the flow through a convergent nozzle. The **elongational viscosity (extensional viscosity, Trouton viscosity, η_E)** is the resistance of a fluid to stretching. For a uniaxial extensional flow, $\eta_E = \sigma_E/\dot{\varepsilon}$ where σ_E is the extensional stress (i.e. the tensile stress) and $\dot{\varepsilon}$ is the rate of extension. The **Trouton ratio** is the ratio of extensional viscosity to shear (dynamic) viscosity, with the value 3 for a Newtonian fluid and generally greater for non-Newtonian liquids. *See also* SHEAR FLOW; VISCOMETRIC FLOW.

extension ratio *See* STRETCH RATIO.

extension spring A tightly-coiled (closed) spring, often having hooks at either end, for use in tension.

extensive property A thermodynamic property of an entire system, including volume, mass, enthalpy, internal energy, and entropy. *See also* INTENSIVE PROPERTY.

extensometer An instrument attached to a tensile testpiece which measures the small displacements that occur over the gauge length with increasing or decreasing load.

external-combustion engine A heat engine such as a steam turbine or Stirling engine in which the fuel is burned outside the system boundary that encloses the engine and heat is transferred into the working fluid across the system boundary. *See also* INTERNAL-COMBUSTION ENGINE.

external convective boiling *See* BOILING.

external energy (Unit kJ) The contributions of potential and kinetic energy to the total energy of a system.

external flow A flow relative to the outer surface of a body immersed in a fluid. *See also* INTERNAL FLOW.

external gear pump A gear pump in which there are two meshed gears, typically of the same size, which rotate in opposite directions. Only one gear is driven. The gears may be circular or lobe-shaped. Flow enters on the side where the gear teeth separate. *See also* INTERNAL GEAR PUMP.

external irreversibility *See* IRREVERSIBILITY.

externally-fired boiler A steam boiler in which the furnace is separate from the boiler itself.

external screw thread *See* SCREW.

external sensor A sensor which measures a variable relating to the environment around a system rather than measuring a variable that forms part of the system. In robotics, a vision system is an external sensor whereas an encoder on a joint is not.

external work 1. The work performed during deformation of a body by the forces acting over its surfaces. *See also* INTERNAL WORK. **2.** In thermodynamics, all the work transfer across the surface of a control volume other than that due to normal forces: primarily shaft or stirring work and electrical work.

extraction turbine (pass-out turbine) A steam turbine in which partially-expanded steam is extracted from one or more stages for heating, process plant, or feedwater heating. *See also* REGENERATIVE RANKINE CYCLE.

extreme-pressure lubricants (EP lubricants) Lubricants employed in highly-loaded gears that operate at high temperatures.

extrudate A product of the process of extrusion.

extruder A machine that performs extrusion.

extrusion The process of forming continuous lengths of rods, tubes, etc. in which a workpiece (usually cylindrical) of metal above its recrystallization temperature, or other solid in a soft state, is pushed from a container by a ram through a steel or ceramic shaped orifice (the die). Solid, hollow, and ribbed sections are made by such 'forward' or 'direct' extrusion. Polymer extrusion is carried out using a screw feed of pellets towards the die where they consolidate. In 'indirect', 'backward' or 'inverted' extrusion, the workpiece is held stationary and the die moves into the container so that the flow of the extrudate is in the opposite direction. *See also* IMPACT EXTRUSION; PULTRUSION.

eye The inlet to the impeller of a centrifugal pump or compressor.

eyebolt (eyescrew) A bolt or screw with a closed loop in place of a head.

fabric A flexible mesh of fibres or threads of natural or artificial materials, including metals, which may be manufactured by knitting, weaving, knotting (in the form of nets) or felting (compacting fibres). *See also* DRAPE; SCREEN MESH.

fabrication 1. The manufacture of components from metal, plastic, silicon, wood, etc. **2.** The building up of components into complete or sub-assemblies.

fabric filter A fabric mounted in a frame for removing particulate material from a flow. *See also* SCREEN MESH.

face 1. *See* TOOTHED GEARING. **2.** The part of a valve that contacts the seat. **3.** Any large flat area of a component such as the surfaces of a pipe-flange. **4.** The dial of a clock gauge.

face cone *See* TOOTHED GEARING.

face gear *See* TOOTHED GEARING.

face mill *See* END MILL.

face plate An attachment for a lathe, consisting of a large circular plate having holes and slots to which workpieces may be bolted.

face velocity *See* DARCY'S LAW.

face width *See* TOOTHED GEARING.

factor of safety (design factor, safety factor, *f*) If σ_A is the actual strength of a component and σ_R is the required strength, the factor of safety $f = \sigma_A/\sigma_R$. The **margin of safety** is $(\sigma_A - \sigma_R)/\sigma_R = f - 1$.

FAD *See* FREE AIR DELIVERY.

Fahrenheit scale (*t*) A largely obsolete, non-SI, relative-temperature scale now defined in terms of the Rankine absolute temperature scale as $t\,(°\text{F}) = T(\text{R}) - 459.67$, where °F is the Fahrenheit degree symbol. The scale was originally defined with the melting point of ice (the ice point) as 32°F and the boiling point of water (the steam point) as 212°F. *See also* CELSIUS SCALE.

fail-safe design The design methodology whereby failure of one part of a component or structure does not lead to failure of the whole assembly. This can be achieved through, for example, redundancy. In control systems, the philosophy is called '**fail soft**'. *See also* DAMAGE-TOLERANT; SAFE LIFE.

fail-safe system A system designed so that the failure of one or more components does not result in danger. For example, air-operated brakes on a heavy-goods vehicle are applied by springs and released by air pressure. A leak causing a pressure loss will thus result in the brakes being applied rather than causing them to become inoperative.

failure The result when a body, component, or structure is incapable of performing the task for which it was designed. The term is often used without reference to what causes failure, such as fracture, buckling, excessive deformation, wear, or erosion. **Failure criteria (failure theories, theories of strength)** are mathematical expressions for the combinations of stress, strain, or strain energy at which materials fail, which are employed in design to dimension components. *See also* DAMAGE; STRENGTH.

failure-assessment diagram *See* R6 FAILURE-ASSESSMENT DIAGRAM.

fairing A streamlined, smooth, thin-walled shell of metal, plastic, or composite material, placed around part of a vehicle, such as an aeroplane, motorcycle, motor vehicle, or train, in order to reduce aerodynamic drag. *See also* COWL.

Falkner–Skan solutions Exact similarity solutions (numerical) to the boundary-layer equations for flows with a variation of free-stream velocity U_∞ according to $U_\infty = Kx^m$ where K is a dimensional constant, x is the distance along the surface, and m is a constant. The Blasius solution corresponds to $m = 0$. *See also* WEDGE FLOW.

falling-ball viscometer (falling-sphere viscometer) An instrument for determining the viscosity of a liquid by measuring the time for a sphere, typically of glass, stainless steel, or tantalum, to fall between two fiduciary lines in a glass tube containing the liquid.

falling-film cooler A type of heat exchanger in which a film of the liquid to be cooled flows down either the inside or the outside of vertical tubes or over flat surfaces while a refrigerant or chilled liquid, such as water or brine, flows on the other side of the cooling surface.

falling-film evaporator A type of heat exchanger in which a thin film of the liquid to be evaporated flows either down the inside of vertical tubes heated by steam or subjected to vacuum, or over the outside of an array of horizontal or vertical tubes through which there is a flow of hot fluid.

fan 1. (blower) A device with vanes or blades attached to a hub on a shaft that rotates to produce an airflow. There are both axial and centrifugal designs. The **fan rating** is the volumetric flow rate of air (i.e. the **fan delivery**, $\dot{Q}$) a fan delivers at a particular rotation speed, and the **fan static-pressure rise** Δp is the average increase in static pressure across the blades of a fan at that speed. The **fan total-pressure rise (fan total-head rise)** is the average increase in stagnation pressure across the blades of a fan at a particular rotation speed, and the **fan dynamic-pressure rise (fan velocity-pressure rise)** the corresponding average increase in dynamic pressure. The overall **fan efficiency (η_E)** is the ratio of the power delivered to the air divided by the power provided by the driving motor P_{SHAFT}, i.e. $\eta_E = \dot{Q}\Delta p/P_{SHAFT}$. The **fan characteristics**, measured in a **fan test**, are curves of Δp, P_{SHAFT} and η_E *vs* $\dot{Q}$. **2.** In a turbofan engine for a civil aircraft, a single-stage compressor at the front of the engine operating on the bypass air stream. For military applications there may be two or three stages.

fan brake A type of dynamometer in which the torque is provided by a fan.

Fanning friction factor (*f*) For pipe or duct flow, a non-dimensional measure of the surface shear stress defined by $f = 2\tau_S/\rho V^2$ where τ_S is the circumferential average surface shear stress at any axial location, ρ is the fluid density, and V is the bulk flow velocity. *See also* DARCY FRICTION FACTOR.

Fanno flow Adiabatic gas flow through a constant-area duct with wall friction. For a perfect gas, tabulated results are available from a one-dimensional analysis assuming a constant friction factor. A **Fanno line** is curve of enthalpy or temperature *vs* specific volume or specific entropy for Fanno flow. *See also* RAYLEIGH FLOW.

fantail A ducted version of a helicopter tail rotor.

farfield In fluid mechanics and acoustics, used to indicate conditions far away from a specified source or disturbance.

fast coupling *See* QUICK COUPLING.

fastener A device for the permanent or temporary joining of components or structures, including bolts, screws, nails, pins, clamps, and circlips.

fast Fourier transform A mathematical technique for producing a Fourier transform from sampled data, i.e. for producing the discrete Fourier transform. It is particularly efficient (i.e. fast) when the number of sampled values is a power of 2.

fast pyrolysis A process in which organic materials, especially biomass, are rapidly heated to between 450°C and 600°C in the absence of air to produce gases, char, and organic vapours. The latter are condensed to yield bio oils. *See also* BIOENERGY.

fathom (fath) A non-SI, imperial unit of length used primarily for measuring water depth. 1 fath = 6 ft.

fatigue A term referring, in components and structures subjected to either random or cyclic periodically-varying loads, to a progressive reduction in strength leading to failure at stresses lower than those that cause failure under monotonic loading. Variable loads arise from out-of-balance machinery and other vibration sources, wind gusts, etc., and a large proportion of service failures is caused by fatigue. Fatigue results from the initiation and slow propagation of cracks. In manufactured components, crack initiation usually occurs at a point of stress concentration. After a period, often of millions of stress cycles, the crack reaches a critical length at which the next peak load causes sudden brittle or ductile fracture. Fracture surfaces resulting from fatigue display characteristic **striations** or **progression marks** emanating from the crack initiation site during the slow crack growth period, with a different surface appearance for the final fracture.

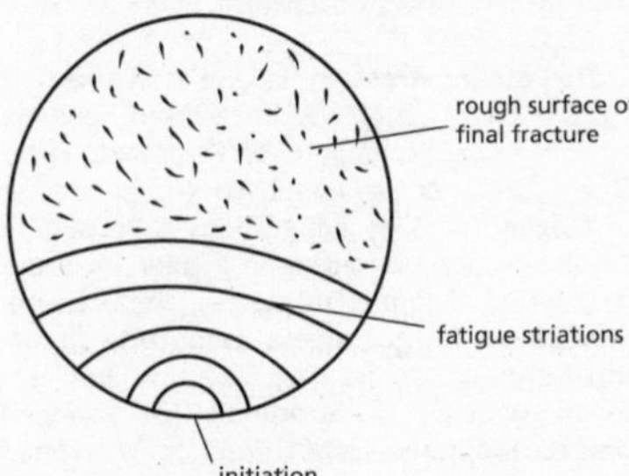

fatigue failure surface

There are various ways of determining the fatigue behaviour of materials. One of the earliest is the rotating cantilever (Wöhler) test, in which a cylindrical specimen is gripped in a chuck and rotated. At its free end the specimen has a bearing below which a dead weight is hung that bends the testpiece.

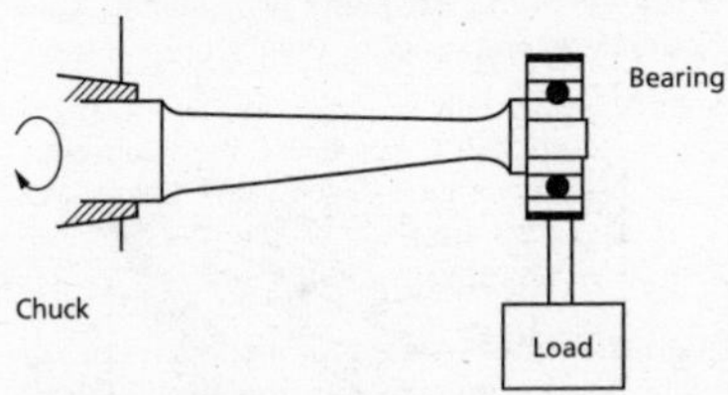

Wöhler testing machine

Surface elements experience the maximum sinusoidal variation of tensile and compressive bending stress (s) as the testpiece rotates, the magnitude of which depends on the hanging load. The machine records the number of cycles N at the instant of fracture. Typical results for alloys of steel and aluminium are plotted as s–N curves, using a log scale on the abscissa.

When s is high, failure occurs in fewer cycles than when s is low (**low-cycle** and **high-cycle fatigue**, respectively). The arbitrary dividing line between the two occurs at about 10^3 or 10^4 cycles. Low-cycle fatigue data are used in the design of components either having relatively short lives or likely to be overloaded. The diagram demonstrates that for $N > 10^6$ cycles

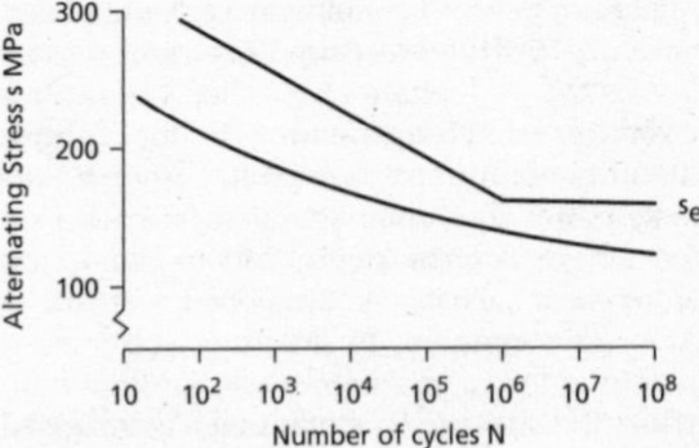

s–N fatigue curve

there is a stress s_e (the **fatigue limit** or **endurance limit**) below which failure does not occur in plain carbon and low-alloy steels, and titanium, no matter how many cycles the testpiece is subjected to. Components designed so that stresses are lower than the endurance limit operate in the so-called **infinite-life regime.** Designs based on the higher fatigue stresses for $N < 10^6$ cycles have **finite lives.** In contrast, aluminium and other non-ferrous alloys, polymers, and composites do not exhibit a fatigue limit, so that for design purposes a **fatigue strength** is defined as the stress at 10^7 or 10^8 cycles. Mandatory requirements are often specified for periodic crack inspection of highly-stressed aluminium components and structures. The **safe fatigue life** is the period of time during which repeated applications of a load to a component or structure are unlikely to result in its failure. Unless stresses are oscillated above 10^4 Hz, frequency has no effect on the s–N curves for metals. This is not true for polymers where localized heating of testpieces occur.

The ratio of the endurance limit to the ultimate tensile strength of the material is called the **endurance ratio** or **endurance-limit, ratio.** For many steels whose UTS < 1.4 GPa, it is found empirically that the endurance ratio is about 0.5 ~ 0.7. The ratio of the fatigue strength to the ultimate tensile strength of a material is called the **fatigue ratio.**

Relations for low-cycle fatigue failure may be expressed in terms of stress or strain amplitudes. **Basquin's** equation is $S = S_f N_f^{-b}$ where S is the stress amplitude (given by $(s_{max} - s_{min})/2$ where s_{max} is the maximum stress in a cycle and s_{min} the minimum, in which compressive stresses are negative), N_f is the number of cycles to failure, S_f is called the **fatigue-strength coefficient,** and b is the **fatigue-strength exponent.** Writing $\Delta\varepsilon_p$ for the applied fluctuating plastic-strain amplitude, given by $(\varepsilon_{max} - \varepsilon_{min})/2$ where ε_{max} is the maximum strain in a cycle and ε_{min} the minimum in which compressive strains are negative, the **Coffin–Manson–Tavernelli** relation is $\Delta\varepsilon_p/2N_f^c = \varepsilon_f'$ where ε_f' is called the **fatigue ductility coefficient** and c is the **fatigue ductility exponent** that ranges from 0.1 for magnesium and some stainless steels to 0.9 for other stainless steels, nickel alloys, and aluminium alloys. A simplified version is $\Delta\varepsilon_p\sqrt{N_f}$ = constant. In structures subjected to pulsating loads, shakedown can be important. When the shakedown stress range is exceeded, the small plastic strains produced in every cycle may accumulate to lead to fatigue failure.

Ordinary screw-driven machines may be used for fatigue testing in which the axial load is arranged to cycle from positive to negative (**push–pull fatigue**); others work like a vibrating tuning fork with the specimen attached to one of the arms. Yet other machines apply fluctuating torsional stresses, or combined stresses. Machines can be programmed to vary the amplitude and frequency of loading during a test, including random loading. Such information is necessary since practical operating conditions for components and structures are rarely the constant-frequency, constant-amplitude patterns of most laboratory testing. In addition to plain samples, specimens containing different sorts of notches may be tested, and even actual machine components rather than material testpieces.

Fatigue data show much scatter, as results are sensitive to surface finish and other factors such as stress concentrations, size of testpiece and the surrounding environment. Specimens are often highly polished to achieve the highest s–N curves (and greatest endurance limits for steels). The endurance limit in reversed-bending and torsion fatigue of round bars depends on the diameter, other factors (such as surface finish) being equal; for non-circular components an equivalent diameter based on the same cross-sectional area has to be found. In **corrosion fatigue**, aggressive chemical environments shorten the crack-initiation period and accelerate crack-growth rates leading to a shorter component life. In consequence there are various empirical correction factors (all < 1) that are applied to the endurance limit of polished-steel samples to arrive at a value for design; they include **surface**, **size**, **temperature**, and **environment factors**. Note that the **fatigue-strength reduction factor K_f** is the ratio of the fatigue limits of notched and un-notched testpieces, and is the stress-concentration factor to employ in fatigue instead of the usual stress-concentration factor K_t employed for monotonic loading. K_f varies with notch or discontinuity geometry. The **notch-sensitivity factor (index** or **ratio)** defined as $q = (K_f - 1)/(K_t - 1)$ indicates the extent to which a body containing a notch is sensitive to fatigue. Notch-sensitive materials have $q = 1$; notch-insensitive materials have $q = 0$.

The **mean stress** in fatigue is defined as $s_{mean} = (s_{max} + s_{min})/2$. In reversed bending, $s_{max} = -s_{min}$, so $s_{mean} = 0$. The **stress ratio** $R = s_{min}/s_{max}$ is also employed as a parameter in fatigue. In many applications a fluctuating stress is superimposed upon a static load and it is found that increasing s_{mean} reduces the permissible **fatigue-stress amplitude (limiting range of stress)**. The diagrams show different versions of **Goodman (or fatigue) diagrams** that show how S and s_{mean} are related to ensure different specified lives (N_e). The locus marked N_f is for an infinite fatigue life for materials having an endurance limit. The dashed regions are of little practical significance for infinite-life design, as s_{max} exceeds the yield stress. In the diagram, $S = s_e(1 - s_{mean}/UTS)$ where s_e for fully-reversed alternating stress is assumed to be ⅔ *UTS*.

The **modified Goodman diagram** sets s_e as the actual limiting stress range. The Goodman relation was preceded by **Gerber's** equation in which the (s_{mean}/UTS) term was squared. Later **Soderberg** proposed an even more conserva-

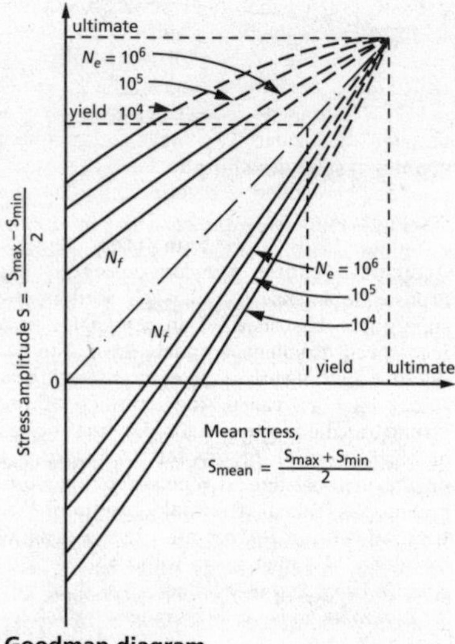

Goodman diagram

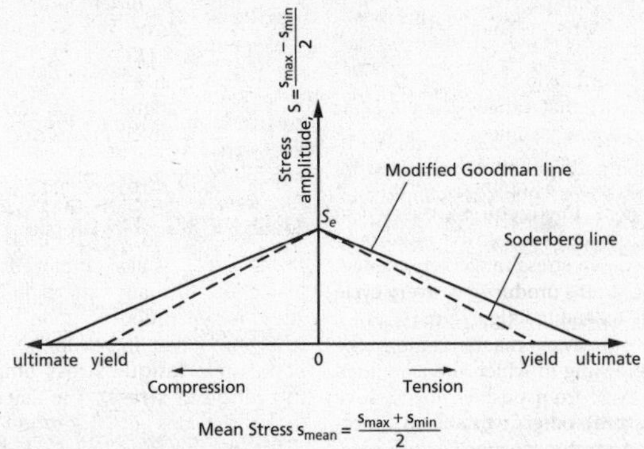

Goodman diagram (alternative version)

tive relationship where UTS is replaced by the yield stress.

Components and structures may be subjected to an accumulation of stress cycles of varying amplitude and varying frequency, or load fluctuations that are random. The **Palmgren–Miner** or **Miner** empirical rule states that $\Sigma_i N_i/N_{fi} = 1$ where N_{fi} is the fatigue life of the component under maximum stress s_i, and N_i ($< N_{fi}$) is the actual number of cycles spent at stress s_i. In practice the constant of unity varies widely, typically between 0.7 and 2, and depends on whether high stress amplitudes occur before or after low. **Rainflow analysis** is employed in the analysis of fatigue data obtained under random loading, in which random stress cycles may be represented in terms of a number of regular sine waves of different amplitude in order that the Palmgren–Miner rule may be applied to determine the life of a component or structure.

Fatigue loads arise from many sources. **Acoustic fatigue** concerns components exposed to intense sound, such as from the supersonic exhaust jet of an aeroengine or a high-speed turbulent boundary layer. The endurance limit of surfaces in cyclic contact (gear teeth, etc.) is called the **contact-fatigue strength**, the safety factor for the design of which is called the **life factor**. Failure of a ductile material due to repeated temperature cycling is called **thermal fatigue**. Two **thermal-fatigue factors** are used to quantify the ability of a material to withstand thermal-strain cycling, and are defined by $\Delta\sigma k/\alpha E$ and $\Delta\varepsilon^P k/\alpha$ where $\Delta\sigma$ is the stress range to which the material is subjected, $\Delta\varepsilon^P$ is the corresponding range of plastic strain, k is the coefficient of thermal conductivity, α is the coefficient of thermal expansion and E is Young's modulus.

Complications in fatigue behaviour arise at high homologous temperatures T_M in metals, and in other materials such as polymers having time-dependent behaviour, leading to **creep fatigue.**

Design based on *s–N* fatigue curves in which a statistical mean life is determined for the expected service load spectrum is **safe-life** design. **Fail-safe** design relies on redundancies in a structure such that failure of one member does not lead to a cascade of fractures and hence complete failure. **Damage-tolerant** design employs fracture mechanics to take into account initial imperfections (starter cracks), crack-growth rates and the conditions of final fracture. The **Paris equation** relates the crack growth per cycle in fatigue da/dN to the range of applied-stress intensity ΔK by $da/dN = C(\Delta K)^n$ where a is crack length and $\Delta K = (s_{max} - s_{min})Y\sqrt{\pi a}$ and in which C and n are material-dependent, experimentally-determined constants. Integration of this relation between $a = a_i$ (the length of a crack found by non-destructive inspection) and $a = a_f$ the critical crack length at which the body fails, gives

$$N_f = \frac{(1/a_i)^{n/2-1} - (1/a_f)^{n/2-1}}{C(s_{max} - s_{min})^n Y^n \pi^{n/2}(n/2-1)}$$

in which Y is assumed to be constant. In this way the fatigue life N_f of a component is determined. As with *s–N* curves, there are modifications to the Paris equation to account for mean stress effects. *See also* FLOW-INDUCED VIBRATION; PITTING; RAINFLOW ANALYSIS; STATIC FATIGUE; THRESHOLD STRESS-INTENSITY FACTOR.

fault monitoring The process of automatically checking the performance of a system to detect the occurrence of faults. Often applied to detect both hardware and software faults in computer-control systems.

fault tolerance The ability of a system to continue correct operation despite the failure of one or more components of the system. *See also* FAIL-SAFE DESIGN.

favourable pressure gradient In boundary-layer theory, a streamwise pressure gradient that accelerates the flow of a fluid, increases the wall shear stress, and delays transition to turbulence. If very strong, it can lead to laminarization of a turbulent boundary layer. *See also* ADVERSE PRESSURE GRADIENT.

Favre averaging A method of time averaging used in the analysis of a turbulent flow in which density variations are significant. For a flow parameter *u*, the Favre average is defined by

$$\tilde{u} = \frac{1}{\bar{\rho}} lim_{T\to\infty} \frac{1}{T}\int_0^T \rho\, u\, dt = \frac{\overline{\rho u}}{\bar{\rho}}$$

where *t* is time, *T* is the averaging time, ρ is the fluid density, and the straight overbar indicates a Reynolds-averaged quantity.

faying surface The part of the surface of a component that is in contact with another surface to which it is joined, for example by welding, brazing, soldering, bonding, or riveting.

FBD *See* FREE-BODY DIAGRAM.

f-chart A computer programme for the analysis and design of solar-heating systems.

feasibility study The determination of the viability and benefits of a proposal with particular emphasis on identifying potential problems.

feathering Adusting a controllable-pitch propeller to a position of minimum drag in the event of power failure.

feather key A metal bar of rectangular cross section with both ends radiused that fits into a keyslot on a shaft to allow torque to be transmitted to a wheel, gear, or other cylindrical component.

feedback The process of taking a signal representing the output of a plant and using this with other signals to determine the input to the plant. In a control system, the **feedback control loop (feedback loop)** is formed by the plant, sensors measuring the plant output, feedback path, summing junction, and controller. The transfer function of the feedback path **(feedback transfer function)** includes the transfer function of the sensor measuring the plant output and the transfer function of the **feedback compensation** (the use of a compensator in the feedback branch, rather than in series with the plant as would be the case in cascade compensation). The signal fed back to the plant input through the **feedback branch (feedback path)** to the summing junction to produce the error signal is the **feedback control signal**. A **feedback control system** is one using one or more closed loops, i.e. where the output from the controlled plant is used by the controller to modify the plant input. **Negative feedback** is where the plant input is reduced by the feedback signal; **positive feedback** is where the plant input is increased. A **feedback regulator** uses feedback control so as to maintain the desired plant output.

feedforward control system A control system that uses one or more sensors to detect disturbances affecting the plant and then applies an additional input to the plant so as to minimize the effect of the disturbance. This requires a mathematical model of the plant so that the effect of disturbances can be predicted and the necessary additional input calculated.

feed gear The means by which, in a machine tool, a tool is fed to the workpiece or vice versa.

feed nut The means by which, in an automatic drill, a drill bit is fed to the workpiece.

feed pipe The connecting pipe from a feed pump.

feed pressure The pressure provided by a feed pump.

feed pump A pump that supplies liquid to a component or machine, such as feedwater to a boiler, fuel or oil to an engine.

feed screw The screw that controls feed mechanisms in a machine tool, such as the drive for the saddle in a lathe. *See also* LEAD SCREW.

feedstock Workpiece material in bulk that is fed into a process, or into the deformation zone of a machine such as an automatic bolt-heading machine, a capstan lathe, etc.

feedwater The water supplied to a steam boiler.

feedwater heater A heat exchanger used to preheat feedwater before it enters a boiler. The energy required is usually derived from the flue gas or steam extracted between the stages of the steam turbine. *See also* ECONOMIZER; EXTRACTION TURBINE.

feeler gauge An array of flat fingers of hardened steel of accurately-known thickness, pinned at their ends like a collapsible fan, used for checking clearances, end float, sparking plug gaps, etc.

FEM *See* FINITE-ELEMENT METHOD.

female fitting (female coupling) The outer part of a connection in which one part fits into another. *See also* MALE FITTING.

female thread *See* SCREW.

FEPA Federation of European Producers of Abrasives.

ferrography A method of analysis of particles in lubricating oil, used to detect wear in gearboxes and such machine components as bearings.

ferrule 1. A metal (literally iron) ring or cap attached to a component, such as the handle of a tool, to protect it. **2.** *See* OLIVE.

FFLD *See* FRACTURE FORMING LIMIT DIAGRAM.

FHP *See* FRICTION POWER.

fibre 1. A natural or man-made filament of material. Natural fibres include wood and asbestos. Man-made fibres include fibreglass and carbon fibres, but the most common are polymer fibres. **2.** A filament, either long (continuous) or short (chopped), used to reinforce matrices in composites.

fibre bridging In fibre-reinforced composites, when intact fibres span across a crack and thereby increase the fracture toughness by pull-out.

fibre content The amount, by volume or mass, of fibres in a composite material. *See also* VOLUME FRACTION.

fibre-optic cable A cable consisting of one or more glass fibres through which light is transmitted.

fibre-optic thermometer A temperature sensor in which a luminescing phosphor is attached to the tip of an optical fibre and excited to luminescence by a pulsed light source. The decay rate of luminescence is temperature-dependent.

fibre reinforcement The increase in stiffness of composites by addition of fibres to the matrix.

fibre stress 1. In composites under load, the stress σ_f in the fibres as opposed to that in the matrix, σ_m. *See also* COMPOSITE MODULUS; COMPOSITE STRENGTH; RULE OF MIXTURES. **2.** A term sometimes used in bending theory to relate to the stresses at different distances from the neutral axis.

fibrous fracture A term used to describe the appearance of fracture surfaces in ductile materials that crack by void initiation, growth, and coalescence. Characteristic surface cavities are nucleated on inclusions, such as manganese sulphide in steels, and elongated before fracture is complete. *See also* CLEAVAGE FRACTURE; CRYSTALLINE FRACTURE; VOID SHEET.

Fick's law The governing equation for mass diffusion due to a concentration gradient. In a binary fluid mixture, the mass flux of species A is given by the equation $\dot{m}''_A = -\rho D \nabla c_A$ where ρ is the mixture density, D is the binary diffusion coefficient and c_A is the mass concentration of A. D can be regarded as being defined by the equation. The corresponding thermal-energy relation is Fourier's law.

fictitious force *See* ACCELERATING FRAME OF REFERENCE.

fiducial temperature *See* FIXED POINT.

fiduciary line A fixed reference line marked on an instrument, such as the crosshairs in the eyepiece or graticule of a microscope.

filament 1. A long fibre, particularly of glass or carbon, employed to reinforce composite materials. **2.** A fine wire, such as the heating element in an incandescent light bulb.

filament winding The manufacture of vessels made from fibre-reinforced composites, by winding the continuous filament or prepreg over a shaped mandrel, thus avoiding joins. Winding may be performed axially, circumferentially, helically, or longitudinally, or as combinations of these.

filar micrometer (bifilar micrometer) An instrument for the measurement of objects viewed under a microscope. It consists of two parallel filaments ('hairs') within the eyepiece, one fixed and one moved by a screw thread.

file A hand-held cutting tool that cuts on the forward stroke only. Files are classified according to the cut (number and size of ridges) and shape (flat, square, round, half-round, triangular, etc). The harder the material being filed, the finer-toothed the file. *See also* DRAW FILING.

file hardness A measure similar to the idea of scratch hardness, whereby a material that cannot be cut by a file must be as hard as, or harder than, the file.

filled-system thermometer *See* FLUID-EXPANSION THERMOMETER.

filler Low-cost material such as wood flour, silica flour, or talc, used to bulk out some polymer-moulding compounds.

filler rod *See* WELDING.

fillet gauge A device used to check whether the dimensions of a fillet weld are within specified tolerances.

fillet weld A weld of triangular cross section used to join two plates at right angles.

film boiling *See* BOILING.

film coefficient *See* HEAT TRANSFER.

film condensation (filmwise condensation) The dominant form of condensation of a vapour onto a surface at a temperature below the saturation vapour temperature and which is wetted and completely covered by the condensate which flows continuously under the action of gravity. *See also* DROPLET CONDENSATION.

film cooling A method of surface cooling in which a cool fluid is blown through either a series of holes or one or more slots in the surface, in a direction nearly tangential with the surface. For a gas flow, the cooling effectiveness can be greatly enhanced by using as the coolant a liquid that evaporates (sweat cooling). *See also* TRANSPIRATION COOLING.

film-cooling effectiveness (adiabatic film-cooling effectiveness, η) If T_{AS} is the temperature of an adiabatic surface for a flow with free-stream temperature T_∞, the film-cooling effectiveness is defined as $\eta = (T_{AS} - T_\infty)/(T_F - T_\infty)$ where T_F is the cooling-fluid temperature.

film pressure The pressure in the lubricant film within a bearing.

film temperature The temperature used to evaluate fluid properties in pipe or boundary-layer flow where there are large temperature differences. It is defined as $(T_{REF} + T_S)/2$ where T_S is the surface temperature, T_{REF} is the bulk temperature in pipe flow and the free-stream temperature for a boundary layer.

film thickness (Unit μm) The thickness of the thin layer of lubricating oil between two surfaces, such as within a bearing or between two contacting gear teeth.

filmwise condensation *See* FILM CONDENSATION.

filter 1. (filter screen) A mesh, gauze, paper, or cloth, usually held by a frame, that removes particulates larger in diameter than the mesh or pore size from a fluid stream passing through it. Metal wires, natural and synthetic fibres are all used for filter construction. *See also* BETA RATIO. **2.** In processing signals, a device that passes a specified range of frequencies and blocks others. Normally used in low-pass, high-pass, or band-pass forms.

filterability The ability of a fluid to pass through a filter without excessive pressure drop.

filter efficiency *See* BETA RATIO.

filter pump An aspirator, such as a jet pump, attached to the outlet side of a liquid filter, for example using a side-arm flask, to apply suction and so increase the filtration rate.

filter screen *See* FILTER (1).

filter thickener A device that increases the proportion of solids in a liquid–solid mixture such as sewage, sludge, or slurry, by filtering off some of the liquid. Applications include wastewater treatment and fruit-juice concentration.

fin 1. *See* FLASH. **2.** *See* COOLING FINS.

final value The steady-state output of a system after transient effects have decayed.

final-value theorem The result, when a Laplace-transformed (i.e. frequency-domain) variable is multiplied by s and evaluated in the limit as $s \rightarrow 0$, is the value of the inverse transform (i.e. time-domain solution) as time $t \to \infty$. Thus: $lim_{t\to\infty} f(t) = lim_{s\to 0} sF(s)$

fin cooling *See* COOLING FINS.

fin effectiveness (fin efficiency) *See* COOLING FINS.

fine particle strengthening *See* PRECIPITATION HARDENING.

fine thread *See* SCREW.

finger gripper A robot gripper using two or more moveable rods, equivalent to human fingers, to pick up components.

finish machining The final stages of manufacture of a component by cutting, where small depths of cut are taken at high speed in order to satisfy dimensional and surface-finish requirements.

finite-difference method A method for the numerical solution of ordinary and partial differential equations in which the solution domain is divided into cells. At the most basic level, a spatial derivative with respect to any quantity is approximated by the difference in that quantity across a cell divided by the cell length, height, or depth. This is termed a **first-order difference**. Temporal derivatives and higher spatial derivatives are also approximated by differences. Special techniques have been developed to improve numerical accuracy and convergence and avoid instability, for example using higher-order differences, as well as forward and backward differences and finer cells, in regions where gradients are steep. The method is applied in many problem areas including elasticity, heat transfer, and fluid mechanics. Steady-flow problems are often solved as the final state of a time-evolving problem. *See also* FINITE-ELEMENT METHOD; FINITE-VOLUME METHOD.

finite-element method (FEM) A numerical method of solving partial differential equations and hence problems in stress analysis, fluid mechanics, heat transfer, etc. In simple terms, the body is divided up into regions (**elements**) of shapes and sizes appropriate to the problem, and the governing equations of every element written down. Elements are then reconnected at nodes, resulting in a set of simultaneous algebraic equations, solution of which solves the problem. By the process of reconnecting nodes, the field quantity of interest (e.g. displacement field, temperature field) becomes interpolated over the whole body in a piecewise fashion by as many polynomial expressions as there are elements. The 'best' values of the nodal field quantities are those that minimize some function such as total energy. In matrix notation, $\boldsymbol{KD} = \boldsymbol{R}$, where $\boldsymbol{D}$ is a vector of the unknown nodal-field quantities, $\boldsymbol{R}$ is a vector of known loads, temperatures etc., and $\boldsymbol{K}$ is a matrix of known constants. In stress analysis, $\boldsymbol{K}$ is called the stiffness matrix.

finite-life fatigue *See* FATIGUE.

finite-strain theory The theories of elasticity and plasticity that employ large-deformation measures of strain, i.e. stretch ratios for rubber elasticity and logarithmic strains for plasticity, rather than incremental strains.

finite-volume method A method for the numerical solution of partial differential equations, widely used in many commercial CFD packages. The solution domain is divided into small but finite volumes surrounding node points on a mesh. The governing equations are formally integrated over each of these volumes. The conservation of any flow variable φ is considered taking into account fluxes of φ due to advection and diffusion across the faces of each volume and the creation of φ within it. The resulting set of algebraic equations is solved iteratively. *See also* FINITE-DIFFERENCE METHOD; FINITE-ELEMENT METHOD.

finned surfaces *See* COOLING FINS.

fin performance *See* COOLING FINS.

fins *See* COOLING FINS.

fire barrier *See* THERMAL BARRIER (2).

firebox The furnace used to heat a fire-tube boiler, such as used on a steam locomotive.

fire bridge (bridge wall) In a reverberatory furnace, a low barrier that separates the fuel from the ore or metal being smelted.

fire damp An explosive gas, primarily methane, found in coal mines.

fire damper A passive fire-protection device installed in air-conditioning and ventilation systems to prevent the spread of fire and smoke.

fired process equipment Equipment in which the heat for the process concerned is provided by combustion of a fuel.

fire pump A pump used to supply water at high flow rates for fire suppression. Fire pumps are usually powered by diesel engines or electric motors, and may be vehicle mounted, portable, or permanently installed in a building, on a ship, etc.

fire-tube boiler (fire-tube shell boiler, shell boiler, shell-and-tube boiler, smoke-tube boiler) A boiler in which the hot combustion gases flow through tubes contained within a steel shell and the water to be heated flows around the tubes within the shell. *See also* DRY-BACK BOILER; ECONOMIC BOILER; WATER-TUBE BOILER; WET-BACK BOILER.

f

firing 1. The process of combustion occurring in a cylinder of an internal-combustion engine or the combustion chamber of a rocket engine. **2.** The process of heating a clay-rich component in a kiln to drive off water and harden the material, including by sintering.

firing order The controlled sequence in which combustion occurs in the cylinders of a multi-cylinder piston engine.

firing rate The rate at which fuel is supplied to a burner or combustion chamber. *See also* TURNDOWN RATIO.

firing stroke *See* POWER STROKE.

first-angle projection *See* PROJECTION.

first law of motion *See* NEWTON'S LAWS OF MOTION.

first law of thermodynamics A statement of the principle of conservation of energy. For a closed system, the change in all forms of energy stored in the system ΔE is given by $\Delta E = Q - W$, where Q is the net addition of energy to the system in the form of heat and W is the net reduction of energy from the system in the form of work. For an open system the law is expressed in terms of energy and work flow rates, and account must also be taken of energy associated with mass crossing the control-volume boundary.

first-level controller The low-level controller for one of the sub-systems when a complicated system is split into sub-systems by plant decomposition.

first normal-stress difference (N_1) (Unit Pa) The difference, for a viscoelastic fluid, between the normal stresses in the x- and y-directions produced by a shear stress τ_{xy}. It is invariably larger than the second normal-stress difference.

first-order difference *See* FINITE-DIFFERENCE METHOD.

first-order system A dynamic system fully described by first-order ordinary differential equations.

first-order transition A discontinuous phase transition, such as melting, in which heat is liberated or absorbed. *See also* SECOND-ORDER TRANSITION.

first tap *See* TAP.

first-yield moment (M_e) (Unit N.m) The bending moment at which the maximum elastic stress in a component just reaches the yield stress σ_{yield}. For a rectangular beam of width $2b$ and depth $2h$, $M_e = 4bh^2\sigma_{yield}/3$. *See also* SHAPE FACTOR.

fir-tree root The serrated wedge-shaped end of a turbine blade used to fix it to a slot of similar design in the shaft.

fishtail burner (flat burner) A burner in which gas is injected through a narrow slot or a line of closely-spaced ports. The flat flame produced may be parallel if the ports are parallel to each other or divergent if they are at angles causing the flame to fan out.

fission *See* NUCLEAR FISSION.

fit *See* LIMITS AND FITS.

fitted bolt A bolt with a plain unthreaded portion immediately beneath the head. *See also* MACHINE SCREW; SHOULDER BOLT.

fittings The small, often standard, parts of a machine or mechanism. *See also* PIPE FITTINGS.

fixed-active tooling A term for powered stationary accessories in a robot system such as a parts feeder or conveyor.

fixed-ended beam *See* FIXITY.

fixed error *See* ERROR (2).

fixed-passive tooling Unpowered stationary accessories, such as jigs and fixtures, in a robot system.

fixed-pitch propeller A propeller for aircraft, marine, or wind-turbine applications that has been optimized for optimum efficiency, low vibration, etc. and cannot be adjusted during operation. *See also* CONTROLLABLE-PITCH PROPELLER.

fixed plug In the manufacture of seamless tubing, a conical plug fixed in position within the drawing die which controls the internal diameter of the pipe. *See also* FLOATING PLUG.

fixed point (fiducial temperature) A temperature assigned to a reproducible equilibrium state on the International Practical Temperature Scale.

fixed-stop robot A primitive robot where each joint moves between mechanical limit stops without precise control between them. The number of positions at which the robot can stop is mechanically determined and cannot be changed by a program.

fixed wave-energy converter A wave-energy system fixed rigidly either to the shore or the seabed. *See also* FLOATING WAVE-ENERGY CONVERTER.

fixing moment The bending moment required at some location in a beam (usually the end) in order to maintain a specified deflexion and rotation (i.e. deflexion slope).

fixity The different ways in which beams or other structural members are kept in position, idealized as **pinned** (having rotation and deflection), **simply-supported** (rotation but no deflection) and **built-in (encastré, fix ended)** (neither rotation nor deflection).

fixture A device for holding a workpiece in a machine tool for mass production of a component. *See also* JIG.

flame A region of hot gas in which mixing and combustion occurs from a solid, liquid, or gaseous fuel. Most flames are visible with a colour dependent upon temperature and chemical composition. The flames from pure methanol and ethanol are invisible.

flame arrester (flame trap) A device that prevents the spread of fire by forcing it through small holes such as in a wire mesh or perforated sheet. The burning gas is cooled by surface contact to the point where the flame is extinguished. An important application is to fuel-supply lines.

flame-ionization detector *See* FLUE-GAS ANALYSER.

flame speed The speed with which a flame propagates through a gaseous fuel. The speed is higher in a turbulent flame than in a laminar flame.

flame spraying A thermal-spraying process in which a consumable, typically a powder or a wire, is melted and propelled onto a substrate to form a coating. The consumables are fed into a jet of gaseous fuel such as acetylene or propane, using air, argon, or nitrogen as the carrier fluid. *See also* ARC SPRAYING; PLASMA SPRAYING.

flame stabilization The creation of a region in a combustion chamber where the local flow speeds are lower than the flame speed, typically a region of recirculating flow. *See also* SWIRL BURNER.

flame trap *See* FLAME ARRESTER.

flange 1. Annular rims at the ends of pipes (**flanged pipe**) or shafts by which they may be coupled together using bolts that pass through holes in the flanges (**flange coupling, flange union**), or by toggle clamps. **2.** An extended rim on a wheel that positions it laterally on a track. **3.** The top and bottom parts of an I-beam. *See also* WEB.

flank 1. The side of a screw thread. **2.** The curved outlines of gear teeth. **3.** The side of a cutting tool that may contact the workpiece when the cutting edge is rounded or profiled.

flank angle 1. The angle made in an axial plane between the flanks of a screw thread and the perpendicular to the axis of the thread. **2.** The angle (part of the tool signature) of the flank of a cutting tool.

flap valve A check valve installed at a pipe outlet consisting of a circular disk hinged at the top, often with a backward-inclined seat to ensure positive closure under gravity. Applications include sewage treatment, water treatment, and water outfalls. The valve body and flap are usually cast iron and the seat bronze, leather, or elastomer.

flare burner A burner consisting essentially of a vertical tube through which flow combustible waste gases, such as the unwanted by-products of refining or other processes, including waste gas from landfill, that mix with air at the tube outlet where they are ignited. The burner may be shrouded by a cylindrical outer tube. The flame at outlet is termed a **flare**.

flared tube A circular tube that increases in diameter at one or both ends.

flash (fin) In drop-forged and moulded components, excess material that has been squeezed into thin fins at the join of closed dies.

flash boiler A boiler in which water flows through tubes at such a high temperature that superheated steam is produced almost immediately without further heat addition.

flashing (flash boiling) The process of producing flash steam.

flash point The lowest temperature at which a volatile liquid can vaporize to produce an ignitable mixture in air. *See also* AUTOIGNITION TEMPERATURE.

flash steam If water at high pressure is allowed to drop to a low pressure, flash steam is produced if the initial temperature is higher than the saturation temperature corresponding to the low pressure. It is commonly produced in a power plant from hot condensate.

flash-steam power plant A steam-power plant supplied with high-pressure brine from a geothermal source at a temperature above about 180°C. The brine is sprayed into a tank at low pressure where it rapidly vaporizes (flashes) before entering a steam turbine. Condensed steam and waste brine from the flash tank are reinjected into the geothermal reservoir. *See also* DRY-STEAM POWER PLANT.

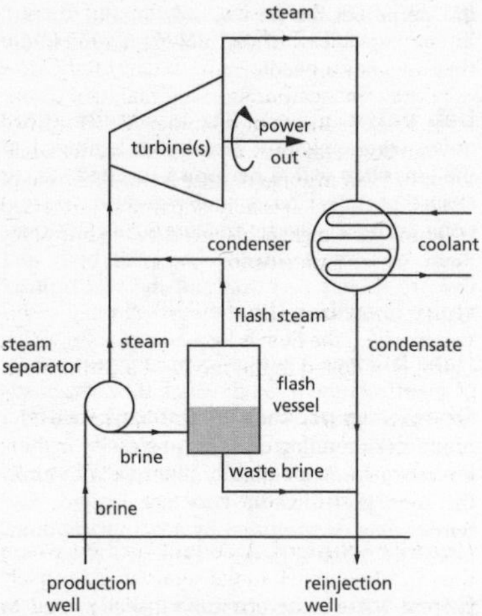

flash-steam power plant

flat belt A belt of rectangular or trapezoidal cross section used to transmit power between pulleys. *See also* TOOTHED BELT.

flat-blade turbine impeller A mixing device suitable for high-viscosity liquids and slurries and consisting of a number of flat radial blades attached to a central driveshaft. The plane of each blade may be parallel to the shaft axis or at an angle to it, often 45°. *See also* PITCHED-BLADE TURBINE IMPELLER; RUSHTON IMPELLER.

flat burner *See* FISHTAIL BURNER.

flat crank (flat crankshaft) A crankshaft in which the throws are 180° apart. Commonly used for high-performance V8 petrol engines.

flat engine *See* VEE ENGINE.

flat-flamed burner *See* FISHTAIL BURNER.

flat-foot follower *See* MUSHROOM FOLLOWER.

flatness 1. *See* SURFACE ROUGHNESS. **2. (flatness factor)** *See* KURTOSIS.

flat-plate air collector *See* SOLAR COLLECTOR.

flat-plate boundary layer (flat-plate flow) A boundary layer that develops in the absence of a streamwise pressure gradient, as it would on a flat surface in a flow of unlimited extent. *See also* BLASIUS PROBLEM.

flat-plate solar collector *See* SOLAR COLLECTOR.

flat-plate water collector *See* SOLAR COLLECTOR.

flat spring A spring in the form of a cantilever.

flaw 1. A defect, imperfection, or blemish in a manufactured component **2.** A visible or hidden crack, void, or discontinuity in a solid material. *See also* FRACTURE MECHANICS; SONIC FLAW DETECTION.

flaw detection Identification of the location and size of a flaw by means of dye penetrant, ultrasonic, or other methods. *See also* SONIC FLAW DETECTION.

FLD *See* FORMING LIMIT DIAGRAM.

Flettner rotor A vertical circular cylinder rotating about its own axis that generates a horizontal lift force perpendicular to a flow of fluid transverse to the cylinder as a consequence of the Magnus effect.

Flettner ventilator A ventilator in which an extraction fan is driven by a Savonius wind turbine, commonly employed on caravans, buses, buildings, etc.

flexibility *See* COMPLIANCE.

flexible coupling A coupling that connects two misaligned shafts where the connexion is made using deformable materials such as rubber or steel springs, as opposed to the stiff metal tongues of an Oldham coupling.

flexible manufacturing system (FMS) A manufacturing system that is not dedicated to the production of a single component, as in mass production, but is capable of producing a variety of components through computer control of a series of linked machine tools. FMS is particularly useful for the production of batches in relatively small numbers, say 50 to 500.

flexible shaft A drive shaft that is capable of transmitting rotary motion around corners. *See also* BOWDEN CABLE.

flexible tooling **1.** A term for common magazines for holding tools that may be used on a number of machine tools. **2.** Fixtures for a machining centre which are capable of positioning and securing a variety of components. **3.** A re-usable tool bed in the forming of composites, in which the shape and curvature of the mould may be altered.

flexometer A device for applying a cyclic load to rubber and elastomeric materials and measuring the resulting temperature increase which gives information about heat generation, anisotropy, time of cure, softening, and stiffening due to flexing.

flexural axis The axis about which a beam or plate is bent.

flexural centre *See* SHEAR CENTRE.

flexural compliance The inverse of flexible stiffness.

flexural rigidity (flexural modulus) (Unit Pa.m^4) In elastic bending, the product EI where E is Young's modulus and I is the second moment of area about the neutral axis in the plane of bending.

flexural shear The shear stresses and strains that exist in bent bodies is caused by non-uniform moments. The shear stresses and strains are zero on the top and bottom surfaces in the plane of bending and are a maximum at the neutral axis. For a rectangular beam of depth h and width b, the distribution of shear stress is parabolic with maximum shear stress given by $3V/2bh$ where V is the shear force. The maximum shear stress is thus 50% greater than the average shear stress. Shear in bending is important in beams of short span.

flexural stiffness (Unit N/m) The ratio of the force on a beam in bending to the resulting deflexion under the load, in particular to the tip deflexion of a loaded cantilever.

flexural strength (modulus of rupture) (Unit Pa) The bending stress at which a beam made of a brittle material fractures. It does not have an absolute value, as it depends on the testpiece and inherent flaws, as explained in fracture mechanics.

flexure *See* BENDING.

flexure theory *See* BEAM THEORY.

flitched beam *See* SANDWICH BEAM.

flitched girder *See* SANDWICH BEAM.

float **1.** An object partially submerged in a liquid as a consequence of a lower average density, used e.g. to act as a float (a **float gauge**) that indicates the position of the surface of a liquid in a tank. **2. (float valve)** A component that operates a needle valve to control the flow of petrol into a carburettor and maintain a constant level in the float chamber. **3. (floating)** The situation whereby an object, such as a balloon, with an average density lower than that of the fluid in which it is fully or partially immersed rises until its weight is balanced by the buoyancy force. *See also* ARCHIMEDES PRINCIPLE.

float chamber (float bowl) A small reservoir in which the float is located and from which petrol is supplied to the throat of a carburettor.

float-glass process (Pilkington process) A method of manufacturing sheet glass by floating molten glass on a bath of molten metal, usually tin. *See also* FOURCAULT PROCESS.

floating control A control method where the rate of approach to the set point is inversely proportional to the error. Particularly used to describe hydraulic and pneumatic valves that can vary the rate of fluid flow rather than simply switching between open and closed.

floating element A small section of a surface, over which there is a fluid flow, isolated from the surrounding surface by a narrow gap and attached to a sensitive load cell or other electrical balance below the surface which measures the tangential force on the element from which the surface shear stress is estimated. A similar technique is employed to determine interfacial tractions between die and metal flow in plasticity.

f

floating-head heat exchanger A shell-and-tube heat exchanger in which the tubesheet that supports the tube bundle is free to move within the shell and so minimize thermal stresses within the tubes.

floating plug A plug used in the manufacture of seamless tubing: both the wall thickness and diameter of the tube are reduced by drawing through a die over either a mandrel or a plug. Plugs floating within the die gap control the inside diameter and remain in position by equilibrium between forces normal to the surface of the cone and the frictional drag along the surface of the cone. *See also* FIXED PLUG.

floating shoes In a drum-brake arrangement, brake shoes that are machined to match precisely the internal radius of the drum and pivoted in such a way that they are free to contact the drum over their entire length.

floating wave-energy converter An offshore wave-energy converter freely floating on the ocean surface but held in position by mooring lines. *See also* FIXED WAVE-ENERGY CONVERTER.

float switch A switch actuated by the rise and fall of a float on the surface of a liquid.

float valve *See* FLOAT (2).

flood To supply so much fuel to a carburettor that it cannot form a combustible mixture with the air available.

flood generation *See* TIDAL ENERGY.

flow 1. The movement of a fluid, such as through pipes or a machine or around an object or of one fluid within another. *See also* FLUID MECHANICS. **2.** The permanent change of shape imparted to a workpiece by plastic deformation, as in rolling, extrusion, etc. *See also* PLASTIC FLOW. **3.** The continuous or intermittent movement of solid material, usually in a specified direction, such as coal or powder on a conveyor belt.

flow chart (flow diagram) A graphical representation using standard symbols of the steps in a process. The technique may be applied to a wide range of processes including major engineering projects, manufacturing processes, chemical-production processes, and computer programs.

flow coefficient 1. For a pump, turbine, compressor, or fan, a non-dimensional measure of volumetric flow rate $\dot{Q}$ defined by $\dot{Q}/\omega D^3$ where ω is the angular rotation speed and D is a characteristic diameter. **2.** For a Bernoulli obstruction flow meter, a factor introduced into the theoretical equation for the flow rate to account for losses.

flow-control valve A valve that regulates the rate or pressure of fluid flow through a pipe system or out of a pressure vessel.

flow curve 1. The true stress–true strain curve of a plastically-deforming solid. **2. (rheogram)** For a non-Newtonian liquid, a graph of shear stress or apparent viscosity *vs* shear rate.

flow energy (convected energy, flow work, transport energy) The work required to cause mass to flow across the boundary of a control volume. Per unit mass it is given by pv where p is the fluid pressure and v is its specific volume. *See also* SPECIFIC PROPERTY.

flow-induced vibration A vibration caused by unsteady, especially periodic, motion of fluid flowing through or around a structure, machine or device. *See also* VORTEX SHEDDING.

flowline The path taken by an element of metal in plastic flow, sometimes being called a streamtube in axisymmetric problems.

flow measurement 1. The process of determining the mass or volumetric flow rate for fluid flow in a pipe, duct, or channel using a **flow meter (flowmeter)**. **2.** The process of determining velocity using instrumentation such as hot-wire and laser-Doppler anemometers and particle-image velocimeters.

flow mixer *See* IN-LINE MIXER.

flow nozzle *See* BERNOULLI OBSTRUCTION FLOW METER.

flow rules The relations, in plasticity theory, between increments of true strain and the current true stresses. There is a different flow rule associated with every yield criterion. For the isotropic von Mises yield criterion, $d\varepsilon_1 = [\sigma_1 - (\sigma_2 + \sigma_3)/2]d\bar{\varepsilon}/\bar{\sigma}$ with corresponding expressions for $d\varepsilon_2$ and $d\varepsilon_3$ by altering the suffices, where $\bar{\sigma}$ is the effective stress and $d\bar{\varepsilon}$ is the effective strain increment, σ_1, σ_2 and σ_3 are the principal stresses, and $d\varepsilon_1$, $d\varepsilon_2$ and $d\varepsilon_3$ are the principal strain increments. In general,

$$d\varepsilon_{ij} = G\left(\frac{\partial f}{\partial \sigma_{ij}}\right)df$$

where f is the yield criterion and G is a scalar that may depend on stress, strain, and history. In total-strain (**Hencky**) plasticity theory, incremental strains $d\varepsilon_1$ etc. are replaced by total strains ε_1 etc. *See also* LÉVY–MISES EQUATIONS; PRANDTL–REUSS EQUATIONS.

flow stress The true stress during plastic deformation.

flow transmitter A system that incorporates a flow meter to measure flow rates for fluid flow through a pipe and provide the output in the form of an electrical signal.

flow variable Any physical parameter that either varies in a fluid flow or influences it, such as velocity, vorticity, shear stress, turbulence intensity, pressure, density, temperature, and enthalpy.

flow visualization Any technique that allows the features occurring in a flow to become visible. In fluid flow, the introduction of smoke into a gas flow, dye or particles into a liquid flow is employed. There are also optical techniques such as shadowgraphy, Schlieren and interferometry that allow density variations in a flow to be made visible. In plastic flow, patterns of grids or circles are marked either on the interior surfaces of split testpieces before deformation, or on the exterior surfaces of sheets, and their distortion determined after flow by direct measurement or using image recognition software. Speckle interferometry is also employed with sheets.

flow work *See* FLOW ENERGY.

fluctuating stresses The shear stresses that arise in unsteady fluid flow, especially turbulent flow, as a consequence of momentum transfer due to velocity fluctuations in orthogonal directions. There are also normal stresses due to fluctuations in a given direction. *See also* FATIGUE.

flue A passage through which mainly gaseous products of combustion, also including any excess oxygen and particulates, from a furnace, oven, or fireplace (**flue gas**), flow to a chimney or pollution-treatment system. A fan (**flue exhauster**) may be used to enhance the gas flow.

flue-gas analyser An instrument used to monitor the composition of the flue gas from a furnace to check whether emissions, including unburned hydrocarbons (UHC), carbon monoxide (CO), oxides of nitrogen (NO_x), sulphur dioxide, and hydrogen sulphide, are within allowable limits, and to analyse furnace efficiency. The analysis methods used include: a flame-ionization detector (UHC); a non-dispersive infra-red detector (carbon dioxide and CO); a chemiluminescent analyser (NO_x). *See also* EXHAUST-GAS ANALYSER.

flue-gas expander (hot-gas expander) A gas turbine used to recover energy from hot, often high-pressure, exhaust or flue gas.

flue-gas stack *See* SMOKE STACK.

flueric element A term sometimes used for a dynamic fluidic device.

fluid A continuous material that cannot support a shear stress when at rest but undergoes unlimited deformation without fracture. Some fluids exhibit a yield stress and so have the characteristics of a solid material at shear stresses below that level. Gases, vapours, liquids, and slurries are all fluids. The physical property that distinguishes any real fluid from a solid is viscosity, the distinction, in materials science, being drawn conventionally at a value of 10^{14} Pa.s. *See also* BINGHAM PLASTIC; DYNAMIC VISCOSITY; IDEAL FLUID; NON-NEWTONIAN FLUID.

fluid amplifier *See* FLUIDICS.

fluid capacitor (fluidic capacitor) A device in which pressure energy can be stored in a compressed fluid. Similar to an accumulator, but usually on a smaller scale.

fluid-controlled valve A valve actuated by pneumatic or hydraulic pressure.

fluid coupling (fluid clutch, fluid drive, fluid flywheel, hydraulic clutch, hydraulic coupling, hydraulic drive, turbo coupling) A type of clutch used in motor- and marine-vehicle drive trains consisting of an impeller driven by the engine and a vaned output rotor coupled to the gearbox, the impeller/rotor pair being in close proximity and enclosed by a casing filled with oil. Torque is transmitted by circulation of the oil between the impeller/rotor pair.

fluid dynamics *See* FLUID MECHANICS.

fluid end The parts of a pump in direct contact with high-pressure fluid, including cylinders, liners, pistons, and valves. An example is the mud pumps used in the drilling of oil and gas wells.

f

fluid-expansion thermometer (filled-system thermometer) An instrument for temperature measurement comprising a fluid-filled bulb connected by a capillary tube to a device that changes shape due to expansion of the fluid caused by a temperature increase, such as a Bourdon tube, diaphragm, or bellows. The fluid may be a liquid, a liquid plus its vapour, or a gas.

fluid-film lubrication *See* HYDRODYNAMIC LUBRICATION.

fluid flywheel *See* FLUID COUPLING.

fluidics (fluidic logic) The use of flow phenomena, such as jet impingement, separation, reattachment, entrainment, the Coanda effect, and vortex motion, to perform analogue or digital operations, such as amplification, rectification, and resistance. **Fluidic devices** are devices, widely used in nuclear applications, with no moving parts that use small pressure differences or a small proportion of the total flow (**control flow**) to control the overall flow, for example by switching a bistable flow from one side to another (**wall-attachment amplifiers**) or by causing a flow to swirl (**vortex amplifiers** or vortex diodes). A **fluidic amplifier** amplifies the difference in pressure across its control ports and increases the flow rate. In a wall-attachment amplifier symmetrical flow through a divergent channel is caused to attach to one of the walls by a control flow applied on one side. A device consisting of a fluidic amplifier with negative feedback that produces an output frequency proportional to flow rate can be used as a **fluidic flow meter (fluidic oscillator meter)**. In a **fluidic diode** there is either much higher resistance to flow in one direction than the other or high resistance is created by the control flow, e.g. by causing the flow to swirl as in a **vortex diode** which consists of a flat circular chamber with the supply flow entering radially at the periphery and the control flow entering tangentially. In the case of a **fluidic cascade diode (fluid-flow rectifier, fluidic rectifier)**, flow in the high-resistance direction is progressively caused to swirl by a series of stators of increasing angle of attack. Pressure ratios approaching 200 are possible. The term **fluidic rectifier** is also used to mean a fluidic device that reduces the fluctuation levels in a periodic flow. Any fluidic device designed such that an increase in back pressure caused by partial or total blockage of an outlet can be used as a proximity sensor **(fluidic sensor)**. The term **fluid network** is a combination of fluidic devices represented by capacitance, resistance, and inductance elements.

fluidity (ϕ) (Unit 1/Pa.s) The reciprocal of dynamic viscosity. A measure of the tendency of a fluid to flow.

fluidization The process created by a fluid flowing upwards through a column of solid particulate material at a sufficiently high rate that the drag force on some or all of the particles is sufficient to overcome their weight. If the resulting **fluidized bed** consists of particles of different densities, the lighter particles will rise to the surface and the heaviest sink to the bottom. The fluid velocity when the bed just becomes fluidized is called the **minimum fluidization velocity**. *See also* AGGREGATIVE FLUIDIZATION; PARTICULATE FLUIDIZATION.

fluidized-bed combustion A combustion technology in which solid fuel particles form a fluidized bed through which air is blown. Limestone or dolomite is added to the fuel to absorb sulphur oxides. At atmospheric pressure the fuel burns at temperatures too low (below about 900°C) for oxides of nitrogen to be formed. Such systems may be used to heat steam boilers. *See also* PRESSURIZED FLUIDIZED-BED COMBUSTION.

fluid-logic system A system of connected elements that utilize orifices, diaphragms, valves, etc. to control the flow in a pneumatic or hydraulic circuit.

fluid mechanics The study of fluids in motion (**fluid dynamics**) or **fluid statics** where there is no relative motion between fluid particles. Fluid statics concerns primarily the variation of pressure with altitude or depth; it includes aerostatics and hydrostatics. Fluid dynamics includes the topics of aerodynamics, gas dynamics, hydraulics, hydrodynamics and many aspects of acoustics, chemical engineering, flight, lubrication, meteorology, non-Newtonian fluid flow, oceanography, power-plant technology, propulsion, and turbomachinery. It involves the application of the laws of mass, momentum, and energy conservation. In general there is a balance between shear stresses, pressure gradients, and inertia represented for a Newtonian fluid by the Navier–Stokes equations, a set of non-linear partial differential equations. The general situation includes fluid properties that depend upon pressure and temperature, imposed unsteadiness or turbulent flow that arises from instabilities. Exact analytical solutions are limited to simple flows such as fully-developed, steady, laminar pipe flow of a constant-property fluid (Poiseuille flow). More generally, even for relatively simple steady, laminar flow, numerical methods must be resorted to. For example, the situation of a developing laminar boundary layer in the absence of a free-stream pressure gradient (Blasius problem) requires numerical solution. Full numerical solution of turbulent-flow problems (DNS) is beyond the scope of any existing computer except for quite low Reynolds numbers, and it is necessary to resort to sophisticated turbulence modelling, in which many terms in the transport equations for turbulent kinetic energy, dissipation, etc. are modelled empirically. Many practical problems are dealt with by making simplifications that allow either analytical or numerical solution. Examples include one-dimensional analysis of compressible gas flow, including shock waves, and laminar boundary-layer theory including pressure-gradient effects. Extensive use is made of dimensional analysis to identify the non-dimensional parameters and variables that characterize a flow and enable analytical or numerical calculations and experimental data to be presented in the most compact form possible. The Reynolds, Mach, and Froude numbers arise most frequently together with the Stanton, Nusselt, and Prandtl numbers for problems involving convective heat transfer. The complexity of fluid-flow analysis is significantly greater for non-Newtonian liquids.

fluid meter *See* FLOW METER.

fluid network *See* FLUIDICS.

fluid particle (material particle) A mass of fluid that can be treated as having infinitesimally small volume, but which contains a sufficient number of molecules for the continuum hypothesis to apply.

fluid pressure *See* PRESSURE.

fluid spring A spring that uses the compressibility of a gas to provide compliance, typically in a piston–cylinder arrangement.

fluid statics *See* FLUID MECHANICS.

fluid stress The stresses that arise in fluid flow are pressure (a normal stress) and shear. *See also* STRESS (2).

flume 1. An open channel, typically rectangular in cross section, for conveying water, other liquids or slurries. *See also* PARSHALL FLUME; VENTURI FLUME. **2.** A recirculating water tunnel with an open section used to study the flow around totally or partially submerged bodies, including marine vehicles and swimmers as well as aircraft, structures, etc. *See also* TOWING TANK.

flush 1. Said of independent adjacent surfaces that are at the same level. **2.** *See* PURGE.

flute A helical groove such as on the inside of a rifle barrel or the outside of a twist drill.

flutter A form of flow-induced vibration in which energy from a flow causes a contiguous elastic structure to vibrate at one of its natural frequencies, resulting in vibration amplitudes that increase the aerodynamic forces causing vibration, until a point is reached where either the structure fails or there is sufficient damping to limit the amplitude. Flutter can affect aircraft wings, chimneys, bridges, valves, etc. *See also* CHATTER.

flutter valve *See* REED VALVE.

flux 1. *See* BRAZING; SOLDERING; WELDING. **2.** In fluid mechanics and heat transfer, the quantity of a substance or property transported per unit area. *See also* DARCY FLUX; HEAT FLUX; MASS FLUX; MOMENTUM FLUX; RADIANT FLUX.

flux density *See* RADIANT FLUX DENSITY.

fly ash Fine, glass-like, particulate matter in the flue gas resulting from the combustion of ground or powdered coal and other solid fuels.

flycutter 1. A milling cutter having only one tooth employed to finish-cut soft materials, fine surfaces being produced as swarf does not get trapped under the tool. **2.** A revolving arm, attached to the end of which is a cutting tool, employed to cut holes of large diameter in sheet material. *See also* TREPANNING TOOL.

flypress A press actuated by a vertical square screw thread, to the bottom of which is attached a punch for stamping out shapes in thin sheets of metal, and at the top of which are attached two heavy balls at the ends of an arm to give a large moment of inertia and hence large energy-storage capacity. It is also used for press fits.

flywheel A device used to store energy and to smooth out motion that is either (a) intermittently driven (e.g. a single cylinder engine) or (b) has energy removed periodically (e.g. a crank press). It is typically a heavy rotating wheel having a large moment of inertia.

FMS *See* FLEXIBLE MANUFACTURING SYSTEM.

foam 1. (froth) A concentrated dispersion of gas bubbles in a detergent or other surfactant solution. Applications include mineral separation from extracted ore (froth flotation), drilling fluids in oil production, and agents to fight oil fires. **2.** A type of cellular polymer or rubber material containing many bubbles, that may be isolated (closed-cell foam) or interconnected (open-cell foam). 'Flexible foams' are deformable; 'rigid foams' are stiff. *See also* SPONGE.

focusing collector *See* PARABOLIC-BOWL CONCENTRATOR; PARABOLIC MIRROR; SOLAR CONCENTRATOR.

foil Thin sheets of metal, such as aluminium, copper, brass, bronze, silver, gold, stainless steel, and nickel (sometimes metal-coated polymer), produced by rolling and employed to make shims and spacers. Other uses include thermal insulation, packaging, shielding against electromagnetic radiation, and electrical components. Some foils (e.g. aluminium kitchen foil) are bright on one side and matt on the other, resulting from cold-rolling two sheets simultaneously, only the surfaces in contact with the rolls being burnished.

follower *See* CAM FOLLOWER.

following error (follower error) The error in a control system when attempting to follow a varying set point.

foot A non-SI, imperial unit of length, now defined as 304.8 mm.

foot valve A form of check valve, often incorporating a strainer, installed at the bottom of a pump suction line to prevent backflow and aid priming.

force (*F*) (Unit N) A force is a push (**compressive force**) or pull (**tensile force**) applied to a body or system of bodies. It is a vector quantity requiring magnitude, direction and point of application for its complete specification. **Statics** is the study of forces in equilibrium applied to a body or system which may cause changes in its shape or size during a transient to a new state of equilibrium. An unknown force may be measured in terms of the final extension it produces in an elastic body, such as a spring (at which extension the load in the spring is in equilibrium with the applied force) compared with that produced by a known force. An unbalanced force applied to a body will move it or tend to move it; if applied to a body already moving, the force will change its motion. *See also* APPLIED MECHANICS; NEWTON'S LAWS OF MOTION.

force-balance meter *See* DIRECT IN-LINE VARIABLE-AREA FLOW METER.

force coefficient (C_F) The ratio, for a wind turbine, of the actual thrust F_A acting on the vanes to the thrust that would be exerted on a disc of the same diameter D as the blading by the dynamic pressure of the wind q, i.e. $C_F = 4F_A/q\pi D^2$.

force-controlled motion The control of a robot so as to apply a controlled force through the end effector rather than a controlled displacement.

forced-air heating A system in which air heated in a furnace is circulated by a fan through ductwork which distributes it to a building.

forced-circulation boiler A water-tube boiler in which the water is circulated by a pump rather than by natural circulation due to density differences.

forced-circulation evaporator A system in which a liquid is pumped through a heat exchanger, where it is heated but does not boil, and then through a flash vessel, where evaporation occurs. It is used for the evaporation of liquids with a high solids content, high viscosity, tendency to foul, and for crystalizers, or as a finishing evaporator for the concentration of a liquid.

forced convection *See* HEAT TRANSFER.

forced-convection boiling (forced-flow boiling) *See* BOILING.

forced draught *See* DRAUGHT.

forced-draught cooling tower A mechanical-draught cooling tower with a fan at the intake to force air through the tower. *See also* INDUCED-DRAUGHT COOLING TOWER; NATURAL-DRAUGHT COOLING TOWER.

forced induction The process of forcing air at a pressure above ambient into the cylinders of an internal-combustion engine using a turbocharger or supercharger. *See also* NATURAL ASPIRATION.

forced lubrication The supply of oil under pressure from a pump to lubricate bearings in engines and other machinery.

forced vibration (forced oscillation, forced response, harmonic response) To maintain steady oscillating motion of a system in the presence of damping, it must be 'forced', i.e. a periodic disturbing influence must be applied. Where the system is linear, the response will be of the same frequency as the input but the phase and amplitude will vary according to the frequency. *See also* FREE OSCILLATION.

forced vortex *See* RANKINE VORTEX.

force feedback Feedback in which the signal is the force applied by the plant. Used, for example, in force-controlled motion.

force fit *See* LIMITS AND FITS.

force polygon *See* FUNICULAR POLYGON.

force pump A reciprocating pump used to pump liquids. Depending upon the valve arrangements, it may be single- or double-acting.

force ratio *See* MECHANICAL ADVANTAGE.

force transducer (force gauge) Any device for measuring an applied force. Modern transducers are usually based upon a load cell.

Forchheimer flow *See* DARCY'S LAW.

forge The workplace, including fire, furnace, and equipment, in which forging is carried out.

forge welding *See* WELDING.

forging 1. A method of component manufacture by hammering metal by hand or by machine (drop forge, press). **2.** A part made by the process of forging.

form cutter *See* TOOTHED GEARING.

form drag *See* DRAG.

form generation *See* TOOTHED GEARING.

forming The process of shaping materials (particularly metals) into separate items such as beverage cans, or the manufacture of feedstock (such as rolled sheet) for further processing. *See also* COLD WORKING; HOT WORKING; WARM WORKING.

forming limit diagram (FLD) In sheet-metal forming, a diagram of major and minor in-plane strains during biaxial loading on which is plotted the locus of experimental strains at which localized necking occurs. Strains are determined from pre-marked circles on the surface of the sheet. Used for tool and die design. *See also* FRACTURE-FORMING LIMIT DIAGRAM.

Fortin barometer A laboratory-standard mercury-in-glass barometer. The mercury reservoir is a flexible pouch beneath which is a screw jack that adjusts the level of the mercury to a fixed point. The height of the mercury in the glass tube is read from a scale using a vernier.

forward extrusion *See* DIRECT EXTRUSION.

forward kinematics The process of determining the position and orientation of a robot end effector from the positions and angles of the joints. It is the opposite of the more useful process of inverse kinematics which determines the joint angles and positions required to achieve a specified end-effector position and orientation.

forward path The path through a control system from the set point input to the plant output, omitting any feedback paths.

forward transfer function *See* OPEN-LOOP TRANSFER FUNCTION.

fossil fuel A naturally-occurring fuel, such as gas, oil, or coal, formed by the decomposition of organisms at great depth below the earth's surface.

Föttinger coupling The original name for a turbo-coupling. *See also* FLUID COUPLING.

fouling factor (fouling resistance, *R*) (Unit m^2K/kW) An empirical factor used to account for the additional thermal resistance to heat transfer caused by deposits on, or corrosion of, a heat-transfer surface.

foundry A place where metals are melted and cast into moulds to manufacture intricately-shaped components.

four-ball tester An instrument, in which one ball is rotated under load against three stationary balls, to assess the performance of lubricants as indicated by the extent of wear scars on the balls.

four-bar linkage (four-bar chain) Four links pinned together to form a trapezium, the motion of all corners of which follows from the motion of adjoining elements. In many applications, one element is fixed, an adjoining link is a crank, and the other arms form part of a mechanism that performs repetitive motion. *See also* KINEMATIC CHAIN; KINEMATIC PAIR.

Fourcault process A method for manufacturing flat glass by drawing a sheet vertically upwards from a pit of molten glass. Now largely superseded by the float-glass process.

four-cylinder engine A piston engine with four cylinders, either in an in-line, vee, or boxer configuration.

475°C embrittlement A change that occurs in ferritic stainless steels when they are held in the temperature range 450–550°C, resulting in a severe loss of toughness and an increase in hardness. Such embrittlement is avoided by rapid cooling after annealing.

Fourier analyser A spectrum analyser that determines the frequency spectrum by applying a fast Fourier transform to the input signal.

Fourier number (Fourier modulus, *F*, *Fo*) *See* HEAT TRANSFER.

Fourier's law of heat conduction (Fourier's heat equation) *See* HEAT TRANSFER.

Fourier transform A mathematical process for the decomposition of a time-domain function or signal into its component frequencies so that the amplitude and phase of each frequency is determined. The process can thus be used to produce the frequency spectrum of a signal with applications such as in a spectrum analyser. *See also* FAST FOURIER TRANSFORM.

four-jaw chuck A chuck, the four jaws of which may be moved independently so as to grip non-circular workpieces.

four-stroke engine An internal-combustion engine operating on a practical four-stroke cycle. For a given cylinder, as shown in the first diagram, each cycle consists of (i) intake of a fuel/air mixture, (ii) compression, ignition and combustion, (iii) expansion, and (iv) exhaust. The dashed lines in the second diagram show the variation of cylinder pressure (p) with cylinder volume ($\mathcal{V}$). For a real spark-ignition engine, the cycle is approximated by the Otto cycle shown by the unbroken lines.

SEE WEB LINKS

- Animation and explanation of the principles of a four-stroke engine

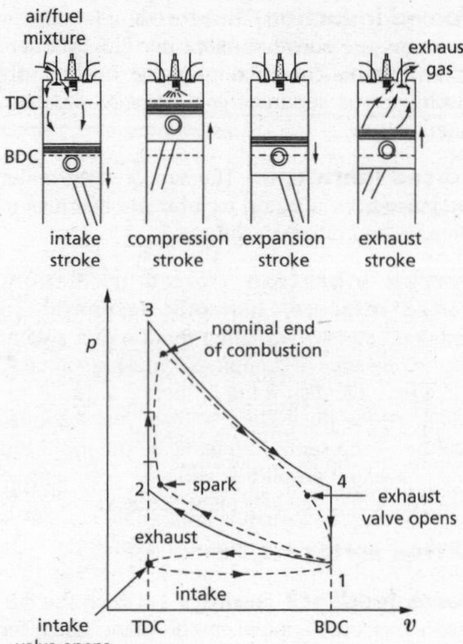

four-stroke engine

four-way valve A two-position flow-control valve, the body of which has four ports A, B, C, and D. In one position, A and B, C and D are connected. In the other A and D, B and C are connected. There is also a fully-closed position.

four-wheel drive A motor vehicle in which power is transmitted to all four wheels, including those steering, usually from a single engine.

fracking *See* HYDROFRACTURING.

fractional pitch *See* SCREW.

fractography The study of the surfaces of broken components in order to establish the type and cause of fracture. *See* CRYSTALLINE FRACTURE; FIBROUS FRACTURE.

fracture (rupture) The separation of materials, components, or structures into two or more parts by the propagation of a crack. Cracking may be globally elastic (brittle) or accompanied by varying degrees of plasticity (ductile).

fracture forming limit diagram (FFLD) A forming limit diagram on which the locus of fracture, as well as necking, strains is plotted.

fracture mechanics The stress analysis of bodies containing cracks. The use of stress-concentration factors and other correction factors of traditional strength of materials is inadequate when flaws are present initially (or develop during loading), since fracture depends not only on stress but also the size of the crack. Traditional stress analysis cannot predict either the safe working stress in the presence of a known flaw, or the critical size of flaw just tolerable with a given working stress.

At the instant of fracture, stresses throughout the body may be smaller than the yield stress σ_y and therefore within the elastic range, so that after fracture the broken parts may be refitted to regain the original size and shape of the cracked body (**brittle fracture**). Alternatively, at fracture the working stresses may exceed the yield stress resulting in regions of plasticity where the body is permanently distorted to varying degrees. In such elastoplastic (**ductile, post-yield**) fracture, it is impossible to refit the broken parts to regain the original size and shape of the fractured component. **Fracture toughness** is the property of a material that determines its resistance to cracking. The basic definition of this rate-, temperature- and environment-dependent mechanical property is the work required to propagate a crack by unit area and given various symbols (R, G_C, J_C), the subscript C meaning 'critical'. Initiation work is performed in the **fracture process zone** at the tip of the crack that, on propagation, forms thin boundary layers contiguous with the crack faces. Values range from 10's of J/m^2 for brittle materials, to kJ/m^2 for ductile materials. An alternative measure of toughness is the critical crack opening displacement (CTOD) with symbol δ_C. Yet another is the critical stress-intensity factor K_C having the peculiar unit N/m$^{3/2}$, i.e. Pa$\sqrt{\text{m}}$. These different ways of defining the same physical property mean that they must be related, and $R = m\sigma_y\delta_C$, where σ_y is the yield stress and $m = 1$ for plane stress and 3 for plane strain; $K_C^2 = ER$ where E is Young's modulus (from which the peculiar unit follows). Note that K_C is also sometimes confusingly called the fracture toughness.

Depending upon conditions, brittle and ductile cracks may variously propagate in tension, in-plane shear and out-of-plane (twisting) shear, such modes of fracture being called by the Roman numerals I, II and III. The fracture toughnesses may differ in magnitude in the different modes and are designated by R_I, R_{II} and R_{III} (similar subscripts are employed with G_C, J_C, K_C and δ_C).

Toughness also depends upon constraint, so that cracks propagating in plane stress (with

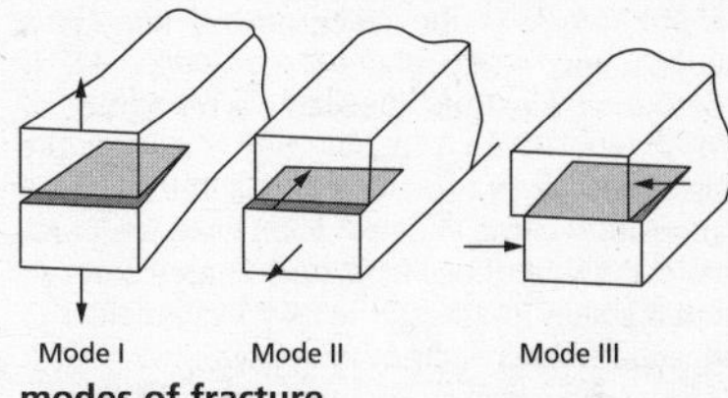

modes of fracture

through-thickness deformation) experience greater resistance than those in plane strain where through-thickness deformation is inhibited and hydrostatic stresses are consequently higher. Values for the plane strain fracture toughness are useful in design. To ensure plane strain deformation, a specimen has to have certain minimum dimensions, in particular that its thickness $B \geq 2.5(K_{IC}/\sigma_y)^2$ where σ_y is the yield strength. Because the plane strain K_{IC} value is not known until after a test is conducted, yet B must be selected beforehand, a preliminary thickness is selected and a tentative toughness value (K_Q) is determined, from which it may be established whether B needs to be thicker.

There are two approaches to the determination of elastic fracture stresses, the one based upon consideration of the complicated stress fields around the tips of cracks and giving rise to K_C; the other based upon energy methods. Both have their parallels in the Theory of Elasticity on the one hand, and Castigliano's theorems on the other. All linear elastic fracture mechanics (LEFM) formulae have the general form $K_C = \sigma_f Y(a/W)\sqrt{\pi a}$ where σ_f is the fracture stress, $Y(a/W)$ is a non-dimensional correction factor to take into account the geometry of different cracked bodies, the way of loading and so on, a is the crack length, and W is a representative dimension of the body.

In elastoplastic fracture mechanics, use is made of the equivalence between non-linear elasticity and total strain (Hencky) plasticity theory to obtain σ_f from the critical value of the J-integral, J_C. While correct for crack initiation, it is not correct for elastoplastic crack propagation during decreasing load, as only part of the work done is recoverable. The use of **J_R crack resistance curves** in ductile fracture mechanics is likely to overestimate the real resistance to propagation and is therefore non-conservative.

High-cycle fatigue crack propagation may be described in terms of fracture mechanics by the Paris equation and its variants: $da/dN = C(\Delta K)^n$

where da/dN is the crack growth per cycle, and C and n are material constants. ΔK is the range of stress intensity factor given by $\Delta K = (\sigma_{max} - \sigma_{min})Y(a/W)\sqrt{\pi a}$ with σ_{max} the biggest stress in a cycle and σ_{min} the smallest. Integration of the Paris equation gives the number of cycles required for a crack to grow from an initial length to a longer length, in particular to the length that results in final failure.

The scale effect in fracture mechanics means that σ_f in large structures can be lower than σ_f in similar small structures, possibly lower than the yield strength of the material, and possibly even smaller than the strength of materials design stress based on the yield stress divided by a factor of safety. Hence there is a danger of brittle fracture of large structures made from materials that behave in a ductile fashion in the laboratory.

Employment of fracture mechanics requires knowledge of crack size (a in the formulae above). In the absence of obvious cracks in brittle materials, a is identified with microstructural features. With ductile materials, cracks are initiated during loading through the formation of voids at inclusions and hard second-phase particles which grow and coalesce. The rate of void growth depends upon hydrostatic stress and plastic strain for which there are various mathematical expressions. In welds, slag inclusions act as starter cracks. *See also* COHESIVE ZONE; CRITICAL CRACK TIP OPENING DISPLACEMENT; DAMAGE MECHANICS; DESIGN METHODOLOGIES; FATIGUE; JOHNSON–COOK FRACTURE EQUATION; KACHANOV'S THEOREM; LEAK-BEFORE-BURST;R6 FAILURE ASSESSMENT DIAGRAM; THRESHOLD STRESS INTENSITY FACTOR.

fracture mechanisms The microstructural mechanisms that cause fracture, such as void initiation, growth and coalescence, cleavage.

fracture strength (fracture stress) The stress at which a material breaks. It is not absolute for a given material, as it depends on the laws of fracture mechanics and is size-dependent. *See also* MODULUS OF RUPTURE.

fracture test **1.** The recording of load-deflexion (or stress–strain) in a specimen loaded monotonically until it breaks in tension, compression, shear, or torsion. **2.** In fatigue, determination of the number of cycles to fracture for a specimen under different patterns of repeated loading. **3.** Experiments to determine fracture toughness.

fracture toughness *See* FRACTURE MECHANICS; SPECIFIC WORK OF FRACTURE.

frame of reference Any set of axes to which points in space, moments in time, etc. may be referred, typified by the Cartesian, polar, cylindrical, and spherical-polar coordinate systems. Frames of reference are usually fixed, but may move. *See also* INERTIAL COORDINATE SYSTEM; ACCELERATING FRAME OF REFERENCE.

framework *See* STRUCTURE; STRUT; TIE.

Francis turbine A high-efficiency (*c.*90%) mixed-flow hydraulic reaction turbine with a spiral inlet scroll (or volute) followed by radial inlet guide vanes and a runner through which the flow turns into the axial direction where it flows to the draft tube and the tailrace.

frangible A qualitative term for ease of fracture.

free-air delivery (*FAD*) (Unit m^3/s) The volumetric flow rate of air $\dot{Q}$ delivered by an air compressor converted to the ambient pressure and temperature, p_A and T_A, using the ideal gas law, i.e. $FAD = p_{ACT}\dot{Q}T_A/p_AT_{ACT}$ where p_{ACT} and T_{ACT} are the actual pressure and temperature of the air at the compressor outlet. The air mass flow rate is given by $\rho_{ACT}\dot{Q} = \rho_A FAD$ where ρ_{ACT} is the actual air density at outlet and ρ_A is the ambient air density.

free-body diagram (FBD) When using force-equilibrium calculations in statics and dynamics, the physical limits of the system under consideration, and the nature of all forces and moments which act upon it, must be identified clearly. Any part of a body may be isolated by means of an imaginary system boundary to give a 'free body', the equilibrium of which is determined solely from the forces and moments acting upon it as shown by the free-body diagram. Application of Newton's third law is particularly important in the correct assignment of internal forces for inclusion on any part-system FBD.

free convection (natural convection) *See* HEAT TRANSFER.

free energy *See* GIBBS FUNCTION.

free enthalpy *See* GIBBS FUNCTION.

free expansion (unresisted expansion) An irreversible process in which a gas expands into an evacuated insulated chamber.

free fall Strictly, the vertically downward motion of an object falling under the influence of gravity, but more frequently used to describe an object falling through the atmosphere. *See also* TERMINAL VELOCITY.

free field In acoustics, an environment free of boundaries so that outgoing acoustic waves are not scattered or reflected.

free flight A term for unpowered flight, such as gliding.

free gyroscope A gyroscope used in aircraft to give an artificial reference horizon during ascent or descent from level flight.

free jet A jet of fluid developing in effectively infinite surroundings.

free-jet amplifier A fluidic amplifier in which a rectangular jet is deflected into downstream collection passages by control nozzles on either side of the jet nozzle.

free joint The joint on a robot arm and wrist that causes the greatest end-effector movement when a specified force is applied to the end effector, i.e. that has the lowest stiffness with regard to the force.

free-molecular flow Gas flow in which the mean-free path between molecular collisions is much larger than a characteristic dimension of a body or flow channel. For such flows the Knudsen number is much greater than unity.

free oscillation *See* FREE VIBRATION.

free outflow The discharge from beneath an underflow gate when the liquid surface immediately downstream is open to the atmosphere. *See also* DROWNED OUTFLOW.

free-piston engine A reciprocating internal-combustion engine in which the pistons move back and forth but are not connected to a crankshaft. Dual piston engines can have separate or common combustion chambers, the pistons being linked for synchronization in the latter case. Such engines are used primarily as gas compressors or gas generators (**free-piston gas generator, gasifier**) to supply hot exhaust gas to drive a gas-turbine.

free-piston gauge *See* DEADWEIGHT TESTER.

free power turbine *See* FREE TURBINE.

free-shear flow A shear flow not in direct contact with a solid boundary.

free shear layer The shear flow that develops between two contacting, parallel fluid streams having different velocities.

free stream In fluid dynamics, the region beyond the outside edge of a boundary layer, jet, wake, or other shear layer.

free-stream turbulence Turbulent fluctuations in the free stream of a boundary layer.

free-stream velocity The flow velocity in the free stream of a shear layer.

free surface 1. The interface between a liquid and a gas, particularly a flowing liquid. **2.** The surface between a supersonic flow, such as a jet, and stagnant surroundings at constant pressure.

free turbine (free power turbine) The turbine in a turboshaft engine that powers the driveshaft, separate from that which powers the compressor. Both extract energy from the same gas stream. *See also* SHAFT TURBINE.

free vibration (free oscillation) A body displaced from its stable position of rest and released will perform free vibrations when acted upon by a restoring force such as that provided by a spring. Under the influence of damping, free vibrations eventually die out. *See also* FORCED VIBRATION.

free vortex *See* RANKINE VORTEX.

freeze drying (lyophilization; vacuum freeze drying) A dehydrating process in which a substance, such as a food product, is deep frozen (typically at −65°C) and then subjected to reduced pressure (below 100 mbar), causing any ice to sublime. The water vapour is then removed.

french coupling *See* GUILLEMIN COUPLING.

frequency (temporal frequency, *f*) (Unit Hz) The number of cycles per second in an oscillation or the repetition rate for a cyclic process.

frequency domain The mathematical representation of a signal or system in terms of the behaviour as a function of frequency, usually obtained through taking Laplace transforms of the time-domain equations. Also used to refer to the representation of a signal through the Fourier transform.

frequency-domain analysis The analysis of a control system using the Laplace transform of its behaviour, e.g. using the transfer function. Also used to refer to analysis using the results obtained from the application of a Fourier transform.

frequency locus The behaviour of the transfer function as a function of frequency. It can be used to determine the describing function for a non-linear system.

frequency response The magnitude (in dB) or phase of the output of a system for a specified input as a function of frequency. It may also be specified as the range of frequencies that are passed by a system within a specified magnitude range. For example, a control system may be specified as having a frequency response of DC to 10 Hz with a specified maximum error. It can be represented as a graph (**frequency-response curve**).

frequency-response function The magnitude and phase of the transfer function of a system as a function of frequency obtained by replacing s with $i\omega$ where s is the Laplace-transform variable, $i = \sqrt{-1}$ and ω is the angular frequency. The frequency-response function is used to produce the Bode plots for a system.

frequency-response trajectory The locus plotted on the complex plane of the frequency-response function as the frequency is changed.

frequency spectrum *See* POWER SPECTRAL DENSITY.

frequency transformation A method for the design of filters based on transforming the behaviour of a low-pass filter into either a high-pass or band-pass characteristic.

frettage The expansion by internal pressure of a thick-walled vessel, such as a large-calibre gun barrel, so that the bore material deforms plastically. On unloading, the inside surfaces are both harder and in a state of residual compression, and therefore may be subjected to higher loads. The same effect produced by shrink-fitting two thick-walled tubes is autofrettage.

fretting The extraneous small-amplitude (<100 μm) oscillatory loading associated with cyclic relative tangential displacement between surfaces notionally in stationary contact.

fretting failure The progressive wear and eventual failure of bodies such as nut-and-bolt assemblies, bearings, and press-fitted joints subjected to fretting.

friction The resistance that a body encounters when sliding over another body, or experiences when a viscous fluid flows over its surface. *See also* AMONTONS FRICTION; COEFFICIENT OF FRICTION; DRAG; DYNAMIC FRICTION; FRICTION FACTOR; ROLLING FRICTION; SKIN-FRICTION COEFFICIENT; STATIC FRICTION; STICTION.

frictional flow The flow of a viscous fluid.

friction angle *See* ANGLE OF FRICTION.

friction bearing *See* BEARING.

friction clutch A clutch, the engagement of which relies on friction between its plates.

friction coefficient *See* COEFFICIENT OF FRICTION; SKIN-FRICTION COEFFICIENT.

friction damping (Coulomb damping) The damping that results from bodies sliding over one another.

friction drag *See* DRAG.

friction drive The coupling of shafts by wheels and discs in contact, operation of which relies on friction that is great enough to prevent slipping.

friction factor *See* DARCY FRICTION FACTOR; FANNING FRICTION FACTOR.

friction gear *See* FRICTION CLUTCH; FRICTION DRIVE.

friction loss The conversion of mechanical energy to heat due to friction within a machine, mechanism, linkage, etc. *See also* LOSS.

friction materials Materials having a high coefficient of friction which, when coupled with a long life, may be employed as brake linings.

friction power (FHP, friction horsepower) (Unit W) The power expended on overcoming frictional losses within an engine, and in pivots, bearings, gears, etc. **Friction torque** (unit N.m) is friction power/engine speed. *See also* LOSS.

friction-stir welding *See* WELDING.

friction-tube viscometer *See* CAPILLARY VISCOMETER (2).

friction vacuum gauge (molecular gauge, viscosity gauge) A device used to measure very low gas pressures.

friction variable *See* WALL VARIABLE.

friction velocity (u_τ, u^*) (Unit m/s) For viscous fluid flow over a solid surface, a scaling velocity particularly used in the analysis of turbulent flows defined by $u_\tau = \sqrt{\tau_S/\rho_S}$ where τ_S is the wall shear stress and ρ_S is the fluid density at the surface. *See also* WALL VARIABLE.

friction welding *See* WELDING.

frigorie (*fg*) An obsolete unit of heat used in refrigeration and cryogenics; approximately equal to one calorie.

frost-point hygrometer A type of hygrometer in which a metal mirror is cooled, typically by a Peltier device, until a temperature called the frost point is reached, at which a film of white frost is observed to appear on it.

froth *See* FOAM (1).

Froude dynamometer *See* HYDRAULIC DYNAMOMETER.

Froude number (*Fr*) A non-dimensional parameter that arises in the study of open-channel liquid flow and is defined by $Fr = V/\sqrt{gH}$ where V is the liquid speed and H is its depth, and g is the acceleration due to gravity. It is an indicator of the relative importance of inertia and gravity forces. The definition can be generalized to characterize flow at the interface between two immiscible fluids of density difference $\Delta\rho$ using a **densimetric Foude number** Fr_D, defined by

$$Fr_D = \frac{V}{\sqrt{gH\Delta\rho/\bar{\rho}}}$$

where $\bar{\rho}$ is the average density of the two fluids. *See also* RICHARDSON NUMBER.

frozen The state of a substance that is normally liquid at STP when caused to become solid, usually by cooling. *See also* MOLTEN.

frozen expansion The expansion of a gas at sufficiently high (hypersonic) Mach numbers that the gas becomes so rarefied that molecular collisions no longer occur, and limiting conditions are reached.

frozen-stress technique A method of analysing 3-dimensional stress distributions using 2-dimensional photoelasticity, in which the model is annealed while under load. This preserves the birefringency after unloading and the model can be sliced into sections, each of which is analysed as a plane problem. To avoid the need for polishing the cut surfaces, the sections are examined in a fluid having a refractive index the same as that of the model material.

fuel 1. A substance from which energy can be extracted: by combustion to give thermal energy in the case of a liquid, gaseous, or solid fossil fuel or biofuel; by nuclear fission to give thermal energy from a radioactive substance. **2.** A substance from which electrical energy can be produced by chemical reaction in a battery or fuel cell.

fuel assembly (fuel bundle, fuel element) A structured group of fuel rods in a nuclear power reactor, or one or more fuel rods in a research reactor.

fuel bed A layer of fuel in a solid-fuel combustion system.

fuel cell A device that converts a fuel and oxidant, such as hydrogen and oxygen, into electricity directly (i.e. without combustion).

fuel consumption (Unit kg/hr) The mass rate at which an engine, furnace or other power-producing device must be supplied with fuel to operate. *See also* SPECIFIC FUEL CONSUMPTION.

fuel element *See* FUEL ASSEMBLY.

fuel injection The supply of pressurized liquid fuel to an engine or furnace by a mechanical pump.

fuel oil Any liquid hydrocarbon (single or blended) that is burned in a furnace, boiler or engine. Most are distilled from petroleum, and they range from low-viscosity products with a low boiling point (around 175°C), such as kerosene, to viscous, heavy fuel oils which boil at about 600°C.

fuel rod A long, slender metal tube containing pellets of fissionable material which provide fuel for a nuclear reactor. *See also* FUEL ASSEMBLY.

fuel system (fuel-supply system) The entire fuel-delivery system for an engine or furnace, which may include: the reservoir in which fuel in liquid or gas form is stored **(fuel tank)**; the **fuel pump(s)** that delivers liquid fuel through a line from a fuel tank; the pipe or tube through which a gas or liquid fuel is delivered (**fuel line**); a main supply pipe that distributes fuel through smaller branch pipes (**fuel

manifold) e.g. to each combustion chamber of a gas turbine; the nozzle arrangement that converts liquid fuel to a spray (**fuel injector**) e.g. as it is injected into the combustion chamber of an engine; a **fuel filter** installed in a fuel line to remove particulates; valves, control units, and heaters. *See also* INJECTION PUMP; PETROL INJECTION.

fulcrum The point at which a lever is supported and turns. *See also* PIVOT.

full admission The admission of working fluid into a turbomachine over its full circumference. *See also* PARTIAL ADMISSION.

full life-cycle analysis *See* LIFE-CYCLE ANALYSIS.

fully-developed pipe flow (fully-developed duct flow) Flow in a long pipe or duct of constant cross section that has reached a location beyond which there are no changes in the distribution of any flow quantity. For turbulent flow, this applies to the time-averaged flow. *See also* DEVELOPING FLOW; POISEUILLE FLOW.

fully-plastic moment *See* PLASTIC HINGE.

fully-rough flow The regime, for turbulent boundary-layer and pipe flow over a rough surface, where the skin-friction coefficient or friction factor is independent of Reynolds number. For pipe flow, this is when the sand-grain-roughness height k exceeds about $70\nu/u_\tau$, ν being the kinematic viscosity of the fluid and u_τ the friction velocity. *See also* HYDRAULICALLY-SMOOTH SURFACE; TRANSITIONAL ROUGHNESS.

functional decomposition The design of a control system by splitting the system into components with specific actions, for example the regulator, adaptive control element, and optimal control element.

functional design A design technique that concentrates on achieving the overall objectives required of the system being designed, without detailed reference to individual components. This is achieved by the assembly of individual sub-systems having specified input-output characteristics which only interact through their inputs and outputs.

functional dimension A dimension on an engineering drawing that directly affects the function of a part or an assembly.

fundamental derivative (*Γ*) In gas dynamics, a non-dimensional measure of the dependence of sound speed c on density ρ, defined by

$$\Gamma = 1 + \frac{\rho}{c}\left(\frac{\partial c}{\partial \rho}\right)_S$$

where the subscript S denotes an isentropic process.

fundamental dimension *See* DIMENSION (1).

fundamental frequency (Unit Hz) The first harmonic of a periodic signal.

fundamental interval The interval between two fixed points on a temperature scale that is divided into units of temperature; 100° for the Celsius scale and 180° for the Fahrenheit scale.

fundamental mode The mode of free oscillation of an oscillatory system with the lowest natural frequency, i.e. at the fundamental frequency.

funicular polygon A graphical construction from which the line of action of a set of coplanar forces may be obtained. In the diagram, a body has forces P_1, P_2, etc. applied to it. The sides 1, 2, etc. of the force polygon are parallel and proportional to the forces P_1, P_2, etc. The closing side 5 of the polygon gives the magnitude and direction of the resultant force P_5. The funicular polygon gives the line of action of P_5. It is constructed by joining an arbitrary point O to the corners of the force polygon, giving lines 1–2, 2–3 etc. and drawing, from an arbitrary point L, LA_1 parallel to 5–1 cutting P_1 at A_1. A_1A_2 is drawn parallel to 1–2, cutting P_2 at A_2, and so on until A_4A_5 cuts LA_1 at A_5 that is a point on the line of action of P_5. If the force polygon is closed the system is not necessarily in equilibrium but may reduce to a couple, revealed by the funicular polygon becoming two parallel forces acting in opposite directions.

furling speed *See* CUT-OUT SPEED.

furnace 1. A type of combustion chamber in which solid, liquid, or gaseous fuels are burned to supply hot gases to a boiler or other process plant. Examples include the firebox, boiler furnace (steam-generating furnace), hot-air furnace, oil-fired furnace, updraught furnace, and water-cooled furnace.
2. A chamber, sometimes having a controlled atmosphere or under vacuum (vacuum furnace), for heating and melting materials. Examples include the blast furnace, direct- and indirect-arc furnaces, the induction furnace

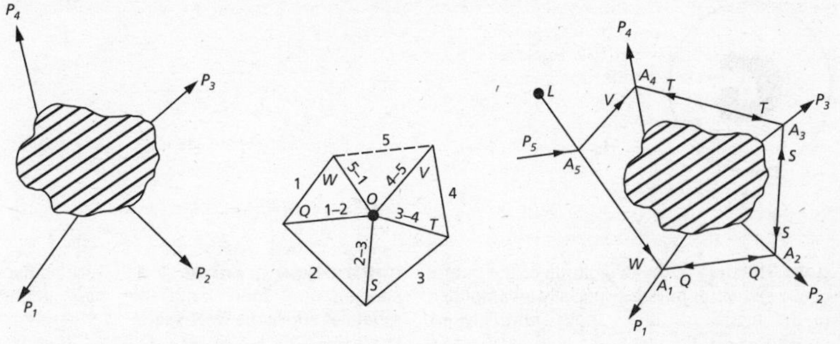

force polygon

funicular polygon

funicular polygon

(high-frequency furnace), muffle furnace, reverberatory furnace, and solar furnace.

fused A term for materials joined together by melting and then allowed to cool, for example glass and metals.

fusibility 1. The ease with which materials can be fused together. **2.** The ease with which a substance can be melted, i.e. the amount of heat required and the temperature to which it must be raised.

fusible plug (safety plug) A safety device used on a pressure vessel, such as an LPG container, based on a metal plug which melts at a predetermined temperature to prevent the buildup of excessively high pressure and allow a controlled release. In the case of a boiler, the plug is designed to melt should the water level drop and the temperature rise to a dangerous level. Steam escapes into the firebox, gives a warning, and damps the fire.

fusion 1. *See* LATENT HEAT OF FUSION. **2.** *See* NUCLEAR FUSION.

fusion welding *See* WELDING.

fuzzy logic An extension of Boolean (i.e. two-valued, true/false) logic to represent situations where variables representing answers to questions may be partially true, i.e. somewhere between completely true and completely false. A set of rules somewhat similar to the human decision-making process can then be applied to the variables to give fuzzy control of the system. **Fuzzy control** is frequently applied to large, complex systems, and to those where the control process replaces human decision-making. A **fuzzy controller** is a control system employing fuzzy logic to determine the input to the plant.

gain The ratio of the magnitude of the output of any system or device, especially an amplifier, to the magnitude of its input, normally expressed in dB. For electronic devices, the gain will usually be the voltage gain (output voltage to input voltage) or the current gain (output current to the input current). For fluidic devices, the gain is determined from the ratio of absolute pressures.

gain asymptotic approximation *See* BODE PLOT.

gain crossover frequency *See* BODE PLOT.

gain margin On Bode or Nyquist plots, the amount by which the gain of the system, at the frequency at which the phase shift is 180°, would have to be increased to reach unity. It is thus a measure of how much the gain could be increased before instability occurs. In decibels, it is: $Gain\ Margin = -20log_{10}|G(i\omega)|_{\phi=180}$ where $G(i\omega)$ is the gain at frequency ω, $i = \sqrt{-1}$ and ϕ=180 indicates that the gain is measured at the frequency where the phase shift ϕ is 180°. *See also* BODE PLOT; NYQUIST DIAGRAM.

gain scheduling A method of controlling a non-linear system in which the controller uses different gains at different times, the gain used being determined by a measurement of the current and previous behaviour of the plant. *See also* DESCRIBING FUNCTION.

Galilean transformation In steady fluid flow, the addition or subtraction of a constant velocity to the entire flowfield, e.g. so that flow can be considered relative to a moving object or flow feature, such as a shock wave.

Galileo number *See* ARCHIMEDES NUMBER.

galvanizing A process for coating steel with zinc to protect against rusting, either by electrochemical deposition or dipping into molten zinc. *See also* HOT DIPPING.

gantry-type robot A rectangular-configuration robot with the base frame mounted above the work space.

Gantt chart A type of bar chart showing the sequence of tasks required to carry out a project, including key steps and intermediate outcomes such as reports and milestones.

gap lathe A lathe having a gap in the bed at the headstock end, allowing larger-diameter workpieces to be accommodated.

gas A highly compressible dry fluid that cannot form a free surface but fills completely any closed container. Any substance at a temperature above the critical isothermal has the characteristics of a gas. *See also* LIQUID; PERFECT GAS; VAPOUR.

gas bearing *See* BEARING.

gas burner A device, often just a tube or a Venturi, through which a combustible gas flows and burns at the outlet. Some air may be mixed with the gas within the tube.

gas calorimeter *See* BOYS GAS CALORIMETER.

gas compressor A machine which sucks in a gas at low pressure and delivers it at high pressure, usually accompanied by a temperature increase. Compressor designs include axial, centrifugal, reciprocating, rotary, and diaphragm.

gas constant *See* PERFECT GAS; UNIVERSAL GAS CONSTANT.

gas-cooled nuclear reactor *See* NUCLEAR FISSION.

gas cycle A thermodynamic cycle in which the working fluid is a gas throughout the cycle.

gas cylinder A metal bottle used to store gas at high pressure (up to 1000 bar) but fitted with a regulator so that the gas can be delivered at lower pressure.

gas dynamics The study of gas flow when compressibility is significant, typically at Mach numbers above about 0.3 including supersonic flow, hypersonic flow, and rarefied gas dynamics.

gas engine A spark-ignition piston engine for which the fuel is natural gas, propane, producer gas, coal gas, or landfill gas. *See also* ALTERNATIVE-FUEL ENGINE; DUAL-FUEL ENGINE.

gas-filled thermometer A fluid-expansion thermometer in which the fluid is a gas.

gas-flow meter 1. An instrument that determines the volume flow rate $\dot{Q}$ of a gas by measuring the pressure drop Δp across a laminar-flow element. Since the flow is laminar, the device has a linear characteristic i.e. $\dot{Q} \propto \Delta p$. **2.** *See* DIAPHRAGM METER.

gas generator 1. (core) The compressor, combustor, and turbine of a gas-turbine engine. **2.** That part of a liquid- or solid-fuel rocket used to produce relatively cool gas to drive the turbopump. **3.** An electrical generator driven by a gas engine.

gas holder (gasometer) A large (typically 50,000 m^3) container in which natural gas or town gas is stored at near atmospheric temperature and pressure. *See also* WATER-SEALED HOLDER.

gasification The process of pyrolysis adapted to maximize the amount of fuel gases produced from solid organic material including coal, wood, biomass, and municipal waste.

gasifier 1. The partial-combustion device used in gasification. **2.** *See* FREE-PISTON GAS GENERATOR.

gasket A shaped flat sheet of cork, rubber, soft metal, or other deformable material, sometimes sandwiched between thin sheets of copper or another metal, inserted between adjoining surfaces to act as a seal against the escape of gas or liquid.

gasket *m*-factor For a gasketed joint of area A in which the clamping force is F, it is essentially that $F \geq mpA$ where p is the pressure that would separate the two halves of the joint and $m > 1$.

gas law Any equation of state that relates pressure, temperature, and density for a gas. *See also* PERFECT GAS.

gas meter Any type of flow meter used to measure the volumetric flow rate of a gas, including diaphragm, gas-flow, rotary, turbine, and orifice meters.

gas mixture A mixture of two or more gases the properties of which can be determined from Amagat's law, Dalton's law, and the Gibbs–Dalton law.

gasohol A fuel for spark-ignition engines consisting of petrol containing at least 26% of ethanol.

gas oil A fuel for internal-combustion engines. An unrefined distillate of crude petroleum after kerosene (i.e. paraffin) fractions have been removed.

gasometer *See* GAS HOLDER.

gas port An inlet or exhaust passage to a cylinder of an internal-combustion engine.

gas power The output of a gasifier calculated assuming isentropic expansion of the gas leaving the cylinder. It is the potential power output.

gas refrigeration cycle (air-cycle refrigeration, Bell–Coleman cycle, reverse Brayton cycle, reverse Joule cycle) An ideal (air-standard) refrigeration cycle consisting, as shown on the diagram of temperature (T) *vs* specific entropy (s), of four internally reversible processes: isentropic compression (in a compressor, 1–2) in which the air temperature increases, isobaric cooling in a heat exchanger (2–3), isentropic expansion (in a turbine, 4–5) in which the air temperature decreases, and isobaric absorption of heat by a heat exchanger from the refrigerated space (5–6). A regenerator can be installed between the two heat exchangers (3–4 and 6–1).

gas regulator 1. A valve fitted to a gas cylinder that maintains constant supply pressure. **2.** A valve in a gas-supply line that maintains either constant pressure or constant flow rate.

gas-shielded arc welding *See* WELDING.

gas spring A piston-cylinder arrangement in which a gas, typically air or nitrogen, in the cylinder supports an imposed load.

gas thermometer *See* CONSTANT-VOLUME GAS THERMOMETER.

gas thread *See* SCREW.

gas-tube boiler *See* WASTE-HEAT BOILER.

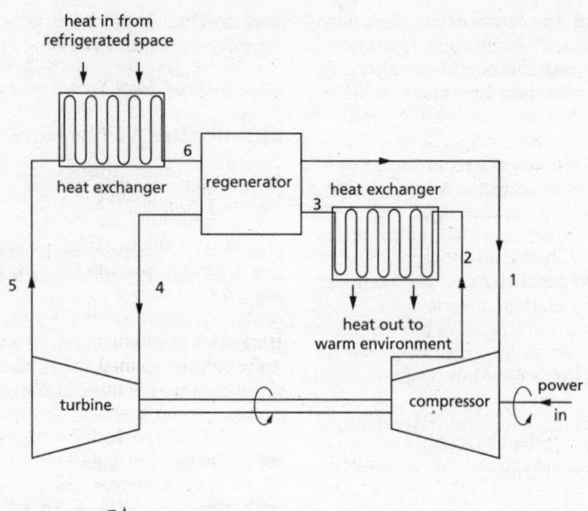

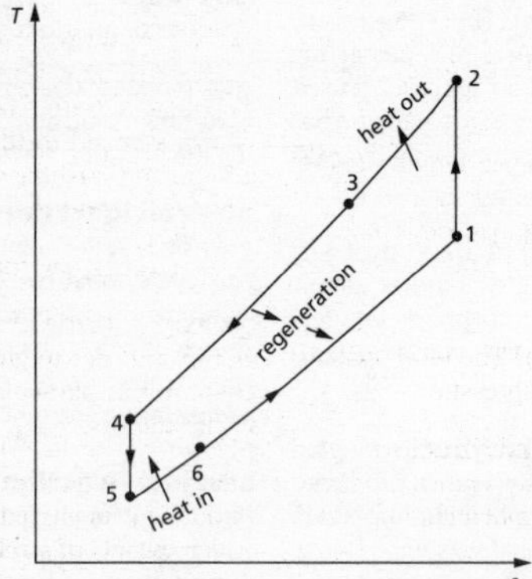

gas refrigeration cycle

gas turbine An engine in which air, compressed by a multi-stage axial compressor and/or one or more centrifugal compressors, flows into a combustion chamber (or chambers) where fuel is burned and the hot gases then drive an axial turbine which powers the compressor. Where the compression takes place in stages, an intercooler between the stages may be used to improve efficiency. Kerosene is the normal fuel used for propulsion applications, while natural gas or waste gas can be used for power production and process applications. The turbofan, turbojet, and turboprop engines are variants designed for propulsion, while the turboshaft engine provides shaft power. *See also* BRAYTON CYCLE.

gas-turbine cycle *See* BRAYTON CYCLE.

gas-turbine nozzle *See* PROPELLING NOZZLE.

gate A barrier that can be partially or fully opened or closed to allow access, flow of a fluid, etc.

gate valve (sluice valve) A stop valve with a circular or rectangular closing element (the gate) that may have a parallel or wedge-shaped cross section and slides in guides normal to the flow.

gauge 1. Any instrument for measuring or checking dimensions, including adjustable gauges such as a micrometer and non-adjustable gauges such as go/no-go limit gauges that indicate the maximum and minimum dimensions allowable in a component during manufacture. **2.** A term often used to mean an instrument, e.g. pressure gauge. **3.** The diameter of a wire or thickness of a sheet according to a standard. *See* WIRE GAUGES. **4.** The distance between the rails of a railway track.

gauge block (Johansson block) A tablet of steel having two parallel faces ground to an accuracy of some 250 nm, a series of which is employed for standard lengths in the toolroom.

gauge diameter *See* SCREW.

gauge factor The multiplying factor by which the change in the electrical resistance of a strain gauge is related to the strain being measured.

gauge glass An externally-mounted transparent tube that indicates the level of water in a boiler or liquid in a tank.

gauge length The reference length against which extensions are measured in a testpiece employed to determine the load-deflexion (or stress-strain) characteristics of a material in tension. *See also* EXTENSOMETER.

gauge pressure (Unit Pa) The level of static pressure above the ambient pressure.

Gaussian velocity distribution The shape of the curve of velocity *vs* distance from the centreline for the far field of a laminar wake flow.

gauze A fine woven mesh of plastic, metal, or other fibres. *See also* FILTER SCREEN.

Gay–Lussac law *See* CHARLES LAW.

gear (gears) *See* TOOTHED GEARING.

gearbox A mechanism consisting of meshing gears which transmit power and torque to an output shaft from an input shaft directly connected to an engine. The mechanism is normally contained within a casing filled with lubricating oil. The diagram shows a simple four-speed **sliding-mesh gearbox** in which gears with different numbers of teeth are slid along parallel splined shafts within the gearbox to mesh with each other and change the ratio of output to input speed (**gear ratio**). A lever with prongs (**selector fork**) engages with grooves cut in the boss of a gear to move it along the shaft. In a **selective transmission** a single lever is used to change from one gear to another. **Synchromesh** is mechanism to ensure gears are rotating at the correct speed prior to meshing. A **baulk ring** prevents engagement until the speed of the gears is the same, thus minimizing crashing noise or 'grinding'. *See also* EPICYCLIC GEARING.

gear cutter *See* TOOTHED GEARING.

geared flywheel A flywheel having teeth on its rim in order to give a large gear ratio. They are found in particular on engines having self-starter motors.

gearless traction The motion transmitted without gears directly from an engine to the driven wheels of a vehicle.

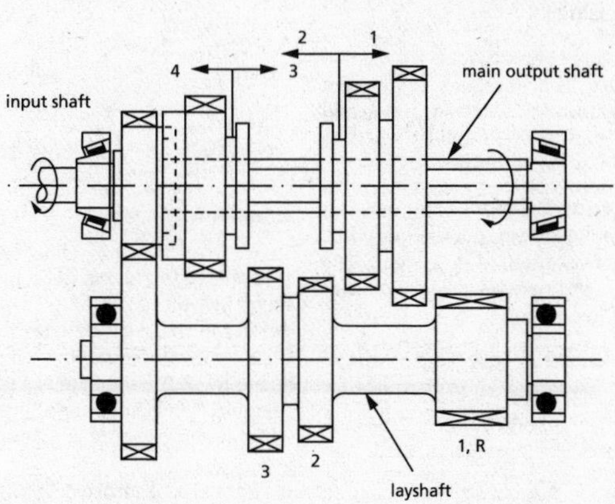

gearbox

gear meter *See* OVAL-GEAR FLOW METER.

gear motor Essentially a gear pump run in reverse, in which high-pressure fluid supplied to one side of the casing causes the gears to rotate.

gear pump A positive-displacement pump in which fluid is pumped by meshing gears, one driven and the other an idler gear, on parallel axes within a closed casing. *See also* EXTERNAL GEAR PUMP; HYDRAULIC PUMP; INTERNAL GEAR PUMP; TOOTHED GEARING.

g

gear ratio 1. For two gears in contact, the ratio of the number of teeth on the driving gear to that on the driven gear. *See also* TOOTHED GEARING. **2.** *See* GEARBOX.

gear teeth *See* TOOTHED GEARING.

gear train (train) Any combination of gear wheels by means of which motion is transmitted from one shaft to another. *See also* TOOTHED GEARING.

gear wheel Any form of toothed wheel, particularly those having conjugate teeth employed in the transmission of motion and power. *See also* TOOTHED GEARING.

Geiringer's equations In plasticity theory, the relations between increments of velocity along plane-strain slip lines (characteristics) given by $du - vd\phi = 0$ along an α-line and $dv + ud\phi = 0$ along a β-line, where the α- and β-lines coincide with the Ox and Oy axes of the problem at the origin.

gel A colloidal suspension of liquid droplets in a solid such as gelatine.

gel coat A thin layer of resin or epoxy, used to protect the surface of a fibreglass component such as a hull against damage from water and ultra-violet solar radiation. It is also used to protect the surface of polyester moulds.

generalized coordinates The smallest number of independent variables required to describe the configuration of a system. The number of such coordinates is called the number of degrees of freedom.

generalized Hooke's law Under triaxial loading in linear elasticity, the expressions for normal strains ε_x, ε_y and ε_z have the form $\varepsilon_x = [(\sigma_x/E) - (\nu/E)(\sigma_y + \sigma_z) + \alpha\Delta T]$ with change of corresponding suffixes for ε_y and ε_z, where σ_x, σ_y, and σ_z are the normal stresses acting in the mutually-orthogonal x, y, and z directions, ν is Poisson's ratio, E is Young's modulus, α is the coefficient of thermal expansion, and ΔT is any change of temperature. The shear strains, that are independent of the axial strains, are given by $\gamma_{xy} = \tau_{xy}/G$, with change of corresponding suffixes for γ_{yz} and γ_{zx}, where τ is the shear stress and G is the shear modulus. The expressions may also be written for stresses with strains as the independent variables. *See also* HOOKE'S LAW.

generalized Newtonian fluid An inelastic non-Newtonian liquid for which the shear rate is not proportional to the shear stress. *See also* SHEAR-THICKENING LIQUID; SHEAR-THINNING LIQUID.

general tolerance *See* TOLERANCES.

generating cutter *See* TOOTHED GEARING.

generating line *See* TOOTHED GEARING.

generator 1. A machine for converting mechanical energy to electrical energy. **2. (generatrix)** A point, line, or surface regarded as moving and so forming a line, surface, or solid, respectively.

Geneva mechanism (Maltese Cross mechanism) A mechanism that gives intermittent motion, consisting of a rotating wheel having a peg that engages in slots in a profiled disc running on a parallel shaft.

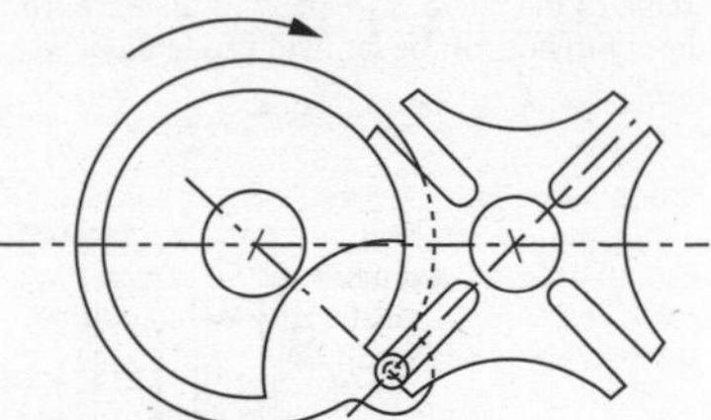

Geneva mechanism

geodesic-isotensoid In the manufacture of filament-wound reinforced-polymer vessels, the dome shape of the ends that gives uniform fibre tension when under pressure. The dome shape by which the filament stresses match the unequal local hoop and meridional stresses when under pressure is a **geodesic ovaloid**.

geometrical factor *See* VIEW FACTOR.

geometric pitch *See* PROPELLER.

geometric similarity *See* PHYSICAL SIMILARITY.

geopressurized brine A high-saline brine containing trapped gas, such as methane, found at depths of typically 4 km with pressures up to 100 MPa. A potentially important geothermal-energy resource.

geothermal energy The energy derived from the enthalpy stored within the Earth's crust, typically at depths of 5 to 10 km, particularly in the form of hot rock and high-pressure water or steam. Below 100°C it is classified as a **low-enthalpy resource**, in the range 100–200°C as a **medium-enthalpy resource** and above about 200°C as a **high-enthalpy resource**. *See also* DRY-STEAM POWER PLANT; FLASH-STEAM POWER PLANT; HYDROTHERMAL ENERGY; ORGANIC BINARY-CYCLE POWER PLANT.

GER *See* EMBEDDED ENERGY.

gerotor pump A type of gear pump in which liquid is pumped by an eccentrically-mounted scalloped rotor rotating within a circular casing. Gerotor is derived from **ge**nerated **rotor**.

GFRP *See* GLASS-FIBRE-REINFORCED PLASTIC/POLYMER.

GHG *See* GREENHOUSE EFFECT.

Gibbs–Dalton law An extension of Dalton's law of additive pressures to include the statement that the internal energy, enthalpy, and entropy of a mixture of gases are equal to the sum of the internal energies, enthalpies, and entropies the individual constituents would have if each existed alone at the same temperature and volume. **Gibbs rule** is that part of the law pertaining to entropy.

Gibbs equations *See* *TDS* EQUATIONS.

Gibbs function (free enthalpy, *g*, Gibbs free energy) (Unit kJ) A thermodynamic property defined by $g = h - Ts$ where h is the specific enthalpy, s is the specific entropy, and T is the absolute temperature. **Helmholtz function (a, *f*, Helmholtz free energy, Helmholtz potential)** is defined by $f = u - Ts$ where u is the specific internal energy. The **Gibbs–Helmholtz equations**, derived from the Gibbs and Helmholtz functions for constant-volume (v) and constant-pressure (p) processes, are

$$\frac{\partial}{\partial T}\left(\frac{a}{T}\right)_v = -\frac{u}{T^2}$$

and

$$\frac{\partial}{\partial T}\left(\frac{g}{T}\right)_p = -\frac{h}{T^2}$$

respectively.

Gibbs paradox A paradox which states that when two different gases mix there is an entropy increase, but no increase if they are identical.

Gibbs phase rule In thermodynamics, if C is the number of chemical species present and P is the number of phases, then the number of degrees of freedom F (i.e. the number of properties of state required to specify a system) is given by $F = C - P + 2$.

giga (G) An SI unit prefix indicating a multiplier of 10^9; thus gigawatt (GW) is a unit of power equal to one billion or one thousand million watts.

gimbal A pivoted support that allows rotation of a supported object about a single axis. Two orthogonal gimbals are used in supports of, for example, compasses. *See also* CARDAN'S SUSPENSION.

girder 1. A rolled steel joist such as an I-beam. **2.** A lattice or other built-up framework.

gland *See* PACKED GLAND.

glass 1. A liquid, when it is cooled far below its freezing point and fails to crystallize (owing to its molecular structure or to a fast rate of cooling), solidifies into an amorphous disordered solid called a glass. Easy glass-formers are substances with molecules that, in the liquid state, have complex shape, or exist as long chains or networks, and are thus very viscous. The distinction between fluids and solids is arbitrarily drawn at a viscosity of 100 TPa.s, and the temperature at which it occurs is called the glass transition temperature. **2.** The substance familiarly called glass, as used for glazing, etc., is derived from silica (SiO_2).

glass fabrics Fabrics woven or knitted from glass fibre and used for reinforcement of polymers.

glass fibre A substance consisting of glass in the form of filaments some 5 to 10 μm in diameter and having a tensile strength of about 1 GPa (of order ten times that of bulk glass).

glass-fibre-reinforced plastic (glass-fibre-reinforced polymer, GFRP, GRE, GRP) A substance that utilizes the high strength of glass fibres by embedding them in a castable polymer matrix.

glass transition temperature (T_g) (Unit °C) The temperature at which the viscosity of a glass-forming material has the value 100 TPa.s. It is a **second-order transition**, and for many materials is about one third of the boiling temperature. For example, $T_g \approx 100°C$ for polymethylmethacrylate (PMMA); for unvulcanized rubber $T_g \approx -80°C$. At temperatures above T_g, materials display increasingly-pronounced time-dependent mechanical behaviour; below T_g they are time-independent.

global solar irradiance (G) (Unit W/m^2) The sum of the direct (I) and the diffuse (D) solar irradiances. If the solar zenith angle is θ, for a horizontal surface $G = I \cos\theta + D$.

global warming *See* GREENHOUSE EFFECT.

globe valve A regulating valve with a generally spherical body within which there is an internal baffle with an opening, the valve seat, which is partially or completely closed by a movable plug or disc.

glowing surface (surface combustion) The phenomenon that results when a material such as wood is subjected to a heat flux which causes decomposition and the release of fuel gases, but the heat flux is too low for flames to appear. The onset is termed **glowing ignition.**

glow plug An electrically-heated, pencil-like plug installed in a diesel-engine cylinder and used to aid cold starting.

glue *See* ADHESIVE.

goal coordination method In a control system where the plant decomposition method has been applied, goal coordination is a technique for adjusting the behaviour of the individual controllers so as to achieve an optimum overall result.

Goodman diagram *See* FATIGUE.

Gorlov helical turbine A vertical-axis tidal-current turbine with a helical-bladed rotor having long blades that run along a cylindrical surface like a screw thread.

Görtler vortices Counter-rotating streamwise vortices, which arise due to centrifugal instability in the boundary layer of flow over a surface which is concave in the streamwise direction. The flow is characterized by the non-dimensional **Görtler number (G)** defined by $G = U_\infty\theta/\nu\sqrt{\theta/R}$ where U_∞ is the free-stream velocity, θ is the boundary-layer momentum thickness, ν is the kinematic viscosity of the fluid, and R is the radius of curvature of the surface.

governor An automatic regulator of the supply of fuel to an engine in order to maintain a constant speed.

Graetz number (Gz) A non-dimensional number that arises in the study of heat transfer in developing pipe flow and is defined by $Gz = RePrD/x$ where Re is the Reynolds number based on the bulk velocity and the pipe diameter D, Pr is the Prandtl number, and x is the distance from the pipe inlet. In some work the definition includes the factor $\pi/4$ and in others the inverse definition is given.

Graham's laws The laws stating that the rates of diffusion and effusion of a gas are inversely proportional to the square root of its density.

grain An individual crystal in a polycrystalline material, the thin contiguous regions between being the **grain boundaries**. In forging and other metalworking processes, the pattern taken up by deformed crystals is the **grain flow**.

gram calorie *See* CALORIE.

granularity In systems engineering, a measure of the extent to which a composite system can be broken down and analysed as a number of separate components.

graphic panel A panel showing diagrammatically the structure of a plant (i.e. each of the components and their interconnections) with, alongside the diagram of each component, one or more displays showing its operation and one or more manual controls to regulate that operation. *See also* MIMIC PANEL.

Grashof number (Gr) A non-dimensional number that arises in the study of natural convection. It represents the ratio of buoyancy to viscous forces and is defined as $Gr = g\beta\Delta TL^3/\nu^2$ where g is the acceleration due to gravity, β is the coefficient of volumetric expansion of the fluid and ν is its kinematic viscosity, ΔT is the temperature difference between the surface and the surroundings, and L

is a characteristic dimension of the heated or cooled surface.

Grätzel cell A photoelectrochemical thin-film solar cell.

gravimetric analysis The mass of each component in a mixture of substances. *See also* MASS FRACTION; MOLAR ANALYSIS; VOLUMETRIC ANALYSIS.

gravitational constant (*G*) *See* NEWTON'S LAW OF GRAVITATION.

gravitational energy *See* POTENTIAL ENERGY.

gravitational force A force of attraction that exists at all points outside the earth, directed toward the centre of the earth and of magnitude GM/r^2 on unit mass, where G is the gravitational constant, $M \approx 5.983 \times 10^{24}$ kg is the mass of the earth, and $r \approx 6.365 \times 10^6$ m is the distance from the centre of the earth. On the surface of the earth the gravitational force on one kg is about 9.81 N.

gravitational system of units A non-SI system of units based on a fundamental unit of weight rather than a unit of mass.

gravity The attractive force by which bodies are drawn towards the centre of earth or other celestial object. Its intensity is measured by the acceleration produced.

gravity-capillary wave A wave on a liquid-gas interface or the interface between two immiscible liquids for which both gravity and surface tension contribute to the restoring force. The dispersion relation is given by

$$\omega^2 = \frac{\sigma}{\rho_H + \rho_L}\left[\frac{2\pi}{\lambda}\right]^3 + \frac{\rho_H - \rho_L}{\rho_H + \rho_L}\left[\frac{2\pi}{\lambda}\right]$$

where ω is the angular frequency of the wave and λ its wavelength, ρ_H is the density of the denser fluid and ρ_L that of the less dense, and σ is the surface tension. *See also* CAPILLARY WAVE.

gravity wave (gravitational wave) A wave, in which the dominant force is gravity, on a liquid surface or liquid–liquid interface or within a density-stratified fluid.

GRE *See* GLASS-FIBRE-REINFORCED PLASTIC.

grease An oily or fatty matter in a semi-solid-state, used as a lubricant.

green design *See* SUSTAINABLE DESIGN.

greenhouse effect The process whereby energy radiated from the Earth is absorbed by the atmosphere and partially radiated back to the Earth, causing a rise in temperature. The gases primarily responsible for increasing the absorptance of the atmosphere are due to emissions from industrial processes, motor vehicles, aircraft etc., primarily carbon dioxide, methane, oxides of nitrogen, and ozone, together with water vapour. As the atmosphere is increasingly polluted by these so-called **greenhouse gases (GHG)**, the temperature gradually increases, a phenomenon known as **global warming**.

grey body A body that emits the same fraction ε_λ of blackbody emission at a given temperature at all wavelengths. *See also* MONOCHROMATIC ABSORPTANCE.

grey iron (gray iron) *See* CAST IRON.

Griffith equation The expression for the remote stress σ_{crack} required to fracture a large plate having a small central through-crack of length $2a$, given by $\sigma_{crack} = \sqrt{ER/\pi a}$ where E is Young's modulus and R is fracture toughness. *See also* FRACTURE MECHANICS.

grinding 1. A process of high-speed multiple scratching of surfaces by a wheel formed from hard grits and a binder which are progressively exposed as the binder wears away. It is a process of micromachining employed particularly in accurate finishing operations on hard materials. **2.** A form of comminution.

SEE WEB LINKS

- Grinding processes terminology

grinding machine A machine tool consisting of a traversing table and a **grinding wheel**, originally a wheel of sandstone (a **grindstone**). Later wheels contained emery, corundum and quartz grit bonded together by, for example, glass or various silicates. Modern grinding wheels employ artificial versions of these ceramics, or polycrystalline diamond.

SEE WEB LINKS

- Primarily the materials used in the manufacture of grinding wheels

grinding stress The residual stresses left in the surface of a component after grinding, owing to non-uniform cooling from the high surface temperatures produced.

gripper An end effector on a robot used to pick up components. A general purpose gripper will have two or more fingers similar to a human hand. *See also* ATTRACTION GRIPPER.

grip vector *See* APPROACH VECTOR.

grit Small hard particles used to abrade or grind a material when bonded to a grinding wheel, disc cloth, paper, belt, etc. Materials used include aluminium oxide, silicon carbide, aluminium oxide–zirconium oxide, and chromium oxide.

grit blasting *See* EROSION.

g

grit size The average size of grit based upon sieving (macrogrit) or sedimentation (microgrit). The Federation of European Producers of Abrasives (FEPA) designations for bonded macrogrits range from F4 (4890 μm) to F220 (58 μm) and for microgrits from F230 (53.0 ± 0.2 μm) to F2000 (1.2 ± 0.3 μm). For coated macrogrits the ranges are P12 (1815 μm) to P220 (68 μm) and for microgrits P240 (58.5 ± 2 μm) to P2500 (8.4 ± 0.5 μm).

Gröber charts *See* HEISLER CHARTS.

grommet A grooved metal, plastic, or rubber ring, often used to prevent fraying of electrical cables passing through a hole in a sheet of metal or other hard material.

groove weld *See* WELDING.

gross calorific value *See* CALORIFIC VALUE OF A FUEL.

gross energy requirement *See* EMBEDDED ENERGY.

gross heating demand The total amount of heat per annum that needs to be supplied to a building.

ground effect The effect on fluid flow over an object in close proximity to a solid surface, such as an aircraft taking off or landing. *See also* HOVERCRAFT.

ground-source heat pump (GSHP) A domestic heating system in which water or a refrigerant is circulated through a buried coil or a pipe in a borehole typically 100 m deep to extract heat from the earth. The liquid passes through a heat pump in the house that transfers heat to indoor air. The process can be reversed in hot weather for cooling.

group velocity *See* PHASE SPEED.

GRP *See* GLASS-FIBRE-REINFORCED PLASTIC/POLYMER.

grub screw A short headless screw with a recess at one end to receive a screw driver or key.

Grüneisen parameter A non-dimensional parameter that arises in the study of thermoacoustics and is defined by $\eta = \alpha c^2/C_P$ where α is the thermal expansivity, c is the sound speed, and C_P is the specific heat at constant pressure.

GSAW *See* WELDING.

GSHP *See* GROUND-SOURCE HEAT PUMP.

guard A mesh or cover placed around a machine to protect workers from accidents.

gudgeon A simple bearing that supports a pintle, such as the hinge for a rudder.

gudgeon pin *See* PISTON.

guide vanes Fixed vanes, which may be just curved sheets of metal or have an aerofoil cross section, that guide fluid flow in a turbomachine, wind or water tunnel, air-conditioning system, etc. *See also* WICKET GATE.

Guillemin coupling (french coupling) A quarter-turn sexless (i.e. symmetrical) coupling, in which a ring with two protuberances at the end of a hose locks with a similar ring on another hose. Widely used for connecting fire hoses.

Gukhman number (*Gu*) A non-dimensional parameter that arises in the study of convective heat transfer with evaporation and is defined by $Gu = (T_\infty - T_{WB})/T_{BP}$ where T_∞ is the ambient temperature, T_{WB} is the wet-bulb temperature, and T_{BP} is the boiling point of the liquid being evaporated. A definition sometimes used is $Gu = (T_{WB} - T_{DB})/T_{DB}$ where T_{DB} is the dry-bulb temperature.

gun burner (gun-type burner) An oil burner in which fuel oil is pumped at high pressure through an atomizing nozzle and into a combustion chamber or furnace.

gun drill A drill used to drill very long holes, such as for gun barrels. A liquid flows through a hole in the shank to cool and lubricate the tip and carry chips back along one or more straight flutes in the shank.

Gurney flap A strip of metal or composite material running along the trailing edge of an aerofoil, protruding either downwards or upwards, which fixes the separation point and increases lift or downforce.

gusset A bracket at an angular join in a framework to increase stiffness.

gust load The transient loading of a component or structure caused by a sudden gust of wind.

gyration A rotating or whirling motion.

gyrocompass A non-magnetic compass in which a continuously-driven gyroscope is so mounted that its axis remains parallel to the earth's axis of rotation.

gyroplane (gyrocopter) *See* AUTOGYRO.

gyroscope (gyro) A rapidly-spinning wheel or disc mounted about an axis that itself can rotate about other orthogonal axes. Because the spin axis tends to maintain the same direction in space, gyros are employed in various devices and instruments to provide stabilization or reference direction, and to determine angular velocity and acceleration.

gyroscopic couple The couple experienced by an object spinning with angular velocity ω when it precesses (i.e. rotates) with angular velocity Ω about an axis perpendicular to the axis of spin. The couple, of magnitude $J\omega\Omega$ where J is the polar moment of inertia of the object about its axis of spin, is exerted about the third perpendicular axis. *See also* POINSOT METHOD.

gyroscopic nutation *See* NUTATION.

gyroscopic precession *See* PRECESSION

hacksaw A hand saw used for cutting metals, plastics, and other materials. Tension in the blade is achieved in smaller models through the use of a stiff frame alone, and in larger models with an adjusting screw that pulls the blade against the frame. A hacksaw blade has limited height, and thus can cut along an arc if the radius is sufficiently large. Standard pitches for large blades are between 14 and 32 teeth per inch (from 2 mm to 0.8 mm between teeth).

Hagen–Poiseuille flow *See* POISEUILLE FLOW.

hair cotter pin *See* RETAINING CLIP.

hairline crack A fine crack observable on the surface of a component. The width is of order 100 μm and so comparable with the diameter of a human hair.

hairpin vortex *See* HORSESHOE VORTEX.

half shaft A final drive shaft in a motor vehicle which is divided into two halves, one on either side of the differential.

half space The region of infinite extent below a plane surface in a continuum.

half-track *See* CATERPILLAR.

Hamiltonian mechanics *See* LAGRANGE–HAMILTONIAN MECHANICS.

hammer shock A shock wave caused by the sudden closure of a valve in a pipeline through which gas is flowing. *See also* WATER HAMMER.

hammer-shock testing A testing method in which a component fixed to an anvil is struck by a pendulum.

hand *See* SCREW.

hand lay-up A process employed in the construction of parts using fibre-reinforced composites, when the object is built up by hand using layers of chopped strand mat and matrix, or prepreg.

Hanning window (Hann window) A technique applied in a Fourier analysis spectrum analyser to reduce the generation of spurious frequencies in the output data.

hardenability The ease with which the extremely hard microconstituent martensite can be formed at different cooling rates in a steel. Plain carbon steels must be quenched in water (and consequently have residual stresses); some alloy steels may be quenched in oil; and there are air-hardening alloy steels. *See also* JOMINY TEST; TEMPERING; TTT CURVES.

hardener *See* EPOXY.

hardness The ability of a ductile material to resist permanent deformation. It is a measure of plastic flow properties determined principally by indentation testing, but scratch hardness and rebound hardness are also employed. *See also* RUBBER HARDNESS.

SEE WEB LINKS

- Types of hardness test and equipment

hardness scales The different measures of hardness given by different tests, such as indentation pressure in the Brinell and in the Vickers hardness tests, and the different Rockwell hardness numbers.

hardness test 1. Any of various tests in which different hard indenters are forced into the surface of a solid under different loads to give permanent impressions, the pressure to cause which being called the hardness. The Brinell test employs a spherical indenter; the Vickers test a square-based pyramid. Originally hardnesses were given in kg/mm^2 but now are often given in Pa. **2.** The tests for scratch and file hardness relate either to one material being able to mark another (Mohs hardness scale for minerals), or to the size of groove produced by a rigid indenter slid under load across a surface. **3.** Rebound hardness concerns the height of rebound of a dropped indenter, or an indenter at the end of a pivoted arm, having struck a surface.

hard soldering *See* BRAZING.

harmonic A frequency within a periodic signal that is an integral multiple of the fundamental frequency which is termed the first harmonic.

harmonic balancer (harmonic damper) *See* DAMPING.

harmonic drive (harmonic speed changer) A reversible reduction drive providing a large reduction ratio (typically up to 320:1) with effectively zero backlash, high torque capability, and high efficiency. The drive has concentric input and output shafts and is of considerably lower mass and volume than comparable drives.

SEE WEB LINKS

- Video of the operation of a harmonic drive.

harmonic motion The simplest form of periodic motion represented by a sinusoidal function of time. *See also* SIMPLE-HARMONIC MOTION.

harmonic order number The ratio of the vibration frequency to the crankshaft rotation speed (both expressed in Hz or rad/s) for vibration of a reciprocating engine.

harmonic oscillation *See* SIMPLE-HARMONIC MOTION.

harmonic oscillator A linear system with inertia which, when displaced from its equilibrium position, experiences a force proportional to the displacement acting in the direction to restore it to equilibrium. Common examples are a mass on a spring, a pendulum experiencing small displacements, or an electrical inductor-capacitor circuit. Without damping, the consequence of an initial displacement is simple harmonic motion; with damping, it is decaying sinusoidal oscillation.

harmonic response *See* FORCED RESONSE.

harmonic speed changer *See* HARMONIC DRIVE.

harmonic synthesizer A device for synthesizing a complex waveform from a number of constituent sinusoidal components, each with a specified amplitude and phase. Unlike Fourier synthesis, the frequencies of the components are not multiples of a fundamental frequency.

harmonic vibration *See* HARMONIC MOTION, SIMPLE-HARMONIC MOTION.

Hartford loop An arrangement of piping between a steam boiler's header and its condensate gravity-return piping that prevents the boiler from running dry in the event of a leak in the return line.

Hartmann–Sprenger tube (Hartmann generator) A research device in which a high-speed jet of gas is directed towards the entrance of a tube which is closed at the other end. Under certain conditions an intense pulsating flow is produced in the tube accompanied by very rapid heating to high temperatures.

hatching Fine lines, typically at 45°, used on an engineering drawing to show solid cross sections.

HAWT *See* HORIZONTAL-AXIS WIND TURBINE.

HAZ *See* WELDING.

H-beam (H-girder) An I-beam in which the flanges are of greater extent than the web.

HDD *See* DEGREE DAY.

head (pressure head, *h*) (Unit m) A pressure difference Δp expressed in terms of the vertical height of a column of liquid, typically water or mercury. If the liquid has density ρ, $h = \Delta p/\rho g$ where g is the acceleration due to gravity.

header 1. An exhaust manifold for a piston engine. **2. (header tank)** A tank in which liquid is stored, located above the equipment it supplies, such as a water tank supplying a boiler.

head loss A loss in stagnation pressure in internal flow due to wall friction and minor losses in fittings, expressed as a head.

head meter *See* BERNOULLI OBSTRUCTION FLOW METER.

headrace An open channel or tunnel that conveys water into a hydraulic turbine. *See also* TAILRACE.

headstock The device on a machine tool (particularly a lathe) which carries the revolving spindle or quill. *See also* TAILSTOCK.

heat (*q*, *Q*) (Unit J) A form of energy that is transferred across the boundary of a system at one temperature to another system (or the surroundings) at a different temperature by virtue of the temperature difference between them. Heat can be identified only as it crosses the boundary. A body can never be said to contain heat which is thus a transient phenomenon. *See also* HEAT TRANSFER; WORK.

heat addition The transfer of heat into a system. *See also* HEAT REJECTION.

heat-affected zone *See* WELDING.

heat balance (heat budget) An energy balance involving only heat transfer across the system boundary and the stored energy.

heat capacity (*C*) (Unit J/K or J/°C) The energy required to raise the temperature of a body by 1K without change of phase. Heat capacity is an extensive thermodynamic property dependent on temperature and pressure. *See also* MOLAR HEAT CAPACITY; SPECIFIC HEAT CAPACITY.

heat-capacity rate (*c*) (Unit W/K) For fluid flow in a duct with mass flow rate $\dot{m}$, the product $\dot{m}C_P$ where C_P is the specific-heat capacity, at constant pressure in the case of a gas.

heat-capacity ratio (c*) **1.** A non-dimensional parameter that arises in the analysis of heat exchangers, defined as the ratio c_{MIN}/c_{MAX} where c_{MIN} is the heat-capacity rate *c* for the fluid with the smaller value of *c* and c_{MAX} the value for the fluid with the larger value of *c*. Different flow rates and specific-heat values give rise to different values for *c*. **2.** *See* SPECIFIC-HEAT RATIO.

heat conduction *See* HEAT TRANSFER.

heat conductivity *See* HEAT TRANSFER.

heat content A term, not recommended, sometimes used to mean specific enthalpy.

heat convection *See* HEAT TRANSFER.

heat duty The heat-transfer rate possible with a given heat exchanger.

heat energy *See* THERMAL ENERGY.

heat engine A continuously operating thermodynamic system in which heat is converted to work and across the boundaries of which flow only heat and work. External-combustion engines are heat engines but, in thermodynamic terms, internal-combustion engines are not because air, fuel and exhaust gas flow across the system boundary. However, the term is often used in the broader sense to include any work-producing device that operates in a mechanical cycle but not in a thermodynamic cycle since the working fluids do not undergo a complete cycle.

heat-engine cycles *See* AIR-STANDARD CYCLES.

heat exchanger (HEX, recuperator) A device in which heat is transferred, by a combination of convection and conduction, between two fluid streams at different temperatures without them coming into contact. There are numerous configurations including counter-flow, cross-flow, parallel-flow, and shell and tube. *See also* REGENERATOR.

heat-exchanger effectiveness (effectiveness, ε) For a given heat exchanger, the ratio of the actual heat-transfer rate $\dot{Q}_{ACT}$ to the maximum possible heat-transfer rate for the given inlet temperatures, flow rates, and specific heats for the two fluids. It can be shown that $\varepsilon = \dot{Q}_{ACT}/C_{MIN}\Delta T$ where C_{MIN} is the lower of the two specific heats and ΔT is the temperature difference between the two fluids at inlet.

heat flow *See* HEAT TRANSFER.

heat-flow line Straight or curved lines, perpendicular to the isotherms in a solid body through which there is heat transfer by conduction, showing the direction of the local heat-flux vector.

heat flux ($\dot{q}''$) (Unit W/m^2) In any heat-transfer problem, a vector quantity being the rate of heat transfer per unit area.

heating degree day *See* DEGREE DAY.

heating value of a fuel *See* CALORIFIC VALUE OF A FUEL.

heat liberation rate The rate at which heat is generated during an exothermic reaction.

heat loss The irreversible heat transfer to the surroundings that occurs during a thermodynamic process.

heat-loss flow meter *See* THERMAL MASS FLOW METER.

heat of ablation A measure of the ability of a material to serve as a heat-protection element by a process of burning off in a severe thermal environment, such as re-entry of a spacecraft into the earth's atmosphere. Approximately equal to $m(h_F + [T_M - T_I]C_P)$ where m is the mass of the material being ablated, h_F is the latent heat of fusion, T_M is the melting temperature, T_I is the initial temperature and C_P is the specific heat.

heat of combustion *See* ENTHALPY.

heat of transformation The heat associated with a phase change, including condensation, crystallization, evaporation (vaporization), melting (fusion), solidification, and sublimation, is generally termed latent heat. For processes such as reaction, including combustion and formation, as well as cooling and compression, the term enthalpy of reaction, etc. is used.

heat of wetting The increase in enthalpy of a substance that occurs when water vapour is adsorbed and condenses. In combustible substances, such as coal or charcoal, this can lead to spontaneous ignition.

heat pipe A metal tube lined with a wick and filled with a liquid in equilibrium with its vapour. The liquid is evaporated at the hot end, the vapour flows along the centre, condenses at the cold end, is absorbed by the wick and transported back to the hot end. Applications include cooling of electronic components, solar power, and heat recovery.

heat pump (thermal pump) A device operating on a refrigeration cycle in which low-grade heat from the surroundings, such as a river or cold ambient air, is used to raise the temperature of a house or building either by heating room air or circulating water.

heat radiation *See* INFRARED RADIATION; HEAT TRANSFER.

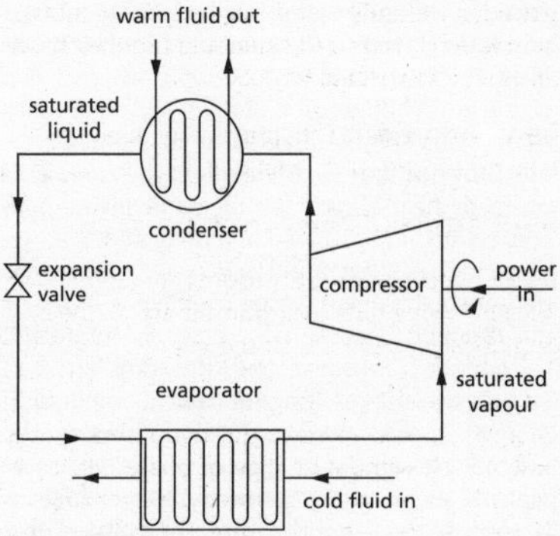

heat pump

heat rejection The transfer of heat out of a system. *See also* HEAT ADDITION.

heat-release rate The rate at which thermal energy is released from combustible materials, as in a fire.

heat seal A join or seal between two sheets of thermoplastic material through the application of heat. Commonly used to seal bags such as vacuum packs.

heat shield *See* ABLATION COOLING.

heat sink (heat reservoir, sink) A thermal-energy reservoir that absorbs energy in the form of heat.

heat source A thermal-energy reservoir that supplies energy in the form of heat.

heat transfer (heat flow, heat transmission, heat transport) The transport of energy due to a temperature difference within a solid object or stationary fluid, between solid objects or between a solid object and a fluid. **Steady-state heat transfer** refers to any heat-transfer process in which the temperature, heat flux, material and flow properties at all points are independent of time. There are three basic modes of heat transfer: conduction, convection, and thermal radiation.

Conduction (thermal conduction) occurs by the translational, rotational and vibrational energy of a molecule being transferred to nearby molecules by collisions. It is the only mode of heat transfer within a solid, but also contributes to heat transfer within a fluid. According to **Fourier's law of heat conduction**, the heat flux $\dot{q}''$ is proportional to the temperature gradient ∇T, i.e. $\dot{q}'' = -k\nabla T$ the constant of proportionality k being the **thermal conductivity (heat conductivity)**, a material property with unit W/m.K. For an isotropic, solid medium, the three-dimensional unsteady heat-conduction equation (**Biot–Fourier equation**) in a Cartesian-coordinate system (x, y, z) may be written as

$$\frac{\partial}{\partial x}\left(k\frac{\partial T}{\partial x}\right)+\frac{\partial}{\partial y}\left(k\frac{\partial T}{\partial y}\right)+\frac{\partial}{\partial z}\left(k\frac{\partial T}{\partial z}\right)+\dot{q}'''=\rho C\frac{\partial T}{\partial t} \quad \text{or} \quad \frac{1}{\alpha}\frac{\partial T}{\partial t}=\frac{\dot{q}'''}{k}+\nabla^2 T$$

where T is the temperature, $\dot{q}'''$ is the rate of thermal energy generated internally, ρ is the density of the medium, C is its specific heat, t is time, $\alpha \equiv k/\rho C$ is the **thermal diffusivity (thermometric conductivity, α)** with unit m^2/s, and ∇^2 is the Laplace operator. A non-dimensional number that arises in the analysis of unsteady heat-transfer problems is the **Fourier number (Fourier modulus, *F, Fo*)** defined as $F = \alpha t/L^2$, where L is a characteristic dimension of the body, sometimes taken as its volume divided by its surface area. The **Biot number (*B, Bi*)** arises in the analysis of unsteady heat-transfer problems involving both convection at the surface of a body and conduction within it. It is defined as $Bi = hL/k$, where h is the convective heat-transfer coefficient. If $Bi << 1$ the thermal response of a body is dominated by surface resistance and the problem can be solved using the lumped-capacity method. If $Bi >> 1$ the response is dominated by internal resistance. In studies of the effects of thermal shock on the properties of materials, B is sometimes called the **Biot modulus**. For a solid plate with thickness Δ and thermal conductivity k, the quantity Δ/k is the **conduction resistance (thermal resistance)**, with unit m^2 K/W. Its reciprocal is the **thermal conductance**.

Convective heat transfer (convection, thermal convection) takes place when a fluid at a different temperature flows over a surface and can be correlated using **Newton's law of cooling** $\dot{q}''_S = h\Delta T$ where $\dot{q}''_S$ is the surface heat flux and ΔT is the temperature difference between the surface and the flowing fluid. The **heat-transfer coefficient (coefficient of heat transfer, conductance, convective heat-transfer coefficient, film coefficient, heat transfer conductance, surface conductance, thermal conductance, thermal-exchange coefficient, *h*)** has unit W/m^2 K and is defined by $h = \dot{q}''_S/\Delta T$. Convection resistance is the inverse of h. Convection involves the combined effects of conduction and fluid motion and is called **forced convection** if the fluid flow occurs due to the action of a fan, pump or movement of a surface. If the fluid motion occurs due to buoyancy effects induced by density differences within the fluid, the convection is called **free** or **natural convection**. The heat-transfer coefficient h is a quantity which depends upon numerous factors, depending upon the physical situation, including the flow and thermal boundary conditions at the surface (e.g. solid with no slip, suction, constant temperature, or constant heat flux). Also important are the free-stream fluid velocity U_∞, a length scale which characterizes the flow geometry, such as a diameter or other length L, the fluid properties density ρ, dynamic viscosity μ, thermal conductivity k, and specific heat C_P. Where the fluid properties vary significantly with temperature, an average value is used to characterize the flow. For high-speed flow, the soundspeed c also plays a role. In the case of forced convection, it is usual to express correlations or plot data for h in non-dimensional form as a Nusselt number Nu $(= hL/k)$ or a Stanton number St defined as $h/\rho U_\infty C_P$. Nu and St depend upon the flow Reynolds number Re $(= \rho VL/\mu)$, Prandtl number Pr $(= C_P\mu/k)$ and, for high-speed gas flows, the Mach number M $(= V/c)$. For natural convection, the usual non-dimensional number which is introduced is the Grashof number Gr $(= g\beta\Delta TL^3/\nu^2)$ where β is the coefficient of volume expansion of the fluid, equal to $1/T$ for an ideal gas, and ν is the kinematic viscosity of the fluid $(= \mu/\rho)$. The **conduction thickness (δ_C)** is a scaling length sometimes used in convective-heat-transfer analysis, defined by $\delta_C = k\Delta T/\dot{q}''_S$. Although numerous empirical correlations exist for convective heat transfer in both laminar flow and turbulent flow, most calculations are now carried out using commercially available or in-house computational fluid dynamics (CFD) packages. Convective heat-transfer problems can also include further complications such as transpiration cooling (involving mass transfer through the surface), phase change of liquid or vapour in contact with the surface (such as boiling or condensation), and evaporation, burning, or ablation of the surface itself. The heat-transfer coefficient can be modified to include the effects of thermal radiation. Closely related to thermal convection is mass transfer associated with diffusion due to near-wall concentration gradients.

The emission of **thermal radiation** from a real surface is usually written as a fraction, the emissivity ε, of the emission of radiation from the surface of a perfect blackbody according to $\dot{q}''_S = \varepsilon\sigma T_S^4$ where σ is the Stefan–Boltzmann constant and T_S is the surface temperature. Because of the T_S^4 dependence, radiation becomes increasingly important as the surface temperature increases. In most practical situations the radiation from one surface to another does pass through a medium, such as a gas, but this merely acts to absorb and re-emit some of the radiation and is not a necessary aspect of the process which basically involves transmission of electromagnetic waves through space between surfaces, the geometry and orientation of which are important factors. The absorptivity of a surface α is the fraction of radiation energy incident on a surface that is absorbed by the surface. According to Kirchhoff's law of radiation, α and ε are equal at the same temperature and wavelength.

SEE WEB LINKS

- Reference book companion website covering many material properties (thermal conductivity, density, thermal expansion coefficients, surface tension, viscosity, etc.)

heat transfer *j*-factor *See* COLBURN *J*-FACTOR.

heat treatment (heat treating) Alteration of the mechanical properties of materials, particularly metals, by different sequences of heating, holding at temperature, and cooling at different rates. *See also* AGEING; ANNEALING; HARDENABILITY; PRECIPITATION HARDENING; TEMPERING.

SEE WEB LINKS

- Heat treatment of steels terminology

heat value (heating value) *See* CALORIFIC VALUE OF A FUEL.

heat wheel *See* ROTARY REGENERATOR.

hecto (h) *See* INTERNATIONAL SYSTEM OF UNITS.

height A dimension of an object measured vertically, or the vertical distance to an object. *See also* ALTITUDE; DEPTH.

Heisler chart A type of chart on which curves show the results of approximate transient heat-conduction calculations for an infinite plate, an infinitely long cylinder, and a sphere, all with convection at the surface. If T_O is the centreline or centre temperature, T_I is the value of T_O at time zero and T_∞ is the temperature of the surroundings, one set of curves shows the non-dimensional temperature difference $(T_O - T_\infty)/(T_I - T_\infty)$ *vs* non-dimensional time in the form of the Fourier number *Fo* with the inverse of the Biot number *Bi* as the curve parameter. A second set shows non-dimensional temperature *vs* $1/Bi$ for different locations within the object. Gröber charts are similar to Heisler charts, but include a third set of curves for heat loss *vs* $FoBi^2$ (non-dimensional time) with *Bi* as the curve parameter.

Hele–Shaw cell An apparatus consisting of two flat plates, typically of thick glass, separated by a narrow parallel gap. Fluid of high viscosity is allowed to flow through the gap around shapes such as circles, rectangles, and aerofoil sections placed within the gap, with dye used to make the streamlines visible. The flow patterns correspond to those for two-dimensional potential flow.

helical-flow turbine (helical turbine) *See* GORLOV HELICAL-FLOW TURBINE.

helical gear *See* TOOTHED GEARING.

helical-rotor pump *See* PROGRESSIVE-CAVITY PUMP.

helical spring A spring manufactured by forming elastic wire into a helix which may be cylindrical, conical, barrel-shaped, or hourglass in overall form. Closed-coiled springs are used in tension; open-coiled in both tension and compression. *See also* WAVE SPRING.

helical winding A process used in the manufacture of fibre-reinforced composite objects, in which the fibres are wound along helical paths.

helicoid *See* SCREW.

helicoil® insert *See* THREAD INSERT.

helicopter An aircraft, typically powered by a gas turbine, for which lift is generated by a bladed rotor rotating in a horizontal or near-horizontal plane.

heliostat A sun-tracking device that uses a plane mirror to reflect solar radiation onto a target such as a receiver on a solar tower. Usually employed in an array of many such mirrors.

helium refrigerator A refrigerator that uses helium as the refrigerant in liquid form for part of the cycle. Temperatures below 5 K are achievable.

helix A curve on the surface of a right-circular cylinder or cone which cuts the generators of the cylinder at a constant angle. A helix is formed on the surface of a rotating cylinder or cone when the point of contact moves axially at constant speed. The relative speeds of rotation and axial movement determine whether the helix angle is large or small, i.e. is 'slow' or 'fast'.

helix angle *See* TOOTHED GEARING.

Helmholtz cavity *See* HELMHOLTZ RESONATOR.

Helmholtz function (Helmholtz free energy, Helmholtz potential, a, *f*) *See* GIBBS FUNCTION.

Helmholtz number (*He*) A non-dimensional number that arises in acoustics defined as $He = 2\pi L/\lambda = kL$ where L is a characteristic length, λ is the acoustic wavelength and k is the acoustic wavenumber.

Helmholtz resonator (Helmholtz cavity) A rigid cavity of volume $\mathcal{V}$, filled with a gas of soundspeed *c*, having a small opening with a short neck of effective length *L* and cross section *A*. The cavity resonates when excitation is applied at the opening: the fundamental natural angular frequency ω is given by $\omega = c\sqrt{A/L\mathcal{V}}$.

hemispherical total emissivity *See* TOTAL EMISSIVITY.

Hencky equations In plane-strain plasticity, the slip lines (orthogonal planes of maximum shear) are labelled α- and β-lines, a common sign convention being that the largest algebraic principal stress σ_1 lies in the first and third quadrants of the α-β system. The Hencky equations relate the hydrostatic (mean) stress *p* and the rotation ϕ of the slip lines. With *p* the compressive mean stress, and positive ϕ an anticlockwise rotation, $p \pm 2k\phi = \text{constant}$ where *k* is the rigid-perfectly-plastic shear yield stress (+ for α-lines; − for β-lines). The constants vary from one slip line to another, but rarely need evaluation. *See also* CHARACTERISTICS.

Hencky total-deformation plasticity theory (Hencky total-strain plasticity theory) A plasticity theory in which initial and final true strains are employed rather than the actual path-dependent incremental true strains.

HEPA filter *See* HIGH-EFFICIENCY PARTICULATE AIR FILTER.

hermetic seal A seal which is airtight. The term is used generally to mean gas tight.

herringbone gear *See* TOOTHED GEARING.

Herschel-type Venturi tube *See* VENTURI.

hertz (Hz) *See* INTERNATIONAL SYSTEM OF UNITS.

Hertzian stresses The stresses set up over the contact patch between two curved surfaces in elastic contact. *See also* CONTACT MECHANICS.

Hertz theory The stress and strain distributions given by elasticity theory for two (singly- or doubly-curved) surfaces under load.

HEX *See* HEAT EXCHANGER.

hex pump *See* SIMPLEX PUMP.

Hg *See* MERCURY.

H-girder *See* H-BEAM.

hierarchical control A control method in which separate controllers operate at two or more levels, some being responsible for low-level tasks and others for tasks at a higher level.

high-cycle fatigue *See* FATIGUE.

high-efficiency particulate air filter (HEPA filter) A filter consisting of a mat of fine fibres designed to remove at least 99.97% of airborne particles with diameters greater than about 0.3 μm.

high-energy forming *See* HIGH-RATE FORMING.

high-enthalpy resource *See* GEOTHERMAL ENERGY.

higher calorific value *See* CALORIFIC VALUE OF A FUEL.

higher pair A mechanism where the elements have line or point (not surface) contact. *See also* KINEMATIC PAIR; LOWER PAIR.

highest useful compression ratio (HUCR) The compression ratio of a piston engine giving the maximum power for a given fuel.

high explosive An explosive material that explodes by detonation. *See also* LOW EXPLOSIVE.

high-frequency furnace (high-frequency heater) *See* INDUCTION FURNACE.

high-head hydroelectric plant A hydroelectric installation where the entire reservoir is located well above (typically by more than 100 m) the turbine. For a **medium-head hydroelectric plant** the reservoir surface is located less than about 100 m above the turbine, as in a typical river-dam installation, and for a **low-head hydroelectric plant** the reservoir is located only slightly above (typically by about 10 m) the turbine, as in a tidal-barrage system.

high-pass filter *See* BAND-PASS FILTER.

high-pressure turbine (HP turbine) 1. (high-pressure cylinder) The turbine in a compound steam engine or steam-turbine system that receives steam directly from the boiler. **2.** The turbine in a two- or three-spool gas-turbine engine that receives high-pressure hot gas directly from the combustor(s). *See also* INTERMEDIATE-PRESSURE TURBINE; LOW-PRESSURE TURBINE.

high-rate forming (high-energy forming) Plastic forming of components by sudden application of load, using impact, explosive, or other means.

high-solidity wind turbine A wind turbine with a large number of blades, typically overlapping, so giving the impression of a solid disc. *See also* LOW-SOLIDITY WIND TURBINE.

high strength low alloy steels (HSLA steels) Steels with various micro-alloying elements such as copper, nickel, chromium, molybdenum, niobium, titanium, and vanadium in small quantities that give improved strength (as high as 900 MPa) and corrosion properties compared with plain carbon steels. Improvements are due to grain refinement and precipitation hardening, better control of the chemistry during steel making, and accurate rolling temperatures. *See also* CONTROLLED ROLLED STEEL; DUAL-PHASE STEEL.

high-technology robot A robot equipped with advanced sensors, such as a vision system, and a sophisticated controller allowing it to adapt to changes in its environment.

high-tensile bolt (high-tension bolt) A bolt manufactured from an alloy steel that has a high tensile strength of about 1 GPa.

Hill's spherical vortex (Hill's vortex) A recirculating viscous flow for which vorticity within a spherical volume varies linearly with radial distance from its centre.

Hilsch tube *See* RANQUE–HILSCH TUBE.

hinge 1. (hinge joint) *See* PIN JOINT. **2.** *See* PLASTIC HINGE.

hipping (HIP) *See* HOT ISOSTATIC PRESSING.

hob *See* TOOTHED GEARING.

hobbing *See* TOOTHED GEARING.

Hodgson number (*Ho*) A non-dimensional parameter used to characterize the damping effect, on an in-line flow meter, of a receiver installed in a pulsating flow system. It is defined as $HO = f\mathcal{V}\Delta p/\dot{\mathcal{V}}\bar{p}$ where f is the fundamental pulsation frequency, $\mathcal{V}$ is the volume of the receiver between the pulsation source and the flowmeter, Δp is the average pressure loss across the flow meter, $\bar{p}$ is the mean static pressure in the receiver and $\dot{\mathcal{V}}$ is the volumetric flow rate.

hodograph 1. (velocity diagram) A graphical construction in plane kinematics for acceleration, in which the velocity in the hodograph is equal to the acceleration in the actual motion. **2.** The name employed in plasticity theory for the vector diagram of particle velocities.

hohlraum A large cavity with a small opening. Radiant energy entering the cavity undergoes repeated reflections so that very little of the incident energy is re-emitted and the cavity has the characteristics of a black surface.

hoist A machine with a winding drum for lifting from directly above the load.

holdup In multiphase-phase flow in a tube or column, the ratio of the cross section of any one component at a particular location to the total cross section, and so effectively a volume fraction.

hole-basis fit *See* LIMITS AND FITS.

hole flanging The forming process in which a punch is passed through an already-existing hole in a sheet to form a circular wall on the far side of the hole.

hole saw *See* TREPANNING TOOL.

Holland's formula An approximate equation used to estimate the rise of a buoyant plume of exhaust gas or vapour from a stack, taking into account the stack height and diameter, the prevailing airspeed, temperature and density, and the effluent velocity and temperature at the stack exit.

hollow punch A thin-walled punch used with softer materials for cutting out discs or making holes.

holography A method of recording and displaying a three-dimensional image (a **hologram**) of an object, usually using coherent radiation from a laser. A hologram is formed on a photographic plate by interference patterns between direct and reflected light, the image being reproduced by illuminating the plate with coherent light. Applications of holographic interferometry include stress and vibration analysis.

Holzer's method A method to determine the natural frequencies of torsional vibration of a built-up system (such as a crankshaft, connecting rods, and pistons) that is idealized as a number of discs on a rotating shaft. It can also be applied to flexural vibrations as an alternative to Stodola's method.

homenergic flow *See* ISENERGIC FLOW.

homentropic flow *See* ISENTROPIC FLOW.

homogeneous **1.** A term for a material whose microstructure and properties do not vary with location, but which may still be anisotropic. **2.** A term used to describe turbulence that on average is uniform in all directions.

homogeneous strain **1.** Strain distributed uniformly throughout a body. **2.** In plasticity theory, homogenous strain is the true strain based on changes in the external dimensions of a body without the inclusion of redundant strains or strains resulting from friction. The plastic work required to achieve homogeneous strain is the minimum possible work required to perform an operation.

h

homogeneous transforms A technique for the kinematic analysis of industrial robots, based on representing each joint and associated link by a 4 × 4 matrix. The matrices are designated A_1 to A_n for a robot with *n* joints numbering from the base frame. The product of the matrices in numerical sequence (i.e. A_1 A_2 A_3 ... A_n) is a 4 × 4 matrix representing the end effector. The first column of this end-effector matrix is the normal vector, the second column the orientation vector, the third column the approach vector, and the fourth column the position vector.

homogenizer A mixing device, such as a blender, used to ensure that the properties of a mixture are uniform throughout.

homokinetic joint (homokinetic coupling) *See* CONSTANT-VELOCITY UNIVERSAL JOINT.

homologous temperature (T_M) The ratio of the absolute temperature of a metal to its melting point. Since alloys (except for those of eutectic composition) do not have melting points, either the liquidus or the solidus temperatures may be used. Creep becomes important if $T_m > 0.5$.

honeycomb A natural or man-made material consisting of thin-walled hollow tubular cells made of paper, metal, polymer, or composites. The cells are often hexagonal in cross section. Honeycomb is used for the cores of sandwich materials. The hexagonal-cell form, usually made of aluminium, is used in wind tunnels as a flow-conditioner to eliminate swirl. *See also* SCREENS.

honeycomb radiator A cross-flow heat exchanger, formerly used for motor-vehicle radiators, in which cooling gas, usually air, flows through a bundle of tubes the ends of which have a hexagonal cross section, while the liquid to be cooled flows in the gaps between the tubes.

honing Finish polishing of surfaces and cutting edges in which very fine abrasive powders, such as rouge, are employed. Honing is usually applied to cylindrical surfaces and lapping to flat, but the usage is not consistent.

Hooke's joint *See* UNIVERSAL JOINT (1).

Hooke's law The deflexion of an elastic body is proportional to the applied load, so **Hookean deformation** is linear elastic behaviour that is reversible and path-independent. In terms of uniaxial stress and strain, Hooke's law is $\varepsilon_x = \sigma_x/E$ where ε_x is the normal strain along the *x*-axis, σ_x is the normal stress and *E* is Young's modulus. *See also* GENERALISED HOOKE'S LAW; STIFFNESS.

hoop strain The strain in the circumferential (tangential) direction in cylindrical and non-cylindrical vessels and rotating bodies.

hoop stress The stress in the circumferential (tangential) direction in cylindrical and non-cylindrical vessels, when loaded by the hydrostatic head of a liquid in a fully- or partially-filled container or by internal or external pressure in a closed vessel. Hoop stresses also arise in spinning bodies. The corresponding strains are the **hoop strains**. *See also* MEMBRANE STRESS; MERIDIONAL STRESS.

hoop winding In the manufacture of cylindrical components from fibre-reinforced composite materials, when the reinforcement is placed only in the circumferential direction.

Hopkinson bar An instrumented device for applying loads to testpieces at very high velocities. A thin testpiece is sandwiched between an incident bar and a transmitter bar, both of which are strain-gauged. A striker bar is suddenly projected by a high-pressure gas gun against the incident bar to send a stress wave through the specimen. Measurement of the strains in the bars enables the stress-strain behaviour of the material to be deduced. Torsional Hopkinson bars give shear stress-strain behaviour.

horizontal-axis wind turbine (HAWT) A wind turbine in which the axis of rotation of the rotor is horizontal. Most wind turbines are of this type. *See also* VERTICAL-AXIS WIND TURBINE.

horizontal boiler A boiler in which the hot flue gases pass through horizontally.

horizontal engine Any piston engine in which the axis (or axes) of the cylinder (or cylinders) is horizontal. *See also* VEE ENGINE.

horizontal-return boiler *See* ECONOMIC BOILER.

horn 1. A device used with loudspeakers and other acoustic transducers so as to direct the sound in a particular direction and, by providing impedance matching, increase the sound pressure level radiated. **2.** An implement which uses ultrasonic vibration for cutting.

hornblock A casting attached to the frames of railway vehicles in which the sprung axlebox is constrained to move vertically.

horsepower (hp) A non-SI unit of power equivalent to 745.7 W.

horsepower transmitted The power at the output of an engine or engine–gearbox combination.

horseshoe vortex 1. (junction vortex) A vortex that forms when a boundary layer interacts with an object, such as a cylinder placed normal to a surface. Several such vortices develop in the junction. **2. (hairpin vortex)** A flow structure used to model near-wall turbulence.

hose Flexible tubing, often made from rubber or polymers, used to carry gases, liquids and slurries. It is sometimes reinforced on the outside to allow high pressures and to protect against wear and abrasion.

hoseclip (hose clamp) A circular metal band, one end of which has a screw mechanism through which passes the other end such that the screw engages with indentations in the band, allowing the clip to be tightened or loosened. Used to hold a hose onto a pipe such as a water outlet.

hot-air engine *See* HOT-GAS ENGINE.

hot-air furnace A heating system in which air passes through a heat exchanger heated by hot gases from a burner.

Hotchkiss drive The system employed on rear-wheel-driven vehicles in which the gearbox is connected to the live rear axle/differential by a propeller shaft having universal joints at both ends. The torque is reacted by the leaf springs from which the rear axle is suspended. *See also* TORQUE TUBE.

hot dipping (hot-dip galvanization) A process of coating an iron, steel, or aluminium object with a thin layer of zinc by passing it through a bath of molten zinc at a temperature of about 460°C. *See also* GALVANIZATION.

hot-dry rock *See* ENHANCED GEOTHERMAL SYSTEM.

hot-film anemometer 1. A device similar to a hot-wire anemometer in which the sensing element is a conducting film on a ceramic substrate. More robust than a hot-wire probe and suitable for use in both liquids and gases. **2.** A surface-mounted device similar to (**1**), used for determining surface shear stress.

hot forging *See* HOT WORKING.

hot-gas engine (hot-air engine) A heat engine in which a gas in a cylinder is alternately heated, causing expansion, and cooled, causing contraction, such that a piston reciprocates and produces mechanical work.

hot-gas expander *See* FLUE-GAS EXPANDER.

hot hardness The hardness of materials at high temperatures, often used to rank the performance of cutting tools.

hot isostatic pressing Sintering of powdered materials at high temperature and high hydrostatic pressure in order to minimize porosity.

hot pressing 1. Sintering of metal or ceramic powder at high temperature. *See also* ISOSTATIC PRESSING. **2.** Formation of components from fibre-reinforced composite material in which bonding is ensured through application of pressure and temperature, often against a mould to give the required shape.

hot rolling *See* HOT WORKING.

hot shortness The reduction of ductility in steels at high temperatures caused by melting of sulphides that wet grain boundaries and spread along them.

hot stamping Hot forging of brass and bronze alloys.

hot strength The yield stress (or sometimes fracture stress) of a material at temperatures above about half its melting point.

Hottel–Willier–Bliss equation An equation used to predict the performance of flat-plate solar collectors: $\dot{q}'' = \tau_{COV}\alpha_P G - U(T_P - T_\infty)$ where $\dot{q}''$ is the heat flux into the plate, τ_{COV} is the transmittance of the cover, α_P is the absorptance of the plate, G is the irradiance on the panel, U is the overall heat transfer coefficient, T_P is the panel surface temperature, and T_∞ is the temperature of the surrounding air.

hot well A collection tank beneath the tubes of a condenser, used to collect the condensate.

hot-wire anemometer An instrument for determining steady or rapidly-fluctuating gas velocity by monitoring the temperature of a fine heated wire typically 5 μm diameter and 1 mm long. Calibration is essential, e.g. using King's law. *See also* HOT-FILM ANEMOMETER.

hot-wire detector *See* KATHAROMETER.

hot working Plastic deformation of a metal, by rolling (**hot rolling**), drawing, forging (**hot forging**) etc., at a temperature above its recrystallization temperature (in commercial alloys above about 40% of their melting points) which results in permanent shape change but no increase in strength or loss of ductility. *See also* COLD WORKING; STRAIN HARDENING; WARM WORKING.

Houdaille damper An untuned torsional damper consisting of a free rotational mass within a cylindrical cavity filled with a viscous liquid.

hovercraft A vehicle that rides over land or sea on a cushion of air supplied by rotors or fans. *See also* AIR-CUSHION VEHICLE.

hp *See* HORSEPOWER.

HP turbine *See* HIGH-PRESSURE TURBINE.

HSLA steels *See* HIGH STRENGTH LOW ALLOY STEELS.

HTU *See* HYDROTHERMAL UPGRADING.

hub *See* BOSS.

HUCR *See* HIGHEST USEFUL COMPRESSION RATIO.

humid heat (C_S) (Unit J/kg.K or J/kg.°C) The specific heat at constant pressure of moist air per unit mass of dry air in the mixture. It is given approximately by $C_S = C_{PA} + C_{PV}H$ where C_{PA} is the specific heat at constant pressure of dry air (1.005 kJ/kg K), C_{PV} is the specific heat of water vapour (1.82 kJ/kg K), and H is the absolute humidity, in mass of water vapour per unit mass of dry air in the mixture. *See also* PSYCHROMETRIC RATIO.

humidify To add water vapour to air or another gas to increase its humidity.

humidistat (humidity element, hygrostat) A device used to monitor and regulate relative humidity and control air-conditioning units, humidifiers, and dehumidifiers.

humidity *See* RELATIVE HUMIDITY; SPECIFIC HUMIDITY.

humidity ratio *See* SPECIFIC HUMIDITY.

Humphries equation An equation for the ratio of the specific heats of moist air: $\gamma_W = (7 + h)/(5 + h)$ where $h = 0.01RHp_{SV}/B$, RH being the relative humidity of the air (%), p_{SV} is its vapour pressure and B is the ambient pressure.

hundred-second creep modulus The constant applied stress divided by the strain reached after 100 s in time-dependent materials such as polymers. *See also* RELAXED MODULUS; TEN-SECOND CREEP MODULUS.

hunting The undesirable oscillation of the output of a controlled system about an average value.

Huygens–Steiner theorem *See* PARALLEL AXIS THEOREM.

H-VAWT A type of vertical-axis wind turbine with vertical straight blades attached to a horizontal support arm pivoted at its centre, thus forming an H. *See also* V-VAWT.

hybrid beam *See* SANDWICH BEAM.

hybrid composite A composite reinforced by more than one variety of filament.

hybrid vehicle A motor vehicle with two or more different engines, usually a piston engine and an electric motor.

hydragas suspension A motor-vehicle suspension system that uses the compressibility of nitrogen as a spring and a liquid, displaced by the nitrogen, for damping. *See also* HYDROPNEUMATIC.

hydraulic An adjective applied to any device in which the working fluid is a liquid, usually water or an oil. *See also* PNEUMATICS.

SEE WEB LINKS

- Hydraulic components terminology

hydraulic accumulator An energy-storage device within a hydraulic circuit in which a liquid is held at high pressure by a spring, weight, or high-pressure gas.

hydraulic actuator A device which uses the pressure of a liquid acting against a piston to produce displacement or force. A typical application is the opening and closing of a valve, a spring often being used as a return element.

hydraulic air compressor 1. A compressor powered by a hydraulic motor. **2.** A device in which air is compressed as a result of being entrained in water flowing in a downcomer pipe. The compressed air is released in a chamber at the bottom of the pipe.

hydraulically-smooth surface (hydrodynamically-smooth surface) A surface for which the scale of any roughness is smaller than the thickness of the viscous sub-layer in turbulent boundary-layer flow such that the roughness has no effect on surface shear stress. The usual criterion is $ku_\tau/\nu < 5$ where k is the roughness height, u_τ is the friction velocity, and ν is the kinematic viscosity of the fluid. *See also* FULLY-ROUGH FLOW; TRANSITIONAL ROUGHNESS.

hydraulic amplifier (hydraulic intensifier) A device which uses a low-pressure hydraulic signal to operate a slave valve and thus control a high-pressure hydraulic source so that a large hydraulic actuator can be controlled from a low-pressure hydraulic signal.

hydraulic brake A brake in which force is applied to a brake shoe or pad by a hydraulic actuator.

hydraulic circuit A connected system of hydraulic components such as actuators, pumps, and valves, used to apply, control, or transmit power.

hydraulic clutch *See* FLUID DRIVE.

hydraulic control system A control system in which the controller and actuator are implemented hydraulically.

hydraulic coupling *See* FLUID COUPLING.

hydraulic cylinder (linear hydraulic motor) A piston-cylinder arrangement in which a force is produced by applying hydraulic pressure to one face of the piston, resulting in linear motion. In double-acting cylinders, hydraulic pressure can be applied to either side of the piston to produce back-and-forth movement.

hydraulic diameter (D_H) (Unit m) An equivalent diameter for a cylindrical tube of any cross section defined by $D_H = 4A/P$ where A is the cross-sectional area and P is the wetted perimeter. D_H is a common choice for the characteristic length in the definition of non-dimensional parameters for flow and heat-transfer in non-circular ducts. *See also* HYDRAULIC RADIUS.

hydraulic drive *See* FLUID COUPLING.

hydraulic dynamometer (Froude dynamometer) A dynamometer in which shaft power is absorbed by a rotor rotating within a liquid-filled casing. The power is determined from the product of the shaft rotation speed and the torque required to prevent rotation of the casing.

hydraulic engine *See* HYDRAULIC MOTOR.

hydraulic fluid A liquid, usually water or an oil, used to transfer force, torque, or power in a hydraulic device. *See also* BRAKE FLUID.

hydraulic forming A method of forming in which hydraulic pressure is used to deform a blank into a die.

hydraulic fracturing *See* HYDROFRACTURING.

hydraulic horsepower *See* HYDRAULIC POWER.

hydraulic intensifier *See* HYDRAULIC AMPLIFIER.

hydraulic jetting The use of a liquid jet, often containing abrasive particles, to clean surfaces or cut a material.

hydraulic jump A spontaneous increase in the level of liquid flow in an open channel when the upstream flow is supercritical. *See also* FROUDE NUMBER.

hydraulic motor (hydraulic engine) A machine powered by high-pressure hydraulic fluid, such as a gear motor, swashplate motor, or vane motor. *See also* HYDRAULIC CYLINDER.

hydraulic nozzle A nozzle designed to convert a liquid flow into a spray. *See also* ATOMIZATION.

hydraulic packing *See* PACKING.

hydraulic power (hydraulic horsepower, *P*) (Unit kW or hp) The power transferred to a fluid by a pump or generated from a fluid flow by a hydraulic turbine. If the pressure change across the machine is Δp and $\dot{Q}$ is the volumetric flow rate, then $P = \dot{Q}\Delta p$.

hydraulic pump A machine, such as a gear pump, peristaltic pump, screw pump, swashplate pump, or vane pump, designed to pump a hydraulic fluid.

h

hydraulic radius (r_H) (Unit m) An equivalent radius for a cylindrical tube of any cross section defined by $r_H = D_H/4$ where D_H is the hydraulic diameter. Note that r_H is not equal to $D_H/2$ as might have been expected.

hydraulic ram 1. A single-acting hydraulic cylinder in which the piston rod almost fills the cylinder bore.
2. A water-powered pump in which there are two connected pistons in cylinders, one of which operates as a water-operated piston engine and the other as a water pump. By using a smaller diameter for the pump piston, a high pressure can be obtained from a low-pressure water supply. They are commonly used for irrigation.

hydraulics The study of water flow in open channels (including canals, rivers, etc.), water-supply, drainage, and irrigation systems.

hydraulic telemotor The part of the steering mechanism of a ship in which hydraulic power is used to turn the rudder in response to rotation of the steering wheel.

hydraulic test Proof testing a pressure vessel by filling it with water and overloading to 1.5–2 times its design pressure.

hydraulic transport The movement of solid material, usually in particulate form, by water flow through a pipe or open channel.

hydraulic turbine (water turbine) A turbomachine, such as a Francis, Kaplan, Pelton, or Turgo turbine, used to produce shaft power from a head of water. *See also* HYDROELECTRIC GENERATOR; HYDROPOWER.

hydraulic valve A device that directs the flow of a hydraulic fluid from its input ports to its output ports, typically by linear movement of a sliding spool that opens and closes the ports.

hydrocarbon An organic chemical compound containing only carbon and hydrogen. Naturally-occurring examples include gases such as methane, propane, and butane, liquids such as hexane, octane, and benzene, and solids such as paraffin wax. Synthetic examples are polymers such as polyethylene and polystyrene.

hydrocolloid *See* HYDROSOL.

hydrocyclone (hydroclone) A device used to separate solids or liquids from gases, or to separate mixtures of liquids of different densities by creating a swirling flow. A hydrocyclone is cylindrical at the top, where flow enters tangentially and the lighter fraction leaves axially at the overflow. It is conical below where the heavier fraction leaves through the underflow.

hydrodynamically-smooth surface *See* HYDRAULICALLY-SMOOTH SURFACE.

hydrodynamic boundary layer *See* BOUNDARY LAYER.

hydrodynamic lubrication (fluid-film lubrication) The use of a film of oil to develop sufficient pressure as a slightly eccentric shaft rotates in a bearing to prevent contact of the surfaces and support a load. *See also* ELASTOHYDRODYNAMIC LUBRICATION; HYDROSTATIC BEARING.

hydrodynamic oscillator A device in which flow kinetic energy amplifies a flow instability and positive feedback leads to sustained oscillation in the form of mechanical vibration or sound.

hydrodynamic roughness *See* FULLY-ROUGH FLOW; HYDRAULICALLY-SMOOTH SURFACE; TRANSITIONAL ROUGHNESS.

hydrodynamics A term sometimes used to mean fluid dynamics, even when the fluid is a gas or a liquid other than water.

hydroelasticity A phenomenon equivalent to aeroelasticity where the fluid is a liquid rather than air.

hydroelectric generator The combination of a hydraulic turbine and an electrical generator.

hydroelectric system An installation comprising a hydroelectric generator, a water source, a penstock, flow and generator controls, and cables for electricity distribution. *See also* HYDROPOWER.

hydrofoil **1.** A wing-like device, usually mounted on struts beneath the hull of a boat, that generates lift when propelled through water. **2.** A vehicle employing hydrofoils to create lift.

hydrofracturing (fracking, hydraulic fracturing) A technique to increase the surface area available for heat transfer from hot dry rock by pumping water at sufficiently high pressure into a borehole to open pre-existing fractures. It is also used to extract shale gas.

hydrogen embrittlement Premature crack growth over time under tensile stress leading to unexpected failure in certain metals, caused by small amounts of hydrogen in the microstructure. Hydrogen may enter steels during melting or heat treating, or during processes such as electroplating.

hydrometer (aerometer) An instrument, typically consisting of a glass cylinder weighted at its lower end and with a graduated stem, used to measure the relative density of a liquid. When the cylinder is submerged, the position of the liquid surface on the stem indicates the relative density.

hydronic heating (radiant heating) A heating system in which a hot liquid, typically water, is circulated through plastic tubing embedded in, or running beneath, a floor.

hydronic separator *See* HYDROSEPARATOR.

hydrophilic A term for a material that has an affinity for water, readily absorbing it or being wetted by it. In the case of a colloid, readily forming or remaining as a hydrosol.

hydrophobic A term for a material that tends to repel, or not absorb water, and is not readily wetted by it. In the case of a colloid, not readily forming or remaining as a hydrosol.

hydrophone A transducer used to measure acoustic pressures in a liquid.

hydropneumatic **1.** A term referring to a fluid power or suspension system combining hydraulic and pneumatic devices. *See also* HYDRAGAS SUSPENSION. **2.** A term referring to a device, such as a pulsation damper, in which the compressive properties of a gas are balanced against the incompressible properties of a liquid.

hydropower (water power) The shaft power generated by a hydraulic turbine from a head of water, either for generating electricity or for direct mechanical purposes. *See also* HYDRAULIC TURBINE; HYDROELECTRIC SYSTEM.

hydroseparator (hydronic separator) A device for separating heavy from light particles using an upward stream of water.

hydrosizer *See* HYDRAULIC CLASSIFIER.

hydrosol (hydrocolloid) A sol in which water is the continuous phase. *See also* AEROSOL.

hydrostatic bearing *See* BEARING.

hydrostatic equation The equation governing the spatial variation of static pressure p (**hydrostatic pressure**) in a liquid in which there is no shearing. For a liquid of density ρ, either at rest or moving with constant speed, the increase of p with depth z, is given by $dp/dz = \rho g$ where g is the acceleration due to gravity. If ρ and g are constant, $p = p_0 + \rho gz$ where p_0 is the pressure at $z = 0$. A fluid satisfying the hydrostatic equation is said to be in **hydrostatic balance**. *See also* FLUID MECHANICS.

hydrostatic modulus *See* BULK MODULUS.

hydrostatic press A press having a small-diameter input cylinder connected to a large-diameter output cylinder. The work done by a small input force acting over a long stroke, becomes a large output force acting over a small distance. *See also* JACK.

hydrostatic pressing *See* ISOSTATIC PRESSING.

hydrostatics *See* FLUID MECHANICS; HYDROSTATIC EQUATION.

hydrostatic stress (σ_H) (Units Pa) In solid mechanics, the stress equivalent to hydrostatic pressure is the mean or hydrostatic stress $\sigma_H = (\sigma_1 + \sigma_2 + \sigma_3)/3$, where σ_1, σ_2, and σ_3 are the principal stresses. σ_H can be tensile as well as compressive. *See also* DEVIATORIC STRESS.

hydrostatic test *See* HYDRAULIC TEST.

hydro-technology The use of falling water (i.e. a head of water) to generate power.

hydrothermal upgrading (HTU) A biofuel conversion technology in which wet biomass is converted at 300–350°C and high pressure to a heavy, hydrocarbon-rich liquid, which can be refined to produce a fuel similar to diesel oil.

hygrograph A recording hygrometer that produces a hygrogram.

hygrometer *See* PSYCHROMETER.

hygrometric chart *See* PSYCHROMETRIC CHART.

hygrometry *See* PSYCHROMETRY.

hygrostat *See* HUMIDISTAT.

hygrothermograph An instrument that records both temperature and relative humidity.

hyperbaric chamber A pressure vessel in which structural and other tests can be carried out at high pressure.

hyperboloid of revolution *See* TOOTHED GEARING.

hypersonic flow (hypervelocity flow) A supersonic flow for which the Mach number M is much greater than unity (typically taken as greater than 5) at which gas ionization and substantial thickening of viscous layers come into play, as in atmospheric re-entry where $M > 25$. Hypersonic flow conditions can be realized in a **hypersonic wind tunnel**.

hyperstatic *See* STATICALLY-DETERMINATE.

hypobaric chamber *See* ALTITUDE CHAMBER.

hypocycloid The curve traced by a point on the circumference of one circle as it rolls around the inside of another circle. *See also* EPICYCLOID.

hypoid axle A final-drive axle equipped with hypoid gearing such that the driveshaft axis is below the centre of the driven ring gear. *See also* TOOTHED GEARING.

hypoid gear *See* TOOTHED GEARING.

hypsometric equation *See* ISOTHERMAL ATMOSPHERE.

hypsometry The science of measuring altitude.

hysteresis Where a change in some property of a physical system lags behind changes in the phenomenon causing it, shown, in a load-deflexion diagram for example, by the formation of a **hysteresis loop** in which the recovered displacement lags the recovered load. The dissipated work is given by the area between the loading and unloading lines. In a stress–strain diagram, the area is work per unit volume. *See also* MECHANICAL HYSTERESIS.

hysteresis damping (hysteretic damping) Damping caused by hysteresis, in which energy is dissipated in a path-dependent, but reversible, system.

Hz *See* HERTZ.

I-beam A beam having a cross-section in the shape of an upper-case letter I, the top and bottom parts of which are called flanges and the vertical part the web. *See also* H-BEAM.

ICAO *See* STANDARD ATMOSPHERE.

ice point A fixed point on some temperature scales that corresponds to the temperature of a mixture of pure water and ice at a pressure of one atmosphere, normally taken as 0°C.

ID *See* INSIDE DIAMETER.

ideal cycle Any thermodynamic cycle in which there are no internal irreversibilities.

ideal fluid *See* PERFECT FLUID.

ideal gas *See* PERFECT GAS.

ideal gas law *See* PERFECT GAS.

ideal-gas temperature scale *See* RANKINE TEMPERATURE SCALE.

ideal liquid *See* PERFECT FLUID.

ideal mixture A mixture of substances in which the effects of dissimilar molecules are negligible.

ideal radiator *See* BLACKBODY.

ideal solid An incompressible solid.

ideal solution A solution in which the effects of dissimilar molecules are negligible.

ideal vapour-compression refrigeration cycle *See* VAPOUR-COMPRESSION REFRIGERATION CYCLE.

identification The estimation of the parameters describing a system from observation of the output(s) over a range of inputs. Normally undertaken using a least-squares method, i.e. a method that adjusts the estimates of the parameters to minimize the sum of the squares of the difference between the actual system output and that predicted using the estimated parameters.

idle (idling, tick over) An unloaded engine slowly running when the throttle is at a preset, almost-closed position is said to be idling. The **idling jet** in a piston-engine carburettor supplies sufficient fuel for the engine to idle.

idler gear (idler wheel) An intermediate wheel in gearing that either reverses rotation so as to ensure that the next gear rotates in the same direction as the driving gear, or fills a gap in the spacing of axles. The shaft on which an idler gear or pulley is attached (**idling shaft, lay shaft, counter shaft**) does not transmit power.

igniter A device, often similar to a spark plug or glow plug, used to initiate combustion in gas-turbine engines, rocket motors and some piston engines.

ignition The initiation of combustion.

ignition delay The time interval, in a compression-ignition engine, between the start of injection and the onset of combustion.

ignition system The electro-mechanical system that controls the voltage, timing, and duration of the spark for a spark-ignition engine. **Ignition timing** refers to the crank angle relative to top-dead centre at which the spark occurs in any cylinder. **Ignition advance (spark lead)** is where the spark is caused to occur earlier in the combustion cycle as the engine speed increases to obtain the highest peak pressure. **Ignition delay (ignition lag)** is the delay between the occurrence of the spark and departure of the pressure trace from that of an unfired cycle.

IHP *See* INDICATED HORSEPOWER.

image table In a programmable logic controller, a part of the memory configured so that signals applied to the input port can be read from these memory locations and data written to these locations appear on the output port.

imep *See* INDICATED POWER.

immediate surroundings In thermodynamics, that part of the surroundings affected by any process taking place within a system. *See also* ENVIRONMENT.

impact Sudden loading of a body or component, as in a vehicle collision or striking by a hammer.

impact energy **1.** The sum of the kinetic energies of all the bodies involved at the instant of collision of two or more moving bodies. **2.** The energy required to fracture a specimen in a Charpy or Izod impact test.

impact extrusion The extrusion of separate components of soft metals whose properties permit the action to be performed quickly.

impact load (Unit N) The load *P* during an impact given by the impulse

$$\int_0^t P dt$$

where *t* is the duration of the impact. The impulse is equal to the change in momentum of the colliding bodies, both of which may be moving or one at rest.

impact modulator An axisymmetric fluidic amplifier involving two coaxial opposed jets.

impact strength **1.** (Unit Pa) The stress to cause failure (by yielding or fracture) under conditions of high strain rate. **2.** A term sometimes used to describe the energy required to fracture a specimen in a Charpy or Izod impact test, even though the units (J) are not those of strength.

impact testing Determination of the mechanical properties of materials under high-rate conditions. Often determined from the behaviour of a testpiece when struck by a weight falling from a known height.

impact tube *See* PITOT TUBE.

impact wrench A pneumatically-or electrically-powered socket wrench used to tighten or loosen nuts through the application of torque in a rapid series of impulses.

impedance The ratio of the potential applied to a system to the flow that results. Typical potentials are pressure, voltage, and temperature, with the resulting flows being mass flow, current, and heat flux, respectively.

impeller An enclosed rotor in a centrifugal turbomachine, such as a turbocharger, pump, or compressor, with radially-oriented vanes that impart kinetic energy to a fluid. The fluid enters the impeller axially and is discharged radially into a diffuser channel. *See also* EYE.

imperfect gas *See* PERFECT GAS.

Imperial Standard Wire Gauge *See* WIRE GAUGES.

imperial ton *See* LONG TON.

impingement **1.** A narrow jet of fluid striking a surface. **2.** A flame striking boiler parts, causing local overheating.

implicit programming The programming of a robot at a high level using a few commands that imply many actions. For example, an implicit program may have statements such as ‘Go to store and pick up component’ whereas an explicit program would require many lines of code for the same task with these lines specifying each movement in detail. Implicit programming is currently a research technique, industrial robots normally being programmed using explicit methods.

impulse **1.** A signal that has a very short (in principle, tending to infinitessimaly short) duration whilst still being time-integrable. It is represented mathematically in continuous time by the Dirac delta function. **2.** The time integral of a transient force *P*, i.e.

$$\int_0^t P dt,$$

that is equal to the change in momentum during an event. Customarily thought of in terms of a rapid application of force, it is equally applicable to slow movements. *See also* IMPACT LOAD.

impulse-reaction turbine A steam turbine or gas turbine in which there are both impulse and reaction stages.

impulse-reaction turbine blade A turbine blade in which the cross section changes from that of an impulse blade close to its root to a reaction blade at its tip. The blade twists from root to tip to compensate for the increase in blade speed with radius and ensure the flow direction relative to the blade has the correct angle of attack.

impulse response The output of a dynamic system as a result of a single impulse input. As the Laplace transform of such an impulse input is unity, that of the resulting output is equal to the transfer function of the system.

impulse theorem *See* MOMENTUM THEOREM.

impulse train A series of impulses equally spaced in time.

impulse-turbine stage A turbine stage in which practically all the pressure drop takes place in one or more stationary nozzles (the stator) and the pressure remains constant as the working fluid passes over the blades of a rotor. Examples include the Pelton, Banki, and Turgo water turbines and the first stages of steam turbines. The diagram shows blading cross sections typical of an impulse steam turbine together with velocity triangles for the flow into and leaving the rotor. *See also* REACTION STAGE.

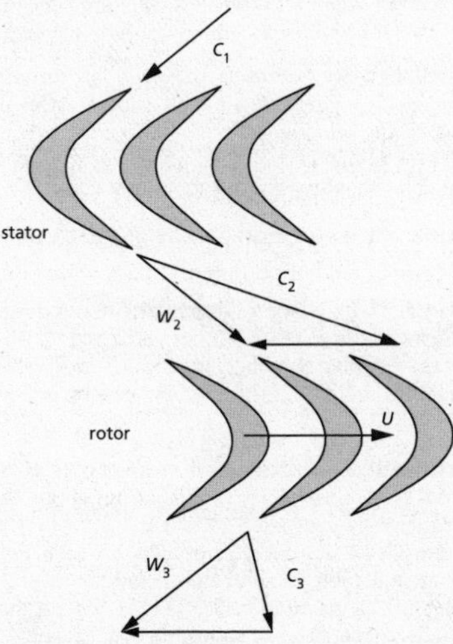

impulse-turbine stage

impulse wheel The rotor of an impulse-turbine stage in which torque is generated as a consequence of momentum transfer from the high-speed working fluid as it changes direction relative to the blading.

impulsive force *See* IMPULSE (2).

impulsive motion The response of a body or system to an impulsive force.

inaccuracy *See* ERROR.

inch A non-SI unit of length now defined as 25.4 mm.

incinerator A furnace or combustion chamber in which waste material is burned to produce heat, flue gas, and ash.

inclined-tube manometer Typically, a liquid-in-glass tube manometer, which can be inclined to the vertical to amplify the displacement (L) of the liquid interface along the inclined tube when a pressure difference ($p_1 - p_2$) is applied. If the vertical height change of the interface is h, $p_1 - p_2 = \rho gh$ where ρ is the liquid density and g is the acceleration due to gravity. If the angle of inclination is α, $h = L \sin \alpha$. Water, paraffin, or mercury is commonly used as the liquid. *See also* WELL-TYPE MANOMETER.

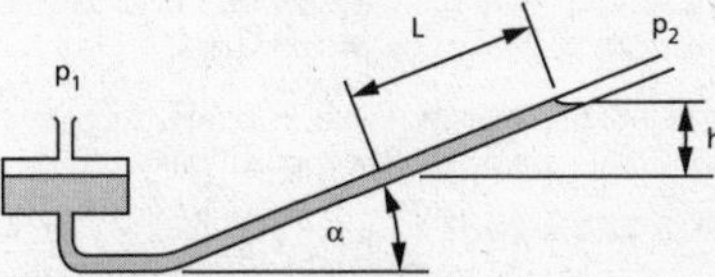

inclined-tube manometer

inclinometer (clinometer) An instrument for measuring the slope of a surface with respect to the vertical.

included angle *See* SCREW.

inclusions Small particles of foreign matter, such as slag and refractory, oxides, etc., found in the microstructure of metals. Fracture often initiates at such sites.

incomplete combustion The partial oxidation of a fuel.

incomplete lubrication *See* STARVED LUBRICATION.

incompressible A material for which the specific volume (or density) does not change with increasing or decreasing pressure, is incompressible, and has an infinite speed of sound. Liquids are practically incompressible. *See also* COMPRESSIBILITY; PERFECT FLUID.

incremental encoder A device for measuring translation or rotation, where the output is a series of pulses which must be counted electronically to determine the actual position or angle.

indentation hardness A measurement of hardness taken by means of an indenter, pressed into a material by a load, which makes a permanent impression. The hardness is given by the load divided by the area of the indentation. Plasticity theory shows that the hardness is about 2.5–3 times the uniaxial yield stress. *See also* HARDNESS SCALES.

indenter The hard, rigid tool employed in hardness testing. Different hardness tests use different types of indenter: a ball for Brinell hardness, a pyramid for Vickers hardness.

independent jaw chuck A four-jaw chuck for a lathe in which each jaw can be moved independently so as to enable workpieces of irregular cross section to be centred accurately.

indeterminate structure A structure in which the loads in the members cannot be evaluated from statics alone. *See also* DETERMINATE STRUCTURE; REDUNDANT STRUCTURE.

indexable teeth Teeth in a tool, such as a saw, that may be replaced individually.

indexable tool A cutting tool having a number of edges that can be used in succession after one or more has worn out.

index centres The two centres on a machine tool about which workpieces are centred and rotated.

index head *See* DIVIDING HEAD.

indexing The process of incrementing, by small fixed amounts, a workpiece in order to present predetermined angles or spaces for cutting tools to work in (as in gear cutting). *See also* DIVIDING HEAD.

index of compression *See* POLYTROPIC PROCESS.

index of expansion *See* POLYTROPIC PROCESS.

index plate A plate having a series of concentric rings of holes, every ring having a different spacing between the holes, used with a dividing head.

indicated power (*IP*) (Unit kW) The theoretical power output calculated, for a piston engine, from the indicated mean-effective pressure *imep* (Pa), and given by $\mathrm{IP} = imepALNn/C$ where A is the cross-sectional area of a piston (m^2), L is the stroke (m), N is the rotational speed (rps), n is the number of cylinders, and the factor C is 2 for a four-stroke engine and 1 for a two-stroke engine. *IP* represents the power output before frictional and other losses are accounted for. The **indicated horsepower (*IHP*)** is the indicated power converted to horsepower. An **indicator diagram** is a plot of the instantaneous cylinder pressure *vs* the instantaneous volume between the piston crown and the cylinder head. For a four-stroke engine, the curve has a power loop (positive area) and a pumping loop (negative area). The **indicated mean-effective pressure** is the average cylinder pressure over a cycle evaluated from an indicator diagram. It is the net positive area within the pressure–volume curve divided by the swept volume. The **indicated thermal efficiency (η_{IT})** is the ratio of the indicated power to the rate at which thermal energy is supplied, and given by $\eta_{IT} = IP/(\dot{m}_F Q_{NET})$ where $\dot{m}_F$ is the mass flow rate of the fuel and Q_{NET} is its net (or lower) calorific value. *See also* BRAKE MEAN-EFFECTIVE PRESSURE; BRAKE POWER.

indirect-arc furnace An electric-arc furnace for high-melting-point metals such as steel in which the arc does not come into contact with the metal that is contained in a refractory-lined hearth. *See also* DIRECT-ARC FURNACE.

indirect extrusion (backward extrusion) *See* EXTRUSION.

indirect-injection diesel engine A diesel engine with a pre-chamber, attached to the main cylinder chamber, into which the fuel is injected and combustion is initiated by a glow plug.

individual distributed numerical control A system in which more than one computer numerically-controlled (CNC) machine tool is controlled using local computers on each machine tool rather than a single computer controlling many machine tools. This is the normal mode of operation for industrial CNC machines.

induced drag *See* DRAG.

induced draught *See* DRAUGHT.

induced-draught cooling tower A cooling tower in which a fan, or fans, situated at the top of the tower draws air upwards against the downward flow of sprayed water. *See also* FORCED-DRAUGHT COOLING TOWER.

induced mass *See* ADDED MASS.

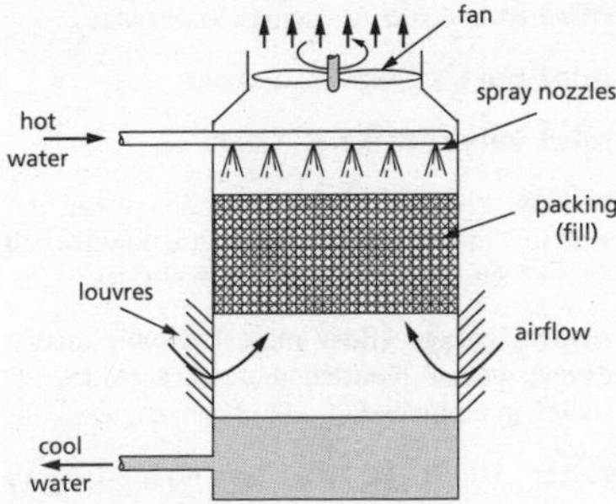

induced-draught cooling tower

inducer A component ahead of the axial-inlet of a pump impeller designed to increase the liquid pressure and avoid cavitation.

induction The flow of the air/fuel mixture into an engine, either through natural aspiration or forced induction.

induction factor *See* INTERFERENCE FACTOR.

induction furnace (high-frequency furnace, high-frequency heater) An electrical furnace in which a conductive material, such as scrap metal, contained in a crucible is melted by induction heating. The alternating magnetic field can also be used to stir the molten material.

induction manifold *See* INTAKE MANIFOLD.

induction port (inlet port) The opening through which the air/fuel mixture enters a cylinder of a piston engine during the **induction stroke** which is a suction stroke with the intake valve open.

induction pump A pump driven by an induction motor.

induction welding *See* WELDING.

inductive resolver A transducer for angular measurement which uses three coils, two fixed and one rotating with the angle being measured. The fixed coils are supplied with alternating current with a phase difference of $\pi/2$ (90°) between the current in the two coils. The output from the rotating coil is shifted in phase with respect to the first input coil by a phase angle equal to the physical angle between the output coil and the first input coil. Measurement of the output phase thus gives the angle. An **inductosyn** is a form of inductive resolver used to measure translational position.

industrial engineering *See* PRODUCTION ENGINEERING.

inelastic Materials that, having been loaded in tension, compression, or torsion, remain permanently deformed after unloading are said to be inelastic.

inelastic buckling Buckling where the strut does not return to its initial unbent state on unloading. *See also* PLASTIC BUCKLING.

inequality of Clausius *See* SECOND LAW OF THERMODYNAMICS.

inertia The property of an object that resists change of angular or linear velocity. *See also* MASS; MOMENT OF INERTIA.

inertial coordinate system (inertial reference frame) A system of coordinates at rest or moving with constant velocity in which Newton's laws apply, i.e. where a body referred to it is either at rest or moves with uniform velocity unless acted on by a force. *See also* ACCELERATING FRAME OF REFERENCE.

inertial force *See* ACCELERATING FRAME OF REFERENCE.

inertial mass (*m*) (Unit kg) According to Newton's second law of motion, the mass of an object that determines its acceleration a due to an applied force F such that $F = ma$. *See also* MASS.

inertia reel The reel in a motor-vehicle seat-belt mechanism that permits the belt to move slowly, but jams under high deceleration owing to inertia.

infeed grinding *See* PLUNGE GRINDING.

inference engine The part of an expert system that makes decisions based on the inputs and the information stored in the knowledge base.

inferential flow meter A flow meter in which the flow rate is determined from the measurement of flow and fluid properties, rather than direct measurement. The majority of flow meters are of this type.

infinite-life regime *See* FATIGUE.

infinitely-variable transmission *See* CONTINUOUSLY-VARIABLE TRANSMISSION.

inflammable *See* COMBUSTIBLE.

inflate To fill a flexible container with a gas, such as air in a tyre, or helium in an airship.

inflexion *See* POINT OF CONTRAFLEXURE.

influence line A diagram showing how, at a fixed location in a member of a structure, the bending moment and shear force change as a live load moves across the structure.

infrared radiation Invisible electromagnetic radiation with wavelengths in the range 0.7 μm to about 1 mm. The near infrared is below 25 μm and longer wavelengths correspond to the far infrared. *See also* THERMAL RADIATION.

i

infrared thermography (thermal imaging) The detection of infrared radiation emitted by an object. An **infrared thermometer (infrared pyrometer, non-contact thermometer)** determines surface temperature by focussing the radiation onto a detector.

ingot A block of cast metal ready for primary working (forging or rolling). *See also* CONTINUOUS CASTING.

inherent damping Damping of vibrations due to the microstructural features of a material. *See also* INTERNAL FRICTION.

inhibitor *See* CORROSION INHIBITOR.

initial tension (spring preload) The internal force that holds the coils of an unloaded extension spring together.

injection *See* FUEL INJECTION.

injection lag The time required, in a diesel engine, to develop injection pressure by the injection pump.

injection moulding Formation of plastic components in which polymer feedstock in granular form is softened by heating and then forced into a water-cooled die cavity under pressure that is maintained until the component has solidified. *See also* VACUUM-INJECTION MOULDING.

injection pump A mechanical pump that delivers a measured amount of fuel to the cylinder of a piston engine. *See also* FUEL SYSTEM.

injector 1. *See* FUEL SYSTEM. **2.** *See* JET PUMP.

inlet manifold *See* INTAKE MANIFOLD.

inlet port *See* INDUCTION PORT.

inlet valve *See* INTAKE VALVE.

in-line engine A multi-cylinder piston engine in which the cylinder axes are parallel and lie in a single plane. *See also* VEE ENGINE.

in-line mixer (flow mixer) A static mixing device, typically consisting of helical blades, installed in a pipeline.

inner dead centre *See* BOTTOM DEAD CENTRE.

inner variables *See* LAW-OF-THE-WALL VARIABLES.

input–output relation The function describing how the output of a system changes as a result of an input to the system. It may be represented in many ways, for example through Laplace transforms as the transfer function.

insensitive time (dead time) The time delay between the application of an input to a dynamic system and the first change in the output.

inside diameter (internal diameter, ID) The diameter of a cylindrical tube or a hollow sphere measured between opposite points on the internal surface. *See also* BORE; OUTSIDE DIAMETER.

insolation (irradiance, solar insolation, solar irradiance) (Unit W/m^2) A measure of the **in**cident **sol**ar radi**ation**, direct plus diffuse, received on a given surface over a specified period of time, typically 24 hours. *See also* GLOBAL SOLAR IRRADIANCE; SOLAR CONSTANT.

inspection gauge Any of various gauges used in manufacturing and quality control to check dimensions, finish, etc.

instability 1. An event or process that cannot be controlled. It depends on the given constraints; for example, crack propagation can often be stable when the external displacement of the body is fixed, but is not when the load is fixed. **2.** In fluid mechanics, the rapid increase in the amplitude of a disturbance due to non-linear effects. In laminar flow this leads to transition to turbulence. **3.** The opposite of stability, i.e. a system is unstable if, when it is subject to a bounded (finite) input or disturbance, the output is not bounded (is infinite). In practice, a system which shows continuing oscillation of the output for a steady input is frequently referred to as unstable even though the output is bounded.

instantaneous centre (instantaneous axis) The position within a system of bodies (such as the links in a mechanism) undergoing combined linear and rotary motion, that has zero velocity at some instant and is therefore the point in space through which an axis passes about which the body may be considered to rotate instantaneously. *See also* BODY CENTRODE, SPACE CENTRODE.

instantaneous strain Strain that appears immediately upon application of a load, as opposed to time-dependent strain.

instroke The stroke of a piston engine in which a piston approaches top-dead centre. *See also* OUTSTROKE.

instrument accuracy *See* ERROR.

instrumentation 1. The field of designing measurement systems. **2.** A system of measuring instruments used to monitor, measure, and usually record variations in physical variables such as acceleration, displacement, force, frequency, pressure, strain, stress, temperature, and velocity, often as functions of time. Modern systems are primarily digital. **3.** Part of a control system.

instrument bandwidth The range of frequencies to which a measuring instrument will respond with a specified accuracy. For example, for an AC voltmeter specified as having an accuracy of ± 1% over a frequency range of DC to 1MHz, the instrument bandwidth would be given as 1 MHz.

instrument error analysis *See* UNCERTAINTY ANALYSIS.

insulation *See* THERMAL INSULATION.

intake manifold (induction manifold, inlet manifold) A system bolted to the cylinder head of a piston engine, typically consisting of a plenum to the inlet of which is bolted the throttle body and from which a series of connected pipes convey air and fuel to the cylinders.

intake stroke *See* SUCTION STROKE.

intake valve (inlet valve) The valve in a piston engine that opens and closes the induction port.

integral-furnace boiler A water-tube boiler in which steam is generated in tubes attached to the furnace wall, thereby also cooling the wall.

integral method A technique for obtaining approximate solutions to the boundary-layer equations by assuming functional forms, such as polynomial, for the distributions of velocity, temperature, etc. that satisfy essential boundary conditions, such as zero gradient at the boundary-layer edge.

integral-mode controller A controller employing **integral control (integral action, integral compensation)** where the signal $x(t)$ applied to the controlled plant is determined from the time integral of the error ε. Hence the plant input is:

$$x(t) = \frac{1}{T_i}\int_0^T \varepsilon\, dt$$

where T_i is the **integral time constant** selected by the designer to achieve the required system behaviour and T is the present time. The **integral network** is the part of an integral-mode controller which determines the time integral.

integral momentum equation *See* MOMENTUM INTEGRAL EQUATION.

integral relations Equations for the conservation of mass, linear momentum, angular momentum, and energy for the flow of fluid through a control volume. *See also* REYNOLDS–TRANSPORT THEOREM.

integral square error The time integral of the square of the error in a control system over a specified period of time. Used as a measure of the performance of the system and in the identification of parameters.

integrated engineering *See* ENGINEERING SCIENCE.

integration The process, in systems engineering, of combining sub-systems so as to produce an overall system having the required performance.

intelligent manufacturing The use of models of the skills and knowledge of human manufacturing experts in intelligent systems and machines so that these can produce high-quality products with little or no human intervention. This may be achieved through expert systems, advanced scheduling methods, CNC control, etc.

intelligent robot A robot capable of performing complex tasks and adapting to changes in its environment without human intervention.

intense explosion An explosion in which a large amount of energy is suddenly released, resulting in the propagation of a strong spherical shock wave. *See also* SHOCK WAVE.

intensity of radiation (*I*) (Unit W/m^2.sr) The rate of radiant energy propagation in a particular direction, per unit area normal to that direction, per unit solid angle about that direction.

intensive property A thermodynamic property, including temperature, pressure, density, and all specific thermodynamic properties, that is independent of the size of a system. *See also* EXTENSIVE PROPERTY; STATE PRINCIPLE.

interaction-prediction method A technique applied in a distributed control system to reduce the effect of interaction between different parts of the plant.

interchangeable gearing *See* TOOTHED GEARING.

intercondenser A condenser between two stages of a multistage steam-jet ejector used to condense the steam from the previous ejector.

intercooler 1. A heat exchanger used to cool the compressed air entering the intake manifold of a turbocharged or supercharged piston engine. **2.** A heat exchanger used to cool the gas between the stages of a multistage compressor, or between a low- and a high-pressure compressor, for example in the gas-turbine Brayton cycle.

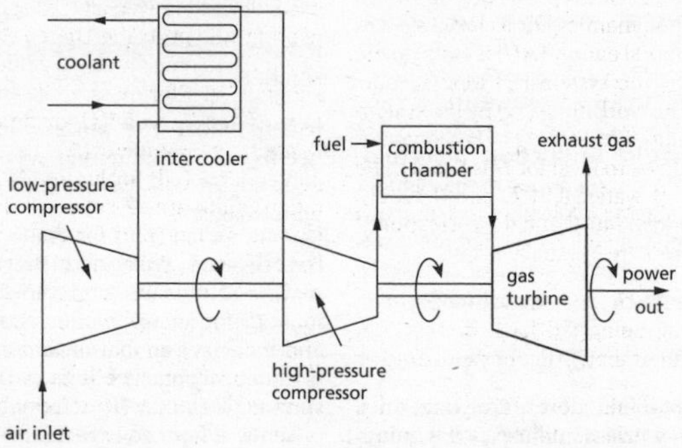

intercooler

interface 1. The permanent or transient region between solid surfaces in contact. **2.** The surface separating two immiscible fluids, either a liquid and a gas or two liquids. *See also* MENISCUS.

interface resistance *See* THERMAL CONTACT RESISTANCE.

interfacial flow Flow on the interface between two immiscible fluids where there is relative tangential movement between the two.

interfacial instability *See* KELVIN–HELMHOLTZ INSTABILITY.

interference When components will not mate owing to dimensional differences. *See also* LIMITS AND FITS.

interference factor (induction factor, perturbation factor, *a*) The fractional decrease in wind speed across the blades of a wind turbine, defined by $a = (V_0 - V_1)/V_0$ where V_0 is the unperturbed wind speed and V_1 is the wind speed in the plane of the blades.

interference fit *See* LIMITS AND FITS.

interferometry A method for determining the density variation in compressible gas flow by analysing the distortion of the interference-fringe patterns produced between two coherent light beams, one of which passes through the flowing gas, the other through the same gas at rest. *See also* HOLOGRAPHY; MACH–ZHENDER INTERFEROMETER.

intergranular fracture Fracture in crystalline materials, where the path of cracking is predominantly between grains along grain boundaries. *See also* TRANSGRANULAR FRACTURE.

interlaminar shear strength The maximum shear stress that can be withstood by a composite material before debonding of the layers occurs.

intermediate-pressure turbine (IP turbine) In a compound steam-turbine system or a three-spool gas-turbine engine, a turbine receiving steam or gas from the high-pressure turbine and delivering it to the low-pressure turbine(s). *See also* COMPOUND STEAM-TURBINE SYSTEM.

intermittency factor (γ) The fraction of time that a transitional or turbulent flow is turbulent at any specified point.

internal Pertaining to the interior of an object or substance.

internal-combustion engine A heat engine in which the fuel is burned within the system boundary that encloses the engine. Usually taken, as in this dictionary, to mean a petrol, diesel, or gas engine, but in principle it also includes the gas turbine, rocket engine, and ramjet. *See also* EXTERNAL-COMBUSTION ENGINE.

internal damping *See* STRUCTURAL DAMPING.

internal diameter *See* INSIDE DIAMETER.

internal energy (U) (Unit J) The sum of all the microscopic forms of energy of a system. It is related to the molecular structure and the degree of molecular activity, and can be viewed as the sum of the kinetic and potential energies of the atoms and molecules. It does not include the macroscopic forms of energy of the system as a whole. The absolute internal energy of a system cannot be measured. According to the first law of thermodynamics, for a closed system the change in internal energy (ΔU) is equal to the heat absorbed by the system (Q) from its surroundings, less the work done (W) by the system on its surroundings, i.e. $\Delta U = Q - W$. For water in all its forms, the zero level for U is taken to be the triple point of water, 0.01°C and 0.006112 bar. For other fluids, −40°C and the corresponding saturation pressure is usually taken.

internal feedback A feedback loop which derives the signal being fed back from within the controller, rather than from the plant output.

internal flow Fluid flow through a duct, such as a pipe, nozzle or diffuser, or within a closed container.

internal force The force acting on a cross section of any loaded component of a structure, mechanism, machine, etc.

internal friction *See* DAMPING; HYSTERESIS.

internal furnace A furnace within a fire-tube boiler from which the hot gases flow directly through the fire tubes.

internal gear *See* TOOTHED GEARING.

internal gear pump A gear pump in which a driven eccentric gear rotates within a rotating ring gear. *See also* EXTERNAL GEAR PUMP; GEROTOR.

internally-fired boiler A fire-tube boiler having an internal furnace, such as a scotch or vertical tubular boiler, or a boiler with a water-cooled plate-type furnace.

i

internal-mix atomizer An atomizer in which the liquid to be atomized is mixed with a gas before being discharged from the atomizing nozzle.

internal screw thread *See* SCREW.

internal stress *See* RESIDUAL STRESS.

internal work *See* STRAIN ENERGY.

International Practical Temperature Scale (IPTS) A standard scale for absolute temperature based in 1990 (IPTS-90) upon 16 fixed points: the triple points for hydrogen, neon, oxygen, argon, mercury, and water; the boiling points of hydrogen at two different pressures; the freezing points of indium, tin, zinc, aluminium, silver, gold, and copper; and the melting point of gallium.

International System of Units (SI system of units, Le Système international d'unités) The standard system of units now used almost universally in science and engineering except in the United States. There are seven basic units: metre (symbol m) for length; kilogram (kg) for mass; mole (mol) for amount of substance; second(s) for time; kelvin (K) for temperature; ampere (A) for electric current; and candela (cd) for luminous intensity. In addition, there are 22 coherent derived units, including radian (rad) for plane angle, steradian (sr) for solid angle, hertz (Hz) for frequency, newton (N) for force, pascal (Pa) for pressure, joule (J) for energy, work, and amount of heat, and watt (W) for power and radiant flux. Recommended practice is to avoid combinations such as N/mm^2, MN/m^2 being preferred.

The SI system also specifies 20 prefixes for the decimal multiples and submultiples of SI units in the range 10^{-24} (yocto, symbol y) to 10^{24} (yotta, Y). For mechanical-engineering purposes, the prefixes normally encountered are: pico (p, 10^{-12}), nano (n, 10^{-9}), micro (μ, 10^{-6}), milli (m, 10^{-3}), centi (c, 10^{-2}), deci (d, 10^{-1}), deca (da, 10^{1}), hecto (h, 10^{2}), kilo (k, 10^{3}), mega (M, 10^{6}), giga (G, 10^{9}) and tera (T, 10^{12}), although the use of centi, deci, deca, and hecto is not common in engineering applications.

SEE WEB LINKS

- Full details of the International System of Units
- Basic information about SI units

interstitial solid solution An alloy in which the solute atoms are much smaller than the solvent atoms and so occupy the interstices in the crystal structure of the solvent. The most important interstitial solutes in metals are carbon, boron, nitrogen, and oxygen, which harden the alloy by distorting the crystal lattice.

interstitial velocity (Unit m/s) For flow through a porous medium, the Darcy flux divided by the void fraction. *See also* SUPERFICIAL VELOCITY.

invar A nickel steel with a very low (*c.*10^{-6} /K) coefficient of linear expansion. *See also* KOVAR.

inverse cam The element corresponding to the follower of a cam mechanism being used as the driver, as in a cam-controlled planetary gear device.

inverse feedback *See* NEGATIVE FEEDBACK.

inverse kinematics The determination through homogeneous transforms of the joint angles and positions (i.e. the **arm solution** or **back solution**) required in a robot to achieve a specified end-effector position and orientation. The opposite of forward kinematics.

inverse Laplace transform The process of converting from the frequency domain (i.e. Laplace-transformed) representation of a signal or function to the time-domain representation, i.e. the opposite process to taking a Laplace transform.

inverse sublimation *See* SUBLIMATION.

inversion 1. (inversion layer) A stable condition in the lower troposphere in which the temperature increases with altitude and so reduces mixing of pollutants which become trapped in a layer. **2.** The process of pushing a tube onto a die that turns the walls inward, thus forming a longer, smaller-diameter tube.

inversion temperature (temperature of inversion) At any particular pressure, the temperature above which a gas cannot be cooled by a throttling process.

inverted engine A piston engine in which the cylinders are below the axis of the crankshaft.

inverted pendulum A pendulum that oscillates stably above its pivot when the pivot is given a vertical simple harmonic displacement having a sufficiently high frequency.

investment casting (lost-wax casting, lost wax technique) A method of casting intricate shapes in which the object is first made in wax that is then covered in a refractory shell by dipping or spraying. The wax is burnt away and the shell mould filled by the casting metal.

inviscid flow The flow of a perfect fluid. *See also* POTENTIAL FLOW.

inviscid fluid *See* PERFECT FLUID.

involute The locus of a point on a straight line which rolls, without slipping, on the circumference of a circle. Equivalently, it is the locus of a point on a taut string which is unwound from a cylinder.

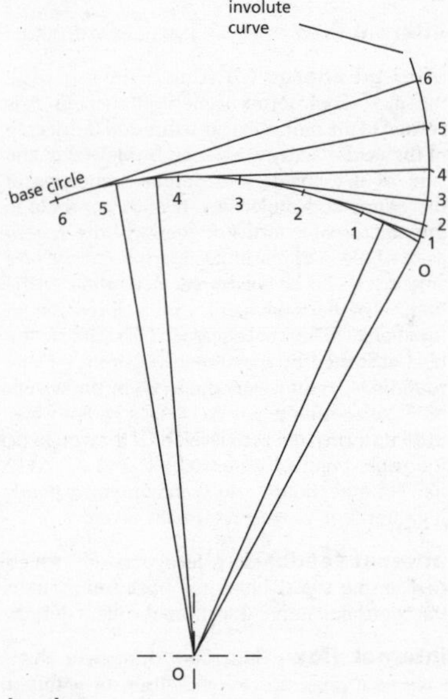

involute

involute gear teeth *See* TOOTHED GEARING.

involute helicoid *See* TOOTHED GEARING.

involute rack *See* TOOTHED GEARING.

inward-flow radial turbine A turbine in which fluid enters the machine radially and is then turned into the axial direction.

ionosphere *See* STANDARD ATMOSPHERE.

IP *See* INDICATED POWER.

Ipsen's method (sequential elimination of dimensions) A method of dimensional analysis in which each of the dimensions is eliminated in turn by successively dividing or multiplying each variable by powers of the initial variables or combinations thereof until all dimensions have been eliminated. *See also* RAYLEIGH'S METHOD.

IPTS *See* INTERNATIONAL PRACTICAL TEMPERATURE SCALE.

IP turbine *See* INTERMEDIATE-PRESSURE TURBINE.

ironing The forming process whereby the wall of a beverage can is thinned by being forced to flow through a circular tapered die in the form of a ring, thus simultaneously lengthening the can.

irradiance *See* INSOLATION.

irradiation (*G*) (Unit W/m^2) The rate at which thermal radiation is incident on a surface from all directions per unit area of that surface.

irreversibility 1. In a thermodynamic process, the difference between the reversible work and the useful work. It is equivalent to the exergy destroyed. Typical causes of irreversibility are combustion (chemical irreversibility), sudden expansion, mixing and friction (mechanical irreversibility), and heat transfer (thermal irreversibility). **2.** In materials mechanics, non-elastic, frictional, hysteretic, and other dissipative processes are said to exhibit irreversibility.

irreversible energy loss In a dissipative-flow process, such as a sudden expansion or shock wave, the conversion of mechanical energy into thermal energy.

irreversible process A thermodynamic process that cannot be reversed in such a way that both the system and its surroundings are returned to their original conditions. *See also* IRREVERSIBILITY; REVERSIBLE PROCESS.

irrotational flow A flow in which fluid particles do not rotate and the vorticity is everywhere zero. *See also* KELVIN'S CIRCULATION THEOREM; POTENTIAL FLOW; ROTATIONAL FLOW.

isenergetic process A thermodynamic process for which the system internal energy remains constant.

isenergic flow (homenergic flow, isoenergetic flow) A flow in which the specific total enthalpy is everywhere the same.

isenthalpic process A thermodynamic process for which the system enthalpy remains constant.

isentrope A curve along which the entropy remains constant.

isentropic atmosphere A model of the atmosphere in which it is assumed the entropy is the same at all altitudes.

isentropic bulk modulus (B_S) (Unit Pa) A measure of the compressibility of a fluid or solid with isentropic conditions imposed, defined by

$$B_S = \rho\left(\frac{\partial p}{\partial \rho}\right)_S$$

where p is the static pressure, ρ is the density and s is the specific entropy. For an ideal gas, $B_S = \rho c^2$ where c is the speed of sound. The **isentropic compressibility (K_S)** is the reciprocal of the isentropic bulk modulus defined by

$$K_S = -\frac{1}{v}\left(\frac{\partial v}{\partial p}\right)_S$$

where v is the specific volume. *See also* COMPRESSIBILITY; ISOTHERMAL COMPRESSIBILITY.

isentropic efficiency 1. (nozzle efficiency, η_N) The ratio, for a nozzle, of the actual kinetic energy of the gas flow at the nozzle exit to the kinetic energy at the exit of an isentropic nozzle with the same inlet state and exit pressure. **2.** (η_T) The ratio, for a turbine, of the actual work output to the output that would be achieved if the process between the inlet state and the exit pressure were isentropic. **3.** (η_C) The ratio, for a compressor, of the work input that would be required to increase the pressure of the working fluid by a specified amount, if the process between the inlet state and the exit pressure were isentropic, to the actual work input.

isentropic index *See* POLYTROPIC PROCESS.

isentropic process A thermodynamic process that is both reversible and adiabatic such that the system entropy remains constant. In an **isentropic flow** the specific-entropy distribution is the same at all times. In a **homentropic flow** the specific entropy is the same throughout the flow.

isobaric process (isopiestic process) A thermodynamic process carried out at constant pressure. An **isobar** is a curve along which the pressure remains constant.

isochoric process (isometric process, isovolumetric process) A thermodynamic process for which the volume of a closed system remains constant.

isoclinic In a photoelastic image, a line joining points of constant inclination of the principal stresses.

isochronous governor A governor that maintains the speed of a prime mover constant under all loads.

isoenergetic flow *See* ISENERGIC FLOW.

isokinetic sampling A method of sampling a gas flow using a hollow probe through which the gas is sucked at such a rate that the gas velocity at the probe tip is the same as the undisturbed gas velocity.

isolated system A closed thermodynamic system for which no energy is permitted to cross the system boundary.

isolation *See* VIBRATION ISOLATION.

isometric drawing *See* PROJECTION.

isometric process *See* ISOCHORIC PROCESS.

isopiestic process *See* ISOBARIC PROCESS.

isopycnic process A thermodynamic process for which the density or specific volume remains constant.

isostatic pressing (hydrostatic pressing) A process in which a more uniform density and higher strength is achieved compared with plain sintering, by placing the powdered material in a sealed bag that is immersed in water or oil under pressure.

isostatics The loci of constant stress in a body under load given by elasticity or plasticity solutions, FEM simulations, or from photoelasticity.

isoteniscope An apparatus for measuring the vapour pressure of a liquid in which one side of a U-tube manometer is attached to a bulb containing the liquid, the other to a vacuum pump and a pressure transducer. Both sides of the U-tube are submerged in a constant-temperature bath. The vacuum pump is used to adjust the pressure sensed by the transducer such that the liquid levels on both sides of the manometer are the same.

isotherm A curve along which the temperature is constant at any instant of time.

isothermal *See* ISOTHERMAL PROCESS.

isothermal atmosphere A model of the atmosphere in which it is assumed the temperature is the same at all altitudes. It is a good approximation for the stratosphere. The **hypsometric equation** gives the variation of pressure with height in the atmosphere assuming constant temperature and acceleration due to gravity.

isothermal bulk modulus (B_T, fluids, K solids) (Unit Pa) A measure of the compressibility of a fluid or solid with isothermal conditions imposed, defined by

$$B_T = \rho\left(\frac{\partial p}{\partial \rho}\right)_T$$

where p is the static pressure, ρ is the density, and T is the temperature. For an ideal gas, $B_T = \rho c^2/\gamma$ where c is the speed of sound and γ is the ratio of specific heats.

isothermal compressibility (K_T) (Unit 1/Pa) For both solids and fluids, the reciprocal of the bulk modulus defined by

$$K_T = -\frac{1}{v}\left(\frac{\partial v}{\partial p}\right)_T$$

where v is the specific volume, p is the pressure, and T is the temperature. *See also* COMPRESSIBILITY; ISENTROPIC BULK MODULUS.

isothermal equilibrium The state in which two bodies have the same temperature, even if they are not in contact. *See also* ZEROTH LAW OF THERMODYNAMICS.

isothermal process A thermodynamic process carried out at constant temperature. In an **isothermal flow** the temperature is everywhere the same at all times.

isothermal transformation *See* TIME–TEMPERATURE TRANSFORMATION DIAGRAM.

isotropic A term used where a material or a physical quantity has the same properties in all directions and no preferred directions. *See also* ANISOTROPY; HOMOGENEOUS.

isotropic turbulence A turbulence field that is statistically invariant under translation, rotation, and reflection.

isovolumetric process *See* ISOCHORIC PROCESS.

iteration The process of finding a solution to an equation, or set of equations, by successive approximations.

IUPAC International Union of Pure and Applied Chemistry. The international authority that defines terminology for chemistry, recommends standardized methods of measurement, atomic weights, etc.

SEE WEB LINKS

- IUPACwebsite

IUPAP International Union of Pure and Applied Physics.

IUTAM International Union of Theoretical and Applied Mechanics.

Izod test A materials impact test in which a pendulum of known energy strikes a notched bar of specified dimensions which is clamped in a fixture as a cantilever, with the notch on the struck side. The loss in energy of the striker is a measure of the impact strength. *See also* CHARPY TEST.

jack A lifting device that exerts large forces over small displacements, achieved by mechanical gearing or hydraulics.

jacket 1. Lagging applied to boilers, pipes, etc. for insulation purposes. **2.** A casing, surrounding boilers, pipes, etc., containing liquid to cool or heat the contents.

jacketed pipe A core pipe through which a process fluid is transported, surrounded by a concentric jacket pipe, with heating or cooling fluid flowing through the annular space between the two pipes. A typical application is to liquids too viscous to flow at ambient temperature. *See also* VACUUM-JACKETED PIPE.

jackshaft *See* COUNTERSHAFT.

Jacobs chuck A self-centring chuck, the three jaws of which are moved simultaneously by a single toothed key via toothed segments on the surrounding holder. Jacobs chucks are attached to their drive shafts by a shaft with a tapered end (**Jacobs taper**).

Jakob number (phase-change number, *Ja*) A non-dimensional number that arises in the study of condensation heat transfer, defined by $Ja = C_P\Delta T/h_{gf}$ where C_P is the specific heat of the condensing fluid, h_{gf} is its latent heat of condensation, and ΔT is the difference between the condensing surface temperature and the saturation-vapour temperature of the fluid.

Jeffery–Hamel flow The radial viscous flow in a wedge geometry due to a line source or sink at the apex.

jerk (Unit m/s^3) The rate of change of acceleration $\boldsymbol{a}$ with time t, i.e. $d\boldsymbol{a}/dt$. It is a vector quantity.

jerk pump A fuel-injection pump for a piston engine consisting of a high-pressure pump that pressurizes the fuel, meters it, and times the injection through an atomizing spray valve or nozzle.

jet The flow of high-momentum fluid out of an orifice, slot, nozzle, pipe outlet, etc., into a surrounding fluid. *See also* OVER-EXPANDED JET; UNDER-EXPANDED JET.

jet compressor A type of ejector that utilizes a jet of high-pressure gas or vapour to entrain a low-pressure gas or vapour, mix the two and discharge at an intermediate pressure. Widely used in the process, paper, petroleum, gas, and power industries. *See also* THERMAL COMPRESSOR.

jet condenser A device in which steam is condensed by direct contact with cooling water sprayed into a steam-filled chamber.

jet engine A device, such as any gas-turbine, rocket, or ramjet engine, that produces thrust by ejecting exhaust gas at high speed into the surroundings. The term is commonly used to mean a turbojet engine.

jet flap *See* BLOWN FLAP.

jet mixer (jet-flow agitator) 1. A mixing device in which liquids, or liquid and powder, are mixed by circulation through a ducted rotor. **2.** A mixing device based upon an ejector in which a high-pressure liquid entrains a low-pressure liquid and the two mix in the exit nozzle.

jet noise Aerodynamic sound produced by the turbulent mixing of a high-speed gas jet and its surroundings. The high-intensity sound produced by a supersonic jet interacting with its surroundings is termed **jet screech**.

jet nozzle A convergent nozzle, usually axi-symmetric, used to increase the speed of a fluid stream. *See also* PROPELLING NOZZLE.

jet pipe In a gas-turbine engine, the duct or pipe downstream of the low-pressure turbine and upstream of the final propelling nozzle. The **jet-pipe temperature** is the temperature of the gas flow in the pipe. The **jet-pipe efficiency (η_{JP})** is defined by $\eta_{JP} = (h_{0_I} - h_E)/(h_{0_I} - h_{E_S})$

where h_{0_I} is the specific stagnation enthalpy at inlet to the jet pipe and h_E is the specific enthalpy at its exit. The subscript S denotes an isentropic expansion.

jet propulsion Propulsion resulting from the thrust created by a jet of gas in the case of an aircraft or water for certain marine craft. *See also* JET ENGINE.

jet pump (eductor jet pump, ejector pump, injector) A device consisting of a convergent-divergent nozzle into which flows a jet of high-speed fluid (motive fluid) upstream of the convergence. If the reduction in fluid pressure due to the Venturi effect is sufficiently low, low-pressure fluid is sucked into the device through an upstream port or ports.

jet screech *See* JET NOISE.

jeweller's vice A portable vice consisting of two bars having pairs of threaded holes in which run hand-driven screws.

***j* factor** *See* COLBURN *j* FACTOR.

jig A device employed in manufacture that accurately locates and guides tools so as to ensure repeatability in dimensions. Also used in the assembly of parts. *See also* FIXTURE.

jig borer An especially accurate vertical milling machine tool for the manufacture of jigs and fixtures.

***J* integral** A parameter of non-linear elastic fracture mechanics given by the line integral on any curve Γ around a crack tip orientated along the x_1 axis, starting from the lower surface and ending on the upper surface in an anticlockwise direction.

$$J = \int_{\Gamma}\left[Wdx_1 - \boldsymbol{T}\left(\frac{\partial u}{\partial x_2}\right)ds\right]$$

where W is the strain energy density, $T \equiv \sigma_{ij}n_{ij}$ is the traction vector on Γ according to an outward unit vector n (with direction cosines n_j) normal to Γ, u is displacement, s is arc length and σ_{ij} is stress. Used to analyse elastoplastic fracture by invoking **Kachanov's theorem**. A critical value of the J-integral (J_C) is required for crack initiation. The **J_R curve** records any changing resistance to crack propagation (in terms of J values at different crack lengths) until a steady value is attained. *See also* FRACTURE MECHANICS.

jockey pulley A pulley in a belt or chain drive which is neither the driving nor driven pulley, but which maintains the drive in tension by its weight or by spring loading. Jockey pulleys transmit no power. *See also* CAM.

Johansson block *See* GAUGE BLOCK.

Johnson–Cook constitutive equation An empirical relation, often employed in large-deformation impact mechanics, between equivalent flow stress σ, equivalent plastic strain ε, non-dimensional equivalent plastic strain rate E ($=\dot{\varepsilon}/\dot{\varepsilon}_0$), and homologous temperature T_M, given by $\sigma = (A + B\varepsilon^n)(1 + ClnE)(1 - T_M^m)$ where $\dot{\varepsilon}$ is the plastic strain rate and $\dot{\varepsilon}_0$ is the quasi-static strain rate at which the material-dependent constants A, B, C, n, and m are determined. The equation does not include the effect of strain-rate workhardening.

Johnson–Cook fracture equation An empirical relation employed in damage mechanics for the effect of hydrostatic stress σ_H, non-dimensional equivalent plastic strain rate E, and homologous temperature T_M, on the equivalent plastic strain at fracture ε_f in a material subjected to large deformations, given by $\varepsilon_f = [X + Yexp(Z\sigma_H)](1 + VlnE)(1 - WT_M)$ where V, W, X, Y, and Z are material-dependent constants and where E is defined in the preceding entry. Fracture is presumed to occur when the microstructural damage given by $\Sigma(\Delta\varepsilon/\varepsilon_f) = 1$ where the $\Delta\varepsilon$ are increments of

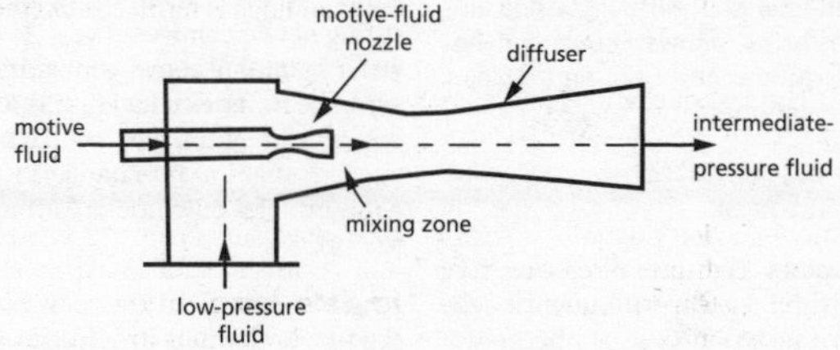

jet pump

equivalent plastic strain along sequential load paths.

Johnson's formula A formula for the failure of a short strut at stresses beyond the proportional limit due to compression without significant buckling. The failure force P is given by

$$P = \sigma_Y A\left[1 - \left(\frac{\sigma_Y}{4\pi^2 E}\right)\left(\frac{L}{k}\right)^2\right]$$

where σ_Y is the compressive yield stress, A is the cross-sectional area of the strut, E is Young's modulus, L is the strut length, and k is the radius of gyration of the strut cross section about the axis of bending.

joint The part of a robot arm permitting either rotational or translational motion. Each joint provides a single degree of freedom, and thus a minimum of six is required if the robot is to be able to position and orientate the end effector anywhere within the workspace. A joint is described by four parameters: the joint angle, joint offset, joint twist, and link length. The **joint angle** is the angle between an extrapolation of the previous link length and the present link length, measured positive anticlockwise in a plane normal to the joint axis. This is variable in a rotational joint and fixed in a translational joint. The **joint offset** is the distance between the link length for the previous link and that for the present link, measured along the joint axis. This is variable for a translational joint and fixed for a rotational joint. The **joint twist** for the n_{th} joint is the angle between the axes of joints J_n and J_{n+1}, measured positive anticlockwise in a plane normal to the link length and viewed from the position of the J_{n+1} joint. The link length is the mutually perpendicular distance between planes passing through the axes of joint J_n and J_{n+1}. Both the joint twist and link length are fixed in all joints.

joint space The description of the position and orientation of a robot in terms of the rotational angles and translational positions of the joints. Kinematic analysis, normally using homogeneous transforms, allows conversion between this description and the end-effector position and orientation in *x,y,z* space.

Joly's steam calorimeter *See* DIFFERENTIAL STEAM CALORIMETER.

Jominy distances The rates of cooling vary along the bar in the Jominy end-quench test, being very high (about 50°C/s) at the water-quenched end and low (about 2°C/s) at the far end. Since thermal conductivities and emissivities do not vary much between different alloy steels, distances along the bar are thought of as being equivalent to particular known cooling rates.

Jominy end-quench test A method to determine the hardenability of steel alloys in which a standard-size bar (25 mm diameter, 100 mm long) is uniformly heated to the normal hardening temperature of that alloy, after which one end is subjected to a metered stream of cold water. The cooling rates vary from quenching at the watered end to air cooling at the far end. After cooling to room temperature, the hardness is measured at a series of Jominy distances along the bar. At the quenched end, the hardness will be high owing to the formation of martensite; at the far end, the hardness will be the lower value of an equilibrium microstructure. Good hardenability is indicated when high hardness is retained at long distances from the quenched end.

Joukowski transformation The conformal transformation $z = \varsigma + 1/\varsigma$ where $\varsigma = \chi + i\eta$ is a complex variable in the original space, and $z = x + iy$ is a complex variable in the new space. The symbol i represents $\sqrt{-1}$. The transformation is used in two-dimensional potential-flow theory. An aerofoil shape (**Joukowski aerofoil**) results from applying the transformation to potential flow around a circular cylinder.

joule (J) The SI unit of energy and work, being the work done by a force of one newton moving its point of application in the direction of the force by one metre, i.e. 1 J = 1 N.m.

Joule cycle *See* BRAYTON CYCLE.

Joule heating The generation of heat in a conductor due to an electric current passing through it. *See also* JOULE'S LAWS.

Joule's equivalent *See* MECHANICAL EQUIVALENT OF HEAT.

Joule's experiment (Joule free expansion) An experiment using an apparatus consisting of two connected vessels submerged in a water bath, one vessel containing a gas at pressure, the other evacuated. It is found that when the cock between the two, initially closed, is opened, there is no change in the water temperature. The outcome is formulated as Joule's law (2).

Joule's laws 1. The rate at which heat is produced $\dot{Q}$ when an electric current I flows through a conductor of resistance R is given by

$\dot{Q} = I^2R = V^2/R = VI$, where V is voltage. **2.** The internal energy of a given mass of an ideal gas is a function of its temperature only. A corollary (**Joule–Thomson law**) is that the specific enthalpy of an ideal gas is also a function of its temperature only.

Joule–Thomson coefficient (μ) (Unit K/Pa) A measure of the change in temperature T with pressure p in an isenthalpic throttling process, defined by

$$\mu = \left(\frac{\partial T}{\partial p}\right)_h$$

where h is the specific enthalpy. For an ideal gas, μ is zero.

Joule–Thomson effect (Joule–Kelvin effect, Joule–Thomson expansion, Joule–Thomson process) The small temperature change observed during adiabatic and isenthalpic expansion of a real gas, such as when flowing at low speed through a porous plug, a partially-open valve, or similar resistance. For most gases the temperature decreases, but for hydrogen and helium it increases, except at very low temperature.

Joule–Thomson inversion temperature For any gas, the temperature for which the Joule–Thomson coefficient is zero at a given pressure.

journal bearing *See* BEARING.

journal box *See* AXLE BOX.

J_R curve *See* *J* INTEGRAL.

judder A vibration at low frequency, such as within a drive train caused by stick-slip friction of a member such as a clutch.

junction vortex *See* HORSESHOE VORTEX (2).

Kachanov's theorem Partly-irreversible elastoplasticity may be analysed as if it were fully-reversible non-linear elasticity provided there is no unloading and that the strain ratios remain constant. *See also* J-INTEGRAL.

Kalina cycle A Rankine cycle in which the working fluid is an ammonia–water mixture.

Kalman filter A method of improving the knowledge of the states of a system by using a combination of measured data and data predicted from a dynamic model of the system. The weightings used for the measured and predicted data are determined according to their likely accuracy.

Kaplan turbine An axial-flow, propeller-type hydraulic turbine, for high-volume flows with relatively low head, in which the blade angles can be adjusted to suit the power demand.

Kármán constant (*κ*) An empirical constant, with a value of about 0.41, in the law-of-the-wall representing the reciprocal of the turbulent-core velocity gradient in semi-logarithmic coordinates.

Kármán integral equation *See* MOMENTUM-INTEGRAL EQUATION.

Kármán–Pohlhausen method *See* POHLHAUSEN POLYNOMIAL.

Kármán–Schoenherr formula An empirical equation, widely used by naval architects, for estimating wall shear stress in a zero-pressure gradient turbulent boundary layer. It may be written as $1/\sqrt{c_f} = 4.15\, log_{10}(Re_x c_f) + 1.7$ where c_f is the skin-friction coefficient and Re_x is the Reynolds number based on the free-stream velocity and distance from the origin of the boundary layer.

Kármán–Tsien equation (Kármán–Tsien rule) An approximate equation that allows the pressure coefficient C_P for a compressible subsonic flow to be obtained from the value for incompressible flow C_{Pi} according to

$$C_P = \frac{C_{Pi}}{\sqrt{1 - M_\infty^2} + \frac{M_\infty^2}{1+\sqrt{1-M_\infty^2}}\frac{C_{Pi}}{2}}$$

where M_∞ is the approach-flow Mach number.

Kármán viscous pump A large flat disk rotating at angular rotation speed Ω in a viscous fluid with kinematic viscosity ν that produces an axial velocity far from the disc of $0.886\sqrt{\nu\Omega}$.

Kármán vortex street In a viscous fluid flow, the alternating sequence of vortices downstream of a cylinder transverse to the flow caused by periodic vortex shedding of the surface shear layer. For a circular cylinder of diameter *D*, shedding occurs in the Reynolds-number range $10^2 < Re < 10^7$ with a Strouhal number $fD/V \approx 0.21$ where *f* is the shedding frequency (in Hz) and *V* is the approach-flow velocity.

katharometer (hot-wire detector) An instrument used in the analysis of gases in which the heat loss from a heated filament of tungsten or platinum in a test gas is compared with that from a filament in a reference gas. The heat loss depends primarily on the thermal conductivity and the specific heat of the gas. *See also* THERMAL-CONDUCTIVITY CELL.

kcal *See* KILOCALORIE.

kelvin (K) *See* ABSOLUTE TEMPERATURE.

Kelvin absolute temperature scale *See* ABSOLUTE TEMPERATURE.

Kelvin equation An equation for the effect of the curvature of a surface on the vapour pressure p_V of a liquid. If σ is the surface tension, ρ is the density, *M* is the molar mass, *T* is the temperature, and p_{VO} is the saturated vapour pressure, for a spherical drop of radius *r*,

$$ln\left(\frac{p_V}{p_{VO}}\right) = \frac{2\sigma M}{r\rho\mathcal{R}T}$$

where $\mathcal{R}$ is the universal gas constant.

Kelvin–Helmholtz instability The instability of an interface, idealized as a vortex sheet, separating two inviscid fluids with a discontinuity in velocity across the interface.

Kelvin material *See* KELVIN–VOIGT MATERIAL.

Kelvin oval In potential-flow theory, a streamline shape produced by the combination of a uniform flow of velocity U_∞ and two counter-rotating point vortices of strength *K*, one above the other, with separation 2a. The parameter on the curves, which represent the streamlines for which the stream function is zero, is $K/U_\infty a$. *See also* RANKINE OVAL.

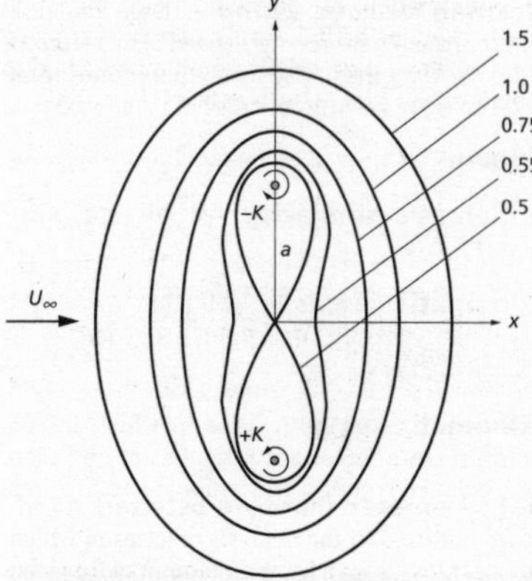

Kelvin oval

Kelvin–Planck statement (Kelvin statement) *See* SECOND LAW OF THERMODYNAMICS.

Kelvin's theorem (Kelvin's circulation theorem) In a homogeneous ideal fluid, the circulation remains constant with time.

Kelvin temperature scale *See* ABSOLUTE TEMPERATURE.

Kelvin–Voigt material (Kelvin material, Voigt material) A mechanical model for a solid or liquid viscoelastic material comprising a dashpot (i.e. a viscous damper with viscosity μ) in parallel with a linear-elastic element (modulus of elasticity *G*). The strains ε and strain rates $\dot{\varepsilon}$ in both elements are identical while the stress τ is the sum of the stresses on the two elements, i.e. $\tau = G\varepsilon + \mu\dot{\varepsilon}$. *See also* MAXWELL MATERIAL.

Kennedy and Pancu circle A technique for representing the modes of a vibrating system on a circular plot and thus determining the natural frequencies.

***k*-ε model** An empirical method of modeling turbulent shear flows based upon the simultaneous solution of two partial differential equations, one for the transport of turbulent kinetic energy *k* and the other for ε, the rate of dissipation of *k*, together with the continuity and momentum equations. Although there are now much more sophisticated and accurate turbulence models with broader applicability, the *k*–ε model is the most widely used in commercial CFD codes.

kerf The actual width of cut made by a saw having splayed (set) teeth.

kerosene (paraffin) A light petroleum distillate with a gross calorific value of 46,400 kJ/kg used as a gas-turbine fuel. *See also* AVIATION FUEL.

KERS *See* KINETIC-ENERGY RECOVERY SYSTEM.

kettle reboiler *See* REBOILER.

Keulegan–Carpenter number (*K*) A non-dimensional parameter that arises in the study of oscillatory fluid flow relative to a stationary object, defined by $K = u_{MAX}T/L$ where u_{MAX} is the velocity amplitude of the oscillatory flow, *T* is its period, and *L* is a characteristic length of the body.

kevlar® The proprietary name for a man-made aramid fibre that has a tensile strength of some 3.5 GPa. It may be woven into fabrics and is used as reinforcement in composite structures.

Kew barometer (Kew pattern barometer, Kew type barometer) A barometer consisting of a fixed mercury cistern, a vertical glass tube with an air trap, and a leather washer to prevent loss of mecury. *See also* FORTIN BAROMETER.

key 1. A bar inserted into a shallow longitudinal slot (**keyway**) cut into a hub and a shaft to prevent relative rotation. Keys have different profiles and sizes depending on the application, the torque transmitted, and whether they have to be periodically removed during maintenance. *See also* FEATHER KEY. **2.** A spanner/wrench used for tightening the jaws in a chuck.

Keyes equation An empirical equation of state relating pressure *p*, absolute temperature *T*, and specific volume *v* for gases. It can be written as

$$p = \frac{\mathcal{R}T}{v-\delta} - \frac{A}{(v+l)^2}$$

where $\mathcal{R}$ is the universal gas constant and $\delta = \beta exp(-\alpha/v)$ The constants A, l, α, and β have been determined for many gases. *See also* BEATTIE–BRIDGMAN EQUATION.

keying fit *See* LIMITS AND FITS.

kiln An industrial oven for burning, baking, or drying.

kilo (k) An SI unit prefix indicating a multiplier of 10^3; thus kilometre (km) is a unit of length equal to one thousand metres.

kilocalorie (Cal, kg-cal, kilogram-calorie, large calorie) An obsolete (i.e. non-SI) unit of energy equal to 1000 cal.

kilogram (kg) The base unit of mass in the SI system. It is equal to the mass of the International Prototype Kilogram, a right cylinder of height and diameter 39.17 mm made of an alloy of 90% (by mass) platinum and 10% iridium.

k

kilogram-calorie *See* KILOCALORIE.

kilogram-force (kg$_f$) A non-SI unit of force equal to the weight of 1 kg where the acceleration due to gravity is taken as 9.80665 m/s^2 so that 1 kg$_f$ = 9.80665 N.

kilomole (kilogram mole, kmol, kmole) The amount of a substance in kilograms numerically equal to its molecular weight.

kilowatt.hour (kW.h) An SI-accepted unit for energy.

kinematically admissible motion Any motion of a system which is geometrically compatible within itself and with the constraints. In plasticity, different kinematically-admissible velocity fields are identified to represent the internal deformation in plastic flow, from which overestimates of the working loads are obtained employing the upper bound theorem. The lowest upper bound of a series of velocity fields will be closest to the true load. *See also* STATICALLY-ADMISSIBLE STRESS FIELD.

kinematic analysis In robotics, the process of relating the joint angles and positions to the end-effector position and orientation.

kinematic pair Two elements or links so connected together that their relative motion is completely constrained. A **kinematic chain** is a combination of kinematic pairs in which each element or link forms part of two pairs. In a **turning pair**, relative motion is limited to rotation. *See also* CLOSED PAIR.

kinematics *See* APPLIED MECHANICS.

kinematic similarity *See* PHYSICAL SIMILARITY.

kinematic viscosity (ν) (Unit m^2/s) The dynamic viscosity of a fluid μ divided by its density ρ, i.e. $\nu = \mu/\rho$.

kinematic wave A wave in which inertial and pressure forces are insignificant and there is a balance between gravitational and frictional forces.

kinetic energy (k) The contribution to the energy of a system made by the bulk motion of its constituent parts. For a body of mass m having velocity V, $k = ½ mV^2$. It is equivalent to the amount of work required to bring the body to rest and dependent on the frame of reference chosen. In fluid mechanics and ther-

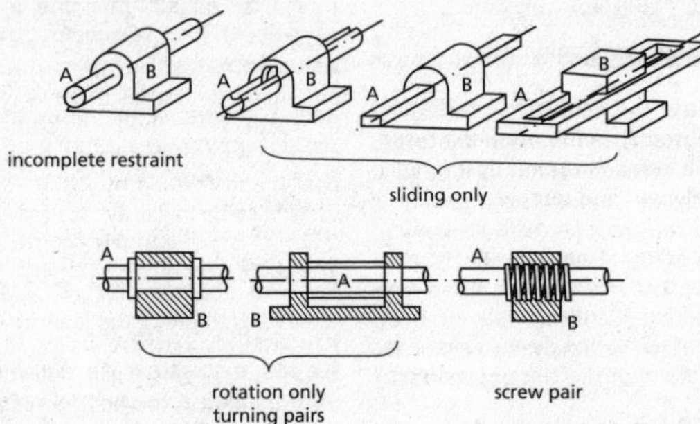

kinematic pair

modynamics, it is considered as contributing to the internal energy of the fluid.

kinetic-energy recovery system (KERS) A system for motor vehicles using a flywheel or battery to store excess energy when braking and release it when extra power is needed for high acceleration.

kinetic equilibrium *See* DYNAMIC EQUILIBRIUM.

kinetic friction (dynamic friction, sliding friction) The sliding resistance to relative motion of two surfaces in contact with each other. *See also* AMONTONS FRICTION; COEFFICIENT OF FRICTION.

kinetic Reynolds number A term for the Reynolds number used in problems where other Reynolds numbers also arise, such as the magnetic Reynolds number.

kinetics *See* APPLIED MECHANICS.

kinetic theory A model that links the temperature of a gas with the velocity of its molecules, which are treated as tiny spheres undergoing perfectly elastic collisions.

kingpin (swivel pin) The nearly vertical pin about which the stub axle of a motor vehicle swivels during steering.

King's law An empirical equation for the convective heat-transfer rate $\dot{Q}$ from a heated wire in a fluid flow of speed V, written as $\dot{Q} = (A + B\sqrt{V})\Delta T$ where A and B are calibration constants and ΔT is the temperature difference between the wire and the fluid. It is the basis for flow measurements using hot-wire and hot-film anemometry, although the square-root term is often replaced by V^n where n is a constant also determined from calibration.

Kirchhoff's law (Kirchhoff's radiation law) The monochromatic emittance of any surface is equal to the monochromatic absorptance at the same wavelength and temperature.

knife-edge bearing A narrow sharp wedge of hard material, such as steel or agate, employed as a low-friction fulcrum about which pivot the beams of a balance, lever arms in instruments, attachment points of extensometers, etc.

knock *See* ENGINE KNOCK.

knock rating The intensity of knock produced by a given fuel compared to that produced by a standard fuel.

Knoop hardness test A hardness test in which the indenter is an elongated diamond pyramid that gives an impression in the form of a parallelogram, in which the longer diagonal is about seven times the shorter. Used for studies of anisotropy in solid materials.

knowledge base The part of an expert system that stores, as a set of logical rules, information on the previous behaviour of the plant being controlled.

knuckle joint A hinged joint between two shafts with a pin passing through a hole in a boss on the end of one, and through holes in a clevis-like piece at the end of the other.

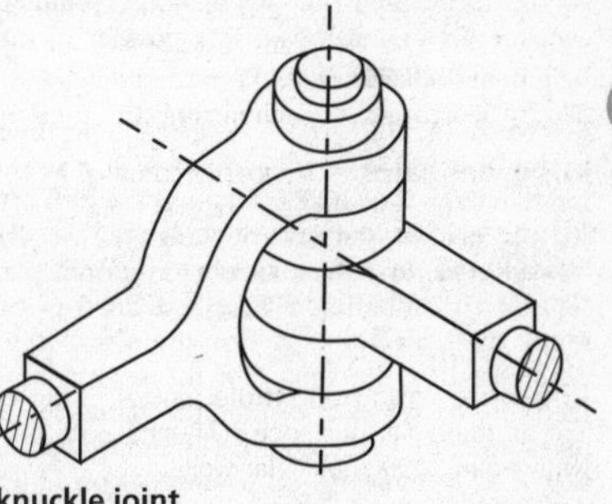

knuckle joint

Knudsen gauge (Knudsen vacuum gauge) An instrument for measuring very low gas pressures (1 μPa to 1 Pa) based upon the momentum transfer by gas molecules between two closely-spaced plates immersed in the gas.

Knudsen number (*Kn*) A non-dimensional parameter that arises in flows where the mean-free path of the fluid Λ is comparable with, or greater than, a characteristic dimension of the flow L and defined by $Kn = \Lambda/L$. Applications include rarefied gas dynamics ($Kn > 10$) and gas flows in microchannels where slip is important ($0.3 > Kn > 0.01$).

Knudsen pump A device in which thermal transpiration causes gas molecules to flow from the cold end to the hot end of a microchannel (<100 nm diameter) subjected to a longitudinal temperature gradient.

knurling A continuous diamond-shape pattern, usually on a cylindrical surface, intended to improve finger grip or as a decorative feature. In some instances the knurling is parallel to an axis of rotation. It is represented in an engineering drawing by fine straight lines on the surface of an object, parallel if the knurling is straight, diagonal at about 30° if a diamond pattern.

Kolmogorov microscales The smallest scales of turbulence in a flow defined in terms of the kinematic viscosity of a fluid ν and the turbulent dissipation rate per unit mass ε: length scale $\eta \equiv (\nu^3/\varepsilon)^{1/4}$, time scale $\tau_\eta \equiv (\nu/\varepsilon)^{1/2}$ and velocity scale $u_\eta \equiv (\nu\varepsilon)^{1/4}$.

Kovar A vacuum-melted iron-nickel-cobalt alloy with thermal-expansion properties similar to those of glasses and ceramics. *See also* INVAR.

Kundt's tube A gas-filled transparent horizontal tube used to demonstrate standing acoustic waves that cause fine powder in the tube to accumulate at nodes.

kurtosis (K, Rku, β_2) A statistical term used in the description of a distribution of a set of random data. It quantifies whether the data are sharply peaked (**leptokurtic**, high K) or flat (**platykurtic**, low K) relative to a normal (i.e. Gaussian) distribution. K may be defined as the fourth moment $\mu_4 =< x_i^4 >$ of the distribution normalized by the square of the second moment μ_2 or $\sigma =< x_i^2 >$ (i.e. the variance or mean square) where x_i is the random quantity and $< x_i^n >$ indicates an average value of x_i^n taken over all values of i with $n = 2$ or 4. The distribution is defined such that $< x_i > = 0$ and then $K =< x_i^4 >/\sigma^2$. Among other applications, kurtosis is used in the description of surface roughness and of turbulence where the term **flatness** (or **flatness factor**) is also used. *See also* SKEWNESS.

Kutta condition In potential-flow theory, the condition that infinite velocities cannot exist at a sharp trailing edge. Circulation is imposed on the flow to ensure that the velocity is finite.

Kutta–Joukowski theorem A vortex of strength (i.e. circulation) Γ in a uniform flow with velocity V experiences a lift force L given by $L = \rho V\Gamma$ where ρ is the fluid density.

kW.h *See* KILOWATT.HOUR.

Kyoto protocol A protocol to the United Nations Framework Convention on Climate Change, adopted in 1997, under which 37 countries committed to the reduction of four greenhouse gases (carbon dioxide, methane, nitrous oxide, and sulphur hexafluoride) and two groups of gases (hydrofluorocarbons and perfluorocarbons) produced by them, by an average of 5.2% of 1990 levels between 2008 and 2012. A new international framework needed to be ratified by the end of the first commitment period in 2012 but to date that has not been achieved in spite of numerous meetings, such as those in Copenhagen (2009) and Cancun (2010).

lab-on-a-chip A microfluidic MEMS device that allows laboratory operations to be conducted at micro- and nano-scales.

laboratory coordinate system An inertial (i.e. non-accelerating) Cartesian-coordinate system, with axes *x*, *y* in the horizontal plane and the *z* axis vertical, which is frequently used to reference experimental observations.

labyrinth seal A low-friction mechanical seal that relies upon an extended path, for example a series of narrow closely-spaced grooves, to minimize fluid leakage. Applications include piston-cylinder arrangements and circular shafts especially to protect bearings.

ladder logic A programming method used in a programmable logic controller, based on the representation of sequential program steps by the opening and closing of relays, the starting and stopping of timers, and the operation of special blocks containing complex functions. A **ladder diagram** is a diagrammatic representation of a program produced using ladder logic. The name is derived from the similarity between the diagram and a ladder.

lag The delay when one event happens after another. A signal is described as lagging when it is delayed relative to another signal. *See also* LEAD; VALVE-TIMING DIAGRAM.

lag compensation *See* LEAD/LAG COMPENSATION.

lagging The thermal insulation material, typically in the form of fabric tape, glass fibre, etc., that is wrapped around hot or cold surfaces such as boilers and pipes. Although still installed for some domestic and industrial applications, the use of asbestos is now illegal because of the risk of serious illnesses that can result from the inhalation of asbestos fibres.

Lagrange–Hamiltonian mechanics Newtonian mechanics expressed in terms of energy rather than forces. Particularly useful for determining the trajectories of objects moving in a gravitational field.

Lagrangian description The analysis of fluid flow and finite-deformation plasticity following the motion of individual fluid particles or material elements as they move along pathlines. *See also* EULERIAN DESCRIPTION.

Lambert's cosine law For a diffuse surface, the intensity i_ϕ of thermal radiation emitted in a given direction is given by $i_\phi = i_n cos\phi$ where i_n is the intensity of radiation emitted normal to the surface and ϕ is the angle between the normal and the direction of the radiation.

Lambert surface (Lambertian surface) *See* DIFFUSE SURFACE.

lamda sensor (lamda probe, oxygen sensor) An electronic device that measures the fraction of oxygen in a fluid flow. Most commonly used to monitor the oxygen level in the exhaust gases from an internal-combustion engine and control the air/fuel ratio of the mixture entering the engine to control pollutants.

Lamé constants (Lamé parameters, λ, μ) (Unit Pa) In linear elasticity for a homogeneous isotropic material, the two parameters that relate stress and strain. For normal stresses $\sigma_i = \lambda e + 2\mu\varepsilon_i$ where σ_i is the normal stress in a given direction, e is the volumetric strain (i.e. dilation), and ε_i is the strain in the same direction as σ_i. For the shear stresses $\tau_{ij} = 2\mu\varepsilon_{ij}$. The first Lamé constant $\lambda = K - 2\mu/3 = 9Ev/[(1-2v)(1+v)]$ where K is the bulk modulus, E is Young's modulus, and v is Poisson's ratio. The second Lamé constant μ is identical to the modulus of rigidity (shear modulus, G) given by $G = E/2(1 + v)$. *See also* GENERALIZED HOOKE'S LAW; HOOKE'S LAW.

Lamé equations For linear-elastic behaviour of a thick cylinder subject to internal or external pressure or rotation, the variation of the radial stress σ_r with radius r is given by $\sigma_r = A + B/r^2$ and the hoop stress σ_θ by $\sigma_\theta = A - B/r^2$ The constants A and B are determined from the boundary conditions and the sign convention employed for stresses.

lamina A plane sheet of zero thickness in theory, but in practice applied to thin layers or sheets of materials.

laminar boundary layer A boundary layer for which the Reynolds number is sufficiently low for the flow to remain laminar. For a flat-plate, zero-pressure-gradient boundary layer, the critical Reynolds number $Re_{x,crit}$ is about 60,000 based upon free-stream velocity and distance from the leading edge, or 162 based upon the momentum thickness. A favourable pressure gradient increases $Re_{x,crit}$ whereas an adverse pressure gradient causes a decrease, as does free-stream turbulence.

laminar flow Highly ordered and deterministic flow of a viscous fluid at Reynolds numbers below the critical value (which is different for every flow) for instability and transition to turbulence to occur. Such flows may be steady or unsteady, and include both periodic and transient behaviour.

laminarization (reverse transition) A phenomenon whereby turbulent fluctuations in a flow can be suppressed, for example in a boundary layer by a strong favourable pressure gradient, and the flow then has many of the characteristics of a laminar flow.

laminate 1. A body made up of bonded layers of thin sheets. **2.** To manufacture a laminate.

laminated spring *See* LEAF SPRING.

lamination An individual layer within a laminate.

Lami's theorem If three coplanar, concurrent forces act on a body to keep it in static equilibrium, each force is proportional to the sine of the angle between the other two.

Lanchester damper A device for damping out linear or torsional vibrations, and thus avoiding resonance, in which the damper spring of an undamped vibration absorber is replaced by a dry friction, or viscous friction, element. For the damping of torsional vibrations (as in the balancing of engines), the device is often referred to as a **Lanchester balancer**.

land 1. *See* TOOTHED GEARING **2.** The space on a piston between two ring grooves. **3.** In tapered dies, any section parallel to the flow of material.

landing gear Those parts of an aircraft, usually retractable wheels, that allow it to take off and land or to move when on the ground or water. The term **undercarriage** is more usual for the floats of seaplanes.

lantern ring An annular ring placed between two sets of packings in a packed gland to allow the introduction of lubricant, cooling, or flushing fluid.

lap joint A riveted, welded, or bonded connection between two plates made by overlapping one plate above the other. Lap joints having plates above and below a butt joint avoid the bending moment associated with a simple lap joint.

Laplace pressure (Unit Pa) The pressure difference due to surface tension across a curved interface between two immiscible fluids as given by the Young–Laplace equation.

lapse rate (γ) (Units °C/m, Pa/m, kg/m^4, etc.) The rate of change with altitude z of a thermodynamic property (temperature T, pressure, density, etc.) in the atmosphere. For temperature, $\gamma = -dT/dz$. *See also* ADIABATIC LAPSE RATE.

Laray viscometer An instrument in which liquid viscosity is determined from the time taken by a weighted rod placed in an orifice, both immersed in the liquid, to fall a fixed distance. Widely used to characterize printing inks.

large calorie *See* KILOCALORIE.

large-eddy simulation (LES) A method of calculating turbulent flows in which the large unsteady features of the flow are resolved while the small-scale dissipative eddies are modelled. *See also* TURBULENCE MODELLING.

large-systems control system *See* HIERARCHICAL CONTROL.

laser annealing Annealing of metals by a moving CO_2 or Nd:Yag laser beam focussed down to a few mm in diameter. A point in the surface of an object is heated above the recrystallization temperature and then cooled to room temperature to remove the effects of workhardening and restore a soft condition.

laser-Doppler anemometer (laser-Doppler velocimeter, laser anemometer, LDA, LDV) An instrument for determining flow or surface velocity by measuring the transit time of small particles in the flow or surface elements as they cross interference fringes formed at the intersection of two laser beams focussed onto the same point.

laser hardening Hardening of steel by a moving CO_2 or Nd:Yag laser beam focussed down to a few mm in diameter. A point in the surface of an object is rapidly heated into the austenite range and then rapidly cooled to form tempered martensite. **Electron-beam hardening** can be employed to achieve similar results.

laser micromachining The use of excimer, Nd:Yag, or CO_2 lasers in the fabrication of microchannels, either directly or using a mask.

laser scribing (laser marking) The use of excimer, Nd:Yag, or CO_2 lasers to mark the surfaces of such materials as ceramic, glass, metals, quartz, and silicon. Principal applications are in the production of semi-conductor devices, light-emitting diodes (LEDs), etc.

laser sintering A manufacturing technique by which parts are built layer by layer (each typically 20 μm thick) from plastic or metal (**Direct Metal Laser Sintering, DMLS**) material in powder form, each layer being sintered by a scanning laser. *See also* ADDITIVE LAYER MANUFACTURING.

laser 2-focus anemometer An instrument for determining flow velocity by measuring the transit time of small particles in the flow as they cross two laser beams focussed to point foci a small, accurately-known distance apart.

laser welding *See* WELDING.

latent heat (heat of transformation, latent energy) (Unit kJ/kg) The amount of energy absorbed or released in a phase-change process. The magnitude depends upon the process, the temperature, and the pressure at which the phase change occurs. The latent heat of a given phase change is often referred to as the enthalpy or heat of that change.

The **latent heat of condensation (h_{gf})** is the amount of energy released in condensing unit mass of a dry saturated vapour. It is equal to the difference between the specific enthalpies of the saturated liquid h_f and the dry saturated vapour h_g, i.e. $h_{gf} = h_f - h_g$. The **latent heat of evaporation or vaporization (h_{fg})** is the amount of energy needed to vaporize unit mass of a saturated liquid (i.e. a liquid at its boiling point). It is equal to the difference between the specific enthalpies of the dry saturated vapour h_g and the saturated liquid h_f, i.e. $h_{fg} = h_g - h_f$, and so equal in magnitude but opposite in sign to the latent heat of condensation.

The **latent heat of freezing, crystallization or solidification (h_{fs})** is the amount of energy released in freezing unit mass of a saturated liquid. It is equal to the difference between the specific enthalpies of the saturated liquid h_f and the solid h_s, i.e. $h_{fs} = h_s - h_f$.

The **latent heat of fusion or melting (h_{sf})** is the amount of energy needed to melt unit mass of a solid (i.e. a solid at its melting point). It is equal to the difference between the specific enthalpies of the solid h_s and the saturated liquid h_f, i.e. $h_{sf} = h_f - h_s$, and so equal in magnitude but opposite in sign to the latent heat of freezing.

The **latent heat of sublimation (h_{sg})** is the amount of energy needed to sublimate unit mass of a solid. It is equal to the difference between the specific enthalpies of the solid h_s and the dry saturated vapour h_g, i.e. $h_{sg} = h_g - h_s$.

latent load The thermal energy required, in an air-conditioning system, to condense water vapour in the air.

lateral extensometer An extensometer that measures displacement perpendicular to the principal loading axis.

lathe A machine tool in which work, gripped in the headstock, is rotated against the cutting tool to produce turned, bored, faced, or threaded components. *See also* CAPSTAN LATHE; CHUCK; FEED SCREW; GAP LATHE; HEADSTOCK; LEAD SCREW; SET OVER; SWING; TAILSTOCK; TURNING; TURRET LATHE.

SEE WEB LINKS

- Basic lathe terminology

lattice–Boltzmann method A class of CFD methods in which, instead of solving the Navier–Stokes equations, fluid flow is simulated using discrete particles that undergo consecutive propagation and collision processes over a discrete lattice mesh.

lattice girder An open framework built of two principal members joined by criss-cross reticulated members.

Laval nozzle *See* DE LAVAL NOZZLE.

law of action and reaction *See* NEWTON'S LAWS OF MOTION.

law of corresponding states *See* PRINCIPLE OF CORRESPONDING STATES.

law of gravitation *See* NEWTON'S LAW OF GRAVITATION.

law of partial pressures *See* DALTON'S LAW.

law of partial volumes *See* AMAGAT'S LAW.

law-of-the-wall variables (inner variables, wall variables, wall coordinates, u^+, y^+, T^+) Non-dimensional variables used in turbulent boundary-layer theory using the friction velocity u_τ together with fluid properties evaluated at or near the surface (the wall). Wall variables are indicated by the superscript $^+$, e.g. for the mean-flow velocity $\bar{u}$, $u^+ = \bar{u}/u_\tau$, for the distance from the wall y, $y^+ = u_\tau y/\nu_S$, where ν_S is the kinematic viscosity of the near-wall fluid, and for the mean temperature $\bar{T}$, $T^+ = u_\tau \rho_S C_P(T_S - \bar{T})\, \dot{q}''_S$ where T_S is the surface temperature, ρ_S is the near-wall fluid density, C_P is the specific heat, and $\dot{q}''_S$ is the surface heat flux. The combination $\mu_S/\rho_S u_\tau$ is termed the viscous length scale, and **wall units** are lengths made non-dimensional by dividing by it.

The **law of the wall** is the variation of mean velocity $\bar{u}$ with distance from the wall y in the near-wall region of a turbulent boundary layer or pipe flow in law-of-the-wall variables. From dimensional analysis it is found that $u^+ = f(y^+)$. In the immediate vicinity of the wall (the viscous sub layer, approximately $y^+ < 5$), $u^+ = y^+$. In the turbulent core (approximately $y^+ > 30$), the **log law** applies,

$$u^+ = \frac{1}{\kappa} ln(y^+) + B$$

where κ is von Kármán's constant and $B \approx 5.5$. The **buffer layer** is the transition region $5 < y^+ < 30$. For the entire wall region, Spalding's formula is widely used. The **law of the wake** is an approximate empirical function that accounts for the deviation in velocity from the law of the wall in the outer region of a turbulent boundary layer. The **overlap layer** is the transition region between the inner layer, where the law-of-the-wall applies, and the outer layer. The log law applies in the overlap layer.

Wall functions are approximate functions, based upon the log law, for the variation of the mean velocity, Reynolds stresses, dissipation, etc. in the near-wall region of a turbulent shear flow. They are used to avoid the expense of high spatial resolution in numerical calculations in near-wall turbulent flows.

laws of thermodynamics *See* THERMODYNAMICS LAWS.

lay 1. The hand (right or left) of twist in a rope or cable. **2.** The main direction of tool marks on a machined surface.

layout blue *See* ENGINEERS' BLUE (2).

lay shaft (layshaft) A secondary or intermediate shaft in a gearbox, running parallel to the main shaft and carrying the paired gear wheels that effect the changes in gear ratio. *See also* COUNTER SHAFT.

lay-up The manufacture of fibre-reinforced composite components using pre-pregs.

lb *See* POUND.

lb_f *See* POUND FORCE.

LDA *See* LASER-DOPPLER ANEMOMETER.

LDR *See* LIMITING DRAW RATIO.

L/D ratio The length-to-diameter ratio of an engineering component. *See also* ASPECT RATIO.

LDV *See* LASER-DOPPLER ANEMOMETER.

lead 1. The occurrence of one event before another. A signal is described as leading when it is in advance of another signal. *See also* LAG; VALVE-TIMING DIAGRAM. **2.** *See* SCREW.

lead angle The angle between a tangent to a helix and a plane normal the axis of the helix.

lead compensation *See* LEAD/LAG COMPENSATION.

leaded petrol Petrol to which tetraethyl lead has been added to increase its octane rating and so improve its knock resistance. Now prohibited in most industrialized countries due to the risk of health and environmental damage.

lead/lag compensation (lead/lag control) In a control system, a network (the **lead/lag network**) introduced in series with the plant having a transfer function given by: $G(s) = (s + z)/(s + p)$ where p and z are positive real constants. When $z < p$, the compensator introduces a phase lead in the output relative to the input. When $z > p$ the opposite is true. A lead compensator lowers the magnitude of the output at low frequencies ($\omega < 10p$) while a lag compensator does so at high frequencies ($\omega > 0.1p$).

leading edge *See* AEROFOIL.

lead screw The master screw running along the length of the bed of a lathe from which all screws threads on that machine are cut.

leaf spring (laminated spring) A beam-like spring made up of thin independently-acting plates placed over one another and held together in a buckle.

leakage flow **1.** The flow of working fluid that occurs in a turbomachine between the tips of the rotor blades and the casing. **2.** The flow of any fluid past a seal. **3.** The escape of fluid through cracks in a pressure vessel.

leakage losses The power losses in a turbomachine associated with the leakage flow that does not do its full complement of work on the rotor.

leakage rate The rate at which fluid leaks past a seal or through cracks in a pressure vessel.

leak-before-burst A fracture-mechanics-based design methodology for pressure vessels, whereby propagation of a crack will result in the pressure being relieved by leakage rather than the vessel exploding.

lean mixture *See* STOICHIOMETRIC MIXTURE.

learning control A control method where the controller continuously uses identification to determine the plant parameters and thus modifies the operation of the controller.

least-energy principle (least-work theory) *See* PRINCIPLE OF MINIMUM ENERGY.

leaving loss The kinetic energy of the working fluid at outlet from the last stage of an axial-flow turbine.

Ledoux bell meter A flow meter in which an inverted weighted bell sealed with mercury moves up and down as the flow rate changes.

Leduc's law *See* LAW OF PARTIAL VOLUMES.

Lee's disc apparatus An apparatus for determining the thermal conductivity of a poor conductor in the form of a thin disc sandwiched between two thick metal discs (typically brass or copper).

LEFM *See* FRACTURE MECHANICS.

left-handed thread *See* SCREW.

lehr A special type of oven or kiln used in glass manufacture for annealing by slow cooling to remove internal stresses.

Leidenfrost effect The phenomenon that occurs when a liquid droplet is in near contact with a horizontal surface at a significantly higher temperature than the liquid's boiling point. A vapour layer is created beneath the bubble causing it to float and evaporate relatively slowly.

Leidenfrost point *See* BOILING.

leptokurtic *See* KURTOSIS.

LES *See* LARGE-EDDY SIMULATION.

Le Système international d'unités *See* INTERNATIONAL SYSTEM OF UNITS.

lever A stiff rod or beam pivoted about a fulcrum with a load at some point along the lever being moved by a force (effort) at a third point. There are three classes of lever defined according to the relative positions of the effort, fulcrum, and load. For class 1, the effort and the load are on opposite sides of the fulcrum (as in a pair of pliers); for class 2, effort and load are on the same side of the fulcrum with the effort further away (as in a wheelbarrow); class 3 is like class 2, but with the load further away (as with tweezers). Classes 1 and 2 provide a mechanical advantage, but class 3 does not.

leverage *See* MECHANICAL ADVANTAGE.

levitation *See* MAGLEV.

Levy–Mises equations (Levy–Mises flow rules) The relationships, in rigid-plasticity theory, between normal and shear plastic strain increments ($d\varepsilon_x$ and $d\gamma_{yz}$ respectively) and the instantaneous deviatoric stresses. There are three equations of the type

$$d\varepsilon_x = \frac{2}{3} d\lambda[\sigma_x - \frac{1}{2}(\sigma_y + \sigma_z)]$$

and three of the type $d\gamma_{yz} = \tau_{yz} d\lambda$ where σ and τ are normal and shear stresses respectively and $d\lambda$ is an instantaneous non-negative constant of proportionality that may vary as the strain changes. *See also* PRANDTL–REUSS EQUATIONS.

Lewis equation *See* TOOTHED GEARING.

Lewis form factor *See* TOOTHED GEARING.

Lewis number (Le) A non-dimensional parameter that arises in problems involving combined convective heat and mass transfer, defined by $Le = \rho C_P D_j/k$ where ρ is the fluid density, C_P is its specific heat, k is its thermal conductivity and D_j is the diffusion coefficient for species j. The inverse $k/\rho C_P D_j$ is sometimes used as the definition.

lexan® An easily-machinable, thermoformable polycarbonate material with high impact strength and optical transmittance. Representative properties are: glass-transition temperature 150°C; Young's modulus 2.5 GPa; yield strength 70 MPa. Glass-filled versions are available with improved properties. Applications include high-performance windscreens, aircraft canopies, bullet-resistant windows, and space helmets.

life-cycle analysis (full life-cycle analysis) An analysis, in terms of money, mass or embodied energy, of all aspects involved in a process and its consequences, including environmental impact, manufacture, mining or recycling and processing of raw materials, energy consumption, and decommissioning, including disposal as scrap.

life factor *See* TOOTHED GEARING.

lift (lift force) The aerodynamic or hydrodynamic force exerted on a body as it moves relative to a fluid, acting in a direction perpendicular to that of the relative motion. *See also* DRAG.

lift check valve A valve similar in configuration to a globe valve but with an automatically-operated cone-shaped plug. The cone is lifted off its seat by flow in the forward direction and returns to its seat if the flow reverses.

lift coefficient (C_L) A non-dimensional quantity defined by $C_L = L/Aq$ where L is the lift force exerted on a body by a fluid flow with dynamic pressure q and A is an appropriate area. In the case of an aerofoil A is taken as the wing planform area, but for other bodies it is usually the projected area in the flow direction.

lifting surface Any surface, such as an aerofoil, a propeller, or a turbomachine blade, that produces a lift force when moving relative to a fluid.

lift pump *See* SUCTION PUMP.

lift valve A valve, such as a poppet valve, that is opened by being mechanically lifted.

limit control (limit switch) A switch used, in devices where motion over a fixed range of displacement or angle is required, to stop the driving motor at a predetermined location.

limit cycling Oscillation of the output of a system through a fixed range between limits.

limited-degree-of-freedom robot A robot with insufficient joints to allow independent control of the six degrees of freedom (translations along x, y, z and rotations about x, y, z) of the end effector.

limited-life fatigue *See* FATIGUE.

limited-rotation hydraulic motor A hydraulic motor or actuator that provides rotary motion over a finite angle.

limited-slip differential A differential that permits one wheel to slip relative to the other by only a limited amount, after which it becomes locked.

limit gauge A 'go/no-go' gauge to check that a component is within specified dimensions.

limiting draw ratio (LDR) In deep drawing of cups, the greatest ratio of circular blank diameter to diameter of the drawn cup before fracture of the cup wall occurs. Allowing for friction and workhardening, LDR ≈ 2.

limiting friction The friction between two bodies in contact when sliding is imminent.

limiting moment *See* PLASTIC HINGE.

limiting range of stress *See* FATIGUE.

limit load *See* COLLAPSE LOAD.

limit of proportionality (proportional limit) The maximum stress up to which a material obeys Hooke's law.

limit of size *See* TOLERANCES.

limits and fits Special tolerance combinations which, for certain applications, have been worked out for successful functioning of particular types of fit: **running fits** for shafts in bearings and pistons in cylinders; **sliding fits** for parts assembled by hand pressure or with the help of a light hammer; keying or **driving fits**, where parts are assembled with a medium hammer; and **press** or **force fits**, for parts such as wheels and hubs which, once assembled, are unlikely to be dismantled. Running and sliding fits are **clearance fits**; press fits, where the shaft is bigger than the hole, are **interference fits**; **keying fits** are transition fits in between the two.

These limits and fits are part of the wider system issued in ISO 286 intended to cover most engineering purposes from the finest to the coarsest. Although written in terms of holes and shafts, the system may be applied to mating parts having other geometries, such as a flat surface where the hole is a slot and the shaft a key. The system is concerned simultaneously with (a) the quality or grade of tolerance (the +/− bandwidth on the limits of size); and (b) the type of fit (the location of the bandwidth with respect to the nominal size). Except for very large sizes (where temperature effects become important), **hole-basis fits** are usually specified, i.e. fits in which the design size for a given hole is the basic size, and variations in the grade of fit for any particular hole are obtained by varying the clearance and tolerance on the shaft. This comes about since it is easier to manufacture and measure the male member of a fit, and only one reamer is required for the hole. *See also* TOLERANCES.

limit state The loading on a structure or a component when plastic collapse is imminent.

limit switch *See* LIMIT CONTROL.

linear acceleration (Unit m/s^2) Acceleration along a straight line. *See also* ANGULAR ACCELERATION.

linear actuator An actuator providing translational motion.

linear control system A control system described by linear differential equations and the resulting transfer functions.

linear elasticity In a material, component or structure, when the applied loads and the resulting deflexions are directly proportional. *See also* ELASTIC; HOOKE'S LAW; LIMIT OF PROPORTIONALITY.

linear expansivity *See* THERMAL EXPANSION.

linear feedback control A linear control system employing closed-loop control.

linear hydraulic motor *See* HYDRAULIC CYLINDER.

linear momentum (Unit kg.m/s) The momentum of a body or fluid in linear motion. *See also* ANGULAR MOMENTUM; MOMENTUM.

linear-momentum conservation equation *See* MOMENTUM.

linear–quadratic-Gaussian control (LQG control) An optimal control method for controlling an uncertain linear system subjected to noise added to its inputs and outputs and with incomplete knowledge of the states.

linear roller bearing *See* BEARING.

linear spring *See* SPRING.

linear system A system in which the output is directly proportional to the input. It is characteristic of such a system that if an input x_1 produces an output y_1 and an input x_2 produces an output y_2, an input $(x_1 + x_2)$ will produce an output $(y_1 + y_2)$.

linear variable differential transformer (LVDT) An inductive sensor which uses a transformer with variable coupling between a single primary coil and a pair of secondary coils to give an output voltage proportional to the translational displacement of the iron core. The core is fitted with a projecting non-magnetic shaft so that the LVDT can be coupled to the displacement to be measured. *See also* ROTARY VARIABLE DIFFERENTIAL TRANSDUCER.

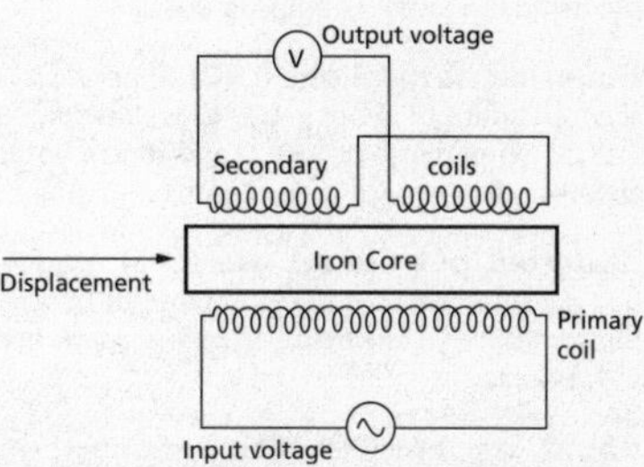

linear variable differential transformer (LVDT)

linear velocity (Unit m/s) The velocity of a body or fluid in linear motion. *See also* ANGULAR VELOCITY.

line load The idealization, in 2-dimensional analysis, that a load applied to a surface acts over zero contact area. *See also* CONTACT MECHANICS; POINT LOAD.

line of action *See* TOOTHED GEARING.

line of condition In a vapour-power cycle, such as the Rankine cycle, the path of the irreversible expansion on a temperature-entropy diagram.

line of contact *See* TOOTHED GEARING.

line oiler *See* AIR-LINE LUBRICATOR.

liner *See* CYLINDER LINER.

line sink An idealized line, in plane potential-flow theory, into which there is a radial inflow of fluid with velocity inversely proportional to the radial distance. There is a singularity at the line itself. A **line source** is the counterpart when there is flow out of the line. *See also* POINT SINK.

link 1. Any connecting member in a mechanism that is pivoted at both ends. **2.** The part of a robot arm running between two joints.

linkage Any combination of links, sliders, pivots, and rotating members forming a mechanism that produces a desired motion.

link length *See* JOINT.

liquefaction The conversion of a gas into a liquid using **liquefier**. This can be achieved by cooling the gas below its critical temperature and compressing it, or by expansion, for example using the Joule–Thompson effect.

liquefied natural gas (LNG) Natural gas at a temperature of about −162°C, well below its critical temperature of −83°C and so in a liquid state.

liquefied petroleum gas (LPG) A liquefied gaseous fuel, normally propane or butane, but sometimes a mix based on 60% propane and 40% butane.

liquid An essentially incompressible wet fluid that forms a free surface in a container. *See also* GAS.

liquid-cooled engine A piston engine in which a liquid, usually water with additives such as antifreeze, is circulated through passages in the engine block and a radiator.

liquid cooling The use of circulated liquid, usually water, to cool engines or other machines.

liquid-in-glass thermometer A bulb thermometer in which the bulb and stem are glass.

liquid-in-metal thermometer A fluid-expansion thermometer in which the expanding fluid is a liquid contained in a metal tube such as a Bourdon tube.

liquidoid line The locus of temperatures of solid-state transformations of a two-component system on a phase diagram (2).

liquid piston compressor A positive-displacement rotary compressor in which a rotor with forward-curved blades rotates within an elliptical casing with the rotor partly filled by a liquid.

liquid-sealed meter *See* CROSSLEY METER.

liquid-sorbent dehumidifier A dehumidifier in which the air to be dried is passed through sprays of a liquid absorbent, such as lithium chloride or glycol solution.

liquid spring A piston-cylinder device dependent upon the compressibility of a liquid, typically silicone-based, to cushion shock loads. Used in heavy-duty applications such as heavy military vehicles and the landing gear of aircraft and the space shuttle. *See also* GAS SPRING.

liquidus line The locus of melting/freezing points of a two-component system on a phase diagram (2).

liquid-vapour dome The two-phase region in a *p-v-T* diagram enclosed by the saturated-liquid and saturated-vapour lines, in which the liquid phase and the vapour phase co-exist in equilibrium.

Lissajous patterns (Lissajous figures) The locus of the resultant displacement of a point moving under the influence of two independent harmonic motions. In the case of two orthogonal simple harmonic motions having the same frequency, the Lissajous figures are a series of ellipses corresponding to possible differences in phase between the two motions.

SEE WEB LINKS

- Interactive simulation of Lissajous figures on an oscilloscope.

lithography The transfer of a pattern to a photosensitive material, typically a photoresist, by selective exposure through a mask to radiation. It is the most important process for fabricating MEMS and other microscale structures. *See also* PHOTOLITHOGRAPHY.

litre (*l*) A non-SI unit of volume defined as 10^{-3} m^3 and approximately equal to the volume of 1 kg of water at 4°C, which was how it was formerly defined.

little end *See* CONNECTING ROD.

live axle An unsprung axle to which wheels are rigidly fixed.

live load The load on a structure caused by vehicles, etc. passing over it, as opposed to the dead weight of the structure itself.

live steam Steam under pressure as supplied direct from a boiler.

Ljungström pre-heater A regenerative air pre-heater for a boiler furnace in which a rotating cylindrical matrix passes alternately through the air and the flue gas. *See also* ROTARY REGENERATOR.

Ljungström turbine An outward-flow steam turbine consisting of concentric rows of blades attached to the opposing faces of two rotor discs that rotate in opposite directions.

LMTD *See* LOG-MEAN TEMPERATURE DIFFERENCE.

LNG *See* LIQUEFIED NATURAL GAS.

load 1. The force applied to a component or structure. **2.** The power drawn from a prime mover.

load-carrying capacity 1. The maximum weight that can safely be carried by a vehicle or structure. **2.** In robotics, the maximum weight that can be carried by the end effector without causing out-of-specification operation, for example, excessive steady-state error.

load cell A device for measuring forces by means of strain gauges, piezoelectric detectors, etc., as opposed to the use of hydraulic pressure or mechanical means. Such devices are sometimes only uniaxial, or simultaneously measure loads along other orthogonal axes.

load control The action of a testing machine when the load applied to the testpiece can be altered in a controlled fashion, the corresponding displacement of the specimen being dependent on the stress–strain characteristics of the specimen.

load-deformation diagram A plot of load *vs* deformation based upon the data given by a testing machine from which stress and strain, and hence mechanical properties, may be calculated.

load distribution factor *See* TOOTHED GEARING.

load factor 1. The ratio of the design working load on a component or structure to the load at which failure would occur. *See also* FACTOR OF SAFETY. **2.** *See* CAPACITY FACTOR.

load inertia The moment of inertia of the load applied to a motor.

loading diagram A schematic of the loads and their points of application to which a component or structure is subjected, often accompanied by the corresponding shear force and bending-moment diagrams.

load intensity (*q*, *w*) (Unit N/m) The value of a distributed load acting on a structure or component.

load rating *See* BEARING.

load shedding 1. The unloading of members of a component or structure following damage or fracture. The loads and energy released are transferred into adjacent members that may become overloaded leading to a cascade of failures. **2.** The process of disconnecting parts of the load on a power station when the demand exceeds the capacity.

lobe 1. The part projecting from a circular profile in a cam, gear or other basically circular component. **2.** A type of rotor or impeller used in some flow meters, pumps, etc.

lobed impeller meter A flow meter in which two lobed impellers rotate in opposite directions within an ovoid housing. A wide range of materials is used for the impellers, including thermoplastics and stainless steel.

lobe pump Similar to a gear pump, but the lobes are prevented from being in contact by external gearing. Commonly used in food handling since they do not damage the material.

local acceleration (Unit m/s^2) In unsteady fluid flow, the time derivative of the vector velocity $\boldsymbol{V}$ at a point, $\partial \boldsymbol{V}/\partial t$. *See also* TOTAL ACCELERATION.

local buckling Buckling of a thin-walled component that is contained within a small region of the surface.

localized necking In sheet-metal forming, a neck that is localized into a thin trough, the orientation of which depends on the applied strain or stress ratio.

local structural discontinuity *See* STRESS CONCENTRATION.

location fit *See* LIMITS AND FITS.

locking differential A differential in which the half-shafts may be locked together if required, thus eliminating differential action so that both wheels rotate at the same speed.

locknut **1.** A thin auxiliary nut tightened against another nut to prevent loosening. **2.** A single nut with special features that prevent loosening.

lock washer *See* TAB WASHER.

locomotive A vehicle on rails, powered by electric, diesel, or diesel-electric motors, used to haul passenger carriages or goods wagons. Vintage steam locomotives are powered by coal-fired boilers.

locomotive boiler A multi-fire-tube boiler with integral firebox.

logarithmic creep *See* CREEP.

logarithmic decrement **(δ)** The natural logarithm of the ratio of the amplitudes of any two successive peaks, x_n and x_{n+1}, in the output of an underdamped oscillatory system, given by $\delta = ln(x_n/x_{n+1})$. It allows the determination of the damping ratio of the system. *See also* DECAY OF OSCILLATION.

l

logarithmic strain *See* TRUE STRAIN.

log law *See* LAW-OF-THE-WALL VARIABLES.

log-mean temperature difference (LMTD, temperature difference, ΔT_M) (Unit °C) If the heat-transfer rate $\dot{q}$ for a simple one-tube pass, one shell-pass heat exchanger, with either parallel flow or counterflow, is expressed as $\dot{q} = UA\Delta T_M$, where U is the overall heat-transfer coefficient and A is the heat-exchanger total surface area, it is found that $\Delta T_M = \frac{\Delta T_1 - \Delta T_2}{ln(\Delta T_1/\Delta T_2)}$ where ΔT_1 and ΔT_2 are the temperature differences between the hot and cold streams at inlet and outlet, respectively. For a more complex arrangement, a correction factor F is introduced such that for the actual heat exchanger $\Delta T_M = F\,\Delta T_{M,CR}$ where $\Delta T_{M,CR}$ is for a simple counterflow heat exchanger.

longitudinal-drum boiler The original type of water-tube boiler operating on the thermo-siphon principle. Cooled feedwater, fed into the drum which is placed longitudinally above the heat source, flows down a header and into the inclined heated tubes that lead back to the drum. *See also* CROSS-DRUM BOILER.

longitudinal wave A wave in which the displacement of the propagating medium is normal to the local wavefront. Examples are plane acoustic waves in fluids and compressional waves in isotropic solids.

long-term repeatability A measure of the ability of a robot to return to the same programmed position over a long period of time.

long ton (imperial ton) A non-SI unit of mass equal to 2240 lb. *See also* SHORT TON; TONNE.

loop gain *See* CLOSED-LOOP GAIN; OPEN-LOOP GAIN.

loop strength (loop tenacity) A mechanical-property test for fibres, in which two interlinked loops are pulled until crushing or cutting occurs under the loading pins.

loss For a machine or system, power that does not perform useful work. The principal losses are due to mechanical or viscous friction windage loss, and leakage loss, for example leakage of working fluid from one stage to the next in a turbine or through seals.

loss modulus (Unit Pa) **1.** **(E'', G'')** A measure of the energy dissipated in a viscoelastic solid, determined by applying an oscillatory stress to the material and recording the strain. E'' corresponds to a tensile stress and G'' to a shear stress. **2.** **(G'')** A measure of the energy dissipated in a viscoelastic liquid, determined by applying an oscillatory shear stress and recording the strain rate. *See also* COMPLEX VISCOSITY; DYNAMIC MODULUS; STORAGE MODULUS.

lost motion The delay in time or in displacement between the driving and driven parts in a mechanism. *See* BACKLASH.

lost-wax casting *See* INVESTMENT CASTING.

lost-wax technique *See* INVESTMENT CASTING.

Love wave *See* SURFACE WAVE.

low-cycle fatigue *See* FATIGUE.

low-E coating (low-emittance coating) A thin metallic or metal-oxide coating on a glass sheet to absorb and reflect infra-red radiation. The coating is applied either by a pyrolytic chemical vapour-deposition process (hard coat), or by sputtering (soft coat).

low-enthalpy resource *See* GEOTHERMAL ENERGY.

lower-bound theorem A statically-admissible stress distribution for a loaded body or structure is one that satisfies equilibrium, the boundary conditions, and is within the yield locus, but does not satisfy displacement compatability. The stiffness of the body according to such a stress field is greater than the exact stiffness. The theorem states that loads producing such stress distributions are equal to, or lower than, the true load that will produce plastic flow. Thus an underestimate of working loads in plasticity (a **lower bound**) is obtained by employing such a field. The highest upper bound of a series of such stress fields will be closest to the true load. *See also* PRINCIPLE OF MAXIMUM PLASTIC RESISTANCE; UPPER BOUND.

lower calorific value *See* CALORIFIC VALUE OF A FUEL.

lower consolute temperature *See* CONSOLUTE TEMPERATURE.

lower pair A mechanism where the elements have surface (not line or point) contact and where one slides over the other during relative motion. *See also* HIGHER PAIR; KINEMATIC PAIR.

low explosive An explosive material that explodes by deflagration. *See also* HIGH EXPLOSIVE.

low-head hydroelectric plant *See* HIGH-HEAD HYDROELECTRIC PLANT.

low heat value *See* CALORIFIC VALUE OF A FUEL.

low-pass filter *See* BAND-PASS FILTER.

low-pressure burner A Venturi device in which gaseous fuel injected at the throat draws in air from the surroundings, the two mix and burn downstream of the divergent section.

low-pressure cylinder The large cylinder, in a compound steam engine or steam-turbine system, that receives steam exhausted from the high- or intermediate-pressure cylinder.

low-pressure steam Steam at a pressure close to atmospheric.

low-pressure turbine (LP turbine) In a compound steam-turbine system or a two- or three-spool gas-turbine engine, the turbine following the high- or intermediate-pressure turbine from which the working fluid is exhausted.

low-solidity wind turbine A wind turbine with a small number (usually no more than three) of narrow blades. *See also* SOLIDITY.

low-temperature solar collector A solar collector operating at temperatures a few degrees (maximum 20°C) above ambient. Swimming-pool heating is a typical application.

LPG *See* LIQUEFIED PETROLEUM GAS.

LQG control *See* LINEAR-QUADRATIC-GAUSSIAN CONTROL.

lubricant Any substance such as oil, grease, or gas under pressure that, when supplied to bearing surfaces in relative motion, reduces friction.

SEE WEB LINKS

- Basic lubrication terminology

lubricating system The system of pipes that takes liquid lubricant from a central reservoir to bearing surfaces within a machine or system.

lubrication The process of providing lubricant to bearing surfaces.

lubrication stability The dependence on the Sommerfeld number, S, of the thickness of the film of lubricant in a bearing. If S exceeds a critical value, the film is thick and stable. Below the critical value the film becomes too thin, is unstable and there is a possibility of metal-to-metal contact.

lubrication theory A theory of viscous fluid flow between two surfaces in close proximity, one moving tangentially relative to the other. The principal application is to bearings.

lubricity (oiliness) A qualitative measure of the effectiveness of a lubricant, high lubricity being desirable to minimize wear.

Lucite® *See* PERSPEX.

Ludwieg–Tillmann drag law A widely used empirical formula for the skin-friction coefficient c_f of a turbulent boundary layer in terms of the shape factor H and the momentum-thickness Reynolds number Re_θ. It is given by $c_f = 0.246\, Re_\theta{}^{-0.268} 10^{-0.678H}$.

Ludwig–Sorét effect *See* SORÉT EFFECT.

Ludwik relation *See* RAMBERG–OSGOOD EQUATION.

Luenberger observer In a control system, a function that uses signals from the feedback sensor output, the plant input, and a knowledge of the mathematical model of the sensor and plant to provide a more accurate feedback signal than that obtained directly from the sensor.

luffing Lifting or lowering the jib of a crane. *See also* SLEWING.

lug *See* EAR.

lumped-capacity method (lumped-system analysis) A method of analysing transient-conduction problems in which it is assumed that a solid body has uniform temperature with convective heat transfer at the surface to or from surroundings at temperature T_∞. If the surface heat-transfer coefficient is h, the surface area is A, the mass of the body is m, and its specific heat is C_P, it can be shown that the variation of temperature T with time t is given by $(T - T_\infty)/(T_0 - T_\infty) = exp(-at)$ where $a = hA/mC_P$ and T_0 is the temperature when $t = 0$. The inverse of α has the dimensions of time and is termed the **thermal time constant (τ)**.

lumped-parameter system A system represented by a number of discrete components such as masses, dampers, and springs, rather than by a continuum. Unlike a continuum model, it has a finite number of states.

LVDT *See* LINEAR VARIABLE DIFFERENTIAL TRANSFORMER.

Lyapunov stability criterion A sufficient criterion for the stability of a system based on the properties of the Lyapunov function for the system. Satisfying the criterion guarantees that the energy in the system cannot increase without limit and thus the output is bounded.

lyophilization *See* FREEZE DRYING.

Lysholm compressor A twin-screw supercharger with improved sealing between the two rotors when compared with a Roots blower. One rotor has thin blades with a thick ridge, while the other has thick teardrop-shaped lobes and a sharp edge.

Mach angle *See* MACH WAVE.

Mach cone *See* MACH WAVE.

machine A mechanism that transmits power in the performance of a useful task. More than one machine in one place, either connected or performing separate functions, is termed **machinery**.

machine bolt *See* BOLT.

machine design *See* MECHANICAL ENGINEERING.

machine element Any individual part used in the design of a machine, particularly a standardized component such as a bearing, fastener, gear, seal, or spring.

machine screw A relatively small screw, usually less than 20 mm in diameter, with the thread running along the whole length up to the head, intended to be screwed into threaded holes. If inserted through plain holes in assembled parts and held together by a nut, an undesirable contact surface of threads bearing against the surface of a hole results. *See also* FITTED BOLT.

machine tool A powered machine, such as a borer, grinder, lathe, milling machine or planer, used for cutting and shaping metal, plastics, composites, etc. (**machining**). A **machining centre** is a CNC machine tool working about several axes, having a stock of tools and automatic tool changing ability, which is capable of diverse machining operations under automated control. *See also* CNC MACHINE TOOLS.

SEE WEB LINKS

- Machine-tool terminology

machining indication *See* SURFACE-TEXTURE INDICATION.

Mach number (*M*) **1.** At a point in a steady flow, the ratio of the fluid velocity *V* to the local speed of sound *c*, i.e. $M = V/c$. **2.** For an object moving through a fluid, the ratio of its speed (relative to the fluid) to the speed of sound. A **Machmeter** is an instrument, typically based on a pitot tube, that determines the Mach number of an aircraft. *See also* BAIRSTOW NUMBER; CAUCHY NUMBER; SUBSONIC FLOW; SUPERSONIC FLOW; TRANSONIC FLOW.

Mach wave (Mach line) An infinitesimally-weak wavefront of expansion or compression in a supersonic flow, caused by an infinitesimal disturbance, usually on a surface. The **Mach angle (θ, μ)**, with unit rad or °, is the angle that a Mach wave makes with the direction of a supersonic flow. If *M* is the Mach number, $\mu = \sin^{-1}(1/M)$. A **Mach cone** is the cone within which are confined the disturbances from an infinitesimal source moving at supersonic speed relative to a fluid. The cone angle is equal to the Mach angle. The two-dimensional equivalent is a **Mach wedge**, where the disturbances are produced by a line. **Mach's construction** is the geometric construction of a mach cone or wedge in which the distance travelled by a source in time *t*, *Vt*, is combined with that travelled by the disturbance, *ct*, *V* being the source speed and *c* is the speed of sound. *See also* PRANDTL-MEYER EXPANSION; SHOCK WAVE.

Mach–Zehnder interferometer An apparatus in which two parallel beams of light from the same source pass through a gas flow, the reference beam in undisturbed flow, the other through a disturbed region of non-uniform density. The two beams are focussed on a screen to produce interference fringes that can be analysed for information about the density variations. *See also* INTERFEROMETRY.

Macpherson strut A type of front suspension for motor vehicles comprising a long strut, with an internal shock absorber, surrounded by a coil spring attached to the body and the strut.

macro A prefix denoting a scale size very much larger than that of atoms or molecules. *See also* MESO; MICRO; NANO.

macromechanics The mechanics of materials at the continuum level, i.e. where the different behaviour of microstructural components is averaged out so that the material can be considered to have homogeneous (but not necessarily isotropic) properties.

macroporous material *See* MICROPOROUS MATERIAL.

macroscopic property Any thermodynamic or material property that can be defined at a scale where the continuum hypothesis is valid.

macrosonics The application of high-intensity sound or ultrasound to such industrial processes as atomization, degassing, mixing, and particle agglomeration.

maglev (magnetic levitation) A system of transportation, usually a train, in which vehicles are supported by contactless magnets and propelled by linear electric motors.

mag meter *See* ELECTROMAGNETIC FLOW METER.

magnetic bearing A lubricant-free, low-friction bearing in which a load is supported by forces generated by electromagnets.

magnetic brake A form of non-contact brake in which a metal fin (typically copper or copper/aluminium alloy) passes between rows of neodymium magnets. Eddy currents are generated in the fin together with a magnetic field that opposes its motion. Commonly used on roller coasters.

magnetic clutch (electromagnetic clutch) A clutch in which an electromagnet is used to magnetize a rotor that attracts one or more friction discs, either directly or through an armature. In an alternative design, torque is transmitted via teeth on the armature and rotor.

magnetic diffusivity (λ) (Unit m^2/s) A quantity analogous to kinematic viscosity or thermal diffusivity that accounts for the diffusion of a magnetic field.

magnetic-disc coupling A non-contacting coupling in which power is transmitted across the air gap between two discs, carried by co-axial shafts, each of which incorporates powerful rare-earth magnets.

magnetic drag dynamometer (eddy-current dynamometer) A dynamometer in which torque is applied to a rotor attached to the input shaft. The torque arises from eddy currents induced in the spinning rotor by the magnetic field created by an electromagnet.

magnetic flow meter *See* ELECTROMAGNETIC FLOW METER.

magnetic levitation *See* MAGLEV.

magnetic powder clutch *See* POWDER CLUTCH.

magneto An electrical generator that produces alternating current for the ignition systems of petrol engines where there is no other electrical supply, such as for lawn mowers, bicycles, and some aircraft engines.

magnetohydrodynamic generator A power-generation device in which a flow of hot ionized gas through a channel is subjected to a magnetic field, thereby generating an electric field. Electrical power can be supplied to an external circuit by electrodes built into the channel walls.

magnetohydrodynamics (MHD) The study of the flow of an electrically-conducting fluid subjected to a magnetic field. Counterparts of the Reynolds and Prandtl numbers which arise are the **magnetic Reynolds number**, defined by VL/λ, where V is a characteristic flow speed, L is a characteristic length, and λ is the magnetic diffusivity of the fluid, and the **magnetic Prandtl number,** defined by ν/λ where ν is the kinematic viscosity of the fluid.

magnetostriction The expansion or contraction of a ferromagnetic material in a magnetic field, e.g. to produce ultrasonic vibration.

Magnus effect The phenomenon whereby a lift force arises on a spinning object in a transverse viscous fluid flow. The force acts in a direction perpendicular to the spin axis and that of the flow.

main bearings In a reciprocating machine, the bearings that locate and support the crankshaft.

main rotor(s) The rotor(s) of a helicopter or other rotorcraft that provide lift and thrust.

major diameter *See* SCREW.

make up The water added to boiler feedwater to compensate for that lost through blowdown, exhaust, leakage, etc.

male fitting (male coupling) The inner part of a connection in which one part fits into another. *See also* FEMALE FITTING; SCREW.

male thread *See* SCREW.

malleable iron *See* CAST IRON.

Maltese-Cross mechanism *See* GENEVA MECHANISM.

mandrel 1. A circular bar, either cylindrical or tapered, that acts as a form on which soft or flexible material can set or be wound or on which partially machined work is supported for further machining. **2.** The driving or headstock spindle of a lathe.

Mangler transformation A method for transforming the axisymmetric boundary-layer equations to the plane (i.e. two-dimensional) boundary-layer equations.

manifold *See* EXHAUST MANIFOLD; FUEL SYSTEM; INTAKE MANIFOLD.

manifold pressure The absolute pressure in the intake manifold of a normally-aspirated piston engine. *See also* BOOST PRESSURE.

manipulator *See* ROBOT.

manometer An instrument used in the measurement of a pressure difference in a fluid (**manometry**). *See also* DIFFERENTIAL PRESSURE GAUGE; INCLINED-TUBE MANOMETER; U-TUBE MANOMETER; WELL-TYPE MANOMETER.

manostat A device for controlling and regulating pressure.

manual control unit A hand-held control panel used to program a robot. *See also* TEACH PENDANT.

manufacturing engineering *See* PRODUCTION ENGINEERING.

maraging A process in which high-nickel alloy steels, which form martensite on air cooling having a strength of about 1 GPa, are reheated to about 500°C for some hours, during which they age to give a room-temperature yield strength of some 2.4 GPa.

Marangoni convection (Bénard–Marangoni convection) A form of natural convection arising as a consequence of thermocapillary instability, in which vertical hexagonal cells occur in a thin horizontal layer of a liquid heated from below. The motion is a consequence of surface-tension gradients in the liquid surface. The flows are characterized by the non-dimensional **Marangoni number (*Ma*)**, defined by $Ma = \Delta\sigma H/\mu\alpha$ where $\Delta\sigma$ is the difference in surface tension due to a temperature difference across a fluid layer of depth H, μ is the dynamic viscosity of the fluid, and α is its thermal diffusivity. *See also* BÉNARD CONVECTION.

marginal stability (neutral stability) A system shows marginal stability if, when given a non-periodic input (for example, an impulse), it shows oscillation that is bounded (i.e. finite) and continues indefinitely.

margin of safety *See* FACTOR OF SAFETY.

Margoulis number *See* STANTON NUMBER.

marine-screw propeller *See* PROPELLER.

Marlborough wheel An extra-wide gear wheel that can mesh with two or more standard-width gear wheels side-by-side.

mass (*m*) (Unit kg) The quantity of matter in a body or system. *See also* INERTIAL MASS.

mass balance weight A mass attached to an aircraft control surface to prevent flutter.

mass-conservation equation *See* CONSERVATION EQUATIONS.

mass density *See* DENSITY.

mass-diffusivity coefficient *See* FICK'S LAW.

mass effect In heat treatment, the rate of cooling of a component during quenching is progressively slower from the outside to the centre. Thus, in the case of steels, martensitic microstructures may be produced on the surfaces of components, but equilibrium microstructures in the centre. *See also* HARDENABILITY.

mass flow meter An instrument, such as a Coriolis flow meter, which measures the mass flow rate of a fluid flowing through a pipe or other duct, rather than its volume flow rate.

mass flow rate ($\dot{m}$) (Unit kg/s) The mass of a material, usually a fluid or powder, that flows across a surface or through a pipe or other duct per unit time. The corresponding **mass flux ($\dot{m}''$)**, with unit kg/s.m^2, is the mass flow across a real surface or through a duct, divided by the surface or cross-sectional area A, i.e. $\dot{m}'' = \dot{m}/A$ or $\dot{m}'' = \rho V$ where ρ is the material density and V is its velocity normal to the surface.

mass fraction The ratio, in a mixture of substances, of the mass of an individual component to the mass of the mixture.

mass moment of inertia *See* MOMENT OF INERTIA.

mass transfer The flow of fluid through a porous surface, to or from another fluid, involving diffusion due to concentration gradients within the fluid and movement of the fluid. The diffusive phenomena occur within the boundary layer. The **mass-transfer coefficient (coefficient of mass transfer, convective mass-transfer coefficient, mass-transfer conductance, *g*)**, with unit kg/s.m^2, is the proportionality factor in the mass-transfer equation $\dot{m}'' = gB$ where $\dot{m}''$ is the mass flux and B is the **mass-transfer driving force.** For mass transfer of component j of a fluid across a porous surface into another fluid, $B = (m_{j\infty} - m_{jS})/(m_{jS} - m_{jT})$ where m_j is the mass concentration of j. The subscript ∞ indicates a value far from the surface in the surrounding fluid, S a value at the surface, and T a value far from the surface in the reservoir of component j.

mass transport In fluid flow, the movement of a substance by advection or diffusion.

master A reference gauge or instrument against which others are compared or calibrated.

master cylinder A piston-cylinder pair attached to a reservoir of hydraulic fluid in a hydraulic circuit, such as in the braking or clutch systems of a motor vehicle, used to transfer fluid to the brake cylinders, clutch, etc.

master/slave manipulator A robot which follows the movement of a manually moved **master arm**, but often with greater force or range of movement. Used for remote operations in hazardous areas, for example nuclear-plant decommissioning. *See also* TELEOPERATOR.

material derivative *See* SUBSTANTIAL DERIVATIVE.

material particle *See* PARTICLE.

materials science The study of the properties, behaviour, and application of solid substances such as metals, ceramics, glasses, polymers, composites, biomaterials, and semiconductors, at all scales from the atomic to the macroscopic. The topic has its origins in metallurgy.

matrix The continuous phase in any reinforced material, which may be polymeric, metal, ceramic, or complex biomaterial.

Matthew effect Applied to such phenomena as crystal or drop growth, whereby large crystals or drops develop at the expense of small ones.

Maupertius principle *See* PRINCIPLE OF LEAST ACTION.

MAV *See* MICRO-AIR VEHICLE.

maximum allowable operating pressure The highest pressure at which any pressure system may be operated, usually 10 to 20% below the maximum allowable working pressure.

maximum allowable working pressure The pressure on which the design of a pressure system is based and the highest pressure at which relief valves should be set. The lowest-rated component in the system typically has a design safety factor of 4.

maximum-and-minimum thermometer *See* SIX'S THERMOMETER.

maximum continuous load (maximum continuous rating) The maximum rate of steam output that a boiler can supply for a specified period, usually 24 hours.

maximum material condition (maximum metal condition) The situation where the volume of a manufactured component corresponds to the upper limit for all toleranced external dimensions, and to the lower limit for all internal dimensions. *See also* MINIMUM MATERIAL CONDITION.

Maxwell material A mechanical model for a solid or liquid viscoelastic material comprising a dashpot (i.e. a viscous damper with viscosity μ) in series with a linear-elastic element (modulus of elasticity G) such that both carry the same load that gives rise to a common stress τ.

The elastic displacement rate is $\dot{\tau}/G$, $\dot{\tau}$ being the rate of change of shear stress with respect to time, the viscous displacement rate (or shear rate) being τ/μ, and the overall displacement rate $\tau/\mu + \dot{\tau}/G$. The shear stress τ is given by $\tau + \lambda\dot{\tau} = \mu\dot{\gamma}$ where $\lambda = \mu/G$ is the relaxation time and $\dot{\gamma}$ is the overall strain rate. *See also* KELVIN–VOIGT MATERIAL.

Maxwell relations (Maxwell equations) A set of equations which thermodynamic properties must satisfy for any thermodynamic system in a state of equilibrium:

$$\left(\frac{\partial T}{\partial v}\right)_S = -\left(\frac{\partial p}{\partial s}\right)_v, \left(\frac{\partial T}{\partial p}\right)_S = \left(\frac{\partial v}{\partial s}\right)_p,$$
$$\left(\frac{\partial p}{\partial T}\right)_v = \left(\frac{\partial s}{\partial v}\right)_T \text{ and } \left(\frac{\partial v}{\partial T}\right)_p = -\left(\frac{\partial s}{\partial p}\right)_T$$

where T is absolute temperature, p is pressure, v is specific volume, and s is specific entropy.

Maxwell's lemma A relation from which the minimum weight of a structure to support a given set of loads may be obtained, given by

$$\mathcal{V}_T\sigma_T - \mathcal{V}_C\sigma_C = \Sigma_i \boldsymbol{F}.\boldsymbol{r}$$

where $\mathcal{V}_T$ is the total volume of members in tension and $\mathcal{V}_C$ that of those in compression, σ_T and σ_C are the failure stresses in tension and compression respectively, and $\boldsymbol{F}$ is the planar vector force applied at the ith node of the truss situated at vector distance $\boldsymbol{r}$ from an assigned origin.

Maxwell's theorem *See* RECIPROCAL THEOREM.

Mayer's formula For an ideal gas, $C_P - C_V = R$ where C_P and C_V are the specific heats at constant pressure and constant volume respectively, and R is the specific gas constant.

McCabe wave pump A wave-energy device consisting of three hinged pontoons pointed parallel to the wave direction, which flex in response to the waves.

M contour On the Nyquist diagram, the locus of the real and imaginary parts of the closed-loop transfer function with unity feedback plotted as a function of frequency.

mean *See* MEAN VALUE.

mean diameter 1. The average of the inside and outside diameters for a helical spring or hollow circular cylinder. **2.** *See* SAUTER MEAN DIAMETER.

mean-effective pressure *See* BRAKE MEAN-EFFECTIVE PRESSURE; INDICATED-MEAN EFFECTIVE PRESSURE.

mean film temperature *See* FILM TEMPERATURE.

mean-free path (Λ) (Unit nm) The average distance that a gas molecule travels before collision with another molecule.

mean stress The average of the maximum and minimum stresses for a material subjected to a stress cycle, as in a fatigue test.

mean temperature *See* BULK TEMPERATURE.

mean temperature difference (ΔT_M) (Unit K or °C) In a heat exchanger, ΔT_M is defined by $\dot{q} = UA_S\Delta T_M$ where U is the overall heat-transfer coefficient, A_S is the surface area associated with U, and $\dot{q}$ is the heat-transfer rate. ΔT_M can be obtained from the log-mean temperature difference (*LMTD*) for a counter-flow heat exchanger and a correction factor F for the specific type of heat exchanger concerned according to $\Delta T_M = F.LMTD$.

mean value (mean) Where a material property is dependent upon either the thermodynamic state (i.e. temperature or pressure) or a variable such as strain (in a solid) or shear rate (in a fluid), an average over the range of the state or variable. For example, the mean stress for a work-hardening solid deformed over a range of strain from 0 to ε would be

$$\sigma_{mean} = \frac{1}{\varepsilon}\int_0^{\varepsilon} \sigma(\varepsilon)d\varepsilon$$

See also BULK DENSITY; BULK MEAN TEMPERATURE; BULK VELOCITY.

measured relieving capacity The actual flow rate through a pressure-relief device measured at its design pressure.

measuring tank (dump tank, metering tank) A calibrated tank used to measure the mass of a volume of liquid that flows into it in a measured period of time. Used to calibrate flow meters.

mechanical admittance (mobility) (Unit m/s.N) The inverse of mechanical impedance.

mechanical advantage (force ratio, leverage) The ratio, for a machine, of the output force (load or resistance) to the input force (applied effort).

mechanical comparator A contact device, such as a dial gauge, in which movements of the indicator are magnified by levers and gear trains. *See also* OPTICAL COMPARATOR.

mechanical damping The damping caused by dry friction in a vibrating system.

mechanical draught *See* DRAUGHT.

mechanical-draught cooling tower A cooling tower in which fans force or draw air through the tower. *See also* FORCED-DRAUGHT COOLING TOWER; INDUCED-DRAUGHT COOLING TOWER.

mechanical efficiency (η) 1. In general for a machine, the ratio of output work to input work. **2.** For a compressor, the ratio of indicated power to shaft power; for a reciprocating engine or an expander, the ratio of shaft power to indicated power.

mechanical energy The sum of kinetic energy and potential energy for an object or a mechanical system, including the energy stored in springs, etc.

mechanical-energy integral equation An equation relating the variation with streamwise distance x of the fluid kinetic energy within a boundary layer to the work done by the shear stress τ. The equation is valid for laminar and turbulent flow. For a flow with constant density ρ, it can be written as

$$\frac{d}{dx}\left[U_\infty \int_0^\infty \bar{u}\left(\frac{1}{2}\rho\bar{u}U_\infty^2 - \frac{1}{2}\rho\bar{u}^3\right)dy\right] = \int_0^\infty \tau\frac{\partial\bar{u}}{\partial y}dy$$

where U_∞ is the free-stream velocity and $\bar{u}$ is the mean velocity a distance y from the surface. The quantity on the right-hand side is termed the **dissipation integral**.

mechanical engineering That branch of engineering concerned with energy conversion, stress analysis, vibration, dynamics, and kinematics, especially applied to design (**machine design, mechanical-engineering design**).

mechanical equivalent of heat (Joule's equivalent, J.) According to the first law of thermodynamics, for a closed system taken through a cycle, the sum of the net work done on the system (or delivered to the surroundings), $\Sigma\delta W$, plus the net heat delivered to (or taken from) the surroundings, $\Sigma\delta Q$, is equal to zero. Thus $\Sigma\delta W$ is proportional to $\Sigma\delta Q$, the proportionality constant being the mechanical equivalent of heat, i.e. $J = -\Sigma\delta W/\Sigma\delta Q$, the value depending upon the units of W and Q. If both are in joules, then $J = 1$.

mechanical hysteresis The lag between stress relaxation and strain recovery that occurs in some materials on unloading. *See also* HYSTERESIS LOOP.

mechanical impedance (Z) (Unit N.s/m) A measure of the resistance to motion. If a force F, varying with angular frequency ω, applied to a mechanical system leads to a velocity $V(\omega)$ at the point of application, the mechanical impedance $Z(\omega) = F(\omega)/V(\omega)$. The **mechanical reactance** is the imaginary part of Z and the **mechanical resistance** is the real part. The **mechanical ohm,** equal to 1 N.s/m, is a non-SI unit of mechanical impedance. *See also* ANGULAR MECHANICAL IMPEDANCE.

mechanical irreversibility *See* IRREVERSIBILITY.

mechanical linkage A mechanism consisting of rigid bars connected at pivot points which can transmit force but not torque.

mechanical pressure (p) (Unit Pa) In a moving fluid, the negative of the average value of the normal component of stress over the surface of an infinitesimally small sphere. Generally taken to be the same as the static pressure. *See also* PRESSURE.

mechanical properties of solid materials. The strength and stiffness properties of solid materials such as fracture toughness, the moduli of elasticity, percent elongation, Poisson's ratio, proof stress, tensile strength, ultimate stress, and yield stress.

mechanical scales A weighing device consisting of one or more pivoted beams along which are moved counterweights until balance is achieved. *See also* ROBERVAL BALANCE.

mechanical seal A device that prevents the leakage of fluid from a high-pressure to a low-pressure region. *See also* O-RING; PACKING.

mechanical separation The partial or total removal of solid particles from a fluid, or liquid droplets from a gas, by cyclone separators, filtration, or gravitational settling.

mechanical units The units of physical quantities, the dimensions of which include mass, length, and time. *See also* INTERNATIONAL SYSTEM OF UNITS.

mechanical vibration The motion of a particle or body which oscillates about a position of equilibrium.

mechanics *See* APPLIED MECHANICS.

mechanism A kinematic chain in which one element or link is fixed so that motion can be transmitted or transformed.

mechanize To convert a hand-operated device into a powered machine.

mechatronics The integration of mechanical engineering, electrical engineering, electronic engineering, and software engineering. *See also* INTEGRATED ENGINEERING.

medium-enthalpy resource *See* GEOTHERMAL ENERGY.

medium-head hydroelectric plant *See* HIGH-HEAD HYDROELECTRIC PLANT.

medium-technology robot A robot having four to six joints and not incorporating a vision system or other advanced sensors. Such robots are typically employed for pick-and-place tasks.

mega (M) An SI unit prefix indicating a multiplier of 10^6; thus megawatt (MW) is a unit of power equal to one million watts.

Melde's experiment An apparatus in which standing waves are excited on a fine thread under tension by vibrating one end either longitudinally or transversely.

melt fracture Instabilities in flow during polymer extrusion, resulting in rough surfaces and variations in dimensions of the extrudate. In extremis, the extrudate fractures.

melting point (Unit K, °C) The temperature at which a solid material undergoes the phase change to a liquid at a specified pressure, usually 1 atm. Pure metals and eutectics have single-valued melting points, while alloys with other compositions melt over a range of temperature such that there is a well-defined start and end to the melting process, but there are states in between where solid and liquid are both present. *See also* LATENT HEAT OF FUSION; LIQUIDUS; SOLIDUS.

melt spinning A method of manufacturing fibres in which molten polymer is pumped through a die having numerous holes about the size of the fibre diameter (a **spinneret**). The fibres are then reeled, during which stretching occurs, giving the fibres orientation.

member A component of a structure such as a beam, column, plate, strut, or tie rod.

membrane 1. A thin sheet of material, usually flexible, but with negligible resistance to bending such that the only stresses are tensile. **2.** A thin sheet of semi-permeable material, often polymeric, used in separation processes such as water purification, reverse osmosis, ultrafiltration, and dehydrogenation of natural gas.

membrane analogy (soap-film analogy) An analogy based upon the fact that the differential equation governing the elastic-stress distribution in a cylindrical bar of arbitrary cross section under torsion is the same as that for the surface slope of a membrane with an outline corresponding to the bar and subjected to a differential pressure. *See also* SAND-HEAP ANALOGY.

membrane pump *See* DIAPHRAGM PUMP.

membrane stresses The in-plane stresses in shell structures such as thin-walled pressure vessels. There are circumferential (hoop) stresses, orthogonal to which are meridional stresses that, in the case of a cylinder, are the axial stresses. **Membrane analysis (membrane theory, thin cylinder theory)** is the analysis of thin shells assuming only in-plane compressive or tensile stresses, with zero bending stresses.

membrane valve *See* DIAPHRAGM VALVE.

MEMS *See* MICRO-ELECTROMECHANICAL SYSTEMS.

MEMS actuator *See* MICROACTUATOR.

MEMS thruster (microthruster) A miniature propulsion device with applications to, for example, nanosatellites. The devices typically have radial dimensions up to 500 μm, are fabricated from silicon, use hydrogen peroxide or hydrazine as propellant, and develop thrust up to about 500 μN.

meniscus The curved portion of a liquid surface where it comes into contact with a solid surface. If the angle of contact is acute (< 90°), as for water, the solid surface is wetted and the meniscus is concave. The opposite is true when the angle is obtuse, as for mercury. *See also* INTERFACE.

mep *See* BRAKE MEAN-EFFECTIVE PRESSURE.

mercury (Hg) A heavy (relative density 13.55 at 20°C) metallic chemical element, liquid in the range −38.8°C to 356.7°C and widely used in bulb-thermometers, barometers, and manometers.

mercury barometer A barometer in which the working liquid is mercury. *See also* FORTIN BAROMETER.

mercury manometer A manometer, such as an inclined or U-tube manometer, in which the working liquid is mercury.

mercury thermometer A bulb thermometer in which the working liquid is mercury.

mercury vapour cycle That part of a binary vapour cycle in which the high-temperature fluid is mercury.

meridional stress *See* MEMBRANE STRESSES.

Mersenne's law The fundamental frequency f of a stretched string of length L, mass per unit length m', and tension T is given by $f = \sqrt{T/m'}/(2L)$.

mesh 1. A pair of rotating gears in contact is said to mesh. **2.** The arrangement of elements in finite-element, finite-difference and finite-volume simulations. **3.** *See* SCREEN MESH; SIEVE MESH.

mesh size *See* SIEVE.

meso A prefix meaning 'in between' other scale sizes; its use depends upon the application.

mesopause *See* STANDARD ATMOSPHERE.

mesosphere *See* STANDARD ATMOSPHERE.

mesoporous material *See* MICROPOROUS MATERIAL.

metacentre For a floating object in the neutral position, a point M vertically above the centre of buoyancy B lying on a line through the centre of gravity G. If M is above G, the situation is stable to a small tilt $\Delta\theta$ but unstable if M is below G. In the diagram, W is the weight of the object and F_B is the buoyancy force. The **metacentric height** is the distance between G and M. It is an important parameter for the stability of any floating object, such as a ship.

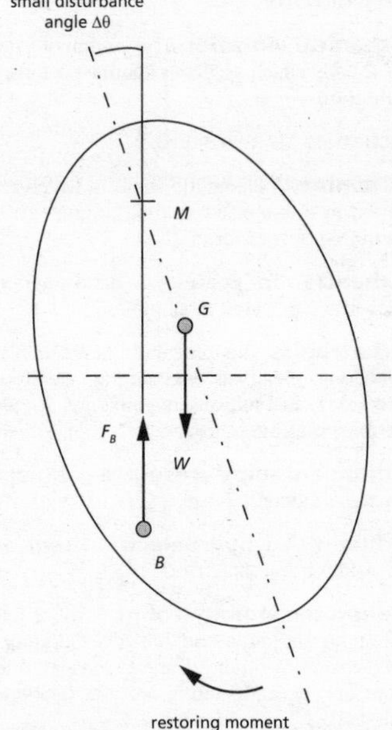

metacentre

metal inert gas welding *See* WELDING.

metallize A non-metallic object, such as a glass mirror, is metallized by coating with a metal, typically by vacuum deposition or thermal spraying.

metallurgical limit The maximum temperature that a highly stressed part of a machine, such as a gas or steam turbine or a reciprocating engine, can withstand before failure occurs.

metallurgy The study of the atomic, crystallographic, microstructural, mechanical, and physical properties of metallic elements and alloys and their applications. Now part of materials science.

metal-matrix composite A composite material in which a reinforcing material, such as carbon fibre, alumina or silicon carbide, is embedded into a metal matrix.

metal spraying The creation of a metal coating using arc, flame, or plasma spraying.

metarheology The rheology of spatially-heterogeneous materials, such as fibre suspensions, that display elements of both particulate and continuous systems.

metastable A state of matter that drops to a lower energy state due to a slight disturbance; for example, the condensation of a supercooled vapour.

meter factor 1. The ratio of the true flow rate passing through a liquid flow meter to the flow rate indicated by the meter. The true flow rate is determined using a **meter prover** whereby an inflatable spheroid is forced through a calibrated section of piping to displace a precise volume of liquid. **2.** The calibration factor for a flow meter, i.e. the output divided by the flow rate.

metering pump A pump that supplies a specified volume of fluid in a specified time period. Types include bellows, diaphragm, piston, and travelling-cylinder pumps. *See also* PROPORTIONING PUMP.

metering rod A type of needle valve consisting of a thin tapering rod used in a carburettor to vary the size of the jet opening and so control the fuel flow rate.

metering tank *See* MEASURING TANK.

method of characteristics A numerical procedure for solving hyperbolic partial differential equations, commonly used for problems in supersonic flow and rigid plasticity.

method of joints A procedure for determining the forces in the individual members of a truss by considering the equilibrium of forces at each pinned joint.

method of sections A procedure for determining the internal forces in the individual members of a statically-determinate structure by making an imaginary cut through the framework and considering the equilibrium of forces on each of the two parts.

metre (m) The base unit of length in the SI system. It is equal to the distance travelled by light in a vacuum in 1/299,792,458 of a second.

metric ton *See* TONNE.

Meyer hardness (Unit Pa) A material hardness similar to the Brinell hardness but based upon the projected area of the impression rather than the surface area.

MHD *See* MAGNETOHYDRODYNAMICS.

Michell turbine *See* BANKI TURBINE.

micro 1. (μ) An SI unit prefix indicating a submultiplier of 10^{-6}; thus micrometre (μm) is the unit of length equal to one millionth of a metre, formerly called the micron. **2.** A prefix usually used to indicate a device or an object with dimensions in the range 1 μm to about 1 mm although now frequently used to indicate a much larger but still small-scale device, such as a micro-air vehicle or micro-CHP. *See also* MACRO; MESO.

microactuator (MEMS actuator) An actuator with dimensions typically in the range 100 to 500 μm and generating a force on the order of tens of μN. Basic principles include constrained bent beams that move due to Joule heating and forces generated electrostatically.

micro-air vehicle (MAV) An unmanned aircraft that may be fixed, flapping, or rotary wing, typically with a scale in the range 50 mm to 200 mm.

microballoon A pneumatic actuator used for turbulence control, in which a silicone-rubber membrane about 120 μm thick is attached to an etched silicon wafer.

microbubbles Bubbles of air or oxygen less than 0.5 μm in diameter. Applications include surface cleaning and drag reduction.

microcalorimeter 1. An X-ray detector used in spacecraft and consisting of an X-ray absorber, a thermistor, and a heat sink. **2.** A calorimeter, fabricated on a silicon wafer and incorporating a heater and a thermometer, used to study the thermal properties of small (typically a few μg) samples.

microcasting *See* SOFT LITHOGRAPHY.

microchannel (microfluidic channel) A channel etched in glass or silicon having depth and width typically in the range 1 to 500 μm. Typical applications include microfluidics, micro heat exchangers, and microreactors.

microchannel reactor *See* MICROREACTOR.

micro-CHP A combined heat and power system, on the scale of a family house or small office building, based upon the use of a Stirling engine.

microcombustor (microburner) A combustor micromachined from silicon, designed to burn a gaseous fuel such as hydrogen and supply hot gas to a microturbine. Such devices are about 10 mm in diameter and generate power on the order of 100 W.

microconstituent In materials science, any identifiable element of a material's microstructure, as seen under an optical microscope. In metal alloys, single phases may be microconstituents but also certain combinations of phases that warrant being assigned separate names, e.g. in steels, the microconstituent called pearlite that is a eutectoid with fixed proportions of iron (ferrite) and Fe_3C (cementite) combined in a characteristic pattern. A **microstructure diagram** is a phase diagram labelled in terms of microconstituents rather than phases. In the case of the silver–copper diagram shown under the entry for phase diagram, the left side of the central $(\alpha + \beta)$ region would be labelled $(\alpha + \varepsilon)$, where ε means eutectic, and $(\varepsilon + \beta)$ to the right of the 71.9% Ag/28.1%Cu eutectic composition.

m

microcooler 1. A solid-state device, typically using the Peltier effect, for cooling semiconductor chips. **2.** A MEMS device for cooling electronics, typically using a micro-capillary pumped loop incorporating an evaporator and a condenser.

microdispenser A device used to distribute liquid volumes in the range 1μl to 1 ml .

micro-electromechanical systems (MEMS, microsystems) Small-scale devices manufactured from silicon, polymer, or glass, including microchannels, micro heat exchangers, microreactors, and micropumps.

microengine (micro gas turbine, microjet) A miniature gas-turbine engine based upon a microcombustor. Applications include micro-air vehicles, experimental and model aircraft.

microexchanger *See* MICRO HEAT EXCHANGER.

microfabrication The fabrication of components from silicon, elastomers, and plastics on scales from fractions of 1 μm to 1 mm. Techniques include photolithography, etching, deposition, moulding, casting, injection, and laser ablation.

microfilter A filter that uses a microporous membrane with pore sizes in the range 0.1 to 10 μm to remove contaminants from a fluid. Applications include drinking-water treatment.

microfluidic channel *See* MICROCHANNEL.

microfluidic mixer *See* MICROMIXER.

microfluidics The study of flow in artificial microsystems, including miniaturized fluidic devices. *See also* MICROCHANNEL.

microfuel cell A fuel cell with dimensions typically a few mm, used for low-power electronic devices. Methanol and hydrogen are commonly used fuels.

micro gas turbine *See* MICROENGINE.

microgeneration *See* MICROPOWER GENERATION.

micrograph An image taken through a microscope or similar optical or electronic device to show a magnified image, for example of metallurgical microstructure.

microgravity (micro-g environment, μg) An environment in which the effective acceleration due to gravity has been reduced to a negligible fraction (typically 10^{-6}) of its normal value on earth (9.81 m/s^2).

microgripper A device designed to handle objects ranging from 1 to 100 μm in size.

microhardness The hardness of a material determined under light loads to give shallow, small-volume indentations, thus enabling the hardness of microconstituents in alloys, thin layers, and brittle materials to be measured. Microhardness machines employ Berkovich, Knoop, or Vickers indenters and usually incorporate a microscope.

micro heat engine A heat engine in which thermocapillary pumping within a microchannel is used to convert heat to power.

micro heat exchanger (microexchanger) A heat exchanger with fluid flowing through microchannels about 100 μm deep, used, for example, to cool electronic components and also in microreactor systems.

micro heat pipe A heat pipe with sufficiently small dimensions (typically < 1mm) that capillary action rather than a wick can be used to return the condensate to the evaporator.

micro-hydroelectricity (micro-hydro power) An imprecise term usually taken to mean a hydroelectric power plant generating less than 100 kW.

microjet *See* MICROENGINE; MICROSYNTHETIC JET.

microlithography *See* PHOTOLITHOGRAPHY.

micromachining A process used to produce components for microsystems with critical dimensions on the order of a few μm. In bulk micromachining, a silicon wafer is selectively etched to produce microstructures. In surface micromachining the microstructures are built up by etching of several layers deposited on the substrate.

micromanipulator A machine used to position an end effector within the view of a microscope, typically to make contact with a semiconductor device during manufacture.

micromanometer *See* BETZ MANOMETER.

micromechanism Any of various crystallographic or microstructural mechanisms, such as dislocation movement, void-initiation, growth, and coalescence, by which plastic flow, fracture, etc. occur.

micrometer (micrometer gauge) A mechanical-contact device for the accurate measurement of the length, width, diameter, etc. of an object, the depth of a hole, the height of a step, etc. The usual arrangement is a spindle that is moved by rotation of a thimble, the distance then being read off a vernier scale.

micromixer (microfluidic mixer) A microchannel-scale mixing device in which mixing occurs either due to molecular diffusion and chaotic advection (passive micromixers) or to externally-imposed effects such as pressure variation, thermal, acoustics, electrohydrodynamics, and MHD (active micromixers).

micromoulding High-precision moulding of miniature plastic components, typically with a mass of a few μg.

micron The former, non-SI, name for 1 μm. *See also* MICRO.

micronozzle A nozzle with a diameter in the range of a few μm to 1mm machined in a thin metal sheet. Applications include ink-jet printing and micropropulsion.

micron rating The ability of a fluid filter to remove particles of a given size. *See also* BETA RATIO.

microporous material (nanoporous material) Porous material with pore diameters less than 2 nm. Pore sizes for micro-, **meso-** (2–50 nm) and **macroporous** (greater than 50 nm) materials follow IUPAC guidelines.

micropower generation (microgeneration) The stand-alone, small-scale generation of low-carbon heat and/or electricity typically using solar photovoltaics, solar thermal water heating, micro CHP, wind turbines, fuel cells, micro-hydro systems, biomass boilers, air-, ground-and water-source heat pumps, and passive flue-gas recovery devices.

micropropulsion Small-scale propulsion, typically for use in outer space (e.g. for satellite positioning), using pulsed-plasma electric propulsion devices.

micropump A pump with internal dimensions in the μm range, typically powered by piezoelectric actuators and incorporating passive check valves. Applications include cooling of electronic components, ink-jet printing, and drug delivery.

microreactor (microchannel rector, microstructured reactor) A device in which flow processes involving chemical reactions take place in microchannels. Especially suited to strongly exothermic and dangerous reactions.

microsensor A miniature sensor, typically using capacitive and piezoresistive effects and manufactured by wet or dry etching of silicon.

microstereolithography A rapid-prototyping process in which an object is formed through the addition of successive layers of a liquid photopolymer that solidifies when exposed to ultra-violet light. *See also* ADDITIVE LAYER MANUFACTURING.

microstrain The measure of strain obtained from a strain gauge and equal to 10^6 x actual strain.

microstructure *See* MICROCONSTITUENT.

microstructured reactor *See* MICROREACTOR.

microsynthetic jet (microjet) A device consisting of a cavity, typically 15 μm deep and 1 to 5 mm long, with a flexible actuating membrane on one side and an orifice, typically 50 to

800 μm wide, on the other, usually machined into a silicon wafer. Movement of the membrane towards the orifice pumps fluid out of the orifice to create a jet. Movement away from the orifice sucks fluid into the cavity.

microsystems *See* MICRO-ELECTROMECHANICAL SYSTEMS.

microtechnology The technology involved in the design and manufacture of microsystems. *See also* MICROFABRICATION.

microthruster *See* MEMS THRUSTER.

microturbine *See* MICROENGINE.

microvalve A MEMS device used to regulate or seal fluid flow in a microfluidic system. Actuation effects include electromagnetic, electrostatic, piezoelectric, bimetallic, thermopneumatic, and shape memory.

mig welding *See* WELDING.

milli (m) An SI unit prefix indicating a submultiplier of 10^{-3}; thus millimetre (mm) is a unit of length equal to one thousandth of a metre.

m

millimetre of mercury A non-SI unit of pressure equal to the pressure exerted by a vertical column of mercury 1 mm high. Approximately equal to 133 Pa or 1 torr.

millimetre of water A non-SI unit of pressure equal to the pressure exerted by a vertical column of water 1 mm high. Approximately equal to 9.81 Pa.

milling A machining process, typically for metals and plastics, in which a multi-tooth rotary cutter removes material to produce flat or profiled surfaces, slots, grooves, etc.

SEE WEB LINKS

- Milling-machine and related terminology

mimic panel A control panel which shows diagrams (mimics) of each part of the plant being controlled, with sensor readings and actuator positions displayed alongside each diagram.

MIMO system *See* MULTIPLE INPUT, MULTIPLE OUTPUT SYSTEM.

Miner's rule *See* FATIGUE.

mini-hydroelectricity (mini-hydro power) An imprecise term, usually taken to mean a hydroelectric power plant generating less than 5MW.

minimal realization The smallest number of states of a system that properly represent the transfer function and allow both observability and controllability.

minimum fluidization velocity *See* FLUIDIZATION.

minimum material condition (minimum metal condition) The situation where the volume of a manufactured component corresponds to the lower limit of all toleranced external dimensions and to the upper limit for all internal dimensions. *See also* MAXIMUM MATERIAL CONDITION.

minimum-phase system A linear system where the poles and zeros have zero or negative real parts. Such systems are thus stable due to the zero or negative real parts of the poles.

minimum potential energy (theorem of minimum potential energy) A rigid body or structure will locate itself so as to have a minimum potential energy when in a position of equilibrium. In a gravitational field, a deformable body or structure will take up the shape and position giving minimum potential energy.

minimum quantity lubrication A term relating to machining with the least possible quantity of coolant/lubricant, as most traditional coolants are ecologically unfriendly, yet dry machining is not acceptable.

minor diameter See SCREW.

minor loop control The placing of an additional control loop around the plant so as to effectively modify the plant transfer function to simplify the design of the main control loop.

misfiring The behaviour of a piston engine when the fuel/air mixture in one or more cylinders fails to ignite. Causes include a non-combustible fuel:air ratio or, in a petrol engine, the absence of spark.

mist Liquid in the form of microscopic droplets (typically 100 μm diameter) following atomization.

mist eliminator *See* DEMISTER PAD.

mist extractor *See* DEMISTER PAD.

mist flow *See* BOILING.

mistuning Small, random, blade-to-blade differences in gas-turbine rotors that lead to increased levels of blade vibration and stress.

Mitchell–Banki turbine *See* BANKI TURBINE.

Mitchell bearing A thrust bearing with wedge-shaped films of oil between pads and a rotating disc.

mitre gears *See* TOOTHED GEARING.

mixed cycle *See* SABATHÉ CYCLE.

mixed-flow heat exchanger *See* SHELL-AND-TUBE HEAT EXCHANGER.

mixed-flow impeller A pump (**mixed-flow pump**), compressor, or turbine (**mixed-flow turbine**) impeller which discharges the working fluid with both an axial and a radial component of velocity.

mixed-mode fracture Fracture resulting from a combination of tensile, shear, or twisting loading, i.e. a combination of modes I, II, and III of fracture mechanics. *See also* MODE OF FRACTURE.

mixed-phase flow *See* MULTIPHASE FLOW.

mixed-pressure turbine A steam turbine in which steam is admitted from two or more independent sources that may be at different pressures.

mixing The process by which two or more fluids become distributed each within the other. **Mixing performance** is a term for how well mixed materials are after a period of times in a mixer. It is generally used in a qualitative sense, but can be quantified in terms of the statistical properties of the concentration distribution of each constituent. Complete mixing means that the mixture properties are uniform throughout.

mixing chamber A chamber within which mixing occurs, typically of fuel and oxidant prior to injection into a combustion chamber.

mixing layer The laminar or turbulent shear layer that develops when there is a tangential velocity difference between two viscous fluids.

mixing length (*l*) (Unit m) The basis of an elementary turbulence model in which it is assumed that the turbulent shear stress τ_T is given by

$$\tau_T = \rho l^2 \left|\frac{\partial \bar{u}}{\partial y}\right| \frac{\partial \bar{u}}{\partial y}$$

where ρ is the fluid density, $\bar{u}$ is the mean velocity a distance y from a reference plane (usually a solid surface), and $\partial \bar{u}/\partial y$ is the gradient of mean velocity. *See also* VAN DRIEST DAMPING FUNCTION.

mixing valve A valve in which two or more miscible fluid streams, which may be different in temperature or composition, are mixed together to produce a fluid stream of desired temperature or composition.

mixtures *See* GAS MIXTURES; SOLUTION.

mixture strength *See* STOICHIOMETRIC MIXTURE.

MKS The obsolete system of units based upon the metre, kilogram, and second that preceded the SI system.

MMC *See* METAL-MATRIX COMPOSITE.

mmscfm An abbreviation for million standard cubic feet per minute; a non-SI unit for volumetric gas flow rate.

mobile robot A robot where the base frame is mounted on a vehicle.

mobility *See* MECHANICAL ADMITTANCE.

modal analysis The study, using transducers such as accelerometers and load cells, of the dynamic response to excitation of structures, objects, systems, etc.

mode (mode of oscillation, mode of vibration) A pattern of vibration of a component or structure corresponding to a natural frequency.

model-following problem The problem of making the behaviour of a real plant accurately match that of a model reference system; that is, the problem whose solution is attempted by model reference adaptive control.

modelling 1. The analysis of complex physical phenomena, such as turbulence or viscoelasticity, based upon simpler systems or processes for which the behaviour is well understood. *See also* KELVIN–VOIGT MATERIAL; MAXWELL MATERIAL. **2.** Sometimes used to mean the process of simulation of results in FEM.

model reduction The process of simplifying the mathematical model of a real system so as to give a model reference system.

model reference system A mathematical model of an idealized plant which shows desirable behaviour. For example, a real robot may experience vibration when a joint is moved, while the model reference system for the joint would show the same basic dynamics without the vibration. A **model reference adaptive-control system** would then attempt to make the behaviour of the real system (that is, the robot) match that of the model and thus avoid the vibration.

model testing The use of small- or large-scale models to determine the behaviour of geometrically-similar full-scale structures and systems using the principles of dimensional analysis and physical similarity.

mode of failure The principal cause of failure in a structure or component, including buckling, creep, fatigue, fracture, impact, and thermal shock.

m

mode of fracture The different types of crack propagation, viz: tensile opening (mode I); in-plane shear (mode II); out-of-plane shear (twisting, mode III). *See also* FRACTURE MECHANICS.

mode of oscillation *See* MODE.

mode of vibration *See* MODE.

moderator A medium used in a nuclear reactor to slow fast neutrons so that a chain reaction with uranium^{235} is possible. Water, heavy water, and graphite are the most commonly used moderators. *See also* NUCLEAR FISSION.

modern control A control method that represents system behaviour by states and the resulting low-order time-domain differential equations, rather than through Laplace transforms. The controller is then designed to achieve optimal control.

modification A change in the design of a component or machine, to correct a fault or deficiency, improve performance, or simplify manufacture.

modified Goodman diagram *See* FATIGUE.

modulation The systematic time variation of the amplitude, frequency, or phase of a periodic signal.

module *See* TOOTHED GEARING.

modulus of compression *See* BULK MODULUS.

modulus of decay (Δ) (Unit s) The time taken for the amplitude of a damped oscillation to decay to 1/e of its initial value, where e is Euler's number, 2.71828183....

modulus of elasticity (Unit Pa) A term that usually refers to Young's modulus of an isotropic solid, although there are also the shear modulus and the bulk modulus.

modulus of elasticity in shear *See* SHEAR MODULUS.

modulus of resilience (u_R) (Unit J/m³ or Pa) The strain energy density in a material stressed to the proportional limit σ_{PL} and given by $u_R = \sigma_{PL}^2/2E$ where E is Young's modulus.

modulus of rigidity *See* SHEAR MODULUS.

modulus of rupture (flexural strength, fracture strength in bending) (Unit Pa) A measure of the breaking strength of a brittle material, typically determined from a three-point bending test in which the test specimen is loaded to fracture. It is not an absolute quantity, as it depends upon fracture toughness and flaw size.

modulus of torsion *See* SHEAR MODULUS.

Mohr–Coulomb fracture criterion A fracture criterion, primarily for brittle materials, according to which failure occurs when the stress at a point in a material falls outside the envelope created by the Mohr's circles for uniaxial tensile strength and uniaxial compressive strength.

Mohr–Coulomb yield criterion A pressure-dependent yield criterion, according to which yielding occurs when the stress at a point in a material falls on the envelope created by the Mohr's stress circles at yielding for various tests such as tension, shear, and compression that have different components of hydrostatic stress.

Mohr's strain circle For a body strained along orthogonal x,y axes by normal strains ε_x and ε_y, and shear strains γ_{xy}, the normal strain ε_θ and shear strain γ_θ on a plane inclined at angle θ to the x-axis are given by $\varepsilon_\theta = \varepsilon_x cos^2\theta + \varepsilon_y sin^2\theta + \gamma_{xy} cos\theta sin\theta$ and $\gamma_\theta = 2(\varepsilon_x - \varepsilon_y)cos\theta sin\theta - \gamma_{xy}(cos^2\theta - sin^2\theta)$

In terms of principal strains ε_1 and ε_2,

$$\varepsilon_x = \frac{1}{2}(\varepsilon_1 + \varepsilon_2) + \frac{1}{2}(\varepsilon_1 - \varepsilon_2)cos2\theta,$$

$$\varepsilon_y = \frac{1}{2}(\varepsilon_1 + \varepsilon_2) - \frac{1}{2}(\varepsilon_1 - \varepsilon_2)cos2\theta$$

and $\gamma_{xy} = (\varepsilon_1 - \varepsilon_2)sin2\theta$.

These relations plot as a circle (Mohr's strain circle) on axes of $\varepsilon,(\gamma/2)$ using the tensor definition of shear strain, i.e. one-half of the engineering shear strain γ. When used for plastic strain increments, simultaneously with the corresponding stress circle, the origin of the strain circle coincides with the value of the hydrostatic stress.

Mohr's stress circle (Mohr's circle) For a body loaded along orthogonal x,y axes by normal stresses σ_x and σ_y, and shear stresses τ_{xy}, the normal stress σ_θ and shear stress τ_θ on a plane inclined at angle θ to the x-axis are given by

$$\sigma_\theta = \frac{1}{2}(\sigma_x + \sigma_y) + \frac{1}{2}(\sigma_x - \sigma_y)cos2\theta - \tau_{xy}sin2\theta$$

and

$$\tau_\theta = \frac{1}{2}(\sigma_x - \sigma_y)sin2\theta - \tau_{xy}cos2\theta.$$

Elimination of θ gives

$$[\sigma_\theta - \frac{1}{2}(\sigma_x + \sigma_y)]^2 + \tau_\theta^2 = \frac{1}{4}(\sigma_x - \sigma_y)^2 + \tau_{xy}^2$$

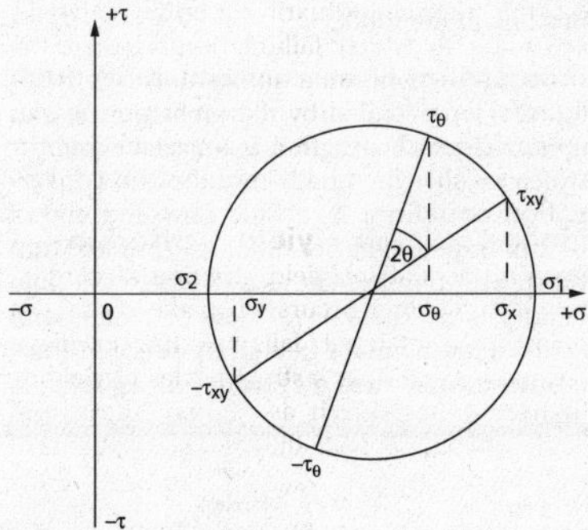

Mohr's stress circle

that, on σ,τ axes, is the equation of a circle having radius

$$\sqrt{\frac{1}{4}(\sigma_x - \sigma_y)^2 + \tau_{xy}^2}$$

centred on the σ-axis at ½ $(\sigma_x + \sigma_y)$, called Mohr's stress circle. For loading in three dimensions, there are three Mohr's circles having diameters $(\sigma_1 - \sigma_2)$, $(\sigma_2 - \sigma_3)$ and $(\sigma_1 - \sigma_3)$ where σ_1, σ_2 and σ_3 are the principal stresses.

Mohs scale A scale of scratch hardness originally developed for minerals. *See also* HARDNESS TEST.

moisture content *See* SPECIFIC HUMIDITY.

moisture separator *See* STEAM SEPARATOR.

moisture separator reheater *See* REHEATER.

molar analysis The number of moles of each component in a mixture of substances. *See also* GRAVIMETRIC ANALYSIS; MOLE FRACTION.

molar gas constant *See* UNIVERSAL GAS CONSTANT.

molar heat capacity (molecular heat capacity, $\tilde{C}$) (Unit J/kmol.K or J/kmol.°C) The energy required to raise the temperature of one kmol of a substance by one degree kelvin without change of phase. For practical purposes, a liquid or a solid is generally regarded as having a single specific heat, whereas for a gas the energy required depends upon how the energy-transfer process is executed and two values are defined: the molar heat capacity at constant volume ($\tilde{C}_V$) and the molar heat capacity at constant pressure ($\tilde{C}_P$). The thermodynamic definitions are $\tilde{C}_V = (\partial\tilde{u}/\partial T)_V$ and $\tilde{C}_P = (\partial\tilde{h}/\partial T)_P$ where the subscript V denotes a reversible non-flow process at constant volume and P denotes a reversible non-flow process at constant pressure, $\tilde{u}$ is the molar internal energy, $\tilde{h}$ the molar enthalpy, and T the absolute temperature. Molar heat capacity is an intensive thermodynamic property dependent on temperature and pressure. *See also* SPECIFIC HEAT.

molar property A form of intensive property obtained by dividing the extensive value of a property by the number of kilomoles present. **Molar mass (*M*)** with unit kg/kmol, is the mass in kg of one kmol of any substance. **Molar**

volume, with unit m^3/kmol, is the corresponding volume, at a given temperature and pressure, and equivalent to the molar mass divided by the substance density. *See also* MOLE; MOLECULAR WEIGHT.

mole (mol) The amount of a substance in grams numerically equal to its molecular weight. *See also* KILOMOLE.

molecular drag pump (molecular pump) A high-vacuum (down to about 10^{-4} Pa) pump in which momentum is transferred to the gas molecules by a rapidly rotating solid surface. *See also* TURBO/DRAG PUMP; TURBOMOLECULAR PUMP.

molecular gauge *See* FRICTION VACUUM GAUGE.

molecular heat capacity *See* MOLAR HEAT CAPACITY.

molecular mass The mass of one molecule of a substance in unified atomic mass units and numerically equivalent to the molecular weight. Unlike molar mass, the isotopic composition of the molecule is taken into account.

m

molecular weight (relative molar mass) The ratio of the mass of a molecule to one unified atomic-mass unit. The molecular formulae and molecular weights of some common substances are: hydrogen H_2, 2.016; helium He, 4.003; carbon C, 12.011; nitrogen N_2, 28.014; oxygen O_2, 31.998.

mole fraction In a mixture of substances, the ratio of the mole number of an individual component to the mole number of the mixture. *See also* MOLAR ANALYSIS.

mole number (Symbol *N*) The number of moles in a given mass m of a substance, equal to m divided by the molar mass M of the substance.

Mollier diagram (Mollier chart, enthalpy-entropy chart) A plot of specific enthalpy *vs* specific entropy, for a substance such as steam, on which are shown isobars, isotherms, lines of constant dryness fraction or quality, and sometimes also density. *See also* T-S DIAGRAM.

Moll thermopile A heat-flux sensor used to measure thermal radiation. It consists of a blackened disc to which are connected thermocouples connected in series with a reference junction maintained at constant temperature. *See also* THERMOPILE.

molten The state of a substance, such as a metal, which is normally solid at STP, when caused to become liquid (usually by heating). *See also* FROZEN.

moment (*M*) (Unit N.m) The tendency of a force to rotate an object to which it is applied. If the force vector is $\boldsymbol{F}$ and the displacement vector is $\boldsymbol{r}$, r being the length of the lever arm from the putative axis or point of rotation to the point at which the vector is applied, $\boldsymbol{M} = \boldsymbol{r} \times \boldsymbol{F}$. If the magnitude of the force vector is F and the angle between F and r is θ, then the magnitude of M is $M = rF\sin\theta$. *See also* BENDING MOMENT; COUPLE; TORQUE.

moment diagram *See* BENDING-MOMENT DIAGRAM.

moment of force *See* COUPLE.

moment of inertia (mass moment of inertia, *I*) (Unit kg.m²) For a particle of mass m rotating about an axis, $I = mr^2$ where r is the perpendicular distance of the particle from the axis. For a body regarded as a system of particles rotating about an axis, $I = \Sigma_i m_i r_i^2$ where particle i has mass m_i and is a distance r_i from the axis. More generally

$$I = \int r^2 dm$$

Similarly for bodies rotating about a point. *See also* SECOND MOMENT OF AREA.

moment of momentum *See* ANGULAR MOMENTUM.

moment sensor A sensor used in robotics that allows the determination of the load carried by the end effector from the moment produced in a link of the robot.

momentum (linear momentum, *M*) (Unit kg.m/s) For a solid body of mass m moving with vector velocity $\boldsymbol{V}$, $\boldsymbol{M} = m\boldsymbol{V}$. It is a vector quantity which satisfies the **linear-momentum conservation equation**. In a fluid flow, the rate of flow of momentum per unit area **(momentum flux)**, with unit kg/ms², is ρV^2 where the fluid density is ρ. The rate of flow of momentum across a given area A in a given direction **(momentum flow rate, $\dot{M}$)**, with unit kg.m/s², is then

$$\dot{M} = \int_A \rho V^2 dA$$

See also ANGULAR MOMENTUM; CONSERVATION EQUATIONS; MOMENTUM THEOREM.

momentum integral equation (integral momentum equation, Kármán integral equation, von Kármán momentum integral equation) An equation that relates the wall shear stress for a boundary layer to the variation of integrals of the mass and momentum flow within the boundary layer, the pressure gradient, and the surface mass-transfer rate $\dot{m}''$. It is conveniently written in terms of the skin-friction coefficient c_f and the displacement and momentum thicknesses, δ^* and θ as

$$\frac{c_f}{2} = \frac{1}{U_\infty^2}\frac{\partial}{\partial t}(U_\infty\delta^*) + \frac{\partial\theta}{\partial x} + (2\theta + \delta^*)\frac{1}{U_\infty}\frac{\partial U_\infty}{\partial x} - \frac{\dot{m}''}{\rho U_\infty}$$

where ρ is the fluid density, U_∞ is the free-stream velocity, t is time and x is the distance along the surface.

momentum source A concept in fluid mechanics in which there is a flow of momentum initially in the absence of mass flow. *See also* ENTRAINMENT; JET.

momentum thickness (momentum-deficit thickness, θ) (Unit m) A thickness that arises in the analysis of a boundary layer based upon the momentum-integral equation. It is related to wall shear stress. From conservation of momentum within the boundary layer,

$$\theta = \int_0^\infty \frac{u}{U_\infty}\left(1 - \frac{u}{U_\infty}\right)dy$$

where $u(y)$ is the distribution of streamwise velocity within the boundary layer, y is the distance from the surface, and U_∞ is the free-stream velocity. *See also* DISPLACEMENT THICKNESS.

momentum theorem (impulse theorem) A form of Newton's second law of motion as used in fluid mechanics, which can be stated as follows: the sum of all the forces $\boldsymbol{F}$ acting on a fixed control volume $\mathcal{V}$ is equal to the time rate of change of momentum $\boldsymbol{M}$ contained in the volume plus the net flow rate of momentum through its surface S. As an equation, the momentum theorem can be written as

$$\boldsymbol{F} = \frac{\partial \boldsymbol{M}}{\partial t} + \int_S \rho \boldsymbol{V}\boldsymbol{V}\cdot\boldsymbol{n}\, dS$$

where t is time, ρ is the fluid density, $\boldsymbol{V}$ is the vector velocity $\boldsymbol{n}$ is the outward-pointing unit vector normal to dS and $\boldsymbol{M} = \int_\mathcal{V} \rho \boldsymbol{V}\, d\mathcal{V}$.

momentum thrust (Unit N) In a jet or rocket engine, if the exhaust mass flow rate is $\dot{m}$ and the exhaust-gas velocity is $\boldsymbol{V}$, the product $\dot{m}\boldsymbol{V}$ is termed the momentum thrust.

momentum transfer The transfer of momentum from one body to another upon impact or from one fluid stream to another upon mixing. In both instances, momentum is conserved. *See also* CONSERVATION EQUATIONS.

MON *See* OCTANE NUMBER.

monatomic gas A gas for which each molecule consists of a single atom. Examples are helium, neon, argon, krypton, xenon, and radon. The specific-heat ratio γ for most monatomic gases at room temperature is 5/3. *See also* DIATOMIC GAS.

monitor To observe, and often record, the output of one or more instruments measuring process or system variables, structural integrity, etc., over a period of time.

monobloc A single-piece piston-engine block incorporating all cylinders and possibly also the crankcase.

monochromatic absorptance (monochromatic absorptivity, α_λ) The ratio of the absorptance of a surface at a given wavelength and temperature to the absorptance of a blackbody at the same wavelength and temperature. For emittance, the corresponding quantity is the **monochromatic emittance (monochromatic emissivity, ε_λ)**.

monochromatic absorption coefficient (α_λ) A measure of the decrease in thermal-radiation intensity at a given wavelength λ due to absorption by a gas. According to Beer's law, $I_{\lambda_x} = I_{\lambda_0} exp(-\alpha_\lambda x)$ where I_{λ_x} is the intensity after absorption by a gas layer of thickness x and I_{λ_0} is the initial intensity.

monocoque structure A structure such as an aeroplane fuselage consisting of a shell that is torsionally very stiff. A **unitized** motor-vehicle **body** has a monocoque structure.

MONO pump *See* PROGRESSIVE-CAVITY PUMP.

Moody chart (Moody diagram) A set of curves of Darcy friction factor *vs* Reynolds number for fully-developed turbulent flow in rough pipes, based upon Colebrook's equation.

Moody friction factor *See* DARCY FRICTION FACTOR.

morse taper A standardized series of tapers employed on the shanks of large-size drills, etc. to mate with internal tapers on machine tool spindles.

Morton number (*Mo*) A non-dimensional parameter that arises in the study of bubbles and drops rising or falling in another fluid and is defined by $Mo = \Delta\rho g\mu^4/\rho^2\sigma^3$ where $\Delta\rho$ is the density difference and σ is the surface tension at the interface between the two fluids, g is the acceleration due to gravity, ρ is the density of the continuous phase and μ is its dynamic viscosity. *See also* BOND NUMBER; EÖTVÖS NUMBER.

motion Any progressive change in the position of a solid object or a fluid particle.

motive fluid A high-pressure fluid used to produce flow of another fluid, as in an ejector.

motive power That which imparts motion to a machine, for example steam, a head of water, and wind.

motor *See* ENGINE; PRIME MOVER.

motor octane number *See* OCTANE NUMBER.

mould A hollow form filled with liquid, pasty, soft, or molten material that is allowed to harden and adopt the shape of the mould in the process of casting.

moulding pressure The pressure required to force material into a mould, particularly highly viscous material.

moulding shrinkage The reduction in the dimensions of a moulded component as it hardens.

mould seam *See* SEAM (2).

mounting The structure supporting a component within a machine or an entire machine, often designed to minimize the transmission of shock or vibration.

movable-active tooling Any tools carried by a robot that require power to operate or which produce sensor signals, for example a gripper or a proximity sensor mounted on the end effector. **Movable-passive tooling** are tools which do not require power, for example a remote centre compliance.

moving frame of reference *See* ACCELERATING FRAME OF REFERENCE.

moving load *See* INFLUENCE LINE.

M thread *See* SCREW.

mud drum A drum at the bottom of a natural-circulation water-tube boiler into which cool water flows from a downcomer, so displacing water that is heated as it flows through a riser leading to the steam drum. Sediment settles in the mud drum and is periodically removed.

muffle furnace A term used historically for a furnace in which the heated object was isolated from the fuel and combustion atmosphere. Nowadays, an oven heated electrically, where the problem does not arise.

multi-fuel engine An engine that can run on a range of petroleum fuels from petrol to diesel.

multilevel control theory A design method for the control of complicated systems, which uses plant decomposition to reduce the complicated control problem to a number of simpler problems, the control of which is coordinated so as to achieve the desired performance of the complete plant.

multi-pass shell-and-tube heat exchanger A shell-and-tube heat exchanger in which the tube-side fluid passes across or through the shell several times. In the cross-flow/counterflow arrangement shown, there are five tube-side passes and one shell-side pass.

multiphase flow (mixed-phase flow) Fluid flow, usually through a duct, in which a single fluid may be present in more than one phase, or a flow in which two or more immiscible fluids or fluids containing solid particles are present.

multiple input multiple output system (MIMO system, multivariable system) A dynamic system having more than one input and more than one output, where the signal on each output is a function of more than one of the inputs.

multiple-loop system A control system with more than one feedback loop.

multiple-start thread *See* SCREW.

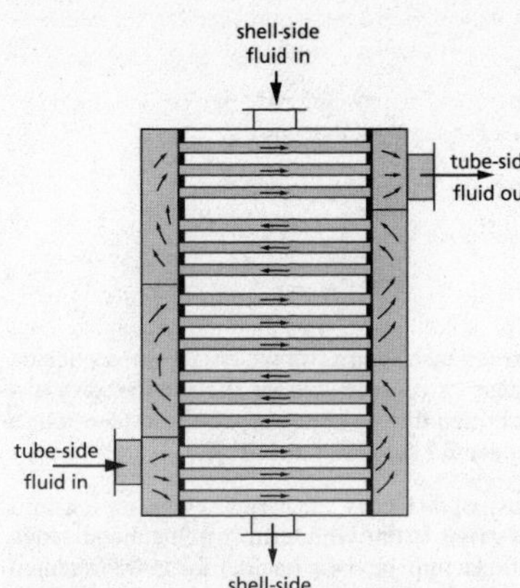

multi-pass shell-and-tube heat exchanger

multi-point injection *See* DIRECT INJECTION.

multiport burner A burner with multiple fuel outlets.

multi-resonant oscillating water columns A wave-energy converter in which columns of water are caused by wave action to oscillate up and down in partially submerged tubes of different lengths. The water columns in turn induce oscillatory flows of air, which can be passed through Wells turbines to generate electrical power. *See also* OSCILLATING WATER COLUMN.

multiscale modelling Modelling applied to physical problems in which significant phenomena occur at different length and time scales.

multistage compressor A compressor in which compression of a gas takes place across two or more stages, with or without intercooling between the stages. In a **multistage pump** the liquid-pressure increase takes place across two or more impellers on one shaft. In a **multistage** steam or gas **turbine**, expansion of the working fluid takes place across two or more stages.

multivariable system *See* MULTIPLE INPUT, MULTIPLE OUTPUT SYSTEM.

Musgrove rotor A type of vertical-axis wind-turbine rotor for which the blades are vertical in normal power generation, but fold about a horizontal beam for control or shutdown. *See also* DARRIEUS WIND TURBINE.

mushroom follower (flat-foot follower) A type of cam follower with a flat surface.

mushroom valve *See* POPPET VALVE.

mutilated gear A gear wheel from which one or more teeth have been removed, in order to provide intermittent motion.

Mylar® A thin (20 to 500 μm thick) polyester film made by biaxially-stretching polyethylene terephthalate (PET) sheet to give a Young's modulus of 4 GPa. Available in transparent and metallized forms with numerous applications, including magnetic recording tape, packaging, and X-ray film.

Nahme–Griffith number (*G*) A non-dimensional parameter that arises in flows, particularly of polymeric fluids, where viscous dissipation is significant, defined by $G = \mu V^2 b/k$ where μ is a typical viscosity of the fluid, b is the temperature coefficient of the viscosity, k is the thermal conductivity, and V is a characteristic velocity. *See also* BRINKMAN NUMBER.

nano (n) An SI unit prefix indicating a sub-multiplier of 10^{-9}; thus nanometre (nm) is a unit of length equal to one billionth of a metre or one millionth of a millimetre.

nanochannel A flow channel with width and depth dimensions in the range 1 nm to 1 μm. *See also* MICROCHANNEL.

nano-fabrication *See* NANOMANUFACTURING.

nanofibre A filament having a diameter in the nm range.

nanofilter A filter with a pore size on the order of a few nm.

nanofluid A colloidal suspension of nanoparticles in a base liquid, such as water.

nanofluidics Fluid flow in devices which have channels, pores, slits, etc. with dimensions typically below 100 nm.

nano machining The machining of silicon chips typically less than 1 nm thick.

nanomanufacturing **1.** The manufacture of nanoscale materials. **2.** The manufacture of components or devices that have critical dimensions of order 1 nm. *See also* NANOTECHNOLOGY.

nanomaterial A material defined by the European Commission as 'A natural, incidental or manufactured material containing particles, in an unbound state or as an aggregate or as an agglomerate and where, for 50% or more of the particles in the number size distribution, one or more external dimensions is in the size range 1 nm–100 nm. In specific cases and where warranted by concerns for the environment, health, safety or competitiveness the number size distribution threshold of 50% may be replaced by a threshold between 1 and 50%.'

nanoparticles **1.** Particles with dimensions of order 1 nm which have either been introduced into microstructures for reinforcement or are present as impurities. **2.** *See* NANOFLUID.

nanoporous material *See* MICROPOROUS MATERIAL.

nano reinforcement Reinforcement of materials using nanometre-size particles, platelets, or filaments.

nanostructure Material microstructure observed at the nm scale.

nanotechnology The science and engineering of materials that have been structured on length scales of 1–100 nm, resulting in modified physical properties owing to changes in the ratio of surface area to volume (atoms on surfaces having different symmetry from those in the bulk) and because many of the fundamental physical processes that underpin the properties of materials have a characteristic length scale of a few nm, so that alteration of microstructure at the nm level alters the bulk properties.

nano tube *See* CARBON NANOTUBE.

nano wire A wire having a diameter in the nm range.

natural aspiration (natural induction) The process by which combustion air is drawn into the cylinders of a reciprocating-piston, internal-combustion engine due to atmospheric pressure acting against the reduced cylinder pressure caused by downward movement of the pistons during an induction stroke. *See also* FORCED INDUCTION.

natural convection *See* HEAT TRANSFER.

natural draught *See* DRAUGHT.

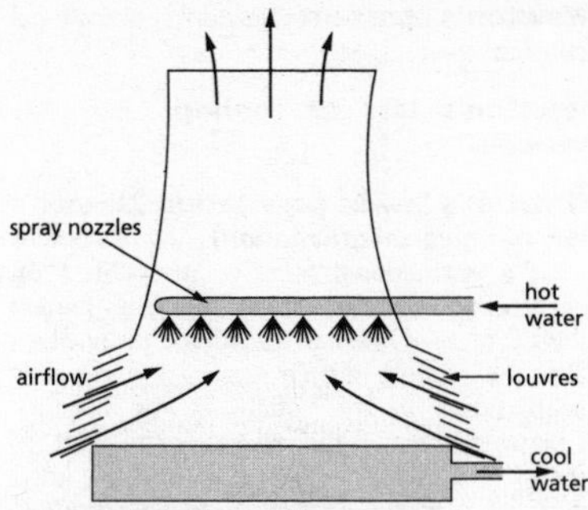

natural-draught cooling tower

natural-draught cooling tower A cooling tower in which the upward flow of air is brought about by its buoyancy. *See also* FORCED-DRAUGHT COOLING TOWER.

natural frequency The frequency of free (i.e. unforced and undamped) oscillations for a given mode of oscillation. *See also* DAMPED NATURAL FREQUENCY.

natural gas A gaseous fuel for both domestic and industrial use consisting primarily of methane, with up to 20% of other gases, primarily ethane. *See also* LIQUEFIED NATURAL GAS.

naturally-aspirated engine (normally-aspirated engine) A piston engine in which air enters the cylinders at atmospheric pressure without the use of a compressor such as a turbocharger or supercharger. *See also* FORCED INDUCTION.

Navier–Stokes equation The momentum equation for a linear viscous (i.e. Newtonian) fluid. In vector form it may be written

$$\rho\frac{DV}{Dt}=\rho g-\nabla p+\frac{\partial}{\partial x_j}\left[\mu\left(\frac{\partial v_i}{\partial x_j}+\frac{\partial v_j}{\partial x_i}\right)+\delta_{ij}\lambda\, div\, V\right]$$

where ρ is the fluid density, μ is its dynamic viscosity and λ is the coefficient of bulk viscosity, $\boldsymbol{g}$ is the acceleration due to gravity, p is the static pressure, $\boldsymbol{V}$ is the vector velocity with components $u\,(i=1)$, $v\,(i=2)$, and $w\,(i=3)$ in the $x\,(i=1)$, $y\,(i=2)$, and $z\,(i=3)$ directions respectively, and δ_{ij} is the Kronecker delta. For a constant-property fluid, the equation reduces to

$$\rho\frac{DV}{Dt}=\rho g-\nabla p+\mu\nabla^2 V$$

See also FLUID MECHANICS; STOKES HYPOTHESIS.

NDE (NDT) *See* NON-DESTRUCTIVE EVALUATION.

near-net-shape A rough-finish blank, component, etc. that is close to the required final shape and size so that minimal forming or cutting is required to finish the object, thus reducing wastage and energy.

neck (necking) 1. (diffuse necking) During tensile plastic flow when deformation becomes localized in a region having a size about the same as the diameter of a round bar testpiece. For a material following $\sigma=\sigma_o\varepsilon^n$, the strain ε_1 at which diffuse necking occurs is given by $\varepsilon_1=n$. **2. (localized necking)** In sheet forming, when the deformation becomes localized into a narrow trough at angle θ to the ε_1-axis given by $tan^{-1}\sqrt{-\rho}$ where $\rho=(\varepsilon_2/\varepsilon_1)$ is the in-plane principal-strain ratio. For uniaxial tension $\rho=-0.5$ and $\theta\approx 55°$. The strain at which localized necking takes place is $\varepsilon_1=n/(1+\rho)$, so in simple tension $\varepsilon_1=2n$.

needle-roller bearing *See* BEARING.

needle valve A regulating valve with a relatively small outlet, a long tapered seat, and a needle-shaped plunger moved axially by a fine-threaded screw.

negative feedback Where the feedback in a closed-loop system is subtracted from the set point to give the error.

negative rake The inclination of the cutting tool in machining in the direction of cutting, so that rake angle, measured perpendicular to the cut surface, is negative. Often used with tools made of brittle materials since loading on the tool is mainly compressive.

Nernst heat theorem The entropy change for a chemical reaction between pure substances approaches zero as the absolute temperature approaches zero.

Nernst theorem *See* THIRD LAW OF THERMODYNAMICS.

net calorific value *See* CALORIFIC VALUE OF A FUEL.

net positive suction head The absolute pressure of a liquid at the inlet to a suction pump less the vapour pressure of the liquid. It is a measure of the likelihood of cavitation.

netting analysis The materials-mechanics analysis of fibre-reinforced materials, assuming that loads are borne entirely by the filaments in tension, neglecting bending and shear, and ignoring any contribution from matrix material.

Neumann–Kopp rule An approximate empirical formula stating that the molar heat capacity of a solid substance is equal to a weighted sum of the heat capacities of its constituents.

neural network (neural net) In artificial intelligence and control, a set of program structures referred to as artificial neurons which are linked in a similar way to the neurons in a human brain. As the neurons operate simultaneously, the computational problem is thus solved in a parallel, rather than sequential, way.

neuroengineering The study of basic and clinical neuroscience employing aspects of mechanical, materials, and robotics engineering with the aim of producing machines that interact in real time with the brain to restore normal function in cases of injury or disease.

neutral axis (neutral fibre, neutral plane, neutral surface) In bending of a beam, that location within its depth where stresses, which are tensile on the outside, change to compressive on the inside, i.e. where the stress and deformation are both zero.

n

neutral equilibrium The condition of a body or system which, having been given a small disturbance, stays in the disturbed state. *See also* STABLE EQUILIBRIUM; UNSTABLE EQUILIBRIUM.

neutral gear The state of gear engagement or disengagement in a gearbox when no power is transmitted from the drive shaft to the driven shaft.

neutral stability *See* MARGINAL STABILITY.

'new' biomass A term used to distinguish purpose-grown energy crops, organic waste, etc. from traditional sources of biomass such as wood, rice husks, and other plant residues.

newton (N) The basic unit of force in the SI system, defined as the force that results in an acceleration of 1 m/s^2 when acting on a 1 kg mass.

Newtonian attraction *See* NEWTON'S LAW OF GRAVITATION.

Newtonian mechanics *See* CLASSICAL MECHANICS.

Newtonian reference frame *See* INERTIAL COORDINATE SYSTEM.

Newton's constant Newton's gravitational constant.

Newton's law of cooling *See* HEAT TRANSFER.

Newton's law of gravitation (Newton's law of universal gravitation) Any two bodies exert a gravitational force of attraction F on each other directed along the line joining their centres of mass, in magnitude proportional to the product of their masses and inversely proportional to the square of the distance r between them. If the masses are m_1 and m_2 then $F = Gm_1m_2/r^2$ where G is Newton's gravitational constant and equal to 6.67 x 10^{-11} $N.m^2/kg^2$. Because G is very small, mutual attractions of bodies on the earth's surface are insignificant compared with the earth's attraction.

Newton's law of viscosity The shear stress between two parallel layers of fluid is proportional to the velocity of one relative to that of the other, and inversely proportional to the distance between them. The constant of proportionality is the dynamic viscosity μ. For a **Newtonian fluid** μ is independent of pressure gradient, shear stress, and strain, but may change with temperature and pressure. *See also* NON-NEWTONIAN FLUID.

Newton's laws of motion The three fundamental laws that are the basis for classical (i.e. non-relativistic and non-quantum) mechanics. They can be summarized as follows:

First law Every body stays at rest or at constant velocity unless acted upon by an unbalanced external force.

Second law The rate of change of momentum $m\boldsymbol{V}$ of a body of mass m and vector velocity $\boldsymbol{V}$ with respect to time t is equal to the magnitude of any unbalanced external force $\boldsymbol{F}$ and is in the same direction, i.e. $\boldsymbol{F} = d(m\boldsymbol{V})/dt$. For a body of constant mass m, the acceleration $\boldsymbol{a}$ $(= d\boldsymbol{V}/dt)$ is directly proportional to $\boldsymbol{F}$.

Third law To every action, there is an equal and opposite reaction.

NFFO *See* NON-FOSSIL FUEL OBLIGATION.

nibbling The production of a curved cut in sheet material by making a series of small overlapping notches or slits by rapid punching.

Nichol's chart A plot of the open-loop gain against the phase shift in a control system. The phase margin and gain margin can be derived from the plot.

nipple A device containing a non-return valve screwed into a lubrication point through which grease may be introduced, for example into a bearing.

nitriding A process of hardening low-carbon steel components by heating in an atmosphere of nitrogen that diffuses into the surfaces to form fine precipitates of hard nitride compounds. *See also* CASE HARDENING; ELECTRON BEAM HARDENING; LASER HARDENING.

nitrogen oxides *See* OXIDES OF NITROGEN.

NLEFM *See* NON-LINEAR ELASTIC FRACTURE MECHANICS.

node In a system of stationary waves, a location where the displacement is zero. *See also* ANTINODE.

nodular cast iron *See* CAST IRON.

noise In instrumentation, unwanted random signals that affect the output of a sensor.

nominal power The average power output of a typical production engine under normal working conditions measured according to SAE standard J 1349/ISO 1585. *See also* INDICATED HORESPOWER.

nominal size The intended size of a component. The actual size will depend on manufacturing tolerances. *See also* LIMITS AND FITS.

nominal strain *See* ENGINEERING STRAIN.

nominal stress *See* ENGINEERING STRESS.

non-adiabatic *See* DIABATIC.

nonblackbody A body emitting thermal radiation for which the emissivity is less than unity.

non-condensable gas A gas not easily condensed by modest cooling. All gases liquefy at sufficiently low temperatures.

non-contact thermometer *See* INFRARED THERMOMETER.

non-destructive evaluation (non-destructive testing, NDE, NDT) Any of various methods, including eddy current, magnetic flux, radiography, ultrasonics, and X-rays, by which the integrity of a material may be assessed without destroying the component.

non-dimensional number (non-dimensional group, non-dimensional parameter, dimensionless number) A combination of physical quantities $A, B, C, D \ldots$ of the form $A^m B^n C^o D^p \ldots$ that has no overall dimensions. The exponents $m, n, o, p \ldots$ can be positive or negative, whole numbers or fractions. In many instances a non-dimensional number represents a ratio of the stresses involved in a flow problem or modes of heat transfer in a heat-transfer problem. Many non-dimensional numbers are given special names, such as the Reynolds, Mach, and Froude numbers. *See also* DIMENSIONAL ANALYSIS.

non-equilibrium thermodynamics The thermodynamics of systems not in thermal equilibrium. *See also* APPLIED THERMODYNAMICS.

non-ferrous metal A term referring strictly to all metals and alloys that do not contain iron, but usually applied to aluminium and copper-based alloys.

non-flow energy equation The increase in internal energy of a closed system is equal to the heat supplied to the system plus the work done on the system. This is a corollary of the first law of thermodynamics that defines internal energy.

non-flow exergy function (A) (Unit kJ) A thermodynamic property defined by $A = U + p_O\mathcal{V} - T_O S$ where U and S are, respectively, the internal energy and entropy of a closed system of volume $\mathcal{V}$, while p_O and T_O are the pressure and temperature of the surroundings. *See also* COMPOSITE PROPERTY.

non-fossil fuel obligation (NFFO) A government requirement in England and Wales for the electricity Distribution Network Operators to purchase energy generated by nuclear reactors and from renewable sources.

non-inertial coordinates (non-inertial reference frame) *See* ACCELERATING FRAME OF REFERENCE.

non-interacting control A technique to allow control of multiple input, multiple output systems using single input, single output control methods.

non-linear elastic fracture mechanics Reversible fracture mechanics where the load-displacement relations are non-linear owing to large displacements or geometric effects.

non-linear mechanics An area of mechanics where elastic forces are not proportional to displacement. Paradoxically, irreversible plastic flow may be analysed by non-linear elasticity providing that applied loads remain proportional, and there is no unloading. *See also* KACHANOV'S THEOREM; TOTAL STRAIN PLASTICITY.

non-linear system A system in which the relation of the output to the input is non-linear.

non-linear vibration Vibration where the restoring forces are not proportional to displacement.

non-minimum-phase system A control system having a transfer function with poles and/or zeros in the right hand, positive real part of the complex plane.

non-Newtonian fluid A fluid, invariably a liquid, which has rheological properties more complex than those of a Newtonian fluid, usually involving a dependence of viscosity on shear rate. Included are viscoelastic, thixotropic, and generalized Newtonian liquids. Most synthetic liquids are non-Newtonian, including polymeric solutions, gels, slurries, and pastes, as are many natural liquids, including blood and synovial fluid.

non-positive displacement pump A pump in which there is substantial leakage of fluid back past the pumping mechanism, such as the impeller of a centrifugal pump or the rotor of an axial pump. *See also* POSITIVE-DISPLACEMENT PUMP.

non-return valve (reflux valve) A valve that prevents reverse flow when the forward flow ceases, by means of parts closing under gravity or by springs.

non-selective radiator A thermal radiator, such as a blackbody or a grey body, for which the emissivity is independent of wavelength.

non-storage calorifier A heat exchanger that uses steam to heat water for heating systems and industrial processes. *See also* STORAGE CALORIFIER.

non-tracking concentrator A solar concentrator that does not track the sun.

non-wetting *See* HYDROPHOBIC.

normal (normal direction) Perpendicular to a line or surface.

normal frequencies (fundamental frequencies) The frequencies of the normal modes of vibration of a system.

normal impact Impact that occurs when a projectile strikes a surface that is perpendicular to the line of flight.

normal-incidence pyrheliometer A solar-tracking device used to measure direct solar irradiance.

normality The principle that the vector sum of plastic strain increments is perpendicular to the yield surface. *See also* FLOW RULES.

normalized variable A physical variable that has been multiplied or divided by fixed quantities, such as length, velocity, and time scales together with material properties, usually to make it non-dimensional and so generalize theoretical (including numerical) solutions or experimental data. *See also* NON-DIMENSIONAL NUMBER.

normally-aspirated engine *See* NATURALLY-ASPIRATED ENGINE.

normal mode of vibration The oscillation of a system where all parts of the system move sinusoidally with the same frequency and phase. Systems with more than one degree of freedom have as many normal modes as degrees of freedom.

normal pitch (circular pitch) *See* TOOTHED GEARING.

normal reaction The perpendicular force between two surfaces in contact. The **tangential reaction** is the corresponding force parallel to the two surfaces.

normal shock (normal shock wave) A shock wave for which the gas flow is perpendicular to the shock front. *See also* OBLIQUE SHOCK; STRONG SHOCK.

normal stress A stress, in a fluid or a solid, that is perpendicular to the surface on which it acts. The surface may be a real external surface or an imaginary internal one. *See also* SHEAR STRESS.

normal temperature and pressure (NTP) Reference conditions usually taken as 20°C and 1 atm. *See also* STANDARD STATE.

normal vector In the kinematic analysis of robots, a 4×4 matrix is obtained representing the position and direction of the end effector.

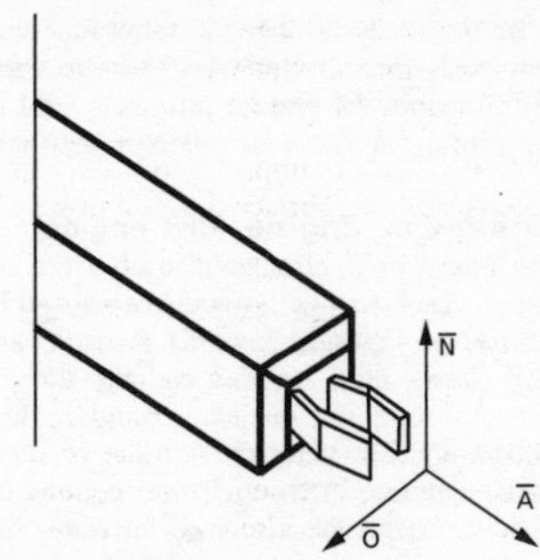

normal vector

The first column of this is the normal vector $\bar{N}$, which is a unit vector running in a direction orthogonal to the approach and orientation vectors, as shown in the diagram.

normal water level (operating water level) The level of water in a steam-raising boiler for proper operation. Combustion is usually halted if the level is too low, and the feedwater supply cut off if too high.

nose The usually rounded or pointed front part of a bulbous or slender body, such as an aeroplane, missile, high-speed train, torpedo, dirigible, or motor vehicle.

no-slip condition The assumption that there is no tangential relative velocity between a viscous fluid and a surface with which it is in contact. The assumption fails if the mean-free path of the fluid is comparable with, or greater than, the smallest dimensions of the flow geometry.

notch A geometric discontinuity, often having a V-shape, that can be microstructural or much larger in size, which acts as a stress concentrator or, depending on its sharpness, a crack. A material that is ductile in a conventional tensile test, but brittle when the testpiece has deep cuts that provide a large constraint to deformation, is said to be **notch brittle**. The load-bearing capacity of a body is given by the product of its cross-sectional area and the working stress. The load-bearing capacity is reduced by the presence of a notch or crack. In **notch-insensitive** materials, the reduction in load-bearing capacity is directly proportional to the reduction in cross-sectional area, but in **notch-sensitive** materials the reduction is disproportionately greater, owing to shear connexion between elements of the body. Fracture mechanics is the analysis of notch-sensitive solids. The effect of notches in fatigue is described in terms of the **notch-sensitivity factor**. *See also* FRACTURE MECHANICS.

notched-bar impact test *See* CHARPY TEST.

NOX (NO_X) *See* OXIDES OF NITROGEN.

nozzle A pipe or tube, usually of varying cross section, through which there is flow. See also CHOKED NOZZLE; CONVERGENT DUCT; CONVERGENT–DIVERGENT NOZZLE; VENTURI.

nozzle blade *See* STATOR BLADE.

nozzle contraction area ratio (ε_C) The ratio of the inlet area to the throat area for a convergent nozzle.

nozzle efficiency *See* ISENTROPIC EFFICIENCY.

nozzle expansion ratio (ε_E) The ratio of the exit area to the throat area for a divergent nozzle.

nozzle guide vane *See* STATOR BLADE.

nozzle-mix gas burner A burner in which the combustion air and gaseous fuel are kept separate until they mix and are ignited in a divergent nozzle.

nozzle throat The section of a convergent or convergent-divergent nozzle where the cross-sectional area is a minimum.

NTP *See* NORMAL TEMPERATURE AND PRESSURE.

NTU *See* NUMBER OF TRANSFER UNITS.

nuclear fission (fission) A reaction in which the nucleus of a heavy atom splits into lighter nuclei accompanied by the release of large amounts of thermal energy. In a **nuclear power plant**, the thermal energy needed to produce steam is derived from controlled nuclear fission in a **nuclear reactor**. The most common fuels are isotopes of uranium and plutonium. Either pressurized water or a gas such as carbon dioxide or helium (**gas-cooled nuclear reactor**) is commonly used to cool the reactor core and generate steam in a boiler. In the **advanced gas-cooled reactor (AGR)** graphite is the neutron moderator, enriched uranium is the fuel, and carbon dioxide gas is the coolant. The gas pressure is about 40 bar and the highest gas temperature 640°C. In a **supercritical-water reactor (SCWR)** the water is at supercritical pressure. In a **boiling-**

water reactor steam is generated within the reactor itself.

nuclear fusion A nuclear reaction in which atomic nuclei of low atomic number fuse to form a heavier nucleus with the release of large amounts of energy. Current research is aimed at realizing sustained fusion in a thermonuclear reactor as a future source of energy for human use.

nucleate boiling *See* BOILING.

nucleation The formation of microcracks in a material.

nucleation sites (nucleation centres) *See* BOILING.

number of transfer units (NTU) In heat-exchanger design, the combination UA/C_{min} where U is the overall heat transfer coefficient, A is the surface area corresponding with U, and C_{min} is the capacity rate (i.e. the product $\dot{m}C_P$ where $\dot{m}$ is the mass flow rate and C_P the specific heat of the fluid) for the fluid experiencing the larger temperature change. For a given heat exchanger, the effectiveness increases with increase in the NTU value, reaching a maximum for NTU $\approx$ 5.

nuclear power plant

Nusselt number (*Nu*) A non-dimensional parameter that arises in problems involving both convective and conductive heat transfer, defined by $Nu = hL/k$ where h is the convective heat-transfer coefficient, L is a characteristic length scale, and k is the thermal conductivity of the fluid.

nut An internally-threaded fastener that works on externally-threaded screws. There is a multitude of shapes for special purposes, and different gripping surfaces for different spanners.

nutating-disc engine (disc engine) An engine based on a circular disc attached to a Z-shaped crankshaft with the disc enclosed in a cylindrical or spherical housing. A fixed radial barrier allows fluid to flow radially into the housing, around the enclosure, and to leave the housing while forcing the disc to nutate without rotating. The nutation rotates the crankshaft. Originally conceived for water flow, there is now an internal-combustion version of the disc engine. A **nutating-disc meter (disc meter)** for the measurement of the volumetric flow rate of a liquid uses the same principle as the nutating-disc engine where the rotation speed of a spindle attached to the disc is proportional to the flow rate. In a **nutating-disc pump (disc pump)**, rotation of the crankshaft causes the disc to nutate and so produce fluid flow.

nutation **1.** An oscillation ('nodding') of the axis of a spinning body which is precessing at a variable rate. **2.** The motion of the surface of an inclined disc rotating about an axis passing obliquely through its centre. *See also* SWASHPLATE.

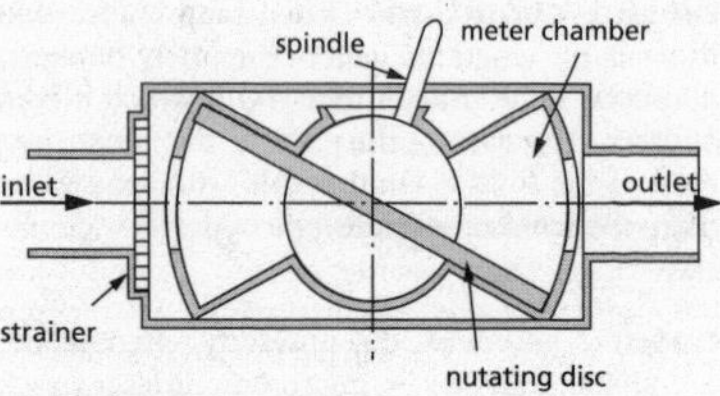

nutating disc meter

nut splitter A tool for removing rusted-on and corroded nuts from bolts. It consists of a stiff steel ring that is placed around the nut. A screw thread passing through the ring bears diametrically on a wedged-shape tip that indents and cuts though a face of the nut.

nylon A group of synthetic thermoplastic polymers (polyamides) that can be extruded and injection moulded. It has a density in the range 1000 to 1850 kg/m^3 and tensile strength in the range 20 to 300 MPa. Engineering applications include fasteners, fibres, and bearings.

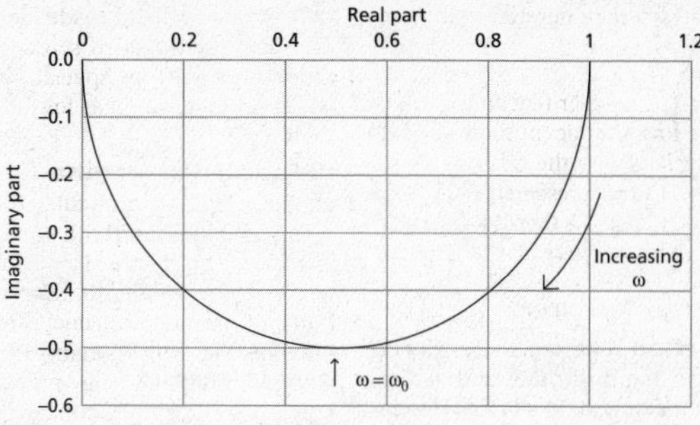

Nyquist diagram

Nyquist diagram (Nyquist plot) A polar plot in the complex plane of the open-loop frequency response of a system. The plot is produced from the transfer function by plotting the gain and phase shift over a range of frequencies where the gain is plotted as the radius to each point and the phase shift as the angle between the abscissa and the radius to the point. For example, a simple low pass filter with a transfer function G(s) and a cut off frequency ω_0 will have the following Nyquist diagram: $G(s) = 1/(\omega s/s_0 + 1)$. The stability of a system can be investigated through the **Nyquist stability theorem**, also known as the **Nyquist stability criterion**. This states that a unity negative feedback closed-loop system will be unstable when the gain of the open loop part of the system is greater than or equal to unity at any frequency where the phase shift in the open loop part is $\pm 180°$. On the Nyquist diagram this corresponds to the contour crossing or encircling the negative real axis at -1. Hence if the contour crosses the negative real axis between the origin and -1, the closed loop system will be stable, whereas if it crosses at a point more negative than -1 the system will be unstable. For example, consider the open loop transfer function $G(s)$ given by: $G(s) = 1/[(s + 1)(s + 2)]$ The contour on the Nyquist plot for this transfer function is shown in the Nyquist stability criterion diagram.

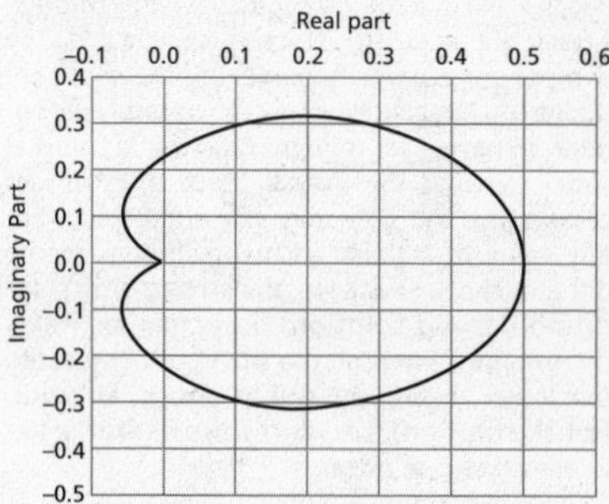

Nyquist stability criterion

By inspection the point (−1,0) lies outside this contour and thus, as would be expected from the transfer function, the system will be stable.

Nyquist–Shannon sampling theorem If a function of time contains no frequency components higher than f_{max} (**Nyquist frequency**), it is completely determined by sampling its value at intervals $\Delta t = 1/(2f_{max})$.

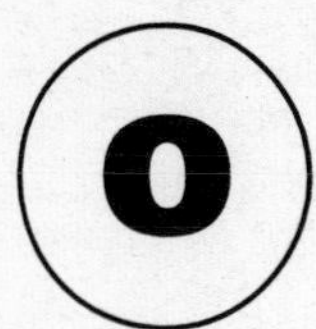

oblique projection A perspective engineering drawing employing horizontal and vertical axes along which a component's actual dimensions are scaled, and an angled axis for receding depth along which the dimensions are often reduced to avoid the impression of distortion.

oblique shock wave (oblique shock) A shock wave that turns a supersonic flow through a 'concave' angle (θ). The shock angle is β. The component of velocity parallel to the shock remains unchanged, whereas the normal component behaves as though crossing a normal shock, so that the overall Mach number decreases but the flow may stay supersonic. For any value of θ there are two solutions to the oblique-shock equations, the **strong-shock solution (strong solution)** being that for which the pressure jump across the shock is greater, the other being the **weak-shock solution (weak solution)**. *See also* PRANDTL–MEYER EXPANSION; SHOCK POLAR.

oblique valve 1. A through-flow valve in which the valve stem and seat are at an angle to the main flow direction. **2.** A valve in which the outlet is inclined to the approach-flow direction.

observability The extent to which the internal states of a system can be estimated by measurement of externally available signals.

observation spillover The parts of the sensor signals from a dynamic system which result from dynamic terms that are not modelled in the observer and thus reduce the accuracy of the state estimates.

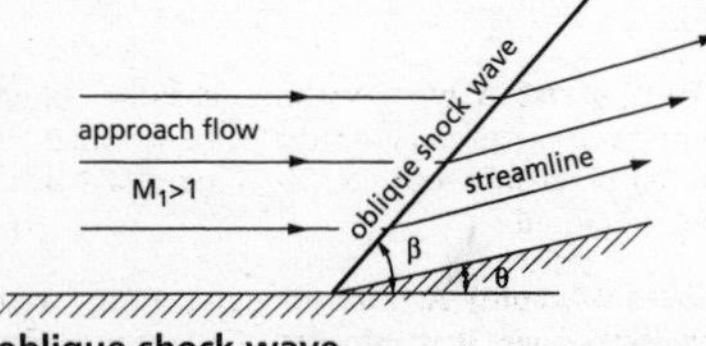

oblique shock wave

observer A mathematical model of a system used to obtain estimates of the internal inaccessible states of the system from those states that are available as external signals.

obstruction flow meter *See* BERNOULLI OBSTRUCTION FLOW METER.

obturator ring A ring of relatively soft material around the outside of a projectile in a gun barrel or piston in a cylinder that expands with heating to improve the seal.

ocean thermal-energy conversion (OTEC) The use of heat engines to utilize the temperature difference (about 20 to 25°C) between the warm near-surface water in an ocean and the colder water beneath the thermocline, and convert some of the available thermal energy into electrical energy. In closed-cycle systems, a low-boiling-point working fluid such as ammonia is used. In open-cycle systems, the working fluid is seawater.

octane number (octane rating) A measure of the anti-knock properties of a fuel taken as the percentage by volume of iso-octane in the fuel. The **octane scale** is an arbitrary scale for the octane number on which iso-octane has an octane number of 100, while heptane is zero. For a piston engine, the **octane requirement (octane number requirement)** is the octane rating required to avoid knock. The **research octane number (RON)** is a guide to the anti-knock performance of a fuel under mild driving conditions, while the **motor octane number (MON)** is relevant to severe driving conditions (full throttle with high ambient temperature). Both RON and MON are obtained from standard comparative tests using a CFR engine. *See also* CETANE NUMBER.

octoid *See* TOOTHED GEARING.

OD *See* OUTSIDE DIAMETER.

ODT *See* OPTICAL DOPPLER TOMOGRAPHY.

offset cylinder A reciprocating compressor, engine, or pump in which the centre lines of the cylinders do not pass through the axis of the crankshaft.

offset yield strength The yield strength at some fixed value of the permanent strain (0.1 or 0.2%) used with materials without a sharply-defined yield point. It is given by the intersection of a line offset from, but parallel to, the elastic loading line and the non-linear stress–strain curve.

offshore energy system A system for generating electrical energy located offshore; in the case of wind turbines, to take advantage of higher wind speeds there. An **offshore wave-energy converter** is designed to exploit the more powerful waves that occur in deep (> 40 m) water.

ogive A pointed, curved, symmetrical surface, typically consisting of intersecting secants, and used as the nose shape for rockets, projectiles, bullets, supersonic wing planforms, etc.

OHC engine *See* OVERHEAD-CAMSHAFT ENGINE.

OHV engine *See* OVERHEAD-VALVE ENGINE.

oil A liquid, used as a fuel, coolant, or lubricant, derived from petroleum, coal, biomass, rape seed, etc. Synthetic silicone oils are widely used as lubricants. *See also* CUTTING FLUID.

oil burner A device that atomizes, vaporizes, and burns fuel oil.

oil-cooled A term for a machine that uses circulating oil to remove heat. Examples include gearboxes, differentials, hydraulic clutches, and some motorcycle engines which use engine oil for cooling.

oil cooler A heat exchanger used to cool an engine's lubricating oil.

oil filter A filter employed in an oil-lubrication or hydraulic system to remove particulates resulting from wear or products of combustion.

oil-fired A furnace or boiler that is heated by burning fuel oil.

oil gallery A small passage cast into an engine block or cylinder head through which lubricating and cooling oil is circulated.

oil grooves Grooves cut into flat or cylindrical bearing surfaces and fed by oilways, through which oil is distributed over the bearing.

oil hardening The formation of martensite before tempering by quenching suitable steels in oil at lower cooling rates than given by water quenching, thus reducing the likelihood of component fracture or severe residual stresses.

oil hole A small hole through which oil may be injected into a bearing.

oiliness *See* LUBRICITY.

oilless bearing A bearing in which solid or liquid lubricants are incorporated into the bearing material, e.g. porous bronze.

oil lift 1. The use of high-pressure oil to support a bearing load in the absence of rotation. *See also* HYDROSTATIC BEARING. **2.** The use in small machines of capillary action to draw lubricant from a reservoir to feed bearings.

oil pump A crankshaft-driven auxiliary pump in a compressor, engine, or pump to pressure-feed oil from the sump to the bearings. *See also* GEAR PUMP.

oil ring *See* SCRAPER RING.

oil seal 1. A device such as an O-ring, gasket, etc. to prevent oil ingress or egress between parts of a system, bearing, etc., or to the exterior. *See also* THROWER THREAD. **2.** A seal in which oil itself prevents passage of gases.

oil sump *See* SUMP.

oilway A passage in an engine or other machine through which lubricating oil flows.

Oldham coupling (double-slider coupling) A device for connecting a pair of misaligned shafts, consisting of a slotted flange on the end of each shaft, between which is a disc held in place by corresponding orthogonal tongues. *See also* DOG CLUTCH.

oleometer A type of hydrometer used with oils.

Oleo strut A type of shock absorber often employed on aircraft landing gear, in which air or nitrogen in a cylinder is compressed by a column of oil.

olive (ferrule) A brass or copper ring with beveled edges. It is the part of a compression

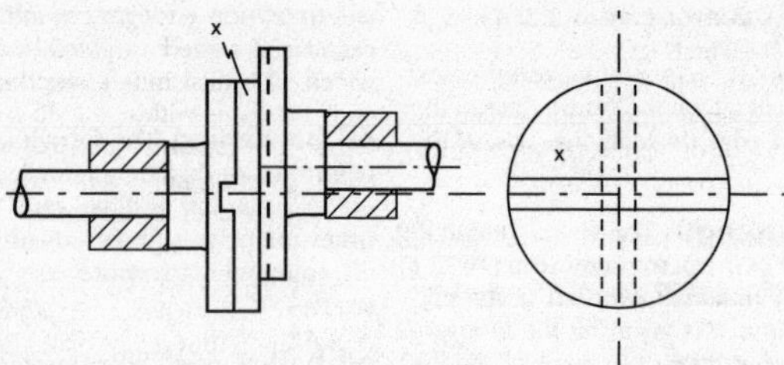

Oldham coupling

fitting that is compressed to seal the space between the pipe, nut and receiving fitting.

once-through boiler A boiler in which water at high pressure flows directly through an economizer, an evaporator, and a superheater tube, usually vertical and which may be formed into a helix. Such boilers are especially efficient at supercritical pressures.

one-dimensional flow An approximation, primarily used for gases, for the analysis of fluid flow in a pipe, duct, or nozzle, assuming the fluid and flow properties are uniform across any cross section but vary in the direction of flow as a consequence of area change, heating, cooling, friction, etc. *See also* FANNO FLOW; FLUID MECHANICS; PLUG FLOW; RAYLEIGH FLOW.

onion diagram A chart used in systems engineering to show the dependence of a process on other processes. The chart comprises a number of concentric circular rings where the processes in a particular ring depend on those processes in rings inside it, but not on those in rings outside.

on-off control (on-off system) *See* BANG-BANG CONTROL.

open-channel flow The flow of a liquid with a free surface through an uncovered channel. The basic force balance is between gravity and friction.

open cycle Any thermodynamic cycle in which there is a continuous flow of fresh working fluid into the system and a corresponding flow of fluid out. In the case of a power plant (**open-cycle plant**), piston engine (**open-cycle engine**), or gas turbine (**open-cycle gas turbine**), the working fluid enters as air and fuel and leaves in the form of exhaust gas.

open-cycle system for OTEC *See* OCEAN THERMAL-ENERGY CONVERSION.

open-die forging Forming of a workpiece by compression between flat top and bottom dies (**open dies**) to make a rough shape of a component. *See* also CLOSED-DIE FORGING.

open feedwater heater *See* REGENERATIVE RANKINE CYCLE.

open kinematic pair A kinematic pair in which contact between elements is not ensured by the constraints, so that a compressive force is required for closure, as in a cam/follower mechanism. *See also* CLOSED KINEMATIC PAIR.

open-loop control system (open-loop system) A control system where the input applied to the plant is independent of the plant output, i.e. the opposite of closed-loop control.

open-loop gain The forward gain of a system with any feedback loops disconnected.

open-loop transfer function (forward transfer function) The transfer function of a system with any feedback loops disconnected.

open pair A kinematic pair in which the constraint between the two bodies is a consequence of an externally applied force or torque, for example, the constraint between a cam and cam follower in an internal combustion engine.

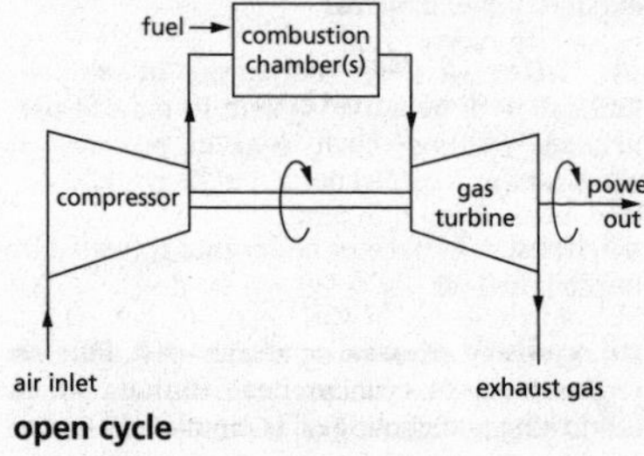

open cycle

open system (control volume) A thermodynamic system for which mass can flow across the boundary. It is a selected region in space which usually encloses a machine or a component, such as a turbine or nozzle, through which mass flows. *See also* CLOSED SYSTEM.

open-web girder *See* LATTICE GIRDER.

operating pressure The pressure at which a system is designed to operate to satisfy either safety or process requirements.

operating stress The stress under which a component, structure, or system is designed to operate.

operating water level *See* NORMAL WATER LEVEL.

opposed-cylinder engine (opposed engine) *See* VEE ENGINE.

opposed-piston engine A piston engine in which two pistons run in each cylinder, alternately moving towards each other and away, with combustion occurring between the two.

optical coherence tomography A technique for obtaining images within translucent or opaque materials.

optical comparator A comparator in which the magnified silhouette of an object is projected onto a screen for comparison with a reference template. *See also* MECHANICAL COMPARATOR; SHADOWGRAPH.

optical Doppler tomography (ODT) An instrument that combines laser Doppler velocimetry with optical coherence tomography to determine the velocity of moving particles in highly scattering fluid suspensions, such as blood.

optical flat A precisely polished flat surface on a transparent block, typically a disc, of glass ceramic or fused silica, used to determine the flatness of another surface.

optical pyrometer *See* DISAPPEARING-FILAMENT PYROMETER.

optical reflectometer An instrument that uses spectral reflectivity to determine such properties as the thickness, roughness, and optical constants of thin films.

optimal control A control method that applies the calculus of variations to maximize a performance measure or minimize a function representing unsatisfactory performance within specified constraints. For example, an optimal controller for an internal-combustion engine may be designed to maximize engine power or torque whilst minimizing fuel consumption and emissions within a specified engine-speed range. The **optimal control element** in the control system applies the optimal control expressed through an **optimizing control function**. Where the optimal control operates using a closed loop this is known as **optimal feedback control**, whereas in **optimal programming** the input to a plant required to give optimal control is determined from the initial states and the optimal control law, and no feedback is used.

optimal smoother A technique for providing the best estimate of a system state at a particular time from measurements of the state before and after that time.

optimistic time In planning using the PERT method, the shortest time in which a process can be undertaken.

optimization A mathematical method which seeks to maximize a performance function or minimize a function representing unsatisfactory performance by updating the set of inputs to the system or process. The term is also applied more generally to the process of improving a design, a process, etc.

optimum design A design in which a particular feature is optimized, such as a structure designed to have minimum weight when satisfying all the specifications.

ORC *See* ORGANIC RANKINE CYCLE.

order The order of a system, in control and dynamics, is equal to the highest-order differential equation required to describe it.

order number *See* HARMONIC ORDER NUMBER.

ordinary gear train A gear train in which the axes of all gears are fixed in position, as opposed to, for example, epicyclic trains.

organic Rankine cycle (ORC) A Rankine-turbine cycle in which the working fluid is an organic liquid, such as pentane or butane, with a lower boiling point than water. Typical applications are in waste-heat recovery, biomass-power generation, solar thermal-power systems, and **organic Rankine cycle plants (binary-cycle power plants)** where, typically (as shown in the diagram), the working fluid in the liquid state is evaporated using hot

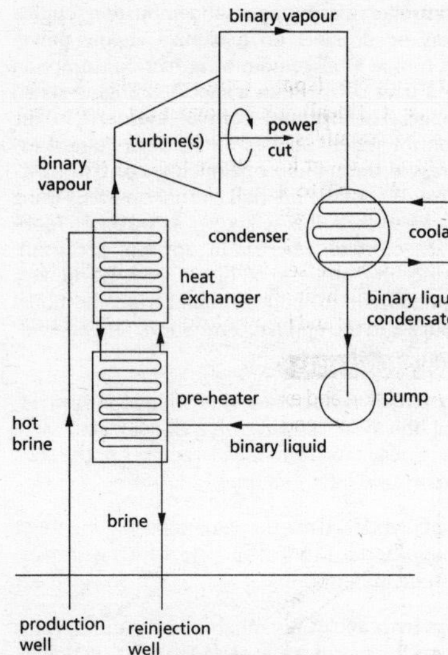

organic Rankine cycle plant

brine from a geothermal reservoir. *See also* DRY-STEAM POWER PLANT; FLASH-STEAM POWER PLANT.

orientation vector In the kinematic analysis of robots, a 4×4 matrix is obtained representing the position and direction of the end effector. The second column of this is the orientation vector $\bar{O}$, which is a unit vector running in a direction orthogonal to the approach and normal vectors, and, where the end effector is a gripper, in the direction of the opening and closing movement of the gripper fingers, as shown in the diagram.

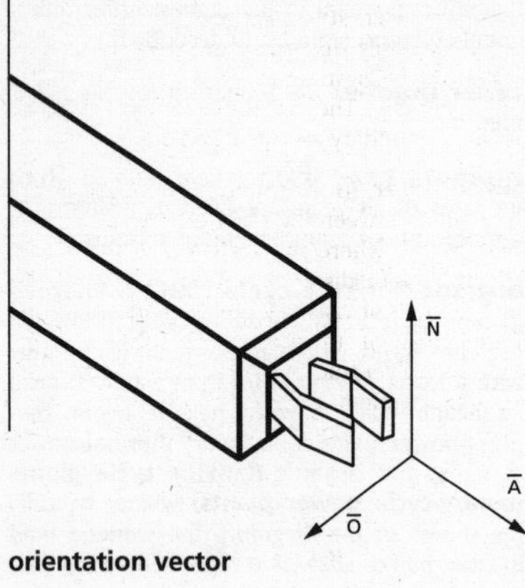

orientation vector

orifice A hole, usually circular, in a plate, tube wall, shell structure, etc., usually to permit fluid flow.

orifice mixer A plate or cylindrical body in which there is one or several holes through which fluids are pumped. The emerging jets are highly turbulent and mix readily with surrounding fluid or, if two or more fluids flow through the same device, each other.

orifice plate A precisely-machined circular hole with beveled edges, machined into a circular disc and used as the basis of a Bernoulli obstruction flow meter **(orifice meter)**.

O-ring (O-ring seal) A toroidal ring of synthetic rubber seated in a groove machined into a flat or cylindrical surface to act as a seal.

Orr–Sommerfeld equation A fourth-order, non-linear, ordinary differential equation for the time variation of an infinitesimal disturbance velocity in a two-dimensional incompressible laminar flow. It is a central element of stability theory. *See also* RAYLEIGH'S EQUATION.

Orsat apparatus A device for analysing the composition by volume of a sample of exhaust gas by successive absorption of CO_2, O_2, and CO. Any remaining gas is assumed to be N_2.

orthogonal At right angles to another line or plane.

orthogonal coordinate system The Cartesian coordinate system of three mutually-perpendicular axes.

orthographic projection *See* PROJECTION.

orthotropic material A material that is anisotropic, but with certain rotational symmetry. For example, a polymer reinforced with longitudinal fibres, or a membrane with different properties in plane and perpendicular to the membrane.

oscillating follower *See* CAM FOLLOWER.

oscillating granulator A machine used to break up solid material by the reciprocating motion of blades or bars above a screen that determines the size of the granulated material. Widely used in the pharmaceutical industry in the production of tablets.

oscillating screen A screen oscillated in a horizontal, or near-horizontal, plane, used to classify small solid objects or to separate large and small objects.

oscillating stresses Stresses that fluctuate in a regular or random manner. They may be all tension, all compression, or mixed tension-compression. *See also* FATIGUE.

oscillating water column (OWC) A wave-energy converter in which a column of water is caused to oscillate up and down a partially submerged tube by wave action. The oscillating water column induces an oscillatory flow of air which passes through a Wells turbine to generate electrical power.

oscillation The time-dependent-variation about a mean value, of a physical quantity, such as pressure, temperature, or displacement. The variation may be periodic (i.e. harmonic) or random. In the case of an object or mechanical system, an oscillation is termed vibration. **Oscillatory motion** is where the oscillating quantity is displacement or velocity, which may be linear or angular.

oscillator A mechanical or electrical device the output of which is a quantity which varies periodically with time, such as the displacement of a component or a voltage.

oscillatory signal Any signal which changes repetitively with time.

oscilloscope An instrument for displaying and measuring a voltage as a function of time.

Oseen theory An *ad hoc* modification to account for deficiencies in Stokes-flow theory far from a body by incorporating a linearized convective term.

osmosis The movement of small molecules through a semipermeable membrane while larger molecules are blocked. A common example is water passing from a solution of low salt concentration to one of high concentration.

osmotic pressure The pressure difference across a semipermeable membrane required to prevent osmosis.

Ossberger turbine *See* BANKI TURBINE.

Ostwald viscometer A type of U-tube capillary viscometer consisting of two bulbs, one higher than the other, linked by a vertical capillary. The time for the liquid level in the upper bulb to fall a calibrated distance is approximately proportional to the kinematic viscosity of the liquid. *See also* UBBLEHODE VISCOMETER.

OTEC *See* OCEAN THERMAL-ENERGY CONVERSION.

Otto cycle (spark-ignition cycle) An air-standard (ideal) thermodynamic cycle that closely resembles the actual operating conditions of a two- or four-stroke spark-ignition piston engine (**Otto engine**), invariably a petrol engine. As shown on a plot of pressure (p) *vs* specific volume (v), it consists of four internally reversible processes: isentropic compression (1–2), isopycnic heat addition (2–3), isentropic expansion (3–4) and isopycnic heat rejection (4–1).

SEE WEB LINKS

- Animation and explanation of the principles of a four-stroke engine

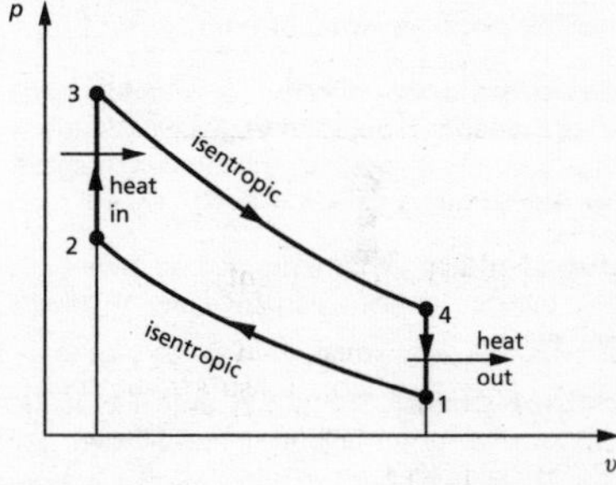

Otto cycle

outer-dead centre *See* TOP-DEAD CENTRE.

outer edge The location of a line representing the locus of points in a shear layer, such as a boundary layer, at which the mean velocity $\bar{u}$ is not appreciably different from the free-stream velocity U_∞. Common choices are $\bar{u} = 0.99\ U_\infty$ and $0.999\ U_\infty$. These values are often used to define a boundary-layer thickness δ.

outer layer In a turbulent boundary layer, the region between the law-of-the-wall and the free stream, where the variation of the mean velocity $\bar{u}$ with distance y from the surface is independent of viscosity but dependent on the fluid density ρ, the wall shear stress τ_W, the boundary-layer thickness δ, and the streamwise pressure gradient dp/dx. It can be shown that the outer-layer velocity profile follows the velocity-defect law:

$$\frac{U_\infty - \bar{u}}{u_\tau} = g\left(\frac{y}{\delta}, \xi\right)$$

where U_∞ is the free-stream velocity, u_τ is the friction velocity and the pressure-gradient parameter:

$$\xi = \frac{\delta}{\tau_W}\frac{dp}{dx}$$

outflow heater A relatively small heater that protrudes into a large storage tank containing a viscous liquid. Only the liquid in the immediate vicinity of the heater is drawn in and raised to the pumping temperature.

outgassing The release of gas dissolved, trapped, frozen, absorbed, or adsorbed in a material. Outgassing is critically important in high-vacuum environments.

outlet port *See* EXHAUST PORT.

out-of-balance A term for rotating parts whose motion is subject to transverse vibrations due to a non-axisymmetric distribution of mass. *See also* DYNAMIC BALANCE; STATIC BALANCE.

out-of-plane A term for forces, moments, etc. that are applied perpendicular to sheets and plates.

output power The power available at the output shaft of an engine or mechanism. *See also* BRAKE POWER.

O

output shaft The driving shaft of an engine from which power is taken.

outside diameter (OD) The diameter of a cylindrical tube or a sphere measured between opposite points on the external surface. *See also* INTERNAL DIAMETER.

outstroke The stroke of a piston engine in which a piston approaches bottom dead centre. *See also* INSTROKE.

outward-flow turbine *See* LJUNGSTRÖM TURBINE.

oval-gear flow meter (gear meter) A positive-displacement rotary flow meter in which oval-shaped gear-toothed rotors rotate within a chamber of specified geometry. As the rotors turn they sweep out an accurately known volume of fluid, the flowrate being calculated from the number of turns in a given time.

oven A thermally-insulated chamber used for heating a substance, for example for drying or heat treatment. *See also* AUTOCLAVE; FURNACE.

overaged A term for precipitation-hardened alloys that have been aged for too long at too high a temperature, so that the fine particles that provide a strengthening mechanism agglomerate and are less effective.

overall heat-transfer coefficient (U-value, *U*) (Unit W/m^2.°C) A coefficient defined, where there is heat transfer at a rate $\dot{q}$ through several thermal resistances separating regions with an overall temperature difference ΔT, by $\dot{q} = UA\Delta T$, where A is a suitable area across which the heat transfer takes place.

overall isentropic efficiency The ratio, for a turbine, of the overall enthalpy decrease to the enthalpy decrease that would occur for an isentropic expansion between the same pressure levels. For a compressor, the ratio of the isentropic enthalpy increase to the actual increase. *See also* STAGE EFFICIENCY.

overall thermal efficiency (overall efficiency) The ratio, for a steam engine, internal-combustion engine (including a gas turbine), or an entire power plant, of the power produced to the product of the latent heat (the enthalpy of combustion) of the fuel supplied, and the fuel mass flow rate.

overdamped response A response in which the output of a system reaches the steady-state value without overshoot, but in a longer time than would have occurred in a critically-damped system.

overdamped system A system which shows an overdamped response.

over-expanded jet A supersonic gas jet exhausting from a nozzle into an environment where the pressure is higher than the pressure in the exit plane of the nozzle. Oblique shocks form at the exit, followed by a series of oblique shock waves and expansion waves confined within the boundaries of a sausage-like jet. *See also* UNDER-EXPANDED JET.

overfire draught *See* DRAUGHT.

overflow The flow of a fluid, usually a liquid, that occurs when an open container is filled beyond its capacity and the fluid passes over the rim. *See also* HYDROCYCLONE.

overgear A gear ratio that results in the drive shaft of an engine rotating faster than the output shaft.

overhaul **1.** To rebuild a used machine, replacing components as necessary, to restore it to its original working condition **2.** When an object is being lowered by means of a nut on a thread, should the thread friction be overcome by the weight of the load so that control is lost, it is said to overhaul.

overhead-camshaft engine (OHC engine) A piston engine in which the camshafts are located above the cylinder head(s) rather than in the engine block.

overhead-valve engine (OHV engine) A piston engine in which the intake and exhaust valves for each cylinder are directly over the pistons. The valves are actuated by pushrods and rocker arms, actuated in turn by a camshaft located in the engine block. *See also* SIDE-VALVE ENGINE.

overlap 1. The amount by which plates extend over one another in a lap joint. **2.** In any sequence of events, where the next event begins before the previous one has ended. For example, in a piston engine, the degrees of crankshaft rotation during which the intake and exhaust valves in any one cylinder are simultaneously open, as at the end of the exhaust stroke and the start of the next intake stroke. *See also* LAG; LEAD; VALVE-TIMING DIAGRAM.

overlap layer *See* LAW-OF-THE-WALL VARIABLES.

override To stop the automatic control of a process, and control the process manually. For example, an emergency-stop switch is an override.

overriding process control A form of distributed control system in which a higher-priority controller can switch off the output of a lower-priority controller and directly control that part of the system that was previously controlled by the lower-priority controller.

overrunning clutch A mechanical device in the engine of a motor vehicle that disengages the starter motor when the engine begins to run. *See also* BENDIX DRIVE.

overshoot A situation in which the output of a system subjected to a change in input exceeds the steady-state value before settling to that value. Hence in the diagram, which shows the response of a second-order system to a unit step input, the output overshoots the steady-state value prior to settling to that value.

overspeed governor A device that limits the speed of an engine, typically by controlling the fuel supply rate.

oversquare engine A piston engine in which the cylinder bore is greater than the stroke.

oversteer The undesirable situation that arises when the set up of the steering gear, suspension, tyres, etc., of a vehicle results in a turn of smaller radius than intended, often accompanied by side slip of the rear wheels. *See also* UNDERSTEER.

overstrain (overstressing, over yielding) Deformation of a component or structure beyond the design level, especially into the plastic range.

overtone A natural frequency of an oscillatory system higher than the fundamental fre-

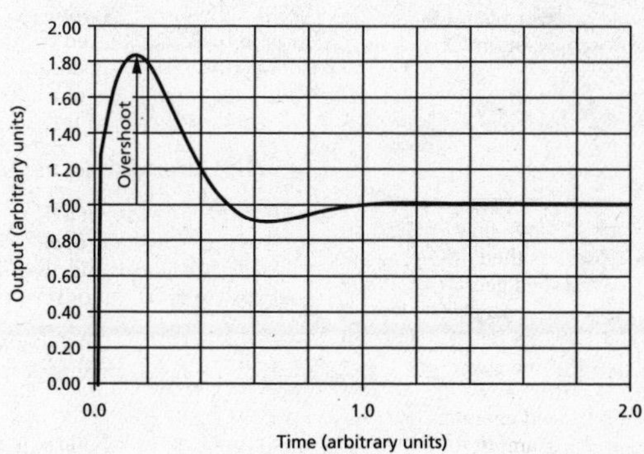

overshoot

o

quency (the first harmonic), but not necessarily an integer multiple of it. *See also* HARMONIC.

OWC *See* OSCILLATING WATER COLUMN.

oxidant (oxidizer) One of the reactants in a chemical reaction in which other reactants are oxidized. *See also* COMBUSTION.

oxides of nitrogen (NOX, NO_X) Gases produced from the nitrogen in the air when hydrocarbon fuels are burned at high temperatures. They include nitric oxide and nitrogen dioxide, and are a principal constituent of photochemical fog.

oxy-acetylene torch (acetylene torch) A torch used for cutting (**oxy-acetylene cutting**) or welding (**oxy-acetylene welding**) ferrous metals in which a hot flame (3200–3500°C) is produced by burning a jet of acetylene mixed with oxygen.

oxygen bomb calorimeter *See* BOMB CALORIMETER.

oxygen lance cutting (oxygen lancing) A process used to cut or pierce thick metal (e.g. steel in excess of 2m thick) by supplying pure oxygen through a consumable steel tube (the lance) to a site pre-heated by other means, such as an oxy-acetylene torch.

oxygen sensor *See* LAMDA SENSOR.

ozone A molecule consisting of three oxygen atoms, chemical formula O_3. It is much less stable than the diatomic allotrope, O_2, normally referred to simply as oxygen.

ozone cracking Cracking occurring in components manufactured from natural rubber and certain synthetic rubbers in which double bonds are attacked by trace amounts of ozone. Cracking occurs in material under tension, gaskets and O-rings being particularly susceptible.

ozone layer A layer mainly in the lower stratosphere, at an altitude between about 15 and 40 km, that contains relatively high concentrations (typically 5 parts per million) of ozone that absorbs about 98% of the ultra-violet radiation received from the sun. A reduction in ozone concentration (**ozone-layer depletion**) is being brought about by pollutants, such as oxides of nitrogen and atomic chlorine and bromine, that result from the decomposition of man-made organohalogens such as chlorofluorocarbons and bromofluorocarbons which used to be widely used as refrigerants, aerosol propellants, and solvents.

packed bed A tube or column in which there is a deep layer of packing material such as small spheres, metal filaments, or a structured matrix.

packed-bed reactor A chemical reactor, often used to catalyse gas reactions, in which the packed bed consists of solid catalyst particles.

packed column A packed bed used to perform separation processes such as distillation and absorption to remove contaminants from a gas stream or volatile components from a liquid stream. *See also* SCRUBBER.

packed gland (gland) A device in which compressed packing is used to seal valve shafts, pump shafts, piston rods, etc. Depending upon the packing, a lubricant such as graphite grease may be necessary.

packing Material tightly packed around a component which rotates or slides within a tube or box to provide a mechanical seal that prevents leakage of water, steam, or other fluids from the high-pressure side to the low-pressure side. Materials include alumina-silica, aluminium mesh, aramid, copper mesh, cotton, flax, graphite fibre, hemp, jute, PTFE, and rubber.

packing ring *See* PISTON RING.

paddle A flat plate at one end of a support arm, the other end being attached to a rotating shaft, used in chemical stirrers or for propulsion through water.

paddle wheel A wheel-like device in which there are paddles at the outer end of the spokes such that each paddle has an axial-radial orientation.

pair Two elements of a kinematic mechanism, each of which constrains the motion of the other, such as a crank and a connecting rod. *See also* CLOSED PAIR; KINEMATIC PAIR; OPEN PAIR; TURNING PAIR; SCREW PAIR.

pallet One of the discs of a chain pump.

Palmgren-Miner rule *See* FATIGUE.

pancake engine *See* VEE ENGINE.

panel 1. A sheet of metal, plastic, or other material held in a frame. **2.** A basic element of a framework in the form of a triangle, square, or other shape. **3.** *See* CONTROL PANEL.

pantograph 1. A mechanical linkage based on a parallelogram used for copying and scaling line drawings. **2.** The rhombus-like device on electric locomotives and trams used to pick up current from overhead wires.

pantometer A device for measuring slopes based on a four-bar chain incorporating a protractor.

PAR An abbreviation for **P**hotosynthetically **A**ctive **R**adiation.

parabolic-bowl concentrator (parabolic collector, parabolic dish) A collector that uses a circular parabolic mirror to focus solar radiation onto a water pipe, steam engine, or Stirling engine.

parabolic flow *See* POISEUILLE FLOW.

parabolic mirror A reflective device the cross section of which has the form of a parabola such that incoming parallel light is focused to a point in the case of a circular mirror, or a line in the case of a two-dimensional trough-shaped mirror. A **parabolic trough collector (parabolic trough concentrator)** uses a mirror of parabolic cross section to focus solar radiation on to a linear absorber. *See also* POINT-FOCUS COLLECTOR.

parabolic shock The approximate form of the detached shock front far from a two-dimensional aerofoil in a supersonic flow.

paraffin *See* KEROSENE.

parallel-axes theorem (Huygens–Steiner theorem) 1. If I_C is the mass moment of inertia of a rigid body of mass m about an axis through its centre of mass, then the mass moment of inertia of the same rigid body about an axis parallel to the first axis is equal to $I_C + md^2$ where d is the distance between the two axes. **2.** If I_C is the second moment of area of a flat surface of area A about an axis through its centroid, then the second moment of area of the same surface about an axis parallel to the first axis is $I_C + Ad^2$ where d is the distance between the two axes. *See also* PERPENDICULAR-AXES THEOREM.

parallel dimensioning A series of dimensions on an engineering drawing, each parallel to the previous one, originating from a common datum.

parallel-flow heat exchanger (parallel-flow recuperator) A heat exchanger in which the two fluid streams enter separately at one end and flow through in the same overall direction.

parallel gripper A robot end effector using two parallel jaws to grip a component.

parallel-motion linkage A system of linked bars, two of which are of equal length and remain parallel in any movement of the connecting link. *See also* PANTOGRAPH.

parallel reliability A system in which there are functionally parallel elements such that the system continues to function even if one of the elements fails.

parameter 1. A quantity that is constant in a particular case under consideration in which independent variables are changed, but which takes a different value in another investigation of the same problem. For example, the variation of cutting force (dependent variable) with depth of cut (independent variable) may be determined at fixed tool rake angle (parameter); then the tool angle is changed and the investigation repeated. **2.** One of the independent variables that determine the values of the dependent variables in a physical problem.

parameter identification A method for determining the governing parameters of a dynamical system from measurements of the observed behaviour of the system.

parametric excitation The process of varying one of the parameters of a dynamical system with time, typically in an oscillatory manner.

parasitic drag *See* DRAG.

Paris equation *See* FATIGUE.

parison *See* BLOW MOULDING.

Parshall flume *See* VENTURI FLUME.

Parsons steam turbine A reaction turbine in which there is pressure drop across both the nozzles and the rotor.

partial admission Admission of working fluid into a turbomachine over less than its full circumference, as in a de Laval turbine. *See also* FULL ADMISSION.

partial bearing *See* BEARING.

partial condensation Condensation to liquid of part of a saturated vapour, brought about by reduction in the temperature or increase in the pressure.

partial fractions The result of splitting a complicated algebraic expression into simpler fraction terms. Used in control-systems design to separate transient and steady-state solutions.

partial pressure *See* DALTON'S LAW.

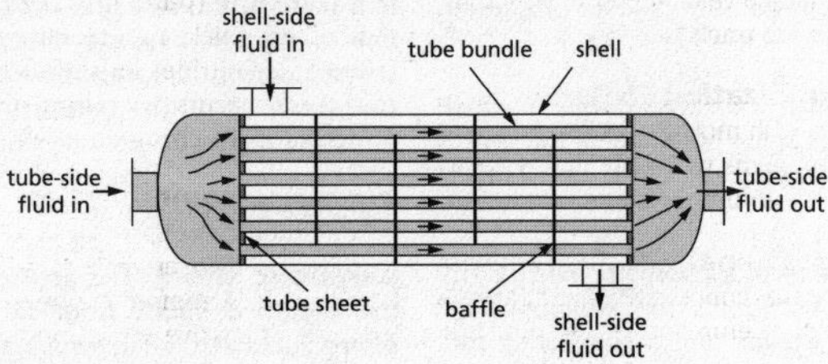

parallel-flow heat exchanger

partial view A view, in an engineering drawing, of part of a component either at enlarged scale or from a different angle.

partial volume *See* AMAGAT'S LAW.

particle (material particle) 1. In theoretical analysis, a volume of solid or fluid material that is infinitesimally small but has finite mass. **2.** A solid piece of material, typically with dimensions of order particle derivative *See* SUBSTANTIAL DERIVATIVE micrometres. *See also* DUST.

particle derivative *See* SUBSTANTIAL DERIVATIVE

particle dynamics (particle mechanics) The study of the motion of one or more particles subjected to a system of forces.

particle energy The sum of a particle's kinetic and potential energies.

particle-image velocimeter (PIV) An instrument for determining the velocities of tracer particles in a flowfield using a digital CCD camera to photograph successive positions of the particles illuminated by light sheets produced by two laser-light pulses a short time apart.

particle-laden gas A gas in which solid or liquid particles are suspended.

particle path The line in a fluid flow followed by an infinitesimally small solid particle. If neutrally buoyant, such a particle can be assumed to move with the local fluid velocity and follow a pathline. *See also* STREAKLINE; STREAMLINE.

particle-size distribution For a mixture of particles covering a range of sizes, the proportion of each size according to number or mass.

particle-sizing anemometer *See* PHASE-DOPPLER ANEMOMETER.

particle-tracking velocimetry (PTV) A form of particle-image velocimetry in which individual particles are tracked.

particulate fluidization The type of fluidization that occurs with most liquid/solid systems and in gas/solid systems when the fluid velocity just exceeds the minimum fluidization velocity.

particulate mass analyser An instrument used to measure the concentration of particulates in an aerosol or emissions from industrial processes.

particulate reinforcement Reinforcement of materials with particles rather than fibres or whiskers. Examples are concrete (stones in mortar), ABS polymers, and cermets.

particulates (particulate matter) Particles of solid or liquid matter suspended in gas or liquid flows and originating either naturally or from domestic and industrial processes, especially combustion.

parting line In mould making the line where two or more parts of a mould meet, evident as a raised line on moulded objects. *See also* FLASH.

parting-off tool A narrow tool, used at the completion of turning on a lathe, for separating the workpiece from the end held in the chuck.

part programming (workpiece programming) The planning and specification of the sequence of steps in the operation of a numerically-controlled machine tool.

parts list A list of parts derived from design drawings that, on assembly, make up a component, a sub-assembly, or an entire product. Sometimes included in the assembly drawing itself.

pascal (Pa) The SI unit of pressure, 1 Pa = 1 N/m^2.

Pascal's law When there is a change in pressure at any point in a confined fluid at rest, there is an equal change at every other point in the fluid volume.

pass 1. One passage through the roll gap in a rolling operation. **2.** The flow of a fluid from one end or side to the other of a heat exchanger.

passband The range of frequencies that can pass through a system without suffering significant attenuation. For example, where a control system is specified as responding to a sinusoidal demand input over a frequency range DC to 10 Hz to within a specified error, the passband is DC to 10 Hz.

passivation (passivity) The natural formation of nm-thick, non-reactive layers of oxide (sometimes nitride) on surfaces to reduce corrosion, as occurs with iron alloys containing more than 12% chromium.

passive accommodation The ability of a robot end effector to move as a result of applied force so as to overcome misalignment with a component. A remote-centre compliance is an example of passive accommodation.

passive capillary effect The movement of fluid, e.g. in a microchannel, due to surface tension.

passive solar heating The absorption of solar energy directly into a building for space heating.

passive solar water heater A system in which differential solar heating of a water storage tank leads to water circulation driven by the thermosyphon effect.

passive vibration suppression *See* VIBRATION ISOLATION.

pass-out turbine *See* EXTRACTION TURBINE.

patent A legal document granted by a government, following expert scrutiny and public disclosure, to protect an invention for a specified period of time.

path 1. In thermodynamics, the series of states a system passes through during a process. **2.** In solid mechanics, the history of loading in terms of stresses or strains.

path computation The determination of the trajectory to be followed by a robot.

pathline In a fluid flow, the actual path followed by a fluid particle. *See also* PARTICLE PATH; STREAKLINE; STREAMLINE.

path of engagement *See* TOOTHED GEARING.

pattern A full-size model of a part to be produced by casting, machining, or any other forming process. *See also* RAPID PROTOTYPING.

pawl A pivoted hook-like component which engages with a ratchet wheel. It is used to prevent reverse rotary motion.

PCM *See* PHASE-CHANGE MATERIAL.

PD control (P+D control) *See* PROPORTIONAL CONTROL.

peaking plants (peaking power plants) Small (typically 10 MW to 100 MW) open-cycle gas turbines which can be run up to full power in less than an hour, or diesel generators of around 1 MW which can be brought on-line in minutes, to meet short-term peak demand for electrical energy.

peak-load power *See* BASE-LOAD POWER.

peak time The time taken for the output of a dynamic system (including a control system) to rise from zero to the first peak value.

peak watts (watt-peak, Wp) The maximum power output that a photovoltaic solar-energy device will produce under ideal conditions measured in an industry-standard light test.

Peaucellier linkage (Peaucellier–Lipkin mechanism) In the linkage shown in the diagram, the pivot point O is fixed, bars OA and OC are of equal length, and bars AB, BC, CD, and DA are all of equal length to form a rhombus. If point B moves along the circumference of the circle centred at E and passing through O, point D moves in a straight line perpendicular to OE extended. Alternatively, if B moves along any straight line not passing through O, point D traces a circle passing through O.

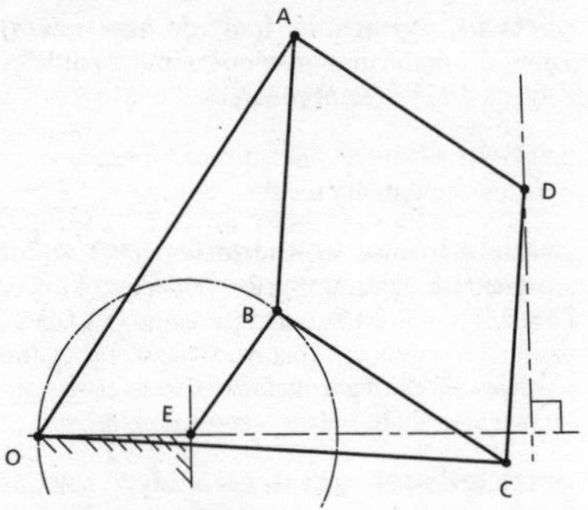

Peaucellier linkage

pebble-bed heater (pebble heater) A heat exchanger in which a bed of refractory pebbles in a column is preheated to the required temperature for a particular process and then heat is transferred to a gas which flows through the bed. They are commonly used to heat the air in a hypersonic wind tunnel.

pebble mill A tumbling mill that grinds or pulverizes materials using balls of steel or ceramic in a cylindrical or conical drum that rotates on a horizontal axis.

Péclet number (*Pe*) A non-dimensional parameter that arises in convective heat and mass transfer problems and represents the ratio of advection to diffusion of either heat or mass: for heat transfer $Pe = VL/\alpha$ and for mass transfer $Pe = VL/D$, where V and L are a characteristic flow velocity and length respectively, α is the thermal diffusivity, and D is the mass-diffusion coefficient. Pe is also equal to $RePr$ or $ReSc$ where Re is the Reynolds number for the flow, Pr is the Prandtl number, and Sc is the Schmidt number.

pedestal robot A robot having a first joint with a vertical axis fixed to a stand so that the workspace is centred around the stand.

peel strength (peel test) The strength of an adhesive bond between two materials as measured by the force required in a given direction to peel apart adhered strips. This force depends on the dimensions and thickness of the adhered strips and their yield stress, together with the fracture toughness of the bond.

peening *See* SHOT PEENING.

peg A projecting pin on which other components can be supported or fixed. *See also* DOWEL.

Pelamis wave-energy converter (Pelamis wave-power device) A semi-submerged device aligned approximately head-on to incoming waves with jointed sections which can move vertically and transversely as a wave passes. Hydraulic rams resist the movement and pump oil through hydraulic motors which drive electrical generators.

Peltier effect The change in temperature produced by passing an electric current through the junction between two dissimilar metals or semiconductors and used e.g. to control the temperature of the plates of a rheometer. *See also* SEEBECK EFFECT.

Pelton turbine An impulse hydraulic turbine in which high-speed water jets, produced from a head of water by flow through nozzles, impinge on buckets around the periphery of a rotating wheel (**Pelton wheel**).

peltric turbo-generator set A very small Pelton turbine driving a generator to produce about 1 kW of power from a head of about 60 m.

PEMFC *See* PROTON EXCHANGE MEMBRANE FUEL CELL.

pencil flame gun A small torch employed instead of a soldering iron.

pendant drop *See* DROP.

Pendulor wave converter A gate hinged at the top fitted at the first antinode of the design wavelength from the back wall of a caisson. A hydraulic pump connected to a generator converts the mechanical energy from the movement of the gate into electrical energy.

pendulum A device consisting of a weight (the **pendulum bob**) at the lower end of a long rod or wire, supported at the top in such a way that the weight can swing freely in a vertical plane. For small-amplitude oscillations, the period T is given approximately by $T = 2\pi\sqrt{(l/g)}$ where l is the length of the pendulum from the point of support to the centre of mass of the pendulum bob, and g is the acceleration due to gravity.

pendulum damper A device for eliminating resonant torsional oscillations in crankshafts. It consists of a small pivoted mass, attached off-centre to the shaft, which becomes a 'centrifugal pendulum' having a frequency proportional to the engine speed, and thus is tuned correctly at all speeds.

pendulum governor An engine-speed governor in which heavy balls at the end of arms spin around a shaft, swing outward and upwards under centrifugal force, thereby forcing progressive closure of the engine throttle. *See also* WATT GOVERNOR.

penetration number An empirical measure of the softness or consistency of a material, such as a grease, obtained by allowing a weighted cone to penetrate into the material for a specified time at a specified temperature.

penetrometer One of a number of instruments used to measure the depth of penetration of different types of indenter into time-dependent materials resulting from the application of a given weight for a given time. *See also* INDENTATION HARDNESS.

p

penstock An enclosed pipe, which may include a gate and a valve, that delivers high-pressure water to the inlet of a hydraulic turbine.

percentage elongation *See* ELONGATION.

percent excess air (percent theoretical air) *See* EXCESS AIR.

percussion drill A machine that drills into a material by simultaneously applying rotation and repetitive impact to a drill bit.

perfect fluid (inviscid fluid) An idealized fluid which has zero viscosity and so cannot support a shear stress. An incompressible perfect fluid is an **ideal fluid**. An **ideal liquid** is incompressible.

perfect gas (thermally- and **calorically-perfect gas** in the nomenclature of aeronautics) A gas for which the specific heat at constant volume C_V is constant and which satisfies the equation of state for an **ideal gas (ideal gas law)**, $p = \rho RT$ where p is the gas pressure, T is its absolute temperature, ρ is its density, and R is the **specific gas constant**. R is different for each gas depending on its molar mass M, such that $\mathcal{R} = \mathcal{R}/M$ where R is the universal gas constant. Also $R = C_P - C_V$ where C_P is the specific heat of the gas at constant pressure. The perfect-gas assumption is a good approximation for real gases of low density (dilute gases). A **thermally-perfect gas (semi-perfect gas)** obeys the ideal-gas equation, but C_P and C_V are functions only of temperature, while for a **calorically-perfect gas**, both are constant in the range of pressure and temperature considered. The departure of the behaviour of a **real gas** from that of a perfect gas **(thermally-imperfect gas)** is accounted for by the compressibility factor being not equal to 1. For a **calorically-imperfect gas**, C_P and C_V, and so also R, are temperature-dependent.

perfect lubrication A complete, unbroken film of lubricant separating two non-contacting surfaces in close proximity moving relative to each other. *See also* STARVED LUBRICATION.

perfectly-plastic yield strength *See* PLASTIC HINGE.

performance curve A graph which illustrates the characteristics of a machine, for example power *vs* rotation speed for an engine, or pressure difference *vs* flow rate for a pump or fan.

p

performance measure The variable that, when optimizing the performance of a device or system, is maximized by adjusting the parameters describing the device or system.

period (Unit s or m) The smallest interval of time or distance over which a periodic oscillation repeats. It is the reciprocal of frequency. For a damped oscillation, it is the interval between alternate zero crossings. *See* also SPATIAL FREQUENCY; TEMPORAL FREQUENCY; WAVELENGTH.

periodic oscillation A steady-state oscillation consisting of repeated identical cycles over either time (**temporally periodic**) or distance (**spatially periodic**). *See also* RANDOM OSCILLATION.

peripheral pump *See* REGENERATIVE PUMP.

peristaltic pump A pump that produces a continuous flow of liquid within a flexible tube using a series of rollers attached to a rotating wheel, each of which rolls over and squeezes the same length of tube. Apart from the tube inner wall, there is no contact between the liquid and parts of the pump and hence minimal possibility of contamination.

permanent gas A gas that has a critical temperature below normal ambient temperature and so cannot be liquefied by pressure alone.

permanent set Permanent plastic deformation of a test specimen, or permanent deflection of a structure, after removal of the applied load. *See also* ELASTIC LIMIT; PROOF STRESS; YIELD STRESS.

permeability (coefficient of permeability, κ) (Unit m^2 or D) A measure of the ability of a porous material to allow fluid flow. It takes into account the pore size, shape, and length. The SI unit of κ is m^2 but the non-SI unit darcy (D) is also used where $1\text{D} = 9.869233 \times 10^{-13}\text{m}^2$ or $1\text{mD} \approx 10^{-15}\text{m}^2$. *See also* DARCY'S LAW.

permeameter An instrument for measuring permeability.

permissible load *See* SAFE WORKING LOAD.

perpendicular-axes theorem 1. If I_x and I_y are the mass moments of inertia of a rigid plane lamina about any pair of orthogonal axes in its plane, then the mass moment of inertia I_z about the z-axis (the polar moment of inertia) is given by $I_z = I_x + I_y$. **2.** If I_x and I_y are the second moments of area of a rigid plane lamina about any pair of orthogonal axes in its plane, then the second moment of area J about the z-axis (the polar second moment of area) is given by $J = I_x + I_y$. *See also* PARALLEL-AXES THEOREM.

perpetual-motion machine A perpetual-motion machine of the first kind produces energy from nothing, thereby violating the first law of thermodynamics. A perpetual-motion machine of the second kind produces mechanical energy by extracting thermal energy from its surroundings, thereby violating the second law of thermodynamics. Because they violate the fundamental laws of thermodynamics, perpetual-motion machines are impossible to construct.

Perspex® (Plexiglas®, Lucite®) The trade names for polymethyl methacrylate, a synthetic polymer of methyl methacrylate, which is an easy-to-machine thermoplastic often used as a lightweight and shatter-resistant substitute for glass. Its glass transition temperature is about 100°C, Young's modulus 2.8 GPa, and yield strength 70 MPa.

perturbation A small change in the magnitude of a quantity, often away from its equilibrium value, imposed to investigate stability or find the solution to an equation.

perturbation factor *See* INTERFERENCE FACTOR.

pet cock (priming valve) A small plug valve for draining condensate from a steam engine or checking the water level in a boiler.

Petroff's law *See* BEARING.

petrol A liquid fuel for spark-ignition engines produced by the fractional distillation of crude oil. Iso-octane, toluene, benzene, etc. are added to increase the octane rating.

petrol engine A spark-ignition internal-combustion engine, using petrol as the fuel, which operates on the Otto four-stroke cycle or a two-stroke cycle.

petroleum (crude oil) A naturally-occurring inflammable liquid consisting of a complex mixture of hydrocarbons and other organic compounds found in rock formations far below the earth's surface.

petrol injection The introduction of petrol into the engine-intake system of a petrol engine using a pump rather than a carburettor. **Petrol pumps** are either driven mechanically from the engine's camshaft or operated electrically.

phase 1. In thermodynamics, a substance can exist as a solid, liquid, or vapour, each of which is termed a phase. At temperatures well above the critical temperature of a fluid, and also at very low pressures, the term gas is used rather than vapour for the phase of a substance. **2.** In solid materials, a phase is any collection of atoms that is homogeneous on a fine scale. *See also* MICROCONSTITUENT; MICROSTRUCTURE.

phase angle (phase, ϕ) If the variation with time t of an oscillatory function $x(t)$ with angular frequency ω and amplitude A is represented by the equation $x(t) = A \sin(\omega t + \phi)$ then ϕ is the phase angle.

phase asymptotic approximation *See* BODE PLOT.

phase change 1. The phenomenon whereby a substance changes from one phase to another, such as the melting of a solid to become liquid, the freezing of a liquid to become solid, the evaporation of a liquid to become vapour, the condensation of a vapour to become liquid, or the sublimation of a solid to become vapour, and vice versa. **2.** In solid-state transformations the term refers to changes in the crystal structure, for example in iron, austenite has a face-centred cubic crystal structure and is the stable phase at temperatures above 912°C but body-centred ferrite is the stable phase below 912°C.

phase-change material (PCM) A substance with a high latent heat of fusion that can store and release large amounts of thermal energy when it changes from solid to liquid and vice versa.

phase-change number *See* JAKOB NUMBER.

phase crossover *See* BODE PLOT.

phase diagram 1. (p–T diagram) In thermodynamics, a diagram of pressure (p) *vs* temperature (T) on which are plotted melting/freezing, vaporization, and sublimation lines separating the solid, liquid, and vapour phases of a pure substance, the three lines meeting at the triple point. *See also* P–V–T DIAGRAM. **2. (equilibrium diagram)** In materials science, a diagram relating the composition of an alloy by mass to its temperature. Changes in ambient pressure are not normally included. It shows the transitions on cooling from an all-liquid state through mixed liquid-solid states to the solid state and shows the phases formed including

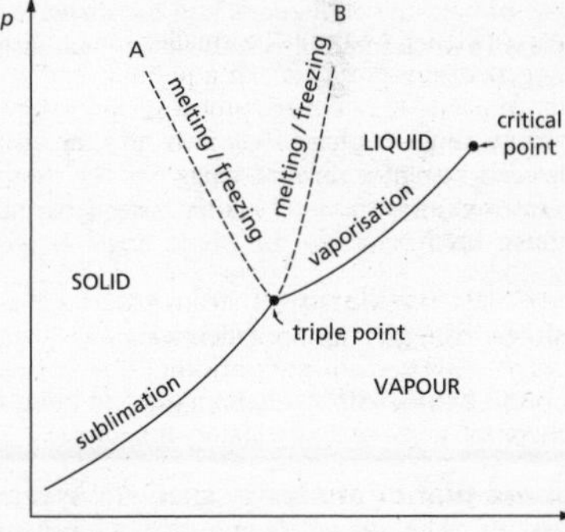

phase diagram

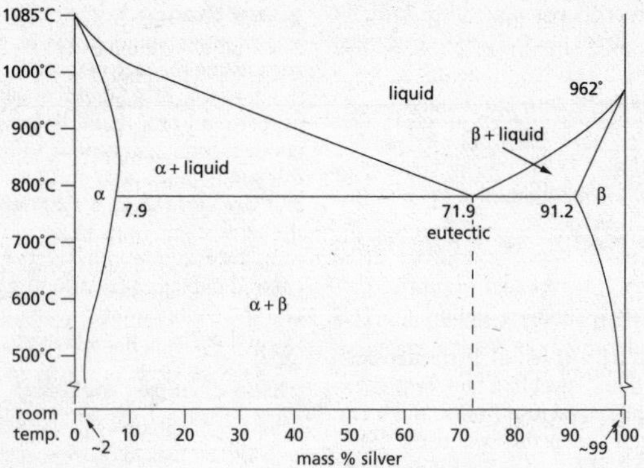

equilibrium phase diagram

any subsequent solid-state transformations as the temperature is slowly lowered. The same phase boundaries usually apply during heating. The example is for alloys of silver and copper. On the left, α is the phase that is almost pure copper: it can contain up to 7.9% silver at 780°C but only about 2% at room temperature. On the right, β is the phase that is almost pure silver: it can contain up to 8.8% copper at 780°C but only about 1% at room temperature. The alloy having the lowest melting/freezing point of 780°C has a composition of 71.9Ag/28.1Cu (the eutectic composition). *See also* EUTECTIC; EUTECTOID; MICROCONSTITUENT; SOLIDUS.

phase-Doppler anemometer (particle-sizing anemometer) An optical instrument used in a multiphase flow to measure the velocities of both the continuous and the dispersed phase (particles, bubbles, or droplets) and also particle sizes.

phase lag The phase angle by which the output of a system is delayed relative to the input measured after transient effects have decayed. **Phase lead** is where the phase angle is advanced.

phase margin *See* BODE DIAGRAM.

phase modulation Slow variation in time of the phase angle of an oscillatory function.

phase-plane analysis A control-system analysis technique where one state of a system is plotted against another state for a range of inputs. For example, in a mechanical system the time derivative of position (i.e. the velocity) would be plotted as a function of position. A **phase portrait** is the graph produced in phase-plane analysis.

phase shift The difference, for a linear system with a sinusoidal input, between the phase angles of the output and the input.

phase speed (wave speed, $v_{ph}^{(x)}$) (Unit m/s) For an obliquely propagating plane-progressive wave with angular frequency ω and wave-number component k_x in the x-direction, the phase speed $v_{ph}^{(x)}$ along the x-axis is given by $v_{ph}^{(x)} = \omega/k_x$. It is the speed traced out along the x-axis by a wavefront. For a plane-progressive wave of wave number k, the **phase velocity (wave velocity)** is $\boldsymbol{n}\omega/k$, where $\boldsymbol{n}$ is the unit vector in the propagation direction. The **group velocity (v_G)** is the velocity with which the shape of a wave propagates through a medium given by $v_G = \partial\omega/\partial k$.

Phillips-head screw A screw having a head with a recess in the form of a cross.

photocell *See* PHOTOVOLTAIC CELL.

photochemical smog A dangerous type of air pollution formed when oxides of nitrogen, aldehydes, peroxyacytyl nitrates, ozone, and volatile organic compounds in the atmosphere react due to exposure to sunlight.

photodetector A sensor that responds to light or some other form of electromagnetic radiation to produce a measurable effect such as a voltage, current, or resistance change.

photoelasticity An experimental method for the determination of the stress distribution in a loaded model constructed from a transparent material that exhibits birefringence (i.e. optical anisotropy). Stress fields can be deduced from fringe patterns viewed through a polariscope. Commonly used with plane models, but three-dimensional stress distributions may be obtained using the frozen-stress method.

photoelectric effect The emission of electrons from a material as a consequence of its absorption of short-wavelength (infra-red to ultra-violet) electromagnetic radiation.

photoelectric liquid-level indicator An indicator of the liquid level in a tank or process vessel using a photodetector to sense the interruption of a light beam as the level rises.

photoelectric pyrometer A pyrometer that uses a photodetector to measure the radiant energy given off by a hot object.

photolithography (microlithography) A microfabrication process in which a photosensitive layer (**photoresist**, **photosensitive resist**) deposited on a substrate of silicon or glass is illuminated, by visible radiation or X-rays, through a mask which is typically a plate of quartz on which chromium has been deposited to form a pattern. *See also* SOFT BAKE.

photometer An instrument used to measure short-wavelength (infra-red to ultra-violet) electromagnetic radiation.

photon An elementary particle that is the basic unit of electromagnetic radiation, including visible radiation. *See also* PLANCK'S CONSTANT.

photosynthesis The chemical process by which plants take in carbon dioxide and water from their surroundings and use solar energy to convert these into carbohydrates, sugars, starches, cellulose, etc., which constitute vegetable matter.

photovoltaic effect The production of a voltage in an inhomogeneous semi-conductor such as silicon, or at a junction between two types of material, by the surface absorption of electromagnetic radiation such as visible light. A **photovoltaic cell (photocell)** is a device which uses the photovoltaic effect to generate electrical energy.

physical realizability *See* REALIZABILITY.

physical similarity The combination of geometric, kinematic, dynamic, and other forms of similarity, applied to a flow or process at a certain scale and a model at a different scale. Two bodies show geometric similarity when all the dimensions of one are the same multiple of the other. Fluid flows with kinematic similarity are those, at different geometric scales, for which the paths followed by fluid particles are geometrically similar, the streamline patterns are the same, and the velocities at corresponding points are in the same ratio. To satisfy the conditions for dynamic similarity, all applicable non-dimensional similarity parameters, such as the Reynolds and Mach numbers, must have the same values for any two physical problems having the same boundary conditions whatever the scale or values of the fluid properties, the only difference being one of geometric scale (Buckingham's similarity rules). The concept also applies to other physical problems and can be extended to include, for example, thermal similarity. *See also* DIMENSIONAL ANALYSIS.

PIANO test (PIANO analysis) A hydrocarbon analysis using gas chromatography to determine the proportions of paraffins (**P**), isoparaffins (**I**), aromatics (**A**), naphthalenes (**N**), and olefins (**O**) present in petrol, aviation fuel, and other hydrocarbon fuels. The individual hydrocarbons determine the octane rating of the fuel. *See also* PONA ANALYSIS.

Piche evaporimeter An atmometer consisting of a vertical graduated tube, part-filled with distilled water, sealed at its upper end and closed at the lower end by a filter paper. The evaporation rate is determined from the rate at which the meniscus level changes.

pick-and-place robot A robot used to transfer components from one location to another.

pickling The removal of grease, scale, salt deposits, etc. from a part by immersion in a bath of, for example, dilute acid. Used during metal processing, such as after the hot rolling of steel before cold rolling; also prior to bonding metals with adhesives, etc.

picnometer *See* DENSITY BOTTLE.

pico (p) An SI unit prefix indicating a submultiplier of 10^{-12}; thus picosecond (ps) is a unit of time equal to one trillionth of a second.

pico-hydroelectricity (pico hydro) Hydroelectric-power generation limited to about 5 kW.

PI control *See* PROPORTIONAL CONTROL.

PID control *See* PROPORTIONAL CONTROL.

piecewise linear system A system in which non-linear behaviour can be approximated by a series of linear functions, each applicable over different intervals.

piercing The punching of holes in metal sheet or plate where, in contrast to blanking, the remaining stock is the required component.

piezoelectric effect The acquisition of electrical charge by certain crystalline materials, such as quartz, barium titanate, lead zirconate, and lead titanate, when compressed, twisted, or otherwise distorted. Used e.g. in **piezoelectric detectors** such as load cells.

piezometer (piezometer tube) An instrument for measuring liquid pressure. The most basic version consists of a vertical, open-top tube, of diameter large enough that capillary effects are negligible, with its lower end immersed in the liquid: the liquid height h within the tube is a measure of the pressure difference Δp compared with the ambient pressure based on the hydrostatic equation $\Delta p = \rho gh$ where ρ is the liquid density and g the acceleration due to gravity.

piezoresistive accelerometer An accelerometer in which a seismic mass exerts a force on a piezoresistive substrate and the resulting change in resistance can be used to estimate the acceleration of the seismic mass.

p

piezoresistive effect In metals, the change in resistance of a material due to changes in its geometry when subjected to stress; in semiconductors, such as germanium and silicon, there is a much larger effect due to changes in the resistivity itself. Some pressure transducers make use of the effect.

piezotropic fluid A fluid for which the density ρ is a single-valued function of pressure p. For example, p/ρ^{γ} = constant which describes the behaviour or of a perfect gas with specific-heat ratio γ undergoing an isentropic process.

pilger mill A machine for producing seamless tubing from pierced metal billets by hot rolling over a mandrel that returns to its starting position after every bite of the non-circular rolls elongates part of the billet. The intermittent process is so named after the pilgrims' chorus in Wagner's opera *Tannhäuser*.

Pilkington process *See* FLOAT-GLASS PROCESS.

pillar drill A self-standing drilling machine.

pilot bit (pilot drill) A small-diameter drill used to drill a **pilot hole** into a material prior to a larger hole being drilled, in order to locate the drill bit and prevent it from wandering.

pilot-controlled valve A valve installed in a flowline, the pressure or flow rate in which is controlled by a pilot valve installed in a by-pass line.

piloted ignition A method of ignition in which the initial energy for combustion to occur is provided by a spark or small flame.

pilot light (pilot flame) A constantly-burning small flame used to ignite a gas burner.

pilot valve (relay valve) A valve controlling pneumatic or hydraulic flow that is operated by pneumatic or hydraulic pressure. Such a valve allows a small pressure or flow of fluid to control a larger pressure or flow, and thus effectively provides amplification.

pin 1. A small-diameter cylinder projecting from a surface, sometimes used as an axle or spindle. **2.** A cylindrical or conical component passing through a shaft and a hub to prevent relative axial movement between them.

pinion *See* TOOTHED GEARING.

pin joint (hinge, hinge joint, revolute joint) A joint in a mechanism or framework, such as a truss, where the connection is a pin about the axis of which both parts can turn without restriction. Forces are transmitted, but not moments.

pinking (pinging) *See* ENGINE KNOCK.

pinned *See* FIXITY.

pint Obsolete (non-SI) British unit of volume, equal to 0.568261500 x 10^{-3} m^3.

pintle 1. A pivot pin, as in a hinge or rudder. *See also* GUDGEON. **2.** The end of the injection spindle which atomizes the fuel in a fuel injector for an internal-combustion engine.

pintle chain A series of open links connected by pins that articulate within the barrels of adjacent links, used to transmit power between sprocket wheels. Typically used in agricultural machinery.

pipe A metal, plastic, or glass tube of circular cross section.

pipe-elbow meter A flow meter in which the pressure difference Δp is measured across a pipe diameter half way round the bend in the plane of the bend. The fluid mass flow rate is approximately equal to to $K R^2 \sqrt{\rho \Delta p}$ where R is the pipe radius, ρ is the fluid density, and K is a non-dimensional constant of proportionality determined by calibration.

pipe fittings Components, usually of plastic, glass, or metal, such as elbows, flanges, couplings, Y-junctions, tees, and reducers, used to connect pipe lengths together or pipes to tanks, pressure vessels, etc. A **pipe tee** has the shape of the letter T with two inline connections and one connection at 90°. A **Y-fitting** connects three pipes in the form of a Y. **Piping** is a system of pipes and pipe fittings used to convey fluids from one location to another. A **pipeline** is a length of pipe incorporating pipe fittings, valves, and other control devices, measuring equipment, etc. used to convey gas, liquids (especially water or oil), vapour, or particulate matter, often over very considerable distances. The **pipe run** is the path or length of a pipeline.

pipe flow The flow of a fluid, slurry, or particulates through a pipe.

pipe scale Deposits of calcium carbonate, magnesium carbonate, barium sulphate, and other minerals on the inner surface of a pipe, which lead to surface roughness and so increased resistance to flow.

pipe thread *See* SCREW.

Pirani vacuum gauge A type of thermal-conductivity vacuum gauge in which an electrically-heated filament is placed within a low-pressure gas. The heat loss from the filament is dependent on the thermal conductivity and the filament temperature.

piston A solid or hollowed cylindrical component that reciprocates within a close-fitting cylinder either due to gas or vapour pressure in an engine or to mechanical force in a pump or compressor. The upper part is the **piston head** and the end face in contact with the working fluid is termed the **piston crown**. The **gudgeon pin (piston pin)** is the short shaft that connects the little-end bearing and the piston. The **piston skirt** is that part below the level of the gudgeon pin. A **slipper piston** is one in which the skirt is cut away on the two sides not used as thrust surfaces to reduce weight and friction. A **slotted piston (split-skirt piston)** has a skirt with an axial slot to allow for thermal expansion. A **piston ring (packing ring, split-ring piston packing)** is a split ring of rectangular cross section fitted into a circumferential groove in the piston which springs outward against the cylinder wall to prevent leakage of hot gases. A **scraper ring** is an auxiliary ring usually fitted on the skirt that scrapes surplus oil from the cylinder wall of a piston engine to prevent it from being burned, so reducing oil consumption and emissions. *See also* CONNECTING ROD.

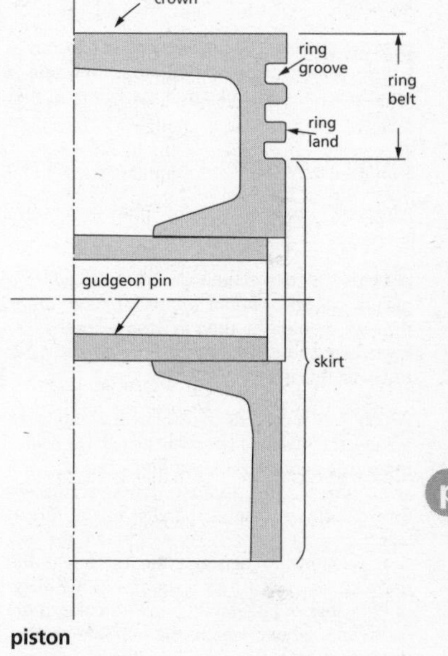

piston

piston displacement *See* SWEPT VOLUME.

piston engine An engine in which power is developed as a consequence of gas or vapour pressure forcing reciprocating movement of a piston within a cylinder, with power being transmitted via a connecting rod to a crankshaft. The most common types are petrol and diesel engines. For multi-cylinder engines, possible cylinder configurations include in-line, vee, flat, and radial. The number of pistons connected to a single crankshaft is essentially unlimited, but the usual range is from one to 24.

piston gauge *See* DEADWEIGHT TESTER.

piston meter *See* ROTARY PISTON FLOW METER.

piston pump A positive-displacement pump in the basic configuration of which a piston reciprocates in a cylinder. On the suction stroke, fluid is drawn into a chamber through the inlet valve, which is then open while the outlet valve is closed. On the delivery stroke the fluid is forced out of the chamber through the outlet valve, which then opens while the inlet valve is closed. Both axial and radial configurations are common.

piston speed The linear speed of a piston in a reciprocating engine which varies from zero at either end of the stroke to a maximum at mid stroke.

piston valve (spool valve) A valve consisting of short pistons (or spools) attached to a rod which slides to open and close ports in the valve body.

piston viscometer A viscometer in which a piston, confined within a circular tube, drops through a fluid, the time to drop a known distance being a measure of the fluid viscosity. *See also* FALLING-BALL VISCOMETER.

pitch 1. In a cascade of turbine or compressor blades, the distance between successive blades measured parallel to the leading edges. **2.** For a screw thread, the distance between adjacent thread forms measured parallel to the thread axis. *See also* SCREW. **3.** For an aircraft, the vertical relationship between the nose and the horizon. The **pitch axis** is an axis in the plane of the wings of an aircraft, perpendicular to the centreline, about which the aircraft rotates. **Pitch motion** is the corresponding up or down movement of the aircraft nose. **Pitch attitude** is the angle between the centreline of an aircraft and the horizontal. The angle is positive when the nose is above its position when the centreline is horizontal. **4.** *See* EFFECTIVE PITCH; GEOMETRIC PITCH.

pitch angle 1. For a bevel gear, the angle between the axis and the pitch-cone generator. **2.** *See* BLADE ANGLE.

pitch circle *See* TOOTHED GEARING.

pitch-circle diameter In a circular component with holes or bolts equally spaced around a circle (the **pitch circle**) centred on the component's axis, the diameter of that circle is the pitch-circle diameter.

pitch cone *See* TOOTHED GEARING.

pitch cylinder *See* TOOTHED GEARING.

pitch diameter *See* SCREW; TOOTHED GEARING.

pitched-blade turbine impeller A flat-blade turbine impeller in which the blades are at an angle to the driveshaft.

pitch element *See* TOOTHED GEARING.

pitching and surging FROG A wave-energy converter which achieves electrical power-generation due to the movement of an inertial mass linked to a generator in a submerged drum.

pitching moment The component of the total aerodynamic moment exerted on an aircraft in the plane containing the lift and drag. It is regarded as positive when it tends to increase the incidence.

pitch point *See* SCREW; TOOTHED GEARING.

pitch ratio *See* PROPELLER.

pitch regulation The continual and optimal adjustment of the blade-pitch angles of a wind turbine.

pitch surface *See* TOOTHED GEARING.

pi theorem (Π theorem) *See* BUCKINGHAM'S Π (PI) THEOREM.

pitot-static tube A device for sensing the difference Δp between the stagnation pressure (p_0) and the static pressure (p) of a fluid flow from which, for an incompressible fluid of density ρ, the flow speed V can be calculated from $V = \sqrt{(2\Delta p/\rho)}$ which is derived using Bernoulli's equation. It consists of a pitot tube, to sense p_0, within a concentric tube that senses p. A **pitot-tube anemometer** consists of a pitot-static tube and a differential pressure transducer.

pitot tube (impact tube, total-head tube) An open-ended tube, placed in a fluid flow with the open end aligned with the flow and facing upstream, which senses the stagnation pressure of the flow. For a supersonic gas flow, the pitot tube senses the stagnation pressure behind the shock wave it causes and to calculate the Mach number requires that the static pressure is also known. *See also* MACHMETER.

p

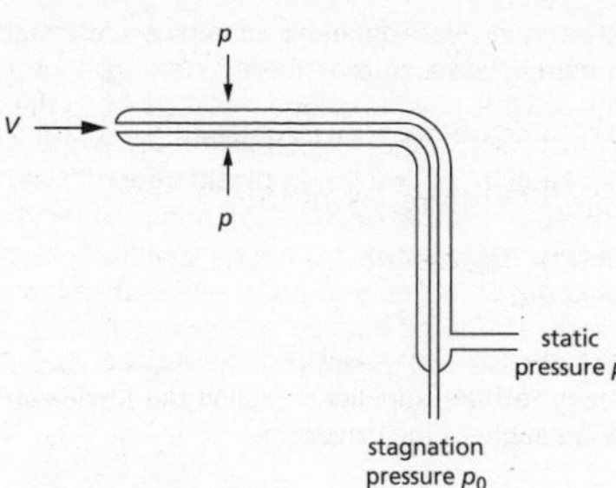

pitot-static tube

pitting A surface fatigue failure which occurs after a number of cycles when two surfaces roll or slide against one another with sufficient force due to the combined influences of the contact stresses, number of cycles, surface finish, hardness, degree of lubrication, and temperature. It is a contributory factor to the wear of gear teeth.

PIV *See* PARTICLE-IMAGE VELOCIMETER.

pivot A short shaft or pin on which a component turns, oscillates, or balances.

pixel An amalgamation of 'picture' and 'element' meaning the smallest controllable element of a digital image.

plain bearing *See* BEARING.

plain milling cutter A cylinder made of tool steel with cutting teeth on its periphery and used to machine a flat surface.

planarization The process of increasing the flatness of the surface of a semi-conductor wafer. The techniques used include oxidation, chemical etching, deposition, resputtering, sacrificial etch back, and mechanical-chemical polishing.

planar linkage A linkage involving movement in a plane.

Planck constant (*h*) Electromagnetic radiation propagates as discrete packets of energy called photons, the energy e of a photon of frequency υ being given by the Planck–Einstein equation $e = h\upsilon$ where $h = 6.62606896 \times 10^{-34}$ J.s (the 2006 CODATA value).

SEE WEB LINKS

- Values of frequently used constants

Planck's function (Planck's law of radiation) The amount of energy $E_{B\lambda}(\lambda,T)$ emitted by a blackbody in a vacuum or gas at absolute temperature T per unit time, per unit surface area, and per unit wavelength about the wavelength λ is given by Planck's function

$$E_{B\lambda}(\lambda, T) = \frac{2\pi c_0^2}{\lambda^5[exp(hc_0/k\lambda T) - 1]}$$

where h is the Planck constant, k is the Boltzmann constant, and c_0 is the speed of light. For a medium of refractive index n, the numerator is replaced by $2\pi hc_0^2/n$. $E_{b\lambda}(\lambda,T)$ is called more precisely the spectral blackbody emissive power and integration of $E_{B\lambda}(\lambda,T)$ over all wavelengths leads to the Stefan–Boltzmann law. *See also* BLACKBODY RADIATION.

Planck's statement *See* SECOND LAW OF THERMODYNAMICS.

plane A flat surface defined by the requirement that any two points in the surface are joined by a straight line that lies entirely within it.

plane lamina A plane sheet which has zero thickness but has mass thanks to a theoretical property called **surface density** σ_S (or ρ_S) defined as the mass per unit area of the sheet (and so units are kg/m^2), i.e. if A represents the sheet area and m is the mass of that area, $\sigma_S = m/A$. If σ_S varies with position, then $\sigma_S = dm/dA$ so that

$$m = \int_A \sigma_S \, dA$$

plane of maximum shear stress *See* MOHR'S STRESS CIRCLE.

plane progressive wave *See* PLANE WAVE.

plane shock A two-dimensional normal or oblique shock wave with zero curvature.

plane strain A strain field in which deformation is restricted to two dimensions. For principal elastic strains $\varepsilon_1 > \varepsilon_2 > \varepsilon_3$, $\varepsilon_2 = 0$, but $\sigma_2 = \nu(\sigma_1 + \sigma_3)$, where σ are principal stresses and ν is Poisson's ratio. The strains under plane strain are the same as those under plane stress if a modified Young's modulus $E/(1-\nu^2)$ and a modified Poisson's ratio $\nu/(1-\nu)$ are used. Plane strain is a useful approximation in elasticity and plasticity (where $\nu = 0.5$) for the loading of bodies that have one dimension (the 2-direction) much greater than the other two.

plane strain fracture toughness *See* FRACTURE MECHANICS.

plane stress A stress field in which the total stress can be reduced to two orthogonal components, the third being zero throughout the stress field. For principal elastic stresses $\sigma_1 > \sigma_2 > \sigma_3$, $\sigma_2 = 0$, but $\varepsilon_2 = -\nu(\varepsilon_1 + \varepsilon_3)$, where ε are principal strains and ν is Poisson's ratio. Plane stress is a useful approximation in elasticity and plasticity (where $\varepsilon_1 + \varepsilon_2 + \varepsilon_3 = 0$), for the loading of bodies that have one dimension (the 2-direction) very small compared with the other two.

planet carrier *See* EPICYCLIC GEARING.

planet gear (planetary gear, planet pinion) *See* EPICYCLIC GEARING.

plane wave A constant-frequency wave for which the wavefronts (surfaces of constant phase) are infinite parallel planes of constant amplitude normal to the direction of propagation. In a **plane progressive wave** the wavefronts propagate in a single direction.

planimeter A mechanical device, consisting of wheels, linkages, and gears, used to determine the area within a plane closed curve by tracing around the curve.

planishing The process of smoothing a metal surface by light hammering or rolling.

p

planoid gears *See* TOOTHED GEARING.

plant In a control system, the system being controlled.

plant decomposition The conceptual division of a complicated plant into sub-processes and thus the grouping of variables with the sub-processes so as to aid control-system design.

plant factor *See* CAPACITY FACTOR.

plasma An inert gas heated, e.g. by an electric arc, to a sufficiently high temperature (10000 to 20000 K is typical) for it to ionize. Among the processes using a plasma to modify the physical and chemical properties of a material's surface is **plasma spraying** in which a consumable in powder form is melted and propelled through a nozzle (a **plasma torch**) onto a substrate to form a coating. The consumables are fed into a plasma jet created using a DC electric arc with argon or nitrogen as the carrier fluid. The process is well suited to materials with high melting points. In **plasma-arc cutting**, a plasma torch is used to melt a metal which is then blown away by the gas jet. **Plasma-arc welding** is a similar application. In **plasma gasification (plasma thermal destruction and recovery)** a plasma torch is used to convert shredded waste material in an oxygen-poor environment into syngas (from the organic constituents) and liquid silicates and metals (the inorganic constituents). *See also* ARC SPRAYING; FLAME SPRAYING; WELDING.

plastic 1. A term used where materials have been loaded beyond the yield point into the **plastic range** of the stress–strain curve so as to be permanently deformed. *See also* PLASTIC DEFORMATION. **2.** *See* PLASTICS.

plastic design In traditional elastic design of structures and components, stresses and strains are maintained within the elastic range, but in the plastic design of plain carbon steel structures the load-carrying capacity (the **limit load** or **ultimate load**) is based on the load at which sufficient **plastic hinges** form so that the structure behaves as a **mechanism** and collapses (**collapse load**). Plastic design leads to more efficient use of material. *See also* SPRINGBACK.

plastic failure Failure of a component or structure by excessive deformation, buckling, or fracture, after loading has caused part, or all, of the body to become plastic.

plastic flow 1. The flow of a yield-stress fluid which results once the maximum shear stress within the fluid exceeds the yield stress. **2.** The permanent change of shape of a body once the applied stresses satisfy the yield criterion.

plastic fracture mechanics *See* FRACTURE MECHANICS.

plastic hinge (hinge) A cross section in a beam, column, or other component where all the bending stresses are above the yield stress and large permanent deformations occur by rotation. The bending moment M_p **(fully plastic moment, limiting moment, plastic moment, plastic moment of resistance, yield moment)** required to form a hinge depends on the geometry of the member and the yield strength. For rectangular beam having width w and depth t in the plane of bending, $M_p = wt^2\sigma_Y/4$ where σ_Y is the **perfectly-plastic yield strength**. A **stationary plastic hinge**

produces increasing rotation at the same location in a structure as the load is increased; in contrast, a **travelling plastic hinge** produces a limited rotation (caused by geometrical constraints) over progressively extending regions of a structure as the load is increased. In a material that work hardens, M_p increases with rotation and stationary hinges spread to adjacent material. *See also* SECTION MODULUS; SHAPE FACTOR.

plasticity theory The mathematical analysis of the plastic flow of materials. *See also* FLOW RULES; FORMING; GEIRINGER'S EQUATIONS; PLASTIC DESIGN; PLASTIC POTENTIAL; YIELD CRITERIA.

plasticizer A liquid in which a polymer is soluble but does not combine with the polymer at the molecular level, resulting in a lowering of the glass transition temperature (i.e. softening) of the polymer. For example, cellulose nitrate plus camphor is used to produce celluloid.

plastic potential A scalar function of stress $g(\sigma_{ij})$, similar to the strain-energy density function, and related to the yield criterion, employed in plasticity theory to relate plastic strain increments $d\varepsilon_{ij}^P$ with the stresses: $d\varepsilon_{ij}^P = (\partial g/\partial \sigma_{ij})d\beta$ where $d\beta$ is a positive constant.

plastic range *See* PLASTIC.

plastics The common name for polymers of high molecular weight with additives to change their characteristics. Thermoplastics, such as nylon, polyethylene, polystyrene, polyvinyl chloride (PVC), and poltytetrafluorethylene (PTFE), soften and melt when heated. Thermosetting plastics, such as vulcanized rubber, Bakelite®, Melamine®, epoxy resins, phenolics, polysters, and polyimides take shape from the liquid state, but char and degrade on heating. *See also* POLYESTER.

SEE WEB LINKS

- Types of plastic material and their properties

plastic state The condition where the stress state over a region of a loaded body or structure satisfies the yield criterion, thereby resulting in some permanent deformation.

plastic strain *See* STRAIN.

plastic zone A circular region at the tip of a crack where local yielding occurs. For conditions of plane stress, the radius of the zone

$$r_p = \frac{1}{2\pi}\left(\frac{K}{\sigma_Y}\right)^2$$

and for plane strain

$$r_p = \frac{1}{6\pi}\left(\frac{K}{\sigma_Y}\right)^2$$

where K is the stress-intensity factor and σ_Y is the yield stress. *See also* FRACTURE MECHANICS.

plate A flat-sheet of material (usually greater than 5 mm thick) produced by rolling or casting.

plate anemometer *See* PRESSURE-PLATE ANEMOMETER.

plate cam A cam with a relatively thin cross section.

plate-coil heat exchanger A heat exchanger made from two embossed metal sheets, which may be flat or curved, held together such that the embossed sunken parts touch and the working fluid flows through the passages surrounding them.

plate-fin heat exchanger A compact cross-flow heat exchanger made of flat metal sheets separated by corrugated sheets to create longitudinal channels with the sheets being sealed along the edges parallel to the corrugations. Separate hot and cold fluid streams flow through alternate channels.

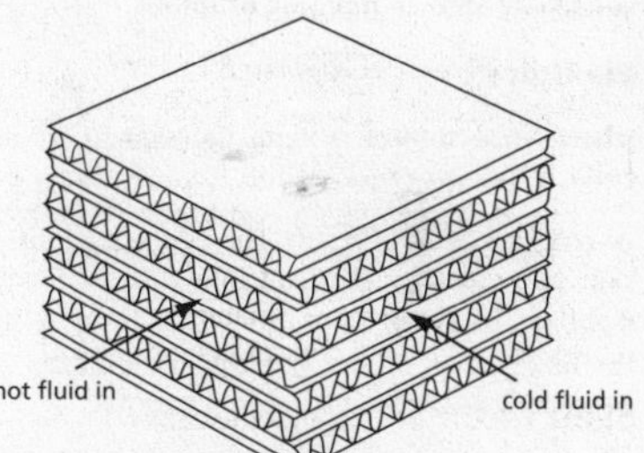

plate-fin heat exchanger

plate heat exchanger A compact heat exchanger consisting of thin corrugated metal plates instead of tubes to separate the hot and cold fluids which alternate between each of the plates. Baffles direct the fluids between the plates.

platen (platten) 1. The component in metal forming that houses the die for forging the required shape. **2.** A hard flat surface used to apply a compressive load to a solid material in a testing machine or a tensile load to a sticky liquid, for example in an extensional viscometer. **3.** The worktable of a machine tool, usually provided with slots for clamping a workpiece.

plate shear test A method used to determine the behaviour of materials such as rubber and honeycomb in shear by bonding the material between two thick parallel steel plates which are subjected to opposing shear forces.

plating *See* ELECTROPLATING.

platinum-resistance thermometer An instrument that determines the temperature by measuring the resistance of a pure platinum wire sensor exposed to that temperature. So-called Pt-100 sensors have a resistance of 100 Ω at 0°C and a sensitivity of 0.385 Ω/°C.

platykurtic *See* KURTOSIS.

play Limited freedom of movement in a bearing or mechanism, which may be intentional (e.g. end float) or the result of wear or poor assembly.

PLC *See* PROGRAMMABLE LOGIC CONTROLLER.

plenum (plenum chamber) A large chamber, for example in a wind tunnel upstream of the working section usually incorporating a honeycomb and screens, in which fluid has very low speed and almost uniform pressure.

p

Plexiglas® *See* PERSPEX®.

plug 1. An object, usually tapered, used to block a hole to prevent fluid from escaping or flowing. **2.** The central moveable part of a valve. **3.** In the flow of a yield-stress fluid through a pipe or duct, the central region often takes the form of a plug with uniform axial velocity. **4.** In thermoforming, a shaping tool that forces a heated plastic sheet into a female mould.

plug-assist forming A thermoforming process in which a plug is used to force a heated plastic sheet into a mould. Once the sheet is in the mould, a vacuum is applied to complete the process.

plug flow Flow of a fluid in a pipe or duct in which the distributions of velocity and fluid properties over the cross section are assumed to be uniform. Viscous effects can be modelled using a friction factor. The term is usually used for liquid flows, while one-dimensional flow is the corresponding term for gases. In the **plug-flow chemical-reactor model** it is assumed that the flow velocity and chemical composition are uniform across every cross section but vary continuously with axial location.

plug gauge A steel gauge used to determine the dimension or dimensions of a hole. It may be straight or tapered, and is typically circular, but in principle can have any cross section. *See also* STEP GAUGE.

plug tap *See* TAP.

plug valve (plug cock) A simple valve which has a plug with a hole in it which is rotated 90° from fully open to closed.

plumb bob (plummet) A weight, usually conical with a sharp point at its lower end, suspended on a string (the **plumb line**) to indicate the vertical direction.

plume A jet-like flow of fluid dominated by buoyancy effects when the initial density of the fluid is different from that of the surroundings. A typical example is the plume of steam emitted from a chimney stack or cooling tower.

plummer block *See* BEARING.

plunge grinding (infeed grinding) A manufacturing process in which a cylindrical grinding wheel is fed directly into the workpiece.

plunger pump A piston pump designed for high-pressure applications.

ply In a composite material, a single layer of a laminate or a single pass of a filament-wound object.

PMC *See* POLYMER MATRIX COMPOSITE.

p_mLAN equation Used to determine indicated horsepower (*ihp*) from an engine indicator diagram, where p_m is the mean effective pressure in psi, L is the piston stroke in inches, A is its cross-sectional area in in^2, and N is the rpm of the engine, then $ihp = p_mLAN/229$.

pneumatic atomizer A device that combines a high-pressure gas stream and a liquid stream at the exit of a nozzle to produce a fine fog of liquid droplets.

pneumatic brake A brake in which the braking mechanism (for example, drum, disc, or band) is air-actuated.

pneumatic controller A device in which the position of a component which controls, for example, fluid flow through a valve, is determined by the force exerted by air pressure acting on a diaphragm.

pneumatic control valve (pneumatic valve) A valve in which the position of the valve stem, which varies the open area, is determined by the net force generated by compressed air acting on a diaphragm operating against the force of a compression spring. Depending upon the arrangement of the spring, in the event of air-supply failure, the valve may open or close.

pneumatic load cell (pneumatic weighing system) A device used to determine the force applied to a sealed, gas-filled chamber by detecting the change in gas pressure it produces.

pneumatics The use of pressurized gas, usually air, to power machines and tools such as a **pneumatic drill**, which is a percussion drill in which compressed air is used to reciprocate a free piston within a cylinder which impacts directly or indirectly on the drill bit. The chisel-shaped bit may rotate by a small amount between each impact. *See also* HYDRAULIC.

SEE WEB LINKS

- Pneumatic components terminology

pneumatic spring *See* AIR SPRING.

pneumatic test A test for leakage or strength of a closed container using high-pressure air. It is less safe than a hydraulic test because compressibility of the gas allows large amounts of energy to be stored.

Pohlhausen polynomial (Kármán-Pohlhausen method) In laminar boundary-layer theory, a fourth-order polynomial which can be used to approximate the velocity profile and used together with the integral-momentum equation to calculate boundary-layer development. The polynomial can be written as $u/U_\infty = 2\eta - 2\eta^3 + \eta^4 + \Lambda\eta(1-\eta)^3/6$ where u is the flow velocity within the boundary layer in the streamwise-direction, $U_\infty(x)$ is the local free-stream velocity, $\eta = y/\delta$, y being the distance perpendicular to the surface, and $\delta(x)$ is the boundary-layer thickness, x being the streamwise location. The **Pohlhausen pressure-gradient parameter** (Λ) is a non-dimensional quantity defined by

$$\Lambda = \frac{\delta^2}{\nu}\frac{dU_\infty}{dx}$$

where ν is the kinematic viscosity of the fluid. Λ is used in boundary-layer theory generally to quantify the streamwise pressure gradient.

Poinsot method A descriptive method of treating the motion of a body about a fixed point O, in which the applied forces do no work and have no moment about O. A good qualitative idea of the motion is given based on the constancy of kinetic energy and angular momentum, both for observers who are fixed in space and those who move with the body. It is useful for gyroscope problems. *See also* CARDAN SUSPENSION.

point-absorber wave-energy device A floating structure with components that move relative to each other due to wave action, the relative motion being used to drive electromechanical or hydraulic energy converters. The device has dimensions which are small relative to the incident wavelengths but draws energy from beyond its physical dimensions.

point-focus collector A solar collector which tracks the sun in two dimensions and uses a parabolic mirror to focus solar radiation onto a small target.

point load (concentrated load, point contact) In 3-dimensional analysis, the idealization that a load applied to a surface acts over zero contact area. *See also* CONTACT MECHANICS; LINE LOAD.

point of contraflexure (point of inflexion) The point in a beam at which the bending moment passes through zero and the beam curvature changes from convex to concave.

points The contact breakers in the ignition system of a spark-ignition engine using a rotary distributor.

point sink In potential-flow theory, a point in three-dimensional space into which fluid flows at a finite rate. If the volume flow rate into the sink is $\dot{Q}$, then the radial inflow velocity v_r at a radius r from the point is given by $v_r = \dot{Q}/4\pi r^2$ As r approaches zero, v_r becomes infinitely large and the point sink is seen to represent a singularity. A **point source** is the counterpart when there is flow out of the point. *See also* LINE SINK.

point-to-point movement Movement of a robot end effector from one position to another without specifying the path between them.

point-to-point programming A robot programming method where during programming the controller records each major change in end-effector position and orientation and then reproduces these when the program is executed.

poise An obsolete unit of dynamic viscosity; 1 poise = 0.1 Pa.s.

Poiseuille flow (parabolic flow, Hagen-Poiseuille flow) Steady, fully-developed, laminar flow of a constant-property fluid, through a circular pipe for which the velocity distribution has the parabolic form

$$u = \left(-\frac{dp}{dx}\right)\frac{R^2}{4\mu}\left(1-\frac{r^2}{R^2}\right)$$

where u is the axial velocity a radial distance r from the pipe axis, R is the pipe radius, μ is the dynamic viscosity of the fluid, and dp/dx is the constant axial pressure gradient.

Poiseuille number (*Po*) A non-dimensional number which characterizes steady, fully-developed, laminar flow of a constant-property fluid through a duct of arbitrary, but constant, cross section and defined by $Po = 2D_H\tau_W/\mu U = c_f Re_H$ where D_H is the hydraulic diameter of the duct, τ_W is the average wall shear stress, μ is the dynamic viscosity of the fluid, U is the bulk velocity of the fluid, c_f is the Fanning friction factor, and Re_H is the Reynolds number based upon U and D_H. For Poiseuille flow, Po =16.

Poisson's ratio (ν, μ) When an isotropic testpiece of a solid is loaded uniaxially within the elastic range, it contracts uniformly as well as extends. The opposite occurs for compression. The ratio of the magnitude of the lateral contraction strain to the longitudinal extension strain is called Poisson's ratio (a positive quantity). For many materials $0.25 < \nu < 0.33$. In general the volume of the body changes, but is constant when $\nu = 0.5$ as occurs in plastic flow of metals and in many rubbers. In non-linear elastic bodies, ν varies with the slope of the strain diagram. In anisotropic materials, ν varies with direction. *See also* AUXETIC MATERIALS.

polar-coordinate robot *See* SPHERICAL-COORDINATE ROBOT.

polarization A property of an electromagnetic wave, such as a light wave, that propagates as a transverse wave, that describes the orientation of its oscillations. If the oscillations are confined to a single plane, the wave is linear polarized. If the oscillations rotate as the wave travels, it is circular polarized.

polar second moment of area (*J*) (Unit m^4) The second moment of area of a lamina about an axis perpendicular to it, given by $J = \int r^2 dA$ where r is distance from the axis and A is area. Polar second moments need not pass through the centre of area, but that is often implied. For an area of uniform surface density ρ_S, the **polar moment of inertia (axial moment of inertia) (I_P)**, with unit kg.m^2, is given by $I_P = \rho_S J$. *See also* PERPENDICULAR-AXES THEOREM.

pole One of the roots of the denominator of the closed-loop transfer function of a control system. Hence the poles are the values of s for which the transfer function becomes infinite. For a transfer function $G(s)$ given by:

$$G(s) = \frac{k(s+z_1)(s+z_2)\ldots..(s+z_m)}{(s+p_1)(s+p_2)\ldots..(s+p_n)}$$

the poles are the constants $p_1, p_2, \ldots p_n$, the **zeros** are $z_1, z_2, \ldots z_m$, and the real constant k is the frequency-independent gain.

pole positioning A control-system design method which explicitly places all of the poles in the left-hand half of the complex plane; that is, it ensures that all of the poles have negative real parts, and thus produces a system that is stable.

pole-zero plot (pole-zero configuration) A plot of the positions of the poles and zeros of the closed-loop transfer function of a control system, with the real parts plotted against the x-axis and the imaginary parts against the y axis of the complex plane. Poles are conventionally plotted as × and zeros as O. If all of the poles lie in the left-hand half (i.e. real part negative) of the plot,

Table: pole zero plot: effect of the pole positions on stability

poles	plot	stability
−0.5, −1.5	+ve imag +2 +1 −ve real +ve real −2 −1 +1 +2 −1 −2 −ve imag	stable
−1.5, +1.5	+ve imag +2 +1 −ve real +ve real −2 −1 +1 +2 −1 −2 −ve imag	unstable
−2 ± j	+ve imag +2 +1 −ve real +ve real −2 −1 +1 +2 −1 −2 −ve imag	stable
± 1.5 j	+ve imag +2 +1 −ve real +ve real −2 −1 +1 +2 −1 −2 −ve imag	marginally stable

the system will be stable. If one or more poles lie in the right-hand half, the system will be unstable. One or more poles on the vertical axis with no poles in the right-hand half indicates marginal stability.

pollutants Man-made waste or naturally-occurring chemicals that contaminate the environment, including the atmosphere, rivers, lakes and oceans, and soil. Air pollution, in both gaseous and particulate (dust) form, originates from internal-combustion engines, external combustion, manufacturing processes, refining, radioactive emissions, volcanic activity, etc. Water pollution originates from manufacturing processes, sewage, chemical deposition from the atmosphere, spillage from oil tankers, etc. The term pollutants sometimes includes light, noise and heat.

pollute (pollution) The introduction of pollutants into the environment.

Polo turbine A tidal-current turbine consisting of a series of vertical-axis water turbines with variable-pitch blades, mounted in a circular rotor which rotates on bearings in a ring system anchored to the sea bed.

polyester A category of polymer, most commonly polyethylene terephthalate (PET), which can be thermoplastic or thermosetting. Applications include the manufacture of plastic bottles, textiles, clothing fabrics, packaging, conveyor belts, tyres, and insulating materials. *See also* SHEET MOULDING COMPOUND.

polygon of forces *See* FUNICULAR POLYGON.

polymer A synthetic or naturally-occurring long-chain molecule in which the individual elements are linked by covalent chemical bonds. Synthetic polymers include synthetic rubber, neoprene, nylon, PVC, polystyrene, polyethylene, polypropylene, polyacrylonitrile, and silicone. *See also* ENGINEERING PLASTICS; PLASTICS.

polymer matrix composite (PMC) A material consisting of a polymer (resin) matrix reinforced with long or short fibres, precipitates, or dispersed phases, resulting in improvement in certain desired properties, but sometimes leading to deterioration of other properties.

polytropic efficiency (small-stage efficiency, $\eta_{\infty c}$) For an axial-flow compressor or turbine, the isentropic efficiency of an infinitesimally small stage. For an expansion,

$$\eta_{\infty e} = \rho \frac{dh}{dp}\bigg|_S$$

and for a compression

$$\eta_{\infty c} = \frac{1}{\rho}\frac{dp}{dh}\bigg|_S$$

where p is the static pressure, ρ is the fluid density, h is the specific enthalpy of the fluid and S denotes entropy.

polytropic process In thermodynamics, an expansion (or compression) process involving an ideal gas in which pv^n = constant where p is the gas pressure, v is the specific volume of the gas and n is a constant called the **index of expansion (or compression).** In the special case where $n = \gamma$, the ratio of specific heats for the gas, the process is isentropic and γ is termed the **isentropic index.**

polyvinyl chloride (PVC) A thermoplastic vinyl polymer having a glass-transition temperature of about 80°C, Young's modulus 2 GPa, and yield strength 25 MPa.

PONA analysis A hydrocarbon analysis using gas chromatography to determine the proportions of paraffins (**P**), olefines (**O**), naphthalenes (**N**), and aromatics (**A**) present in petrol, aviation fuel, and other hydrocarbon fuels. *See* also PIANO ANALYSIS.

pool boiling *See* BOILING.

pop action In some types of steam safety valve, when the steam pressure exceeds the set pressure of the valve it starts to lift, exposing a greater surface area to the steam. This leads to a sudden increase in the force on the valve, which then pops to the full-open position.

pop-off valve A safety relief valve fitted to an LPG tank designed to prevent rupture in the event of excessive pressure build-up.

Popov's stability criterion A frequency-domain method for determining stability in non-linear systems.

poppet valve (mushroom valve) A circular disc connected to a valve stem which reciprocates axially. The side of the disc attached to the stem has a contoured surface, shaped like the top of an onion, which seals against a seat of similar shape. Poppet valves are commonly used to open and close the inlet and exhaust ports of a piston engine. Mechanical force is applied by a cam to the end of the valve stem to open the valves which close due to spring or direct (desmodromic) force.

pop rivet A hollow rivet that enables a connexion to be made from one side only of an assembly. *See also* BLIND RIVET.

pore diameter The actual, average, or effective diameter of the pores in a membrane, screen, filter, or other porous device. *See also* POROUS MEDIUM.

porosimeter An instrument in which gas pressure is applied to a specimen submerged in a liquid until the first bubble of gas passes through the material. The pore diameter is then calculated from the surface tension of the liquid, the contact angle of the liquid, and the applied pressure difference.

porosity (ϕ) A measure of the void space in a porous medium defined by $\phi = \mathcal{V}_V/\mathcal{V}_T$ where $\mathcal{V}_V$ is the total volume of the void space and $\mathcal{V}_T$ the total or bulk volume of the material including both the solid and void components.

porous bearing *See* BEARING.

porous-media flow *See* DARCY'S LAW.

porous medium (porous material) A solid, often called the frame or matrix, permeated by an interconnected network of pores or voids which are filled with a gas or liquid. Rock, soil, and bone are naturally-occurring porous materials. Cement, open-cell foams, synthetic sponges, and ceramics are man-made porous materials.

port An opening in an engine, pump, hydraulic cylinder, etc. through which fluid enters (an inlet port) or leaves (an outlet port), usually controlled by a valve.

portal frame A frame consisting of two uprights connected by a cross member at the top.

positional-error constant (position-error constant) (K_P) The value of the open-loop transfer function of a system as $s \to 0$. This is finite for a type 0 system and infinite for higher system types. Hence for a unity-feedback type 0 system responding to a step input, the steady state error ε_{ss} is $\varepsilon_{SS} = 1/(1 + K_P)$ and zero for higher system types.

positional servomechanism A closed-loop control system designed to control the position of a mechanical system.

position contouring system A CNC machine-tool control system that provides smooth movement along a tightly specified path.

position control A control system designed to control the position of a mechanical system.

position feedback (proportional feedback) A term used to describe the normal situation in which the feedback signal is the output of the controlled plant. Although the term originally referred to feeding back the position of a mechanical system, it is now applied more generally to the output of any plant.

positioning action In robotics, the movement of a robot arm and wrist so as to position the end effector.

position sensor (position transducer) Any device, such as an LVDT or RVDT, used to measure position.

position vector In robot forward kinematics, a 4×4 matrix is obtained representing the position and orientation of the end effector. The fourth column of this is the position vector that represents the *x*, *y*, and *z* coordinates of the end effector.

positive clutch A clutch that consists of two co-axial mating elements, with teeth or square jaws, which lock together when engaged, preventing slip between the two. *See also* DOG CLUTCH.

positive-displacement compressor Any type of compressor, including piston and rotary-screw types, that delivers a fixed volume of gas at high pressure per unit time. A **positive-displacement pump** delivers a fixed volume of fluid, usually a liquid, per unit time.

positive-displacement flow meter An instrument of high accuracy that determines volumetric flow rate by dividing the flowing fluid into successive fixed volumes, and measuring their times of passage through the meter.

positive-displacement machine A machine that incorporates pistons, valves, etc. to ensure positive admission and delivery of a working fluid and prevent undesired reversal of flow. This class of machine incorporates all reciprocating compressors and expanders and some types of rotary compressor such as the Roots blower, gear pump, lobe pump, and vane pump. *See also* NON-POSITIVE-DISPLACEMENT MACHINE.

positive draught *See* DRAUGHT.

positive feedback Feedback in which the plant output is added to the reference input, rather than subtracted from it as in negative feedback. It is likely to lead to instability.

postforming The bending of flat thermoplastic polymer sheets into curves and surfaces after local heating.

post-yield fracture *See* FRACTURE MECHANICS.

potable water (drinking water) Water of sufficiently high quality that it can be consumed by humans without risk.

potential energy The energy an object or system possesses as a result of its elevation in a gravitational field. If g is the acceleration due to gravity, an object or system of mass m has potential energy mgz where z is the elevation of its centre of gravity relative to an arbitrarily selected reference plane. In fluid mechanics and thermodynamics, the potential energy per unit mass is gz and the potential energy per unit volume is ρgz where ρ is the third density.

potential flow An idealized flow which is everywhere irrotational so that $\nabla \times \boldsymbol{V} = 0$ where ∇ is the gradient operator,

$$\nabla = \boldsymbol{i}\frac{\partial}{\partial x} + \boldsymbol{j}\frac{\partial}{\partial y} + \boldsymbol{k}\frac{\partial}{\partial z},$$

and $\boldsymbol{V}$ is the vector velocity, $\boldsymbol{V}=\boldsymbol{i}u + \boldsymbol{j}v + \boldsymbol{k}w$. Such flows are also inviscid. A velocity potential ϕ exists such that $\boldsymbol{V} = \nabla\phi$, which, when combined with the continuity equation, leads to Laplace's equation for ϕ, i.e. $\nabla^2\phi = 0$. A **potential-flow analyser** solves potential-flow problems using either an electrolytic tank or conducting paper, based upon the fact that the phenomena involved in all cases are governed by Laplace's equation.

potentials In thermodynamics, specific enthalpy, the Helmholtz function, and the Gibbs function. *See also* PLASTIC POTENTIAL.

potential temperature 1. *(θ)* In meteorology, the temperature an unsaturated parcel of dry air would have if brought isentropically from a state at absolute temperature T and pressure p to a standard pressure p_0, typically 1 bar. It can be shown that

$$\theta = T\left(\frac{p_0}{p}\right)^{\frac{R}{C_P}}$$

where R is the specific gas constant for air and C_P is its specific heat at constant pressure. **2.** In oceanography, the temperature a water sample would have if raised adiabatically to the surface.

potential vortex *See* RANKINE VORTEX.

potting The process of embedding an electronic assembly in an epoxy resin or thermosetting compound for resistance to impact and vibration as well as insulation from moisture and corrosive chemicals.

pound (lb, lb$_m$) A non-SI, British (i.e. imperial) unit of mass legally defined as 453.592338 x 10^{-3} kg. *See also* POUND FORCE.

poundal An obsolete, non-SI, British (i.e. imperial) unit equal to the force required to accelerate a one pound mass (1 lb$_m$) by 1 ft/s^2, i.e. 1 poundal = 1 lb$_m$ft/s^2.

pound force (pound, lb$_f$) An obsolete, non-SI British (i.e. imperial) unit equal to the force required to accelerate one pound mass (1 lb$_m$) by an acceleration equal to the acceleration due to gravity, i.e. 1 lb$_f$ = 32.1740 lb$_m$ft/s^2 = 4.4482216152605 N.

pound per square inch (psi) An obsolete British unit of pressure corresponding to one pound force applied uniformly to an area of one square inch, i.e. 1 psi = 1 lb$_f$/in^2. Used for both absolute, differential and gauge pressure.

pour test (pour-point test) Cooling of a liquid petroleum product by discrete temperature steps to determine its **pour point**, the lowest temperature at which it can be handled without excessive quantities of wax crystals forming out of solution, and so an index of its utility for certain applications.

powder blasting A micromachining erosion process in which fine particles (typically 10–30 μm) of alumina and other materials are blasted at high speed (typically 100–200 m/s) against a substrate to remove material.

powder clutch (magnetic powder clutch) A proportional-torque device containing a magnetic powder, such as a mixture of powdered iron and graphite or iron, cobalt and nickel. An electric current is passed through a coil to produce a magnetic field which changes the property of the powder from free flowing, at zero excitation, to solid at full excitation.

power (*P*) The rate at which work is done on or by a system or device. It is also the rate at which energy is converted from one form to another or transferred from one place to another. The basic SI unit is the watt (W), but in engineering applications, kilowatt (kW), megawatt (MW), and gigawatt (GW) are used, depending upon the scale of the application.

power-assisted steering *See* POWER STEERING.

power brake A braking system for motor vehicles, in which the force applied to the braking system by the driver is amplified by mechanical, pneumatic, or hydraulic power.

power coefficient *See* WIND ENERGY.

power control valve A control valve operated by an electric motor rather than hydraulic or pneumatic pressure.

power cycle A thermodynamic cycle in which heat is received by the working fluid at a high temperature and rejected at a low temperature, while a net amount of work is done by the fluid. *See also* REVERSE CYCLE.

power cylinder A linear actuator consisting of a piston in a cylinder driven by an electric motor or by pneumatic pressure or hydraulic fluid at high pressure.

power-law approximation 1. In turbulent pipe or zero-pressure-gradient boundary-layer flow, the near-wall mean-velocity distribution can be represented approximately by $u = ay^n$ where u is the streamwise velocity a distance y from the pipe wall or surface, a is a dimensional constant, and n is the power-law exponent, often taken as $\frac{1}{7}$. **2.** An approximation for the true stress–true strain σ *vs* ε curve in large-deformation plastic flow, given by $\sigma = \sigma_0\varepsilon^n$, in which $0 < n < 0.5$ from experiments. *See also* RAMBERG-OSGOOD RELATION.

power plant (power station, thermal power plant) 1. An installation, in which the primary energy source is coal, natural gas, solar, waves, wind, or nuclear, that generates electrical energy using gas turbines, wind turbines, diesel engines, hydraulic turbines, wave-energy devices, and steam turbines. Power plants are frequently referred to in terms of the primary fuel source, thus coal plant, nuclear plant, natural-gas plant, etc. The overall **power-plant efficiency ($\eta_{overall}$)** is the ratio of the net electrical power output $P_{net,overall}$ to the rate of the fuel-energy input, given by $\eta_{overall} = P_{net,overall}/\dot{m}_F HHV$ where $\dot{m}_F$ is the mass flow rate of fuel and *HHV* is the higher heating value of the fuel. **2.** An internal-combustion engine for motor-vehicle or aircraft applications.

power pool A centrally-controlled mix of generating plant, owned by different companies, that supplies electricity to consumers via the National Grid in the UK.

power rating The output power of an internal-combustion engine, electric motor, jet engine, etc. Torque and engine rotation speed are also usually included.

power spectral density (frequency spectrum, power spectrum) For a steady-state signal, the power contained within a signal per unit frequency. It can be determined analytically from Fourier analysis of the signal, or experimentally using a spectrum analyser.

power steering (power-assisted steering) A system that directs a fraction of the engine power to augment the torque exerted by the driver on the steering wheel of a motor vehicle, such as a car, lorry, or bus, in order to turn the vehicle. Most common are hydraulic systems, but electric and electro-hydraulic systems are also employed.

power stroke (firing stroke) That part of the cycle of a piston engine during which a piston is caused to move and deliver power by high-pressure combustion gases or steam.

power tower A part of a solar electricity-generation system in which a field of heliostats reflect the sun's rays on to a boiler or collector, which operates at high temperatures (typically 500–700°C), on top of a central tower.

power train The engine, clutch, gearbox, drive shaft, differential, and final drive which transmit power to the wheels of a motor vehicle.

power turbine *See* FREE POWER TURBINE.

practical entropy (virtual entropy) The value of absolute entropy, less that due to nuclear spin and frozen rotational energy.

Prandtl-Meyer expansion (expansion fan, Prandtl-Meyer fan) In compressible gas flow, an isentropic, centred **expansion wave** which progressively turns a supersonic flow around a sharp convex corner through an angle θ. The fan consists of an infinite number of straight Mach waves centred on the corner, across each of which there is an infinitesimal decrease in the pressure and the Mach angle μ and a corresponding increase in the Mach number M. *See also* OBLIQUE SHOCK.

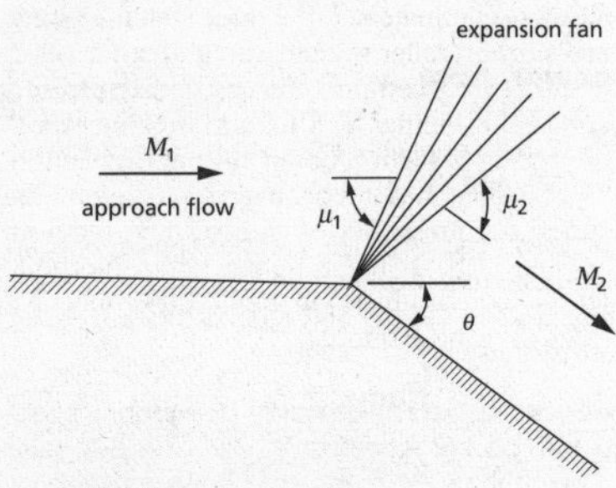

Prandtl-Meyer expansion

Prandtl-Meyer function (ν) The angle through which an initially sonic gas flow (i.e. $M = 1$) must turn through a Prandtl–Meyer expansion to reach a final Mach number M. It can be shown that

$$\nu(M) = \sqrt{\frac{\gamma+1}{\gamma-1}} \tan^{-1}\sqrt{\frac{\gamma-1}{\gamma+1}(M^2-1)} - \tan^{-1}\sqrt{M^2-1}$$

where γ is the ratio of specific heats for the gas. For a flow with initial Mach number M_1 and final Mach number M_2, the turning angle θ is given by $\theta = \nu(M_2) - \nu(M_1)$. The maximum possible turning angle (at which the Mach number reaches infinity) ν_{MAX} from $M = 1$ is given by

$$\nu_{MAX} = \frac{\pi}{2}\left(\sqrt{\frac{\gamma+1}{\gamma-1}} - 1\right)$$

Prandtl number (*Pr*) In convective heat transfer of a viscous fluid, a non-dimensional number that characterizes the relative influences of the fluid's specific heat capacity C_P, dynamic viscosity μ, and thermal conductivity k. It is defined by $Pr = C_P\mu/k$. It can be shown that $Pr = \nu/\alpha$ where $\nu \equiv \mu/\rho$ is the kinematic viscosity, ρ being the fluid density, and $\alpha \equiv k/\rho C_P$ is the thermal diffusivity.

Prandtl relation For an oblique shock wave, the velocity components w_1 and w_2 normal to the shock are related by

$$w_1 w_2 = c_*^2 - \frac{\gamma-1}{\gamma+1}v^2$$

where the subscripts 1 and 2 refer to conditions upstream and downstream of the shock respectively, γ is the ratio of specific heats, v is the flow velocity parallel to the shock, and c_* is the sound speed corresponding to sonic conditions, which is known in terms of the upstream conditions according to

$$c_*^2 = \frac{2}{\gamma+1}c_1^2 + \frac{\gamma-1}{\gamma+1}u_1^2$$

where c_1 is the sound speed and u_1 is the flow speed.

Prandtl–Reuss equations In elasto-plasticity theory, the relationships between total (elastic plus plastic) strain increments and the instantaneous stresses. There are three equations of the type

$$d\varepsilon_x = \tfrac{2}{3}d\lambda[\sigma_x - \tfrac{1}{2}(\sigma_y+\sigma_z)] + \frac{1}{E}[d\sigma_x - \nu(d\sigma_y + d\sigma_z)]$$

and three of the type $d\gamma_{yz} = \tau_{yz}d\lambda + d\tau_{yz}/2G$ where ν is Poisson's ratio, E is Young's modulus and G is the shear modulus. $d\lambda$ is an instantaneous positive constant of proportionality that may vary as the strain changes. *See also* LEVY–MISES EQUATIONS.

pre-admission In a piston engine, the admission of working fluid into a cylinder near the end of a stroke to allow full pressure at the beginning of the return stroke.

precessing vortex core (PVC) In intensely swirling flows which undergo vortex breakdown, the vortex core develops a spiral structure which precesses around the central axis of the flow at a rotation rate different from that of the surrounding fluid. This motion can excite acoustic modes of a flow system such as in a combustion chamber where swirl is introduced to stabilize the flame zone.

precession The rotation of the spin axis of a spinning object when a torque is applied to tilt the spin axis away from the position at which the object can spin stably. The spin axis generates a cone. A spinning top, for example, precesses with angular velocity $\omega_P = mgL/I\omega$ where m is the mass of the top,

g is the acceleration due to gravity, L is the distance from the centre of gravity of the top to the point of contact with a surface, I is the mass moment of inertia of the top about its axis of symmetry, and ω_P is the angular velocity of the top about its spin axis. In this case, the torque is exerted by its own weight. As the top slows down its precession rate increases, but the motion becomes unstable, the top begins to nutate and eventually falls over. *See also* GYROSCOPE.

prechamber *See* PRECOMBUSTION CHAMBER.

precipitation hardening (fine particle strengthening) Increasing the strength of a metallic element or alloy by the formation of very small second-phase particles (about two or three orders of magnitude greater than atomic dimensions) that interfere with dislocation motion. Well-known illustrations of the effect are found at the aluminium-rich end of the aluminium-copper system (**duralumin**®); and the nickel-rich ends of both nickel–aluminium and nickel–titanium stainless steels.

precision 1. The ability of a machine tool or manufacturing process to produce a component to close tolerances. **2.** *See* ERROR.

precision grinding A manufacturing process in which a workpiece is machined by grinding to achieve specified dimensions and low tolerances.

precombustion chamber (prechamber) A small chamber attached to the cylinder of a piston engine into which fuel and air are injected and ignited prior to injection into the cylinder itself, where the bulk of combustion takes place. Primarily used on indirect-injection diesel engines and stratified-charge engines.

precooler An evaporative cooling module used to cool the air entering a heat exchanger or refrigeration unit to improve efficiency.

preferred fit A system based on the basic-hole system for specifying the clearance, transition, or interference fit of two components, one of which must fit within the other. *See also* LIMITS AND FITS.

preform *See* NEAR NET SHAPE.

preheater A heat exchanger used to heat a substance before another process, such as the combustion air entering a boiler to recover waste heat from the flue gas, thereby increasing thermal efficiency. *See also* ECONOMIZER.

pre-ignition The undesirable ignition of the air/fuel mixture in the cylinder of a spark-ignition engine before the spark plug fires, typically due to hot spots in the combustion chamber.

preloading 1. For a bearing, the removal of internal clearance by applying a thrust load. The aim is to eliminate play, increase system rigidity, and prevent roller skidding under high acceleration. **2.** For a compression or extension spring, applying an initial load to compress or extend the spring a slight amount during installation. The preload has to be overcome before the spring compresses or extends further under an applied load. **3.** For a bolted joint, the initial tensile load produced in the bolt by applying a torque during assembly.

pre-mixed flame A flame where the fuel and oxidant are mixed together prior to ignition and combustion. *See also* DIFFUSION FLAME.

pre-mix gas burner A combustion chamber in which a gaseous fuel and oxidant are mixed together to produce a pre-mixed flame.

prepreg Pre-impregnated composite fibres in the form of unidirectional sheets or woven fabrics, containing the matrix material used to bond the fibres and layers together and to other components after curing.

preprogrammed robot A robot that is not equipped or programmed to show accommodation and thus must follow a programmed sequence of movements and operations.

pre-selector gearbox A gearbox in which the selection of a particular gear ratio takes place before it is required.

press A machine which applies a compressive force to a workpiece to cut, shape or compress it, to insert one part into another, or to cause a material to flow through an orifice or die.

press bonding A process of manufacturing laminated components and structures where adhesion takes place under pressure.

press fit An interference or force fit. *See also* LIMITS AND FITS.

pressing (stamping) A sheet-metal component made using a press.

pressure (*p*) (Unit Pa) In thermodynamics and fluid mechanics, the compressive force exerted by the fluid per unit area. The pressure exerted by a fluid on a surface acts normal to the surface. *See also* ABSOLUTE PRESSURE; GAUGE PRESSURE; HYDROSTATIC PRESSURE; MECHANICAL PRESSURE; PASCAL'S LAW; VACUUM PRESSURE.

pressure altimeter (barometric altimeter) An aneroid barometer that converts atmospheric pressure into indicated altitude using standard atmospheric pressure-height relations.

pressure angle *See* TOOTHED GEARING.

pressure bar *See* HOPKINSON BAR.

pressure chamber A chamber in which components or devices can be subjected to high or low fluid (liquid or gas) pressure. *See also* HYPERBARIC CHAMBER; HYPOBARIC CHAMBER.

pressure coefficient (C_P) In fluid mechanics, a non-dimensional measure of the difference between the static pressure p (or stagnation pressure p_0) at a point in a flow and a reference pressure p_{REF}. Defined as

$$C_P = \frac{p(\text{or } p_0) - p_{REF}}{\frac{1}{2}\rho V^2}$$

where ρ is the fluid density and V is the flow speed at the location of p (or p_0) or a reference point. *See also* CAVITATION NUMBER; EULER NUMBER.

pressure-compounded steam turbine (Rateau turbine) An impulse steam turbine in which the steam pressure is progressively reduced by passing through a series of nozzle/rotor stages. The nozzles are carried in diaphragms which separate the stages. *See also* DE LAVAL TURBINE; VELOCITY-COMPOUNDED STEAM TURBINE.

pressure control valve A valve used to set the pressure level in a pressure vessel or piping system.

pressure deflection The deflection of the sensing element in a pressure gauge, such as the tube in a Bourdon gauge, when pressurized by the process fluid.

pressure difference (pressure differential, Δp) (Unit Pa) The difference between two pressures, one of which may be a reference pressure such as barometric pressure. In many flow processes, the pressure difference is more important than the absolute pressure level.

pressure drag *See* FORM DRAG.

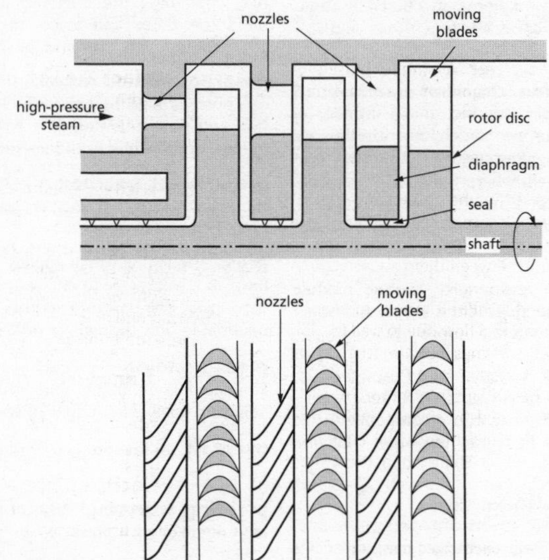

pressure-compounded steam turbine

pressure forming A thermoforming process that combines pressures above and below ambient to force plastic material into and around a mould.

pressure gauge An instrument used to measure absolute or gauge pressure. The sensing element may be a tube which deflects when pressurized, as in a Bourdon gauge, a bellows as in an aneroid barometer, a piezoelectric crystal, a piezoresistive element, etc. *See also* DIFFERENTIAL PRESSURE GAUGE; MANOMETER.

pressure-gradient parameter In boundary-layer theory, a non-dimensional measure of the streamwise pressure gradient dp/dx, variously defined as

$$\frac{L}{\tau_W}\frac{dp}{dx}, \quad \frac{L}{\rho U_\infty^2}\frac{dp}{dx} \quad \text{or} \quad \frac{L^2}{\mu U_\infty}\frac{dp}{dx}$$

where p is the static pressure at the surface, U_∞ is the free-stream velocity, x is the streamwise distance along the surface, ρ is the fluid density and μ is its viscosity, L is a length scale, and τ_W is the wall shear stress. Choices for L include the displacement thickness δ^*, the momentum thickness θ, the boundary-layer thickness δ, and a length defined in terms of the kinematic viscosity ν and the friction velocity u_τ, i.e. ν/u_τ. *See also* POHLHAUSEN POLYNOMIAL.

pressure head *See* HEAD.

pressure-jet burner A simple burner in which high-pressure liquid or gaseous fuel is injected through an orifice into a furnace or other type of combustion chamber. In the case of a liquid fuel, the orifice acts as an atomizer.

pressure line 1. A tube connecting a pressure tapping to a pressure gauge. **2.** For gears in contact, the line representing the direction of the resultant force between them.

pressure loss (Unit Pa) The loss in stagnation pressure in internal flow due to wall friction and minor losses in fittings. *See also* HEAD LOSS.

pressure plate A plate in a motor vehicle's clutch which is pushed by springs against the clutch disc and flywheel to lock the engine to the transmission input shaft when the clutch pedal is released.

pressure-plate anemometer (plate anemometer) A crude anemometer in which the drag force on a suspended plate is used to estimate wind speed.

pressure rating The internal pressure at which a pressure vessel, boiler, tank, piping, etc. is designed to operate safely.

pressure ratio One pressure divided by another which may be a reference pressure. In many flow and thermodynamic processes, the pressure ratio is more important than the absolute pressure levels. For example, in compressible gas flow the Mach number is determined by the ratio of the stagnation pressure to the static pressure.

pressure recovery The progressive increase in static pressure for unseparated flow through a diverging nozzle or diffuser.

pressure regulator (pressure-regulating valve) A device installed in a pneumatic or gas system to maintain the downstream pressure at the required level.

pressure-relief valve A valve that limits the maximum pressure in a pressure vessel or fluid-power system to a specified level.

pressure storage tank (pressure vessel) A closed container for storing gases or volatile liquids, such as liquefied gases, at pressures significantly above atmospheric pressure. Such tanks are commonly cylindrical with domed ends, spherical, spheroidal, torispherical or hemispherical. *See also* AIR RECEIVER; RESERVOIR (2).

pressure surface The high-pressure surface of an aerofoil, turbine, or compressor blade. In normal aircraft applications, this is the lower surface of a wing. For applications, such as to high-performance motor vehicles, where the aerofoil is inverted to generate downforce, it is the upper surface. *See also* SUCTION SURFACE.

pressure tap (pressure tapping, static-pressure tap) A small hole in the wall of a pipe or pressure vessel to which is attached a tube, the other end of which is connected to one side of a pressure transducer.

pressure transducer An instrument (pressure gauge) for measuring absolute or differential pressure, usually with an electrical output signal.

pressure-velocity compounded steam turbine An impulse steam turbine consisting of two or more Curtis stages on the same shaft.

pressurize To increase the pressure within a closed vessel above that of its surroundings, or

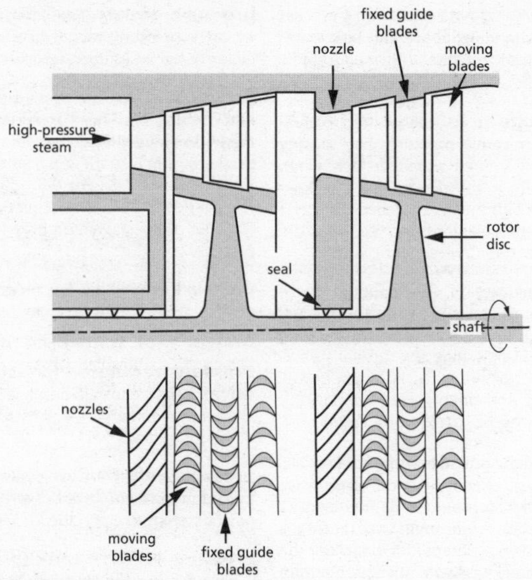

pressure-velocity compounded steam turbine

to increase the pressure of the surroundings above that within a closed vessel.

pressurized fluidized-bed combustion A process of fluidized-bed combustion operating at elevated pressures, producing a hot high-pressure gas stream to drive a gas turbine or heat boiler tubes immersed in the fluidized bed to generate steam to supply a steam turbine.

pressurized-water reactor *See* NUCLEAR FISSION.

presswork Components manufactured on a press.

pre-stress To induce stresses into a component or structure before it is subjected to operating loads. *See also* RESIDUAL STRESS.

pre-tinning *See* SOLDERING.

primary air In a combustor, the fraction (typically 15–20%) of the total air required for combustion which is introduced at the same time as the fuel.

primary couple *See* BALANCING.

primary creep *See* CREEP.

primary energy Energy that has not been converted or transformed. Primary-energy sources include fossil fuels, nuclear fuels which are found in nature, solar radiation, wind, tides and waves, and geothermal.

primary instrument A scientific instrument the output of which depends upon its inherent physical properties, and which requires no calibration. Examples include the equal-arm balance, the U-tube manometer, and the laser Doppler anemometer.

primary shear The shear stress directly attributable to the shear force acting on a component, not including any contribution from applied moments.

prime mover (motor) An engine or mechanism that transforms thermal, chemical, electrical, pressure, or any other source of energy into mechanical energy.

priming The filling of a pump intake with liquid to expel air so that the pump can impose suction on the liquid.

priming pump A pump to supply fuel to an internal-combustion engine during starting.

priming valve *See* PET COCK.

primitive The simplest instruction in robot programming.

principal axes of inertia The x-y-z axes for which the products of inertia vanish. *See also* PRINCIPAL MOMENTS OF INERTIA.

principal axes of strain and stress The particular set of axes in which either shear-strain components vanish leaving the normal principal strains ε_1, ε_2, and ε_3 or shear-stress components vanish leaving the normal principal stresses σ_1, σ_2, and σ_3. In isotropic bodies these axes are coincident. *See also* MOHR'S STRAIN CIRCLE; MOHR'S STRESS CIRCLE; STRAIN.

principal moments of inertia (*I*) It is always possible to choose rectangular axes *Oxyz* such that the moment of inertia of a body about a line L through O is given by $I = A\alpha^2 + B\beta^2 + C\gamma^2$ where α, β and γ are the direction cosines of L relative to O*xyz*. The axes are the principal axes of inertia at O and A, B and C are the principal moments of inertia. Used in the analysis of the motion of a body about a point (as opposed to an axis).

principal strain One of the three normal strains ε_1, ε_2, or ε_3 on the faces of an element of a loaded body in the special case where all the shear strains on those faces are zero. *See also* MOHR'S STRAIN CIRCLE.

principal stress One of the three normal stresses σ_1, σ_2, or σ_3 (where $\sigma_1 > \sigma_2 > \sigma_3$) on the faces of an element of a loaded body in the special case where all the shear stresses on those faces are zero. In two dimensions,

$$\sigma_1 = \frac{1}{2}(\sigma_x + \sigma_y) + \sqrt{\frac{1}{4}(\sigma_x - \sigma_y)^2 + \tau_{xy}^2}$$

and

$$\sigma_3 = \frac{1}{2}(\sigma_x + \sigma_y) - \sqrt{\frac{1}{4}(\sigma_x - \sigma_y)^2 + \tau_{xy}^2}$$

where σ_x and σ_y are the normal stresses on faces of an element aligned along orthogonal x, y axes, and τ_{xy} are the shear stresses on those planes. σ_1 and σ_3 are algebraically the greatest and least normal stresses in the system. They act on planes inclined at angle θ with the x-axis where $\tan 2\theta = 2\tau_{xy}/(\sigma_x - \sigma_y)$. *See also* MOHR'S CIRCLE.

principle of corresponding states (law of corresponding states) The compressibility factor for all gases is approximately the same at the same reduced pressure and temperature.

principle of dynamic similarity *See* PHYSICAL SIMILARITY.

principle of least action A variational principle stating that the integral with respect to time of $(T-V)$, where T is the kinetic energy and V is the potential energy of a body, is always smaller for its actual motion than for any other possible motion. *See also* LAGRANGE–HAMILTONIAN MECHANICS.

principle of maximum plastic resistance 1. In plasticity theory, for any plastic strain increment, the state of stress actually occurring gives an increment of work that equals or exceeds the work which would be done by that strain increment and any other state of stress within or on the yield locus. **2.** The former name of the Lower Bound theorem.

principle of optimality An optimal policy has the property that whatever the previous decisions (i.e. control actions) have been, the remaining decisions must constitute an optimal policy with regard to the state resulting from the first decision. The principle can be applied in many areas including economics, dynamic programming, and control systems. For example, a control system that is divided into parts labeled $p_1, p_2, \ldots p_n$ can be described as satisfying the principle provided that the performance of all parts $p_2, p_3, p_4, p_5, \ldots p_n$ is optimal (i.e. achieves the best control action), irrespective of the output of part p_1.

principle of superposition *See* SUPERPOSITION PRINCIPLE.

principle of virtual work *See* VIRTUAL WORK.

principle of work and energy *See* WORK-KINETIC-ENERGY THEOREM.

prism joint A sliding (i.e. translational) joint on a robot.

probe-type liquid-level detector A device to sense or measure the level of a conductive liquid in a container using an immersed electrode.

process 1. A controlled sequence of continuous or discrete events or actions designed to produce a pre-specified end result which may be a product. **2.** *See* THERMODYNAMIC PROCESS.

process control The monitoring and adjustment of process variables, usually automatically, to ensure they are within pre-specified ranges.

process-reaction-curve *See* ZIEGLER AND NICHOLS METHOD.

producer gas *See* WOOD GAS.

product design Design of components or objects where the emphasis is on styling and ergonomics rather than engineering principles.

production engineering (industrial engineering, manufacturing engineering) The organization of a factory so that products are fabricated and assembled to the specified design, having specified tolerances etc., at the required rate, from appropriate materials, at the least cost on the equipment available.

product of inertia With respect to Cartesian-coordinate axes x, y, and z, the three products of inertia for a rigid body of mass m are

$$I_{xy} = \int xy\,dm, I_{yz} = \int yz\,dm \text{ and } I_{xz} = \int xz\,dm$$

Products of inertia are required in expressions for the kinetic energy and momentum of a rigid body rotating about a point (as opposed to an axis).

product re-engineering Changes made in the detail design of a component, or assembly of components making up a product, in order to permit manufacture by a better, or more modern, method while maintaining the function of the part, e.g. forming rather than cutting.

products of combustion The chemical species, in gaseous, vapour, and/or solid form, which result from the exothermic reaction between a fuel and an oxidizer. For a hydrocarbon fuel reacting with air, the products of combustion typically include carbon dioxide, carbon monoxide, water vapour, nitrogen, oxides of nitrogen, and soot. Sulphur-containing fuels also produce sulphur dioxide while solid fuels, such as coal or wood, also produce ash.

profile The shape of a cross section of an object, such as an aerofoil or cam.

profiled path The programmed path of a CNC machine tool or robot where the position is specified at all times rather than as a series of discrete steps as in point-to-point movement.

profile projector See SHADOWGRAPH.

profiling machine (profile milling machine) A milling machine in which the cutter is guided by the contour of a model having the required shape.

profilometer An instrument used to quantify the roughness of a surface. For a contact profilometer, a diamond stylus sweeps across the surface along a series of parallel lines. Non-contact profilometers use optical techniques to map the surface irregularities.

program control A control system where the set point is derived from a programmable controller, and can thus be programmed to change with time.

programmable logic controller (PLC, programmable controller) A computer used for the automation of a process. Originally designed to replace controllers based on relays, modern programmable-logic controllers can be programmed in high-level languages (for example, BASIC, C++) and are capable of implementing complex control equations. They have digital and analogue inputs and outputs and are designed to be robust, both against mechanical damage and excessive voltage inputs.

programming by demonstration A method of programming a robot by moving it, or a lightweight model, through the task to be performed whilst the robot controller records the movement.

programming panel The panel on a computer control system, including a CNC controller, used to enter the program. This is typically a keyboard and screen.

program scan The time taken for a programmable-logic controller to execute a full program cycle including all instructions.

progression marks *See* FATIGUE.

progressive bonding Curing of a resin adhesive by applying first heat, then pressure.

progressive-cavity pump (helical-rotor pump, MONO pump, progressing-cavity pump) A positive-displacement pump with a rotor in the form of a single helix which seals tightly against a stator to form a series of fixed-shape cavities. The rotor is usually steel, the stator hard rubber, steel, or plastic. As the rotor rotates, fluid trapped in the cavities is moved through the pump. *See also* SCREW PUMP.

progressive spring A spring for which the spring rate increases progressively with spring deflection.

progressive wave *See* TRAVELLING WAVE.

projected area The area which corresponds to the shape produced by projecting a three-dimensional object on to a plane surface.

projection A representation of the features of a three-dimensional object on a two-dimensional engineering drawing. There are two versions of the most common projection, the **orthographic projection.** In **first-angle orthographic projection (first-angle projection)** the side view of an object as seen from the left-hand side is shown to the right of the front view and the back view to the right of that, the side view as seen from the right-hand side is shown to the left of the front view, the plan view is shown directly below the front view and the bottom view is placed directly above the front view. In **third-angle orthographic projection (third-angle projection)** the side-view positions are the opposite of those in first-angle projection, as are the plan and bottom views. The back view is again on the far right. First-angle projection is preferred in the UK and most of the rest of Europe, while third-angle projection is used in the USA, Australia and India. An **oblique projection** employs horizontal and vertical axes along which a component's actual dimensions are scaled, and an angled axis for receding depth along which the dimensions are often reduced to avoid the impression of distortion. In an **isometric drawing (isometric projection)**, one of the three mutually-perpendicular axes of the body coincides with a vertical axis and the two others are at $\pm 120°$ with respect to it. There is no perspective in such projections and lengths retain their true scaled values.

projection lines On an engineering drawing, parallel lines used to show the dimensions of an object. A projection line should start just clear of the outline of the feature being dimensioned, and extend just beyond the dimension line.

pronate The positioning of a robot end effector with the gripper fingers pointed downwards.

prony brake An absorption dynamometer where the torque of an engine is balanced by applying friction to a drum on the output shaft.

proof load The test load in quality control to which different components made of different materials must be subjected without failing in order to perform properly.

proof strength (proof stress) That stress on an engineering stress *vs* engineering strain curve determined by the intersection of the initial elastic loading line offset by an arbitrary amount such as 0.1, 0.2, or 0.5% permanent strain (the **proof strain**). Used for materials, such as light alloys, which do not exhibit a clearly-defined yield point. *See also* R_p.

prop A short column acting in compression employed to limit deflexions of structural items such as beams.

propellant 1. The chemicals used to generate thrust in a propulsion system, typically a liquid or solid fuel and an oxidizer. **2.** The pressurized gas in an aerosol spray can.

propeller (prop) A rotor consisting of a number of shaped blades attached to a shaft and designed to produce thrust when rotated in a fluid. A **propeller blade** is an arm with the cross section of an aerofoil (for an airscrew) or hydrofoil (for a marine screw). The chord and thickness of the blade both vary with distance from the root and the cross section also twist. The component by which the blades of a propeller are attached to the propeller shaft is the **propeller boss (propeller hub).** An **airscrew** is a propeller that operates in air. A **marine-screw propeller** consists of a number of blades, typically three, four, or five, of helicoidal form attached to a central boss and used to provide propulsive thrust for a ship. The thrust is produced by the propeller adding momentum to the fluid in which it is rotating. An **adjustable-pitch propeller** is a propeller the pitch of whose blades can be altered when the engine is not running, in contrast to a **controllable-pitch propeller (variable-pitch propeller),** for which the pitch can be altered when rotating. In the case of a **constant-speed propeller** the pitch is automatically adjusted so as to maintain constant rotational speed. **Propeller efficiency (η)** is the ratio of the thrust power produced by a propeller to the shaft power provided by the engine, given by the equation $\eta = FV/T\Omega$ where F is the thrust, V is the airspeed, T is the torque produced by the engine, and Ω is the rotation speed. **Geometric pitch** is the distance an element of a propeller would advance in one revolution in the absence of slip, while the **effective pitch (advance)** is the actual forward distance (less than the geometric pitch due to slip) moved by a propeller-driven vehicle for one revolution of the propeller. The **advance ratio (advance coefficient, *J*)** is the ratio of the distance moved forward in one revolution to the propeller diameter D. If V is the forward speed and N the rotation speed in revolutions per second, $J = V/ND$. For a screw propeller, **propeller slip** is given by $1 - J/P$ where P is the **pitch ratio**, the ratio of geometric pitch to diameter. The **propeller tip speed** is the peripheral speed V of a propeller tip given by the equation $V = \omega R$ where ω is the angular velocity of the propeller shaft and R is the radial distance from the axis to the tip.

propeller anemometer (windmill anemometer) An anemometer comprising a streamlined body, pivoted to move about a vertical axis, with a propeller at one end and a wind vane at the other. The wind speed is approximately proportional to the propeller rotation speed.

propeller fan An axial-flow fan using a propeller to produce an airflow.

propeller meter *See* TURBINE FLOW METER.

propeller pump *See* AXIAL-FLOW PUMP.

propeller shaft (Cardan shaft, prop shaft) 1. In the power train of a motor vehicle, the shaft that transmits power from the gearbox to driven rear wheels. **2.** *See* PROPELLER.

propeller turbine An axial-flow hydraulic turbine with a rotor (or runner) similar in design to a marine propeller. *See also* FRANCIS TURBINE; KAPLAN TURBINE.

propelling nozzle (exhaust nozzle) The nozzle at the rear end of a turbojet engine, rocket engine, or ramjet, which is normally choked. For subsonic flight the nozzle is convergent, but may be convergent-divergent for supersonic flight.

property A macroscopic, observable, physical quantity that characterizes the state or behaviour of a substance or system. Examples include pressure, temperature, mass, volume, elastic moduli, Poisson's ratio, and coefficients of thermal expansion. Many thermodynamic material properties, such as density and enthalpy, are defined in terms of other properties and also depend upon temperature and pressure. *See also* EXTENSIVE PROPERTIES; INTENSIVE PROPERTIES.

propfan A form of propeller with 8 to 10 very wide swept-back blades, designed to enable turboprop aircraft to cruise at a Mach number of about 0.8.

propfan engine *See* UNDUCTED FAN ENGINE.

proportional control A form of control where the controller output $x(t)$ and thus the plant input is proportional to the error ε, i.e. $x(t) = k\varepsilon$ where k is the **proportional gain**. The **proportional band** is the range of output of a proportional controller over which proportionality is maintained. For example, a controller driving an actuator cannot produce a force greater than the capability of the actuator, and proportionality can be maintained only up to this maximum force. More sophisticated control methods involve the time derivative or time integral of ε as follows: (a) **proportional-plus-derivative control (PD control, P+D control)** with

$$x(t) = k\varepsilon + T_d \frac{d\varepsilon}{dt}$$

where T_d is the derivative time constant; (b) **proportional-plus-integral control (PI control, P+I control)** with

$$x(t) = k\varepsilon + \frac{1}{T_i}\int \varepsilon\, dt$$

where T_i is the integral time constant; (c) **proportional-plus-integral-plus-derivative control (PID control, P+I+D control)** with

$$x(t) = k\varepsilon + \frac{1}{T_i}\int \varepsilon\, dt + T_d \frac{d\varepsilon}{dt}$$

proportional dividers Dividers with two legs of equal length, pointed at each end, held together by a movable pivot. The separation of the points at either end are in the same ratio as the distances from pivot to point on either side. Proportional dividers are used to transfer measurements from one scale to another.

proportional feedback *See* POSITION FEEDBACK.

proportional limit *See* LIMIT OF PROPORTIONALITY.

proportioning probe A leak-testing device in which the ratio of tracer gas to an inert diluting gas (dry air or nitrogen) can be varied without any substantial change in the total flow from the probe.

proportioning pump A pump that supplies two liquids, usually water and a concentrate, in the exact ratio needed for a process. *See also* METERING PUMP.

proprioceptor In robotics, a sensor that detects the internal movement, forces, or torques in a robot. An encoder or resolver measuring the angle of a robot joint is a proprioceptor.

prop shaft *See* PROPELLER SHAFT.

propulsion Any method of applying a force or torque to an object which causes it to move, usually in a forward direction. In the case of motor vehicles, this is through an engine which applies torque to the driving wheels. In the case of aircraft, it is through thrust provided by a propeller or jet engine. Space vehicles are propelled by rocket engines. *See also* PROPELLANT; PUMP JET.

propulsion efficiency The ratio of the thrust power (i.e. the product of thrust and airspeed) of a jet or rocket engine to the rate of production of propellant kinetic energy.

proton exchange membrane fuel cell (PEMFC) A fuel cell operating at temperatures below 100°C in which hydrogen is the fuel and the electrolyte is a solid polymer membrane in the form of a thin plastic film.

prototype A preliminary example, usually full size, of a machine such as a motor vehicle or aircraft used to evaluate design and performance.

protractor A circular or semi-circular plate, usually transparent, used to measure or construct angles by reference to radial lines marked on the plate.

proud A component that projects above the nearby level in a surface is said to be proud of that surface.

proving ring A type of load cell in the form of an alloy-steel ring that, when loaded elastically across a diameter, deforms in proportion to the load. The capacity depends upon the inside and outside diameters and its thickness. Sometimes used for the calibration of testing machines.

proximal Mounted close to the base frame of a robot arm. For example, the actuator driving the waist joint of an articulated robot would be described as proximal as it is mounted on the fixed structure at the base frame.

proximity sensor A sensor that operates without making contact with the object being sensed. For example, a capacitance sensor that detects the presence of a component by observing the change in electrical capacitance between the sensor and ground.

pseudo force *See* ACCELERATING FRAME OF REFERENCE.

pseudoplastic liquid *See* SHEAR-THINNING LIQUID.

pseudo-random signal A digitally-programmed test signal that obeys criteria for randomness. It can be used for the frequency-response testing of dynamic and control systems. *See also* RANDOM SIGNAL.

pseudo-reduced specific volume (reduced specific volume) The ratio of the specific volume of a substance to the value of the specific volume at the critical point, calculated from the ideal gas equation. *See also* REDUCED PROPERTY.

psi *See* POUND PER SQUARE INCH.

psychrometer (hygrometer) An instrument used to measure relative humidity.

psychrometric chart (hygrometric chart) A chart showing the thermodynamic properties of moist air at a constant pressure in which the specific humidity ω is plotted *vs* dry-bulb temperature. The chart includes lines of constant specific enthalpy h, constant wet-bulb temperature, and constant specific volume v, as well as curves of constant relative humidity ϕ. **Psychrometric tables** present the same data in tabular form.

psychrometric ratio (r) The ratio of the convective heat-transfer coefficient h to the product of the convective mass-transfer coefficient h_M and the humid heat c_S, given by the equation $r = h/h_M c_S$.

psychrometry (hygrometry, psychrometrics) The study of steam–air mixtures, of importance to heating-and-ventilating systems.

***p–T* diagram** *See* PHASE DIAGRAM (1).

PTFE *See* TEFLON.

PTV *See* PARTICLE TRACKING VELOCIMETRY.

pulley A free or driven wheel on a shaft with an appropriately shaped rim to carry a flat belt, vee belt, notched belt, rope, or chain. Used to transmit power or motion. *See also* JOCKEY PULLEY.

pulley block (block) A wooden or metal frame which carries pulleys placed side by side. *See also* BLOCK AND TACKLE.

pull out The contribution to fracture toughness in composites made by fractured and debonded filaments being pulled out of the holes they occupied.

pulsating flow A confined fluid flow in which the pressure and flow rate vary periodically, often due to fluctuations in the output of the pump(s) or compressor(s) which drive(s) the flow. A **pulsation damper**, typically using a bladder or bellows to separate the process fluid from gas stored in a closed chamber, is used to control and minimize flow pulsations.

pulsation The periodic increase and decrease in the magnitude or intensity of a physical quantity such as pressure, flow rate, volume, or light intensity.

pulse An increase or decrease in the magnitude of a physical quantity, such as pressure, voltage, or force, with a time scale short compared with other time scales in a process, after which there is a return to the original level.

pulse-jet engine (athodyd, pulse jet) A rocket-like jet engine with no rotating parts, in which air and fuel are drawn into a combustion chamber where they burn to generate a high-pressure, hot gas which exhausts through the engine's tailpipe. In a **valved pulse jet**, the airflow is controlled by a one-way inlet valve which opens and closes in response to the combustion-chamber pressure. In a **valveless engine**, the airflow is controlled by the fluctuating chamber pressure.

pulse-transfer function The output–input relationship of a system expressed using z-transforms, i.e. using sampled-data methods. It is equivalent to the transfer function determined using Laplace transforms for a continuous-time system.

pultrusion A continuous manufacturing process in which continuous-filament reinforced composite materials of uniform cross section are produced by being pulled through a die.

pulverization The process of crushing and grinding materials into powder.

pump A machine designed to cause a liquid, gas, vapour, or slurry to flow due to the reciprocating motion of pistons, rotation of vanes, or rotation of an impeller.

pumpability test A test to determine the limiting conditions at which a liquid can be pumped. For example, the lowest temperature at which a petroleum product remains pumpable or the solids content for a slurry. *See also* POUR TEST.

pump-down time The time required to reduce the pressure of a pressure vessel of a given volume to a specified level, using a vacuum pump of known characteristics.

pumped hydrostorage system (pumped storage) A hydroelectric power plant in which, at times of low demand, surplus power is used to pump water to a high-level reservoir. When demand is high, the water flows through the turbine(s) to generate electrical power. In practice, the same machine, usually a Francis turbine, is used for both pumping and generating, and the electrical generator is also used as an electric motor. For 100 m head, the potential energy stored is 1 MJ/m^3.

pumped solar water heater A domestic solar-power system in which a pump circulates water through a roof-mounted solar panel and a heat exchanger at the bottom of a storage tank.

pump hydraulic efficiency (η_P) The fraction of work done on a fluid by a pump that results in an increase in the mechanical energy stored in the fluid, defined by $\eta_P = \dot{m}\Delta p_0/\rho\Omega T$ where $\dot{m}$ is the mass flow rate through the pump, Δp_O is the rise in total pressure across the pump, ρ is the fluid density, Ω is the angular rotation speed of the pump impeller, and T is the torque.

pumping loss (pumping work) In a four-stroke piston engine, the negative work that occurs during the induction and exhaust strokes.

pump jet (water jet) A marine-propulsion system in which a pump is used to create a high-speed jet of water to provide thrust.

punch 1. A hand tool with a sharp point, used to mark the position of a hole centre in a workpiece prior to drilling. **2.** A flat-ended tool with sharp edges, used to shear out a hole in a plate that is supported underneath on a die having a slightly larger same-shaped orifice. *See also* HOLLOW PUNCH.

puncture A rupture, often very small, in the wall of a pressure vessel or tyre, which allows the contained fluid to escape.

purchase The ability of one component to grip another without slipping.

pure bending Flexure of a component or structure under a constant bending moment with zero shear force.

pure shear The result when an element of material is subjected to shear stresses that result only in shear strains. For example, when a circular bar, either solid or hollow, is subjected to torsion, a small thin element removed from between two parallel cross sections is in a state of pure shear.

purge (flush) To remove contaminants or unwanted material from piping or a pressure vessel by pumping through a cleaning liquid or inert gas such as nitrogen.

pusher A propeller-powered aircraft in which the propeller is mounted behind the engine. Thrust is created in the same way as when the propeller is conventionally mounted.

push fit A tight sliding fit between a shaft and a hole.

push–pull fatigue *See* FATIGUE.

push rod A rod which opens and closes a valve via a valve rocker in an overhead-valve piston engine. The rod is actuated by a camshaft located in the crankcase.

PVC 1. *See* POLYVINYL CHLORIDE. **2.** *See* PRECESSING VORTEX CORE.

***p–v–T* diagram** A two-dimensional graphical representation of the phase changes experienced by a pure substance as its pressure (p), specific volume (v), and temperature (T) change. The changes are shown with p as the ordinate and v as the abscissa and include lines of constant T. The same information can be shown with temperature as the ordinate, isobars then being included.

***p–v–T* surface (thermodynamic surface)** A three-dimensional graphical representation of the phase changes experienced by a pure substance as its pressure (p), specific volume (v), and temperature (T) change. The changes are shown on a graph with p as the vertical axis, T and v as orthogonal horizontal axes. Isotherms and isobars may also be included.

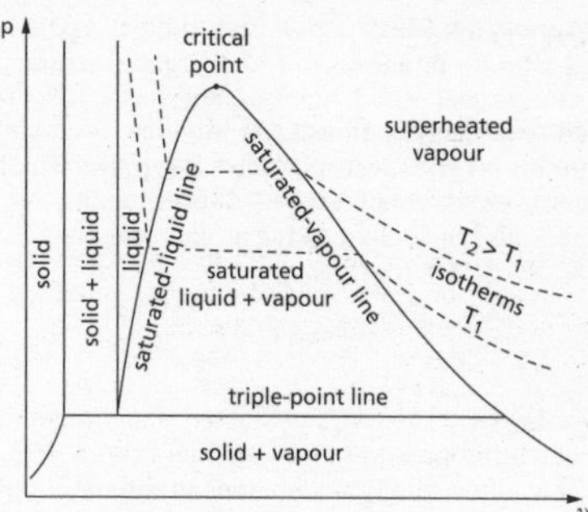

***p–v–T* diagram**

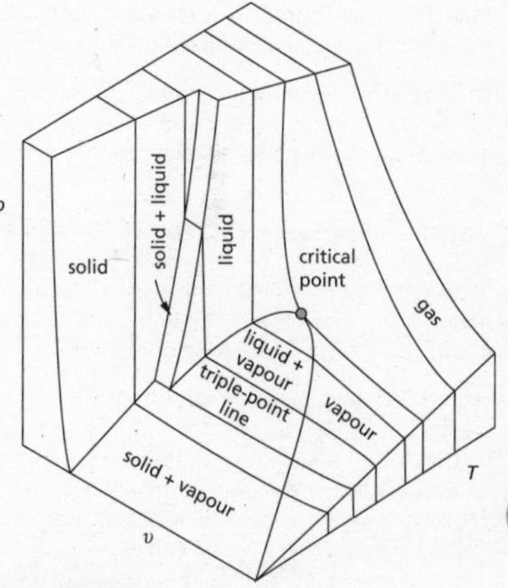

***p–v–T* surface**

pycnometer *See* DENSITY BOTTLE.

pylon 1. The strut assembly used to connect an engine to an aircraft wing. **2.** A tower, typically of lattice-frame construction, used to support overhead electrical-power lines.

pyranometer *See* SOLARIMETER.

pyrgeometer An instrument that measures the net long-wave (4.5 μm to 100 μm) radiation balance between the infra-red radiation received from the atmosphere and that emitted by the earth's surface.

pyroheliometer (pyrheliometer) An instrument that measures the intensity of direct-beam solar radiation. *See also* WATER-FLOW PYRHELIOMETER.

pyrolysis (destructive distillation) A general term for processes in which organic material, such as coal, wood, biomass, and waste, is heated or partially combusted to produce secondary fuels and chemical products. The products include gases, condensed vapours as liquids, tars, and oils, and solid residue as charcoal and ash. *See also* GASIFICATION.

pyrometer 1. (radiation thermometer) A non-contact instrument that measures an object's surface temperature by detecting the electromagnetic radiation (infra-red or visible) emitted by the surface. The two main types are the optical pyrometer and the infra-red pyrometer. **2.** A thermometer used to measure very high temperatures.

Q-factor (quality factor, sharpness of resonance) A non-dimensional parameter that describes, for a given resonant mode, the sharpness of the peak in the frequency response of a lightly-damped linear oscillator: $Q = \omega_{RES}/\Delta\omega$ where ω_{RES} is the resonance frequency and $\Delta\omega$ is the half-power bandwidth of the resonance. It is inversely related to the damping such that a broad peak corresponds to high damping and a narrow peak to low damping.

quadplex pump *See* SIMPLEX PUMP.

quadrant 1. A lever that can swing through 90°, as used to provide signals on railway systems. **2.** A slotted link in the form of a quarter circle.

quadratic performance index A measure of the performance of a control system that evaluates the time integral of a quadratic function of the error. Unlike many other performance measures, this can be determined in the frequency domain (i.e. from the transfer function) without having to revert to the time domain.

quadruple thread *See* SCREW.

qualification test A test that a component or system has to pass before being put into service.

quality *See* DRYNESS FRACTION.

quality assurance (quality-control test) The use of statistical methods to check whether mass-produced products satisfy design and performance criteria.

quality factor *See* Q-FACTOR.

quarl The divergent conical part of a swirl burner.

quartz fibre A fine quartz filament made from high-purity (typically 99.95%) silicon dioxide, which has a very low coefficient of thermal expansion, can be heated to about 1000°C, and has excellent resistance to thermal shock. The fibres can be spun and woven.

quartz-resonator transducer A device that utilizes the piezoelectric properties of crystalline silicon dioxide **(quartz crystal)** in the measurement of force, pressure, or temperature, all of which when applied to a quartz crystal cause its mechanical resonant frequency to change. **Quartz thermometers** are based upon the near-linear temperature dependence of the resonant frequency. **Quartz pressure transducers** are particularly suited to rapidly-varying pressure and pressure transients.

quasi-linear feedback control system A control system which shows substantially linear behaviour despite the incorporation of non-linear control elements.

quasi-linear system A non-linear system whose behaviour can be approximated using linear functions.

quasi-static process A mechanical event that is so slow that it can be analysed without regard to inertial forces.

quench (quenching) 1. The process of rapid-cooling by plunging an object into a bath of water, oil, salt, molten metal, or other media. It is a method of heat treatment used particularly to form martensite preparatory to tempering steels. The bath temperature is the **quench temperature. 2.** The suppression of combustion. **3.** In a piston engine, the cooling of a fraction of the gases during combustion, typically by reducing the clearance between the piston crown and the cylinder head.

quench-tank extrusion Extrusion of plastic film into a quenching medium.

quick coupling (quick disconnect, fast coupling) A hose connection allowing rapid assembly or disassembly and comprising a socket and a plug incorporating a spring-loaded locking mechanism. *See also* BAYONET CONNECTOR; GUILLEMIN COUPLING.

quick return A mechanism in which the return stroke in a reciprocating machine is faster than the power stroke. A typical design involves a slotted arm pivoted at one end and caused to move back and forth by a pin, moving along the slot, attached to a rotating wheel.

quill A hollow shaft into which another shaft is inserted so that both relative axial and rotational movement is possible. A gear mounted on a quill is a **quill gear.**

Quimby screw pump A screw pump with two meshing screws, each having a right-hand and a left-hand screw. Liquid enters at either end and is discharged from the middle.

quintuplex pump *See* SIMPLEX PUMP.

q

race 1. Either of the inner or outer hard-steel rings in a ball or roller bearing. **2.** *See* HEADRACE; TAILRACE.

rack (rack gear) A straight or curved bar having equidistant teeth to engage with a spur pinion. It is equivalent to a spur gear having an infinite pitch radius. *See also* TOOTHED GEARING.

rack and pinion A system comprising a rack and a spur pinion by which rotary motion may be converted into linear or rotary motion and vice versa. Typical applications include rack railways. *See also* TOOTHED GEARING.

rack-and-pinion steering In motor vehicles, steering achieved by rotation of a pinion on the end of the steering column which meshes with a rack to move it left or right.

rack cutter *See* TOOTHED GEARING.

rack differential A linear-displacement differential in which two parallel racks moving in opposite directions engage with opposite sides of a pinion. The resultant displacement of the axis of the pinion is half the algebraic sum of the displacements of the two racks.

rack gear *See* RACK.

radial acceleration *See* RADIAL MOTION.

radial bearing *See* BEARING.

radial engine A piston engine in which the cylinders are arranged radially around the crankshaft, a design commonly used in early aircraft engines.

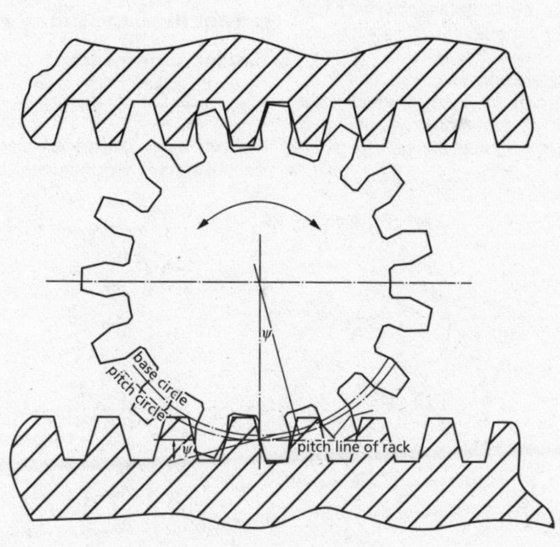

rack differential

radial flow Fluid flow for which the direction of flow is either radially inward or outward.

radial-flow compressor A compressor in which the working gas enters the machine axially and is compressed as it flows radially out through the impeller.

radial-flow turbine 1. A turbine, such as the Francis hydraulic turbine, in which the working fluid enters the machine through a volute at its periphery, as shown in the diagram, flows radially inwards through guidevanes into a runner, and exits axially. **2.** A turbine such as the Ljungström turbine in which the working fluid enters the machine close to its axis and is expanded as it flows radially outwards through the blading.

radial gate *See* TAINTER GATE.

radial load 1. Centrifugal loading induced in a rotating body. **2. (through-wall-thickness load)** The radial component of loading arising from internal or external pressurization of a closed vessel. There will also be axial and hoop loading.

radial motion Motion along any straight line emanating from a fixed reference point. For a particle or body moving relative to a fixed point a distance r away, the **radial velocity (v_r)**, with unit m/s, is the component of velocity equal to the rate of change of r with time, i.e. $v_r = dr/dt = \dot{r}$. The **radial acceleration (a_r)**, with unit m/s^2, is the component of acceleration equal to the rate of change of the radial velocity dr/dt with time, $d^2r/dt^2 = \ddot{r}$, plus the centripetal acceleration, i.e.

$$a_r = \frac{d^2r}{dt^2} + r\dot{\theta}^2 = \ddot{r} + r\dot{\theta}^2$$

where $\dot{\theta}$ is the instantaneous angular velocity. *See also* ACCELERATING FRAME OF REFERENCE; TANGENTIAL ACCELERATION; TANGENTIAL VELOCITY.

radial tyre (radial-ply tyre) A type of tyre construction in which the cords that make up the carcass are laid across the tyre at 90° to the direction of rotation.

radial wave equation A differential equation describing the transmission of a wave in a system with radial symmetry.

radian (rad) A coherent derived SI unit defined as the plane angle subtended at the centre of a circle by an arc having a length equal to the radius. Thus 2π radians are equivalent to 360° and 1 rad $\approx$ 57.3°. *See also* STERADIAN.

radiant emissive power (radiant exitance) *See* EMISSIVE POWER.

radiant energy *See* HEAT TRANSFER.

radiant flux (radiating power, Φ_e) (Unit W) The total rate of radiation emitted or received by a body.

radiant flux density (RFD, ϕ_e) (Unit W/m^2) The radiant flux per unit area.

radiant heating *See* HYDRONIC HEATING.

radiant superheater In a boiler, a superheater installed directly within the furnace and heated primarily by thermal radiation.

radiant-type boiler A boiler in which thermal radiation is the predominant mode of heat

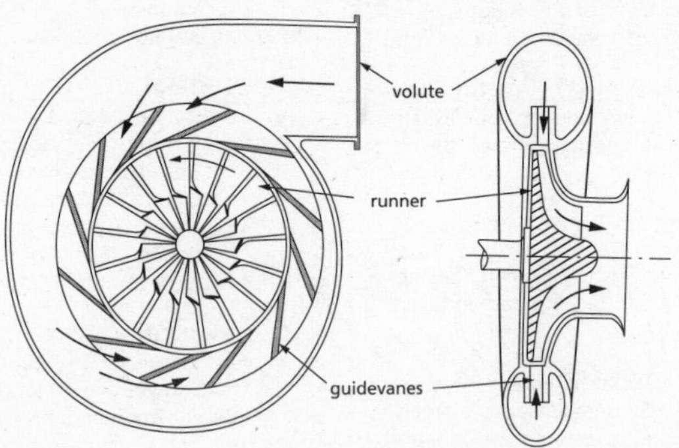

radial-flow turbine

transfer within the furnace. *See also* WATERWALL BOILER.

radiating power *See* RADIANT FLUX.

radiation *See* HEAT TRANSFER.

radiation pressure (Unit μPa) The pressure exerted on a surface exposed to any form of electromagnetic radiation. If the radiation is absorbed, it is equal to the power-flux density divided by the speed of light. For example, the power flux density of solar radiation at the earth's surface is 1370 W/m^2 and the corresponding radiation pressure is 4.6 μPa.

radiation pyrometer (radiation thermometer) Any type of sensor that determines temperature by measuring thermal radiation, usually at a single wavelength.

radiator A heat exchanger used to transfer thermal energy from one fluid to another for heating or cooling purposes. Despite the name, the principal mode of heat transfer is convection rather than radiation. In motor vehicles, water circulated through the engine block is cooled as it flows through the tubes of an air-cooled heat exchanger. In domestic radiators, hot water from a boiler is circulated through a heat exchanger with a large surface area which transfers heat to the surrounding air.

radioactive heat Thermal energy released from the nucleus of an atom such as uranium^{235} by fission due to the absorption of a neutron. Heat is also produced by radioactive decay.

radiosity (*J*) (Unit W/m^2) The total radiation leaving a given surface per unit area, including emitted, reflected, and transmitted radiation.

radius arm An arm used to locate a front- or rear-suspension component in a motor vehicle.

radius of gyration (*k*) (Unit m) The moment of inertia I of any body may be expressed as $I = mk^2$ where m is its mass and k is the radius of gyration.

rainflow analysis *See* FATIGUE.

rake angle The angle between the cutting face of a cutting tool and the perpendicular to the machined surface.

ram A piston, usually hydraulically or pneumatically powered, used to apply a precisely-controlled force. Applications include testing machines and extrusion presses.

ram-air turbine A propeller driven by the airflow relative to an aircraft used to generate power in emergency situations.

Ramberg-Osgood equation An empirical relation between stress σ and strain ε during elastoplastic deformation, given by

$$\varepsilon = \frac{\sigma}{E} + \frac{\alpha\sigma_Y}{E}\left(\frac{\sigma}{\sigma_Y}\right)^N$$

where E is Young's modulus and σ_Y is the yield strength, and α and N are empirical constants. N is the reciprocal of the parameter n in the **Ludwik relation** where $\sigma = \sigma_o\varepsilon^n$, σ_o being a dimensional constant.

ram effect The increased static pressure experienced at the intake to an engine due to the airflow being brought to rest. *See also* AIR SCOOP.

ramjet (ramjet engine) A simple form of supersonic jet engine (operating typically in the Mach-number range 3 to 6) with no moving parts, in which air is compressed subsonically then mixed with fuel which burns in a combustion chamber downstream of a flameholder. The exhaust gases pass through a choked nozzle and are exhausted to the atmosphere at supersonic speed. *See also* SCRAMJET.

ramp A signal which shows a proportional increase with time t, i.e. $y = \alpha t$ where y is the ramp signal and α is the slope.

random error *See* ERROR (1).

random molecular energy *See* SENSIBLE ENERGY.

random reinforcement In filament-reinforced composites, when the fibres are discontinuous and not orientated.

random signal A signal ideally having equal power at every frequency that is thus suitable for frequency-response testing of dynamic and control systems. In practice such signals are difficult to generate and thus a practical random signal will have equal power at every frequency within the frequency range of interest. *See also* PSEUDO-RANDOM SIGNAL.

random vibration Unpredictable, non-harmonic vibration caused by random excitation, e.g. by ocean waves or the passage of a vehicle over a rough or irregular surface.

Rankine *See* ABSOLUTE TEMPERATURE.

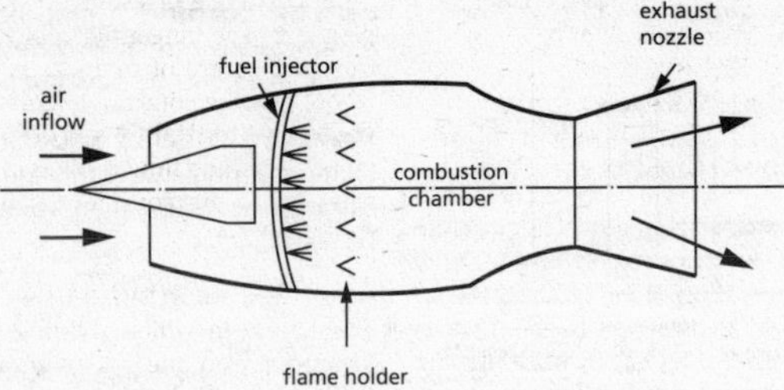

ramjet

Rankine criterion *See* FAILURE.

Rankine cycle The ideal thermodynamic vapour-power cycle consisting, as shown in the diagram, of isentropic pressurization (or compression) in a pump (1–2); isobaric heat addition in a boiler (2–3); isentropic expansion in a turbine (3–4); and isobaric heat rejection in a condenser (4–1). In the case of a steam turbine, water enters the pump as saturated liquid, where it is raised to boiler pressure. It leaves the boiler as wet, dry saturated, or superheated vapour and passes into the turbine. The steam leaves the turbine as a saturated liquid–vapour mixture at reduced temperature and pressure and then passes into the condenser, which it leaves as saturated liquid. *See also* KALINA CYCLE.

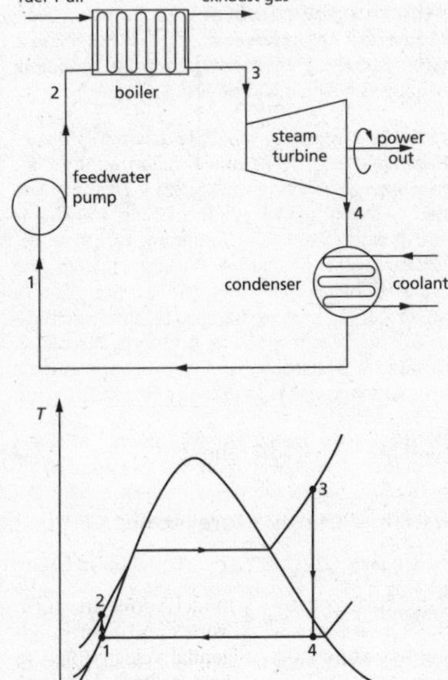

Rankine cycle

Rankine cycle with reheat A Rankine cycle, in which the temperature of the working fluid is increased, typically at constant pressure, part way through the expansion process. In a steam-power plant, steam leaving the high-pressure turbine re-enters the boiler, or a separate heater, before passing into the intermediate- or low-pressure turbine.

Rankine–Hugoniot equation An equation relating thermodynamic quantities on either side of a normal shock wave. In terms of the specific enthalpy h, the density ρ and the static pressure p:

$$\frac{h_2 - h_1}{p_2 - p_1} = \frac{1}{2}\left(\frac{1}{\rho_2} + \frac{1}{\rho_1}\right)$$

where the subscripts 1 and 2 refer to conditions ahead of and behind the shock, respectively. The equation also applies to situations where there is heat release due to the shock, as in a detonation wave. According to the Rankine-Hugoniot equation there should be a **rarefaction shock** with a reduction in static pressure. The corresponding entropy reduction violates the second law of thermodynamics, thereby demonstrating that a steady-state expansion shock cannot exist in practice.

Rankine oval In potential-flow theory, a streamline shape produced by the combination of a uniform flow and a line-source-line-sink pair aligned parallel to the uniform flow. *See also* KELVIN OVAL.

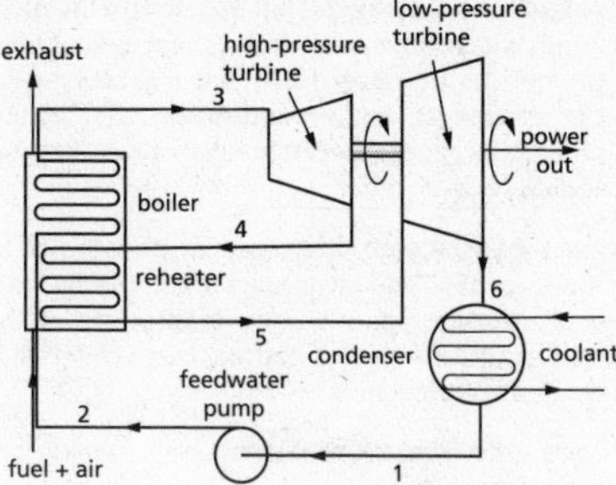

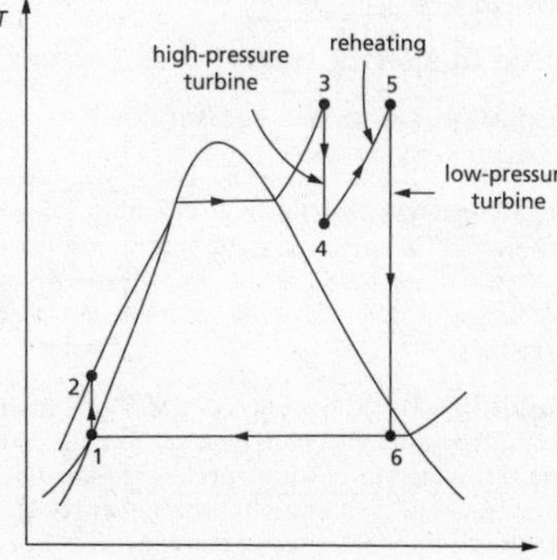

Rankine cycle with reheat

Rankine temperature scale *See* ABSOLUTE TEMPERATURE.

Rankine vortex A model vortex in which there is an inner forced vortex and a surrounding free vortex, the tangential velocity of each being matched at a given radius. A **forced vortex** is an idealized rotational swirling fluid motion like that of a solid body in which there is no shear. The tangential velocity v, a distance r from the centre of rotation, is given by $v = \omega r$ where ω is the constant angular velocity. A **free vortex (potential vortex)** is an idealized irrotational swirling fluid motion in which $v = \Gamma/2\pi r$ where Γ is the constant circulation or vortex strength.

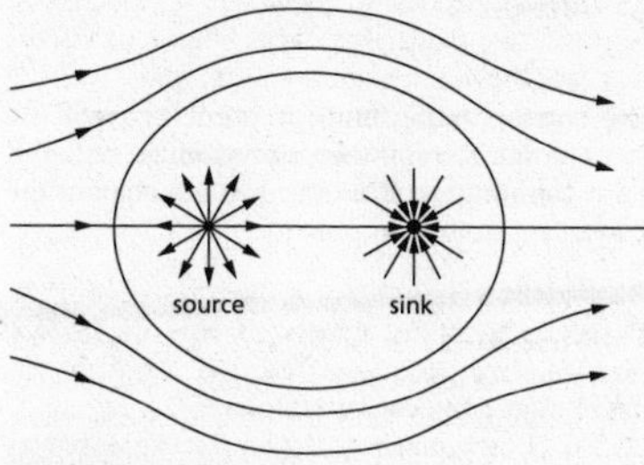

Rankine oval

Ranque–Hilsch tube (Hilsch tube) A cylindrical tube in which swirling flow introduced tangentially on one side of an internal constriction leads to high-temperature outflow on one side and low temperature on the other.

RANS equations *See* REYNOLDS DECOMPOSITION.

rapid prototyping The production of alternative prototypes during the design stage by building up layers of material using a CNC machine, or by sintering, etching, etc. *See also* ADDITIVE-LAYER MANUFACTURING.

RAPS *See* REMOTE AREA POWER SYSTEM.

rarefaction Reduction in the density of a substance, usually a gas. It is the opposite of compression. *See also* EXPANSION.

rarefaction shock *See* RANKINE–HUGONIOT EQUATION.

rarefaction wave A progressive wave or wavefront that causes expansion of the medium through which it propagates. *See also* COMPRESSION WAVE; PRANDTL–MEYER EXPANSION.

rarefied gas dynamics (rarefied flow) A flow in which the mean-free path of the gas molecules is not small compared with the smallest characteristic dimension of the flow geometry, i.e. the Knudsen number is not negligibly small.

ratchet (ratchet wheel) A wheel or ring with inclined teeth that engage with a pawl, resulting in one-way motion with reverse motion being prevented.

ratchet coupling A joint employing a ratchet system between two shafts, so that not only does the driven shaft run in one direction only, but also the driven shaft can, if necessary, run more quickly than the driving shaft.

Rateau turbine *See* PRESSURE-COMPOUNDED STEAM TURBINE.

rate controller *See* VELOCITY CONTROLLER.

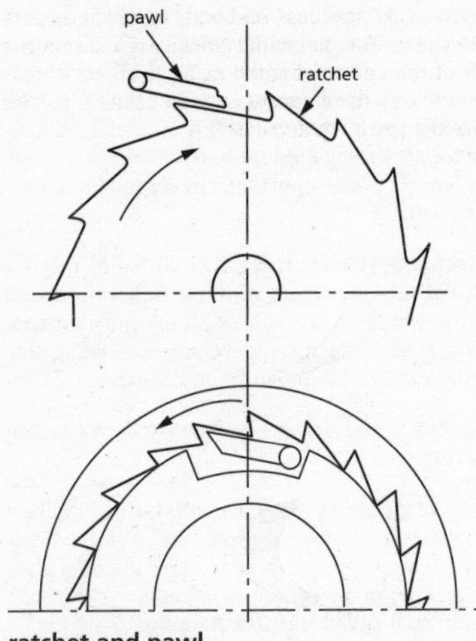

ratchet and pawl

rated capacity 1. *See* RATED POWER. **2.** (Unit kg) For machines such as cranes, the maximum load that can be lifted safely. **3.** (Unit kg/hr) For a steam boiler, the mass of steam that can be produced per hour for a given feedwater temperature and pressure. In some cases, a kilowatt rating is given.

rated engine speed (Units rpm, rps) For a piston engine, the rotation speed at which rated power is obtained.

rated flow (Unit m^3/s) In a hydraulic system or machine, the flow rate a manufacturer specifies for a component as the maximum desirable for it to function as designed.

rated load (Unit N) The maximum load that a structure or component has been designed to carry.

rated power (rated capacity, rated horsepower) (Unit hp or kW) The maximum power output that can be sustained continuously for any power-producing machine or system, such as an engine or a power plant. *See also* DECLARED NET CAPACITY.

rated relieving capacity That portion of the measured relieving capacity used as the basis for the application of a pressure-relief device, determined from the applicable code or regulation.

rated wind speed (Unit m/s, kph) The minimum wind speed for a wind turbine to generate the maximum power for which it is designed. Overspeeding at higher wind speeds, that would potentially generate even greater power, is prevented by a governor.

rate gyroscope A gyroscope mounted in a single gimbal and restrained by a spring that measures the rate of change of direction of an axis by means of the torque exerted on the bearings of the gimbal.

rate servomechanism *See* VELOCITY SERVOMECHANISM.

rating life *See* BEARING.

ratio of specific heats *See* SPECIFIC HEAT.

Rayleigh–Bénard convection *See* BÉNARD CONVECTION.

Rayleigh damping In a vibrating system, where the damping matrix $\boldsymbol{C}$ is proportional to both the mass matrix $\boldsymbol{M}$ and the stiffness matrix $\boldsymbol{K}$, such that $\boldsymbol{C} = \alpha\boldsymbol{M} + \beta\boldsymbol{K}$ where α and β are constants.

Rayleigh flow Friction-free gas flow through a constant-area duct with heat addition or cooling. Tabulated results for a perfect gas are available based upon a one-dimensional analysis. A **Rayleigh line** is a curve of enthalpy or temperature *vs* specific volume or specific entropy for Rayleigh flow. *See also* FANNO FLOW.

Rayleigh instability (Rayleigh–Plateau instability) An instability that occurs due to surface tension on the surface of a liquid jet surrounded by gas.

Rayleigh number (*Ra*) A non-dimensional parameter that arises in the study of combined convection and conduction in a fluid. For a vertical plate of length L, $Ra = g\beta(T_S - T_\infty)L^3/\nu\alpha$ where g is the acceleration due to gravity, β is the thermal-expansion coefficient for the fluid, α is its thermal diffusivity, ν is its kinematic viscosity, T_S is the surface temperature, and T_∞ is the free-stream temperature. In some problems, a local Rayleigh number is defined with L replaced by x, the distance from the leading edge of the plate.

Rayleigh's method A method of dimensional analysis in which, if the dependent physical variable in a problem is y and the independent physical variables are x_1, x_2, $x_3\ldots$ etc., it is assumed that y can be expressed as an infinite sum with terms of the form $x_1^a x_2^b x_3^c \dots\dots$, the exponents *a, b, c*... being selected to ensure dimensional homogeneity. *See also* IPSEN'S METHOD.

Rayleigh's stability equation The form assumed by the Orr–Somerfeld equation for inviscid flow.

Rayleigh step bearing A slider bearing in which there is a step decrease in the clearance in the direction of travel. This bearing has the highest load capacity of all slider shapes.

Rayleigh's theorem *See* RECIPROCAL THEOREM.

Rayleigh–Taylor instability An instability that occurs on the interface between two immiscible fluids at rest, one above the other, the upper fluid having the higher density. *See also* KELVIN–HELMHOLTZ INSTABILITY.

Rayleigh wave A surface wave whose propagation velocity is given by $c_S = f(v)c_2$ where $c_2 = \sqrt{G/\rho}$ is the shear-wave velocity, G is the shear modulus, ρ is the density, and $f(v)$ is a function of Poisson's ratio v having value 0.9194 when $v = 0.25$ and 0.9953 when $v = 0.5$. $c_S < c_2$ because at the free surface of a solid, normal stresses are zero. Rayleigh waves are strong only within a depth of about one wavelength below the surface. Ultrasonic inspection is based upon Rayleigh-wave propagation. *See also* SURFACE WAVE.

R-clip *See* RETAINING CLIP.

reaction 1. The force, or force and couple, required to maintain a body, structure, or system in place. *See also* NEWTON'S LAWS OF MOTION. **2.** *See* DEGREE OF REACTION. **3.** *See* CHEMICAL REACTION.

reaction chamber A chamber within which a controlled chemical reaction occurs, for example a bomb calorimeter or the combustion chamber of an engine.

reaction injection moulding (RIM) Moulding of plastic parts by the confluence of two streams of reactants into the die. *See also* INJECTION MOULDING.

reaction stage An axial-flow turbomachine stage consisting of a rotor and a stator (or nozzle row) with the enthalpy or pressure change divided between the two. If the division is equal, the degree of reaction is 50%. The diagram shows blading cross sections typical of a reaction stage together with velocity triangles for the flow into and leaving the rotor. A **reaction turbine** consists of a series of reaction stages. *See also* IMPULSE–TURBINE STAGE.

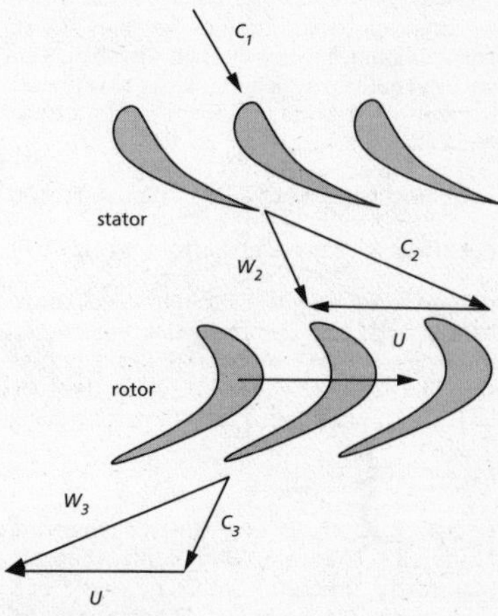

reaction stage

reactor A vessel for containing and controlling a biochemical, chemical or nuclear reaction process.

reactor core (core) That part of a nuclear reactor where the fuel assemblies are located and where the heat-producing nuclear reactions occur.

real gas *See* PERFECT GAS.

realizability (physical realizability) The possibility, for a given theoretically-determined transfer function, of constructing a physical system showing that transfer function. A theoretically-determined transfer function which cannot be realized by constructing a system showing that transfer function has no practical value.

reamer A cutting tool having longitudinal or helical flutes for final finishing of drilled holes to achieve the specified diameter to high accuracy. A taper reamer produces holes for tapered pins. *See also* EXPANDING REAMER.

r

reattachment The phenomenon in which a separated boundary layer regains contact with a surface. Recirculating flow occurs within the reattachment region.

reboiler (kettle reboiler) A type of shell-and-tube heat exchanger, usually horizontal, in which steam passes through the tubes and vaporizes liquid flowing through the shell. They are used in the petroleum industry to vaporize a fraction of the bottom product from a distillation column.

rebound hardness *See* DYNAMIC HARDNESS.

receiver *See* AIR RECEIVER; RESERVOIR.

receptivity Any mechanism by which disturbances, such as those originating from surface roughness and free-stream turbulence or pressure fluctuations, enter a laminar boundary layer and provide initial conditions for instability growth and ultimately transition to turbulence.

reciprocal theorem (Betti's theorem, Maxwell's theorem, Rayleigh's theorem, reciprocity theorem) A linearly-elastic body or structure, loaded in two different ways, will have different displacements and rotations under the different loads and couples. The theorem states that the work done by the forces and couples of the first loading system on the corresponding displacements and rotations of the second loading system is equal to the work done by the forces and couples of the second loading system on the displacements and rotations of the first loading system. That is, the displacement (or angle of rotation) at a point A due to a load (or couple) acting at B is equal to the displacement (or rotation) at B due to the same load (or couple) acting at A. The theorem may be extended to bodies in motion or in vibration by including inertia forces. It can be useful in solving problems in elasticity, and in the construction of influence lines in the theory of structures. The theorem was discovered independently by Maxwell and by Mohr, and generalized by Betti and Rayleigh.

reciprocating machine A machine in which pistons connected to a crankshaft move back and forth in cylinders. In the case of an engine, the piston is driven by the varying pressure of the working fluid, whereas for a compressor or expander, the working fluid is compressed or expanded by the pistons which are driven by the crankshaft.

recirculating-ball steering A steering mechanism consisting of a worm and nut, interspersed with recirculating balls. *See also* BALL SCREW AND NUT.

recirculating flow A flow, such as occurs downstream of a sudden step, in which fluid particles travel around a closed path, i.e. for part of the time the local flow direction is reversed.

recovery The return to the original dimensions on unloading a body. It may be instantaneous or time-dependent. *See also* SPRINGBACK.

recovery factor (temperature-recovery factor, $\mathcal{R}$) A parameter that characterizes the frictional heating of a surface in high-speed gas flow, defined by $\mathcal{R} = (T_{AS} - T_\infty)/(T_{0\infty} - T_\infty)$ where T_{AS} is the adiabatic wall temperature **(recovery temperature)**, $T_{0\infty}$ is the free-stream stagnation temperature, and T_∞ is the free-stream static temperature. For laminar flow $\mathcal{R} = \sqrt{Pr}$; for turbulent flow $\mathcal{R} \approx \sqrt[3]{Pr}$ where Pr is the Prandtl number of the gas.

recrystallization The process of forming new grains when cold-worked metal alloys are heated to above about a half of their absolute melting temperatures.

rectangular-configuration robot *See* CARTESIAN-COORDINATE ROBOT.

rectifier *See* FLUIDICS.

rectilinear motion Motion in a straight line.

recuperative air heater A recuperator, usually a counterflow type, in which waste heat is recovered from flue gas and used to pre-heat combustion air in power plants, chemical plants, etc.

recuperator *See* HEAT EXCHANGER.

redrawing When a single deep drawing operation does not produce a part of the required dimensions, the drawing process is repeated one or more times to achieve them.

reduced modulus (E_r) (Unit Pa) Plastic buckling may be analysed using the reduced modulus, $E_r = 4EE_t/(\sqrt{E} + \sqrt{E_t})^2$ where E_t is the tangent modulus and E is Young's modulus, in place of E in the Euler buckling formulae. *See also* SECANT MODULUS.

reduced property In thermodynamics, the ratio of the value of any property of a substance, such as absolute temperature (**reduced temperature**), pressure (**reduced pressure**) or specific volume (**reduced specific volume, pseudo-reduced specific volume**), to the value of that property at the critical point. *See also* PSEUDO-REDUCED SPECIFIC VOLUME.

reducing valve A valve used to reduce the pressure of fluid flowing through a pipeline.

reduction gearing A set of meshed gears, as in a gearbox, for which the output rotation speed is lower than that of the input, with a concomitant increase in torque.

redundancy A term for a structure that has more members than required to make it simply stiff, and so statically indeterminate.

redundant system A system in which critical components are duplicated so that the system will continue operating within specification despite the failure of a critical component. Where very high reliability is required, critical components may be triplicated.

Redwood viscometer A type of viscometer primarily used to determine the viscosity of oils and other liquid petroleum products. It is based upon the time for a given quantity of liquid to flow through a short capillary tube.

reed valve A valve in which one or more elastic cantilevers permit flow in one direction only. Applications include two-stroke petrol engines, pulse jets, and some compressor designs.

re-entrant Any sections that point inwards in a body having a generally outwards-pointing shape.

reference circle *See* TOOTHED GEARING.

reference cylinder *See* TOOTHED GEARING.

reference input *See* SET POINT.

reference temperature 1. In dimensional metrology, the reference temperature is 20°C. **2.** *See* STANDARD REFERENCE STATE.

reflectance (reflectivity, ρ) The fraction of radiant flux incident upon a surface that is reflected by that surface. The term also applies to the transmission of radiation by a volume of fluid. *See also* ABSORPTANCE; TRANSMITTANCE.

reflux valve *See* NON-RETURN VALVE.

refrigerant The two-phase working fluid in a refrigeration cycle.

refrigeration condenser *See* VAPOUR-COMPRESSION REFRIGERATION CYCLE.

refrigeration cycle Any thermodynamic cycle designed to transfer heat from lower-temperature regions to higher-temperature ones. The reversed Carnot cycle is the basic liquid-vapour cycle. The reverse Brayton cycle is the basic gas refrigeration cycle. Practical refrigeration cycles include the vapour-compression and vapour-absorption cycles.

refrigerator (refrigeration system) A cooling device operating on a refrigeration cycle.

regelation The process in which, for a substance such as ice that expands upon freezing and for which the melting point decreases with increasing pressure, a small region melts under high pressure and then refreezes when the pressure is reduced. The process may be illustrated by the passage through a block of ice of a weighted wire made of conducting material.

regeneration A thermodynamic process in which heat is transferred to a thermal-energy storage device (a regenerator) during one part of a cycle and back to the working fluid during another. Both the Stirling cycle and the Ericsson cycle involve regeneration, which can also be used to increase the efficiency of the Brayton cycle.

regenerative braking *See* DYNAMIC BRAKING.

regenerative cooling 1. In a liquid-propellant rocket engine, the use of the fuel and/or oxidant to cool the combustion-chamber and nozzle walls before being injected into the combustion chamber. **2.** A two-stage method for cooling compressed gas by first expanding it, then passing it through a heat exchanger.

regenerative pump (regenerative turbine pump, peripheral pump) A pump with a double-sided impeller having a large number of radial blades. The pressure of the pumped liquid increases progressively over several revolutions of the impeller. It is particularly suited to producing large heads at small flow rates without cavitation.

regenerative Rankine cycle A Rankine cycle in which the feedwater is heated before it enters the boiler. As shown in the diagram, an **open feedwater heater (direct-contact feedwater heater, regenerator)** mixes the feedwater with steam extracted just downstream of

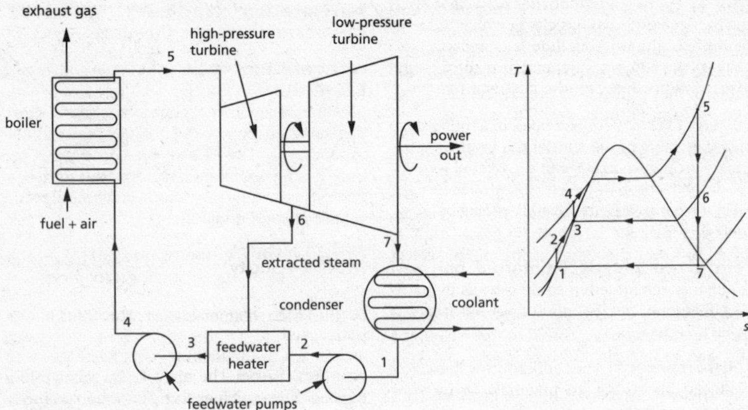

regenerative Rankine cycle

the high-pressure turbine. A **closed feedwater heater** uses extracted steam to heat the feedwater in a heat exchanger without mixing. The thermodynamic cycle is shown on a plot of temperature (*T*) *vs* specific entropy (*s*).

regenerator 1. A type of heat exchanger in which the hot and cold fluids alternately occupy the same space within the exchanger core. In the arrangement shown in the diagram, this is accomplished by switching the paths of the two fluids using valves. *See also* ROTARY REGENERATOR.
2. In a thermodynamic cycle, a device that takes energy from the working fluid during one part of the cycle and returns it during another. For example, in the Brayton cycle, hot, low-pressure gas leaving the turbine is used to increase the temperature of the air leaving the compressor.

register A ring of angled turning vanes designed to introduce swirl into the airflow entering a combustion chamber to improve flame stabilization.

regulator *See* STARTING VALVE.

regulator problem A control problem where the requirement is to maintain the plant output at a steady value despite disturbances.

reheat Increasing the enthalpy of the exhaust gases of a turbojet engine by the combustion of additional fuel injected into the afterburner.

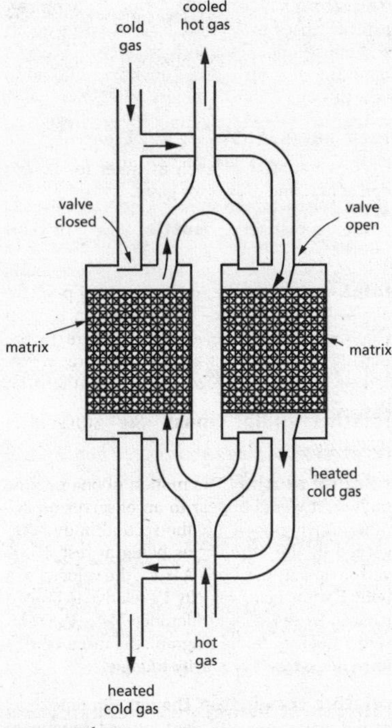

regenerator

reheat cycle A thermodynamic cycle such as the Rankine or Brayton cycle, in which the temperature of the working fluid is increased, typically at constant pressure, part-way through the expansion process. *See also* RESUPERHEATING.

reheat factor (*RF*) The ratio, in a multi-stage steam turbine, of the sum of the isentropic enthalpy drop in each stage to the overall isentropic enthalpy drop. If the efficiency of each stage η_S is the same and the overall efficiency is η_O, then $RF = \eta_O/\eta_S$.

reinforced plastic (reinforced polymer) A polymer reinforced in some way, as by particles, filaments, etc. *See also* GLASS-REINFORCED PLASTIC/POLYMER.

reinforcement The addition of particles, platelets, fibres, etc. to materials to increase stiffness or strength.

***Rel* (R_{el})** The symbol for yield point.

relaminarization (reverse transition) The suppression of turbulent fluctuations in a strongly accelerated turbulent boundary layer such that the boundary layer has many of the characteristics of a laminar flow. Relaminarization can also occur in a diffuser if the local duct Reynolds number falls to a low enough value.

relative density (specific gravity) The ratio of the density of a substance to that of a reference substance, such as water for liquids and solids, and dry air for gases.

relative-density bottle *See* DENSITY BOTTLE.

relative humidity (ϕ) The ratio or percentage of the actual mass of moisture in a given volume of air at a given temperature to the maximum possible mass of moisture at the same temperature. *See also* SPECIFIC HUMIDITY.

relative molar mass *See* MOLECULAR WEIGHT.

relative motion The motion of one moving body as it would appear to an observer on another moving body, i.e. the second body is regarded by the observer as being at rest. If the vector velocity of a body A is $\boldsymbol{V_A}$, the velocity of a body B with vector velocity $\boldsymbol{V_B}$ relative to body A is given by the vector difference $\boldsymbol{V_B} - \boldsymbol{V_A}$ (**relative velocity**). In the diagram, the three vectors form the sides of a velocity triangle.

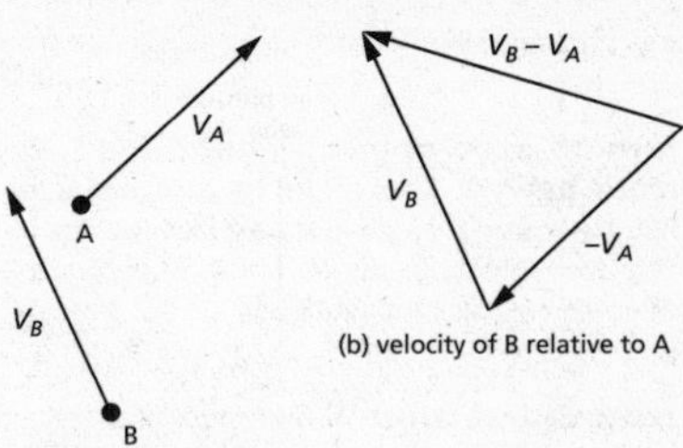

(a) absoute velocities

(b) velocity of B relative to A

relative velocity

relative roughness The ratio, in pipe flow, of the actual or equivalent surface roughness to the pipe diameter. *See also* COLEBROOK EQUATION.

relative wind The air velocity relative to a moving blade, such as that of a wind turbine.

relaxation The phenomenon in which the stress in a body that is kept at a fixed displacement (strictly fixed strain) reduces with time. In a **relaxation test** a specimen is rapidly strained to a chosen level, after which the reduction in stress at constant strain is monitored. *See also* CREEP.

relaxation method A method of successive approximation applied to difference equations. *See also* FINITE DIFFERENCE.

relaxation process A phenomenon physical process in which a body, flow, or system adjusts to sudden changes in the imposed conditions. The **relaxation time** is the time taken to approach a new equilibrium state.

relaxed modulus (Unit Pa) In rate-dependent solid materials, any of the elastic moduli determined under quasi-static conditions. *See also* UNRELAXED MODULUS.

relay valve *See* PILOT VALVE.

reliability The extent to which a component or system will function repeatedly as intended under the design operating conditions.

relief (relief angle) The clearance behind and beneath the cutting edge of a tool.

relief valve *See* SAFETY VALVE.

remote-centre compliance A device attached to the end effector of a robot that allows a component held by a gripper to translate sideways without rotation. It is used to compensate for a small positioning error when fitting one part into another. *See also* PASSIVE ACCOMODATION.

remote area power system (RAPS, remote power) A term for power-generating plants, typically small-scale and using renewable technology, in remote areas of a country, often operating independently of the national grid.

removed section A representation, in an engineering drawing, of the cross section of a component or assembly removed from, and often to a different scale to, the main drawing. *See also* REVOLVED SECTION.

renewable energy (sustainable energy) Energy obtained from naturally-occurring renewable-energy sources (**renewables**) such as the sun (solar power), tides and ocean waves, hot rocks (geothermal), and wind. It may also include biomass and biofuels.

repeatability The RMS difference between successive results when a robot or other plant is repeatedly moved to the same set point.

repeatable Experimental results which, within experimental uncertainty, are the same every time a given experiment is performed are said to be repeatable.

representative strain *See* EFFECTIVE STRAIN.

representative stress *See* EFFECTIVE STRESS.

research octane number *See* OCTANE NUMBER.

reservoir 1. In thermodynamics, a system that can exchange heat, work, or matter with another system without changes in its properties. A **reservoir of heat (thermal reservoir)** is that part of the surroundings that exchanges energy with a system at a different temperature and is of sufficient capacity that, however much heat crosses the boundary, the reservoir temperature is unchanged. *See also* SINK. **2.** A container for storing liquid, sometimes with an open surface, or a closed container for pressurized gas (**receiver**).

residence time 1. In a chemical process, the average time a particle spends within the reactor. **2.** The average time a fluid element passes through a deformation field, such as lubricant in a rolling mill.

residual strength The strength of a damaged body containing defects induced by microcracking, thermal shock, etc.

residual stress (internal stress) An internal-stress system found in components that have experienced elastic unloading from non-uniform plastic-strain fields during manufacture.

resilience 1. The ability, on unloading, of a body to recover and spring back from its displaced condition. **2.** The energy per volume stored elastically when loaded. *See also* STRAIN ENERGY.

resin Any synthetic solid or liquid organic polymer, thermoplastic or thermosetting, that can form the main ingredients of a plastic material.

resin-transfer moulding *See* TRANSFER MOULDING.

resist *See* PHOTOLITHOGRAPHY.

resistance curve An increase in fracture toughness with crack propagation shown by rising plots of toughness *vs* growing crack length. *See also* FRACTURE MECHANICS.

resistance thermometer A thermometer in which the sensing element is basically a metal wire, usually of platinum, for which the variation of resistance with temperature is known accurately. A **resistance pyrometer** is a resistance thermometer intended for use up to about 1000°C.

resistance welding *See* WELDING.

resisting moment The distribution of bending stresses within a body set up in reaction to an externally-applied moment.

resolution The smallest difference that can be measured by any instrument.

resolver An inductive device for the measurement of translation or rotation.

resonance In a mechanical, acoustic, or electrical system, the circumstance when a forcing frequency coincides with one of the system's natural frequencies, leading to a peak in the amplitude (theoretically infinite without damping) of oscillation. A **resonant frequency (critical frequency, resonance frequency)** is any frequency at which resonance occurs.

resonator **1.** Any device that exhibits resonance at one or more frequencies. *See also* HELMHOLTZ RESONATOR. **2.** The part of a motor-vehicle silencer system that assists the main silencer in reducing noise at certain frequencies.

response The behaviour, as a function of time, of the output of a system under specified conditions or caused by a specified input. *See also* CRITICALLY-DAMPED RESPONSE; FORCED RESPONSE; HARMONIC RESPONSE; IMPULSE RESPONSE; STEP RESPONSE; UNDAMPED RESPONSE.

restrictor An orifice plate introduced into a duct or inlet to restrict the flow rate, such as the airflow into the turbocharger of a high-performance piston engine.

resultant force The single force obtained by vector addition of all forces acting on a body. *See also* FUNICULAR POLYGON.

resultant moment The single moment obtained by vector addition of all moments acting on a body.

resuperheater *See* STEAM REHEATER.

retaining clip (R-clip, hair cotter pin) A wire clip roughly with the shape of the letter R. The straight part of the clip passes through a hole drilled into a shaft or rod such that a wheel is held in position on the shaft, or the shaft is prevented from moving axially.

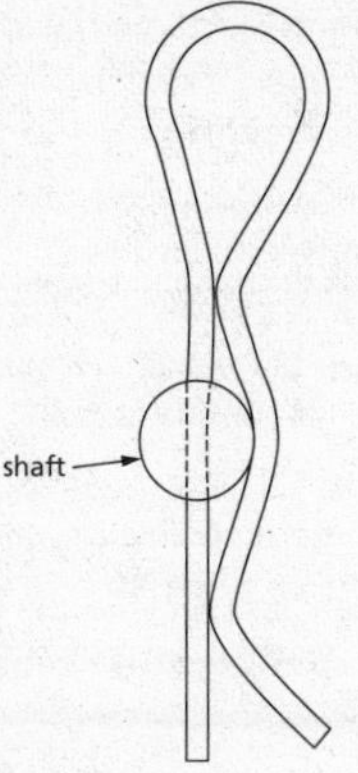

retaining clip

retarder The generic term for a hydrodynamic braking system; basically, a fluid coupling in which the turbine wheel is stationary.

return-flow oil burner An atomizing oil burner in which part of the oil supplied to the atomizer is returned to the oil reservoir or the supply line.

reverberatory furnace A process furnace in which the material being processed comes into contact with the combustion gases but not the fuel. *See also* FIRE BRIDGE.

reverse Brayton cycle (reversed Brayton cycle) *See* GAS REFRIGERATION CYCLE.

reverse Carnot cycle A refrigeration or heat-pump cycle using wet vapour as the working fluid. As shown in the diagram, the vapour is compressed isentropically (1–2), passed through a condenser (2–3), then expanded isentropically (3–4) and finally evaporated at constant pressure back to its original state (4–1). In a refrigeration cycle, heat is rejected from the condenser to the surroundings and the evaporation extracts heat from the cold space. In a heat pump, the condenser rejects heat to a hot-water tank.

reverse cycle (reversed heat-engine cycle) A thermodynamic cycle in which heat is received at a low temperature and rejected at a high temperature, while work is done on the fluid. Reverse cycles are the basis for heat pumps and refrigeration cycles. *See also* POWER CYCLE.

reverse engineering The disassembly of a machine, mechanism, system or device, measurement of its component parts, and identification of the materials used so that if required, a functioning replica can be produced.

reverse-flow annular combustor A compact gas-turbine combustion chamber design in which the hot gases are turned through 180° before passing through the turbine blades. A typical application is to helicopter turbo-shaft engines.

reverse Joule cycle *See* GAS REFRIGERATION CYCLE.

reverse osmosis A filtration process used to extract a solvent, usually water, from a solution such as salt water. High pressure, in excess of the osmotic pressure, is applied to the solution on one side of a semi-permeable membrane to force the solvent into low-solute solution on the other side.

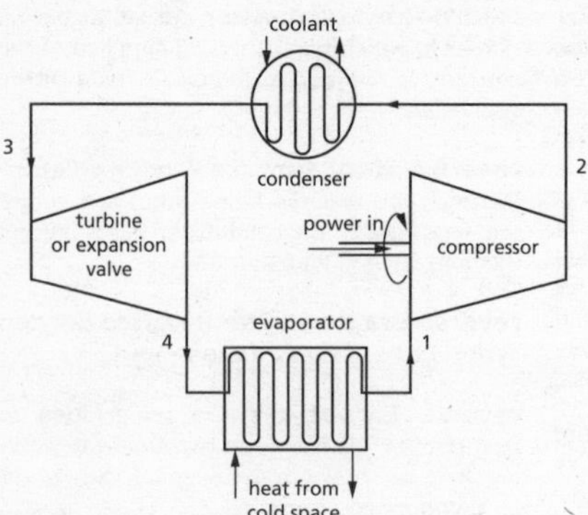

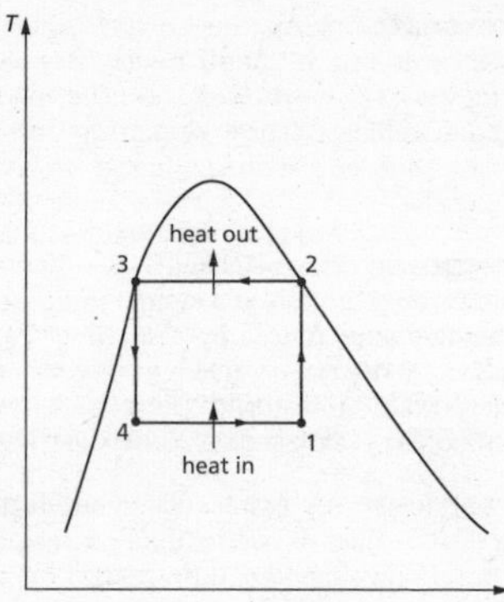

reverse Carnot cycle (Carnot refrigerator)

reverse pitch The pitch of a variable-pitch propeller that produces negative thrust.

reverse-return system A closed-loop heating or cooling system in which the working liquid is pumped through a series of loads, such as heat exchangers, in parallel and returned to the boiler or evaporator after the load furthest from the pump. *See also* DIRECT-RETURN SYSTEM.

reverse sublimation *See* SUBLIMATION.

reverse transition *See* RELAMINARIZATION.

reversibility *See* REVERSIBLE PROCESS.

reversible cycle A reversible thermodynamic process in which the initial and final states are identical.

reversible heat engine A heat engine operating on the Carnot cycle.

reversible path A line representing each stage in a reversible process on a diagram of one thermodynamic property plotted *vs* another, such as pressure *vs* specific volume.

reversible-pitch propeller A propeller for which the pitch can be adjusted to produce zero or negative thrust.

reversible-pitch turbine A hydraulic turbine, typically installed in a tidal barrage, in which the blade pitch can be reversed to allow both ebb and flood power generation.

reversible process In thermodynamics, a system undergoes a reversible process if it takes place as a succession of equilibrium states and the reverse process restores both the system and its surroundings to their original states. The net heat and work exchange between the system and the surroundings must be zero. **Reversibility** is an unachievable idealization because either friction or heat transfer are always present. *See also* ISENTROPIC PROCESS.

reversible pump A machine that pumps fluid either forwards or backwards, depending upon the sense of rotation of the blades.

reversible pump turbine A machine in a pumped-storage system that pumps water up to a reservoir and subsequently acts as a turbine to generate power.

reversible work **1.** In thermodynamics, the maximum amount of useful work that can be produced, or the minimum work that needs to be supplied, as a system undergoes a process between specified initial and final states. *See also* EXERGY. **2.** In elasticity, the elastic strain energy.

reverted gear train Gearing where the input and output shafts are co-axial.

revolute configuration robot *See* ARTICULATED ROBOT.

revolute joint **1.** *See* PIN JOINT. **2.** *See* ROTATIONAL JOINT.

revolution (revolving, rotary motion, rotation) Motion around an axis or centre.

revolved section In an engineering drawing, similar to a removed section but not removed from the main drawing, although often broken away on both sides.

Reynolds analogy The assumption, for convective heat transfer in a turbulent boundary layer, that the eddy diffusivities for heat and momentum transfer are equal, i.e. the turbulent Prandtl number is equal to unity.

Reynolds decomposition In a turbulent flow, the separation of any fluctuating quantity, such as a velocity component, temperature, enthalpy, and density, into a mean (steady) part and a fluctuating part, which on average is zero. **Reynolds averaging** is a method of averaging products of the fluctuating parts, with respect to time. Applying Reynolds decomposition to the unsteady Navier–Stokes equations and then Reynolds averaging, results in the set of equations that describes the time-averaged behaviour of a turbulent flow, the **Reynolds-averaged Navier–Stokes equations (RANS equations, Reynolds equations)**. The averaging introduces such terms as the Reynolds normal and shear stresses that must be modelled in order to predict flow behaviour.

Reynolds lubrication equation An equation for the spatial variation of pressure p in laminar flow through a narrow channel, such as that between two surfaces moving relative to each other. It is fundamental to the analysis of many hydrodynamic lubrication problems. For flow of a liquid lubricant of constant viscosity μ, the usual form of the equation is

$$\frac{\partial}{\partial x}\left(h^3\frac{\partial p}{\partial x}\right)+\frac{\partial}{\partial y}\left(h^3\frac{\partial p}{\partial y}\right)=6\mu U\frac{dh}{dx}$$

where $h(x)$ is the variation of the channel width with longitudinal distance x, y is the distance from one side of the channel, and U is the relative velocity between the two surfaces. *See also* SOMMERFELD NUMBER.

Reynolds normal stress *See* TURBULENT FLUCTUATION.

Reynolds number (*Re*) A non-dimensional parameter that arises in the study of viscous fluid flows, defined by $Re = \rho VL/\mu = VL/v$ where ρ is the density, μ is the dynamic viscosity and v is the kinematic viscosity of the fluid, V is a characteristic velocity of the flow and L is a length scale for the flow geometry. *Re* represents the relative importance of the inertial and viscous stresses in the flow. For a Newtonian fluid in which there are property variations, such as in heat-transfer problems, v or ρ and μ are chosen at a specific location, such as the surface or free stream, or corresponding to an average temperature. The values of V and L are chosen according to the problem. For example, in duct flow, V is typically the bulk velocity and L is the hydraulic diameter. For flow over an aerofoil, V is typically the relative speed between the aerofoil and the approach flow and L is the chord length.

Reynolds shear stress *See* TURBULENT FLUCTUATION.

Reynolds transport theorem An equation for the rate of change with time t of a fluid property of a system B_{SYST}, such as mass, momentum, energy, or angular momentum, in terms of the amount of that property contained within a control volume (CV) taking account of fluxes across the control surface (CS). A compact form of the equation is

$$\frac{d}{dt}(B_{SYST})=\frac{d}{dt}\left(\int_{CV}\beta\rho d\mathcal{V}\right)+\int_{CS}\beta\rho(\boldsymbol{V}.\boldsymbol{n})dA$$

where $SYST$ refers to the system, β is the intensive property corresponding to B, ρ is the fluid density, $\boldsymbol{V}$ is the vector velocity on CS, $\boldsymbol{n}$ is an outward unit vector on CS, dA is an element of area of CS, and $d\mathcal{V}$ is an element of volume within CV.

RF *See* REHEAT FACTOR.

R-factor (r-factor, R-value) The ratio of width true strain to through-thickness true strain in the plastic region during the tensile testing of sheet materials. In an isotropic material, the R-factor is unity. In practice, it varies with orientation in the plane owing to anisotropy and this is important in drawability. *See also* EARING.

RFD *See* RADIANT FLUX DENSITY.

rheocasting Casting of metal alloys in the partially-solidified pasty state that occurs between the liquidus and solidus temperatures.

rheogram *See* FLOW CURVE (2).

rheology The science of the deformation and flow of matter, especially the flow of non-Newtonian liquids and their properties.

rheometer An instrument for determining the viscous properties of a liquid. Sophisticated rheometers, such as a **rheogoniometer (Weissenberg rheogoniometer)**, can be used to measure elastic properties, such as the first- and second-normal stress differences, and oscillatory properties (storage and loss moduli).

rheostatic braking *See* DYNAMIC BRAKING.

rib A thin protruding strip of material attached to a surface, usually to stiffen it.

riblets Streamwise parallel ribs or grooves (typically 10 to 50 μm high and 10 to 500 μm apart) on a surface exposed to a turbulent boundary layer, designed to modify the near-wall turbulence structure and so reduce surface friction.

Ricardo–Cussons HYDRA engine A modular-design, direct-injection, single-cylinder, naturally-aspirated, water-cooled piston engine widely used for research into engine design, fuel, oil, and additive performance.

Richardson number (*Ri*) A non-dimensional-parameter that arises in flow problems where buoyancy and inertia forces are important. If the flow involves both density and velocity gradients, the usual definition is

$$Ri = -g\frac{\partial \rho}{\partial z}/\rho\left(\frac{\partial u}{\partial z}\right)^2$$

where g is the acceleration due to gravity, ρ is the fluid density at height z, and u is the horizontal velocity component at height z. In the case of an interface between two immiscible fluids with a density difference $\Delta\rho$ between them and a mean density $\bar{\rho}$, Ri can be defined as $g\Delta\rho L/\bar{\rho}V^2$ where L is a characteristic vertical length and V is a characteristic velocity. *See also* BRUNT–VÄISÄLÄ FREQUENCY; FROUDE NUMBER.

Richardson's annular effect The occurrence of a velocity peak in the near vicinity of the wall of a pipe in which there is oscillatory flow.

rich mixture *See* STOICHIOMETRIC MIXTURE.

right-handed thread *See* SCREW.

rigid body An idealized body that has invariant size and shape irrespective of the forces and displacements applied to it.

rigid-body dynamics *See* APPLIED MECHANICS.

rigid-body mode **1.** The degrees of freedom of a body deforming elastically comprise rigid-body displacements and elastic displacements. In vibrations of a structure or component mounted on low stiffness (soft) springs, the supported body may behave, for all practical purposes, in rigid-body mode with all oscillations confined to the supports. **2.** Rigid-body mode is required for accurate FEM simulations of the deformation of bodies, as rigid-body motion must be absent in the mesh. It is achieved by prescribing degrees of freedom (displacements and rotations) at particular nodal points.

rigid-body rotation *See* STRAIN.

rigid-body translation *See* STRAIN.

rigid coupling A coupling in which two shafts are locked together, for example by mating flanges bolted together. It is essential that the two shafts are precisely collinear.

rigidity The quality of being undeformable.

rigidity modulus *See* SHEAR MODULUS.

RIM *See* REACTION INJECTION MOULDING.

rim clutch A clutch in which the contacting surfaces engage on its rim.

rim-generator turbine *See* STRAFLO TURBINE.

ring gauge A ring of precise internal diameter used to check the diameter of a cylindrical component.

ring gear **1.** A gear, usually of large diameter, where the teeth are formed on the inside or outside of an annulus. **2.** A gear formed by teeth on the periphery of the flywheel of a piston engine with which the pinion of the starter motor engages. **3.** A gear within the differential of a motor-vehicle transmission, used to divide the driveshaft torque between two output shafts that may rotate at different speeds.

ringing High-frequency oscillation in a system excited at a lower frequency than the oscillation itself.

ripple *See* CAPILLARY WAVE.

riser A pipe used to convey fluid to a higher level, such as oil from a subsea well.

rise velocity The speed with which a gas or vapour bubble travels upwards through a liquid. *See also* TAYLOR BUBBLE.

rivet A short rod with a head on one end that is inserted through aligned holes in plates to be joined, after which a second head is made on the protruding shank by hammering or forming.

The most common head shapes are flat, domed, and inverse conical. In an array, the **rivet pitch** is the distance between the centres of adjacent rivets. Failure may occur by different mechanisms or modes. For example, a riveted lap joint of width w and where both plates have the same thickness t may fail by (a) shear of a rivet requiring load $F_{shear} = (\pi d^2/4)(\tau_Y)_{rivet}$ (b) tearing of the plate across the remaining ligament at $F_{tear} = 2(w - d)t(\sigma_Y)_{plate}$ or (c) high bearing stresses in the hole at $F_{bearing} = td(\sigma_Y)_{plate}$ where $(\tau_Y)_{rivet}$ is the shear yield stress of the rivet and $(\sigma_Y)_{plate}$ is the yield strength of the plate. These forces vary with d as shown schematically in the diagram. Greatest load-carrying capacity occurs (in the case illustrated) at the transition between rivet shear and tearing. Such considerations permit an optimum size and layout of rivets. *See also* BLIND RIVET; FAILURE; FRACTURE MECHANICS; POP RIVET.

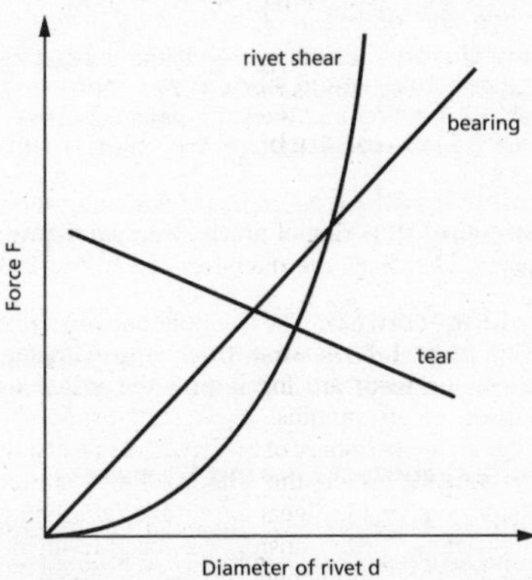

rivet

Rm **(R_m)** The symbol for ultimate tensile strength.

RMS *See* ROOT-MEAN SQUARE.

Roberval balance A weighing machine consisting of a centrally-supported parallelogram linkage with pans above. Correct measurements can be made irrespective of the position of the object and the weights in the pans.

robot The International Standards Organization (ISO) defines a robot as 'an automatically controlled, reprogrammable, multipurpose manipulator programmable in three or more axes, which may be either fixed in place or mobile for use in industrial automation applications' (Standard ISO 8373).

SEE WEB LINKS

- Website of EUROP, the European Robotics Technology Platform
- Website of EURON, the European Robotics Research Network
- Website of the Robotic Industries Association
- Website of the Japan Robot Association

robot classification The description of the types (i.e. rotational or translational) of the joints of a robot and the directions of the axes of these joints. There are five classifications: articulated (revolute), Cartesian, cylindrical, SCARA, and spherical.

robotics The study of the design, manufacture, control, operation, and integration of robots.

robot kinematics The analysis of the motion of a robot without reference to the forces or torques causing that motion (**robot dynamics**). The kinematics are normally analysed using homogeneous transforms where each joint and associated link is represented by a 4×4 matrix and the relationship between the base frame and end effector obtained as the product of these matrices.

robust control The design of a control system so as to show stability and performance within specification despite external disturbances, measurement noise, and modelling errors. Robust controllers are not adaptive.

rocker arm (valve rocker) A centrally-pivoted lever by which motion from a push rod or cam is transmitted to a valve stem in a piston engine.

rocker box The cover enclosing the valve-operating mechanism in a piston engine.

rocket (rocket vehicle) Any kind of flying vehicle, such as a missile or a spacecraft, powered by a **rocket engine (rocket motor),** which is a jet engine that produces thrust by ejecting a high-velocity gas stream through a nozzle. The gas is usually produced by the combustion of liquid or solid propellant with an oxidant. As shown in the diagram, liquid fuel can be used to cool the walls of the propelling nozzle (**regenerative cooling**).

Rockwell hardness test A direct-reading hardness test based on depth of indentation, in which a minor load is applied before the major.

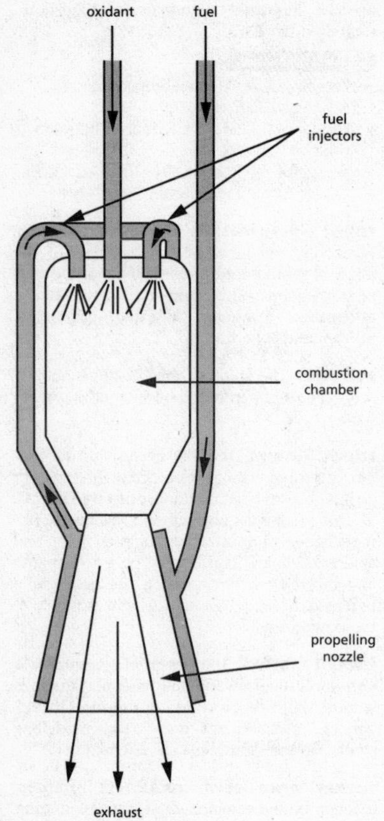

r

rocket engine

There are different scales using conical or spherical indenters for different ranges of hardness.

rod-climbing effect *See* WEISSENBERG EFFECT.

roll **1.** A cylinder employed in a mill to roll materials. **2.** Rotation of a body in motion about its longitudinal axis (**roll axis**).

roll acceleration Angular acceleration of a body during the action of rolling.

rolled-steel joist (RSJ) An I-beam formed by hot rolling.

rolled thread A thread formed on a circular bar by plastic flow in a **roll-threading machine** rather than by cutting with a die, so that it stands proud with a slightly greater diameter than the shank.

roller bearing (rolling-contact bearing) *See* BEARING.

roller cam follower A follower having a rotatable wheel at the end in contact with the cam.

roller chain A drive chain made up of a series of connected inner and outer links. The transverse pins which connect the links carry hardened hollow rollers. For heavy-duty applications, two (duplex) or three (triplex) chains may be assembled in parallel.

roller clutch (roller drive) A clutch that permits free relative rotation of the driven element (the rotor in the diagram) in one direction only. When the direction of rotation is reversed, sprung rollers in angled recesses in the rotor wedge between it and the outer race, preventing relative rotation and allowing power transmission. *See also* SPRAG CLUTCH.

roller levelling A process of flattening sheet by pulling it through a series of staggered rollers.

rolling contact The motion between two contacting surfaces when there is no slipping, i.e. their relative velocity at the point of contact is zero.

rolling-contact bearing *See* BEARING.

rolling friction The tangential force which opposes motion between bodies in rolling contact, such as a wheel on a road surface.

rolling mill A machine employing rolls to thin material. If the rolls are profiled, sections of various shapes can be made.

rolling road A band of material passing over a flat support and stretched around two parallel, rotating cylindrical rollers. Used in a wind tunnel to simulate the relative movement between a motor vehicle and the road surface.

roll straightening A process for straightening long bars and tubes emerging from a mill or drawbench.

roll-threading machine *See* ROLLED THREAD.

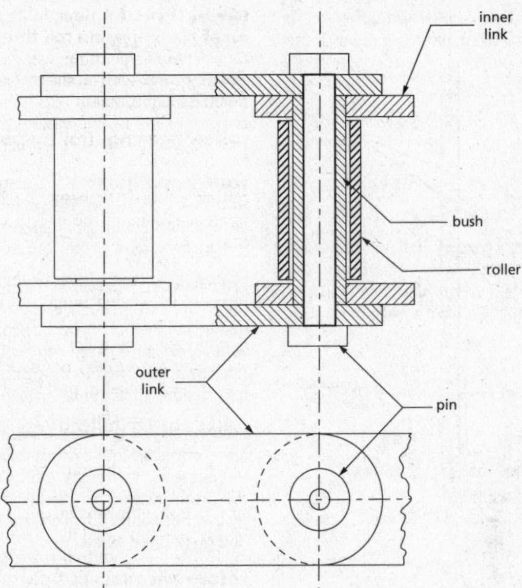

roller chain

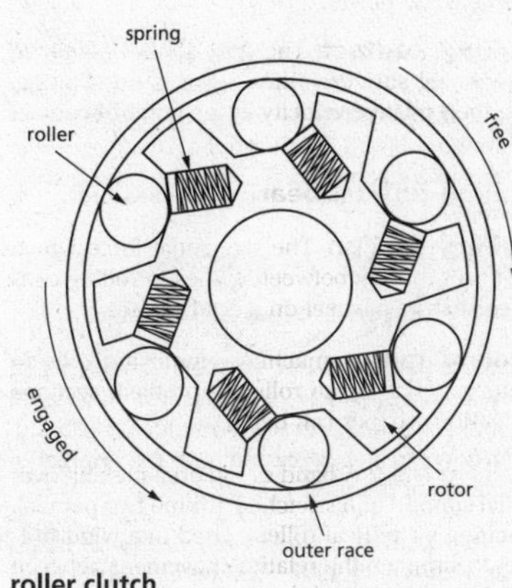

roller clutch

RON *See* OCTANE NUMBER.

root 1. The base region of a gear tooth or screw thread where adjacent flanks join. **2.** *See* FIR-TREE ROOT.

root diameter The diameter of the circle passing through the roots of teeth in gear teeth or screw threads.

root locus analysis A technique for investigating the behaviour of a control system by plotting the position of the poles of the transfer function on the complex plane, that is, plotting the real part of each pole against the abscissa and the imaginary part against the ordinate. The locus followed by the poles as a parameter (for example, the gain) is changed shows how the system behaviour is influenced by changes to that parameter.

root-mean square (RMS, σ) The square root of the mean of the sum of the squares of n observations of a variable x_i with $i = 1$ to n,

i.e. $$\sigma = \sqrt{\frac{\sum_{i=1}^{n} x_i^2}{n}}$$

It is commonly applied to obtain an average value for a quantity fluctuating spatially or temporally, especially where the quantity changes sign. It is also frequently applied to the estimation of uncertainty due to random error where the variable squared is now the difference between each observation and the mean $\bar{x}$ of all observations,

i.e. $$\sigma = \sqrt{\frac{\sum_{i=1}^{n}(x_i - \bar{x})^2}{n}}$$

See also VARIANCE.

Roots blower (rotary-piston blower) A positive-displacement gas compressor with two meshing lobed rotors that rotate within a close-fitting casing. Commonly used as a supercharger.

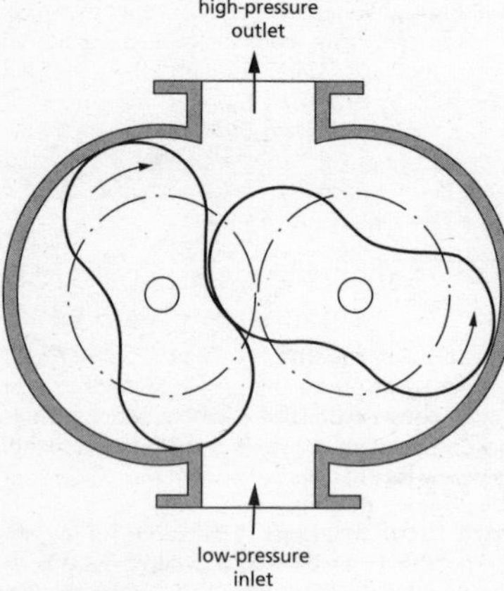

Roots blower

rope brake A band brake in which rope is employed rather than a flat strip.

Rossby number (*Ro*) A non-dimensional parameter that arises in flows where there is both rotation and translation such that Coriolis forces arise. It is defined by $Ro = V/\Omega L$ where V is a characteristic velocity, L is a characteristic length, and Ω is the angular velocity.

rotameter A type of variable-area flow meter in which fluid flows up through a vertical diverging tube containing a float. The float settles at an equilibrium position when the drag on it just balances its apparent weight.

rotaplane *See* AUTOGYRO.

rotary air heater *See* ROTARY REGENERATOR.

rotary blower (rotary compressor) A positive-displacement gas compressor which may be of the sliding-vane type, meshing-lobe type (the Roots blower), or helical-screw type.

rotary-combustion engine (rotary engine) A non-reciprocating internal-combustion engine such as the Wankel engine in which there is a rotor but no pistons.

rotary-cup burner A burner in which fuel oil is injected through a central tube onto the inside surface of a rapidly-rotating (typically 5000 rpm) cone from which it emerges as a fine spray into surrounding primary, secondary, and tertiary airstreams.

rotary dryer A device in which material to be dried is loaded into a drum that is rotated while hot gas passes through. On an industrial scale, the drum is often inclined slightly with respect to the horizontal to allow moisture to collect and be drained away.

rotary encoder *See* SHAFT ENCODER.

rotary hydraulic actuator A device that converts hydraulic power into rotational mechanical power.

rotary meter A gas flow meter similar in design to a Roots blower, in which the rotors are caused to rotate by the gas. The flow rate is determined from the number of rotations of the rotors in a given time. *See also* OVAL GEAR FLOW METER.

rotary motion *See* REVOLUTION.

rotary piston blower *See* ROOTS BLOWER.

rotary piston flow meter A rotary liquid flow meter in which an eccentrically mounted cylindrical rotor is caused to rotate by the liquid. The flow rate is determined from the number of rotations of the rotor in a given time.

rotary pump A positive-displacement pump that pumps a liquid by rotation of internal components, such as a gear pump, lobe pump (similar to a Roots blower), or progressive-cavity pump. *See also* SEMI-ROTARY PUMP.

rotary regenerator (heat wheel, rotary air heater) A regenerator with a core in the form of a wheel rotating within a housing with heat-absorbing material (the matrix) between the hub and the wheelrim. At any instant, hot gas flows through one half of the core and cold through the other. As the wheel rotates, heat is transferred from the hotter to the colder stream. *See also* LJUNGSTRÖM PRE-HEATER.

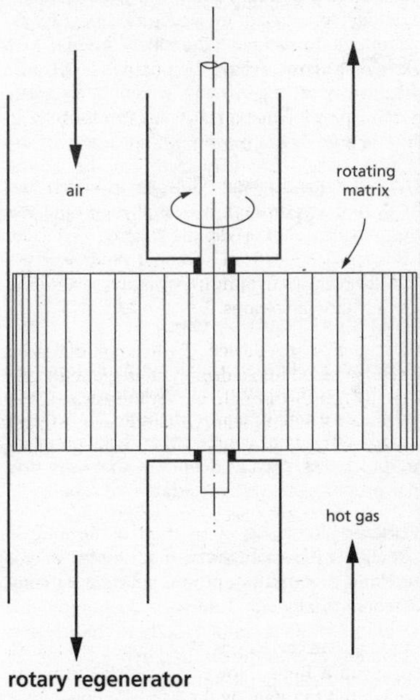

rotary regenerator

rotary swager A device in which hammers are arranged to deliver radial blows to reduce the diameter of ductile bars.

rotary valve (rotating valve) A cylindrical or conical plug in which there is a transverse hole through which fluid can flow when the hole is aligned with the adjacent piping.

rotary-vane compressor *See* SLIDING-VANE COMPRESSOR.

rotary variable differential transformer (RVDT) An inductive sensor which provides an output voltage proportional to the angular displacement of the core of the sensor. *See also* LINEAR VARIABLE DIFFERENTIAL TRANSFORMER.

rotating beam test *See* FATIGUE.

rotating follower *See* CAM FOLLOWER.

rotation 1. *See* REVOLUTION. **2.** *See* RIGID-BODY ROTATION. **3.** The angle between the tangents (slopes) at any two sections along a loaded beam is called the rotation of the beam between these two locations.

rotational flow A fluid flow in which the vorticity is non-zero, i.e. the fluid particles rotate. *See also* IRROTATIONAL FLOW.

rotational inertia *See* MOMENT OF INERTIA.

rotational joint (revolute joint) In robotics, a single degree-of-freedom joint where the controlled variable is the joint angle.

rotational kinetic energy (rotational energy) (Unit J) The kinetic energy of a rigid body rotating with angular velocity ω about an axis is given by $I\omega^2/2$ where I is the moment of inertia about that axis. When rotating about a fixed point it is given by $(A\omega_1^2 + B\omega_2^2 + C\omega_3^2)/2$ where A, B, and C are the principal moments of inertia about the Cartesian axes 1, 2 and 3, and ω_1, ω_2, and ω_3 are the component angular velocities about those axes.

rotational momentum *See* ANGULAR MOMENTUM.

rotational moulding (rotational casting) A method of producing plastic components in which polymer powder is spread over the interior surfaces of a heated mould by tumbling. The powder then melts to form the object that is subsequently cooled and removed.

rotational stability A state in which the angular displacement of a body is always opposed by a restoring couple.

rotational stiffness *See* TORSIONAL RIGIDITY.

rotational viscometer (Couette viscometer, torque-type viscometer) An instrument for measuring the dynamic viscosity μ of a Newtonian liquid consisting of two vertical concentric cylinders of length L and almost equal radii, the inner of which (the bob, radius R_1) is rotated at angular velocity ω while the outer (the cup, radius $R_2 > R_1$) is stationary and maintained at constant temperature. The shear rate $\dot{\gamma}$ within the liquid is given approximately by $\dot{\gamma} = \omega R_1/(R_2 - R_1)$ and the torque T on the inner cylinder by $T = 2\pi R_1^2 L\mu\dot{\gamma}$, which allows μ to be determined. In **controlled-strain mode,** ω is specified and T is measured, while in **controlled-stress mode** T is specified and ω is measured. In some instruments, the bob is stationary and the outer cylinder rotates. *See also* RHEOMETER.

rothalpy The rotational stagnation enthalpy in a swirling flow.

rotodynamic machine A machine in which energy is exchanged between a flowing fluid and a spinning rotor or impeller. Examples include hydraulic, wind, steam and gas turbines, centrifugal pumps, and compressors.

rotor A part of a machine that rotates on a shaft (**rotor shaft**) about its own axis, such as the blade-carrying discs of a turbine, the blades of a helicopter, or the rotating parts of a Roots blower.

rotorcraft Any aircraft, such as a helicopter or autogyro, that uses rotors, both powered and unpowered, to generate lift.

rotor-disc area *See* DISC AREA.

rotor moment of inertia *See* POLAR SECOND MOMENT OF AREA.

roughing pump A pump used to reduce the pressure in a vacuum system down to about 0.1 Pa, below which high-vacuum pumps are used.

rough machining (roughing) The removal of material from a workpiece using deep cuts and slow speeds to get down almost to the required size that is finally achieved by fine cuts at high speed.

roughness *See* SURFACE ROUGHNESS.

router A machine tool having a high-speed cutter for making slots and profiles, especially in wood.

Routh–Hurwitz stability criterion (Routh test) A technique for determining the stability of a control system from the coefficients of s^0, s^1, s^2, ..., s^n of the n terms in the denominator of the transfer function. A necessary condition for stability is that all of the coefficients are positive and non-zero. A sufficient condition for stability is that all elements are positive in the first column of a table constructed from the coefficients using a system of rules.

Rp **(R_p)** The symbol for proof stress. The notation $Rp^{0.2}$ or $R_{p0.2}$ is employed for the 0.2% proof stress.

RP *See* REINFORCED PLASTIC.

rpm (*N*) An abbreviation for revolutions per minute. It is a widely used non-SI unit for rotational speed. The corresponding angular velocity ω in rad/s is given by $\pi N/30$. **rps (revolutions per second)** is also used, the angular velocity then being $2\pi N$.

R6 failure assessment diagram A methodology, employed in structural-integrity assessment, to estimate the safety margin of a flawed structure, where the possibility of elastic fracture is set against the possibility of widespread plastic yielding. It takes the form of an interaction diagram with *Kr*, the ratio of the applied stress intensity factor to the critical stress intensity factor, plotted against *Sr*, the ratio of the applied load to the plastic collapse load. *See also* FRACTURE MECHANICS.

RSJ *See* ROLLED-STEEL JOIST.

RTM *See* TRANSFER MOULDING.

rubber **1.** Natural rubber is made from the latex of various plants by vulcanization, which produces a tough, elastic material. **2.** Artificial (synthetic) rubber consists of artificial polymeric substances, such as neoprene, that have similar properties to vulcanized natural rubber.

rubber forming A method of forming in which a rubber plug within a tubular ductile workpiece, surrounded by a split die, is compressed axially. As rubber is incompressible (Poisson's ratio is about 0.5), there results lateral loading against the die, and hence the formation of intricate parts is possible.

rubber hardness The indentation hardness of rubber-like materials in which the depth of indentation under load is measured. This is a measure of elastic moduli rather than yield stress as with ductile materials, since there is little, if any, permanent impression remaining on unloading rubbery materials.

rule of mixtures The properties of two or more materials in combination are given by the weighted average of the individual components, e.g. the hardness H of hypo-eutectoid plain carbon steels is given by $H = vf_\alpha H_\alpha + vf_P H_P$, where vf_α and vf_P are the volume fractions of ferrite and pearlite respectively, and H_α and H_P are their individual hardnesses. *See also* COMPOSITE MODULUS.

runner **1.** The rotor of a water turbine. **2.** A channel through which molten polymer is introduced to the mould in plastics moulding.

running The term used to indicate that a machine such as an engine is operating.

running fit *See* LIMITS AND FITS.

run on *See* DIESELING.

rupture *See* FRACTURE.

rupture disc *See* BURSTING DISC.

Rushton impeller (Rushton turbine) A flat-bladed impeller, used for mixing, in which the blades are attached to a disc mounted on a shaft.

rusting The corrosion of ferrous metals due to the formation of a surface layer of iron oxides (**rust**).

RVDT *See* ROTARY VARIABLE DIFFERENTIAL TRANSFORMER.

r

s The Laplace transform complex variable of the form $s = a + i\omega$ used when converting between the time and frequency domains, where $i = \sqrt{-1}$ (the symbol j is also used for $\sqrt{-1}$).

Sabathé cycle (dual-combustion cycle, mixed cycle, Seiliger cycle) An air-standard cycle for a high-speed diesel engine which is a combination of the Otto and diesel cycles. The five steps are shown on a pressure (p) *vs* specific volume (v) diagram: isentropic compression (1–2), isopycnic compression (2–3), and isobaric expansion (3–4), both with heat addition, isentropic expansion (4–5) and isopycnic cooling (5–1).

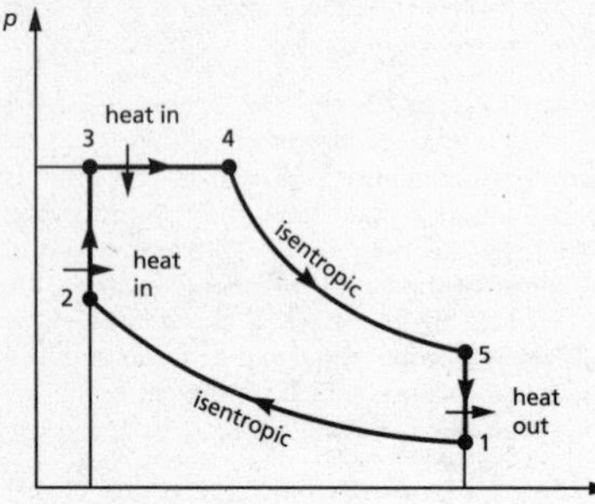

Sabathé cycle

sacrificial anode Material used to protect a buried or submerged metal structure from corrosion, comprising an anode of a metal with a more negative electro-potential than that being protected, and to which it is electrically bonded. The sacrificial anode corrodes in preference to the protected metal, and needs to be replaced from time to time.

sacrificial layer A layer deposited during fabrication of microstructures such as microfluidic channels, which is subsequently removed. The sacrificial material can be a thermoplastic, a wax, an epoxy, a phosphorous silicate glass, etc.

saddle 1. The part of a lathe that carries the cross slide. **2.** A support with the basic shape of a riding saddle. Used, for example, to guide a rope or hawser.

SAE number A code defined by the US Society of Automotive Engineers which indicates the viscosity of transmission, crankcase, and differential lubricants and other motor oils.

safe fatigue life *See* FATIGUE.

safe-life design *See* DESIGN METHODOLOGIES.

safety coupling A friction coupling designed to slip at a given torque to protect the driving or driven device from overload.

safety engineering Assessment of the risk of accidents in the operation of equipment, components, structures, etc. under both normal and fault conditions. *See also* STRUCTURAL INTEGRITY ANALYSIS.

safety factor *See* FACTOR OF SAFETY.

safety hoist A hoist which stops running when tension is released. A **safety stop** prevents the load from falling.

safety plug *See* FUSIBLE PLUG.

safety valve (safety-relief valve) A mechanical valve, fitted by law to all pressure vessels (e.g. steam boilers), which opens to prevent the internal pressure exceeding the maximum safe value for that vessel. The valve closes again once the pressure reduces to a safe level.

safe working load The steady or unsteady load against which a component or structure is designed for normal operation. It is lower than that which would cause failure by buckling, fracture, or yielding, so as to accommodate uncertainty, possible fault or accident conditions.

Saint-Venant criterion *See* THEORIES OF STRENGTH.

Saint-Venant's principle Strains that result from the application, to a small area of a body's surface, of a system of forces that are statically equivalent to zero force, zero moment, and zero torque and become negligible at distances that are large compared with the dimensions of the area. That is, when studying the long-range effects of applied forces, the local details at the loading points may be replaced by the simplest statically-equivalent conditions.

salinity The concentration of common salt (sodium chloride) in a liquid. It can be measured with a **salinometer** based upon an electrical conductivity meter or a hydrometer graduated to show salt concentration directly (**salimeter**).

Salter duck A wave-power device designed to extract energy from water waves. As the buoyant 'duck' bobs up and down on the waves, an internal mechanism, such as a pendulum, spins an electrical generator to generate electrical energy.

salt-gradient solar pond *See* SOLAR POND.

salt-velocity meter A volume flow meter based on detecting the transit time for a small quantity of salt or radioactive isotope in a flow by measuring electrical conductivity or radiation level.

sample 1. A single observation of a variable in a sampled-data system. **2.** A group of components taken at random on which quality assurance procedures are carried out.

sample and hold A device used to precede an analogue-to-digital (A–D) converter which switches rapidly between track mode, where the output follows the input, and hold mode, where the output is fixed. In the hold mode, the device applies a **zero-order hold** to maintain the output constant. The process of switching from track to hold occurs each time the A–D converter makes a conversion, so that the input to it does not change during the conversion. This process is sampling, i.e. the conversion of a signal changing continuously in time to a number of discrete values equally spaced in time.

sampled-data system A control or other system in which the values of variables are only known at discrete, usually constant, intervals of time, and where there is no knowledge of the variables between these times, that is, between samples. A continuous input signal is converted into discrete values by a **sampler**.

sampling The process of measuring a series of values of a time-varying quantity.

sampling frequency (sampling rate) The number of measurement samples per unit time taken from a continuous analogue signal to produce a discrete signal. It is the inverse of the time between successive measurements **(sampling interval, sampling period,** or **sampling time)**.

sampling theorem *See* NYQUIST–SHANNON SAMPLING THEOREM.

sandblasting A process in which small particles of an abrasive material, such as sand or grit, are blown against a surface to clean or roughen a smooth surface or smooth a rough surface, e.g. by removing dirt, rust, scale, or surface irregularities.

sand-grain roughness An empirical measure of surface roughness for turbulent flow in pipes, quantified in the Moody chart.

sand-heap analogy An extension of the membrane analogy to a bar made of rigid perfectly plastic material, in which dry sand is heaped upon a plate having the same cross-section as the twisted bar. The constancy of the angle of repose of the sand replicates the constant shear yield stress, k. The volume of the sand heap is proportional to the torque required for plastic collapse, which, in the case of a cylinder of radius r, is $2\pi kr^3/3$.

sandwich beam (sandwich construction) 1. (flitched beam, flitched girder) A beam constructed of structural timbers bolted together with a continuous steel plate or plates between them or on the sides, in the plane of bending. **2.** A composite construction of metal, plastics, wood, etc. consisting of a foam or honeycomb layer bonded between two outer sheets. *See* ALSO LAMINATE.

Sankey diagram A diagram in which flows of energy, money, material, etc. are represented by arrows the width of which are proportional to the quantity of flow.

satellite solar-power system (SSPS) A system in which power is generated by very large (greater than 50 km^2) arrays of photovoltaic cells in geostationary orbit, and transmitted to earth as microwave radiation.

saturated liquid A solution containing enough of a dissolved solid, liquid, or gas that no more will dissolve at a given temperature and pressure.

saturated steam *See* DRY SATURATED STEAM.

saturated vapour A vapour at such a temperature and pressure that any increase in pressure without change in temperature, or any reduction in temperature without change in pressure, would cause it to condense. A **saturated liquid** is a liquid at such a temperature and pressure that any reduction in pressure without change in temperature, or any heat addition without change in pressure, would cause it to vaporize (i.e. boil). A **saturated mixture** is a mixture of the two phases of a substance which co-exist in equilibrium at the saturation pressure corresponding to a given temperature.

saturated-vapour pressure (saturation pressure, saturation vapour pressure) At a given temperature, the pressure at which a pure liquid substance vaporizes or a vapour condenses, i.e. changes phase from liquid to vapour and vice versa. The **saturation temperature** is the temperature, at a given pressure, at which a pure liquid substance vaporizes or a vapour condenses. The relation between the saturation temperature and the saturation pressure is represented by the vapour-pressure curve.

saturation specific humidity The value of the specific humidity of saturated air at a given temperature and pressure. It is a thermodynamic function of state.

S

Saunders air-lift pump A pump used to raise water from a well by introducing compressed air below the water level in the well.

Sauter diameter (*SD*, D[3,2], d_{32}) (Unit mm) For a single non-spherical particle of volume $\mathcal{V}$ and surface area A, the diameter of a sphere having the same $\mathcal{V}$ and A, i.e. $SD = 6\mathcal{V}/A$. The usefulness of this definition is limited by the difficulty of determining A, for example where the particle surface has re-entrant elements. The **Sauter mean diameter (d_{sm}, d_{32}, *D*[3,2], *SMD*], surface mean diameter)** is the arithmetic mean of several measurements of the Sauter diameter for a single particle or, for a distribution of particles of different size, particularly droplets in a spray, an equivalent diameter based upon the total droplet volume divided by the total surface area assuming that each particle is spherical, i.e. $SMD = \sum\mathcal{V}_i/\sum A_i$ where $\mathcal{V}_i$ and A_i are respectively the volume and area of particle i and the summation is taken over all particles. For the definition to reduce to the correct diameter when all particles are of the same size, the factor 6 must be introduced.

Savonius rotor A wind-turbine rotor comprising two or more semi-circular offset vanes rotating about a vertical axis. A **Savonius–Darrieus wind turbine** is a Darrieus wind turbine with one or more Savonius rotors added so that the turbine is self-starting.

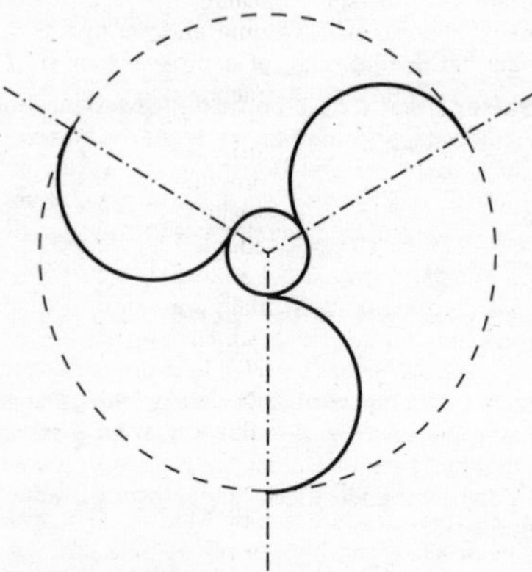

Savonius rotor

Saybolt Universal viscometer An instrument for determining the Saybolt Universal Viscosity for a low-viscosity liquid, which is the time in seconds it takes 60 ml of the liquid at a given temperature to flow through a calibrated vertical capillary tube. The **Saybolt–Furol Viscometer** is a high-viscosity version in which the capillary tube is replaced by a tube of larger diameter. *See also* CAPILLARY-TUBE VISCOMETER.

scalar A quantity which has magnitude but not direction e.g. length, time, area, volume, speed, and temperature. *See also* VECTOR.

scale 1. A series of marks and numbers on a measuring device to indicate the value of the quantity being measured. **2. (scaling)** To change the magnitude of a physical variable by multiplying it by a constant factor (**scaling factor, scaling ratio**), such as the ratio of a dimension of a model to the same dimension for a full-size version. **3. (scaling)** To generalize a physical problem by converting the variables to non-dimensional form using quantities which characterize the problem, such as a length scale, time scale, velocity scale, and temperature difference, together with material properties such as density. *See also* DIMENSIONAL ANALYSIS. **4.** A weighing instrument. **5.** A deposit of calcium and magnesium minerals (limescale) which adheres to a surface used to heat water. **Scaling** is its removal.

scale effect The change in behaviour with size of a material, component, or structure. For example, cube-square scaling where one element changes with volume and another with area, as in the range of a projectile or ship, droplet formation, and fracture mechanics.

scale height (Unit m) The altitude H at which the atmospheric pressure has decreased to a certain fraction of its value at sea level B. It depends on the model adopted for the atmosphere: for an isothermal atmosphere, the pressure decreases exponentially with altitude. If H is taken as the altitude at which the pressure has fallen to B/e, then $H = RT/g$ where R is the gas constant for air, T is the absolute temperature T, and g is the acceleration due to gravity. *See also* LAPSE RATE.

scale-up The process by which data from a small-scale experiment or pilot plant is used to design a large-scale (e.g. industrial scale) unit.

scaling factor *See* FOULING FACTOR.

scaling parameter A quantity, such as a characteristic length (**scaling length**), characteristic velocity (**scaling velocity**), e.g free-stream velocity, or a material property or combination of properties, used to non-dimensionalize a physical variable or an equation representing a physical situation or process. A scaling length may also be derived from the parameters of a problem. For example, in turbulent flow near a solid surface, $\nu\sqrt{\rho/\tau_S}$ is found to be a useful scaling length, ν being the kinematic viscosity of the fluid, ρ is its density, and τ_S is the wall shear stress. *See also* DIMENSIONAL ANALYSIS.

scalping The removal of hardened outer layers of barstock by passing through a die in order to improve machinability and surface finish.

scanning electron microscope (SEM) A type of microscope which images a solid surface by scanning it with a focused beam of high-energy electrons that interact with the surface atoms and produce a signal which provides information about surface topography, texture, chemical composition, and crystalline structure. In contrast to a transmission electron microscope (TEM), a perspective image is obtained, but the magnification ($< 5.10^5$) and resolution (typically 50 to 100 nm) are lower.

SEE WEB LINKS

- Tutorial featuring virtual scanning electron microscopy

scanning laser beam A laser beam which is moved repetitively and continuously over a controlled path.

SCARA *See* SELECTIVE COMPLIANCE ASSEMBLY ROBOT ARM.

scarf joint A joint between two components, made by tapering the ends of each with identical taper angles and overlapping the ends so the cross section is maintained. The joint is secured by a method best suited to the materials being joined: bonding, fastening, brazing, welding, etc.

scatter The departure of data from an average or correlation such as a trend line. A **scatter diagram (scatter graph, scatter plot)** is a two-dimensional plot of a measured dependent variable against an explanatory variable to determine whether there is a correlation between the two. If the variables are correlated, the plotted points will cluster around a line or curve. The higher the degree of correlation, the lower the scatter. A scatter diagram can, in principle, be extended to three dimensions to include a second explanatory variable.

scavenge pump (scavenging pump) A suction pump in the dry sump system of lubrication which returns used oil from the crankcase of an engine to the main oil tank.

scavenging The removal of gaseous combustion products from the cylinder of an internal-combustion engine, and their replacement by a charge of fresh air.

scavenging stroke *See* EXHAUST STROKE.

schematic diagram (schematic) An illustration that represents the essential elements of a machine, experimental facility, industrial plant, system, process, etc. Details inessential to understanding are omitted and a schematic can range from a simplified scale drawing to a symbolic representation. *See also* BLOCK DIAGRAM.

Schlieren method An optical technique for observing fluid flow by detecting the changes in refractive index of an optically transparent flowing medium brought about by changes in density (gases) or shear (certain liquids). A ray of light passing through the fluid is deflected by the refractive-index gradient normal to the ray.

Schmidt number (*Sc*) A non-dimensional parameter which arises in the analysis of flow processes involving simultaneous momentum and mass transfer. It relates viscous diffusion to mass diffusion and is defined as $Sc = \nu/D$ where ν is the kinematic viscosity and D is the coefficient of mass diffusivity.

Schrader valve (American valve) An air check valve used in motor-vehicle and bicycle tyres, which consists of a valve stem into which is threaded a poppet valve assisted by a spring (the valve core).

scientific notation The representation of real decimal numbers, in the form $a \times 10^n$, where a is a number with magnitude between 1 and 10 and n is an integer. *See also* ENGINEERING NOTATION.

scleroscope An instrument which determines the rebound hardness of a material by recording the height to which a diamond-tipped hammer or a standard ball rebounds when dropped onto a surface from a standard height. *See also* SHORE HARDNESS; SHORE SCLEROSCOPE.

score To scratch the surface of a material.

scoring 1. Surface damage caused by lubrication failure in bearings. **2.** The formation of a groove where the path of fracture is intended, as in the tops of beverage cans.

scouring 1. The process of finishing or cleaning a surface using an abrasive material. **2.** Damage caused to a surface by erosion, etc.

scragging machine 1. A machine to test springs by impulsive loading **2.** A machine to increase the service life of coil springs by compressing them to their minimum solid length before use, thereby inducing favourable residual stresses.

SCRAMJET (Supersonic Combustion RAMJET) A ramjet in which the combustion takes place in a supersonic airstream.

scrap Surplus material which remains after a machining or forming operation to produce a component.

scraper ring *See* PISTON.

scratch hardness The resistance of a material, such as a metal, alloy, plastic, or mineral, to scratching by a much harder indenter, such as a diamond stylus, moved slowly across the surface. The scratch hardness number is computed from the loads and the dimensions of the residual scratch.

screen mesh 1. A wire network or cloth, mounted in a frame and used e.g. for removing particulate material from a fluid or flow, or classifying the particulates using a screen of known mesh size. *See also* SIEVE. **2.** In wind tunnels, several screens (1) are used to reduce the turbulence intensity. *See also* HONEYCOMB.

screened windmill A windmill where a barrier screens the sails from the wind during the backward part of the windmill's cycle.

screw A fastener with a **screw thread** cut into its cylindrical or conical shank, intended either to cut its own thread (as in a wood screw) or engage in a threaded hole. A **self-tapping screw (sheet-metal screw, tapping screw)** has a sufficiently hard thread that it cuts an internal thread in thin sheet metal or a soft material when driven into a hole in the material. The **screw head** is the part of a screw used to apply torque to the screw. Common designs are round with a diametral slot or cross, or hexagonal and recessed with a cross or hexagon.

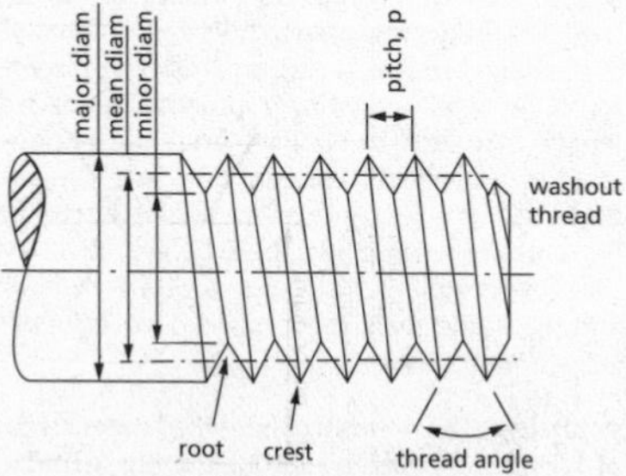

screw thread: right-handed thread

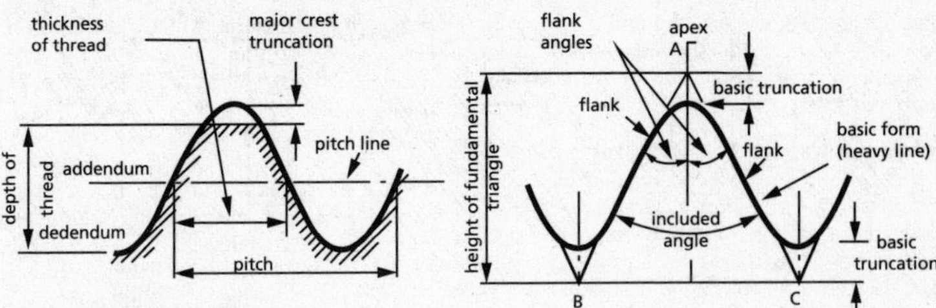

screw thread nomenclature

A **screw thread** is one or more continuous helical grooves of uniform section along either the exterior surface (**male thread**) or the interior surface (**female thread**) of a cylindrical or conical body. The three-dimensional shape that results when the thread cross section is rotated and axially advanced along an axis is called a **helicoid**, the angle that the thread makes when projected onto the axis being called the **angle of inclination**. Threads are employed in fasteners such as bolts, nuts, and screws; location and measuring instruments; in power drives; in some electrical fittings (**Edison thread**); and on the ends of crankshafts to suppress oil leakage (**thrower thread**). **Parallel threads** are formed on cylinders; **tapered threads** on cones, typically with a taper rate of 1:16. A screw with a **right-handed thread** appears to move away from the observer when turned clockwise. All standard screws, bolts and nuts have right-hand threads, but **left-handed threads** are sometimes employed. The axial distance between corresponding points on adjacent threads is called the **screw pitch** or **screw rate** and, for a single continuous helical groove (a **single-start** thread), is the same as the change in axial distance (the **lead**) between a nut and the head of a bolt during one revolution, the number of **thread forms per mm** then being the reciprocal of the pitch. For the same screw diameter, **coarse threads** have fewer threads per mm than **fine threads**. A **multiple-start** screw thread (usually coarse, see later) consists of two or more identical threads running simultaneously along its axis so as to provide greater bearing area and greater velocity ratio. The starts are separated by 180° (double start), 120° (triple start), 90° (quadruple start) etc., depending upon the number of threads. Thus, in double-start threads, the lead is twice the pitch; and so on. The axial distance between corresponding points on two adjacent threads in a multiple start thread is called the **divided pitch**. The position on a screw thread where there is equal distance between the flanks on the solid part of the thread and in the space between the threads is the **pitch point**, the associated diameter of which is termed the **pitch diameter**. The basic nomenclature for threads, some of which is common with that for toothed gearing, is shown in the diagram.

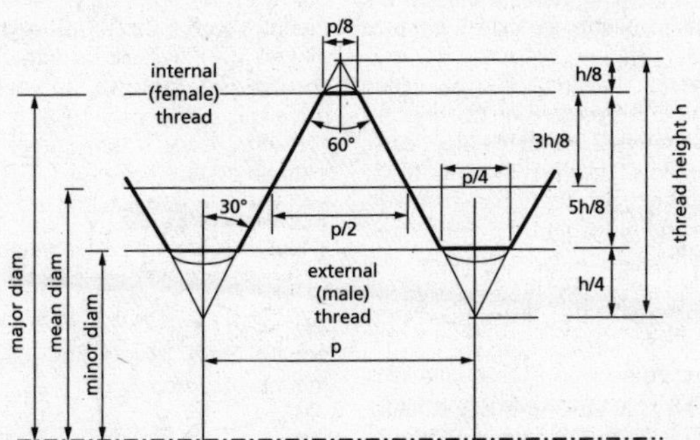

M threads

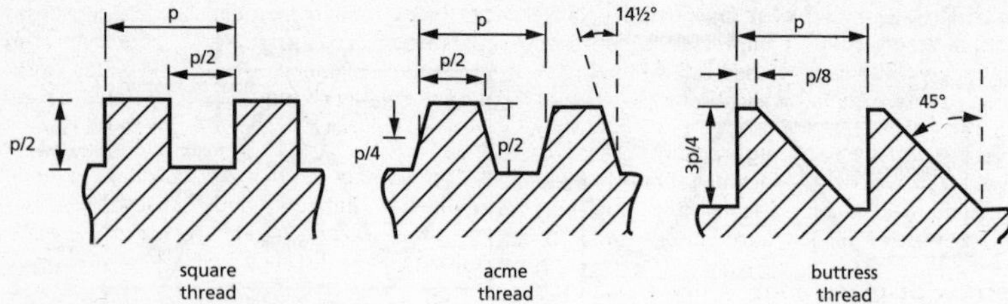

square/acme/buttress threads

There is a variety of standard (basic) **design thread forms** (the ideal shape and size of the groove in an axial plane, without regard to practical manufacturing tolerances), the form of ISO metric 60° threads (**M threads**) being shown in [ISO 724:1993 'ISO general purpose metric screw threads – Basic dimensions'.] Standard metric threads are designated by the nominal **major diameter** and the **pitch**, so that a 20 mm diameter coarse metric thread having a pitch of 2.5 mm becomes M20×2.5 and the corresponding fine thread version is M20×1.5. Now-obsolete V-shaped thread (**V-thread**) systems include **British Association (BA)**, **British Standard Whitworth coarse (BSW)** and **fine (BSF)** threads. Some Imperial threads are still in use, such as the **British Standard Pipe (BSP)** thread (also known as the **gas thread**). **Unified screw threads**, such as **unified coarse (UNC)** and **unified fine (UNF)** that are a merger of the Whitworth and American Standard thread systems, are still used in the USA and Canada. **Acme**, **buttress** and **square threads** have greater load-carrying capacity than V-threads, and are used for power drives, breech block mechanisms, and in other heavily-loaded situations. Threads that do not conform to any standard are called **bastard threads**.

On an engineering drawing, the cross section of a screw thread is represented by two concentric circles: for an internal thread the inner circle is continuous while slightly less than 90° of the outer circle is omitted; for an external thread the outer circle is continuous and the inner circle broken. A lengthwise section of a threaded part is represented by two parallel lines corresponding to the inner and outer extremities of the thread.

In addition to production by taps and dies, screw threads may be manufactured by progressively cutting on a lathe with a single-point tool having the profile of the thread, the pitch being set from rotation of the **lead screw** by the relative rotational speed of the workpiece and feed of the tool. When the pitch of a thread is an integral multiple, or sub-multiple, of the pitch of the lead screw, it is called **even pitch**; otherwise it is **fractional pitch**. A **screwing** or **threading machine** is a mass-production lathe in which threads are made with taps and dies. **Thread rolling** is a cold-forming plastic-deformation process in which an external thread is produced by pressing dies having the profile of the thread on to either side of a rotating rod. In contrast to other methods of manufacture, thread rolling produces no waste swarf. Very precise screw threads may be finished using a profiled grinding wheel. The incompletely-formed parts of a screw thread at the beginning and end of the thread are called **washout threads**.

Screw-thread gauges, to measure the pitch, major, and minor diameters, and thread angles, include special micrometers and gauges of various designs, such as thin metal strips having the zig-zag form of a particular thread along an edge, a series of which is often hinged together like a penknife or feeler gauges, or a cylindrical bar with an accurately-machined thread. In quality control **thread-plug gauges** are employed to measure internal threads, and **thread-ring gauges** to measure external threads.

See also ALLEN SCREW; BENDIX DRIVE; BOLT; THREAD INSERT.

SEE WEB LINKS

- Nut, bolt and screw terminology
- Thread terminology

screw area The area of a circle described by the tips of a propeller.

screw blank *See* BOLT BLANK.

S

screw compressor A positive-displacement rotary compressor in which gas is progressively compressed by two intermeshing, counter-rotating, helical screws.

screw conveyor (auger, spiral conveyor, worm conveyor) A machine for bulk handling of semi-solid materials, consisting of a helical screw which rotates in a trough or casing.

screw displacement A rotation of a rigid body about an axis accompanied by a translation of the body along the same axis.

screw extractor A device for removing broken-off screws from threaded holes. It is rather like a drill with, for broken right-hand threads, a fast (long-pitch) left-hand thread that is driven into a hole drilled in the broken screw, thus untwisting it.

screw feeder A mechanism for handling bulk materials in which a rotating helicoid screw moves the material axially forward. Similar to a screw conveyor, but required to discharge material at a controlled rate very accurately. It operates with the screw completely full.

screw gear *See* TOOTHED GEARING.

screw head *See* SCREW.

screwing machine *See* SCREW.

screw jack A jack consisting of a nut and square-threaded shaft where rotation of the nut raises or lowers the jack.

screw machine A machine tool for high-volume manufacture of small turned components from rod or bar.

screw pair Two links connected together to form a kinematic pair, in which the contacting surfaces are screw threads, so that their relative motion consists of rotation and sliding.

screw press A press having the ram driven by a screw mechanism.

screw propeller A marine or aeroplane propeller consisting of a streamlined hub and two or more blades which rotate and advance as they do so.

screw pump A positive-displacement pump that uses one or more helical rotors rotating within a casing to transfer liquids or slurries. *See also* PROGRESSIVE-CAVITY PUMP.

scriber A sharp-pointed, hard-steel tool for marking off, i.e. drawing fine scratch lines on the surface of metal, plastic, and other workpieces. Engineers' blue may be painted on the workpiece surface to provide contrast.

scroll diode An axisymmetric fluidic diode in which a bluff centrebody causes the formation of an intense annular vortex in the high-resistance flow direction.

scroll gear *See* TOOTHED GEARING.

scrubber (spray chamber, spray tower) An air-pollution control device for the removal of liquid droplets, undesired gas, or particulates from an industrial exhaust stream. *See also* DRY SCRUBBER; EXHAUST SCRUBBER; WET SCRUBBER.

scuffing wear Frictional wear caused by rubbing between two surfaces. *See also* FRETTING.

SCWR *See* NUCLEAR FISSION.

SD *See* SAUTER DIAMETER.

seal A component which controls or prevents leakage of fluids into or out of parts of a machine. *See also* BELLOWS SEAL; GASKET; LABYRINTH SEAL; O-RING SEAL; PACKED GLAND; PACKING.

seam 1. A line, groove, or ridge showing where components have been joined mechanically or by welding. **2. (mould seam)** A mark on a moulded or cast object caused by parting of the mould.

seam welding A series of overlapping spot welds. *See also* WELDING.

seat *See* VALVE SEAT.

seating A surface for the support of another part of an assembly, such as a bearing or seal.

S

secant modulus (Unit Pa) The straight line joining the origin of a stress–strain curve to any point on the curve. For non-linear materials that stiffen, the secant modulus is smaller than the tangent modulus at that point; the secant modulus is greater for materials that soften.

second (s) The SI base unit of time equal to 9192 631 770 periods of the radiation corresponding to the transition between the two hyperfine levels of the ground state of the cesium-133 atom.

secondary air 1. Fresh air introduced into a combustor after the main combustion zone to enhance completeness of combustion. **2.** Fresh

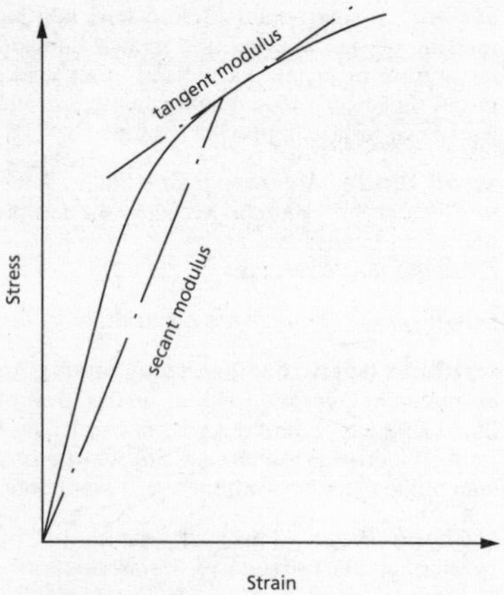

secant modulus

air injected into the exhaust stream of an engine to ensure more complete combustion and assist the performance of the catalytic converter. *See also* ASPIRATED AIR INJECTION

secondary creep *See* CREEP.

secondary flow (crossflow) Flow within a boundary layer with a velocity component parallel to the surface which deviates from the direction of the free-stream velocity. *See also* SKEWED BOUNDARY LAYER.

secondary hardening The process whereby some low-alloy steels, quenched to produce martensite, produce fine precipitates when tempered above 550°C, which inhibit dislocation motion and reverse the trend towards lower strength at higher tempering temperatures.

secondary stresses Stresses different from those induced by the major loading but nevertheless resulting from the major loads; for example, the hoop stresses that occur around the circumference of barrelled compression testpieces and forgings.

second coefficient of viscosity *See* BULK VISCOSITY.

second law of motion *See* NEWTON'S LAWS OF MOTION.

second law of thermodynamics It is impossible to construct a system which will operate in a cycle, extract heat from a reservoir, and do an equivalent amount of work on the surroundings (**Planck statement**). The second law is based upon observation and cannot be proved from any other laws. A corollary related to refrigerators and heat pumps is that it is impossible to construct a device that operates in a cycle and produces no effect other than the transfer of heat from a lower-temperature body to a higher-temperature body (**Clausius statement**). For example, a refrigerator will not operate unless its compressor is driven by an external power source. The **Kelvin–Planck** or **Kelvin statement** is that it is impossible for any device that operates on a cycle to receive heat from a high-temperature reservoir and produce a net amount of work. The working fluid must also exchange heat with a low-temperature reservoir. No heat engine can have an efficiency of 100%.

A consequence of the second law of thermodynamics is the **Clausius inequality** (**Clausius theorem, inequality of Clausius**): when a closed system operates in a cycle,

$$\oint \frac{dQ}{T} \leq 0$$

where Q is the amount of heat transferred to or from the system and T is the absolute temperature. The equality applies to reversible cycles; the inequality to irreversible cycles.

second moment of area (I_C) (Unit m^4) The sum of the products of each element of a plane area, δA, and the square of its distance x from an axis contained within the plane, to give

$$I_C = \int x^2 dA$$

See also POLAR SECOND MOMENT OF AREA; PRODUCT OF INERTIA; MOMENT OF INERTIA.

second normal-stress difference (N_2) (Unit Pa) For a viscoelastic fluid, the difference between the normal stresses in the y- and z-directions produced by a shear stress τ_{xy}. It is invariably smaller than the first normal-stress difference.

second-order system A dynamic system, such as a mass-spring-damper system, in which the behaviour is described by a second-order differential equation.

second-order transition temperature *See* GLASS TRANSITION TEMPERATURE.

second tap *See* TAP.

sectional view The outline, in an engineering drawing, of an object at the cutting plane, together with all visible outlines behind the cutting plane.

section modulus (*Z*) (Unit m^3) The ratio of the second moment of area of the cross section of a beam to the distance from the neutral axis of the element where the stress is greatest. The maximum bending stress is given by M/Z where M is the bending moment. For a rectangular beam of breadth b and depth h, $Z = bh^2/6$; for a circular cross section of diameter d, $Z = \pi d^3/32$.

sector gear (mutilated gear, segmental gear) A component resembling a gear wheel from which one or more teeth have been removed, employed in intermittent-motion mechanisms.

sedimentation The tendency of dense solid particles to drop through a fluid of lower density due to the influence of gravity. *See also* SETTLING.

Seebeck effect The generation of a voltage V due to a difference in temperature ΔT between two junctions of dissimilar metals in the same circuit. It is the basis of the thermocouple and quantified by the **Seebeck coefficient** (unit μV/K) defined by $\alpha = V/\Delta T$. *See also* PELTIER EFFECT.

seepage velocity *See* DARCY FLUX.

segmental gear *See* SECTOR GEAR.

segmental meter A variable-head flow meter in which the orifice plate has an opening in the shape of a semi-circle.

SEGS The acronym for **S**olar **E**nergy **G**enerating **S**ystem.

Seiliger cycle *See* SABATHÉ CYCLE.

seismic mass (test mass) The mass in an accelerometer, the inertia of which results in deflexion of the supporting spring when the substrate to which the spring is attached experiences acceleration. *See also* ACCELEROMETER; PIEZORESISTIVE ACCELEROMETER.

seizing up (seizing, seizure) Abrasive damage to one or both metal surfaces which rub against each other, due to partial welding caused by frictional heating. In severe cases, relative movement between the surfaces may be impossible. It occurs e.g. in bearings, or the cylinder of a piston engine where the clearance is too small or there is insufficient lubrication.

Selective Compliance Assembly Robot Arm (SCARA) A robot with a rotational joint, joint angle θ_1, above the base frame; a rotational joint, joint angle θ_2, with axis parallel to the first joint; and a translational joint, joint offset d_3, with axis parallel to the second joint. The robot has high vertical stiffness and is thus particularly suited to assembly tasks. The diagram shows an idealized selective compliance assembly robot arm.

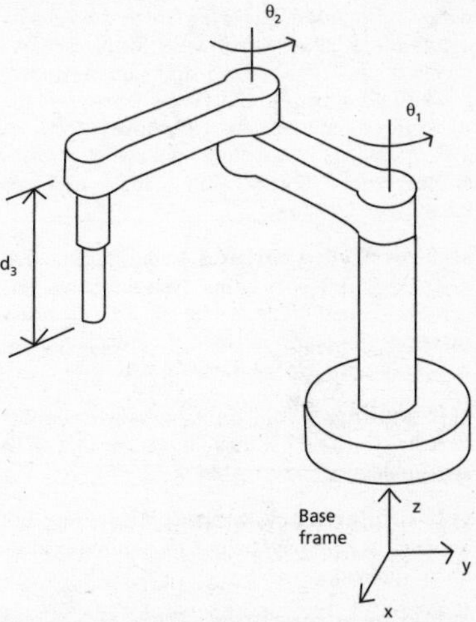

Selective Compliance Assembly Robot Arm

selective emitter A material whose monochromatic emissivity varies with wavelength, angle of incidence, or surface temperature.

selective surface A solar-collector surface for which the energy gain is maximized and the energy loss minimized by optimizing the absorptance (α) and the emittance (ε), both quantities being weighted averages of the monochromatic absorptances and emittances for the surface.

selective transmission *See* GEARBOX.

selector fork *See* GEARBOX.

self-adapting system A control or other system which can adapt to external changes.

self-aligning ball bearing *See* BEARING.

self-excited vibration (self-induced vibration) The vibration of a mechanical system which arises from non-oscillatory motion of the system. Examples are the shimmy of car wheels and the flutter of an aircraft wing.

self-locking nut A nut with an inherent locking action which minimizes loosening due to vibration. A **self-locking screw** locks itself in place without the need for a separate self-locking nut or lock washer.

self-preserving flow (self-similar flow, self-similar motion) A flow, such as a free jet, wake, or boundary layer, for which the velocity distributions at all streamwise locations have precisely the same shape and can be represented in the form $u/U = f(y/\delta)$ where u is the streamwise flow velocity a distance y from an axis or wall, U is a scaling velocity, and δ is a scaling length. Both U and δ may vary with streamwise location.

self-rectifying turbine A wind turbine that can accept airflow in either axial direction. The aerofoil-shaped blade profile must be symmetric about the plane of rotation, untwisted, and with zero pitch. *See also* WELLS TURBINE.

self-sealing A fluid container, such as a fuel tank or tyre, lined with a substance that seals any small puncture or rupture.

self-similar crack propagation Fracture in which the propagating crack continues in its initial direction.

self-tuning regulator (STR) An adaptive control system in which the parameters in the controller, for example the gain and time constants, are automatically adjusted by a second control loop as a consequence of observed performance of the system.

sem An abbreviation for pre-as**sem**bled washer and screw: a machine screw with a washer, or washers, permanently attached under the screw head.

S

SEM *See* SCANNING ELECTRON MICROSCOPE.

semi-closed cycle gas turbine A gas turbine in which less than 100% of the working fluid is recycled.

semi-diesel engine A piston engine which uses heavy oil as a fuel but operates at a lower pressure than a conventional diesel engine. The fuel is either sprayed against a hot surface or spot, or ignited by precombustion or supercharging of part of the charge.

semi-elliptic spring (leaf spring, carriage spring) A spring consisting of one or more layers of arc-shaped metallic plates of progressively decreasing length, joined to act as a single unit.

semi-floating axle An axle that transmits torque and carries wheel loads at its outer end.

semi-inverse method A method of obtaining the solution of a problem in elasticity in which certain assumptions are made at the outset. If it can be shown that, with these assumptions, the equations of equilibrium and the boundary conditions are satisfied, it follows from the uniqueness of solutions in the equations of elasticity that the solution obtained is exact. This method was first used by Saint-Venant, to solve the problem of torsion of prismatic bars.

semi-perfect gas *See* PERFECT GAS.

semi-permeable membrane 1. An idealized membrane which is permeable to a single gas in a mixture of gases. **2.** A membrane that is permeable to molecules of the solvent but not the solute in osmosis.

semi-rotary pump A form of self-priming pump, often hand-operated, suitable for pumping water and light oils such as diesel oil and

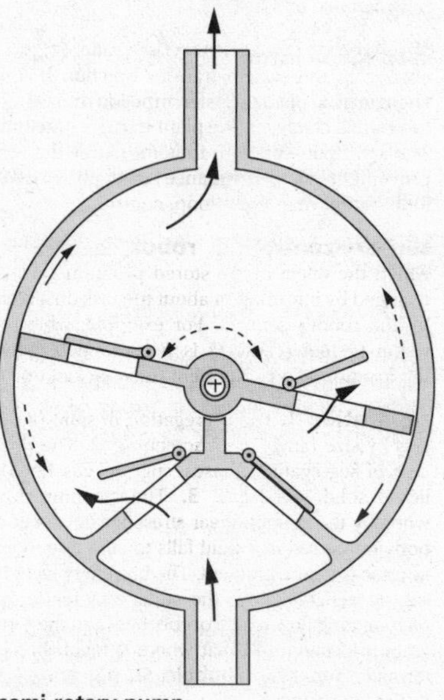

semi-rotary pump

petrol. As shown in the diagram, liquid is sucked into one side of the pump through flap valves and simultaneously ejected from the other side on one stroke. The sequence is reversed on each successive stroke.

sense A qualitative indication of direction or position. For example, clockwise and anti- (or counter-) clockwise rotation.

sensible energy (random molecular energy, sensible internal energy) The portion of the internal energy of a system associated with the kinetic energies of the molecules due to translation, rotation, and vibration, together with translation and spin of electrons and nuclear spin.

sensible enthalpy (sensible heat) The enthalpy change of a substance due to temperature change not accompanied by a change of phase.

sensitivity **1.** The relative importance of different input variables to an output. **2.** For a linear transducer, the ratio of the output magnitude to the input magnitude at a particular frequency. **3. (spread)** The difference between research and motor octane numbers which measures the effect of a change in severity of engine-operating conditions on the anti-knock performance of a fuel.

sensitivity function The ratio of the change in the system transfer function to the change of a plant transfer function measured for a small change in the plant transfer function. The sensitivity function thus measures the improvement in performance through closed-loop, rather than open-loop, control.

sensory-controlled robot A robot in which the effect of the stored program can be changed by information about the task observed by the robot's sensors. For example, where a vision system (camera) is used to determine the location of a component to be picked up.

separation **1.** The segregation of solid particles by size range, as in screening. **2.** The process of segregating phases, such as gas–liquid, liquid–solid, solid–gas. **3.** The phenomenon whereby the surface shear stress for flow over a body immersed in a fluid falls to zero due to an adverse pressure gradient. The boundary layer is said to separate from the surface. A region of recirculating flow which occurs between the separation location and that where it reattaches is termed a **separation bubble**. *See also* STALL (1). **4.** *See* FRACTURE.

sequential elimination of dimensions *See* IPSEN'S METHOD.

serrations A row of notches or tooth-like projections on an edge or surface. On an engineering drawing, serrations on the surface of a circular component are shown over about 60° of arc.

service The installation, maintenance, repair, etc. of a machine, plant, or instrument.

servomechanism (servo, servo system) A control system for the position and time derivatives (for example, velocity) of a mechanical system using closed-loop control to reduce steady-state error and improve dynamic response. A **servo loop** is a single feedback loop within a servo system. A **servomotor (servo)** is a rotational or translational motor to which power is supplied by a **servo amplifier** and which serves to apply torque or force to a mechanical system, such as an actuator **(servo actuator)** or brake **(servo brake)**. Where the actuator or brake is hydraulically operated, a hydraulic valve **(servo valve)** is used in which a low-powered signal controls the rate of liquid flow through the valve.

setover A device which is used to move a lathe tailstock or headstock laterally on its base, enabling a taper to be machined on a turned workpiece.

set point The value selected to be maintained by a control system. This is also known as the demand input to the controller.

set pressure The pressure at which a relief valve or safety valve is set to open, corresponding to the relevant code or standard which applies to the pressure vessel being protected.

set screw (set bolt) A screw with a pointed, square, or otherwise shaped end, used to secure a pulley, gear, or other component on a shaft.

setting The precise adjustment of a mechanism such as an engine valve or the zero of an instrument.

settling The tendency of dense solid particles or immiscible droplets to drop through a fluid of lower density due to the influence of gravity. The **settling velocity** is dependent on the particle's size and density and on the density and viscosity of the fluid. *See also* SEDIMENTATION.

settling time (τ) (Unit s) The time taken for the output of a control system to reach and remain within a specified percentage of the

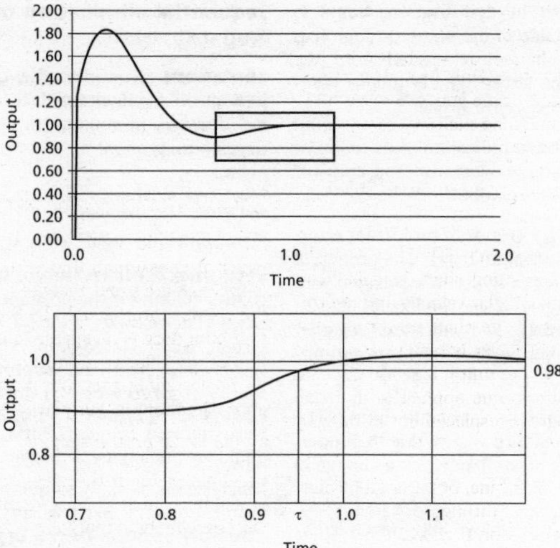

settling time

final (steady-state) value, typically 2%. As an example, in the upper diagram the system output is shown achieving a steady-state value of unity. The lower diagram, which is an enlargement of the highlighted part of the upper diagram, shows the settling time τ when the output has reached 0.98, i.e. 2% below the steady state value.

sextuplex pump *See* SIMPLEX PUMP.

SFC *See* SPECIFIC FUEL CONSUMPTION.

SFEE *See* STEADY-FLOW ENERGY EQUATION.

shackle A U-shaped metal fastening with a pin or bolt passing through holes at the ends of the U, similar to a clevis.

shadowgraph (optical comparator, profile projector) A metrological instrument in which parts such as screw threads are lit from below and the silhouette projected at large magnification onto a ground glass screen. Dimensions may be measured and compared with specifications to check whether a component is within tolerances.

shadowgraph method A technique in which flow structures can be visualized if there are associated differences in the density and hence the refractive index of the fluid. A ray of light passing through the fluid is bent towards a region of higher refractive index so that if a parallel beam of light is passed through the fluid, the shadowgram can be viewed on a screen behind the flow. *See also* SCHLIEREN METHOD.

shaft A revolving machine component, usually cylindrical or conical and supported by bearings, which carries gear wheels, pulleys, etc. and is used to transmit power. For the same volume of material, a hollow shaft has a higher torque-carrying capacity than a solid shaft of the same length. *See also* CRANKSHAFT; HALF SHAFT; SPINDLE.

shaft alignment The positioning of the rotational axes of two or more shafts such that they are co-linear.

shaft angle *See* TOOTHED GEARING.

shaft balancing The process of redistributing the mass of a rotating body, or adding corrective masses to it, to reduce to zero the product of inertia with respect to the shaft axis and an axis in the plane of symmetry of the shaft. The shaft is then statically and dynamically balanced and vibrations due to centrifugal effects are eliminated.

shaft encoder (rotary encoder) An electromechanical device that converts the angular position of a rotating shaft into an electrical signal which can, if needed, be differentiated to give the rotation speed.

shaft horsepower (Unit hp) The output horsepower of any prime mover or the input horsepower to a machine such as a compressor, pump or generator.

shafting A system or assembly of shafts for transmitting power.

shaft power (Unit W) The power being transmitted by a rotating shaft, calculated as the product of the angular velocity and the torque. The product of the shaft power and the time during which power is being transmitted is the **shaft work** with unit J. It is also equal to the product of the torque applied to the shaft and the total angular displacement in radians. *See also* BRAKE POWER.

shaft turbine A turbine, or part of a turbine, which transmits power through a shaft not connected to the compressor, such as a free turbine to drive a propeller or helicopter rotor.

shaft whirl Rotation of the plane made by a shaft which bows out as it rotates in the line of centres of the bearings. The phenomenon results from such causes as mass unbalance, hysteresis damping in the shaft, gyroscopic forces, and fluid friction in the bearings. *See also* SYNCHRONOUS WHIRL.

shakedown A state of self-stress in a component or structure which, when added to a variable applied stress, gives a total stress that satisfies the equilibrium conditions and does not at any time exceed the yield stress anywhere, i.e. the response is purely elastic to the variations of applied stress, even though the component or structure may earlier have been deformed plastically to get into the shakedown state. The possibility of a shakedown state existing depends upon the range of variation in applied stress. *See also* FATIGUE.

shakedown test An equipment test of a machine or plant after first assembly or installation, to detect defective components or incorrect assembly.

shake table (shaker table) A machine for subjecting an object or assembly to controlled mechanical vibration to determine its tolerance to vibration. An important application is to simulate the excitation of buildings by earthquakes.

shaking screen A screen vibrating in a horizontal plane, used to separate particulate or granular material from material with dimensions larger than the mesh size of the screen.

shale gas Natural gas extracted from shale, typically using hydraulic fracturing to create extensive fractures around a borehole drilled into the shale.

shallow-water approximation A simplification of the Navier–Stokes equations for an incompressible fluid with a free surface, in which it is assumed that vertical accelerations of the fluid are small, that the fluid is inviscid, so the viscous force exerted by the horizontal bottom boundary is negligible, and that the variation of the velocity in the horizontal plane can be neglected in the continuity equation. The approximation is fairly well satisfied for the waves termed *tsunami*. Surface waves on the liquid are analogous to density waves in a compressible gas and the approximation leads to equations which represent the flow of a gas with specific heat ratio of two.

shank 1. The stem of a tool, such as a broach, drill bit, reamer, or tap, which fits into a holder such as a chuck. **2.** The stem of a rivet or the unthreaded part of a screw or bolt. **3.** The shaft of a tool connecting the tip and the handle.

shape factor 1. In plastic theory of beams, the ratio of the fully-plastic moment to the first yield moment. **2.** (***H***) The ratio of the displacement thickness δ^* to the momentum thickness θ, i.e. $H = \delta^*/\theta$, for a boundary layer. It is an indicator of the velocity profile shape within the boundary layer. **3.** *See* VIEW FACTOR.

shape-memory alloy (SMA) An alloy that changes shape reversibly when subjected to a particular range of temperature, unlike a bimetal strip that deforms on any change of temperature. *See also* SMART MATERIALS.

shape number (*S*) A non-dimensional measure, closely related to specific speed, of the operating conditions of a hydraulic turbine defined by $S = P^{1/2}\omega/\left[\rho^{1/2}(gH)^{5/4}\right]$ where P is the turbine output power, ω its rotation speed (rad/s), ρ is the water density, H is the water head, and g is the acceleration due to gravity.

shaping A machining process in which the cutter reciprocates along the workpiece which moves in a direction orthogonal to the cutter after every stroke.

sharpness of resonance *See* Q-FACTOR.

shaving 1. A finishing operation in which a thin layer is machined from the surface of a workpiece. *See also* SCALPING. **2.** The process of trimming uneven edges from forgings, stampings, and tubing.

shear A form of deformation of a fluid or solid in which adjacent planes in a material are displaced in opposite directions.

shear angle *See* SHEAR STRAIN.

shear centre (centre of flexure, centre of twist, flexural centre) When planes of bending are not principal planes, beams may twist as well as bend. The shear centre is the point on any plane which includes a cross section where a transverse load produces only a bending deflection, and no twist of the section.

shear energy The strain energy due to shear forces acting on a beam or other component.

shear failure Failure from shear loading of a structural element, such as where a rivet may deform excessively or fracture (**shear fracture**).

shear flow (shear layer) The flow of a viscous fluid in which shear stresses are confined to a narrow region, such as a jet, wake, or boundary layer. The flow may be laminar or turbulent. *See also* FREE SHEAR LAYER.

shear force (shearing force) A force, determined from a free-body diagram, which acts tangentially to a surface which may be a real external surface or a defined surface within a material, such as the cross section of a beam. A **shear-force diagram (shear diagram)** is a plot of the shear force *vs* distance from one end of a beam or other component. *See also* BENDING-MOMENT DIAGRAM.

shearing blade (shearing punch) A blade or punch that cuts material by shear; for example, a guillotine.

shearing strain *See* SHEAR STRAIN.

shearing stress *See* SHEAR STRESS.

shear layer *See* SHEAR FLOW.

shear legs A triangular framework used for lifting.

shear loading Loading that is parallel to the surface over which it acts, as in a lap joint fastened by rivets or bolts, when the two connected components are subjected to equal but opposite tensile or compressive forces, and the fasteners are subjected to shear loading. In such overlapped joints there is a bending moment given by $F(t_1 + t_2)$ where F is the load and $2t_1$ and $2t_2$ are the thicknesses of the two plates. To avoid bending, a third plate may be added on the outside, making a joint having the appearance of a fork. *See also* FAILURE.

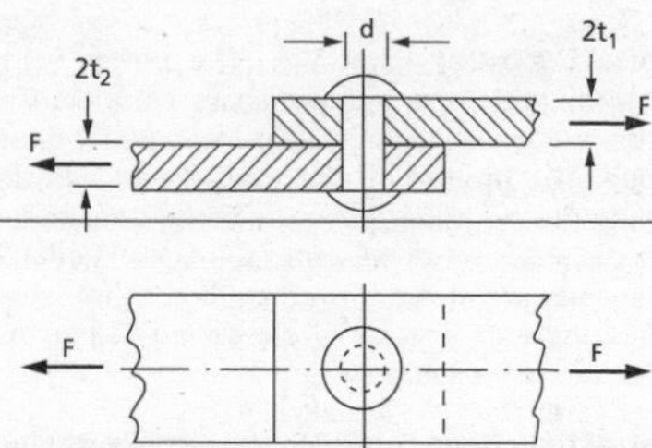

shear loading

shear modulus (modulus of rigidity, *G*) (Unit Pa) The modulus of elasticity in shear given by $G = \tau/\gamma$ where τ is the shear stress and γ is the corresponding shear strain. For an elastic material, the shear modulus can be determined from a torsion test on a cylindrical bar of length L in which applied torque T and angular twist θ are measured, since $G = TL/J\theta$ where J is the polar second moment of area of the bar section.

shear number A non-dimensional parameter for oscillatory viscous flow in a pipe, given by $\sqrt{\nu/\omega R^2}$ where ω is the angular frequency of the oscillation, R is the internal radius of the pipe and ν is the kinematic viscosity of the fluid. Equivalent is $\delta/(R\sqrt{2})$ where $\delta = \sqrt{(2\nu/\omega)}$ is the viscous penetration depth. It is the inverse of the **Womersley number**.

shear pin A safety device in a power transmission system designed to fail in shear if the design torque of the transmission is exceeded by a specified factor.

shear plane A confined zone along which shear deformation is concentrated, as in the primary shear zone in metal cutting.

shear rate ($\dot{\gamma}$) (Unit 1/s) (sometimes referred to as reciprocal seconds). For a viscous fluid, the shear rate $\dot{\gamma}_{ij}$ in a three-dimensional flowfield is the sum of velocity gradients $\partial u_i/\partial x_j + \partial u_j/\partial x_i$. *See also* DEFORMATION (2).

shear spinning The process of forming a rotating flat disc into an axisymmetric contoured part by progressive deformation onto a mandrel. In contrast to spin forging, the thicknesses of the starting blank and the final product remain approximately the same.

shear strain (γ) A rectangular element of a material when under load may be distorted by the sliding of parallel layers, resulting in shear strains as well as normal strains. The shift undergone by two parallel planes unit distance apart is the magnitude of the shear strain. An equivalent small-deformation definition is the angular deformation produced in a solid circular rod by a torque (pure shear). *See also* STRAIN.

shear strain rate ($\dot{\gamma}$) (Unit 1/s) In the deformation of solids, the rate at which shear strain is applied in a test or changes with time in a loaded component or structure.

shear strength (Unit Pa) The maximum shear stress that can be withstood by a material either (a) before plasticity occurs or (b) before rupture.

shear stress (shearing stress, tangential stress, τ) (Unit Pa) **1**. The stress which acts parallel to any plane within a solid material. It can arise due to a bending moment, a shear force, or torque applied to the body. **2**. The stress corresponding to velocity gradients within a flowing viscous fluid. *See also* STRESS; SHEAR RATE.

shear-stress multiplication factor (K_S) The maximum stress τ_{MAX} in the wire of an axially loaded helical spring is given by the maximum stress arising in the spring due to the torque *(FD/2)* induced in the spring multiplied by K_S, i.e. $\tau_{MAX} = K_S 8FD/\pi d^3$ where F is the axial load acting on the spring, D is the mean diameter of the spring, and d is the wire diameter. τ_{MAX} occurs at the inner fibre of the spring.

shear tearout Failure of a bolted or riveted joint by shear between holes.

shear test *See* TORSION TEST.

shear-thickening liquid (dilatant liquid) A non-Newtonian liquid whose apparent viscosity increases as the shear rate increases. Custard and other starchy liquids are examples. In a **shear-thinning liquid (pseudoplastic liquid)** the apparent viscosity decreases as the shear rate increases. Many synthetic liquids, including most polymers, are shear thinning, as is blood. *See also* THIXOTROPIC LIQUID.

shear viscosity *See* DYNAMIC VISCOSITY.

shear wave A transverse wave which occurs in an elastic solid when subjected to periodic shear. The propagation velocity of a shear wave is given by $\sqrt{(G/\rho)}$ where G is the shear modulus of the material and ρ is its density.

sheave A grooved pulley or wheel used to support ropes, including wire ropes, flat belts, and V-belts. *See also* BLOCK AND TACKLE.

sheet forming The permanent (plastic) deformation of flat sheet into complicated shapes by either (a) pressing and drawing between tools and dies, or (b) hydraulic (bulge) forming, or rubber forming, against a cavity.

sheet gauges *See* WIRE GAUGES.

sheet-metal screw *See* SCREW.

sheet moulding compound (SMC) A ready-to-mould fibre-reinforced polyester material manufactured by dispersing long (> 25 mm) strands of chopped glass fibres in a bath of polyester resin. It is used primarily in compression moulding.

shell An open or closed container or structure, having walls that are thin compared with its other dimensions. *See also* MEMBRANE ANALYSIS; MEMBRANE STRESS.

shell-and-tube boiler *See* FIRE-TUBE BOILER.

shell-and-tube condenser A shell-and-tube heat exchanger in which the shell-side fluid enters in the vapour phase and leaves as liquid. In the diagram, cooling water is shown as the tube-side fluid. *See also* SURFACE CONDENSER.

shell-and-tube heat exchanger (tubular exchanger) A design of heat exchanger where one fluid flows within the tubes of a tube bundle (the tube-side fluid) and the other (the shell-side fluid) flows around the tubes which are contained within a surrounding shell. Transverse baffles are used to increase the residence

shell-and-tube condenser

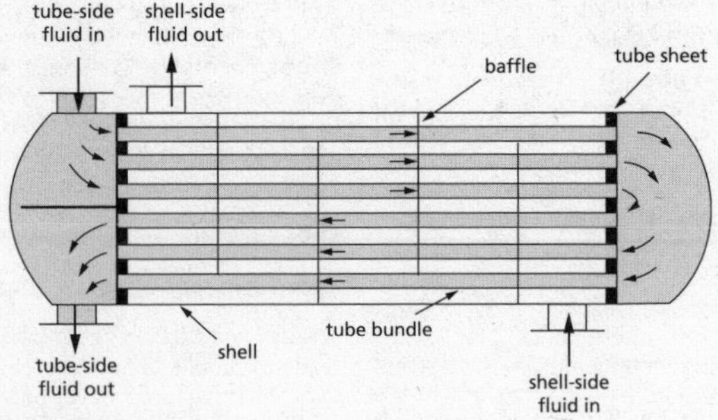

shell-and-tube heat exchanger

time of the shell-side fluid. A shell-and-tube heat exchanger is a mixed-flow type, since there is a mixture of parallel-, cross- and counterflow. *See also* MULTI-PASS SHELL-AND-TUBE HEAT EXCHANGER.

shell bearing *See* BEARING.

shell boiler *See* FIRE-TUBE BOILER.

shell element *See* FINITE-ELEMENT METHOD.

shell theory *See* MEMBRANE STRESSES.

Sherwood number (*Sh*, N_{SH}) A non-dimensional parameter used to characterize mass transfer, analogous with the Nusselt number for heat transfer, which represents the ratio of convective mass transport to diffusive mass transfer. It can be defined as $Sh = k_T L/D$ where k_T is the mass-transfer coefficient, L is a characteristic length, and D is the diffusion coefficient.

shielded bearing *See* BEARING.

shielded thermocouple A thermocouple, for use in very hot gas flows, in which the thermocouple sensor is surrounded by a concentric radiation shield through which the hot gas can flow freely.

shim A thin piece of material, such as metal of accurately-known thickness, placed between two surfaces to ensure they are the correct distance apart.

SHM *See* SIMPLE HARMONIC MOTION.

shock **1.** The sudden application of a load or temperature change to a component or structure. **2.** A transient motion characterized by a very rapid change in displacement, velocity, or acceleration. **3.** *See* SHOCK WAVE.

shock absorber A device, such as a frictional spring or hydraulic damper, which absorbs the energy associated with mechanical shock loading and dampens vibration.

shock adiabat The relationship between the pressure ratio across a shock wave and the ratio of specific volumes. For a perfect gas,

$$\frac{p_2}{p_1} = \left(\frac{\gamma+1}{\gamma-1} - \frac{v_2}{v_1}\right) \Big/ \left(\frac{\gamma+1}{\gamma-1}\frac{v_2}{v_1} - 1\right)$$

where γ is the ratio of specific heats, p is static pressure, v is specific volume, and the subscripts 1 and 2 represent conditions before and after the shock, respectively.

shock angle (β) For an oblique shock, the angle of inclination between the shock wave and the upstream flow direction.

shock isolator (shock mount) A device to eliminate or minimize the transmission of mechanical shock to a system or item of equipment.

shock polar (shock polar hodograph) For an oblique shock wave, a plot of the components V_Y *vs* V_X of the velocity downstream of the shock V_2, where V_X is the component in the direction of the upstream velocity V_1 and V_Y is the component perpendicular to V_X. In the diagram, θ is the turning angle and β is the shock angle. Analysis shows there can be a weak shock or a strong shock with different values of V_2 and other flow properties.

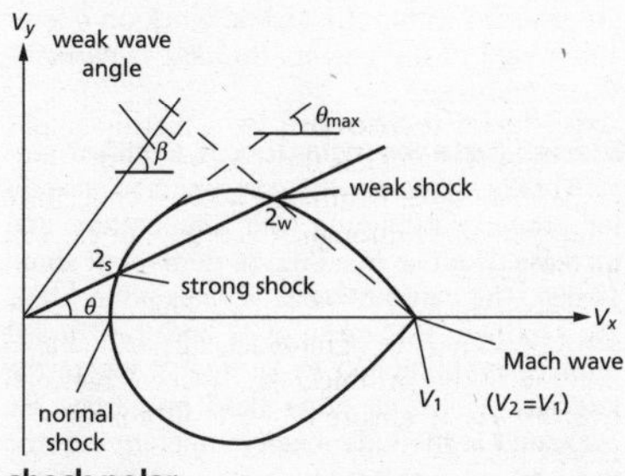

shock polar

shockproof (shock resistant) A mechanism or device which has very high resistance to the effects of mechanical or thermal shock.

shock resistance The ability of a component or structure to avoid failure by any mechanism when subjected to mechanical or thermal shock.

shock tube A cylindrical metal tube in which a high-pressure driver gas is separated from a low-pressure driven gas by a diaphragm, which breaks when the driver-gas pressure is sufficiently high. A shock wave separating the two gases propagates into the driven gas. Shock tubes are used to study compressible gas dynamics and gas-phase chemical reactions.

shock tunnel A hypersonic wind tunnel in which a shock wave generated in a shock tube ruptures a second diaphragm in the throat of a nozzle at the end of the tube, causing gas to emerge at high Mach number (greater than 6) into a vacuum tank. Shock tunnels are used to produce a high-enthalpy gas flow in investigations of aerodynamic heating and re-entry physics.

shock wave (shock) A thin zone which separates an upstream supersonic gas flow from a subsonic downstream gas flow and across which rapid changes in temperature, pressure, gas, and flow properties occur, depending upon the upstream Mach number. The front surface is the **shock front**. For most practical purposes, a shock wave may be regarded as a discontinuity since the **shock thickness (Δ_M)** for shocks of moderate strength amounts to a few times Λ, the molecular mean free path for the gas downstream of the shock. The actual **shock structure** is the detailed variation of temperature, pressure, and gas properties within it. Approximate continuum theory leads to the result $\Delta_M = 12\Lambda\gamma p_1/[(\gamma + 1)\Delta p]$ where γ is the ratio of specific heats, p_1 is the gas pressure ahead of the shock, and Δp is the pressure increase across the shock. The **shock strength (Π)** is the non-dimensional pressure jump across a shock wave defined as $\Pi = \Delta p/\rho c^2$ where ρ is the density of the gas ahead of the shock and c is its soundspeed. For a **strong shock** $\Pi \gg 1$, for a **weak shock** $\Pi \ll 1$. *See also* INTENSE EXPLOSION; SHOCK ADIABAT.

shoe brake A type of brake in which a retarding torque is applied to a rotating system by one or more **shoes**. In simple versions, solid shoes bear onto the steel wheels of, for example, railway vehicles. Alternatively drums are fitted to axles or driving shafts, and shoes (to which renewable friction linings are attached) bear on the inside or outside cylindrical surfaces (internally-expanding, or externally-contracting, shoe brakes). The **shoe pressure** is the pressure acting on the friction material. *See also* BAND BRAKE; DISC BRAKE.

Shore hardness *See* DUROMETER.

Shore rebound scleroscope A dynamic hardness tester in which a small indenter in a glass tube falls from a standard height of 250 mm on to the surface of a specimen and the rebound height is noted. The height of fall is divided into 140 equal divisions and this scale is calibrated against Brinell hardness. It is used for rubber, plastics, and metals.

short column A column with a slenderness ratio too low for Euler buckling theory to apply, which fails in compression rather than buckling.

short-term repeatability The difference between repeated positions and orientations of a robot end effector where the robot is programmed to repeatedly make the same movement over a short period of time.

short ton A non-SI unit of mass equal to 2000 lb; used in USA. *See also* LONG TON; TONNE.

short-wave radiation Radiation in the near infra-red, visible, and near ultra-violet wavelengths, ranging from about 0.1 μm and 5 μm, though there are no standard cutoffs. *See also* SOLAR RADIATION.

shot peening A cold-working process for hardening a metal surface by blasting small hard metallic, glass, or ceramic balls against it. Compressive residual stresses are induced in surface layers which improve fatigue life.

shoulder The portion of a shaft, stepped component or flanged component, where a change in diameter or other dimension occurs.

shoulder bolt A bolt for which the unthreaded cylindrical section between the thread and the head is of larger diameter than the threaded section, precisely machined to length and diameter and hardened. *See also* FITTED BOLT.

S

shoulder joint The second joint on an articulated robot, which has a horizontal axis and is analogous to the human shoulder.

shrink fit A tight interference fit between two components resulting from heating an outer part and/or cooling an inner part to allow easy assembly. The outer component contracts on cooling, while the inner part expands on warming to ambient temperature, thus gripping the two parts together.

shrink ring A ring, which is expanded by heating, is placed around an assembly of parts, and then contracts upon cooling to hold the assembly in place.

shrink wrapping A method of packaging using a thin plastic film which is applied hot around a component and shrinks tightly upon cooling.

shroud 1. A protective guard, often of sheet metal, partially surrounding a machine with parts which are moving, emitting hot gas, releasing liquid, or are electrically live. **2.** Part of the tip of a turbine blade which interlocks with adjacent blades: designed to increase damping and so minimize blade flutter.

shrouded propeller A propeller surrounded by a stationary shroud or duct. *See also* DUCTED FAN.

shroud ring A peripheral ring designed to prevent the escape of gas past the blade tips of an axial-flow turbine.

shunt flow meter *See* BYPASS FLOW METER.

shut down windspeed The windspeed at which a wind turbine ceases power generation and shuts down to prevent damage to the turbine from overspeeding. Also called cut-out speed or furling speed. *See also* CUT-IN SPEED; RATED SPEED; START-UP SPEED.

side leakage *See* BEARING.

side-valve engine A piston engine in which the intake and exhaust valves (**side valves**) are situated within the engine block on one or either side of the engine. *See also* OVERHEAD-VALVE ENGINE.

Sieder–Tate correlation A modified version of the Dittus–Boelter correlation to account for property variations with temperature and increase its range to a Prandtl number of about 17 000. The constant 0.023 is changed to 0.027, the exponent n is taken as ⅓ and the Nusselt number is multiplied by $(\mu/\mu_W)^{0.14}$ where the viscosity μ and all other fluid properties are evaluated at the bulk mean temperature of the flow, except for μ_W, which is evaluated at the wall temperature.

SI engine *See* SPARK-IGNITION ENGINE.

sieve A meshed sheet with apertures of uniform size used for sizing or separating granular materials. The standard opening in a sieve (**sieve mesh**) or screen is defined by four boundary wires (warp and weft). Mesh openings are usually square and the **mesh size** is the number of parallel wires per unit width. The **sieve diameter** is the size of a sieve opening through which a particle of known diameter will just pass. **Sieve analysis (sieve classification, sieving)** is the size distribution of solid particles determined using a series of standard sieves of decreasing size, the **sieve fraction** being the fraction which passes through a sieve of a given number and is retained by a **sieve** of a higher **number.** A **sizing screen** is a mesh sheet with standard-size apertures used to separate granular material into classes according to size. *See also* SCREEN MESH.

sight feed A transparent section of pipe or tubing through which dripping or flowing liquid can be observed.

sight-feed lubricator A drip-feed lubricator with a small glass tube supplying oil from a reservoir in drops which can be observed.

sight glass A glass section on a process line or vessel through which the liquid level can be observed or measured.

sighting tube A tube, often ceramic, inserted into a hot chamber such as a furnace in which the temperature is to be measured. An optical pyrometer is sighted into the tube to observe its interior end and obtain a temperature reading.

signal Any measurable physical quantity can be regarded as a signal, though usually it refers to the electrical voltage or current output from a transducer.

signal analyser An instrument to measure and analyse the output(s) of a dynamic system, usually as functions of both time and frequency, for a range of inputs to that system.

signal-to-noise ratio (S/N, SNR) The ratio of the amplitude of a measured signal to the amplitude of the noise signal, often expressed in decibel. For random noise the root-mean square value is usually used for the amplitude.

sign convention A convention for physical quantities which have a sense of direction as to which is positive and which is negative. For example, clockwise rotation is regarded as positive, anti-clockwise rotation as negative. In stress analysis, a tensile stress is taken as positive and a compressive stress as negative, while a shear stress is positive if it is directed along a positive axis when it acts on a face whose outward normal points along a positive axis.

significant wave height (H_S) (Unit m) Four times the root-mean square height of a wavy sea surface above the average surface height. It is used to estimate available wave power, which is proportional to H_S^2.

silencer (exhaust silencer) An expansion chamber attached to the exhaust pipe of a piston engine, designed to lessen the noise transmitted to the surroundings.

silent chain A drive chain having inverted teeth, where the working faces of each link come gradually into contact with the faces of the sprocket teeth. Used for heavily-loaded, high-speed drives.

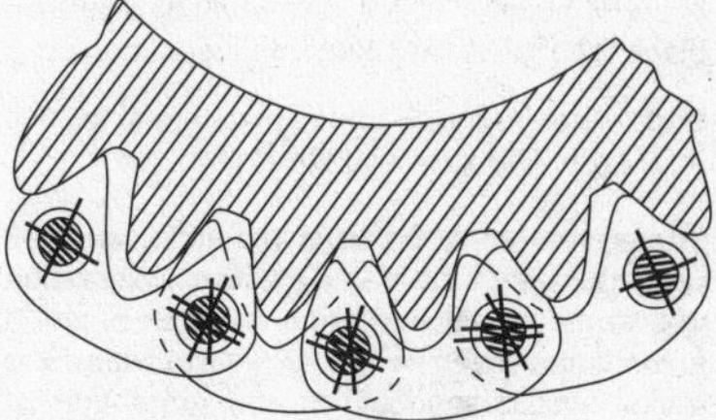

silent chain

silver soldering *See* BRAZING.

similarity (similitude) *See* PHYSICAL SIMILARITY.

similarity parameter *See* PHYSICAL SIMILARITY.

similarity variable A combination of physical quantities which arises when the equations which govern a physical problem are non-dimensionalized in such a way that self-similar solutions are possible. An example is the variable η, defined as $\eta \equiv y\sqrt{U_\infty/\nu x}$ such that the variation of velocity u with distance from the surface y within a laminar boundary-layer where $U_\infty \propto x^n$ can be shown to satisfy the form $u/U_\infty = f(\eta)$. U_∞ is the free-stream velocity, x is the distance measured along the surface, n is a constant and ν is the kinematic velocity of the fluid. *See also* BLASIUS SOLUTION; FALKNER–SKAN SOLUTIONS.

simple engine An engine, such as a **simple steam engine**, in which expansion of the working fluid occurs without change of phase and the fluid is then exhausted to the surroundings.

simple harmonic motion (SHM) The motion of a mass-spring system in which the acceleration of the mass m is proportional to its linear displacement x and in the opposite direction. The governing differential equation

S

is $md^2x/dt^2 + kx = 0$ where k is the spring constant and t is time. The solution is $x = C_1 sin(\omega t) + C_2 cos(\omega t)$ where $\omega \equiv \sqrt{(k/m)}$ is the natural circular frequency and C_1 and C_2 are constants determined from the initial conditions of the motion.

simple machine An elementary machine such as a pulley, lever, wheel and axle, inclined plane, or screw. All mechanical machines are made up of such simple machines.

simple pendulum A pendulum consisting of a bob attached to a weightless inextensible string of length l which can swing in a vertical plane. The governing differential equation is

$$\frac{d^2\theta}{dt^2} + \left(\frac{g}{l}\right)\sin\theta = 0$$

θ being the angle between the string and the vertical at any time t and g is the acceleration due to gravity. The solution leads to a period of oscillation given by $T = 4K\sqrt{(l/g)}$ where K is a constant dependent on the amplitude of oscillation. For small-amplitude oscillations, $T = 2\pi\sqrt{(l/g)}$.

simple shear *See* DIRECT SHEAR.

simplex pump A single-cylinder reciprocating piston or plunger pump, which may be single or double acting. Multiple simplex pumps driven by a common crank are often combined into a single unit with a common outlet. A **duplex pump** is essentially two simplex pumps arranged so that as one piston is on its upstroke, the other is on its downstroke, and the combined output is practically continuous. **Triplex pumps** (three simplex pumps) with the strokes 120° apart are the most common type of pump used to pump drilling mud in oilwell-drilling operations, but have many other applications for pumping all types of liquid including hydrocarbons, glycol, salt water, slurries, abrasive and corrosive liquids. There are also **quadplex** (90° separation), **quintuplex** (72°) and **sextuplex (hex pump)** (60°) variants.

simply-supported *See* FIXITY.

simulate To reproduce some or all of the characteristics of one system with a dissimilar system, either real or virtual, using either analogue- or digital-computer models. A **simulator** simulates a system or condition and the effect on it of applied changes. Simulators are commonly used for training purposes, e.g. racing-car and flight simulators.

simulation block diagram A block diagram used to simulate the behaviour of a dynamic system numerically.

simultaneous Two events are simultaneous if they take place at precisely the same time.

simultaneous engineering *See* CONCURRENT ENGINEERING.

sine wave A wave in which the particles move in simple-harmonic transverse motion.

sine-wave response *See* FREQUENCY RESPONSE.

singing Undesirable oscillation and a form of instability of the output of a system under closed-loop control, typically at high frequency.

singing margin The margin between the normal operating gain of a control system and the gain that will result in singing.

single-acting actuator A translational pneumatic or hydraulic actuator where the pressure is applied only to the cylinder on one side of the piston. It can produce force only in one direction, with an external force required to return the piston.

single-acting engine A reciprocating engine in which the working fluid acts on one side of the piston only, as in most piston engines.

single-acting pump A pump which delivers liquid in alternate strokes only. *See also* DOUBLE-ACTING PUMP.

single-axis gyroscope A gyroscope, such as a rate gyroscope, suspended in a single gimbal whose bearings form its output axis.

single-bladed wind turbine A low-solidity, horizontal-axis wind turbine with a single blade.

single-curve teeth *See* TOOTHED GEARING.

single-cylinder engine 1. A piston engine with only one cylinder; used for some motorcycles and research purposes. **2.** A steam engine with only one steam cylinder.

single-degree-of-freedom gyroscope A gyroscope for which the spin reference axis is free to rotate about only one of the orthogonal axes, usually the input or output axis

single-degree-of-freedom system A dynamical system that requires only one variable and its time derivatives to define it at any instant.

single-entry compressor A rotary compressor in which the working fluid is admitted to one side of the impeller or at one end of the rotor only.

single-flash steam power plant *See* FLASH-STEAM POWER PLANT.

single gear *See* TOOTHED GEARING.

single input, single output (SISO, two-port system) A control system having a single input port and a single output port

single-loop feedback A control system, such as a **single-loop servomechanism,** in which there is a single feedback path from the output to the summing junction at the input.

single-phase flow The flow of gas, vapour, or liquid, separately but not in any combination. *See also* MULTIPHASE FLOW.

single-plate clutch An axial clutch with a single friction plate.

single point incremental forming (SPIF) A manufacturing process, used in rapid prototyping and small batch production, in which a part is formed from sheet metal or polymer in a stepwise fashion by a CNC-controlled rotating spherical tool, without the need for a supporting die.

single-stage compressor A compressor which compresses a gas or vapour over a single stage between inlet to and outlet from the machine.

single-stage pump A pump in which flow and head are developed over a single stage between inlet to and discharge from the machine.

single-stage turbine A turbine in which energy is extracted from the working fluid over a single stage between inlet to and discharge from the machine.

single-start thread *See* SCREW.

singlet A modern type of precision-machined, nozzle-diaphragm design for impulse steam turbines, which reduces diaphragm distortion.

singularity In continuum modelling of material behaviour, where a quantity tends to an infinite value, such as the elastic stress at the tip of a sharp crack having zero crack tip radius. In practice, the radius cannot be smaller than two atoms across, so the stress is bounded. Furthermore in ductile solids, the stress is limited by yielding. An example in potential-flow theory is the infinite velocity that is predicted to occur when flow emerges from a source or enters a sink.

sink 1. *See* HEAT SINK. **2.** *See* LINE SINK; POINT SINK.

sink mark A shallow depression or dimple on the surface of an injection-moulded part caused by local internal shrinkage.

sinter (sintering) A solid-state diffusion densification process for the production of objects, particularly porous objects, from raw material in powder form by heating to a temperature below the melting point until the particles adhere. It is used for ceramic materials and metals. *See also* ISOSTATIC PRESSING.

siphon A tube, pipe, or hose in the form of an inverted U with one end lower than the other. If the device is completely full of liquid and its higher end is immersed below the surface of a liquid, liquid will flow up the short leg and down the longer leg.

siphon barograph A recording siphon barometer.

siphon barometer A simple mercury barometer consisting of a U-shaped tube in which one end is closed and the other end open. If the closed leg of the U-tube is filled with mercury and the device set in the vertical position, a partial vacuum will form in the space above the mercury in the closed side. The height difference between the menisci in the two legs is a measure of the barometric pressure.

siren A device for generating a whistling sound by the mechanical interruption of a jet of gas, usually air, by a perforated disc or cylinder.

SISO *See* SINGLE INPUT, SINGLE OUPUT.

SI system of units (SI units) *See* INTERNATIONAL SYSTEM OF UNITS.

six-axis system A robot with three translational and three rotational degrees of freedom.

size dimension The specified value of a diameter, length, width, etc. of a feature required to specify the finished form of a component or assembly. *See also* DIMENSIONING.

size factor *See* FATIGUE.

sizing 1. A finishing operation to ensure the specified dimensions and tolerances for a component are met. **2. (size classification)** Separating an aggregate of mixed particles into groups according to size using a series of screens.

sizing screen *See* SIEVE.

sizing treatment A surface treatment applied to glass fibres used in reinforced plastics in order to avoid surface damage and consequent reduction in strength.

skew bevel gears *See* TOOTHED GEARING.

skewed boundary layer A boundary layer within which the velocity vectors deviate from the direction of the free stream. If the angle of deviation is constant throughout the boundary layer, the skewing is unidirectional. If it varies, the skewing is bidirectional. Flow with a velocity component parallel to a surface but at an angle to the free stream is termed secondary flow.

skewness (γ_1) A statistical term used in the description of a distribution of a set of random data. It quantifies the degree of asymmetry in the distribution of a random variable. γ_1 may be defined as the third moment $\mu_3 =< x_i^3 >$ of the distribution normalized by the $\frac{3}{2}$-power of the second moment μ_2 or $\sigma =< x_i^2 >$ (i.e. the variance or mean square) where x_i is the random quantity and $< x_i^n >$ indicates an average value of x_i^n taken over all values of i with $n = 2$ or 3. The distribution is defined such that $< x_i > = 0$. Thus $\gamma_1 =< x_i^3 > /\sigma^{3/2}$ Among other applications, skewness is used in the description of surface roughness and turbulence. *See also* KURTOSIS.

skin-friction coefficient (friction coefficient, c_f) A dimensionless quantity used in boundary-layer theory and defined by $c_f = 2\tau_W / \rho U_\infty^2$ where τ_W is the wall shear stress, U_∞ is the free-stream velocity, and ρ is the fluid density. It is analogous to the friction factor for pipe and duct flow.

skin-friction drag *See* DRAG.

skirt *See* PISTON.

slab analysis An approximate method of analysis in plasticity in which stresses on plane or spherical surfaces perpendicular to the flow direction are assumed to be uniform principal stresses.

slack (slackness) The looseness or play in a mechanism due to wear or lack of precision in manufacture.

slack side The untensioned side of a driving belt or chain.

slag 1. A vitreous by-product of smelting ore to extract metal, such as iron, copper, lead, and aluminium. It is typically a mixture of metal oxides and silicon dioxide. *See also* BLAST FURNACE. **2.** *See* WELDING.

slave An actuator that replicates the movement of another component.

slave arm A robot arm that follows the motion of a master arm in real time, rather than being previously programmed. This process is also known as **teleoperation**, and is frequently applied in medical robotics and in operations in hazardous environments such as nuclear plant decommissioning.

slave cylinder A hydraulic or pneumatic cylinder and piston, so connected that the movement of the piston replicates that of a piston in a second, master, cylinder. Such systems are frequently employed to transmit force from a pedal for the operation of the clutch in a motor vehicle.

sleeve A thin cylinder, usually machined to very fine tolerances inside and out, which fits into the bore of another cylinder, such as in a piston engine, or into which a piston is inserted. *See also* LINER.

sleeve bearing *See* BEARING.

sleeve burner A simple type of burner consisting of a perforated cylinder in which oil is vaporized, then burned. They are commonly used in kitchen ranges.

sleeve coupling (sleeve joint) A hollow cylinder which fits over the ends of two concentric shafts, pipes, wires, or cables to join them.

sleeve valve A machined sleeve which fits between the piston and cylinder wall of a piston engine, where it slides or rotates so that openings in the sleeve align with the inlet and exhaust ports at appropriate times in the engine cycle.

slender body A body with width or diameter very much smaller than its length.

slenderness ratio The ratio of the length of a column *L* to the minimum radius of gyration of the cross section *k*.

slewing 1. Alteration of the alignment of the track of a conveyor or railway. **2.** The motion of a crane in which the jib is swung round. *See also* LUFFING. **3.** Rotational movement about a single axis.

slew rate The maximum rate of change in output of a system that can be achieved with a specified finite control input. Hence, for example, the maximum velocity of a robot actuator or the maximum rate of change of the voltage or current supplied by an amplifier.

slide A component which moves over a flat or curved surface between guides.

slide blocks *See* GUIDE BLOCKS.

slider bearing A simple bearing consisting of two almost parallel flat surfaces with a narrow gap filled with a lubricant. *See also* STEP BEARING.

slider coupling *See* OLDHAM COUPLING.

slider-crank chain (sliding-block linkage) A mechanism in which a crank is connected by a link to a sliding crosshead. Rotation of the crank causes the crosshead to reciprocate.

sliding fit A close fit between two components, permitting them to slide relative to each other. *See also* LIMITS AND FITS.

sliding friction *See* KINETIC FRICTION.

sliding gear (sliding pinion) A pinion which moves along a shaft.

sliding-mesh gearbox *See* GEARBOX.

sliding pair Two links connected together to form a kinematic pair, in which the relative motion is one of sliding only. *See also* CLOSED PAIR.

sliding pulley A pulley wheel free to slide along its supporting shaft.

sliding-vane compressor (rotary-vane compressor, vane compressor) A rotary compressor in which gas is compressed as the spaces between spring-loaded sliding vanes held in an offset rotor reduce as the rotor revolves within a cylindrical housing.

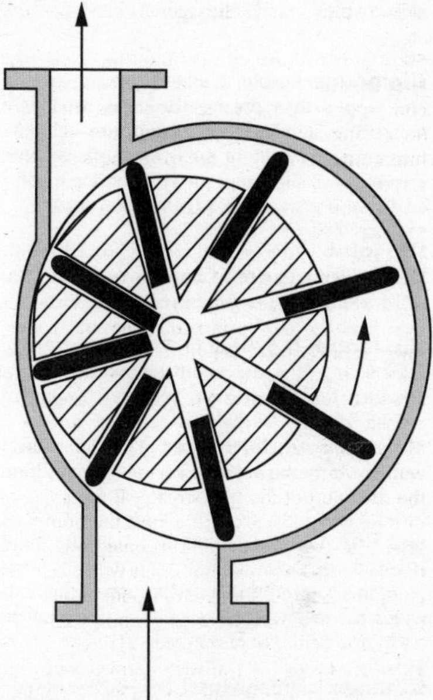

sliding-vane compressor

slinger A spinning disc used to give radial motion to a liquid, such as a lubricant to be sprayed onto sliding parts or the tip of a cutter.

sling psychrometer A psychrometer in which a wet- and a dry-bulb thermometer (**sling thermometer**) are mounted in a frame attached at one end to a handle which allows the instrument to be whirled rapidly in the air by hand for simultaneous measurement of the wet- and dry-bulb temperatures. The whirling ensures evaporation of the liquid, usually water, which coats the wet bulb, and also that the dry bulb is at ambient temperature.

slip 1. The slipping of a belt or clutch, or in general any two surfaces in contact when there is insufficient frictional grip. A **slip-friction clutch** is designed to slip when a certain torque is exceeded. **2.** In rarefied gas dynamics where the Knudsen number is much greater than 1, the gas flow close to a surface does not satisfy the no-slip condition. **3.** *See* PROPELLER SLIP. **4.** Plastic flow in crystals by relative sliding (gliding) of certain planes in directions determined by the lattice structure (body-centred cubic, face-centred cubic, hexagonal close-packed, etc.)

slip gauge A tablet of steel, precision-ground and lapped to a precise thickness used as a measuring standard, for example to calibrate measuring equipment such as a micrometer. Two or more slip gauges may be wrung together. *See also* JOHANSSON BLOCK; WRINGING.

slip joint A mechanical connection which allows limited endwise relative movement of two components such as pipes, rods, and ducts.

slip length (L_S) (Unit m) For a flow with slip velocity u_S, L_S is the depth below the surface at which the extrapolated velocity component parallel to the surface u reaches zero, i.e. $L_S = u_S/(\partial u/\partial y)_S$ where $(\partial u/\partial y)_S$ is the velocity gradient at the surface and y is the distance from the surface.

slip line A line tangent to the maximum shear stress at any point on which plastic flow takes place in a continuum. The collection of slip lines within the region of plastic deformation is called a slip-line field. *See also* PLASTICITY.

slippage (slippage loss) The leakage of fluid between a piston and cylinder wall.

slipper brake A brake for vehicles running on rails in which retardation is achieved by pressing a flat shoe against the rails, rather than curved blocks against the wheels.

slipper piston *See* PISTON.

slip plane The surface over which plastic flow takes place by shear. *See also* SLIP.

slip ratio For a screw propeller, the ratio of actual advance to theoretical advance, given by the product of pitch and rotation speed.

S

slipstream Fluid motion in the wake of an object moving through a fluid. Relative to the object, the flow velocity is reduced compared with the relative velocity between the object and the free stream.

slip velocity 1. In multiphase flow, the difference in velocity of the fluids on either side of an interface between the two. **2.** The velocity of the fluid immediately adjacent to a surface for flows where the no-slip condition does not apply. *See also* SLIP (2).

slitting A process of cutting thin metal, plastic, paper, cloth sheet, etc., using a rotary knife.

slot A groove machined into a component, e.g. to allow for thermal expansion, or in which the tongue or tip of another component can fit or slide. *See also* KEYWAY.

slotted piston *See* PISTON.

slug 1. A starting workpiece for forging and similar operations, such as a length of wire or rod to make a bolt blank on which a thread can be rolled or cut. **2.** An obsolete British unit of mass equal to 14.59390294 kg. **3.** A large-scale flow structure which occurs in the transition from laminar to turbulent flow in a pipe.

slug flow The regime of two-phase pipe flow, involving a gas–liquid mixture or a vapourizing liquid, characterized by large gas- or vapour-filled voids (**slugs**) which form when the gas or vapour fraction exceeds the level at which discrete bubbles occur but is still below the level for annular flow. *See also* BOILING; BUBBLY FLOW.

sluice valve *See* GATE VALVE.

SMA *See* SHAPE-MEMORY ALLOY.

small calorie *See* CALORIE.

small end *See* CONNECTING ROD.

small-scale CHP generation plant *See* MICRO-CHP.

small-scale hydropower (small-scale hydroelectricity, SSH) Power generation using hydraulic turbines with capacity limited to between about 5 MW (UK) and 30 MW (US), but there is no formally-agreed range. *See also* MICROHYDROELECTRICITY.

small-scale wind turbine There is no formally-agreed range, but typically a wind turbine which generates between 100 W and 6 kW, suitable for uses ranging from charging car batteries to providing the energy needs of a small public building.

small-stage efficiency *See* POLYTROPIC EFFICIENCY.

smart materials Materials which have one or more properties that can be changed significantly in a controlled fashion by external stimuli such as temperature, stress, electric or magnetic fields. Examples include shape-memory alloys (**smart alloys, smart metals**), piezoelectric materials, electrorheological materials, and magnetorheological materials.

smart sensor A microsensor integrated with signal-conditioning electronics that can process information or communicate with an embedded microprocessor.

smart structure A structure which is capable of sensing and reacting to its environment in a predictable and desirable manner through the integration of sensors, actuators, power sources, signal processors, etc. A smart structure may alleviate vibration, reduce acoustic noise, monitor its condition and environment, perform alignments, and change its shape or mechanical properties.

smart tool A robot end effector which uses sensory information from the task being performed, for example force or position measurement, to control the movement of the robot or an actuator in the end effector.

SMC *See* SHEET MOULDING COMPOUND.

SMD *See* SAUTER DIAMETER.

smelting The process of extracting metals, including copper, iron, lead, tin, and zinc, from ore. The basic process uses heat and a reducing agent, typically a source of carbon such as coke, together with a flux such as limestone. Electrolytic reduction of aluminium is also referred to as smelting. *See also* BLAST FURNACE.

Smith predictor A form of controller for a plant with a pure transport lag that predicts the effect of changes in the plant input on the future plant output.

smog Atmospheric pollution consisting of smoke and fog, originally associated with smoke and sulphur dioxide resulting from coal burning, but now more with photochemical smog.

smoke A visible contributor to air pollution consisting of solid and liquid particulates, typically in the range 0.01 to 5 μm, and gases which are the products of combustion or pyrolysis.

smokebox A chamber external to a boiler for trapping unburned products of combustion.

smoke point **1.** The maximum flame height in millimetres at which paraffin and other aviation-turbine fuels tested under standard conditions will burn without smoking. It is related to the hydrocarbon composition of such fuels, and to the potential radiant heat transfer from their combustion products. **2.** The temperature at which an oil begins to break down to free fatty acids and glycerol and subsequently acrolein which is a component of bluish smoke irritating to the eyes.

smoke stack (chimney, flue-gas stack, stack) A tall vertical pipe, typically of brick or metal, for the discharge into the atmosphere of steam and flue gases from industrial furnaces, steam-generating boilers, gas-turbine combustors, etc. If the combustion system is dependent upon natural rather than forced draught, the smoke stack is designed to maximize the effect.

smoke-tube boiler *See* FIRE-TUBE BOILER.

smoke washer A device for removing particulates from smoke by passing it through a water spray.

snap ring *See* CIRCLIP.

S–N curve *See* FATIGUE.

S–N diagram *See* FATIGUE.

snifter valve (snifting valve) A relief valve on a pump to release liquid or gas when a certain pressure is reached.

SNR *See* SIGNAL-TO-NOISE RATIO.

S/N ratio *See* SIGNAL-TO-NOISE RATIO.

snubber A hydraulic or mechanical device to limit the acceleration of a system subjected to a shock load.

soap-film analogy *See* MEMBRANE ANALOGY.

socket A recess (often cylindrical) into which a component is intended to fit, for purposes either of location or gripping.

socket wrench A form of spanner with an internally ridged steel socket to fit a nut or the head of a bolt or screw. *See also* BALL-AND-SOCKET JOINT; IMPACT WRENCH.

Soderberg line *See* FATIGUE.

soft bake In photoresist processing, which is an important aspect of lithography, resist coating of a component such as a wafer is followed by a soft bake at between 65°C and 100°C to evaporate solvent.

softening spring *See* STIFFENING SPRING.

soft flow The free-flowing characteristics of a plastic material under conventional moulding conditions.

soft lithography (microcasting) A family of techniques for fabricating or replicating structures using elastomeric stamps, moulds, and conformable photomasks, the term *soft* arising from the use of elastomers such as polydimethylsiloxane (PDMS).

soft solder *See* SOLDERING.

software A general term, shortened from computer software, which incorporates digitally stored data such as computer programs, word processors, computer operating systems, and data-processing programs.

soft water Water containing little or no calcium or magnesium ions. It occurs naturally, or is produced through a deionization process such as reverse osmosis. *See also* HARD WATER.

soft-wired numerical control *See* COMPUTER NUMERICAL CONTROL.

sol A colloidal suspension of solid particles in a continuous liquid medium. Examples include blood and pigmented ink. *See also* HYDROSOL.

solar air heater *See* SOLAR COLLECTOR.

solar battery A connected array of solar cells.

solar car A car driven by electric motors using electrical power generated by a panel of solar cells.

solar cell A device which uses the photovoltaic effect to convert the radiant energy of sunlight incident on the earth's surface directly into electrical energy. The **solar-cell efficiency (conversion efficiency)** is the percentage of solar radiation falling on the surface of a solar cell that is converted into electrical energy.

solar chimney A chimney in which air is heated at ground level, for example in a greenhouse or by solar collectors, creating an updraught that drives a turbine at its base to generate electricity. There are economies of scale to be had, but the overall efficiency remains low, at around 1%.

solar collector (flat-plate solar collector, flat-plate water collector) A basic solar collector, for low-temperature solar-energy collection, consists of a black metallic absorber plate of high thermal conductivity to which are attached water pipes or an air duct, or which is perforated to allow air to flow through (**solar air heater**). The absorber plate has a transparent cover and is backed by insulating material.

solar concentrator (concentrating solar collector, concentrator) An arrangement of mirrors or lenses, usually mounted on a tracking system, to concentrate sunlight onto a receiver, such as a solar furnace, where the radiation is absorbed and converted to another energy form. The **concentration ratio** is the ratio of the projected area of the concentrator facing the solar beam to the area of the receiver. *See also* NON-TRACKING CONCENTRATOR; SOLAR POWER TOWER.

solar constant (total solar irradiance) The rate at which solar energy is incident on a surface normal to the sun's rays at the outer edge of the atmosphere when the earth is at its mean distance from the sun (about 1.5×10^8 km) and equal to about 1.367 kW/m^2. *See also* INSOLATION.

solar cooling Any cooling system or device, such as an absorption refrigerator, which is powered by electricity generated from solar radiation.

solar day The period of time during which the earth makes one complete revolution on its axis relative to the sun. The duration of the mean solar day is defined as precisely 24 hours.

solar distillation A process in which solar radiation is used to evaporate sea water or less saline brackish water to produce pure water, free of sodium chloride and other dissolved salts. In a typical arrangement, the water vapour rises, condenses on the lower surface of a sloping glass cover, the condensate runs down, and is collected.

solar energy Energy derived from the solar radiation which reaches the earth.

solar furnace A device onto which mirrors or lenses focus solar radiation to create temperatures up to about 4000°C.

solar gear *See* EPICYCLIC GEAR TRAIN.

solar heating (solar space heating, SSH) The conversion of solar radiation into heat, using a solar collector, for industrial or domestic purposes or space heating.

solar heat storage The storage of solar energy in water, brick, concrete, eutectic salt, etc.

solarimeter (pyranometer) An instrument for measuring the intensity of combined direct and diffuse solar radiation incident on the earth's surface.

solar insolation (solar irradiance) *See* INSOLATION.

solar panel An array of photovoltaic cells covering a flat or near-flat surface.

solar photovoltaics The conversion of solar energy into electrical energy by a semiconductor device in which the active material is one of three types: silicon (single-crystalline, polycrystalline, or amorphous); thin films of polycrystalline copper indium diselenide, cadmium telluride, etc.; thin films of single-crystal gallium arsenide, etc.

solar pond A large, salty lake which acts as a flat-plate solar collector, in which solar energy is absorbed by dense salt water at the bottom of the pond.

solar power Electrical and other forms of power derived from solar radiation using solar cells, solar concentrators, or flat-plate solar collectors.

solar-power tower A tower on which a receiver is mounted to receive radiation from a solar concentrator.

solar radiation The electromagnetic radiation and particles, such as electrons and protons, emitted by the sun. A small fraction reaches the earth's surface both as direct and diffuse, or scattered, radiation. The maximum flux density for short-wave radiation, in the wavelength band 0.3 to 2.5 μm, is about 1 kW/m^2. *See also* SOLAR CONSTANT.

solar sea power *See* OCEAN THERMAL ENERGY CONVERSION.

solar space cooling *See* SOLAR COOLING.

solar space heating (SSH) *See* SOLAR HEATING.

solar thermal-electric power system A system in which solar energy is concentrated to provide a high-temperature heat source for generating electricity. *See also* SOLAR CONCENTRATOR.

solar thermal energy The technology for collecting solar radiation and using it for heating and electricity generation.

solar thermal engine An extension of active solar heating using more complex collectors to produce temperatures high enough to drive a steam turbine.

solar water heater **1.** A domestic hot-water system comprising a flat-plate solar collector mounted on the pitched roof of a house, a water storage tank, and a pumped circulation system. **2. (thermosyphon solar water heater)** A similar system to (1) in which the storage tank can be mounted outdoors, the hot water circulates by natural convection, and the circulation pump can be dispensed with.

soldering Bonding of parts without melting the mating surfaces, using a thin film of molten solder that adheres to the contacting surfaces. Solders melt at less than 450°C; brazing alloys above this temperature. A successful joint requires the parent metals to be wetted and the gap to be filled by capillary action. **Fluxes** contain chemical compounds that combine with the oxide of the parent materials to form a new compound, having a lower melting point and lower density than the filler material. The new compound floats on the molten metal that now readily wets the surfaces of the parent materials. A **sweated joint** is a soldered joint of large contact area, made by coating both surfaces with solder (**pre-tinning**), holding in place, and subjecting to uniform application of heat. Although solder is soft and weak, joints can be strong when adequate contact area and a thin layer of filler are used. A **cold joint** is an inadequately-heated, and therefore subsequently weak, soldered connection in which only some solder flows, wets, and adheres to the contacting surfaces. Similarly, a **dry joint** has regions unwetted by the molten solder, with a consequently high electrical resistance and reduced strength. Until recently, **soft solder** having the eutectic composition 65%/35% tin/lead and a single melting point of 183°C was used in electrical applications, and soft solder of non-eutectic composition (about 30%/70% tin/lead) and therefore a melting range between about 250 and 183°C was used in plumbing. However, to avoid ecological problems, the use of lead-free solder is now a legal requirement in electronics and for capillary fittings on water systems, e.g. the EU Waste Electronic Equipment (WEE) regulations, Restriction of Hazardous Substances (RoHS) directive, and various water supply by-laws. Thus lead-free soft-solders are employed, typically tin/copper alloys with melting points around 200°C–250°C. *See also* BRAZING.

soldering iron A tool consisting of a tip for applying heat to melt solder at a joint. It may be heated either electrically or indirectly.

solenoid A coil of electrically-conducting wire wrapped around a metal core, typically iron, to produce a magnetic field and hence a force on the core when an electric current passes through the coil. Solenoids are widely used to produce linear movement to actuate valves **(solenoid valves)** and other devices.

sol-gel technique deposition A microfabrication process in which a polymer compound dissolved in a solvent is spin-coated onto a substrate surface to form a gelatinous network (the gel) on the surface. Subsequent removal of the solvent solidifies the gel. The process can also be used to deposit various ceramics.

solid-body rotation Rotation of a body or mass of fluid about a fixed axis in which the tangential velocity varies in proportion to radial distance from the axis. *See also* FORCED VORTEX.

solid coupling A coupling forged in a solid piece with its shaft.

solid damping *See* STRUCTURAL DAMPING.

solid head A piston engine in which the cylinder head and cylinder block are cast in one piece and so cannot be separated.

solid-injection system A fuel-injection system for diesel engines in which liquid fuel at very high pressure is pumped through a fuel line, and then atomizing nozzles, into the combustion chamber.

solidity (σ) **1.** For a screen, the ratio of the blockage area of the screen wires to the total area of the screen. **2.** For a wind turbine, propeller, turbine, or compressor rotor, the ratio of the projected area of the blades to the total disc area.

solidity ratio For a cascade of turbine or compressor blades, the solidity ratio is equal to the blade chord divided by the pitch.

solid mechanics The elastic, plastic, or time-dependent behaviour of solid materials subjected to stress, temperature change, displacements, etc.

solidus The locus of temperatures on a phase diagram below which a given substance is completely solid (crystallized) and above which melting of the substance begins. Above the **liquidus temperature**, the material is homogeneous and liquid at equilibrium. Between the solidus and liquidus temperatures, mixed liquid-solid microstructures exist in alloys. When solid-state transformations occur, the corresponding names are solidoid and liquidoid.

soliton A self-reinforcing wave packet or pulse that maintains its shape while propagating, for example on a water surface, at constant speed. The permanency of form is a consequence of the non-linear effects that prevent dispersion.

solid-solution strengthening The process by which metal alloys have increased strength compared with the elements of which they are composed, owing to the presence of interstitial or substitutional atoms.

solution A homogeneous mixture of two or more substances. The **solute** is dissolved in the **solvent**, both of which may be gaseous, solid, or liquid, but it is usual for the solvent to be a liquid. *See also* GAS MIXTURES.

Sommerfeld number *See* BEARING.

sonar The acronym for **so**und, **na**vigation, and **r**adar. It is an underwater system which uses reflected sound waves at sonic or ultrasonic frequencies to detect and locate submerged objects.

sonic A term for anything pertaining to the speed of sound.

sonic anemometer An instrument for measuring a component of airspeed V based on the time of flight t of sound between a transmitter and a receiver a known distance s apart. V is found from $V = s(1/t_1 - 1/t_2)/2$ where t_1 is the time in one direction and t_2 the time in the opposite direction. The soundspeed c can be found from $c = s(1/t_1 + 1/t_2)/2$. The total airspeed and its direction can be determined if three sets of measurements are made with the receiver–transmitter pair successively positioned at the corners of a triangle.

sonic area (A^*) (Unit m^2) The cross-sectional area for isentropic flow of a perfect gas through a duct at which the gas velocity equals the speed of sound at that location. For a choked convergent–divergent nozzle, the sonic area occurs at the throat (**sonic throat**). In other situations, it may not be a physically realized area but simply a reference area.

sonicator A high-frequency (*c.*20 kHz) sound generator used to agitate liquids. Applications include ultrasonic cleaning, agitation of suspended particles, dispersion of nanoparticles, and degassing.

sonic boom The sudden noise associated with the shock waves caused by an object flying through the lower atmosphere at supersonic speed. The shock waves create an N-shaped overpressure profile at ground level, with a sudden initial pressure increase, a gradual pressure decrease to below ambient, and a sudden return to normal pressure.

sonic cleaning Cleaning of contaminated surfaces immersed in a liquid by the action of intense high-frequency sound waves. *See also* ULTRASONIC CLEANING.

sonic depth finder *See* ECHO SOUNDER.

sonic flaw detection The process of detecting flaws in components such as wheels, turbine blades, welds, or solid-material test specimens by observing internal reflections of sound waves or a variation in their transmission.

sonic line An imaginary line or surface in the flow behind the detached bow shock wave in front of a supersonic blunt body, or the conical shock wave attached to the tip of a supersonic conical body, which separates the region of subsonic flow from the region of supersonic flow.

sonic thermometer A thermometer which uses the principle that the absolute air temperature T can be determined from the sound speed $c = \sqrt{\gamma RT}$, where for air the specific heat ratio $\gamma = 1.4$ and the specific gas constant $R = 287$ J/kg.K. c can be obtained from the sonic-anemometer measurements.

sonic throat *See* SONIC AREA.

sonic velocity *See* SOUNDSPEED.

soot (black carbon) Black particles, of submicron-size, composed primarily of carbon produced by incomplete combustion or pyrolysis of coal, oil, wood, and other hydrocarbon-containing materials. Diesel engines produce a major fraction of air pollution due to soot. *See also* ALBEDO.

soot blower A system of steam or air jets used to remove ash and slag from heat-transfer surfaces.

Sorét effect (Ludwig–Sorét effect, thermal-diffusion effect, thermodiffusion effect, thermophoresis) Diffusion of a species within a mixture due to a temperature gradient. It is characterized by the non-dimensional **Sorét number (*Sr*)**. *See also* DUFOUR EFFECT.

sound barrier A term introduced in the 1940s and given to phenomena associated with the fact that, as the speed of an aircraft approaches that of sound, shock waves develop, leading to a significant increase in drag force, loss of lift, and loss of control. It was thought, erroneously, at the time that sound speed could not be exceeded by an aircraft.

sound energy *See* ACOUSTIC ENERGY.

sound intensity (*I*) (Unit W/m^2) The acoustic power of sound per unit area related to sound pressure by $I = p^2/\rho c$ where p is the root-mean-square of the fluctuating air pressure caused by the passage of sound waves, ρ is the air density, and c is the sound speed. *See also* SOUND-PRESSURE LEVEL.

sound pressure (Unit μPa) The instantaneous deviation of pressure from the ambient pressure caused by the passage of a sound wave, which can be measured by a microphone in air and a hydrophone in water.

sound-pressure level (sound level, SPL) (Unit decibel (dB)) A logarithmic (i.e. non-dimensional) measure of the effective sound pressure p_{RMS} relative to a reference value p_{REF} usually taken as 20 μPa, which is considered to be the threshold of human hearing, i.e. $SPL = 10 log_{10}(p^2_{RMS}/p^2_{REF}) = 20 log_{10}(p_{RMS}/p_{REF})$.

sound spectrograph An instrument used to analyse the spectral composition of audible sound and produce an energy-*vs*-frequency spectrogram of the result.

soundspeed (acoustic velocity, sonic velocity, speed of sound, c) (Unit m/s) The speed with which a sound wave propagates through a medium. (a) For a perfect gas, $c = \sqrt{\gamma RT}$ where γ is the ratio of specific heats for the gas, R is the specific gas constant, and T is the absolute temperature of the gas. For an unbounded liquid, $c = 1/\sqrt{\beta_a \rho}$ where β_a is the adiabatic compressibility and ρ is the density. *See also* FUNDAMENTAL DERIVATIVE. (b) For solids, both shear and longitudinal waves can be propagated. For an isotropic solid, the velocity of shear waves within an extensive medium is $c_S = \sqrt{G/\rho}$ and of longitudinal waves is $c_L = \sqrt{(K + 4G/3)/\rho}$ where G is the shear modulus and K is the bulk modulus. When waves travel near a surface, the velocities are different. In a long slender bar, the elastic wave velocity $c = \sqrt{E/\rho}$ where E is Young's modulus, and for plastic deformation $c = \sqrt{(d\sigma/d\varepsilon)/\rho}$ where $d\sigma/d\varepsilon$ is the local slope of the nominal stress–strain plot. When a rod strikes a rigid body with velocity V, the strain is that value such that the integral of the wave velocities over the corresponding strains is equal to the impact velocity, i.e. $V = \int c(\varepsilon)d\varepsilon$. For an elastic impact it can be shown that the maximum stress is $\sigma = EV/c$.

source 1. (heat reservoir) A reservoir that supplies energy in the form of heat. **2.** *See* LINE SOURCE; POINT SOURCE.

space centrode The path traced by the instantaneous centre of a rotating body relative to an inertial frame of reference.

space frame (space structure, space truss) A statically-determinate rigid structure constructed from pin-jointed interlocking members in a geometric pattern (of which the triangle and square are the usual basic elements) in which all members carry loads in tension or compression. Practical frameworks are often bolted or welded (and are therefore not statically-determinate), but may be analysed in approximately the same way.

space heating The heating required to heat the habitable parts of a building to a specified level.

space structure *See* SPACE FRAME.

space truss *See* SPACE FRAME.

Spalding number 1. (B_M) A non-dimensional number which characterizes the rate of diffusion of fuel from a burning fuel droplet; defined by $B_M = (Y_{FS} - Y_{F\infty})/(1 - Y_{FS})$ where Y_{FS} is the mass fraction of fuel at the surface of the droplet and $Y_{F\infty}$ the fuel mass fraction of fuel in the gaseous oxidant surrounding the droplet. **2.** **(*Sp*)** A non-dimensional parameter that arises in convective heat transfer through a pipe, defined by $Sp = StPr/\sqrt{f/2}$ where *St* is the Stanton number, *Pr* is the Prandtl number, and *f* is the Fanning friction factor.

Spalding's formula An empirical equation for the entire law-of-the wall for a turbulent boundary, given by

$$y^+ = u^+ + e^{-\kappa B}\left[e^{\kappa u^+} - 1 - \kappa u^+ - \tfrac{1}{2}(\kappa u^+)^2 - \tfrac{1}{6}(\kappa u^+)^3\right]$$

where u^+ and y^+ are the law-of-the-wall variables, κ is von Kármán's constant and $B \approx 5.5$.

spallation (spalling) A process in which small fragments of material (**spall**) are removed in a brittle fashion from a surface by impact or a suddenly applied stress. *See also* BEARING.

spalling resistance index *See* THERMAL SHOCK.

span A dimension measured between the extremities of a body or structure, such as between the tips of a wing or the supports of a beam.

spanner A tool for applying torque to tighten or loosen a nut, bolt, or screw. Grip is applied by a serrated ring or an open U-shape at the end of the tool. On imperial-size spanners, the BSW and BSF marking refers to the diameter of the screw thread; similarly for BA spanners. Spanners for unified and metric threads are marked with the distance across the flats ('A/F') of the nut or head of the bolt.

sparger (sparge pipe) A length of pipe or tubing with holes at spaced intervals along the length which can be used to spray one fluid into another, such as liquid into gas or gas into liquid.

spark A visible, and usually audible, short-duration electric discharge due to ionization and heating of the gas between two electrodes when a sufficiently high voltage is applied across them.

spark erosion (spark machining) *See* ELECTRO-DISCHARGE MACHINING.

spark-ignition cycle *See* OTTO CYCLE.

spark-ignition engine (SI engine) A piston engine in which combustion of the air-fuel mixture in each cylinder is initiated by a spark from a spark plug, i.e. **spark ignition.** *See also* DIESEL ENGINE; OTTO CYCLE.

spark knock *See* ENGINE KNOCK.

spark lead *See* IGNITION SYSTEM.

spark plug (sparking plug) A circular plug screwed into the cylinder head of a spark-ignition engine which incorporates two electrodes separated by a small gap across which a voltage difference is applied to create a spark within the combustion chamber.

spatial frequency (ν) (Unit cycles/m) For any quantity which varies periodically with spatial position, the spatial frequency is the number of cycles per unit of distance. It is the inverse of the wavelength λ, i.e. $\nu = 1/\lambda$. The wave number $k = 2\pi\nu$. *See* ALSO TEMPORAL FREQUENCY; WAVELENGTH.

spatial linkage A linkage which involves movement in all three spatial directions.

spatially periodic *See* PERIODIC OSCILLATION.

special tolerance *See* TOLERANCES.

specifications (specs) A document giving all relevant technical information about a device, machine, system, etc., for example dimensions, weight, power output, torque, emissions levels, load-carrying capacity, fuel consumption, fuel capacity, and lubrication requirements.

specific dynamic capacity *See* BEARING.

specific energy **1.** (E_S) (Unit m^2/s^2) The combined potential and kinetic energy of an object divided by its mass. *See also* SPECIFIC PROPERTY. **2.** *See* HIGHER CALORIFIC VALUE.

specific enthalpy *See* SPECIFIC PROPERTY.

specific entropy *See* SPECIFIC PROPERTY.

specific exergy *See* SPECIFIC PROPERTY.

specific fuel consumption (SFC) 1. (brake specific fuel consumption) (Unit kg/kW.s or kg/kW.h) The mass flow rate of fuel consumed by an engine per unit power output and defined by $SFC = \dot{m}_F/bp$ Where $\dot{m}_F$ is the fuel mass flow rate and bp is the brake power. **2.** Unit kg/N.s) For an aircraft engine, SFC is the fuel flow rate $\dot{m}_F$ divided by the thrust T i.e. $\dot{m}_F/T$.

specific gas constant *See* PERFECT GAS.

specific gravity (sp gr) *See* RELATIVE DENSITY.

specific-gravity bottle *See* DENSITY BOTTLE.

specific-gravity hydrometer A hydrometer which is used to measure the relative density of a liquid relative to that of water at the same temperature.

specific heat (specific-heat capacity, *C*) (Unit J/kg.K or J/kg.°C) The energy required to raise the temperature of unit mass of a substance by one degree (K) without change of phase. For practical purposes, a liquid or a solid is generally regarded as having a single specific heat whereas for a gas the energy required depends upon how the energy-transfer process is executed and two values are defined: the specific heat at constant volume (C_V) and the specific heat at constant pressure (C_P). The thermodynamic definitions are $C_V = (\partial u/\partial T)_V$ and $C_P = (\partial h/\partial T)_P$ where the subscript V denotes a reversible non-flow process at constant volume and P denotes a reversible non-flow process at constant pressure, u is the specific internal energy, h is the specific enthalpy, and T is the absolute temperature. Specific heat is an intensive thermodynamic property dependent on temperature and pressure. The **specific-heat ratio (adiabatic index, isentropic index, ratio of specific heats, γ, κ, *k*)** for a gas, defined by $\gamma \equiv C_P/C_V$, is an important quantity in compressible flow. Although there is a slight temperature dependence, for many diatomic gases, including air, $\gamma \approx 1.4$ while for monatomic gases $\gamma \approx 1.677$. *See also* MOLAR HEAT CAPACITY; PERFECT GAS; POLYTROPIC PROCESS.

specific humidity (absolute humidity, humidity ratio, moisture content, vapour concentration, ω) The mass of water vapour per unit mass of dry air in a given volume of the mixture. *See also* RELATIVE HUMIDITY.

specific impulse (I_{sp}) (Unit s) The impulse per unit weight of propellant consumed by a rocket engine. If m_p is the total mass of propellant consumed over time t, then $I_{sp} = I/m_p g$ where I is the impulse given by

$$I = \int_0^t F dt$$

with F being the thrust of the rocket engine, and g the acceleration due to gravity.

specific internal energy *See* SPECIFIC PROPERTY.

specific modulus (Unit m^2/s^2) An elastic modulus per unit density, such as E/ρ where E is Young's modulus for the material and ρ is its density. It is used to rank materials for stiffness/weight. If relative density is used instead of density, the unit becomes Pa. *See also* SPECIFIC STRENGTH.

specific propellant consumption For a rocket engine, the ratio of the total mass of propellant consumed to the total impulse delivered. *See also* SPECIFIC IMPULSE.

specific property In thermodynamics, any extensive property of a substance per unit mass of that substance, i.e. an intensive thermodynamic property. **Specific volume (sp vol, *v*)**, with unit m^3/kg, is the volume of a substance per unit mass. If $\mathcal{V}$ is the volume of the substance and m is its mass, $v = \mathcal{V}/m$. It is the reciprocal of density. **Specific internal energy**

(*u*), with unit J/kg or J/kmol, is the internal energy per unit mass, i.e. $u = U/m$ where U is the internal energy of a system and m is its mass. **Specific enthalpy (*h*)**, with unit kJ/kg or kJ/kmol, is the enthalpy per unit mass, i.e. $h = H/m$ where H is the enthalpy of a system. If p is the static pressure then $h = u + pv$. The product pv is termed the **flow energy**. **Specific stagnation enthalpy (stagnation enthalpy, h_0)** is the specific enthalpy a flowing fluid with velocity V would have if brought adiabatically to rest, i.e. $h_o = h + ½ V^2$ and **specific total enthalpy (total enthalpy, h_T)** for a fluid flow is the sum of h, the kinetic energy per unit mass $½ V^2$ and the potential energy per unit mass gz, i.e. $h_T = h + ½ V^2 + gz$ where g is the acceleration due to gravity and z is the height above a datum level. **Specific entropy (*s*)** is entropy per unit mass, i.e. $s = S/m$ where S is the entropy of a system. Entropy cannot be measured directly but can be determined using $T\,ds = dh - v\,dp$ where T is the absolute temperature. **Specific exergy (ϕ, ψ)**, for a closed system: $\phi = u - u_0 + p_0(v - v_0) - T_0(s - s_0) + \frac{1}{2}V^2 + gz$ the subscript 0 referring to the environment; for a flowing fluid: $\psi = \phi + (p - p_0)v$ *See also* EXERGY.

specific speed (N_S) A non-dimensional quantity used to characterize turbomachinery performance. For a pump or compressor, $N_S = \Omega \dot{Q}^{1/2}/(gH)^{3/4}$ and for a hydraulic turbine, $N_S = \Omega P^{1/2}/(gH)^{5/4}$ where Ω is the angular rotation speed (rad/s), $\dot{Q}$ is the volumetric flow rate , P is the shaft power, H is the head, and g is the acceleration due to gravity. In the case of compressors, the flow rate is determined using an average density for the fluid. Care has to be taken to ensure that consistent units are used: in some applications g is omitted and Ω replaced by the rotation speed in rps or rpm, so that the quantity N_S is no longer non dimensional.

specific steam consumption (SSC, steam consumption) (Unit kg/kW.h) For a steam turbine, the mass flow rate of steam required per unit power output.

specific strength (Unit m²/s²) The yield strength, or ultimate tensile strength, divided by density, used to rank materials for strength/weight. If relative density is used instead of density the unit is Pa. *See also* SPECIFIC MODULUS.

specific total enthalpy *See* SPECIFIC PROPERTY.

specific volume *See* SPECIFIC PROPERTY.

specific weight (γ) (Unit N/m³) The weight per unit volume of a substance: if W is the weight of a volume $\mathcal{V}$ of the substance, ρ is its density and g is the acceleration due to gravity, $\gamma = W/\mathcal{V} = \rho g$.

specific work of fracture (fracture toughness) (Unit J/m²) In fracture, the work per unit area required to propagate a crack. *See also* FRACTURE MECHANICS.

specific work output (Unit kW.h/kg) For a gas turbine, the shaft work produced per unit mass of gas flowing through the turbine. It is equal to the shaft power divided by the mass flow rate of gas.

specs *See* SPECIFICATIONS.

spectral A term indicating wavelength (λ) or frequency dependence.

spectral absorptance The ratio of the radiant flux absorbed by a body to that incident upon its surface at a specific wavelength. *See also* TOTAL ABSORPTANCE.

spectral blackbody emissive power ($E_{B\lambda}$) (Unit W/μm.m²) The amount of radiation emitted per unit time, per unit surface area, and per unit wavelength about the wavelength λ by a blackbody at an absolute temperature T. According to Planck's law,

$$E_{B\lambda}(\lambda, T) = \frac{C_1}{\lambda^5[exp\ (C_2/\lambda T) - 1]}$$

where $C_1 = 2\pi hc_0^2 = 3.742 \times 10^8 \text{W.μm}^4/\text{m}^2$, $C_2 = hc_0/k = 1.439 \times 10^4 \text{μm.K}$, c_0 is the speed of light, $k = 1.38065 \times 10^{-23}$J/K is the Boltzmann constant, and $h = 6.6256 \times 10^{-34}$J.s is Planck's constant. This equation for $E_{B\lambda}$ is valid for a surface in vacuum or a gas. For another medium, it needs to be modified by replacing C_1 with C_1/n^2 where n is the refractive index for the medium.

spectral density *See* POWER SPECTRAL DENSITY.

spectral directional emissivity (ε_λ) The emitted radiation intensity $I_{\lambda\theta}(\theta,\phi,T)$ from a surface in a specified direction θ,ϕ at a given wavelength λ and temperature T divided by the radiation intensity of a blackbody at the same temperature and wavelength $I_{b\lambda}(T)$, i.e. $\varepsilon_\lambda = I_{\lambda\theta}(\theta, \phi, T)/I_{b\lambda}(T)$. *See also* PLANCK'S DISTRIBUTION.

spectral distribution (spectral power distribution) The power per unit frequency in a random signal, plotted as a function of frequency and averaged as time tends to infinity.

spectral hemispherical emissive power (E_λ) (Unit W/m^2 .µm) The rate at which radiation of wavelength λ is emitted in all directions from a surface per unit wavelength $d\lambda$ about λ and per unit surface area. It is the spectral heat flux associated with radiation into a hypothetical hemisphere centred on a point on the surface.

spectral intensity The thermal radiation emitted by a body in a small wavelength interval around a single wavelength. *See also* TOTAL INTENSITY.

spectrum analyser A device for plotting in decibel the power in a signal, normalized by the bandwidth over which the power is measured, as a function of frequency. Such an analyser is thus a method of obtaining a plot of power spectrum density from a signal. Some spectrum analysers may include a second channel monitoring the input to a system under test, in which case both the power and phase of the output can be plotted.

specular reflection A reflection process in which the angles of incidence and reflection are equal. *See also* DIFFUSE REFLECTION.

speed (s) (Unit m/s) Instantaneous speed is a scalar quantity which is the magnitude of instantaneous velocity. The average speed of a moving object is defined as the distance travelled by the object in a given time, divided by that time. For a sequence of uniform speeds s_1, s_2, s_3 . . . over distances L_1, L_2, L_3 . . . , the average speed s_{av} is not given by the rule of mixtures weighted average

$$s_{av} = (L_1/L_{tot})s_1 + (L_2/L_{tot})s_2 + (L_3/L_{tot})s_3 + \ldots$$

where $L_{tot} = L_1 + L_2 + L_3 + \ldots$, but rather by the weighted harmonic mean

$$1/s_{av} = (L_1/L_{tot})(1/s_1) + (L_2/L_{tot})(1/s_2) + (L_3/L_{tot})(1/s_3) + \ldots$$

speed cone(s) (speed pulley(s)) A cone-shaped pulley or a set of pulleys of increasing diameter forming a single stepped cone. *See also* STEPPED-CONE PULLEYS.

speed of light (c, c_0) The speed at which electromagnetic radiation, including light, travels. In a vacuum, c = 2.997 924 58 x 10^8 m/s. When light travels through any medium its speed is reduced by a factor equal to n, the absolute refractive index of that medium.

speed of rotation (N) (Unit rps or rpm) For an object rotating about a fixed axis, the speed of rotation is the number of turns (revolutions) of the object per unit time. It is equal to the angular velocity Ω (or ω) divided by 2π.

speed of sound *See* SOUND SPEED.

speed pulley *See* STEPPED-CONE PULLEYS.

speed reducer A train of gears with gear ratios such that the rotational speed of the final gear in the train is lower than that of the first.

sp gr *See* RELATIVE DENSITY.

spherical-coordinate robot (polar-coordinate robot, spherical-configuration robot, spherical-polar robot) A robot having a rotational joint, joint angle θ_1, with a vertical axis above the base frame (i.e. a waist joint), a rotational joint, joint angle θ_2, with a horizontal axis at the end of the first link (i.e. a shoulder joint) and a translational joint, joint offset d_3, with axis normal to the axis of the shoulder joint. The workspace is thus a hollow sphere centred on the base frame. The diagram shows an idealized spherical coordinate robot.

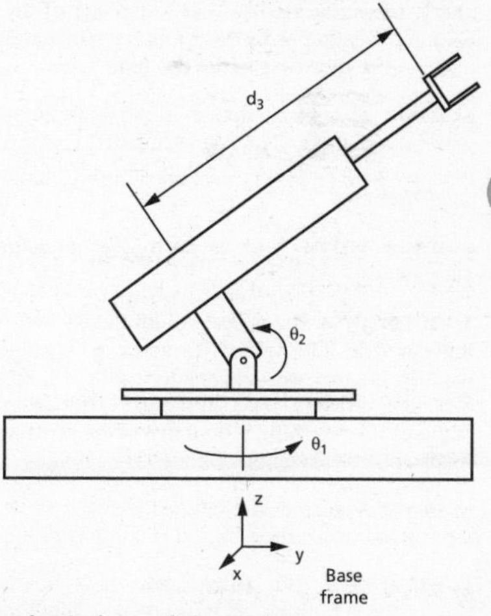

spherical-coordinate robot

spherical plain bearing *See* BEARING.

spherical roller thrust bearing *See* BEARING.

spherical stress A term occasionally used for hydrostatic or mean stress.

spheroidal cast iron *See* CAST IRON.

spherometer An instrument used to measure the radius of curvature of a spherical surface.

spider 1. In a universal joint, the cross-shaped component pivoted between the forked ends of two shafts which are normally not aligned. **2.** A set of radial arms, much like the spokes of a wheel, attached to a shaft and used to support cylindrical components, such as planet gears, at their outer ends.

SPIF *See* SINGLE POINT INCREMENTAL FORMING.

spiffing The manufacturing process of single point incremental forming.

spillway An open channel used for the controlled release of excess water flow e.g. from a dam.

spin Rotation about an axis.

spin coating A microfabrication process in which a solution containing a material to be deposited is dripped onto a substrate surface which is spun at high speed (typically 5000 rpm). The surface tension and viscosity of the solution together with the spinning speed determine the thickness of the coating.

spindle 1. A short, slender shaft, sometimes with pins at the ends. **2.** The threaded component in a valve which is rotated to adjust the valve opening.

spindle valve A screw-down stop or gate valve or stopcock.

spin forging The process of forming a rotating flat disc into an axisymmetric contoured part by progressive deformation onto a mandrel. In contrast to shear spinning, the thickness of the product is thinner than that of the starting blank.

spinner A co-axial streamlined fairing enclosing the hub of a propeller and rotating with it.

spinneret A die with many fine holes, through which plastic melt or polymer solution is forced to produce long filaments for spinning into yarn.

spinning 1. The twisting together of fibres to form a continuous length of thread or yarn. **2.** A manufacturing technique having a localized deformation zone used to produce axisymmetric components by working sheet-metal over a profiled mandrel as the workpiece is rotated. *See also* SHEAR SPINNING; SPIFFING; SPIN FORGING.

spin welding *See* WELDING.

spiral angle *See* TOOTHED GEARING.

spiral bevel gears *See* TOOTHED GEARING.

spiral conveyor *See* SCREW CONVEYOR.

spiral flow *See* SWIRL FLOW.

spiral helical spring An extension spring in the form of a cone such that the turns increase in diameter with distance along the spring. When compressed flat, the spring forms a spiral.

spiral-plate heat exchanger A heat exchanger in which two metal sheets are rolled into spirals, one within the other, with a gap between the two. The ends are capped, as are the outer ends of the curved paths. Inlet and outlet ports are provided for the two fluid streams, which can pass through the exchanger either axially or spirally.

spiral pressure gauge *See* BOURDON GAUGE.

spiral spring A spring of circular or flat cross section wound into a spiral. Commonly used to power clockwork mechanisms.

spiral thermometer A thermometer based on a bimetallic strip in the form of a spiral which, as the temperature increases or decreases, opens or tightens causing a pointer to move across a calibrated scale.

spiral-tube heat exchanger A shell-and-tube heat exchanger in which one or more metal tubes are coiled into spirals with a manifold at either end and this assembly is fitted into a compact shell.

spirit level A device for checking whether a surface is horizontal, vertical, or at some specified angle by ensuring a small bubble in a sealed glass tube almost full of ethanol or similar liquid is centred.

spirit thermometer A thermometer consisting of a closed glass capillary tube having a bulb at one end filled with coloured alcohol, which rises up the capillary tube as the bulb temperature increases. The temperature is indicated by the position of the alcohol meniscus on a scale marked on the outside of the capillary tube.

spiroid gear *See* TOOTHED GEARING.

SPL *See* SOUND-PRESSURE LEVEL.

splash lubrication 1. In an engine, a system in which the big end of the connecting rod dips into a pool of oil within the crankcase and splashes lubricant onto the cylinder wall and other parts. **2.** In an enclosed gear drive, gear teeth partially submerged in oil splash lubricant onto other gears.

spline Narrow grooves resembling long gear teeth machined into a shaft (a **splined shaft**) or hole (using a **spline broach**) and represented on drawings in simplified form as shown in the first diagram. As shown in the second diagram, the teeth may be either straight-sided or of involute form. Torque and rotation are transmitted between splined pairs which also permit sliding motion.

split bearing *See* BEARING.

split crankcase An engine crankcase split transversely close to the centreline of the crankshaft.

split-disc check valve A check valve similar to a swing check valve in which the flap is split and hinged down its centre, the two halves being free to swing in one direction only.

split gear 1. A gear wheel in two halves which are bolted together during installation into a machine. **2.** *See* ANTIBACKLASH GEAR.

split mould (split cavity mould) A mould in which the cavity is formed from two or more parts, called **splits**, which are held together by an outer chase.

split pin *See* COTTER PIN.

split ring 1. A flat ring consisting of between one and two overlapping turns of a helix formed from circular or flat wire. **2.** A ring with a small gap or overlapping joint, such as a piston ring. *See also* CIRCLIP.

split-ring piston packing *See* PISTON.

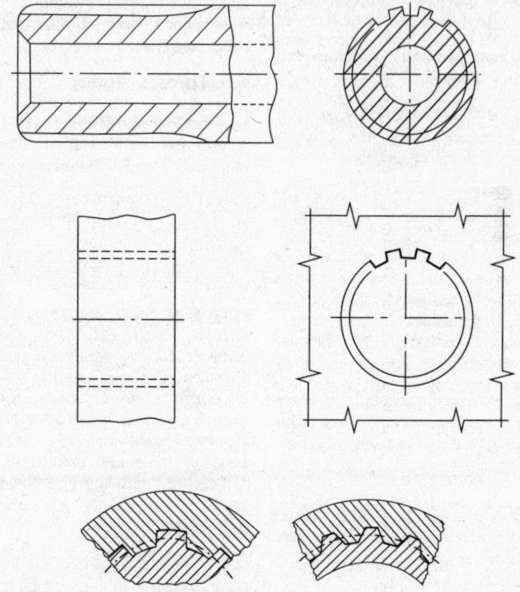

spline 1

spline 2

split-skirt piston *See* PISTON.

splitter (splitter plate, splitter vane) A thin sheet installed in a duct to divide flow through it into two streams, e.g. to minimize the possibility of flow separation in a diffuser or to suppress flow unsteadiness. *See also* GUIDE VANES.

spoiler An aerodynamic device attached to a motor vehicle designed to improve aerodynamic efficiency. Front spoilers, often called air dams, deflect airflow around the vehicle and reduce flow underneath it to increase downforce on the front wheels, while wing-like rear spoilers add downforce to improve rear-wheel traction. *See also* AEROFOIL.

spoke The connection between the hub and rim of a wheel, usually in the form of a bar, rod, or wire, but which may also be cut or cast from material such as steel or aluminium alloy. Spokes having cross-sections large enough to resist buckling are usually radial. Wire spokes, as in a bicycle wheel, are pre-tensioned and also tangential to the hub (thus crossing) in order to give torsional stiffness.

sponge A soft, absorbent, porous material, typically made from cellulose wood fibres or foamed plastic polymers such as polyester that mimic a natural sponge. Non-absorbent sponges are made from low-density polyether.

spontaneous combustion (spontaneous ignition) *See* AUTOIGNITION.

spontaneous process In thermodynamics, a process which occurs due to the inherent properties of a system, with no external input.

spool 1. The revolving drum of a hoist. **2.** The movable part of a cylindrical slide-type valve. **3.** A cylinder which may have flanged ends onto which cable, rope, chain, etc. can be wound for transport, distribution, or storage. **4.** In a gas turbine, a compressor and turbine mounted on a single shaft. *See also* THREE-SPOOL TURBOFAN; TWO-SPOOL TURBOFAN; TWO-SPOOL TURBOJET.

spool up The term used to describe a jet engine, supercharger, or turbocharger beginning to spin more rapidly and increase engine power.

spool valve *See* PISTON VALVE.

spot drilling A process of drilling a small hole or indentation in the surface of a workpiece as a centring guide for subsequent machining. *See also* PILOT HOLE.

spot welding *See* WELDING.

sprag clutch A clutch similar in appearance to a roller bearing, but instead of rollers there are asymmetric pieces (often 'figure-of-eight'-shaped) which allow relative rotation between the races in one direction, but tilt and lock the races together in the other. *See also* ROLLER CLUTCH.

spray A dispersed stream of liquid droplets produced either by forcing the liquid through a spray nozzle, by subjecting a jet of liquid to ultrasound, or by directing a liquid jet onto a spinning disc. *See also* ATOMIZATION; MIST.

spray chamber *See* SCRUBBER.

spray drying A process for producing solid particles by spraying a solution or slurry into hot gas to evaporate the liquid component.

spray nozzle A nozzle designed to atomize a liquid stream as it is forced at high pressure through one or more fine holes in the nozzle tip. In some designs, the droplet size and flow rate are controlled by a needle.

spray tower *See* SCRUBBER.

spray-up An open-mould process used to produce composite components by spraying chopped fibreglass reinforcement, catalysed resin, and, if required, filler material on to the mould surface.

spread *See* SENSITIVITY.

spreading coefficient (S) (Unit N/m) The factor which determines whether a drop of oil placed on a water film on a solid surface within ambient air, or another gas, spreads or ruptures the film (termed **dewetting**). $S = \gamma_{gw} - \gamma_{go} - \gamma_{ow}$ where γ is the surface tension and the subscripts indicate the gas-water interface, the gas-oil interface, and the oil-water interface, respectively.

spring An elastic component which stores mechanical energy and exerts a force when deformed. The slope of the curve of applied force F to the deflexion of a spring x, dF/dx, is termed the **spring rate** or **spring modulus** k with unit N/m. For a **linear spring**, the $F(x)$ curve is proportional and k is called the **spring constant**. *See also* SPRING REPRESENTATION; STIFFENING SPRING; TORSION SPRING.

SEE WEB LINKS

- Spring types and associated terminology

springback The elastic recovery of plastically-deformed sheet parts on removal from press tooling, so that the final shape and size is not what was intended.

spring balance (spring scale) An instrument for determining force by measuring the deflection of a calibrated spring.

spring calliper (spring-joint cauiper) A calliper in which the measuring legs are pushed against an adjusting nut by a spring, placed either between the legs or at the hinge.

spring clip A clip made of a material such as spring steel which grips an inserted part.

spring index (*C*) For a helical coil spring of mean diameter D and wire diameter d, $C = D/d$.

spring load A load exerted on a component by bending, compressing, extending, or twisting an attached spring. Mechanical energy stored in the spring can be released if any constraints on the component are removed.

spring-loaded regulator A pressure-regulator valve in which opening or closing of the valve is determined by a spring.

spring-loaded variable-area flow meter A variable-area inline flow meter in which the flow rate through a pipe is determined by measuring the pressure difference across a spring-loaded cone located within an orifice plate. *See also* DIRECT IN-LINE VARIABLE-AREA FLOW METER.

spring materials Mainly metals including plain-carbon and corrosion-resisting steels, phosphor bronze, spring brass, beryllium copper and nickel alloys, all of which in their work-hardened states permit large reversible elastic strains without permanent deformation. *See also* SPRING STEEL.

spring preload *See* INITIAL TENSION.

spring representation In an engineering drawing a helical spring is represented in one of three ways: a side or end view, a sectional view, or in simplified form as a zig-zag line.

spring scale *See* SPRING BALANCE.

spring shackle A U-shaped swinging support for leaf springs which allows the springs to straighten out when loaded.

spring steel High-carbon or alloy steel including piano wire, oil-tempered wire, hard-drawn wire, chrome vanadium alloy, and chrome silicon alloy steels.

spring surge Propagation of a compression wave back and forth along a helical compression spring subjected to a sudden load, which may cause the spring to cease contact with its end supports or suffer failure. *See also* VALVE BOUNCE.

sprocket wheel (sprocket) A wheel, with a thickness:diameter ratio usually much less than one, with teeth around its periphery which engage with a **sprocket chain** for the transmission of mechanical power, usually to another sprocket with which it is aligned.

sprue 1. In casting, a channel through which molten material flows into a mould. **2.** In injection moulding, the channel through which liquid plastic flows into a die. **3.** Material which solidifies in the channel of (1) or (2).

sprung axle The rear axle of a motor vehicle supported by springs.

sprung mass (sprung weight) (Unit kg (or N)) That part of the mass (or weight) of a vehicle which is supported by the suspension springs. *See also* UNSPRUNG MASS.

spur gear (spur wheel) *See* TOOTHED GEARING.

sputtering A process for depositing a thin film of metal onto a glass, metal, plastic, or other surface in a vacuum by using high-energy gaseous ions to eject atoms from the surface of a solid metal target.

sp vol *See* SPECIFIC PROPERTY.

square 1. A regular quadrilateral, i.e. a polygon with four equal sides and four right angles. **2.** An instrument to check angles of internal and external surfaces and flatness.

square-edged orifice An orifice plate in which the hole is cylindrical with no chamfer or rounding.

square engine A piston engine in which the cylinder bore and stroke are equal.

square jaw clutch *See* DOG CLUTCH.

square key A key for which the cross section, which may be uniform or tapered, is square.

square mesh A wire or textile mesh for which the mesh count is the same in both directions.

square thread *See* SCREW.

square wave A signal with rectangular pulses which alternate between two fixed levels for equal lengths of time.

squish The radially inward, or transverse, gas motion that occurs toward the end of the compression stroke in the cylinder of a piston engine when the piston face and cylinder head approach each other closely.

SSC *See* SPECIFIC STEAM CONSUMPTION.

SSH **1.** *See* SOLAR HEATING. **2.** *See* SMALL-SCALE HYDROPOWER.

SSPS *See* SATELLITE SOLAR-POWER SYSTEM.

stability **1.** A term relating to whether a body, structure, system, or flow is in a state of stable, unstable, or neutral equilibrium. **2.** In numerical analysis, whether numerical errors, such as roundoff or input-data errors, are damped, propagate, or cause divergence of the solution as the numerical method proceeds. **3.** A control system is often described as stable when a step change in input results in a steady output without continuing oscillation. This definition of stability is frequently used in practice as real systems cannot show the infinite output necessary to be considered unstable under the BIBO stability definition.

stability criterion A necessary and sufficient condition, such as the Nyquist or Routh–Hurwitz criteria, for a control system to be stable.

stability margin The gain and/or phase margin(s) of a control system.

stabilizer A device that restores a system or object to a state of stable equilibrium.

stabilizing feedback *See* NEGATIVE FEEDBACK.

stable equilibrium The state of a body, structure, or system such that it will return to its original position after being subjected to an infinitesimal displacement. It corresponds to minimum potential energy. *See also* NEUTRAL EQUILIBRIUM; UNSTABLE EQUILIBRIUM.

stable flow A laminar flow in which infinitesimal disturbances are damped out by viscosity. *See also* UNSTABLE FLOW.

stack *See* SMOKE STACK.

stack damper *See* DRAUGHT.

stack effect *See* DRAUGHT.

stacking sequence The ply ordering in a laminated structure which determines flexural properties and interlaminar stresses.

stactometer (stalagmometer) A device for determining surface tension by measuring the mass of a liquid drop suspended from a capillary tube.

stage **1.** In an axial-flow compressor or turbine, a ring of fixed guide vanes or nozzles (a stator row) and a ring of moving blades or buckets (a rotor row) across which energy is extracted from, or delivered to, the working fluid. *See also* CURTIS STAGE. **2.** In a radial-flow turbine, either a rotor-stator pair or a pair of contra-rotating rotors.

stage efficiency (η_S) The ratio of the actual change in specific enthalpy Δh across one stage of a compressor or turbine to the specific-enthalpy change Δh_S for isentropic flow through the stage, i.e. $\eta_S = \Delta h/\Delta h_S$. For a turbine, Δh and Δh_S are both negative, whereas for a compressor, both are positive. *See also* OVERALL EFFICIENCY; REHEAT FACTOR.

stagger angle *See* BLADE ANGLE.

stagnation conditions The thermodynamic state which would exist at a point in a fluid stream (**stagnation point**) if the flow at that point was brought isentropically to rest. In practice, a stagnation point is a point in a flow where the fluid velocity is zero, typically where a streamline terminates on a solid object in the flow. The **stagnation pressure (p_0)**, with unit Pa, is the static pressure p_0 at a stagnation point. From Bernoulli's equation, for an incompressible fluid of density ρ, $p_0 = p + \rho V^2/2$ where V is the flow speed and p is the static pressure. For an ideal gas with speed of sound a and specific-heat ratio γ,

$$p_0 = p\left(1 + \frac{\gamma - 1}{2}M^2\right)^{\frac{\gamma}{\gamma-1}}$$

where $M = V/a$ is the Mach number. The **stagnation temperature (T_0)**, with unit K, is the static temperature T_0 that a flowing fluid would have if brought adiabatically to rest: $T_0 = T + V^2/(2C_P)$ where T is the static temperature. The corresponding specific enthalpy is the stagnation enthalpy. For an incompressible fluid C_P is the specific heat, while for an ideal gas it is the specific heat at constant pressure. *See also* SPECIFIC PROPERTY.

stagnation enthalpy *See* SPECIFIC PROPERTY.

stainless steels (stainless irons) Iron-based alloys containing at least 13% chromium, which are resistant to many corrosive environments. Hardenable stainless steels, which may be quenched and tempered, contain up to 0.6% carbon. Stainless irons have virtually no carbon, and are always ferritic. The addition of nickel to stainless steels promotes the formation of austenite at room temperature, makes the alloy non-magnetic, and further improves corrosion resistance. The most common austenitic stainless steel is '18/8', having 0.1% C, 18% Cr and 8% Ni as the principal alloying elements.

stalagmometer *See* STACTOMETER.

stall 1. The condition where the lift on a wing, aerofoil, or other lifting surface begins to decrease with increase in the angle of attack beyond the **stall angle (critical angle of attack)**. It is a consequence of the flow on the suction surface (usually the upper surface) being dominated by flow separation.
2. In an axial compressor, rotational stall occurs when a small number of blades experience aerofoil stall. Axi-symmetric stall or compressor surge occurs when stall extends to all blades. **3.** The sudden stopping of a piston, usually due to a sudden increase in load or running in too high a gear. **4.** The condition that occurs in a steam-heated heat exchanger when the steam pressure falls below the condensate back pressure (i.e. the pressure in the steam trap).

stall control (stall regulation) A method of preventing damage by limiting the power output of a wind turbine above a certain wind speed by designing the turbine blades to stall at that speed. *See also* PITCH CONTROL.

stall torque The torque available from an engine running at close to zero rotation speed.

stamping 1. A sheet-metal component made using a press. **2.** The term for a forging made in brass or other non-ferrous metal.

stanchion 1. An upright bar designed to carry a guard rail or to constrain goods on a vehicle. **2.** A structural steel member designed to resist compressive loads.

stand-alone energy system *See* AUTONOMOUS ENERGY SYSTEM.

standard (standard specification) A set of specifications for components, machines, materials, or processes intended to achieve uniformity, efficiency, and a specified quality. In the UK, standards issued by the British Standards Institution (BSI) are now generally those of the International Organization for Standardization (ISO).

standard atmosphere (atmosphere) The layer of gases, primarily air, surrounding the Earth. Its total mass is about 5.15×10^{18} kg. The **I**nternational **C**ivil **A**viation **O**rganization (**ICAO**) model of the earth's atmosphere attempts to represent average atmospheric conditions in temperate latitudes based upon dry air with a zero-altitude temperature of 15°C and pressure of 1 atm by a tabulated set of temperature, pressure, and density values at altitudes up to 80 km. The model represents the atmosphere as a series of concentric spherical layers in which the pressure falls monotonically to about 0.1 kPa at 50 km, at which altitude the density has dropped to about 10^{-3} kg/m^3. The first major layer, between the earth's surface and the tropopause, is the **troposphere**, in which the lapse rate is constant at about −6.5°C/km. It extends from the earth's surface to the **tropopause** at about 11 km at which altitude the temperature is −56.5°C. The second major layer is the **stratosphere**, within which the temperature remains constant at −56.5°C up to about 20 km but heating by ultra-violet radiation from above leads to a temperature increase to about −2.5°C at the **stratopause** (47.35 km). Commercial jet aircraft cruise at altitudes in the range 9 to 14 km so close to the troposphere. Between 51.4 km and 86 km is the **mesosphere**, within which the temperature drops to −86.2°C at the **mesopause** due to decreasing solar heating and cooling due to radiation emitted by CO_2. The **ionosphere** is a part of the upper atmosphere incorporating the outer part of the mesosphere, the **thermosphere**, and the **exosphere** (to an altitude of about 1000 km). It is a shell of particles ionized by solar radiation in which the temperature increases with altitude from about 300 K to over 1000 K. *See also* OZONE LAYER.

standard conditions for temperature and pressure *See* STANDARD STATE.

standard deviation (σ) The square root of variance, a measure of the spread of data about the mean value. *See also* ROOT-MEAN SQUARE.

standard fit The fit of a component machined or otherwise manufactured to standardized clearances and tolerances. *See also* LIMITS AND FITS.

standard gauge A highly-accurate reference gauge against which to check working gauges.

standard gravity (g_0, g_n) A nominal value for the acceleration due to gravity *g* at sea level, defined to have precisely the value 9.808665 m/s^2, which is about 0.09% higher than the average value of *g* at sea level.

standard heat of formation (standard enthalpy of formation, ΔH_f^0) (Unit kJ) The enthalpy change which accompanies the reaction when one mole of a compound is formed from its elements, the reactants and products being in a given standard state, usually 1 bar and 25°C. The heat of formation of a pure element is zero by definition. *See also* ENTHALPY OF FORMATION.

standard hole A hole in a workpiece bored to a specified tolerance where clearance with a shaft is accomplished by allowance on the shaft. A **standard shaft** is machined to a specified tolerance where clearance with a hole is accomplished by allowance on the hole. *See also* LIMITS AND FITS.

standardization 1. National and international agreements for design, manufacture, materials, performance, practices, requirements, strength, etc. which ensure common results wherever an item is made and wherever used. **2.** The manufacture of components so that interchangeability of parts during assembly of new, or repair of old, items is possible without 'fitting'.

standard leak A precisely controlled flow of a tracer gas to allow calibration and adjustment of a leak detector.

S

standard specification *See* STANDARD.

standard state (standard reference state) In thermodynamics, a reference state used to specify the properties of a pure substance, mixture, or solution. The International Union of Pure and Applied Chemistry recommends a standard pressure p^o = 1 bar although 1 atm is sometimes used. A standard temperature T^o = 298.15 K (i.e. 20°C) or 273.15 K (0°C) is commonly used. Property values at the reference state are indicated by the superscript °, e.g. h° and u°. **Standard temperature and pressure (standard conditions for temperature and pressure, STP)** refers to any of a number of reference conditions for experimental measurements, generally close to 1 bar and 0°C or 20°C. Some also include relative humidity. *See also* NORMAL TEMPERATURE AND PRESSURE.

Standard Wire Gauge (SWG) *See* WIRE GAUGE.

standing wave A wave fixed in space produced by interference between two travelling waves of the same amplitude and frequency, propagating in opposite directions.

Stanton number (Margoulis number, *St*) A non-dimensional parameter used in the study of forced convection and defined by $St = h/\rho C_P V$, where *h* is the heat-transfer coefficient, ρ is the density of the fluid, C_P is its specific heat at constant pressure, and *V* is a characteristic fluid velocity.

Stanton tube A device used in the measurement of the difference Δp between the dynamic pressure very close to a surface over which there is turbulent fluid flow and the static pressure. Through a calibration, it is possible to infer the wall-shear stress from Δp.

star gear *See* EPICYCLIC GEAR TRAIN.

Stark number *See* STEFAN NUMBER.

starting friction (limiting friction, static friction, stiction) The force required to initiate movement between two bodies in contact.

starting taper The taper on the end of a reamer or tap which aids in starting the cut.

starting torque The torque developed by a motor at zero rotational speed in order to initiate rotation of the applied load.

starting valve (regulator) A valve which admits steam from the boiler to the cylinder(s) of a steam engine.

starved lubrication (incomplete lubrication) A condition between bearing surfaces where there is insufficient lubricant present to provide a full fluid film of lubricant.

state 1. In control theory, the minimum vector of values (the state variables) which characterizes a system at a given time and thus determines the control action to be applied. The number of values required is typically equal to the order of the system's defining differential equation. **2.** *See* THERMODYNAMIC STATE.

state equations 1. The set of first-order differential equations that fully models the behaviour of a control system. They allow a system modelled by a differential equation of order 'n' to be modelled equally well by 'n' first-order differential equations. The variables on which these equations operate are the state variables. Although the equations are in general non-linear and time-dependent, for many classes of system they are both linear and time-invariant, thus simplifying their expression in matrix form, which in turn simplifies the design of controllers for multiple input, multiple output systems. **2.** *See* EQUATION OF STATE.

state feedback A feedback-control method where the control action is a function of the state variables at the present time and is not influenced by the time history of these variables.

state of strain 1. The nine components of strain at any point in a solid, as defined by the three normal strains ε_{ii} and the six shear strains ε_{ij} (i = 1, 2, 3). Because $\varepsilon_{ij} = \varepsilon_{ji}$ (complementary shear strains), only six of the nine strain components are independent. Again, like stress, the strain at a point may be split up into hydrostatic and deviatoric parts, namely $\varepsilon_{ij} = \varepsilon_m \delta_{ij} + (\varepsilon_{ij} - \varepsilon_m \delta_{ij})$ where $\varepsilon_m = 1/3(\varepsilon_{11} + \varepsilon_{22} + \varepsilon_{33})$ with δ_{ij} the Kronecker delta (the change of sign in comparison with the equivalent expression for stress is because p is a negative stress). For large deformations, true strains are employed. *See also* MOHR'S CIRCLE; PLANE STRAIN; PRINCIPAL STRESS; STRAIN. **2.** *See* DEFORMATION.

state of stress *See* STRESS.

state principle For a pure substance consisting of a single chemical species, specifying any two independent intensive thermodynamic properties uniquely determines all the remaining intensive thermodynamic properties.

state-space analysis In control-system analysis using the state-space method, the behaviour of a system is fully described by the set of **state variables** (the **state vector**) which describe the inputs to the system, the outputs from the system, and any internal variables, and also by the state equations which operate on these variables. These equations are a matrix of first-order differential equations which, for many classes of system, can be linear and time-invariant. By modelling the system behaviour through this matrix method, the analysis of multiple-input, multiple-output systems is simplified.

statically-admissible loads (statically-admissible stresses) Any stress distribution that is in equilibrium internally in a body, and externally with the applied loads. *See also* LIMIT ANALYSIS; LOWER-BOUND THEOREM.

statically-admissible stress field A guessed stress field which is in equilibrium with the external loading on a body, but which pays no regard to displacements. The stiffness of the body according to the guessed stress field is greater than the true stiffness. Employed with the lower bound theorem to give underestimates of working loads.

statically-determinate (hyperstatic) Bodies, structures, or systems in which the loads in the members can be determined from the equations of equilibrium alone are said to be statically determinate. **Statically-indeterminate** is where the equations of equilibrium alone are insufficient to determine the loads in the members. Their determination requires the stress–strain relations and considerations of displacement compatibility.

static balancing The process of adjusting the weight distribution, usually by adding small balance weights, of a wheel or other rotatable mechanism such that it is in static equilibrium in any position. *See also* BALANCE; DYNAMIC BALANCING.

static coupling The joining of parts of a vibrating system by elements that have stiffness, such as a spring. *See also* DYNAMIC COUPLING.

static equilibrium A solid body, structure, or physical system (or any subdivided part) will be in static equilibrium, either at rest or moving with constant velocity, if the resultants of all external forces and moments acting on it are zero. *See also* D'ALEMBERT'S PRINCIPLE.

static fatigue Delayed fracture caused by environmental effects under dead-weight loading of materials such as glass. The name is unfortunate, as cyclic loads are not involved and glass is not susceptible to fatigue.

static friction *See* STARTING FRICTION.

static line A tube which connects a device such as a manometer or pressure transducer to a static-pressure tapping.

static load (dead load) A force or moment (**static moment**) having unchanging magnitude, direction, and point or points of application.

static position error *See* POSITIONAL-ERROR CONSTANT.

static pressure (pressure) In fluid mechanics, the pressure of a moving fluid which is associated with its state, not its motion. It is the pressure which would be sensed by a pressure probe moving with the fluid. *See also* DYNAMIC PRESSURE; STAGNATION CONDITIONS; TOTAL PRESSURE.

static-pressure tapping (static-pressure tap) A small flush hole, typically with diameter of order 1mm, in a surface in contact with a fluid which may be at rest or in motion. When the tapping is connected by a tube to a pressure transducer, the pressure measured is the static pressure of the fluid in the immediate vicinity of the surface.

static-pressure tube (static tube) For subsonic flow, a smooth tube with a rounded nose which has pressure tappings about 10 tube diameters from the nose. To measure the static pressure of the flowing fluid, the tube is pointed into the flow and is aligned with it. For supersonic flow, the nose of the tube is normally conical or ogival in form and sharply pointed.

static reaction *See* REACTION.

statics *See* APPLIED MECHANICS.

static seal *See* GASKET.

static stiffness *See* STIFFNESS.

static temperature (temperature) In fluid mechanics, the temperature of a moving fluid which is associated not with its motion, but with its state. It is the temperature which would be sensed by a temperature probe moving with the fluid. *See also* STAGNATION CONDITIONS; TOTAL TEMPERATURE.

static tube *See* STATIC-PRESSURE TUBE

stationary plastic hinge *See* PLASTIC HINGE.

stationary system In thermodynamics, a closed system for which the velocity and mass remain constant during a process. The change in its total energy is identical with the change in its internal energy.

statistical thermodynamics The application of probability theory to study the thermodynamic behaviour of systems of a large number of particles.

stator A ring of non-rotating blades or nozzles in a compressor, turbine, or other turbomachine, which directs gas flow into an adjacent rotor. *See also* STAGE.

stator blade (nozzle blade, nozzle guide vane) A blade with aerofoil cross section fixed to the casing of a compressor or turbine.

stay A tensioned cable, wire rope, chain, or slender rod used to support a structure, such as a cantilever bridge, a suspension bridge, or a mast.

steadiness 1. A measure of the smooth motion of a robot arm or end effector. It is the absence of vibration and jerkiness such as is caused by stick–slip friction. **2.** The extent to which any motion is free from high-frequency variations in velocity magnitude and direction.

steady flow (steady-state flow) A fluid flow in which the flow velocities and fluid properties may vary spatially but not temporally. For duct flows, the mass flow rate is constant.

steady-flow energy equation (SFEE) A form of the first law of thermodynamics applied to the steady flow of fluid through a fixed control volume. If the work associated with electric, magnetic, surface-tension, and viscous effects is negligible, the SFEE can be expressed as

$$\dot{Q}_{net\ in} + \dot{W}_{shaft,\ net\ in} = \textstyle\sum_{out}\dot{m}(h + V^2/2 + gz) - \sum_{in}\dot{m}(h + V^2/2 + gz)$$

where the summations are taken over all inlet streams (subscript *in*) and all outlet streams (subscript *out*), $\dot{Q}_{net\ in}$ represents the net rate of heat transfer into the control volume and $\dot{W}_{shaft,\ net\ in}$ the net rate of work (i.e. power) input, $\dot{m}$ is the mass flow rate, h is the specific enthalpy of the fluid, V is the average flow velocity for a given stream, z is the altitude, and g is the acceleration due to gravity. The three terms within parentheses represent internal plus flow energy (from $h = u + p/\rho$, u being the internal energy of the fluid, ρ is its density, and p is the pressure), kinetic energy, and potential energy.

steady pin A dowel, key, or pin that prevents a pulley from turning on its shaft.

steady state A process or motion which is independent of time. Sometimes extended to include processes or motions that vary periodically with time in a regularly repeating manner.

steady-state creep *See* CREEP.

steady-state error The error in a stable control system that remains after transient behaviour has decayed; that is, as time tends to infinity.

steady-state flow *See* STEADY FLOW.

steady-state heat transfer *See* HEAT TRANSFER

steady-state vibration Vibration of a system in which the velocity and displacement at all points are periodic in time.

steam The vapour phase of the chemical substance water (H_2O). At any temperature and pressure for which water is liquid, it is in equilibrium with its vapour phase. When the temperature equals the saturation temperature for a given pressure, or the pressure equals the saturated vapour pressure for a given temperature, the water boils and changes to the vapour phase in its entirety. With further increase in temperature or reduction in pressure, the steam becomes superheated. *See also* P–V–T DIAGRAM.

steam accumulator A device in which pressurized water is stored at its saturation temperature. The water flashes into steam when the pressure is reduced. Steam accumulators are a more economical way of meeting peak demand than storing steam at high temperature.

steam-atomizing oil burner A burner which incorporates a supply line for a jet of steam which assists in atomizing the fuel oil.

steam attemperation *See* STEAM DESUPERHEATING.

steam boiler (boiler) A heat exchanger where heat, originating from combustion of fossil fuel, nuclear fission, or other sources, is transferred to water essentially at constant pressure to produce dry saturated steam. In a shell-and-tube boiler, the water flows into and out of a pressure vessel while the hot gas flows through tubes within the liquid. In a water-tube boiler, the water flows through tubes while the gas flows into and out of the pressure vessel. A **steam generator** is a steam boiler together with a separate heat exchanger, or a separate bank of tubes, in which additional heat is added to the saturated steam leaving the steam drum to produce superheated steam (**steam superheater**). A small boiler may be heated electrically. *See also* WATER-TUBE BOILER.

SEE WEB LINKS

- Extensive information on steam engineering

steam calorimeter (throttling calorimeter) An instrument utilizing an adiabatic, steady-flow throttling process, in which the fluid enthalpy is constant, to determine the dryness fraction of wet steam.

steam chest A chamber in a steam engine through which steam, controlled by a slide or piston valve, is delivered from the boiler to a cylinder.

steam cock A stop valve in a steam line fitted to steam engines and steam boilers.

steam consumption *See* SPECIFIC STEAM CONSUMPTION.

steam condenser *See* SURFACE CONDENSER.

steam cycle (steam-plant cycle) A thermodynamic cycle, based upon the ideal Rankine cycle, in which steam is produced in a boiler (2–3), expanded in a turbine (3–4), condensed back to the liquid state (4–1), then pumped through a feedwater heater back into the boiler (1–2). The diagram shows ideal and actual steam cycles on a temperature (T) *vs* specific entropy (s) plot. The actual cycle includes entropy increases due to irreversibilities. Variations on the basic cycle include reheat and regeneration cycles as well as tandem- or cross-compound, multi-cylinder arrangements that incorporate high-, intermediate- and low-pressure turbines.

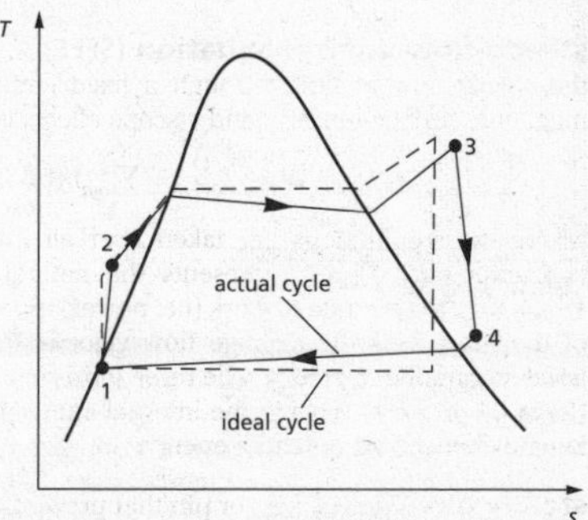

steam cycle

steam desuperheating (steam attemperation, steam conditioning) The injection of water into a flow of superheated steam to reduce the steam temperature.

steam drum A reservoir at the top end of the heated tubes in a steam boiler, in which water and saturated steam are separated before the steam flows to the superheater and the water is returned to the feedwater drum.

steam engine A reciprocating piston engine designed to produce work by expanding steam, formerly used in locomotives and stationary engines, but now as an alternative to a turbine in small (< 1 MW) steam plants. As shown in the diagram, steam is admitted through an inlet port at constant pressure to a steam chest and then passes to a cylinder with the piston initially at outer-dead centre. The steam expands during the admission stroke and is then blown 'down' through the exhaust port, either to the atmosphere or to a condenser. After the exhaust port closes, any remaining steam is compressed on the return stroke and mixes with the next charge. Most steam engines are double acting in that a piston or slide valve admits high-pressure steam alternately into either end of the cylinder so that the steam acts alternately on both faces of the piston. *See also* COMPOUND ENGINE.

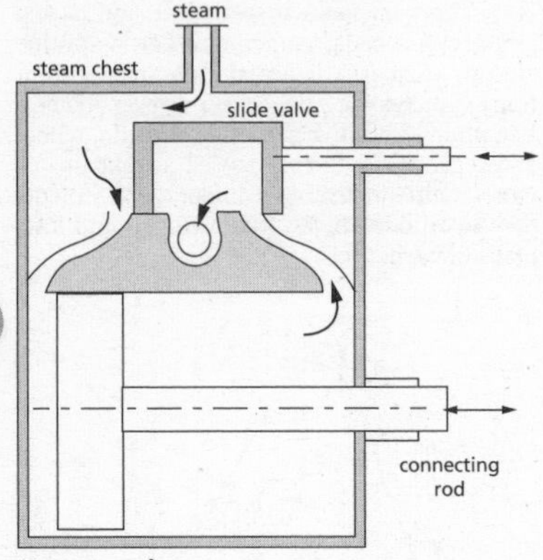

steam engine

steam-engine cycle A sequence of thermodynamic processes in which work is produced by a reciprocating piston engine from steam initially at high pressure and temperature. The five steps in the cycle are shown on a diagram of pressure (p) *vs* cylinder volume ($\mathcal{V}$).

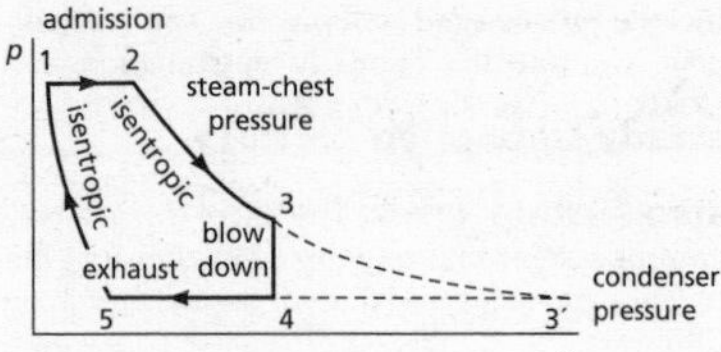

steam-engine cycle

steam gauge A pressure gauge used to measure gauge pressure in a line, boiler, cylinder, or other device operating with steam.

steam-generating furnace *See* BOILER FURNACE.

steam generator *See* STEAM BOILER.

steam-heated evaporator A heat-exchange device in which steam is used to heat a surface in contact with a liquid, such as a liquid gas or organic liquid of low boiling point, which then evaporates.

steam heating The use of steam in radiators for space heating, or in heat exchangers for industrial processes.

steam injector A device, submerged in a water tank, such as a steam accumulator, in which cold water is drawn in by and mixed with steam.

steam jacket A casing or jacket supplied with live steam wrapped around a vessel to heat the contents

steam-jet ejector A jet pump in which the motive fluid is high-pressure steam.

steam-jet refrigeration A system in which medium-pressure steam flows through a steam-jet ejector which is used to reduce the pressure and temperature of warm water sprayed into a reservoir and remove the vapour produced. Typically used in air conditioning.

steam line 1. A pipe through which there is a flow of steam. **2.** *See* DRY-SATURATED STEAM LINE.

steam loop A closed recirculation system in which steam is generated in a boiler and distributed to its point of use, where heat transfer from the steam occurs and the steam condenses, the condensate being returned to the boiler.

steam nozzle A convergent-divergent nozzle in which high-pressure steam is expanded through a choked throat.

steam-plant cycle *See* STEAM CYCLE.

steam point The temperature at which pure water boils on the Celsius thermodynamic scale at a pressure of 1 atm, assigned the value 100°C. *See also* ICE POINT.

steam port An opening which supplies steam to, or exhausts steam from, the cylinder of a steam engine.

steam purifier *See* STEAM SEPARATOR.

steam reheater (resuperheater) In a compound steam-turbine system, moist steam leaving the high-pressure turbine is superheated in a reheater, usually to the original superheat temperature, before entering the low-pressure turbine. The reheater is usually incorporated into the steam boiler. *See also* MOISTURE SEPARATOR REHEATER.

steam separator (moisture separator, steam purifier) A component in a steam pipeline containing a series of plates or baffles which removes water droplets from a steam-water mixture to produce dry steam. *See also* BAFFLE-TYPE SEPARATOR; COALESCENCE-TYPE SEPARATOR; CYCLONIC-TYPE SEPARATOR.

steam superheater *See* STEAM BOILER.

steam tables A set of tabulated values of the intensive properties of steam (specific enthalpy, specific internal energy, specific entropy, and specific volume) over a range of temperatures and pressures to include the critical state, saturated liquid, saturated vapour, saturated solid, superheated conditions, and phase change. Similar tables exist for other fluids, particularly refrigerants such as ammonia, dichlorodifluoromethane, and tetrafluoroethane.

steam tracing (trace heating) The prevention of freezing, thickening, or condensing of fluid flowing through a process line by the use of small-bore steam-heated tubing wrapped around the line or running alongside it, the whole being surrounded by insulation. Heat-conductive paste may be used between the tracer pipes and the process pipe.

steam trap A device which removes condensate, air, and other incondensable gases from a steam line.

steam-tube dryer A device for drying free-flowing granular solids, including foodstuffs, pharmaceuticals, and minerals, which pass through a rotating cylindrical drum incorporating a bundle of steam-heated tubes which run the length of the drum.

steam turbine A rotodynamic machine in which steam, initially at high pressure and temperature, expands through a series of stages which convert thermal energy into shaft power. Efficiencies in excess of 40% are achieved by coupling (tandem compounding) high-, intermediate- and low-pressure turbines running on a single shaft and delivering up to 1500 MW. The diagram shows the cross section of an intermediate-pressure turbine with central inflow and double outflow. High-pressure turbines are usually single-flow machines and low-pressure turbines often double-flow to minimize axial thrust loads.

steel-cable conveyor belt (steel-cord conveyor belt) A heavy-duty conveyor belt made of rubber reinforced with steel cables.

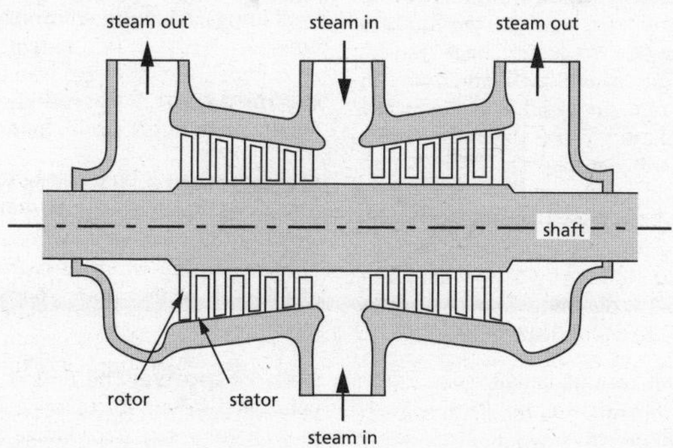

steam turbine

steel-wire rope A rope or cable made up of many steel strands wrapped helically about an axis, each strand being made of metal wires twisted together like a helix.

steelyard (steelyard balance) A pivoted straight-beam balance with arms of unequal length, the load to be weighed being suspended from the short end and a known counterbalance weight being moved to the position of balance on the longer calibrated arm.

steering arm That part of the steering mechanism of a motor vehicle consisting of a short arm which carries the stub axle, pivoted at one end and linked to the tie rod at the other. *See also* ACKERMAN LINKAGE.

steering axis A line which runs through the upper and lower steering pivots of a steered wheel. *See also* ACKERMAN LINKAGE.

steering gear The combination of components, including wheel supports, shafts, gearing, tie rods, and other linkages, which control the direction of a ship, the steered wheels of a motor vehicle, or other steerable device.

Stefan–Boltzmann constant (Stefan constant, σ) The fundamental physical constant of proportionality in the Stefan–Boltzmann law of thermal radiation, with a value of 5.670400×10^{-8} $W/m^2.K^4$.

Stefan–Boltzmann law (Stefan's law) The underlying law of thermal radiation: the emissive power of a blackbody $\dot{E}_B$ is directly proportional to the fourth power of its absolute temperature T and given by the equation $\dot{E}_B = \sigma T^4$ where σ is the Stefan–Boltzmann constant.

Stefan number 1. (Stark number) A non-dimensional parameter used in the study of combined conduction-radiation heat transfer, defined by $k\alpha/4\sigma T^3$ where k is the thermal conductivity of the radiating solid, σ is the Stefan-Boltzmann constant, α is the absorption coefficient of the radiating surface, and T is its temperature. Sometimes the inverse definition is used. **2.** A non-dimensional parameter used in the analysis of heat-transfer problems involving melting of a solid, defined by $\Delta h/L$ where Δh is the specific enthalpy difference between the two phases and L is the latent heart of melting.

stem 1. A part of a component, such as a valve plug or disc, in the form of a cylindrical rod used to move the component along its axis. **2.** The calibrated capillary tube along which the liquid moves in a liquid-in-glass thermometer.

stem correction A correction which must be made to the reading of a thermometer, such as a mercury- or alcohol-in-glass thermometer, if part of the stem of the thermometer is at a different temperature to the bulb.

step An instantaneous change in the set point input to a control system between one steady value and another. *See also* UNIT STEP.

step bearing (Rayleigh step bearing) A slider bearing in which one of the bearing surfaces has a small step to produce a reduction in clearance in the direction of sliding. For an optimized bearing, the load-carrying capacity W is given by $W = 6\mu VL^2/h^2$ where μ is the viscosity of the lubricant, V is the sliding velocity, L is the length of the bearing, and h is the final clearance.

step gauge 1. A plug gauge with two or more gauging diameters to check go and no-go tolerances simultaneously. **2.** A device used to measure steps in a surface by monitoring the movement perpendicular to the surface of a ball, roller, or blade.

stepless transmission *See* CONTINUOUSLY-VARIABLE TRANSMISSION.

stepped-cone pulleys (speed cones, speed pulleys, step pulleys, stepped pulleys) A pair of pulleys, each with a set of steps of increasing diameter, mounted on parallel shafts such that the smallest and largest diameters of one are aligned with the largest and smallest diameters of the other, thereby allowing a wide range of speed ratios to be achieved by shifting a belt from one end of the pair to the other.

stepped gear A gear with two or more sets of teeth with differing pitch-circle diameters.

stepper motor (stepping motor) An electric motor that does not rotate continuously and smoothly, but in a series of small angular steps. Each step occurs when the current in one or two of the coils of the motor stator is switched on or off.

step response The output behaviour of a controlled system in response to a step change in set point or to a step change in a disturbance.

steradian (*sr*) A coherent derived SI unit defined as the solid angle subtended at the centre of a sphere of radius R by a circular cap of area R^2. *See also* RADIAN.

stereolithography A three-dimensional process used in rapid prototyping in which a component is produced from a CAD drawing by depositing thin layer upon thin layer of material, often a liquid resin which is hardened by a scanning laser beam.

stern tube The tube which supports a ship's propeller shaft and its bearings.

stick–slip friction Friction between two contacting surfaces that are alternately at rest and sliding one over the other.

stick–slip motion Intermittent motion resulting when a body slides over a surface where the instantaneous static friction is locally higher than the kinetic friction.

stiction *See* STARTING FRICTION.

stiffening spring A spring for which the spring constant increases with increasing deflection of the spring. The opposite of a **softening spring**.

stiffness (*k*) (Unit N/m) For an elastic body, the rate of change of force F with linear displacement δ, i.e. $k = dF/d\delta$. Usually defined under quasi-static conditions but sometimes under dynamic loading. *See also* TORSIONAL RIGIDITY; SPRING.

stiffness matrix *See* FEM; GENERALIZED HOOKE'S LAW.

stimulus Any applied signal that affects the output from a controlled system.

sting A long slender beam used to support models placed in a wind or water tunnel, usually from behind and aligned with the body to minimize flow disturbance.

stir casting A method for the manufacture of metal-matrix composites, in which the particle or short-fibre reinforcement is stirred and dispersed into a molten matrix prior to conventional casting.

Stirling boiler (bent-tube boiler) A boiler in which cold feedwater enters an upper drum, then descends under gravity through heated boiler tubes to a lower water drum. From the lower drum the water rises due to buoyancy through heated boiler tubes to one or more upper drums. Steam produced within the boiler tubes is taken off through a manifold.

Stirling cooler A Stirling engine run in reverse to produce very low temperatures.

Stirling cycle The ideal cycle is a four-step, reversible, regenerative, closed thermodynamic cycle with a gaseous working fluid. As shown in the plot of pressure (p) *vs* specific volume (v), in step 1 (states 1–2), an amount of heat Q_H is supplied from an external source to the gas during isothermal expansion at temperature T_H; step 3 (3–4) is an isothermal compression at temperature T_L (with $T_L < T_H$) during which an amount of heat Q_S is rejected; the two isothermal processes are connected by step 2 (2–3) in which the gas undergoes reversible isopycnic expansion from T_H to T_L at specific volume v_2 and step 4 (4–1) in which there is a reversible isopycnic compression at specific volume v_4 from T_L to T_H. In step 2 the gas is cooled as it transfers heat to a regenerator and in step 4 passes back through the regenerator where the same amount of heat is recovered. The net work per cycle W is given by $W = Q_H - Q_S$ and for a perfect gas of mass m with specific gas constant R, this can be written in terms of T_L and T_H as

$$W = mRln(r)(T_H - T_L)$$

with r being the compression ratio v_2/v_4.

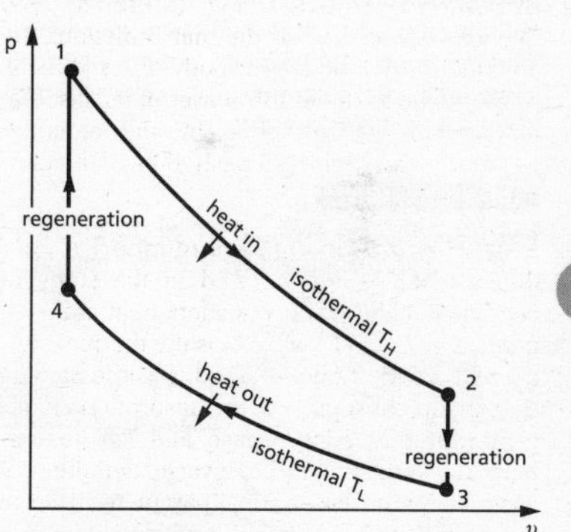

Stirling cycle

Stirling engine (hot-air engine) A reciprocating gas engine operating on a practical approximation to the ideal Stirling cycle. There are several variants including the **alpha type**, shown in the first diagram, which has two

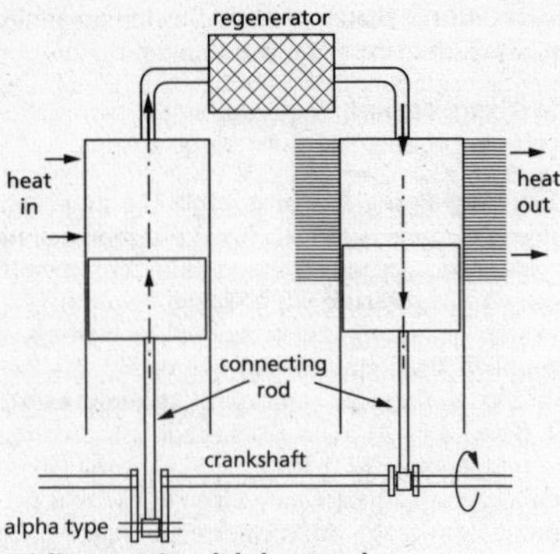

Stirling engine (alpha type)

cylinders connected by a regenerator. The pistons in each cylinder are connected to the same crankshaft but operate 90° out of phase. One cylinder is heated, typically from external combustion, while the other is cooled. As the pistons reciprocate, the gas is moved back and forth through the regenerator. In the **beta type** (second diagram), two pistons reciprocate with a 90° phase difference within a single cylinder, one piston acting as the power piston and the other as a non-power **displacer piston** to move the gas at constant density between the heated and cooled ends of the cylinder. Air, helium, nitrogen, and hydrogen are typical working fluids. Applications include powering submarines and solar power. *See also* THERMOACOUSTIC STIRLING ENGINE.

SEE WEB LINKS

- Animations and explanations of the principles of the alpha- and beta-type Stirling engines

stirred-flow reactor An open or closed tank within which chemicals mix and react, the mixing being enhanced by a rotating impeller.

stochastic control theory A control technique which aims to minimize the effect of random errors in the state variables and in the model of the system by optimizing the controller design.

stochastic error *See* ERROR (1).

Stodola method An iterative method of calculating the fundamental transverse frequency of vibration of a beam (historically of steam-turbine rotors). A deflexion curve and frequency are first assumed for the shaft or beam. From the known masses m_i along the rotor or beam, the assumed frequency ω, and deflection y_i, the dynamic (inertia) loads are calculated using $m_i y_i \omega^2$. These forces are then used to predict another deflexion curve y'_i by the usual methods of statics. To a first approximation $\omega'/\omega = \sqrt{y_i/y'_i}$ where ω' is the improved estimate for the true frequency. With reasonable guesses for y_i, the accuracy can be good, but further refinement is possible by repeating the calculation with y'_i as the assumed starting curve.

stoichiometric coefficients (υ_i) In a chemical reaction, the number of molecules (or moles) of each reactant and product. For example, in the reaction between nitrogen (N_2) and hydrogen (H_2) to produce ammonia (NH_3), $N_2 + 3H_2 \rightarrow 2NH_3$, the stoichiometric coefficients are $\upsilon_{N_2} = -1, \upsilon_{H_2} = -3$ and $\upsilon_{NH_3} = +2$, it being the convention to assign negative coefficients to the reactants.

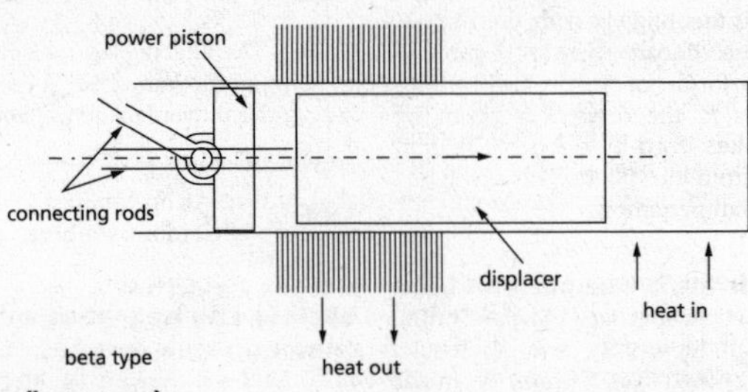

Stirling engine (beta type)

stoichiometric mixture A term used in combustion to refer to a mixture of oxidant (usually air) and fuel which contains just enough oxygen for complete combustion of the combustible elements in the fuel. More generally, in any chemical reaction it is the mixture for which all of the reactants are consumed with no residue or shortfall. In a **lean mixture (weak mixture)**, the air/fuel ratio is greater than stoichiometric. In a **rich mixture** it is lower. The **equivalence ratio** is the ratio of the actual fuel-air ratio to the stoichiometric fuel-air ratio. It is the inverse of **mixture strength**. **Excess air** is the amount of air in excess of stoichiometric, usually expressed as per cent excess air or percent theoretical air, i.e.

$$\frac{\textit{actual A/F ratio} - \textit{stoichiometric A/F ratio}}{\textit{stoichiometric A/F ratio}}$$

where *A/F ratio* is the air-fuel ratio.

stokes (*St*) An obsolete (non-SI) unit of kinematic viscosity, 1 St = 10^{-4} m^2/s.

Stokes first problem The flow produced by an infinite flat plate which suddenly moves at velocity U in its own plane within an infinite body of fluid of kinematic viscosity ν initially at rest. The velocity u a distance y from the plate after time t is given by $u/U = 1 - erf(y/2\sqrt{\nu t})$ where *erf* is the error function.

Stokes flow (creeping flow, Stokes approximation) A viscous fluid flow at a very low Reynolds number (« 1) in which inertia (i.e. acceleration) terms in the Navier–Stokes equations are negligible and the flow is determined by a balance between pressure gradient and viscous stresses. For an incompressible Newtonian fluid, the continuity and momentum equations reduce to $\nabla . \boldsymbol{V} = 0$ and $\nabla p = \mu \nabla^2 \boldsymbol{V}$ where $\boldsymbol{V}$ is the vector velocity, p is the static pressure, and μ is the fluid viscosity. For simple geometries, these linear equations can be solved in closed form: for Stokes flow past a sphere of radius R, the drag D is given by $D = 6\pi\mu RU$ (**Stokes drag law**) where U is the flow velocity far from the sphere. This result has applications to sedimentation.

Stokes hypothesis The assumption that the viscosity μ and the coefficient of bulk viscosity λ for a Newtonian fluid are related by the equation $\lambda + \frac{2}{3}\mu = 0$, which ensures that the mean pressure in a deforming viscous fluid is equal to the thermodynamic pressure.

Stokesian fluid A non-Newtonian fluid in which the stress is a continuous function of the deformation. A Newtonian fluid can be regarded as a linear Stokesian fluid.

Stokes number (*S*, *Sk*) A non-dimensional parameter that arises in the study of unsteady flow: (a) For oscillatory flow of a fluid with kinematic viscosity ν past a solid body with a characteristic dimension d, $S = \omega d^2/\nu$ where ω is the angular frequency of the motion. S represents the ratio of inertia forces to viscous forces, i.e. a form of Reynolds number. (b) For a suspended particle, of density ρ_P, with characteristic dimension d in a fluid flow with velocity U, $S = \tau U/d$ where τ is the relaxation time of the particle which can be taken as $(\rho_P - \rho)d^2/18\mu$, ρ being the fluid density and μ its dynamic viscosity.

Stokes paradox In two-dimensional Stokes flow, such as that past a circular cylinder, no solution to the governing equation can be found that matches to the boundary conditions both at the solid surface and at infinity.

Stokes second problem The flow produced by an infinite plate which oscillates in its own plane within an infinite body of fluid of kinematic viscosity ν with velocity $U = U_0 \cos(\omega t)$ where U_0 is the amplitude of the oscillation, ω is the angular frequency of the oscillation and t is the time. The velocity u a distance y from the plate is given by

$$\frac{u}{U_0} = exp\left(-y\sqrt{\frac{\omega}{2\nu}}\right) cos\left(\omega t - y\sqrt{\frac{\omega}{2\nu}}\right)$$

where *exp* is the exponential function.

Stokes stream function A stream function used to describe an axisymmetric flow.

stone (*st*) An obsolete (i.e. non-SI) British unit of mass. 1 st = 14 lb = 6.350293180 kg.

stop 1. The final position reached by a robot in point-point movement. **2.** A metal block or pin to stop the movement of a machine or mechanism.

stop valve An on-off valve in a fluid line, such as a **stop cock** in a water line.

storage calorifier A hot-water storage vessel heated by a heating coil, often using steam. *See also* NON-STORAGE CALORIFIER.

storage modulus (Unit Pa) **1. (*E′*, *G′*)** A measure of the elastic energy stored in a viscoelastic solid, determined by applying an oscillatory stress to the material and recording the strain. E' corresponds to a tensile stress and G' to a shear stress. **2. (G')** A measure of the elastic energy stored in a viscoelastic liquid, determined by applying an oscillatory shear stress and recording the strain rate. *See also* COMPLEX VISCOSITY; DYNAMIC MODULUS; LOSS MODULUS.

storage of energy An intermediate step between energy generation and its use. Examples of materials in which energy can be stored are fossil and nuclear fuels, pressurized air and other gases, heated water and steam, rock, refractory bricks, water in reservoirs at elevated locations, hydrogen–oxygen fuel cells as well as a wide variety of batteries such as lead–acid, nickel-cadmium, lithium ion, zinc–carbon, zinc–chloride, and alkaline (zinc–manganese dioxide). The energy is often released through chemical reaction, including combustion, or expansion through a turbine. Mechanical devices such as flywheels and springs are also used for energy storage.

storage retrieval machine A machine used to retrieve components from a storage system and move these to a transfer station.

stored-program numerical control *See* COMPUTER NUMERICAL CONTROL.

STP *See* STANDARD STATE.

STR *See* SELF-TUNING REGULATOR.

straddle milling The simultaneous milling of more than one face of a workpiece by using more than one cutter mounted on a common arbor.

Straflo turbine (rim-generator turbine, straight-flow turbine) A tidal-flow turbine in which the turbine runner is situated at the throat of a convergent water channel and acts as a spider to support the electrical generator around its periphery but out of the water flow.

straight bevel gear *See* TOOTHED GEARING.

straightening vanes A number of flat or curved sheets, often formed from sheet metal, installed within a duct with the intention of producing a unidirectional, non-swirling flow downstream of the vanes.

straight-line mechanism Various linkages that produce exact straight-line motion without the use of a sliding pair. Other mechanisms that produce an approximately straight path are mostly based on the four-bar chain. *See also* PEAUCELLIER LINKAGE; WATT'S MECHANISM.

straight-line motion A method of moving a robot such that the end effector follows a straight line. Where the robot uses rotational joints, this implies the use of kinematic analysis to determine the joint angles and angular velocities required to achieve the desired straight-line motion of the end effector.

strain 1. A non-dimensional geometric measure of distortion in a body loaded in tension, compression, shear, or combinations of these. The axial engineering, or nominal, tensile strain e of a uniformly-stretched body is given by $(l-l_0)/l_0$ where l is the stretched length and l_0 the starting ('gauge') length; similarly for the reduced height in compression. Simple shear strain is given by the tangent of the angle γ.

More generally, a particle P having coordinates x_i $(i = 1,2,3)$ with respect to a set of fixed Cartesian axes independent of the body, moves to position P' having coordinates $(x_i + u_i)$ when deformed, where u_i are the components of the displacement. When u_i is constant, there is no strain, only a rigid translation of the whole body. Uniform strain occurs when $u_i \propto x_i$, i.e. for infinitesimal strains $e = du/dx$, etc. In general, $u_1 = e_{11}x_1 + e_{12}x_2 + e_{13}x_3$, etc., i.e. $u_i = e_{ij}x_j$ where $i,j = 1, 2, 3$ and using the repeated-suffix convention. The terms $e_{ii} = \partial u_i/\partial x_i$, are tensile/compressive strains along the i-axis; $e_{ij} = \partial u_i/\partial x_j$ and $e_{ji} = \partial u_j/\partial x_i$ together produce both shear strain and rigid-body rotation, which may be separated by writing $e_{ij} = (\varepsilon_{ij} + \omega_{ij})$ and

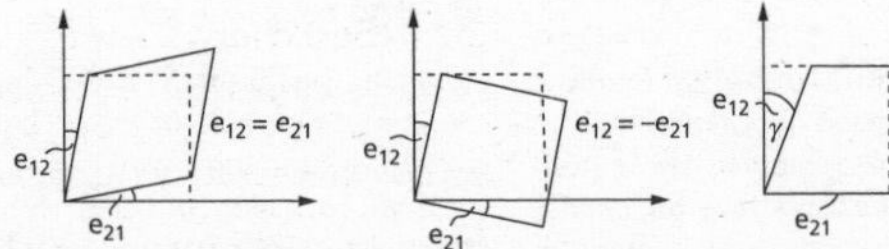

strain

$e_{ji} = (\varepsilon_{ji} + \omega_{ji})$ in which $\varepsilon_{ij} = (e_{ij} + e_{ji})/2$ and $\omega_{ij} = (e_{ij} - e_{ji})/2$, etc. This gives $\varepsilon_{ij} = \varepsilon_{ji}$ that represents the shear strain produced by e_{ij} and e_{ji}, and $\omega_{ij} = -\omega_{ji}$ that represents rigid-body rotation. Simple shear is when $\varepsilon_{ij} = \omega_{ij} = \gamma$. Because $\varepsilon_{ij} = \varepsilon_{ji}$, the strain tensor given by ε_{ij} is symmetric and has only six independent terms. The total change of angle is $(e_{12} + e_{21}) = (\varepsilon_{12} + \varepsilon_{21}) = 2\varepsilon_{21}$ Hence γ the angle of shear (sometimes called the tangent strain) is twice the corresponding component of the strain tensor. This is important in transformation of strain using Mohr's circles, where $\gamma/2$ has to be plotted along the ordinate. There are related expressions for strain in cylindrical and spherical polar coordinates.

In problems involving large deformation, when strains are not small compared with unity, incremental principal strain is defined as $d\varepsilon = dl/l$ where l is the current length (height) rather than l_0. This gives rise to the concept of true (or logarithmic) strain, $\varepsilon = ln(l/l_0) = ln(1 + e)$ employed in plasticity, which not only automatically gives a sign for strain (tension positive, compression negative) but also makes physical sense, in that doubling a length in tension corresponds with halving a height in compression. *See also* DEVIATORIC STRAIN; STRETCH RATIO. **2.** *See* DEFORMATION.

strain amplitude *See* FATIGUE.

strain axis *See* PRINCIPAL AXIS OF STRAIN.

strain circle *See* MOHR'S STRAIN CIRCLE.

strain energy *See* ELASTIC STRAIN ENERGY.

strain-energy density (Unit J/m^3) The strain energy per unit volume, given by the area under the stress–strain curve between the initial and final strains. In plasticity, the strain-energy density is the irreversible work done which is the same numerically as the working stress for homogeneous deformation.

strain gauge A small device consisting of very thin metal wire, metal foil, or semi-conductor filament bonded to a backing sheet that is glued to the surface of a body. Changes in the electrical properties of the device when the body is loaded produce changes in the electrical signal which are proportional to the strain experienced in that region. A **strain-gauge rosette** is a set of three strain gauges fixed closely together on the surface of a body at known relative angles, readings from which may be combined to give the principal strains at that point, and hence the principal stresses using the generalized Hooke's law. The algebra is simplified when the gauges are at 0, 45° and 90°, or 0, 60° and 120° (a 'delta' rosette). A temperature-compensated Wheatstone bridge circuit (**strain-gauge bridge**) is employed to take measurements. *See also* GAUGE FACTOR.

strain hardening (workhardening) The increase in stress above the initial yield stress required to deform a metal in the plastic range at temperatures below the recrystallization range (cold working). Unloading after some strain or work input gives a harder and stronger material. *See also* MECHANICAL WORKING.

strain rate In solid mechanics, the rate at which strain is applied to a body or material, given by $d\varepsilon/dt$ where ε is strain and t is time. For a testing machine with a constant cross-head velocity V, the rate of true strain is V/l where l is the instantaneous length of the specimen. *See also* DEFORMATION; SHEAR RATE; SHEAR STRAIN RATE; VOLUMETRIC STRAIN RATE.

strain-rate tensor The strain tensor written out in terms of strain rates in place of strains. *See also* STRAIN.

strain tensor *See* STRAIN.

strain-time diagram See CREEP.

strake A helical cable or ridge around the outside of a tall chimney, intended to prevent vortex shedding and so avoid wind-induced transverse oscillations.

strand A number of wires, threads, or filaments twisted or braided together to form a cable, electrical conductor, or rope.

stratification The tendency in a fluid in which the density varies, for layers to form with the denser fluid below the lighter fluid. The density variation may be due to variation in temperature, concentration, or immiscibility of fluids with different densities. *See also* BRUNT–VÄISÄLÄ FREQUENCY.

stratified atmosphere An atmosphere in which the potential density decreases with altitude. *See also* LAPSE RATE.

stratified-charge engine A piston engine in which the air–fuel mixture is introduced into the combustion chamber such that the mixture is readily ignitable in the vicinity of the spark plug, and weaker (normally non-ignitable) further away. This controlled inhomogeneity or stratification is often achieved by utilizing a prechamber that contains the spark plug, and controlled by the timing and rate of fuel injection. Such engines run with lower peak combustion temperatures, reduced emissions and improved fuel economy.

stratified flow The flow of a fluid with density that increases with depth. *See also* STRATIFICATION.

stratosphere *See* STANDARD ATMOSPHERE.

streakline In a fluid flow, the locus of all fluid particles that have previously passed through a particular fixed point. *See also* PATHLINE.

stream function (ψ) In fluid flow, a function defined such that it satisfies the continuity equation. For steady, two-dimensional compressible flow in the *xy* plane, $\rho u = \partial\psi/\partial y$ and $\rho v = -\partial\psi/\partial x$ where ρ is the fluid density, u and v are the velocity components in the y- and x-directions, respectively. Physically, ψ can be interpreted as a flow rate.

streaming flow A steady flow created by an oscillating solid body immersed in a viscous fluid. *See also* ACOUSTIC STREAMING.

streamline A continuous line within a fluid, of which the tangent at any point is in the direction of the fluid velocity at that point. In a steady flow, streamlines, particle paths, and streaklines are identical. *See also* PATHLINE.

streamlining The shaping of a solid body over which there is relative fluid flow to reduce frictional drag.

streamtube 1. A tubular region within a fluid flow bounded by streamlines. **2.** *See* FLOWLINE.

streamwise direction The main direction of a fluid flow.

strength The maximum stress, in tension, compression, shear, or combinations thereof, that may be monotonically applied to a material, component, or structure before failure (defined as fracture, yielding, buckling, etc. as appropriate). *See also* CREEP RUPTURE STRENGTH; FACTOR OF SAFETY; FAILURE; FATIGUE STRENGTH; THEORIES OF STRENGTH.

strength of materials A confusing name for what is really stress analysis, reflecting older engineering design based upon elasticity and 'theories of strength' for different materials, without regard to cracks.

strength of shock wave *See* SHOCK WAVE.

strength of vortex *See* VORTEX STRENGTH.

stress 1. Stress 'at a point' (i.e. over a volume of material that is very small compared with that of the component or structure) is the load per unit area for every face of an infinitesimal cube surrounding the point. There are nine components σ_{ij} $(i, j = 1, 2, 3)$: σ_{ii} are uniaxial tension (positive) or compression (negative) 'normal' stresses; σ_{ij} are shears acting on the faces whose normals are along the i-axis, in the direction of the j-axis. Because $\sigma_{ij} = \sigma_{ji}$ (complementary shear stresses), only six of the nine components are independent. Again, like strain, the stress at a point may be split into hydrostatic and deviatoric parts, namely $\sigma_{ij} = -p\delta_{ij} + (\sigma_{ij} + p\delta_{ij})$ where $p = -(\sigma_{11} + \sigma_{22} + \sigma_{33})/3$ with δ_{ij} the Kronecker delta (the change of sign in comparison with the equivalent expression for strain is because p is a negative stress). Note that where different symbols are employed for normal and shear stresses (σ and τ), single suffixes are often used, with σ_x meaning the normal stress in the x-direction. For large deformations, the instantaneous areas A over which loads act change from their starting values A_0. For small deformations $A \approx A_0$, and engineering or nominal stress s uses A_0, but for large deformations true stress σ employs A. When elastic volume changes are neglected, $\sigma = s\ exp\varepsilon$, where ε is true strain, so $\sigma > s$. *See also* DEVIATORIC STRESS. **2.** In fluid flow the stress tensor σ_{ij} is the sum of an isotropic part $-p\delta_{ij}$, δ_{ij} being the Kronecker delta, having the same form as the stress tensor for a fluid at rest, p being the static pressure, and the non-isotropic, deviatoric stress tensor d_{ij} which is due entirely to the fluid motion. The deviatoric stress tensor includes the shear stresses and can be written for a Newtonian fluid with dynamic viscosity μ as $d_{ij} = 2\mu(e_{ij} - \Delta\delta_{ij}/3)$ where Δ indicates the rate of expansion, $\nabla.\boldsymbol{u}$ where $\boldsymbol{u}$ is the vector velocity, and

$$e_{ij} = \frac{1}{2}\left(\frac{\partial u_i}{\partial x_j} + \frac{\partial u_j}{\partial x_i}\right)$$

is the rate of strain. *See also* DEFORMATION.

stress amplitude *See* FATIGUE.

stress circle *See* MOHR'S STRESS CIRCLE.

stress concentration (**stress raiser**) The rapid increase in stress in the vicinity of a concentrated load or a sharp change in the geometry of a component.

stress-concentration factor **1.** **(K_t)** In monotonic loading, the ratio of the maximum stress at a discontinuity in the cross section of a component to the average stress at that cross section. **2.** **(K_f)** The ratio of fatigue strengths measured on un-notched and notched specimens. *See also* FATIGUE.

stress corrosion (stress-corrosion cracking) Corrosion failure accelerated by local stresses (particularly tensile) which enhance cracking in both monotonic and fatigue loading.

stress difference The algebraic difference between the largest and least principal stresses in a loaded body, equal to twice the greatest shear stress in the system. For $\sigma_1 > \sigma_2 > \sigma_3$, it is given by $(\sigma_1 - \sigma_3)$.

stress distribution The manner in which tensile, compressive, and shear stresses are distributed within a loaded body, indicated by loci of constant stress.

stress function In elasticity problems of plane stress and plane strain, the equations of compatibility, stress equilibrium (without body forces), and generalized Hooke's law may be combined to give

$$\frac{\partial^4 \phi}{\partial x_1^4} + \frac{2\partial^4 \phi}{\partial x_1^2 \partial x_2^2} + \frac{\partial^4 \phi}{\partial x_2^4} = \left(\frac{\partial^2}{\partial x_1^2} + \frac{\partial^2}{\partial x_2^2}\right)^2 \phi = 0$$

where ϕ is the stress function, and in which

$$\sigma_{11} = \frac{\partial^2 \phi}{\partial x_2^2}, \quad \sigma_{22} = \frac{\partial^2 \phi}{\partial x_1^2}, \text{ and } \quad \sigma_{12} = \frac{-\partial^2 \phi}{\partial x_1 \partial x_2}.$$

ϕ is a biharmonic function, different forms of which satisfy the boundary conditions of particular problems. As elastic moduli do not enter the biharmonic equation, solutions are valid for any isotropic, Hookean solid. Moduli are required later to determine the strains from the stresses.

stress-intensity factor *See* FRACTURE MECHANICS.

stress lines *See* PHOTOELASTICITY.

stress raiser *See* STRESS CONCENTRATION.

stress range *See* FATIGUE.

stress ratio *See* FATIGUE.

stress relaxation The reduction in stress with time when the strain is kept constant in a time-dependent material. *See also* CREEP.

stress relieving **1.** The elimination of residual stresses in as-quenched martensite by heating in the range 150–260°C. *See also* TEMPERING. **2.** The elimination of residual stresses in weldments by heating to about 650°C. **3.** In glass, reduction of stresses induced by working processes (called annealing in the glass industry).

stress rupture diagram *See* CREEP RUPTURE STRENGTH.

stress sensor Any device that measures stress (strictly load) either directly, as with a piezo crystal, or indirectly, as with strain gauges. Can be applied in robotics to provide force feedback from a gripper to allow handling of fragile components.

stress–strain curve (stress–strain diagram) A plot of stress *vs* strain for a material obtained from a test in which a specimen is loaded in tension, compression, shear, or combinations thereof. Diagrams of true stress *vs* true strain deviate from those of nominal stress *vs* nominal strain only at large strains where appreciable changes in cross section of testpieces occur.

stress–strain equation (stress–strain relationship) *See* CONSTITUTIVE EQUATION (1).

stress tensor *See* STRESS.

stress-time diagram A plot, derived from creep curves of stress *vs* time at constant strain and temperature for materials that creep or relax.

stress trajectories Contours of constant stress for given problems of loading, obtained either theoretically as closed-form algebraic solutions or numerical simulations, or experimentally by photoelasticity or holography.

stress wave An unbalanced force, such as an impact, applied at one location in a body sets the material there into vibration that is transmitted to adjacent elements and ultimately to all parts of the body in the form of stress waves or wave packets. The velocity of propagation is given by $\sqrt{E/\rho}$, where E is Young's modulus and ρ is density, with the size and shape of the wave depending on the initial disturbance. For many engineering materials, the wave speed is in the region of 5 km/s. *See also* SOUND SPEED.

stress-wave emission *See* ACOUSTIC EMISSION.

stretch forming A manufacturing process in which sheet-metal components are physically stretched over a forming die using a press, punch, or edge clamps. Stretching combined with bending reduces springback. *See also* DEEP DRAWING.

stretch ratio (extension ratio, λ) A measure, used for highly-extensible materials such as rubber, of the extensional or normal strain of a component subjected to tension and defined as the ratio of the final length l, for a given load, to the initial or gauge length l_0, where the coordinate axes are those of principal strain. It is related to the engineering strain e by $\lambda = 1 + e$.

striations *See* FATIGUE.

stroboscope A device that illuminates a rotating, reciprocating, oscillating, or vibrating object with flashes of light at a controlled frequency such that it appears to be stationary. The speed of rotation of a **stroboscopic disc**, which is marked with several concentric rings, each of which is divided into light and dark segments, can be determined from the stroboscope frequency which makes one of the rings appear to be stationary. A **stroboscopic tachometer** is an instrument incorporating a stroboscope and a scale, from which the rotation speed of the object can be read directly when the flashing frequency makes the object appear to be stationary.

stroke 1. The linear distance between top-dead centre and bottom-dead centre of a piston in a reciprocating engine or mechanism. **2.** The movement of a piston or plunger in a reciprocating machine to execute a particular function; for example, the exhaust stroke of an engine in which the exhaust gases are expelled from a cylinder.

S

stroke-bore ratio The ratio of the stroke (1) to the bore diameter of the cylinder in a piston engine.

strong shock *See* SHOCK WAVE.

strong-shock solution *See* OBLIQUE SHOCK WAVE.

Strouhal number (*St*) A non-dimensional number characterizing oscillatory fluid flow with a frequency ω, defined by $St = \omega L/U$ where L and U are a characteristic length and velocity scale, respectively.

structural analysis (structures) The determination of the forces, displacements, stresses, and strains in a given structure.

structural damping (internal damping, solid damping) The energy dissipated within a cyclically-stressed material, approximately independent of frequency and proportional to the square of the vibration amplitude.

structural deflection The displacement of any point of a structure due to applied load.

structural engineering A branch of civil engineering concerned with the design of buildings, dams, bridges, and other large structures.

structural frame The underlying load-bearing framework of beams, columns, bracing, etc., of steel, concrete, brick, stone, timber etc., required to support the static and dynamic forces a building or component is subjected to, including its weight and that of its contents.

structural-integrity analysis Assessment of components and structures for likelihood of failure by buckling, fracture, yielding, etc. under conditions of normal and accident loading.

structures *See* STRUCTURAL ANALYSIS.

strut A structural component designed to carry compressive forces along its length. Short struts fail by crushing or yielding, long struts by buckling. *See also* EULER BUCKLING FORMULAE.

stub axle A short axle, such as is used to support an undriven wheel.

stud (studding) A length of rod threaded at one end or both ends.

stuffing Packing material used to create a pressure-tight seal in the annular space between a rod and cylinder.

subassembly An assembled set of components forming part of a larger machine or device.

subcooled boiling Pool boiling in which vapour bubbles form at the hot surface, but the bulk of the liquid is below saturation temperature.

subcooled liquid A liquid at a temperature lower than the saturation temperature for the liquid's pressure. *See also* COMPRESSED LIQUID.

sub-critical flow 1. Open-channel liquid flow for which the Froude number $V/\sqrt{gh}$ is less than unity, V being the liquid velocity, h

the liquid depth, and g the acceleration due to gravity. **2.** A swirling flow in which inertial waves can propagate upstream. **3.** A viscous flow in which a non-dimensional parameter, such as the Reynolds, Taylor or Rayleigh number, is below the value at which the flow becomes unstable. *See* SUPERCRITICAL FLOW; UNSTABLE FLOW.

sub-critical crack growth Slow crack propagation in a component or structure before final failure.

sub harmonic An integer submultiple of the fundamental frequency f of vibration of an oscillatory system, i.e. $f/2$, $f/3$, $f/4$, etc.

sublayer *See* VISCOUS SUBLAYER.

sublimation The thermodynamic process by which a substance can transform (**sublime**) directly from the solid phase to the gas or vapour phase without passing through a liquid phase. It occurs at temperatures and pressures below a substance's triple point. **Sublimation cooling** is a consequence of the need to supply thermal energy to a subliming solid equal to the latent heat of sublimation (**sublimation energy**). The **sublimation pressure** is the vapour pressure of a solid substance at a given temperature and a **sublimation curve** is a plot of sublimation pressure *vs* temperature. The temperature at which the sublimation pressure of a substance is equal to the pressure of the vapour phase in contact with it is the **sublimation point**. It is analogous to the boiling point of a liquid. **Reverse sublimation (deposition, inverse sublimation)** is the opposite of sublimation, i.e. the thermodynamic process by which a substance can transform directly from the gas or vapour phase to the solid phase without passing through a liquid phase.

submerged arc welding *See* WELDING.

submerged combustion The practice of heating a liquid by bubbling through it hot combustion gases from a submerged burner.

submerged-combustion vaporizer (submerged-combustion evaporator) A liquid-evaporation system in which submerged combustion is used to heat the liquid. Where the liquid to be vaporized is itself combustible (e.g. LNG), it is passed through tubes submerged in a bath of water or other non-inflammable liquid heated by submerged combustion.

submerged outflow *See* DROWNED OUTFLOW.

submersible pump A pump, including its drive motor, designed to operate when completely submerged in a liquid.

subsonic flow A fluid flow for which the Mach number M is less than one, i.e. the flow speed V is below the speed of sound for the fluid a. For a perfect gas, $a = \sqrt{\gamma RT}$ where γ is the ratio of specific heats for the gas, R is the specific gas constant, and T is the gas temperature. Thus if T varies then so does a, and this has to be accounted for in determining whether $M > 1$ (supersonic flow) or $M < 1$.

subsonic inlet An orifice or entrance, such as a jet-engine intake, within which the Mach number of the flowing fluid is less than one.

subsonic leading edge A wing with leading-edge sweep where the Mach number is such that the Mach-wave angle is greater than the sweep angle. *See also* SUPERSONIC LEADING EDGE.

subsonic nozzle A nozzle through which a fluid, usually a gas, flows at a speed below the speed of sound in the fluid.

substantial derivative (material derivative, particle derivative total derivative, total-time derivative, *D/Dt*, *d/dt*) In a fluid flow, the total-time derivative which follows a particle, thereby taking into account both spatial and temporal changes in a variable, e.g. in the case of pressure p,

$$\frac{Dp}{Dt} = \frac{\partial p}{\partial t} + (\mathbf{V}.\nabla)p$$

where t is time, $\mathbf{V}$ is the vector velocity, and ∇ is the gradient operator. *See also* CONVECTIVE DERIVATIVE.

substrate The underlying supporting surface upon which a film, generally very thin, of another material is deposited, e.g. by chemical or physical vapour deposition, chemical and electrochemical techniques, sputtering, spraying, or epitaxy.

subsystem Part of a larger system having a well-defined function and characteristics.

subtractive manufacturing A term for those traditional manufacturing processes that remove material from the workpiece to attain the desired shape and size of a component, in contrast to additive manufacturing.

suction 1. The removal of fluid flowing over a porous surface when the wetted surface is at higher pressure than the pores. **2.** The reduction of pressure by a pump to a level below that of its surroundings.

suction box The lower chamber of a suction pump into which liquid is drawn during the upward stroke of the piston.

suction cup A dome-shaped cup of rubber or other flexible material in which sub-atmospheric pressure is created when a force, applied to the exterior causing distortion of the cup and expelling air from inside, is released. If placed on a flat, non-porous surface, the cup attaches itself to the surface and can be used to lift objects, such as sheets of glass. A bulb-shaped cup attached to a tube can be used to suck liquid into the tube, which can then be released in a controlled way. Applications include droppers and pipettes.

suction head (suction lift) The pressure reduction Δp compared to its surroundings which a pump creates at its inlet converted to the height h of a vertical column of liquid of density ρ, i.e. $h = \Delta p/\rho g$ where g is the acceleration due to gravity.

suction line A pipe or tube through which fluid is drawn into a pump, compressor, turbocharger, supercharger, blower, etc.

suction pump (lift pump) A pump which reduces the pressure at its inlet below that of the surroundings. If the pump is at the surface of a body of water but liquid is drawn in through a pipe with its inlet at depth, the maximum possible lift is determined by the vapour pressure of the liquid, since at this pressure the liquid will cavitate.

S

suction pyrometer (suction thermocouple) A thermocouple, typically of platinum-rhodium, protected from chemical attack by a sintered alumina sheath and surrounded by a number of concentric radiation shields, with hot gas being drawn at high velocity between the shields and over the sheath.

suction stroke (intake stroke) 1. The piston stroke that draws a fresh charge of air and, for a non-injected engine, fuel, into the cylinder of a piston engine. **2.** The piston stroke that draws fluid into the cylinder of a reciprocating compressor or pump.

suction surface *See* AEROFOIL.

suction thermocouple *See* SUCTION PYROMETER.

summing junction The part of a closed-loop controller that determines the difference (the error ε) between the set point and an observation of plant output. The output of the summing junction is $\varepsilon = i - o$, where i is the set point and o is the plant output.

sump (oil sump) A tank which acts as a reservoir for lubricating oil at the lowest part of an engine or other machine. *See also* DRY-SUMP ENGINE; WET-SUMP ENGINE.

sun gear *See* EPICYCLIC GEAR TRAIN.

supercentrifuge A centrifuge designed to operate at very high rotation speeds.

supercharge method A method for determining the octane quality of spark-ignition aviation fuel using a standardized single-cylinder, four-stroke, indirect-injected, liquid-cooled CFR engine under supercharged rich-mixture conditions.

supercharging The process of increasing the mass flow rate of air (or air/fuel mixture) into the cylinder(s) of a piston engine using a compressor driven from the crankshaft (the **supercharger**). The power output is increased compared with a naturally-aspirated engine of the same capacity. The process also increases the air pressure and density to greater than ambient. In the past, most superchargers were mechanically driven from the engine's crankshaft, but these have been largely superseded by turbochargers. *See also* LYSHOLM COMPRESSOR; ROOTS BLOWER.

supercompressibility factor *See* COMPRESSIBILITY FACTOR.

supercooling (supersaturation) A metastable (i.e. non-equilibrium) state involving the reduction, without a phase change, of the temperature of a substance to a level below the temperature at which a phase change would be expected. Examples are a superheated vapour expanded isentropically to a temperature below the saturated vapour line, and a liquid cooled to a temperature below that at which it would normally freeze to a solid. *See also* DEGREE OF SUPERCOOLING.

supercritical aerofoil A transonic aerofoil designed to minimize wave drag.

supercritical flow 1. Open-channel liquid flow for which the Froude number is greater than unity. **2.** A swirling flow in which inertial waves propagate at a lower speed than the flow speed. **3.** A viscous flow in which a non-dimensional parameter, such as the Reynolds, Taylor, or Rayleigh number, is above the value at which the flow becomes unstable. *See also* SUBCRITICAL FLOW; UNSTABLE FLOW.

supercritical fluid A fluid at a temperature and pressure above the values at its thermodynamic critical point. Such a fluid has the properties of both a gas and a liquid.

supercritical pressure and temperature The pressure and temperature of a substance which are above the vapour-liquid critical point of that substance.

supercritical steam cycle A steam-power cycle in which the boiler pressure is supercritical.

supercritical-water oxidation A high-efficiency thermal-oxidation reaction process which takes place at supercritical temperatures and pressures, used to treat aqueous waste materials.

supercritical-water reactor *See* NUCLEAR FISSION.

superficial expansivity *See* COEFFICIENT OF SUPERFICIAL EXPANSION.

superficial velocity (U_o) (Unit m/s) For fluid flow through a pipe or duct, with cross-sectional area A, the total volumetric flow rate $\dot{Q}$ divided by A, i.e. $U_o = \dot{Q}/A$. U_o takes no account of any packing within the pipe, for example in a packed column, and in such situations is not a physically real velocity. It is widely used in the analysis of flow through porous media, and also multiphase flows where one phase may be moving at a different velocity to another. *See also* INTERSTITIAL VELOCITY.

superheat *See* DEGREE OF SUPERHEAT.

superheated vapour (superheated steam) A vapour at a temperature above that of dry saturated vapour at a given pressure. *See also* DEGREE OF SUPERHEAT.

superheater A part of a steam-generating system, which may be incorporated into the boiler or separated from it, in which steam is heated above the temperature of dry saturated steam at a given pressure.

superplasticity Deformations of over 90% reduction in area which can occur over long times at high homologous temperatures in materials (such as 80%Zn/20%Al alloy and Pb–Sn of eutectic composition) which display great strain-rate sensitivity, a characteristic that is the antithesis of a good creep-resistant material. Formability limits are much increased under such slow processing, and in **superplastic forming** components are manufactured by extensively-deforming sheet material.

superposition principle (principle of superposition, superposition theorem) For any linear system or equation, if $y_1(t), y_2(t), \ldots y_n(t)$ are the responses or solutions to inputs $x_1(t), x_2(t), \ldots x_n(t)$, then the responses and solutions can be combined by addition such that $y_1(t) + y_2(t) + \ldots + y_n(t)$ is the response or solution to the input $x_1(t) + x_2(t) + \ldots + x_n(t)$ The principle finds application in problems involving e.g. linear elasticity, thermal conduction, control-system analysis, and wave propagation.

supersaturation *See* SUPERCOOLING.

Supersonic Combustion RAMJET *See* SCRAMJET.

supersonic compressor A gas compressor in which the tip speed of the rotor blades is greater than the local speed of sound.

supersonic diffuser A convergent nozzle which, for supersonic flow, leads to a decrease in velocity and an increase in pressure.

supersonic flow A fluid flow for which the Mach number is greater than one, i.e. the flow speed is greater than the speed of sound for the fluid. *See also* SUBSONIC FLOW.

supersonic leading edge A swept wing where the Mach number is such that the Mach-wave angle is less than the sweep angle. *See also* SUBSONIC LEADING EDGE.

supersonic nozzle *See* CONVERGENT–DIVERGENT NOZZLE.

supersonic turning The process of changing the direction of a supersonic flow by continuous compression through a series of compression waves, discontinuous compression via an oblique shock, or continuous expansion through a series of expansion waves. *See also* PRANDTL–MEYER EXPANSION.

supervisory expert control system Software implemented using an expert system of knowledge base and inference engine to adapt a control system to improve performance.

supination The orientation and motion of a robot component with its unprotected side facing upward and exposed.

supply flow The main flow which is to be controlled in a fluidic device. *See also* FLUIDICS.

suppressed-zero instrument A measuring instrument in which the zero is outside the range of the marked scale.

surface The boundary of a solid object or liquid mass which separates it from any substance (including a vacuum) with which it is in contact.

surface analyser (profilometer, surface meter) An instrument that quantifies surface texture or roughness, either mechanically by moving a stylus over the surface, or optically by scanning the surface with a laser beam, typically with μm-level resolution.

surface burning An exothermic reaction between a solid fuel and an oxidant, in which thermal energy is released by convection and radiation in the absence of a flame.

surface combustion Pre-mixed fuel and air burning in or near a permeable medium surface layer made of ceramic fibres.

surface condenser (steam condenser) A water-cooled shell-and-tube condenser operated at sub-atmospheric pressure, used to condense a vapour such as exhaust steam from a steam turbine. The condensed steam is pumped back into the boiler as feedwater. *See also* CONDENSER VACUUM.

S

surface conductance *See* HEAT TRANSFER.

surface-contact stresses *See* CONTACT MECHANICS.

surface density *See* PLANE LAMINA.

surface durability The ability of a surface to maintain its as-manufactured dimensions and properties, tailored to the application, without damage by erosion, wear, corrosion, etc.

surface finish The texture and roughness of a surface following machining, shot peening, sand blasting, grinding, sanding, etc., and including any flaws, coatings, etc. *See also* SURFACE ROUGHNESS.

surface free energy (γ) (Unit J/m^2) Within bodies, chemical bonds are matched and balanced by those of surrounding atoms. On surfaces, the exposed bonds and those of the surrounding vapour are not in equilibrium, and this gives rise to the thermodynamic concept of surface free energy, being the energy of unit area. It is a short-range parameter having magnitudes of no more than a few J/m^2 for all solids. *See also* FRACTURE MECHANICS.

surface gauge **1.** An instrument for determining the heights of points on a surface, with respect to a reference plane. **2.** A tool, adjustable for height, employed to mark metal components with a scriber.

surface gravity wave A wave that propagates along the free surface of a liquid under the influence of fluid inertia and gravity. The phase speed c in a liquid layer of depth h for a small-amplitude wave is given by

$$c = \sqrt{\frac{g}{k} tanh(kh)}$$

where k is the wavenumber and g is the acceleration due to gravity. *See also* SURFACE WAVE.

surface hardness (superficial hardness) Indentation hardness on, and for some distance below, the surface of a component, as opposed to its bulk hardness. Depending on the resolution required, micro- or nano-hardness testing machines are employed.

surface ignition Unwanted ignition in the combustion chamber of an internal-combustion engine, caused by the air:fuel mixture coming into contact with an excessively hot surface, which may result in knock.

surface mean diameter *See* SAUTER DIAMETER.

surface meter *See* SURFACE ANALYSER.

surface micromachining The production of small-scale (typically μm-size) structures by the deposition and etching of thin layers of polysilicon and silicon dioxide.

surface plate A rigid, accurately-flat plate used to test the flatness of other surfaces, or to provide a datum surface for marking off work for machining with a surface gauge.

surface roughness (roughness) The small-scale, irregular peaks and troughs in a solid surface which are quantified using a surface analyser (**profilometer** or optical interferometer). Roughness may be a consequence of wear and corrosion, or of the manufacturing process. It leads to increased friction in solid-solid contact or fluid flow over a surface. Errors of form and waviness are excluded, but **surface texture** includes roughness and waviness. **Surface topography** incorporates the small-scale, three-dimensional geometry of a surface including surface roughness, machined, and etched features, typically at sub-mm scales. *See also* CENTRE-LINE AVERAGE; KURTOSIS; SKEWNESS.

surface tension (σ, γ) (Unit N/m) At the interface between a liquid and a gas or two immiscible liquids, unbalanced cohesive forces acting on the liquid molecules at the interface lead to the property surface tension which causes tensile forces to develop as if it were a skin or membrane. Along a line of length δl in such an interface, the surface-tension force δF within the interface and perpendicular to it is given by $\delta F = \sigma \delta l$. The pressure difference Δp across an interface between two fluids is given by the **Young-Laplace equation**, $\Delta p = \sigma(1/R_x + 1/R_y)$ where R_x and R_y are the principal radii of curvature of the interface. Surface tension is responsible for the spherical shape of small bubbles and drops, and for causing liquid in a small-diameter capillary tube to be pulled up or down and for the interface to adopt a spherical shape.

SEE WEB LINKS

- Reference book companion website covering many material properties (thermal conductivity, density, thermal expansion coefficients, surface tension, viscosity, etc.)

surface-tension-driven flow Flow due to differences in surface tension within a liquid-gas interface, for example due to the surface tension being lowered by the addition of a detergent to some part of the interface.

surface-texture indication (machining indication) On an engineering drawing the basic symbol to indicate surface texture is a tick, the two legs of which are inclined at approximately 60° to the line representing the surface. If machining is required, a horizontal bar is added from the tip of the shorter leg to a point on the longer leg. If removal of material is not permitted, an inscribed circle is added. Numerical values are added to show permitted maximum, and if required, minimum roughness values in μm, together with any features required such as the preferred lay on the machined surface (indicated by the symbol = when the lay is parallel to the projection plane in the drawing; when perpendicular by ⊥; when criss-cross by ×).

surface topography *See* SURFACE ROUGHNESS.

surface treatment Any process, including chemical, electrochemical, mechanical, and thermal, designed to protect a surface against corrosion and wear or to alter its mechanical properties.

surface wave **1.** A mechanical wave, combining longitudinal and transverse motion, that propagates along the interface between two immiscible fluids, the one having the lower density above the other, the restoring force being due to either gravity or surface tension. In the case of a liquid of density ρ and surface tension γ in contact with a gas of negligible density, the angular frequency ω of small-amplitude waves of wavenumber k is given by $\omega = \sqrt{(gk + \gamma k^3/\rho)\tanh(kh)}$ *See* ALSO SURFACE-GRAVITY WAVE. **2. (Rayleigh wave)** A mechanical wave, combining longitudinal and transverse motions, that propagates along the surface of an elastic solid. Its velocity is given by $c_s = f(\nu)\sqrt{G/\rho}$ where G is shear modulus, ρ is density, and $f(\nu)$ is a function of Poisson's ratio ν, equal to 0.9194 when $\nu = 0.25$, and to 0.9953 when $\nu = 0.5$. Such waves are strong only within a depth of about one wavelength from the surface. **3. (Love wave)** A surface mechanical wave that, like a Rayleigh wave, also involves transverse displacements, and is important in seismology owing to stratification of the earth's surface.

surfactant A surface-active agent, or wetting agent, that lowers the surface tension of a liquid

6.3

surface-texture markings

S

or lowers the interfacial tension between two immiscible liquids.

surge 1. In a compressor, overall variations in mass flow with time due to the compressor changing from unstalled to stalled and back again. **2.** An unstable pressure buildup in a flowline.

surge device A wave-energy converter in which incoming waves run into shore-based tapered channels where the wave height increases, spills over into a reservoir, and creates sufficient head for a water turbine.

surroundings The entire mass or region outside a thermodynamic system. *See also* ENVIRONMENT; IMMEDIATE SURROUNDINGS; SYSTEM BOUNDARY.

suspended-level viscometer *See* CAPILLARY VISCOMETER.

suspension (suspension system) 1. The system of springs, shock absorbers, struts, axles, and linkages which connect the chassis and wheels of a vehicle; designed to damp vibration and reduce the effect of shock loads. **2.** A fine wire or coil spring that supports the moving element of certain measuring instruments. *See also* BIFILAR SUSPENSION; TRIFILAR SUSPENSION.

sustainable design (green design) A philosophy of design applied to objects, machines, industrial processes, the built environment, etc., aimed at ecological, resource, economic, and social sustainability.

sustainable energy *See* RENEWABLE ENERGY.

sustained oscillation Continued oscillation in the output of a controlled system as a consequence of marginal stability.

Sutherland's viscosity law (Sutherland's formula) An equation for the dynamic viscosity μ of a gas as a function of absolute temperature T, based upon kinetic theory but including a correction factor, according to which $\mu = KT^{3/2}/(T + C)$ where K and C are constants characteristic of the gas.

swaging A metal-working process for shaping barstock, using radially-oscillating hammers positioned around an opening into which the barstock is inserted. The process is also used to flare out the end of a tube, or to reduce its diameter.

swarf Material, in the form of shavings, turnings, chips, or powder, that is removed from a workpiece during a machining operation.

swashplate (wobble plate) A circular plate mounted obliquely on a rotating shaft which imparts a reciprocating motion to axially-mounted components in contact with it, or is itself rotated by axially-mounted reciprocating components in contact with it. In a **swashplate engine (swashplate motor, barrel engine)**, a circular array of reciprocating pistons held in a barrel causes rotation of a swashplate attached to an output shaft. In a **swashplate pump**, reciprocating axial pistons are driven by a swashplate. *See also* NUTATION.

sweat cooling Cooling of a component by evaporation of a liquid forced through pores in its surface; applications include rocket engines and turbine blades. *See also* TRANSPIRATION COOLING.

sweated joint *See* SOLDERING.

swept volume In the cylinder of a piston engine or other device, the volume between top- and bottom-dead centre positions of a piston; equal to $\pi R^2 S$ where R is the piston radius and S is the stroke.

swept wing (sweptback wing) An aircraft wing for which the angle between the leading edge and the centreline of the fuselage (the sweep angle) is less than 90°. Highly swept wings are designed to reduce wave drag from shock waves. *See also* DELTA WING.

SWG *See* WIRE GAUGE.

swing The radial clearance between the chuck and bed of a lathe, which determines the maximum radius of workpiece that may be machined.

swing axle Part of a simple independent suspension for motor vehicles in which each half shaft is attached to the differential by a universal joint, allowing the axle to move vertically.

swing separator *See* TRENNSCHAUKEL.

swing-type check valve A non-return valve installed in a pipe, in which a circular disc hangs down from a hinge pin to contact a circular seat in the valve body. The valve opens when the fluid pressure exerts a sufficiently high force on the disc to overcome the closing moment due to its weight.

swirl burner (vortex burner) A gas or oil burner in which swirl is added to the combustion air and sometimes also the fuel, typically using an annular array of angled guide vanes (the **swirler**), to enhance mixing between the air and fuel and create a region of recirculation for flame stabilization. A quarl is usually located just downstream of the guide vanes. *See also* ROTARY-CUP BURNER.

swirl chamber A container in which swirling flow occurs, such as in the cylinder of a piston engine to promote mixing, a cyclone separator, vortex amplifiers, and vortex diodes.

swirl flow (swirling flow) A fluid flow in which the fluid particles have a tangential component of velocity about an axis which, when combined with an axial component of velocity, produces a helical or spiral flow.

swirl flow meter A flow meter in which a swirler creates a swirling flow and a sensor then detects oscillations in the flow which have a frequency proportional to the volumetric flow rate.

switching flow *See* FLUIDICS.

switching surface In a bang-bang feedback controller, the surface in state space which separates regions of maximum positive and maximum negative control effort.

swivel (swivel joint) A coupling component which allows an attached object to turn freely.

swivel block A block incorporating a sheave which can turn freely on its shackle or other support.

swivel coupling A connection between two pipes or pipe fittings which allows one to be rotated relative to the other.

swivelling propeller A propeller attached to a shaft which can be turned so that thrust can be delivered in any direction. *See also* THRUST VECTORING.

swivel pin *See* KINGPIN.

sylphon bellows A thin-walled axisymmetric metal bellows.

symbols on drawings Standard symbols used in engineering drawings to show diameter, machining and surface texture, projection system, tapers, welding, etc., as specified in ISO 128.

symmetry axis *See* AXIS OF SYMMETRY.

synchromesh *See* GEARBOX.

synchronous A term for events occurring simultaneously.

synchronous belt *See* TIMING BELT.

synchronous vibrations Vibrations that have the same frequency and phase, but may differ in amplitude.

syngas (synthesis gas) A gas mixture, primarily hydrogen and carbon monoxide, produced by gasification of coal, biomass, and waste matter, and in steam reforming of natural gas to produce hydrogen. It can be used as a gaseous fuel, or as feedstock to produce liquid fuels. *See also* PLASMA.

system 1. In control engineering, a control system is a collection of components designed to cause a plant to produce an output that follows accurately a desired behaviour. **2.** In systems engineering, a system is an aggregation of end products and enabling products to achieve a given purpose. **3.** In thermodynamics and solid mechanics, a system is a quantity of matter or a region in space selected for study. The matter or space outside the system is called the surroundings. The real or imaginary surface which separates the system from its surroundings is called the boundary. *See also* CLOSED SYSTEM; ISOLATED SYSTEM; OPEN SYSTEM.

system analysis The determination of how a set of interconnected components will respond in total given a knowledge of the behaviour of the individual characteristics of the components.

systematic error *See* ERROR (1).

system bandwidth In a control system, the bandwidth is the input frequency at which the output has dropped by −3dB relative to the output for a same-magnitude low-frequency input.

system boundary The real or imaginary surface which separates a thermodynamic or solid mechanics system from its surroundings.

system design The design of a complex engineering product by the selection of individual sub-systems, based on knowledge of the behaviour of each sub-system without reference to their detailed design. For example, a motor, gearbox, and ball screw could be combined to make a product that gives translational motion. A system-design approach would match the three sub-systems based on the input–output

S

relationships of each, without considering their internal design or operation.

systems engineering (system engineering) A methodology which integrates all disciplines and specialty groups into a team effort, forming a structured development process that proceeds from concept to production to operation.

system type (type) The highest order of the time-domain differential equation required to properly describe a system. This is equal to the highest power of s in the denominator of the transfer function.

S

tab washer (lock washer) A washer with one or more protruding tabs which can be bent in such a way that a nut or bolt head is prevented from becoming loose.

tachometer An instrument for measuring the rotation speed of a shaft, wheel, disc, etc.

tackle *See* BLOCK AND TACKLE.

tailrace An open channel or tunnel that conveys water away from a hydraulic turbine. *See also* HEADRACE.

tail rotor A rotor, with its axis perpendicular to the vertical plane dividing the fuselage of a helicopter, that provides side thrust to counteract the torque of the main rotor and give directional control. *See also* FANTAIL; TORQUE REACTION.

tailshaft *See* DRIVESHAFT.

tailstock The workpiece in a lathe is held and driven at the headstock. The tailstock is used at the far end of a workpiece, either to support it along its axis of rotation in a centre, or to hold tools to drill or bore the workpiece. It is also part of a milling or grinding machine. *See also* HEADSTOCK.

tainter gate (radial gate) A spillway gate in the form of a section of a circular cylinder supported on radial arms. The gate rotates about a horizontal axis on its downstream side to control water flow, e.g. through a dam.

Tait equation An empirical equation widely used to represent the variation of liquid density ρ with pressure p according to

$$\frac{\rho - \rho_0}{\rho} = C\ log_{10}\left(\frac{p+B}{p_0+B}\right)$$

where B and C are fitted parameters and the subscript 0 refers to a low-pressure value, usually 0.1 MPa or the saturation pressure.

tandem-compound steam-turbine system A multi-cylinder steam-turbine arrangement in which several machines running on the same shaft are connected in series to a generator. The diagram shows tandem compounded high- (HPT), intermediate- (IPT) and low-pressure (LPT) turbines. *See also* CROSS-COMPOUND STEAM-TURBINE SYSTEM.

tangent screw A worm screw used to adjust measuring instruments such as sextants and vernier callipers.

tangential acceleration (a_θ) (Unit m/s^2) For a particle or body moving relative to a fixed point a distance r away, the component of acceleration corresponding to the angular acceleration $r\ddot{\theta}$ plus the Coriolis acceleration, i.e. $a_\theta = r\ddot{\theta} + 2\dot{r}\dot{\theta}$ where $\dot{\theta}$ is the instantaneous angular velocity. *See also* ACCELERATING FRAME OF REFERENCE; ANGULAR ACCELERATION; RADIAL ACCELERATION.

tangential load (Unit N) The component of load applied to an object, such as a gear, that tends to cause or resist rotation.

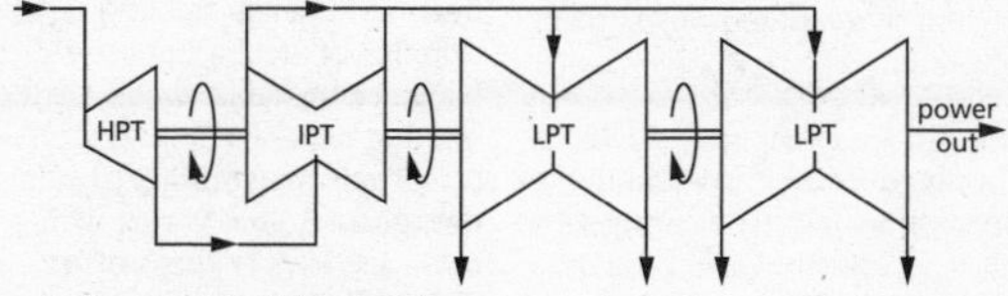

tandem-compound steam-turbine system

tangential reaction *See* NORMAL REACTION.

tangential stress *See* SHEAR STRESS.

tangential velocity (azimuthal velocity, v_θ) (Unit m/s) For a particle or body moving relative to a fixed point a distance r away, the component of velocity corresponding to the angular velocity ω (or $\dot{\theta}$), i.e. $v_\theta = \omega r = r\dot{\theta}$. *See also* RADIAL MOTION.

tangent modulus (E_t) (Unit Pa) The slope of a stress (σ)-strain (ε) curve at a point above the proportional limit, i.e. $E_t = d\sigma/d\varepsilon$. Plastic buckling can be analysed using E_t in place of E in the Euler formulae. *See also* REDUCED MODULUS; SECANT MODULUS.

tap 1. A threaded hard metal plug for cutting screw threads in holes. It has longitudinal grooves for the clearance of chips while cutting. Progressive cutting is achieved by using three taps in succession: taper, second, and plug. First and second taps have tapers to help start the thread along the axis of the hole; the plug tap has no taper and is used on the final cut in a blind hole. **2.** A valve in, or at the end of, a pipeline.

TAPCHAN A wave-energy converter in which ocean waves grow in amplitude and propagate through a **tap**ered **chan**nel into a reservoir within which the water level is increased. The water then flows back out through a low-head Kaplan turbine to generate electrical power.

taper A gradual, often linear, reduction in cross section or shape. *See also* MORSE TAPER.

tapered thread *See* SCREW.

taper key A key with parallel sides but tapering cross section along its length.

taper pin A pin or peg of circular cross section that tapers along its length.

taper reamer (tapered reamer) *See* REAMER.

taper roller bearing *See* BEARING.

taper symbol On an engineering drawing, the axial cross section of a tapered object takes the form of a trapezium.

tappet A sliding rod in a machine driven in one direction by a cam or rocker arm and in the other (return) direction usually by a spring. In a piston engine, used to operate valves. *See also* DESMODROMIC VALVE.

tapping 1. Cutting a screw thread with a tap. **2.** *See* PRESSURE TAPPING.

tapping drill (tapping-size drill) A drill bit used to drill a hole which is to be threaded with a tap. For metric sizes, the drill size is the thread diameter minus the pitch.

tapping screw *See* SCREW.

tap wrench A lever with an adjustable square hole at its centre into which is fitted the shank of a tap so that torque can be applied to cut a thread.

tar A phenolic resin produced by destructive distillation of wood or biomass under pyrolysis. It can be used as a fuel, with a heat of combustion of about 22 MJ/kg, or a source of chemicals.

target flow meter *See* DRAG-BODY FLOW METER.

Taylor–Aris dispersion (Taylor dispersion) An approximate theoretical analysis of the dispersion of a passive tracer in fully-developed laminar flow through a small-diameter tube.

Taylor bubble A long gas bubble that rises through a liquid in a vertical tube. It has a rounded nose and flat tail. If the liquid is stationary, the bubble's rise velocity is approximately $0.35\sqrt{gD}$ where g is the acceleration due to gravity and D is the tube diameter. **Taylor flow** is a two-phase flow in which the gas forms Taylor bubbles.

Taylor–Couette flow The flow of a fluid contained in the annular gap between two concentric cylinders where there is relative rotation between the two, either in the same sense or in the opposite sense. The flow is characterized by a non-dimensional parameter, the **Taylor number (*Ta*)**. If the inner cylinder is rotating at angular velocity Ω_I, the outer at Ω_O, with a fluid of kinematic viscosity ν in the annular gap of width δ, $Ta = R_I\delta^3(\Omega_I^2 - \Omega_O^2)/\nu^2$ where R_I is the radius of the inner cylinder. There are other definitions of *Ta*, but all represent the ratio of centrifugal to viscous forces. If the inner cylinder is rotating faster than the outer cylinder or in the opposite direction, the flow becomes unstable above a critical Taylor number that depends upon the cylinder radius ratio (the **Taylor instability**) leading to the formation of a toroidal vortex with a roughly square cross section (**Taylor vortex**). A **Taylor cell** is the region occupied by a Taylor vortex.

Taylor's hypothesis The assumption that the advection of a turbulence field is due entirely to the mean flow.

Taylor vortex A vortex flow for which the variation of tangential velocity v_θ with radius r and time t is given by

$$v_\theta = \frac{H}{8\pi}\frac{r}{vt^2}exp\left(\frac{-r^2}{4vt}\right)$$

where v is the kinematic viscosity of the fluid and H is a constant equal to the magnitude of the angular momentum in the vortex.

***Tds* equations (thermodynamic-property relations)** Thermodynamic relations between specific entropy s, specific enthalpy h, specific volume v, pressure p, and static temperature T according to: $Tds = du + pdv = dh - vdp$. The first *Tds* equation is referred to as Gibbs equation. The second *Tds* equation is also sometimes called a Gibbs equation, as are the equations $sdT = -da - pdv = vdp - dg$ where g is the Gibbs function and a is the Helmholtz function. The equations take identical forms when written in terms of the corresponding extensive properties.

teach To program a robot for a task. In **teach-by-doing** a robot is programmed by manually moving it, or a model of it, through the required path. A **teaching interface** is the hardware used by an operator to program a robot; for example, a **teach pendant (teach gun)**, which is a handheld device with switches or joysticks to allow control of the positions of the robot joints or, through inbuilt inverse kinematics, of the end effector position and orientation. The program is built up by storing each set of positions as a program step **(teach-by-driving)** so that when the program is run the robot follows each of the steps in turn.

tear strength (Unit N/m) The tensile force divided by the sheet thickness required to tear a pre-split material in sheet form at a specified rate. Particularly used for plastics, rubber, fabrics, and elastomers. Since the units are not those of stress but equivalent to J/m^2, the units of toughness, the term 'strength' is strictly incorrect.

technical atmosphere (at) An obsolete unit of pressure in the MKS system equivalent to 9.80665 x 10^4 Pa.

technical drawing *See* ENGINEERING DRAWING.

technical potential *See* ACCESSIBLE RESOURCE.

teetered rotor A bladed rotor, used in some helicopters and wind turbines, in which the blades are hinged at the hub.

teeth The projecting elements on gears, cutting tools, etc. *See also* TOOTHED GEARING.

Teflon® (PTFE) A white synthetic fluoropolymer of tetrafluoroethylene (polytetrafluorothylene) used as a low-friction, non-stick surface. Its glass transition temperature is about 120°C, Young's modulus 400 MPa, and yield strength 25 MPa.

Teledeltos® paper A black, high-resistance conducting paper that can be used to generate analogue solutions to problems, such as those governed by Laplace's equation, by plotting equipotential lines when a potential difference is applied between surface electrodes.

teleoperation *See* SLAVE ARM.

teleoperator (telechir) A robot under continuous human control (a **telerobot**), for example where a robot is used to perform a task in an area unsafe for the human operator such as in nuclear decommissioning. The robot may incorporate feedback to the human operator, for example force feedback, to aid in the performance of the task. Their specification, design, manufacture, installation, and use is termed **telechirics**.

TEM *See* TRANSMISSION ELECTRON MICROSCOPE.

TEMA standard A periodically-updated guide to all aspects of shell-and-tube heat-exchanger design, issued by the Tubular Exchanger Manufacturers Association, Inc.

temper *See* TEMPERING.

temperature A quantitative measure of the molecular kinetic energy of a substance and so how hot or cold it is. *See also* ABSOLUTE TEMPERATURE; ZEROTH LAW OF THERMODYNAMICS.

temperature boundary layer *See* THERMAL BOUNDARY LAYER.

temperature coefficient of expansion *See* THERMAL EXPANSION.

temperature coefficient of sensitivity A measure of the sensitivity to temperature changes of the calibration of any type of measuring instrument or transducer.

temperature compensation The inclusion in a measuring instrument of a means to compensate, usually automatically, for changes in the ambient temperature. For example, for a strain gauge, the inclusion in adjacent legs of the Wheatstone-bridge circuit of an unstrained 'dummy' gauge bonded to a piece of the same material as the test piece and at the same temperature.

temperature difference *See* LOG-MEAN TEMPERATURE DIFFERENCE.

temperature drift The tendency of the calibration, and hence the output, of a measuring instrument or transducer to change if the ambient temperature changes.

temperature–entropy diagram (*T–s* diagram) A diagram showing the variation of specific entropy with temperature for a pure substance. Included are isobars and, in the case of a substance where phase change occurs, curves separating the various phases. *See also* MOLLIER DIAGRAM.

temperature gradient (∇T) (Unit K/m) A vector quantity being the spatial gradient of the temperature at any point in a material. For a three-dimensional temperature distribution, in a Cartesian-coordinate system, the components are $\partial T/\partial x$, $\partial T/\partial y$, and $\partial T/\partial z$.

temperature inversion *See* INVERSION.

temperature of inversion *See* INVERSION TEMPERATURE.

temperature-recovery factor *See* RECOVERY FACTOR.

temperature scale A scale of numbers between two or more fixed points with divisions between them. *See also* INTERNATIONAL PRACTICAL TEMPERATURE SCALE.

temperature transducer (temperature sensor) A device that produces a single-valued electrical signal directly related to its temperature.

temper embrittlement The anomalous reduction in toughness of martensitic stainless steels when tempered in the range 370–600°C.

tempering (temper) The re-heating of steel with a martensitic structure that is very hard but brittle to introduce some ductility. The greater the tempering temperature, the lower the hardness, but the greater the ductility. *See also* STRESS RELIEVING.

template A pattern in the form of a thin plate used to mark out a workpiece or check the outcome of a machining operation.

temporal averaging An average of a quantity varying with time *t*. For a continuously-varying quantity *f(t)*, the average over a period of time *T* is given by

$$\bar{f} = \frac{1}{T}\int_0^T f(t)dt$$

temporal decomposition The splitting of a control problem into parts with relatively short and long time constant so that different controllers may be used for the different parts.

temporal frequency *See* FREQUENCY.

temporally periodic *See* PERIODIC OSCILLATION.

tenon A component located at the tip of a steam-turbine blade that holds the shroud.

ten-second creep modulus The constant applied stress divided by the strain reached after 10 s in time-dependent materials such as polymers. *See also* RELAXED MODULUS; HUNDRED-SECOND CREEP MODULUS.

tensile force *See* TENSION (2).

tensile modulus *See* YOUNG'S MODULUS.

tensile strength (ultimate strength, ultimate tensile strength, UTS) (Unit Pa) The nominal or engineering stress given by the maximum load in a tension test divided by the original cross-sectional area of the specimen. The maximum load could be the fracture load for a brittle material, but for a ductile material it is usually taken as the load at which necking begins, beyond which the load falls.

tensile stress (Unit Pa) A stress that tends to stretch a component, or local region of a component, when under load. In simple tension it results from the applied axial load, but tensile stresses occur in bending, torsion and other forms of loading.

tensile test (tension test) A test in which a precisely-machined **tensile specimen (tensile testpiece)**, typically circular or rectangular in cross section with large end sections that are gripped in the testing machine, is subjected to an increasing tensile load, usually to the point of fracture, to produce a stress–strain curve from which such material properties as modulus of elasticity, limit of proportionality, proof stress, yield point, and ultimate tensile stress can be determined. The relevant standard is ISO 6892-1.

tensimiter An instrument used to determine the vapour pressure of a liquid, consisting of two sealed bulbs, one containing the liquid of interest and the other a liquid with known vapour pressure. The unknown vapour pressure is deduced by measuring the pressure difference between the bulbs.

tensiometer An instrument used to measure surface tension.

tension 1. The condition in a bar, belt, cable, spring, string, wire, etc. that is being pulled from either end. **2. (tensile force)** The force associated with tension as in (1). It is measured by a **tension meter**.

tensometer A bench-top device, used to perform tension and compression tests.

tera (T) An SI unit prefix indicating a multiplier of 10^{12}; thus terawatt (TW) is a unit of power equal to one trillion watts.

terminal temperature difference (Unit K) The temperature difference between the two fluid streams at either inlet or outlet of a heat exchanger.

terminal velocity (Unit m/s) The velocity reached by an object in free fall through a fluid when the upward drag and buoyancy forces just balance the object's weight. For an object with a density lower than that of the fluid, the terminal velocity is vertically upwards.

terminator wave-energy device A wave-energy converter in which the principal axis of the moving (i.e. power-generating) element is parallel to the incident wave fronts.

tertiary air Air pre-heated to a high temperature, added to a furnace or other combustion chamber downstream of the main combustion zone to improve the combustion of exhaust gases and reduce pollution.

tertiary creep *See* CREEP.

testing machine A machine used to apply either a steady or oscillatory or impact load to a testpiece. The load may be tensile, compressive, shear, bending, or torsional.

test mass *See* SEISMIC MASS.

testpiece A precisely-machined object used in a testing machine for the determination of material properties. *See also* TENSILE TEST.

tethered-buoy wave-energy converter (tethered wave-energy converter) A system in which the wave-induced vertical movement of a buoy floating on the surface is used to pump water through a turbine, either submerged or on land.

tex A means of indicating the size of fibres, being the mass in grams of one kilometre of the filament.

texture The distribution of the crystallographic orientations within a polycrystalline material. A fully random distribution has no texture, and the material has isotropic properties. A single crystal has anisotropic properties. *See also* R-FACTOR; SURFACE TEXTURE.

theorem of minimum potential energy *See* MINIMUM POTENTIAL ENERGY.

theoretical air In a combustion process, the amount of air corresponding to the stoichiometric air:fuel ratio.

theoretical relieving capacity The theoretical loss-free flow rate through a fully open pressure-relief valve.

theories of strength Criteria for failure by yielding or rupture of structures and components under complex conditions of loading based on, for example, the attainment of a maximum normal stress (**Rankine criterion**) or maximum shear stress irrespective of other stresses present (**Tresca failure theory**); maximum normal strain (**Saint-Venant criterion**) or maximum shear strain; or maximum strain energy density (**Beltrami criterion**). With a factor of safety, and account being taken of stress concentrations, they were used in traditional engineering 'strength of materials' elastic design. In the presence of cracks or other defects, the approach is inadequate. *See also* FRACTURE MECHANICS; STRENGTH; VON MISES YIELD CRITERION; YIELD CRITERION.

therm An obsolete (non-SI) British unit of heat and energy, equal to 10^5 Btu or approximately 105.5 MJ.

thermal 1. A term for anything concerning heat. A **thermal process** is any physical or chemical process primarily dependent upon heat; for example, distillation, heat treatment, sterilization, steam generation, or vulcanization. **2.** A buoyant mass of hot gas rising in the atmosphere or other large body of gas, or a dense mass of liquid descending through a liquid, such as at a sewage outfall.

thermal advantage (Φ) For a power station, the ratio of the heat used for district heating to the extra heat used in the network to replace the lost electricity.

thermal arrest The constant-temperature section of a heating or cooling curve for a pure substance, or for alloys having eutectic or eutectoid compositions, that occurs when there is a phase change. It is applied particularly to solidification.

thermal barrier (thermal barrier coating) **1.** A coating, consisting of a metallic bond coat, an oxide layer, and a ceramic top layer, applied to metallic surfaces, such as gas-turbine parts, to insulate them from large and prolonged heat loads. **2. (fire barrier)** An insulating coating applied over polyurethane foam to reduce the rate of heat transfer during a fire.

thermal boundary layer (temperature boundary layer) The thin region close to a surface that is heated or cooled by a fluid flow where the temperature is different from that of the free stream. *See also* BOUNDARY LAYER.

thermal break A layer of material of low thermal conductivity introduced into an assembly to reduce the rate of conduction heat transfer.

thermal bubble A bubble formed by electrically heating a liquid to create a vapour which forces some of the liquid through a small (*c.*100 μm) orifice. Applications include inkjet printers.

thermal bulb The fluid-filled bulb in a thermal-expansion thermometer.

thermal capacitance (thermal-energy capacity, thermal mass) (Unit J/K or J/°C) The product of the specific heat C_P of a substance and its mass. Used in the lumped-capacity analysis of unsteady heat-conduction problems. Sometimes defined as the product of C_P and the density of a substance.

thermal capacity *See* HEAT CAPACITY.

thermal checking The pattern of cracks that appears on the surface of components having undergone thermal fatigue or thermal shock.

thermal compressor 1. (thermocompressor) A jet compressor used to boost the pressure of low-pressure waste or exhaust steam. **2.** A device consisting of an absorber, a generator, a pump, and a throttling device, used instead of a mechanical vapour compressor in an absorption-cooling refrigeration system.

thermal conductance *See* HEAT TRANSFER.

thermal conduction *See* HEAT TRANSFER.

thermal conductivity *See* HEAT TRANSFER.

thermal-conductivity cell An apparatus used to determine the thermal conductivity of a substance. Typically, for a fluid it consists of a sealed glass cylinder containing the fluid with an electrically-heated wire passing through it. For a solid, the material is clamped between a heated plate and a heat sink. *See also* KATHAROMETER.

thermal-conductivity vacuum gauge (thermocouple vacuum gauge) An instrument in which a thermocouple is used to measure the temperature of a heated surface immersed in a gas at very low pressure. Heat transfer is by conduction through the gas, and there is a linear relationship between the thermal conductivity of a gas and its pressure. *See also* PIRANI GAUGE.

thermal-contact resistance (interface resistance) The resistance to conduction heat transfer when two solid surfaces are in close contact. Air or any other fluid trapped in the spaces between the two surfaces acts as insulation.

thermal convection *See* HEAT TRANSFER.

thermal-diffusion effect *See* SORÉT EFFECT.

thermal diffusivity *See* HEAT TRANSFER.

thermal dissociation *See* DISSOCIATION.

thermal efficiency The ratio, for a heat engine or thermodynamic cycle, of the net work output to the net heat input into the system. *See also* OVERALL THERMAL EFFICIENCY; WORK RATIO.

thermal effusion (thermal transpiration) Effusion of a gas due to a temperature difference across a small hole.

thermal emissivity *See* EMISSIVITY.

thermal energy (heat energy) The sensible and latent forms of internal energy.

thermal-energy capacity *See* THERMAL CAPACITANCE.

thermal-energy integral equation *See* ENERGY INTEGRAL EQUATION.

thermal-energy reservoir A body with a sufficiently large thermal capacitance that it can supply or absorb finite amounts of heat without undergoing any temperature change.

thermal equation of state *See* EQUATION OF STATE.

thermal equilibrium *See* ZEROTH LAW OF THERMODYNAMICS.

thermal equilibrium constant *See* DISSOCIATION CONSTANT.

thermal-exchange coefficient *See* HEAT TRANSFER.

thermal expansion (expansivity) The increase in the dimensions, area, or volume of an object due to an increase in its temperature, usually expressed in terms of a **coefficient of thermal expansion** with unit 1/°C or 1/K. The **coefficient of linear expansion (linear expansivity, α_L)** for a solid is the proportional increase in length L corresponding to a 1°C rise in temperature T. In general,

$$\alpha_L = \frac{1}{L}\frac{dL}{dT}$$

and for small increases in temperature ΔT, the increase in length $\Delta L \approx \alpha_L L \, \Delta T$. The **coefficient of area expansion (coefficient of superficial expansion, α_A)** for a flat surface of area A is the proportional increase in A corresponding to a 1°C increase in temperature T. In general,

$$\alpha_A = \frac{1}{A}\frac{dA}{dT}$$

and for a small increase in temperature ΔT, the increase in area $\Delta A \approx \alpha_A A \, \Delta T$. For a fluid, the **coefficient of volumetric expansion (coefficient of cubical expansion, volume thermal expansivity, volume thermal expansion coefficient, $\alpha_{\mathcal{V}}$, β)** is the fractional change in volume $\mathcal{V}$ at constant pressure p caused by a change in temperature T. In general,

$$\alpha_{\mathcal{V}} = \frac{1}{\mathcal{V}}\left(\frac{\partial \mathcal{V}}{\partial T}\right)_p$$

and for a small increase in temperature ΔT, the increase in volume $\Delta \mathcal{V} \approx \alpha_{\mathcal{V}} \mathcal{V} \, \Delta T$. α_v can also be defined in terms of specific volume v as

$$\alpha_v = \frac{1}{v}\left(\frac{\partial v}{\partial T}\right)_p$$

or density ρ as

$$\alpha_v = -\frac{1}{\rho}\left(\frac{\partial \rho}{\partial T}\right)_p$$

For a solid, the constant-pressure constraint is unnecessary. *See also* BULK MODULUS.

SEE WEB LINKS

- Reference book companion website covering many material properties (thermal conductivity, density, thermal expansion coefficients, surface tension, viscosity, etc.)

thermal fatigue *See* FATIGUE.

thermal flame safeguard (combustion safeguard) A safety control used in combustion systems, such as boilers and ovens, which senses the presence or absence of flame. In the event of flame failure, the fuel supply is cut off, typically within 4 seconds.

thermal flux *See* HEAT FLUX.

thermal gradient (Unit °C/m or K/m) A term synonymous with temperature gradient, but sometimes specifically meaning the increase in temperature with depth below the Earth's surface.

thermal hysteresis A phenomenon in which temperature-dependent physical properties are different when a substance is heated through a defined temperature interval and cooled through the same interval. The final properties may differ from the initial properties.

thermal imaging *See* INFRARED THERMOGRAPHY.

thermal impedance *See* THERMAL RESISTANCE.

thermal insulation (insulation) A material that may be a low-conductivity solid, or a porous solid or fabric in which there are voids containing air or another gas, resulting in overall low thermal conductivity. When applied to the surface of a hot or cold object, surface heat transfer is generally reduced. *See also* CRITICAL INSULATION THICKNESS; LAGGING.

thermal irreversibility *See* IRREVERSIBILITY.

thermally-imperfect gas *See* PERFECT GAS.

thermally-perfect gas (thermally- and calorically-perfect gas) *See* PERFECT GAS.

thermal mass *See* THERMAL CAPACITANCE.

thermal mass flow meter (heat-loss flow meter) A gas flow meter in which the measured temperature difference ΔT produced by a known rate of heat transfer $\dot{Q}$ to a gas flowing through a pipe is used to determine its mass flow rate $\dot{m}$ from $\dot{Q} = K\dot{m}C_P\Delta T$, K being a calibration constant and C_P is the specific heat at constant pressure.

thermal-penetration depth (Unit m) In unsteady heat conduction, a length scale for the penetration of temperature variations into a material. For temperature variations with angular frequency ω, it is equal to $\sqrt{2\alpha/\omega}$ where α is the thermal diffusivity of the medium.

thermal power plant *See* POWER PLANT.

thermal process *See* THERMAL (1).

thermal pump *See* HEAT PUMP.

thermal radiation (heat radiation, radiant energy, radiation) *See* INFRARED RADIATION; HEAT TRANSFER.

thermal reactor An enlarged exhaust manifold bolted directly to the cylinder head of a piston engine, in which the oxidation of carbon monoxide and unburned hydrocarbons in the exhaust gas is enhanced.

thermal recharge The natural process, in a geothermal-energy system, of heating the ground in summer to replace heat extracted in winter.

thermal reservoir *See* RESERVOIR OF HEAT.

thermal resistance (thermal impedance, *R*) (Unit K/W) A term primarily used in the analysis of one-dimensional heat-transfer problems, defined by $\Delta T/\dot{Q}$ where ΔT is the temperature difference and $\dot{Q}$ is the heat-transfer rate. For conduction through a plane slab of material, $R_C = L/Ak$ where L is the slab thickness, A is the surface area, and k the thermal conductivity. For convection from the face of a plane slab, $R_H = 1/hA$ where h is the heat-transfer coefficient. For radiation heat transfer between two surfaces at temperatures T_1 and T_2, $R_R = \Delta T/[AF_{1-2}\sigma(T_1^4 - T_2^4)]$ where $F_{1\text{-}2}$ is a factor taking account of the emittances and geometries of the surfaces and σ is the Stefan-Boltzmann constant. For combined modes of heat transfer, the resistances are additive. *See also* FOULING FACTOR.

thermal resistivity (Unit m.K/W) The inverse of thermal conductivity.

thermal shock The partial or complete fracture of a brittle material as a result of a temperature change, usually sudden cooling. The critical shock temperature difference ΔT on the surface of a body to induce cracking depends on $k\sigma_f/E\alpha$ where k is the thermal conductivity, σ_f is the fracture strength, E is Young's modulus, and α is the coefficient of expansion. The parameter $k\sigma_f/E\alpha$ is sometimes called the **spalling resistance index** and has values of less than 1 kW/m for glasses, less than 10 KW/m for ceramics, but 100 KW/m for refractory metals such as molybdenum and nickel-based superalloys, with even larger values for high-strength steels. The extent of crack growth depends upon σ_f/K_C where K_C is the dynamic fracture toughness in the shock environment.

thermal spraying *See* ARC SPRAYING; FLAME SPRAYING; PLASMA SPRAYING.

thermal strain (ε_T) The strain that arises in a body due to heating or cooling. For a straight rod prevented from expanding or contracting that has its uniform temperature changed by an amount ΔT, $\varepsilon_T = \alpha\Delta T$ where α is the coefficient of linear expansion. The thermal strain is added to strain due to tensile or compressive stress σ, i.e. total strain $\varepsilon = \sigma/E + \alpha\Delta T$ where E is Young's modulus. **Thermal stresses** result if thermal strain is resisted and can cause cracking (**thermal-stress cracking**).

thermal time constant *See* LUMPED-CAPACITY ANALYSIS.

thermal transducer A sensor, such as a thermocouple, thermistor, or resistance thermometer, that changes a property, such as resistance, in response to a temperature change. It is the basis of many instruments used to measure temperature.

thermal transpiration *See* THERMAL EFFUSION.

thermistor A resistor, typically of sintered semi-conductor material formed into a small bead, the resistance of which varies strongly with temperature, usually decreasing. Widely used for temperature measurement in the range −90°C to 300°C with an uncertainty as low as 0.1°C.

thermoacoustic engine A device in which heat input into standing or travelling sound waves in a tube generates high-amplitude acoustic power.

thermoacoustic refrigerator A type of heat pump in which heat is extracted from high-amplitude standing or travelling sound waves in a tube.

thermoacoustics **1.** Thermal-energy exchanges driven by sound. **2.** The combination of thermal effects and sound.

thermoacoustic Stirling engine A Stirling engine in which the external heat input is provided by high-amplitude travelling sound waves.

thermocapillarity effects Fluid-flow phenomena caused by the decrease in surface or interfacial tension with temperature. Examples include Bénard convection and Marangoni convection.

thermochemical calorie An obsolete non-SI unit of energy, equal to 4.184 J.

thermocline A transition layer in a large body of water, such as a deep lake, a fjord or an ocean, between the mixed surface layer and the deep-water region in which the temperature decreases rapidly with depth. The extent is typically from 200 m to 1000 m.

thermocompression evaporator A steam-jet ejector used to raise the pressure of low-pressure steam.

thermocompressor *See* THERMAL COMPRESSOR (1).

thermocouple A temperature-measuring device in which two wires of dissimilar metals are joined together. The Seebeck effect causes a voltage difference to be produced that depends upon the temperature of the junction. Commonly-used metal pairs include copper/constantan, iron/constantan, and platinum/rhodium. *See also* THERMOELECTRIC THERMOMETER; THERMOPILE.

thermocouple pyrometer (thermoelectric pyrometer) A pyrometer in which a thermocouple is used to measure either gas or surface temperature.

thermocouple vacuum gauge *See* THERMAL CONDUCTIVITY VACUUM GAUGE.

thermodiffusion effect *See* SORÉT EFFECT.

thermodynamic cycle A linked series of thermodynamic processes undergone by a system such that the final and initial states are identical. *See also* CLOSED CYCLE; OPEN CYCLE.

thermodynamic efficiency Sometimes used to mean thermal efficiency.

thermodynamic engine *See* HEAT ENGINE.

thermodynamic equation of state *See* EQUATION OF STATE.

thermodynamic equilibrium A state in which no further changes occur within a system when it is isolated from the surroundings in such a way that no heat or work crosses the boundary. All properties must be uniform throughout the system.

thermodynamic function of state *See* THERMODYNAMIC VARIABLE.

thermodynamic potentials The thermodynamic properties Gibbs function, Helmholtz function, and specific enthalpy.

thermodynamic process A change in equilibrium state undergone by a quantity of matter or a system.

thermodynamic properties Temperature, pressure, and the intensive properties that define the state of a working fluid: specific internal energy, specific enthalpy and specific entropy. *See also* PROPERTY.

thermodynamic-property relations *See* TDS EQUATIONS.

thermodynamics *See* APPLIED THERMODYNAMICS.

thermodynamics laws The zeroth, first, second, and third **laws of thermodynamics**.

thermodynamic state The condition of a system or working fluid according to its properties. *See also* CRITICAL STATE; REDUCED STATE; STANDARD STATE.

thermodynamic surface *See* P–V–T SURFACE.

thermodynamic system *See* SYSTEM (3).

thermodynamic temperature scale A temperature scale defined in terms of a reversible power cycle operating between a high temperature T_H and a low temperature T_L for which it can be shown $T_L/T_H = Q_L/Q_H$, where Q_L is the quantity of heat added and Q_H is the amount of heat rejected. Although impractical, the defini-

t

tion has the virtue of being independent of the properties of any substance. *See also* INTERNATIONAL PRACTICAL TEMPERATURE SCALE.

thermodynamic variable (thermodynamic function of state) Any thermodynamic property.

thermoelectric converter (thermoelectric generator) A device consisting of series-connected alternate n- and p-type semiconductor elements sandwiched between two ceramic plates. Due to the Seebeck effect, electrical power is generated when a temperature difference is maintained across the plates. Due to the Peltier effect, a thermoelectric converter can act as either a **thermoelectric heater** or a **thermoelectric cooler (thermoelectric refrigerator)** by passing an electrical current through it. The **thermoelectric figure of merit (*Z*)**, with unit 1/K, is a dimensional parameter on which the efficiency of a thermoelectric device is primarily dependent, defined by $Z = \sigma\alpha^2/k$ where α is the Seebeck coefficient, σ is the electrical conductivity, and k is the thermal conductivity. The product of Z and a temperature is nondimensional.

thermoelectric junction (thermojunction) A welded, soldered, or twisted connection between the ends of two wires of dissimilar metals, as in a thermocouple.

thermoelectric material A material in which any of the thermoelectric phenomena, such as the Peltier, Seebeck, and Thomson effects, are especially strong.

thermoelectric pyrometer *See* THERMOCOUPLE PYROMETER.

thermoelectric thermometer Two thermocouples connected as shown, with two junctions held at a reference temperature T_{REF}, typically ice-water (0 °C), and the third at an unknown temperature T_U which can be determined from the voltage difference ΔV. Branches A, B, and C are made of dissimilar metals, typical choices being copper for A, platinum-rhodium for B, and platinum for C.

thermofluids A conflation of thermodynamics and fluid mechanics, a term also taken to include heat and mass transfer.

thermoforming A manufacturing process in which extruded plastic sheet is heated and formed over a heated die, often using vacuum. *See also* PLUG-ASSIST FORMING.

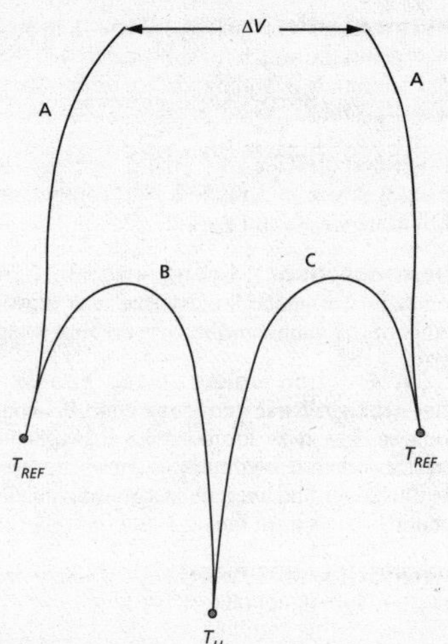

thermoelectric thermometer

thermojunction *See* THERMOELECTRIC JUNCTION.

thermometer Any instrument used to measure temperature according to a scale, such as the Celsius or absolute (kelvin) scales.

thermometric conductivity *See* HEAT TRANSFER.

thermometry The science and practice of temperature measurement. A **thermometric property** is any physical property of a material, such as the coefficient of resistance, the value of which is strongly related to temperature, making it suitable for use in thermometry. In the case of **thermometric fluids**, such as alcohol, mercury (liquids), hydrogen and helium (gases), it is the volumetric expansion properties which are used in thermometry.

thermonuclear fusion *See* FUSION.

thermophoresis *See* SORÉT EFFECT.

thermopile A device consisting of several (*N*) thermocouples connected in series, such that all reference junctions are at the same temperature and all hot junctions are at the same temperature, used for both temperature and heat flux measurement. The arrangement increases the measurement sensitivity of a single thermocouple by the factor *N*. *See also* MOLL THERMOPILE.

thermoplastic (thermosoftening plastic) *See* PLASTICS.

thermoregulator *See* THERMOSTAT.

thermoset plastic (thermosetting plastic) *See* PLASTICS.

thermosphere *See* STANDARD ATMOSPHERE.

thermosyphon A method of liquid circulation using the tendency of hot liquid to rise and cooler liquid to descend.

thermosyphon solar water heater A solar-heating system consisting of an insulated water-storage tank located above a collector panel to which it is connected. Hot water rises up the panel and into the tank, while cooler water descends from the tank into the collector.

thermostat (thermoregulator) A device that monitors temperature, in a room, a car, a building or other confined space, and switches a heating or cooling system on and off at set temperatures.

thickness gauge An instrument used to measure the thickness of a sheet of material or a coating. Micrometers and ultrasonic or eddy-current gauges are often used.

thin-cylinder theory *See* MEMBRANE STRESSES.

third-angle projection *See* PROJECTION.

third law of motion *See* NEWTON'S LAWS OF MOTION.

third law of thermodynamics (Nernst theorem) For a pure substance in thermodynamic equilibrium, the entropy approaches zero as the temperature approaches absolute zero.

thixotropic liquid A non-Newtonian liquid for which the viscosity depends upon the time of shearing as well the shear rate.

Thoma cavitation constant (σ_{TH}) A non-dimensional parameter used in pump design, defined as $(p_0 - p_V)/\rho gH$ where p_0 is the total pressure at the centreline of the pump impeller, p_V is the saturation vapour pressure of the liquid being pumped (usually water), ρ is the liquid density, g is the acceleration due to gravity, and H is the total pump head. *See also* CAVITATION NUMBER.

Thomson effect A thermoelectric effect in which heat is either generated or absorbed by a current-carrying conductor along which the temperature varies. For a cylindrical conductor of cross-sectional area A, the heat production rate is given by

$$\dot{Q} = -\frac{\mu I}{A}\frac{dT}{dx}$$

where μ is the Thomson coefficient, I is the current, and dT/dx is the temperature gradient. Joule heating is an additional heat-generation mechanism.

thread *See* SCREW.

thread forms per mm *See* SCREW.

thread gauge *See* SCREW.

thread grinding A method for producing or finishing very precise screw threads using a profiled grinding wheel.

threading die *See* DIE.

threading machine A machine used to cut an external thread on a rod, tube, bolt, etc. or an internal thread in a hole, tube, nut, etc.

thread insert (threaded bushing, helicoil®) Either a thin cylinder with an internal thread (and sometimes also an external thread) or a helical coil of wire, pressed or screwed into a hole to accept a bolt or screw. Used in material too soft or a component too thin to be threaded, to change one form of thread to another or to repair a damaged thread.

thread-plug gauge *See* SCREW.

thread representation *See* SCREW.

thread-ring gauge *See* SCREW.

thread roll A circular tool in which a thread has been machined into the outer surface.

thread rolling *See* SCREW.

thread system and size *See* SCREW.

three-degrees of freedom gyroscope A gyroscope which, when mounted in three gimbals (universally mounted), has three degrees of freedom: the spin axis, tilt axis, and veer axis.

three-dimensional flow Fluid flow or plastic flow in which there are mean velocity components in all three directions.

three-jaw chuck A self-centring chuck, the three jaws of which are moved simultaneously by a single key via spiral segments on the back of each jaw.

three-spool turbofan A turbofan engine with three concentric shafts; for example, one to carry the fan and low-pressure turbine, another the intermediate-pressure compressor and turbine, and a third the high-pressure compressor and turbine.

three-way spool valve *See* SPOOL VALVE.

threshold stress-intensity factor The stress-intensity factor below which short cracks either do not propagate, or soon arrest if they do. *See* FRACTURE MECHANICS.

threshold value The smallest change in the error that results in a change in the output of a control system.

throat The region of minimum cross section in a convergent-divergent nozzle, such as a Venturi or a convergent nozzle.

throttle A mechanism that controls the rate of flow of fuel into an engine.

throttle valve **1.** The butterfly valve of a petrol-engine throttle. **2.** **(throttling valve, expansion valve)** Any type of flow-restricting device, such as a valve or porous plug, that causes a significant pressure drop in a fluid flow (**throttling**). The process is essentially isenthalpic. Used to control the flow of refrigerant in a refrigerator. *See also* JOULE–THOMSON EXPANSION.

throttling calorimeter A device used to determine the dryness fraction of a wet vapour by measuring its temperature and pressure after throttling by an orifice in an insulated chamber.

through-wall-thickness load *See* RADIAL LOAD.

throw **1.** The linear motion of an eccentric. **2.** The swing of a lathe. **3.** *See* CRANKSHAFT.

thrower thread *See* SCREW.

thrust (thrust force) (Unit N) **1.** A compressive force in a mechanism or structure exerted by one component on another. **2.** The reaction force exerted on an aircraft, rocket, marine craft, structure, etc. due to fluid flow, especially a jet, caused by a gas turbine, rocket motor, propeller, etc.

thrust bearing A bearing designed to support a compressive axial load.

thrust collar A shoulder on a shaft that transmits an axial load to a thrust bearing.

thrust factor (*K*) A non-dimensional measure of the radial thrust F developed by a centrifugal pump, defined by $K = F/\Delta pDw$ where Δp is the pressure rise across the impeller, D is the impeller diameter, and w is the impeller width at discharge.

thrust vectoring (thrust vector control, vectored thrust) A method of changing the direction of the thrust from a propulsion device, especially a jet or rocket engine, to control the attitude of an aircraft, rocket, etc.

Thwaites' method An approximate integral method used to calculate laminar boundary-layer development for a prescribed pressure distribution.

tick over *See* IDLE.

tidal energy (tidal power, tidal range power, tidal stream power). Electrical power generated from tidal flow using a structure similar to a low dam built across an estuary and incorporating hydraulic turbines (**tidal barrage**). Power can be generated as the incoming tide flows through the turbines (**flood generation**), or the incoming tide is allowed to pass through the sluice gates with the turbines idling, the water being trapped behind the barrage by closing the sluices at high tide, and then passed back through the turbines on the outgoing ebb tide to generate power (**ebb generation**), or power is generated on both the incoming and outgoing tides (**two-way generation**). *See also* LOW-HEAD HYDROELECTRIC PLANT.

tie (tie rod) A rod in tension, as in a light framework or motor-vehicle suspension.

tight fit *See* LIMITS AND FITS.

tig welding *See* WELDING.

time constant *See* TRANSIENT RESPONSE.

time domain Analysis of a system as a function of time. Such an analysis is converted to the frequency domain when Laplace transforms are taken.

time-invariant system A system where the transfer function is constant with time such that the same input applied at different times causes the same output. In a **time-varying system** the transfer function varies with time and the same input applied at different times causes different outputs.

time of flight The time taken for a particle being transported in a fluid stream to traverse a prescribed distance, for example a number of interference fringes in a laser Doppler anemometer.

time-of-flight flow meter *See* ULTRASONIC FLOW METER.

time-temperature transformation diagram (TTT curves) A plot of temperature *vs* log time, with superimposed different-rate continuous-cooling curves on which are marked the start and finish times of different solid-state transformations at different rates, such as austenite to coarse, medium, and fine pearlite, and martensite in steels. Used to assess hardenability. Since cooling and holding at different temperatures is easier to perform experimentally than continuous cooling, isothermal transformations are often plotted instead.

timing The sequencing of events related to the operating cycle of a piston engine, in terms of the crankshaft angle, such as the opening and closing of valves, injecting fuel and, for a spark-ignition engine, igniting the mixture with a spark. *See also* IGNITION TIMING; VALVE-TIMING DIAGRAM.

timing belt (synchronous belt) A toothed belt, typically of a rubber material, that connects the crankshaft of a piston engine to the cam shaft(s) to ensure the valves open and close at precisely the right times in the engine cycle. An alternative is a **timing chain** which engages with sprocket wheels on the crankshaft and camshaft(s) of an engine, as well as any intermediate gears. The system of belts or chains and gears is termed the **timing gear**. *See also* OVERHEAD CAMSHAFT.

tip angle *See* TOOTHED GEARING.

tip cone *See* TOOTHED GEARING.

tip speed (Unit m/s) The tangential speed with which the tip of a propeller, rotor, or impeller moves, given by ΩR where Ω is the angular velocity and R is the tip radius.

tip-speed ratio (λ) The ratio of the tip speed of the blades of a wind turbine to the wind speed.

tip vortex The trailing swirling flow that occurs at the wing tip of a lifting wing due to the pressure difference between the pressure and suction surfaces.

toggle (toggle clamp) A device consisting of two pinned levers that are almost in line between two end points in the same plane. Bringing the levers into line generates large forces between the ends. Used to obtain a large mechanical advantage or as a locking mechanism.

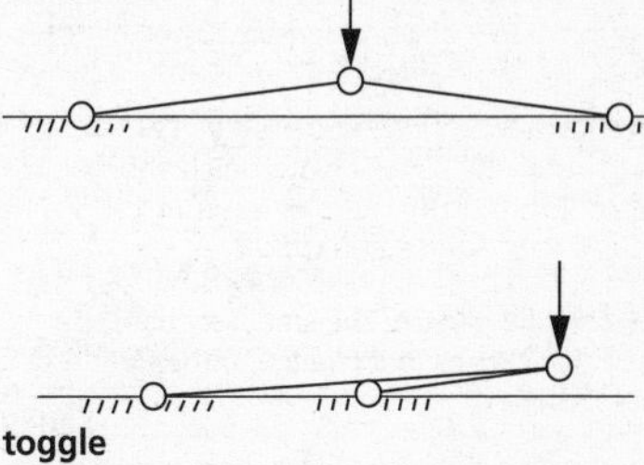

toggle

tolerances In batch, or mass, production using automatic machines, parts are intended to be completely or partially interchangeable and in the first case should ideally fit without the need for selection. In consequence, variations from exactness, termed tolerances are assigned to what is permissible on the basic size. In practice, due to tool wear, variations in raw materials, thermal effects, etc., it is impossible for mass-produced parts to have exactly the same size. In consequence, tolerances are assigned to what is permissible on the basic size. The maximum or minimum size of a dimension, shown on an engineering drawing, is termed the **limit of size**. When both are shown, the larger limit is placed above the smaller. The tolerance is the difference between the limits of size. The limits are the high and low tolerance values that give the maximum and minimum extremes of size of a component or its position in an assembly. A **unilateral tolerance** allows a deviation in only one direction, either plus or minus, from the specified dimension, while a **bilateral tolerance** specifies the amounts by which a manufactured part is permitted to vary above and below a nominal dimension, e.g. 1.234 +0.010/−0.005 mm means that the relevant dimension may be between 1.244 mm and 1.229 mm.

Designers specify tolerances to ensure proper functioning of a component. Manufacturing engineers are concerned with achieving economical production (cost of manufacture increases with greater accuracy); production engineers need to ensure components can be assembled into a unit without difficulty (cost of assembly increases with lower accuracy). It follows that there is an optimum tolerance for least overall cost. When statistical distributions of the variation in manufactured components are taken into account, including the possibility of the largest example of one part coinciding during assembly with the smallest of another, coarsening of tolerances is possible.

For non-mating parts, **general tolerances** control shape, weight, etc., and prevent fouling of other parts of an assembly. **Special tolerances** are required to control the fits between mating parts, including screw threads and gearing, or to maintain desired clearances. The degrees of accuracy that can be reasonably expected range from a few mm in the case of sand casting or flame cutting down to a few μm for reference gauges.

In addition to tolerances on limits of size or relative location, there are geometrical tolerances called errors of form, including straightness, flatness, parallelism, squareness, angularity, roundness, concentricity, cylindricity, and symmetry. *See also* LIMITS AND FITS; MAXIMUM MATERIAL CONDITION; MINIMUM MATERIAL CONDITION.

Tollmien–Schlichting instability (Tollmien–Schlichting wave) A viscous instability that develops in a laminar boundary layer above a critical Reynolds number. It is the first stage in the transition to turbulence.

tonne (metric ton) An SI-accepted, but non-SI, unit of mass equal to 1000 kg. *See also* LONG TON; SHORT TON.

tool An implement, which may be powered, that can be used for cutting, fastening, moving, shaping, etc. Examples are drills, hammers, planes, saws, and screwdrivers. *See also* ROBOT END EFFECTOR.

tool-centre point The position in a robot end effector or tool which defines the end of the robot, i.e. the position of the distal end of the position vector. Similarly in CNC machining, the programmed position of the tool without tool-length compensation.

tool changer In CNC and robotic machines, the device that changes cutting tools at prescribed stages in the manufacture of a component.

toolhead The slide-rest of a capstan lathe on which the various tools to be used sequentially are mounted.

tooling Any of the different types of cutting tools, dies, jigs, and fixtures employed in the manufacture of a component.

tool-length compensation (tool-radius compensation) The process of determining where the tool centre point must be positioned in a CNC machine tool or robot controller so that the part of the tool or end effector in contact with the work is correctly placed, i.e. compensating for the distance from the tool centre to the point of contact.

toolmaker's vice An accurately-made vice as used in a toolroom.

tool post The device by which a cutting tool is positioned and clamped on the slide rest of a lathe.

tool signature The various angles to which the cutting edges of a tool are ground, such as side and back rake angles and side and back relief or clearance angles.

toothed belt A flat belt, typically of a reinforced-rubber material, with transverse teeth that engage with teeth on a wheel or pulley.

toothed gearing In principle, friction between circular discs in tangential contact could be used to transmit rotation and power between shafts, but the power level would be very limited and the velocity ratio affected by slip. To make the drive positive, gear wheels are used instead in which **teeth** extend above and below the diameters (the **pitch-circle diameters**) of discs in ideal rolling contact. The diagram shows the nomenclature for **spur gears**, the simplest type of gearing between parallel shafts, which are cylindrical in shape with straight teeth parallel to the axis of rotation.

When two gears are in mesh, the smaller is termed the **pinion** and the larger the **spur** or **wheel**. Since motion is transmitted continuously through contact between the teeth, their profiles have to ensure no interference on engaging and disengaging, and rolling contact in between with uniform velocity ratio (**conjugate action**). **Conjugate teeth profiles** must be such that the common normal at the point of contact passes through the **pitch point P** (the point of contact between the pitch circles of two gears). Within limits it is possible to choose an arbitrary shape for one of the profiles and then determine the matching profile that satisfies the normality requirement. In practice an **involute** tooth profile is most often employed, where every tooth profile curve is an involute to the **base circle,** but **cycloidal** and other forms are possible depending upon the method of manufacture. The profile of **cycloidal gear teeth** is formed by the trace of a point on a circle rolling without

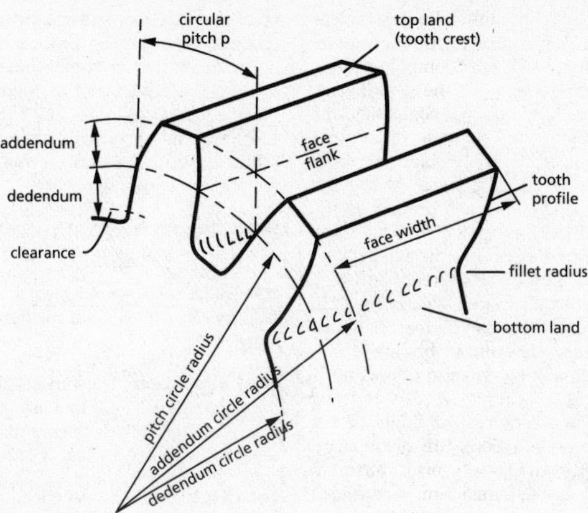

spur gear nomenclature

slipping on the outside or inside of the pitch circle. When in mesh, the faces of one are epicycloids and the flanks of the other are hypocycloids (*see* CYCLOID; INVOLUTE).

In the diagram showing the geometry of meshing teeth, $\mathbf{I_1I_2}$ is tangential to the two base circles from which the tooth profile is generated, and passes through pitch point **P**. It is the **common normal** of all points of contact of the teeth during engagement and, in the absence of friction, is the line along which the contact force between mating gear teeth acts. $\mathbf{I_1I_2}$ is variously called the **generating line**, the **line of action**, **line of contact**, or the **pressure line**. It lies at a **pressure angle** ϕ to the common tangent at the pitch point. The first and last points of contact, **A** and **B**, of mating teeth are where the addenda circles cut $\mathbf{I_1I_2}$. The path of contact **AB** is divided into **AP**, the **arc of approach**, and **PB**, the **arc of recess**.

ϕ is a design choice, but standard gears usually have ϕ equal to 20 or 25°, while 14½° was once common. As the shape of the involute depends solely on the radius of the base circle, the distance between the centres of spur gears may be changed within limits without affecting the motion transmitted, except that the pressure angle is altered. Since a pair of involute gears is in contact for only a small part of one revolution of either gear, continuous motion requires that a second pair of profiles come into contact before engagement of the first pair ends. The face of a tooth (the curved profile of the flank of a gear tooth) has involute form only beyond the pitch circle so that tooth contact below the base circle at entry and exit of engagement is non-conjugate, leading to **interference**. Use of a larger pressure angle, a greater number of teeth, and manufacture of the gear by **form generation** (see below) reduces interference.

The **module m** is the ratio of the pitch-circle diameter in mm to the number of teeth, and indicates the size of a tooth. With inches in place of mm, the reciprocal of m gives the number of teeth per inch and is called the **(transverse) diametral pitch** $\boldsymbol{p_d}$. The **circular pitch (normal pitch)** $\boldsymbol{p}$ is the distance between corresponding points on teeth measured around the pitch circle, and is related to the diametral pitch by $p_d = \pi/p$. Gears of different size having the same diametral pitch will mesh correctly. The **contact gear ratio** (**contact ratio**) is the ratio of the length of the **path of engagement** between two gears to the base pitch. There are standard proportions (in particular for m and ϕ) so that not only are gears interchangeable but also the number of cutters required for manufacture is minimized.

In place of straight-cut spur gears, the teeth may be cut on a helix (**screw gears**) to transmit motion between parallel shafts. To mate, the helices must be of different hand on each gear, but both have the same **helix angle**. For an involute tooth profile, the resulting gear surface is an **involute helicoid**. The **transverse (circular) pitch** of helical gearing is the distance

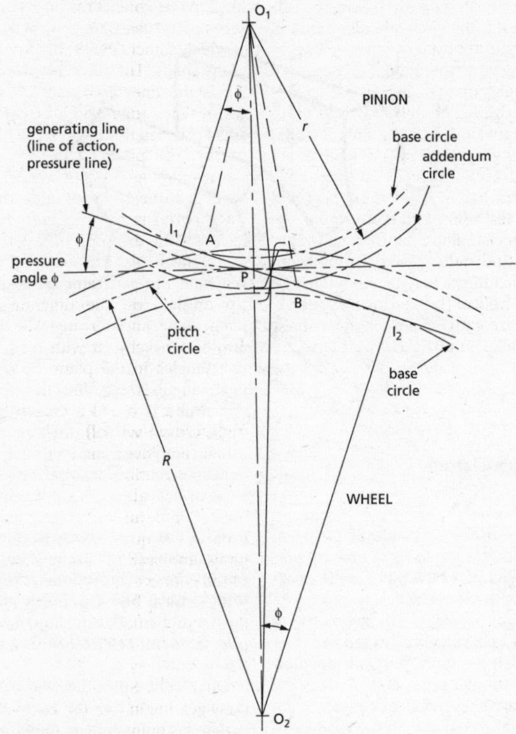

geometry of meshing teeth

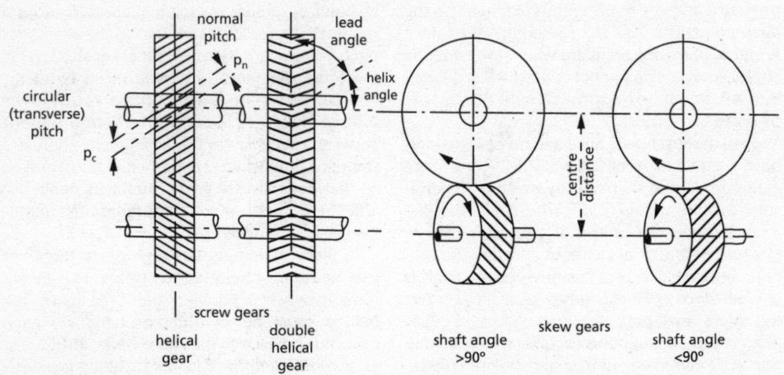

helical gears

between adjacent teeth measured along the **reference circle** that is the circle of intersection of the **reference cylinder** by a transverse plane. In contrast to spur gears, where tooth engagement can be noisy, engagement of helical gear teeth is more gradual, with smooth tooth-to-tooth transfer of load making them suitable for high speed, heavy-duty applications. **Helical gears** often have a pressure angle of 20° and helix angles no greater than 45°. Since the teeth are not parallel to the axis of rotation, there are thrust forces directed along the shaft as well as circumferential and radial forces during tooth engagement. Such forces may be eliminated by using a **double-helical** (**herringbone**) **gear.**

Bevel gears are employed to connect shafts that are not parallel, but whose axes intersect.

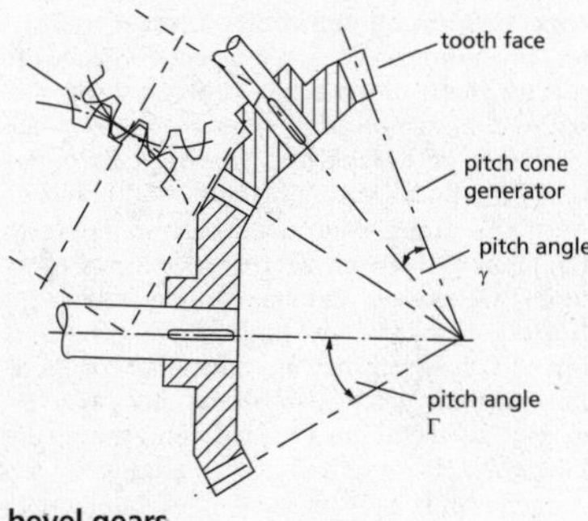

bevel gears

Ideal rolling contact results when the **pitch surfaces** of the gears are conical frustra, the apices of which coincide with the point of intersection of the shafts. The **pitch element** of bevel gearing is the line of contact between the **pitch cones** when rolling and the **pitch angle** is the angle between the axis and the **pitch-cone generator**. The **pitch circle** is the circle passing through the outermost contact point where the bevel gear meshes with another. The **tip cone (face cone)** is the imaginary surface that coincides with the tops of the teeth of a bevel gear. The angle between the axis of a bevel gear and the generator of the tip cone is the **tip angle**, and the difference between the pitch angle and tip angle is the **addendum angle.** A bevel gear with teeth that project at right angles to the plane of the gear, i.e. the pitch angle is 90°, has the appearance of a crown and is called a **crown gear (contrate gear, crown wheel)**. Involute tooth forms are usually employed and teeth may be straight cut as in spur gearing, or spiral cut (the bevel equivalent of helical gearing), the latter again being quieter and having greater load-transmitting capacity. A quasi-involute tooth form sometimes employed in bevel gearing is the **octoid**.

To connect non-parallel, non-intersecting shafts, **spiral bevel gearing** is used in which the teeth (usually involute) are not radially-orientated but angled as in cylindrical-type helical gears; they may also have sides formed of circular arcs. Spiral gearing differs from true skew gearing in that the contact between pitch surfaces is point contact, not line contact.

The **spiral angle (ψ)** is the local slope of the spiral at the mean radius of the gear and is typically about 35°. **Worm gearing** is a type of

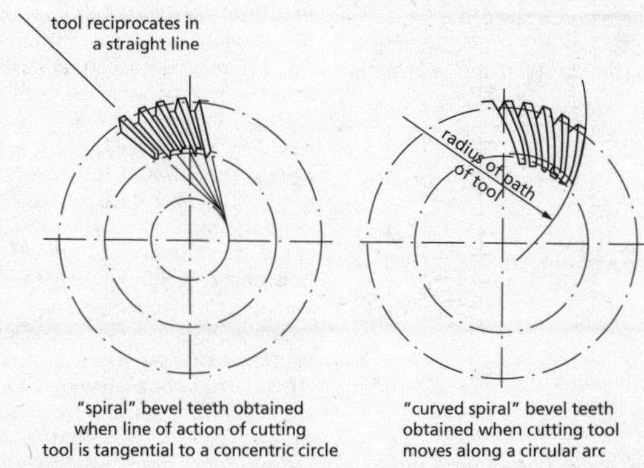

spiral bevel gearing

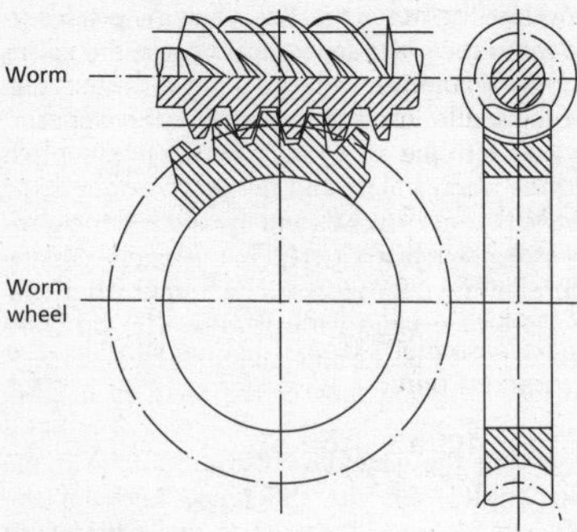

worm gearing

spiral gearing in which the teeth of the **driving wheel** (the **worm**) have a **convex involute helicoid** form and the **worm wheel** tooth profiles are concave.

The worm is a helical gear whose tooth extends over more than one revolution and resembles a multiple-start screw; the worm wheel is like an axial section of a long nut bent into a cylinder. Worm gears have a high **velocity ratio** as the worm is of small diameter relative to the wheel. Terminology for gears connecting non-parallel, non-intersecting shafts which are similar to bevel gearing, and their relationship to spiral and worm gears, is given in the diagram.

For small shaft offsets, the name **hypoid gear** is employed, so-called because the pitch surfaces are **hyperboloids of revolution**; for larger offsets where the driving pinion looks like a tapered worm, the name is **spiroid gear**.

From the driving torque T applied to a gear, the load F along the line of action is given by

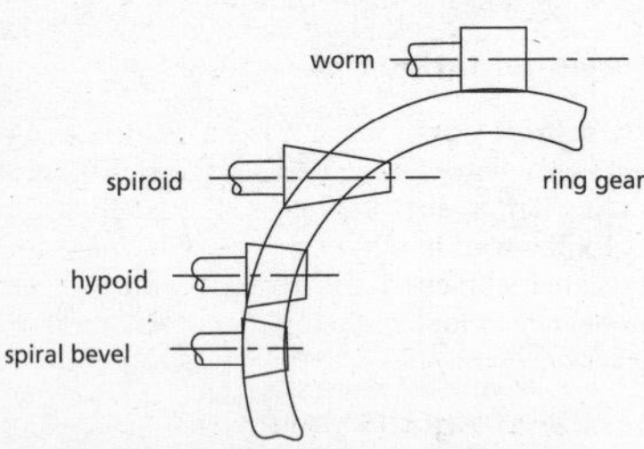

terminology for gears connecting non-parallel non-intersecting shafts

$F = 2T/d$ where d is the pitch-circle diameter; in turn, F may be resolved into tangential F_t, radial F_r, and axial F_z components. Bending and contact stresses in loaded gear teeth may be determined using empirical modifications of simple tip-loaded cantilever-beam theory and Hertz contact-stress formulae. The **Lewis equation** gives Td/Yf for the maximum bending stress in a tooth, where Y is the Lewis form factor that depends upon the number of teeth, the pressure angle, and the depth of the gear face, and f is the face width of gear. Gear teeth may then be proportioned for adequate bending stiffness and strength, contact stiffness and strength, surface wear and fatigue resistance using various empirical parameters (**life factor**, **load distribution factor**).

Gears are manufactured by a variety of methods, including casting, extrusion, rolling, or metal cutting from gear blanks with either **form cutters** or **generating cutters**. Form cutters (including profiled milling cutters) have the precise shape of the space between teeth, and cut with a reciprocating motion parallel to the axis of the gear blank. The process is discontinuous, the blank having to be turned about its own axis through an angle corresponding to the pitch of the teeth after completion of every tooth. Generating cutters (among which are **hobs** that look like a worm in which the convex involute helicoid having the same pitch as the required gear is separated into a series of cutting teeth) have a different shape but nevertheless continuously produce the proper profile by combined linear and rotary motion, and also eliminate interference as the non-involute region below the base circle is removed by **undercutting**. Since a **rack** is a spur gear of infinite radius, the sides of an **involute rack** are straight and inclined at the pressure angle ϕ to a normal to the rack. An involute rack can therefore be manufactured with great accuracy, a feature taken advantage of in gear-generating machines. Grinding is employed as a finishing operation in gear manufacture to correct any distortion of profiles after heat treatment. Depending upon the application, the dimensional accuracy of gears may be improved by shaving, burnishing, and lapping. *See also* ANTI-BACKLASH GEAR; BOLTS; COG; CONSTANT-MESH GEARBOX; CONTACT STRESSES; CONTINUOUSLY-VARIABLE TRANSMISSION; COUNTER SHAFT; ELLIPTIC GEAR (ELLIPTICAL GEAR); EPICYCLIC (PLANETARY) GEARING; FATIGUE; GEAR RATIO; GEAR TRAINS; LAY SHAFT; MARLBOROUGH WHEEL; MUTILATED GEAR; NUTS; SCREWS; SECTOR GEAR; STEPPED GEAR.

SEE WEB LINKS

- Website of the American Gear Manufacturers Association
- Website of the British Gear Association
- Website of the ESDU (Engineering Sciences Data Unit), a UK-based organization that provides validated engineering design data, methods, and software
- Gear terminology
- Gearing terminology

top-dead centre (outer-dead centre) The position of the crank in a reciprocating compressor, engine, or pump when the piston is closest to the cylinder head. *See also* BOTTOM-DEAD CENTRE.

top land *See* TOOTHED GEARING.

topping cycle *See* BINARY VAPOUR CYCLE.

torque (*T*) (Unit N.m) The twisting moment of a force or couple about an axis which results in torsion.

torque arm A bar fixed at one end used to resist torque applied at the other, for example by an electric drill. *See also* TORSION BAR.

torque coefficient 1. **(K_T)** For a propeller of diameter D, a non-dimensional parameter defined by $K_T = T/\rho\Omega^2D^5$ where Ω is the rotational speed (in rad/s) and ρ is the fluid density. **2.** **(C_T)** An alternative to (1) for a wind turbine, defined by $C_T = 2T/\rho V^2AR$ where V is the wind speed, A is the swept area of the blades, and R is the blade radius. **3.** **(K)** An empirical coefficient used to calculate the torque T required to achieve an axial load F in a bolt according to $T = KFd$ where d is the nominal bolt diameter.

torque control A system of optimizing rotor speed for a variable-speed wind turbine by controlling the torque demands of the generator.

torque converter A turbomachine used for torque amplification consisting of an impeller, a turbine, and a reaction member. Applications include motor-vehicle transmissions.

torque dynamometer *See* DYNAMOMETER.

torque meter An instrument for measuring torque using spring-loaded, piezoelectric, or strain-gauged devices.

torque motor A type of induction motor that can apply a steady torque even when prevented from rotating.

torque reaction The torque needed to counteract an applied torque. For example, in a helicopter with a single main rotor, the tendency of the fuselage to rotate in the opposite direction to the rotor. *See also* TAIL ROTOR.

torque tube In a rear-wheel drive motor vehicle, a rigid tube, through which the drive-shaft passes, that connects the rear-axle housing to the gearbox or chassis to counteract the torque transmitted to the differential. *See also* HOTCHKISS DRIVE.

torque-type viscometer *See* ROTATIONAL VISCOMETER.

torque wrench A socket wrench or ring spanner that can be set to allow a specific torque to be applied to a nut or bolt head.

torr A non-SI unit of pressure commonly used in vacuum systems; defined by 760 torr = 1 atm so that 1 torr = 133.3223684 Pa, and approximately equivalent to the pressure corresponding to 1 mm of mercury.

Torricellean barometer A vertical glass tube with its upper end sealed and the lower end submerged in a pool of mercury. The space above the mercury is evacuated. The height to which the mercury rises up the tube is a measure of the barometric pressure.

torsion The twisting of an object about an axis due to an applied couple (torque).

torsional angle (ϕ) The angular deflexion between two locations on a straight bar subjected to a torque.

torsional compliance (torsional flexibility) (Unit rad/N.m) The reciprocal of torsional stiffness, being the angular displacement resulting from the application of unit torque.

torsional damper *See* TORSIONAL VIBRATION DAMPER.

torsional fatigue *See* FATIGUE.

torsional pendulum A mass, suspended by an elastic wire, that undergoes angular oscillations when twisted and released. If the rotational moment of inertia of the mass is J and the torsional stiffness of the wire is K, the angular frequency ω for small-amplitude oscillations is given by $\omega = \sqrt{K/J}$.

torsional rigidity (Unit Pa.m^4) For a shaft subjected to torque, the product GI_P where G is the shear modulus of elasticity and I_P is the polar second moment of area of the cross section. The angle of twist per unit length, θ', produced by a torque T is given by $\theta' = T/GI_P$. The **torsional stiffness (rotational stiffness)** (with unit N.m/rad) is the torque required to produce a unit angle of rotation of a shaft and equal to GI_P/L where L is the length of the shaft.

torsional vibration Angular oscillations in a rotating shaft caused by shaft eccentricity, unbalance mass distribution, oscillatory torque, misalignment, etc.

torsional-vibration damper *See* LANCHESTER DAMPER.

torsion bar A metal bar designed to act as an elastic spring when torque is applied. *See also* TORQUE ARM.

torsion-bar suspension A type of motor-vehicle suspension in which one end of a torsion bar is rigidly fixed to the chassis while the other carries a lever to which are attached the components that carry a wheel. In a **torsion-beam suspension**, a beam connects the wheels on either side of the vehicle.

torsion damper *See* LANCHESTER DAMPER.

torsion spring **1.** A spring in the form of a torsion bar. **2.** A helical spring to which torque can be applied at the ends.

torsion test A test used to determine the shear modulus of elasticity G of a material by subjecting a cylindrical test specimen with circular cross section to torque T and measuring the torsional angle ϕ. G is then given by $G = T/I_P\phi$ where I_P is the polar second moment of area of the test specimen. Other information obtainable from such a test are the yield point and fracture data.

tortuosity A measure of the extent to which a curve or passage twists and turns; for example, connected pores in a porous medium.

total absorptance The ratio of the radiant flux absorbed by a body to that incident upon its surface, measured over all wavelengths. *See also* SPECTRAL ABSORPTANCE.

total acceleration In a fluid flow, the total-time derivative of the vector velocity $\boldsymbol{V}$ taking into account both spatial and temporal changes, i.e.

$$\frac{D\boldsymbol{V}}{Dt} = \frac{\partial \boldsymbol{V}}{\partial t} + (\nabla \cdot \boldsymbol{V})\boldsymbol{V}$$

where t is time and $\boldsymbol{\nabla}$ is the gradient operator. *See also* CONVECTIVE ACCELERATION.

total combustion air **1.** The combination of the stoichiometric flow of air required for combustion together with any excess air. **2.** The flow of fresh air into a boiler plus any flue gas recirculated.

total derivative *See* SUBSTANTIAL DERIVATIVE.

total emissivity (hemispherical total emissivity, total emittance) The emissivity over all wavelengths and directions (i.e. for a flat surface over a solid angle of 2π steradians) at a given temperature.

total energy (Unit J) The sum of all forms of energy associated with a system, including kinetic, potential, internal, magnetic, chemical, and electrical energy.

total-energy plant *See* COMBINED HEAT AND POWER PLANT.

total enthalpy *See* SPECIFIC PROPERTY.

total head (h_T) (Unit m) Total pressure p_T expressed in terms of the vertical height of a column of liquid, typically water or mercury, i.e. $h_T = p_T/\rho g$ where g is the acceleration due to gravity and ρ is the liquid density.

total-head tube *See* PITOT TUBE.

total heat An alternative term for enthalpy. It is inappropriate as it suggests that it includes kinetic energy in the same way as total enthalpy, but it does not. Also, heat and enthalpy are quite different quantities, as their definitions show.

total-loss lubrication A system in which the lubricating oil for an internal-combustion engine is burned together with the fuel. *See also* TWO-STROKE ENGINE.

total pressure (p_T) (Unit Pa) The sum of the static pressure p, the dynamic pressure $\rho V^2/2$, and the potential energy per unit volume ρgz for a fluid flow, i.e. $p_T = p + \rho V^2/2 + \rho gz$ where ρ is the fluid density, V is the flow speed, g is the acceleration due to gravity, and z is the height above a datum level. *See also* BERNOULLI'S EQUATION; STAGNATION PRESSURE.

total resource *See* AVAILABLE RESOURCE.

total solar irradiance *See* SOLAR CONSTANT.

total strain plasticity theory *See* HENCKY TOTAL DEFORMATION PLASTICITY THEORY.

total temperature *See* STAGNATION TEMPERATURE.

total time derivative *See* SUBSTANTIAL DERIVATIVE.

touch feedback Feedback from the fingers of a robot end effector used to control the force applied to an object being picked up.

touch sensor A sensor which detects contact between a robot end effector and the work, or measures the force of that contact.

toughening mechanisms Various methods of increasing the resistance to crack initiation and propagation in materials. They include transformation toughening, in which the microstructure around the crack tip alters so as to slow down or arrest cracks; deflection of cracks; various ways of de-sharpening crack tips; fibre bridging of cracks; and fibre pull-out.

toughness 1. The ability of a material to resist crack initiation and propagation. **2.** The ability of a material to absorb strain energy without fracturing. *See also* FRACTURE TOUGHNESS; RESILIENCE.

tow 1. To cause a vehicle to move by pulling with a rope, chain, bar, etc. **2.** A bundle of fibres employed in reinforcement.

towing tank A long, open tank of water, typically rectangular in shape, in which models of marine craft can be pulled to allow study of the flow around the hull, and from which the power to drive prototype craft can be estimated. *See also* FLUME.

trace heating *See* STEAM TRACING.

track rod A bar connecting the ends of the steering arms in an automotive vehicle.

tracking problem A control problem where the set point changes with time and thus the plant output must follow the changing set point. The opposite of a regulator problem.

traction (tractive force) A tangential force transmitted between two objects resulting in motion, including acceleration or deceleration, or the transmission of torque and power. *See also* AMONTONS FRICTION.

traction-control system A computer-controlled system in a motor vehicle, which prevents the driven wheels from losing traction on a slippery surface. *See also* ANTILOCK BRAKING SYSTEM.

tractive force *See* DRAWBAR PULL.

trailing edge *See* AEROFOIL.

trailing vortex A vortex downstream of a lifting wing, formed by the combination of the tip vortex and distributed vortices shed from the trailing edges of a wing.

train *See* GEAR TRAIN.

training data The data entered into a robot controller when teaching the robot a task.

trajectory The path followed by a projectile moving through the atmosphere or outer space.

trajectory control A form of continuous path control using a model reference method.

tramp elements Elements in metal alloys which are not of the intended composition but are present in very small quantities (e.g. owing to the recycling of scrap). They may affect the properties of the material.

transaxle A combined transmission and axle in a motor vehicle.

transducer A sensor that converts an input signal into an output signal of a different form. The output signal is usually electrical.

transfer efficiency The ratio of the output power from a solar collector to the net heat flowrate into the absorbing plate.

transfer function The Laplace transform from zero initial conditions of the ratio of a control-system output to the input causing that output. A **transfer matrix** is a matrix of transfer functions representing the output/input relationships for a system with multiple inputs and multiple outputs.

transfer moulding (resin-transfer moulding, RTM) A method of compression moulding polymers in which the dies are closed before the operation starts.

transfer ratio (transfer constant) A complex variable representing the ratio between the output of a transducer and the input causing that output.

transformation toughening The improvement of fracture toughness of a material by stress-induced transformation of the microstructure.

transgranular fracture Fracture in crystalline materials where the path of cracking is predominantly across grains. *See also* INTERGRANULAR FRACTURE.

transient 1. A signal of short duration. **2.** A term for non-steady conditions during the change from one steady-state (equilibrium) condition to another.

transient creep *See* CREEP.

transient response The output of a system as a function of time before the steady state is reached. Where the output is described by a linear first-order differential equation, the **time constant** is the time taken for the output to reach 63% of the final steady-state value following a rising step change in input, or 37% of the initial value following a falling step change. *See also* THERMAL TIME CONSTANT.

transition The process by which small disturbances in an unstable laminar flow are amplified and, as a result of non-linear effects, the flow becomes turbulent. In the case of a boundary layer, this occurs over a certain distance. For an internal flow, the entire flow changes from laminar to turbulent with increase in Reynolds number. *See also* DEFORMATION TRANSITIONS; DUCTILE–BRITTLE TRANSITION.

transitional flow A flow in which transition to turbulence occurs.

transitional fit *See* LIMITS AND FITS.

transitional roughness For turbulent boundary-layer and pipe flow over a rough surface, the regime where the skin-friction coefficient or friction factor has a slight dependence on Reynolds number. For pipe flow, this is when the sand roughness height k is in the range $5\nu/u_\tau$ to $70\nu/u_\tau$ where ν is the kinematic viscosity of the fluid and u_τ is the friction velocity. *See also* FULLY-ROUGH FLOW; HYDRAULICALLY-SMOOTH SURFACE.

transition temperature (transition point) 1. The temperature at which the mechanism of fracture in metal alloys having a face-centred cubic crystal structure changes from ductile void growth to brittle cleavage. *See also* GLASS TRANSITION TEMPERATURE. **2.** The temperature at which a material changes from one crystal state to another.

translating follower *See* CAM FOLLOWER.

translation (translational motion) Any change in the position of an object or particle excluding rotation.

translational joint A robot joint where the controlled variable is straight line movement.

transmissibility The ratio of the transmitted force to the disturbing force for a system subjected to a vibratory disturbance. The ratio may also be defined in terms of displacements, velocities, or accelerations.

transmission The system that transmits power and torque from a power source; for example a shaft, belts and pulleys, or a gear train. In the case of a motor vehicle, it includes the gearbox, clutch, propeller shaft, differential and final drive shafts.

transmission dynamometer A dynamometer in which the power transmitted by a shaft is obtained from the product of its rotation speed and the torque determined from measured shear strains along the shaft.

transmission electron microscope (TEM) A form of microscope in which a focussed electron beam passes through a metal specimen less than 50 nm thick and forms an image on a fluorescent screen. Magnifications above 5.10^5 are possible.

transmission error *See* LOST MOTION.

transmittance (transmissivity, τ) The fraction of incident radiant flux that passes through a semi-transparent surface. The term also applies to transmission of radiation by a volume of fluid. *See also* ABSORPTANCE; REFLECTANCE.

transonic A flow for which the Mach number M is close to unity at some location and both subsonic and supersonic regions occur. For flow over an object, it begins when the highest local Mach number reaches unity and ends when the lowest local Mach number reaches unity.

transpiration cooling Cooling of a surface exposed to very high temperatures, such as gas-turbine blades or the interior walls of a combustion chamber, by forcing a cool gas or liquid through pores in the surface into the hot region.

transport delay (transport lag) A delay in a system due to the time taken for material to move from one position to another. For example, when the temperature of the strip in a rolling mill is being controlled, the heater is necessarily prior to the rolls. There is thus a delay between a change in heat input and the resultant change in the temperature of the strip at the rolls. This effect can be modelled in a simplified way as a pure delay τ determined by the distance from the heater to the rolls L and the velocity of the strip V, i.e. $\tau = L/V$. Such a delay can be detrimental to the stability of a system to control the temperature.

transport energy *See* FLOW ENERGY.

transport of energy The flow of thermal energy through any heat-transfer mechanism, or of kinetic or potential energy by advection.

transport properties The properties of a material that account for heat, mass, and momentum transfer on a molecular scale (diffusion), i.e. thermal conductivity, diffusion coefficient, and kinematic viscosity respectively.

transverse base pitch *See* TOOTHED GEARING.

transverse direction A direction perpendicular to another direction considered to be the main direction.

transverse stability The tendency of a floating object, such as a ship, to return to the upright position when rolled by waves or other phenomena to one side. For this to happen, the metacentre must be above the centre of gravity.

transverse vibration Vibration in a plane, perpendicular to the direction of wave travel.

transverse wave *See* SHEAR WAVE.

travel The amplitude of angular or linear movement; for example, of a piston or traverse.

travelling microscope A microscope in which the optical system can move, for example along a rail, allowing the distance between points on a surface to be measured accurately.

travelling plastic hinge *See* PLASTIC HINGE.

travelling wave (progressive wave) A wave that propagates in one direction. *See also* PLANE PROGRESSIVE WAVE.

traverse A mechanism used to move a measuring probe, often under computer control, through a series of angular or linear positions at which measurements are made. The steps are usually small and precise.

tread The grooved part of a tyre that contacts the road surface.

trennschaukel (swing separator) An experimental device used to measure the thermal diffusion factor in binary gas mixtures.

trepanning tool A cutting tool designed to cut a circular groove in a material, often in the form of a circular cylinder with teeth at the end that moves axially while rotating.

Tresca criterion (Tresca yield criterion) The assumption that plastic yielding occurs when the maximum shear stress in a loaded body reaches a critical value, irrespective of other stresses present.

Tresca failure theory *See* THEORIES OF STRENGTH.

triangle of forces *See* FUNICULAR POLYGON.

tribology The study of lubrication, friction, and wear.

trifilar suspension The suspension of a body by three fine vertical wires of equal length. The polar moment of inertia of the body about a vertical axis passing through its centre of mass can be determined by measuring the frequency of small rotational oscillations about the axis. *See also* BIFILAR SUSPENSION.

triple point The point on a pressure-temperature diagram (i.e. a phase diagram) for a pure substance at which the liquid, vapour, and solid phases co-exist in equilibrium. *See also* P–V–T DIAGRAM.

triplex pump *See* SIMPLEX PUMP.

trip wire A wire stretched across the surface of a wind-tunnel model to fix transition at a chosen position.

tropopause *See* STANDARD ATMOSPHERE.

troposphere *See* STANDARD ATMOSPHERE.

trough The lowest point of a wave or of a screw thread. *See also* CREST.

Trouton ratio *See* EXTENSIONAL FLOW.

Trouton's rule An empirical observation that for many liquids the entropy of vaporization has a value of about 88 J/K.mol.

Trouton viscosity *See* EXTENSIONAL FLOW.

true strain (logarithmic strain, ε) The increment of true strain is given by $d\varepsilon = dL/L$, where L is the current gauge length of a test piece. This leads to $\varepsilon = ln(L_1/L_0)$, where L_1 is the final length and L_0 is the initial length. If n is the engineering strain given by $(L_1-L_0)/L_0$, then $\varepsilon = ln(1+n)$.

true stress (σ) (Unit Pa) The applied load divided by the current cross-section area over which it acts. $\sigma = s\ exp\varepsilon$ where s is the engineering stress given by the applied load divided by the original cross-section area over which it acts, and ε is the true strain.

truing 1. Adjusting a wheel such that its rim is concentric with its axis and its faces are in a plane perpendicular to the axis. **2.** Grinding or otherwise machining a surface to ensure it is flat.

trunnions Two protruding, coaxial, cylindrical supports, one on either side of an object, about which it can pivot.

truss A statically-determinate framework comprising one or more triangular units connected at the ends by pin joints such that the individual members are in tension or compression but carry no moments. *See also* METHOD OF JOINTS; METHOD OF SECTIONS.

***T–s* diagram** *See* TEMPERATURE–ENTROPY DIAGRAM.

tsunami A wave, or series of waves, generated in a large body of water by the sudden vertical displacement of the underlying floor. *See also* SHALLOW-WATER APPROXIMATION.

TTT curves *See* TIME-TEMPERATURE TRANSFORMATION DIAGRAM.

t

tube A long hollow cylinder of metal, plastic, or glass, which may be rigid or flexible, used to convey fluids or provide protection for cables. *See also* PIPE.

tube bundle An assembly of a number of parallel straight tubes, typically supported by plates perpendicular to the tubes, as in a heat exchanger, condenser, or boiler.

tube hydroforming Hydroforming in which a tube workpiece is fed axially into the die cavity in the manufacture of complex shapes.

tubular exchanger *See* SHELL-AND-TUBE HEAT EXCHANGER.

Tubular Exchanger Manufacturers Association, Inc *See* TEMA STANDARD.

tubular turbine A tidal-power turbine in which water flows through a turbine runner situated within a duct and connected by a long shaft to an electrical generator situated outside the duct. *See also* BULB TURBINE; STRAFLO TURBINE.

tuft A strand of wool, cotton, or other thread attached to a surface over which there is gas flow in order to reveal the near-surface flow pattern, i.e. a flow-visualization technique. Ideally, there is complete freedom of movement of the tuft at the attachment point such that the near-surface flow direction and large-scale unsteadiness are revealed.

tumbling 1. A surface-finishing process in which components together with abrasive particles are turned over in a rotating drum, causing them to rub together. **2.** A gyrocompass is said to be tumbling when it loses the ability to sense departures from a set direction owing to movement of the outer gimbal.

tuned damper A vibration damper designed to be effective at frequencies close to resonance.

tuned resonator (tuned cavity) A resonator, such as a Helmholtz resonator, designed to give a high impedance to an acoustical control signal at a specific frequency so that the device is selective.

tungsten inert-gas welding *See* WELDING.

tup The heavy metal head of a power hammer, as used in drop forging.

turbidity A reduction in transparency in an inherently transparent material due to light scattering by tiny suspended or trapped particles of solid, gas bubbles, or liquid droplets.

turbine A turbomachine in which a rotor (**turbine wheel**) or impeller is caused to rotate and convert flow energy into shaft power or thrust. *See also* GAS TURBINE; STEAM TURBINE.

turbine blades (turbine buckets) 1. The aerofoil-shaped vanes that form the rotor and stator of a gas, steam or hydraulic turbine. *See also* COMPRESSOR BLADES **2.** The cup-shaped vanes of a Pelton wheel.

turbine flow meter (axial flow meter, propeller meter) An in-line flow meter in which the rotation speed *N* of a propeller or rotor is a measure of the mass flowrate $\dot{m}$. Calibration is always necessary but with appropriate design $\dot{m}$ is closely proportional to *N* over a wide range.

turbine pump (turbopump) An axial or centrifugal pump driven by a turbine. Typically used to supply fuel to the combustion chamber of a rocket engine such as on the space shuttle.

turboblower A centrifugal or axial compressor or fan.

turbocharging A method of supercharging in which the hot exhaust gas from a piston engine is used to drive a turbine which powers the supercharging compressor. A **turbocharger** is the turbine/compressor combination. The compressor is usually of radial outflow design while radial, axial, and mixed-flow turbines are all in use. The **wastegate** is a valve that reduces the flow of exhaust gas into the turbine to limit the boost produced or overspeeding. Not shown in the diagram are bearings, oil passages, etc.

turbo coupling *See* FLUID COUPLING.

turbo/drag pump A high-vacuum (down to about 10^{-8} Pa) pump which combines turbomolecular and molecular-drag pumping elements on a single shaft, with the turbo stages pumping at pressures below 10^{-6} Pa.

turbofan engine An adaptation of a turbojet engine in which some of the shaft power is used to drive one or more large diameter fans at the front of the engine to provide a bypass stream that gives additional thrust. The diagram shows a triple-spool arrangement incorporating high-, intermediate-, and low-pressure axial turbines, an intermediate- and high-pressure axial compressor in addition to the fan. Turbofan engines for combat aircraft typically incorporate an afterburner. *See also* UNDUCTED FAN ENGINE.

turbo generator (turboset) The combination of a steam or gas turbine and an electrical generator with a single shaft or connected coaxial shafts.

turbojet engine (jet engine) A gas-turbine engine in which the exhaust gases pass through

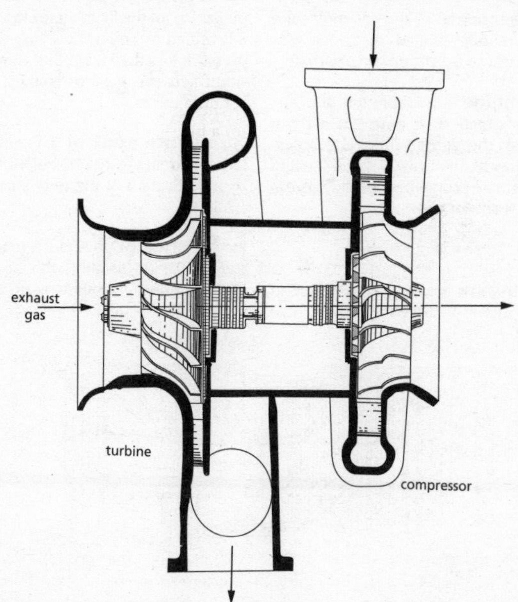

turbocharger

turbofan engine **Diagram courtesy of Rolls-Royce plc**

a jetpipe and a propelling nozzle to provide thrust. The diagram shows a twin-spool bypass arrangement. Turbojet engines for combat aircraft typically incorporate an afterburner.

turbomolecular pump A high-vacuum (down to about 10^{-8} Pa) pump in which momentum is transferred to the gas molecules by a rapidly rotating bladed disc. *See also* MOLECULAR DRAG PUMP; TURBO/DRAG PUMP.

turboprop engine An adaptation of a turbojet engine in which shaft power is used to drive a propeller, through an integral gearbox, that provides most of the thrust, a small fraction also coming from the propelling nozzle. *See also* PROPFAN; UNDUCTED FAN ENGINE.

turbopump *See* TURBINE PUMP.

turboshaft engine A gas-turbine engine in which the compressor is driven by one turbine and shaft power is generated by a second so-called free turbine. The diagram shows a twin-spool arrangement. Applications include electrical-power generation, ship, tank, and helicopter-rotor drives.

turbulence amplifier A fluidic device that makes use of the fact that for the same flow rate, a confined turbulent flow experiences a higher pressure loss than a laminar flow, and the transition between the two is achieved by a small control flow.

turbulence control The reduction of surface shear stress in a turbulent boundary layer by modification of the near-surface turbulence structure.

turbulence grid A mesh made up of wires, rods, or bars, usually with square openings, that generates turbulence in a flow passing through it.

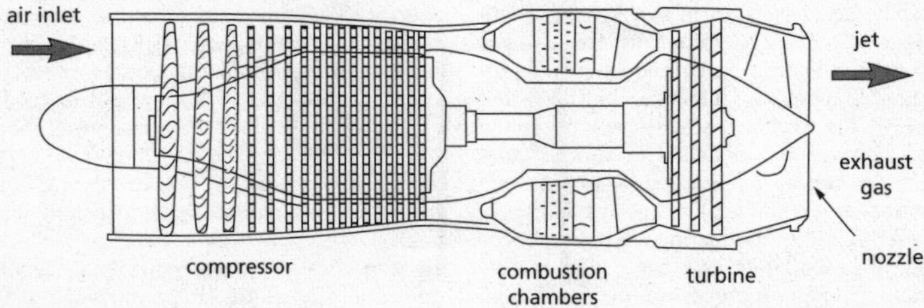

turbojet engine **Diagram courtesy of Rolls-Royce plc**

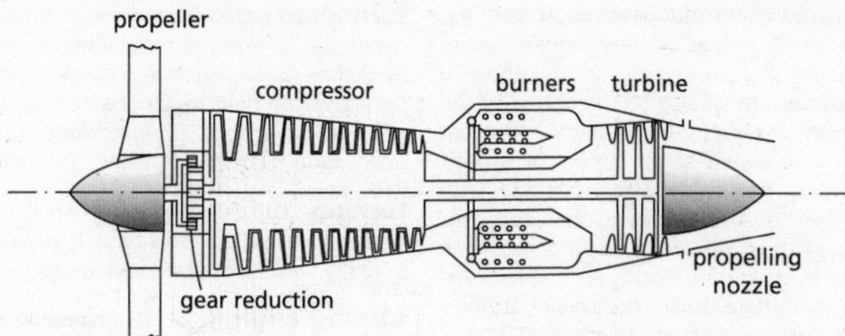

turboprop engine Diagram courtesy of Rolls-Royce plc

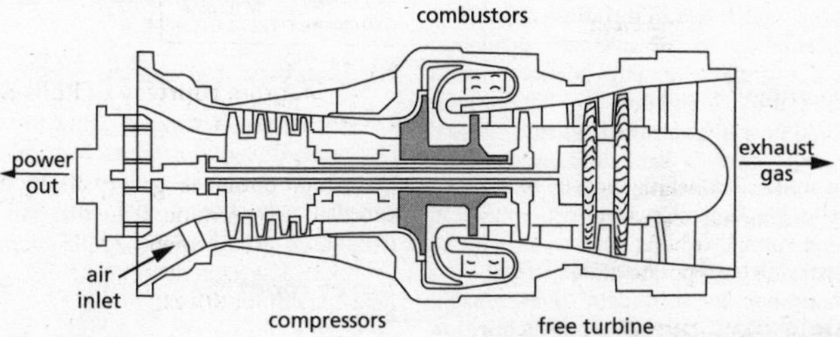

turboshaft engine Diagram courtesy of Rolls-Royce plc

turbulent diffusivity *See* EDDY VISCOSITY.

turbulent flow (turbulence) Fluid motion characterized by disorderly, rotational, (i.e. vortical) three-dimensional velocity fluctuations covering a wide range of frequency and length scales. The pressure, temperature, and other fluid properties also fluctuate and the diffusion of heat, mass, and momentum is greatly enhanced. Although governed by the Navier-Stokes equations, turbulence is so complex that it is generally accepted that complete analysis and quantification, even numerically, will never be achieved even in such simple situations as fully-developed pipe flow, except at low Reynolds numbers. As a consequence, the production, dissipation, diffusion, and pressure strain of the Reynolds stresses and turbulent kinetic energy are modelled in various ways which are largely empirical and approximate (**turbulence modelling**). Constants that arise are established by comparing calculated quantities with experiment as is the overall performance of models, a process termed validation. *See also* DIRECT NUMERICAL SIMULATION; FLUID MECHANICS; LARGE-EDDY SIMULATION.

turbulent fluctuation The seemingly random variation in time of a variable in a turbulent flow, such as velocity (u), temperature, or pressure. The **turbulence intensity** is the root mean square value, e.g. if $u' = u - \bar{u}$, u being the instantaneous value and $\bar{u}$ the time-average value, the turbulence intensity is

$$u'_{RMS} = \sqrt{\frac{1}{T}\int_0^T u'^2\,dt}$$

where t is time and T is the interval over which the average is taken. The **turbulent kinetic energy (*k*)**, with unit m^2/s^2, is the kinetic energy per unit mass associated with the intensities of the three orthogonal turbulent velocity fluctuations, u', v', and w', defined by $k = (u'^2 + v'^2 + w'^2)/2$. A conservation equation for k, including the rate of dissipation (**turbulent dissipation, ε**) due to the work done against viscous stresses by fluctuations at the smallest scales of turbulence, can be derived from the Navier-Stokes equations. Although conventionally referred to as an energy, k must be multiplied by the fluid density for it to properly represent a specific energy. The **turbulent**

normal stress (Reynolds normal stress, σ_x, σ_y and σ_z), with unit m^2/s^2, is the mean-square value of any of u', v', and w', i.e. $\overline{u'^2}$, $\overline{v'^2}$, and $\overline{w'^2}$. These quantities must also be multiplied by the fluid density for them to properly represent stresses. An apparent shear stress (**turbulent shear stress, Reynolds shear stress**) with unit Pa arises in turbulent shear flow as a consequence of the average correlation between any two of u', v', and w'. For a two-dimensional flow, the dominant turbulent shear stress is $-\rho\overline{u'v'}$, a positive quantity because u' and v' instantaneously are more likely to be of opposite sign than the same sign.

turbulent heat flux (Unit W/m^2) An apparent heat flux that arises in turbulent shear flow as a consequence of the average correlation between a turbulent velocity fluctuation and the temperature fluctuation T'. For a two-dimensional flow, the dominant turbulent heat flux is $\rho C_P\overline{v'T'}$ where ρ is the fluid density, C_P is its specific heat, and v' is the velocity fluctuation normal to the surface.

turbulent Prandtl number (Pr_t) A Prandtl number for turbulent heat-transfer, analogous to that for laminar convection but based upon the eddyconductivity k_t and the eddy viscosity μ_t. It is defined by $Pr_t = C_P\mu_t/k_t$ where C_P is the specific heat at constant pressure of the fluid.

turbulent shear flow *See* SHEAR FLOW.

turbulent viscosity *See* EDDY VISCOSITY.

turbulator **1.** A device, typically consisting of small baffles, angular or spiral metal strips, or coiled wire, used to promote turbulence in tubes, such as in a firetube boiler, to enhance the heat transfer rate. **2.** A device, such as a tripwire or row of vortex generators, used to promote transition to turbulence in a boundary layer, such as on an aerofoil to delay separation.

Turgo turbine A development of the Pelton turbine for medium-head applications, in which the double buckets are replaced by single shallower ones with water entering on one side and leaving on the other.

turnbuckle A device for adjusting the tension in a cable, rope, structural element, etc. It consists of two threaded eyelets screwed into opposite ends of a metal loop. The threads at opposite ends have opposite hands, so that rotating the loop either tightens or loosens the item under tension.

turndown ratio (turndown) **1.** The ratio of maximum to minimum flowrates over which a fluid-flow device, such as a steam desuperheater, valve, or flow meter, can operate. **2.** The ratio of maximum to minimum firing rates over which a boiler or burner can operate.

turning The rotation of a workpiece held against a cutting tool in a lathe to produce components with a circular cross section.

turning angle (θ) The change in direction experienced by a gas flow passing through an oblique shock wave or Prandtl–Meyer expansion fan.

turning-block linkage *See* WHITWORTH QUICK-RETURN LINKAGE.

turning pair *See* KINEMATIC PAIR.

turret lathe A capstan lathe in which the capstan is driven mechanically or hydraulically to provide assigned tool paths to perform a specified sequence of operations in the repetitive production of parts.

turret robot *See* CYLINDRICAL-COORDINATE ROBOT.

tuyere A duct or nozzle through which pressurized air is supplied to a blast furnace or hearth, such as in a forge.

twin-camshaft engine A piston engine equipped with separate camshafts to operate the inlet and exhaust valves of a bank of cylinders.

twin-cylinder engine A piston engine with two cylinders powering a single crankshaft; commonly used for motorcycles.

twin-shaft turbojet *See* TWO-SPOOL TURBOJET.

twist The helix produced in a cylindrical component, such as a shaft, wire, tensioned cable, or rope, when one end is rotated relative to the other. Measured either as the number of turns per unit length, or by the helix angle (**twist angle**).

twist drill A hardened-steel drill bit having one or more helical flutes running from a conical tip to the smooth part of the shank.

two-cycle engine *See* TWO-STROKE ENGINE.

two-degrees-of-freedom gyro A gyroscope mounted in two gimbals, providing two axes about which it can precess.

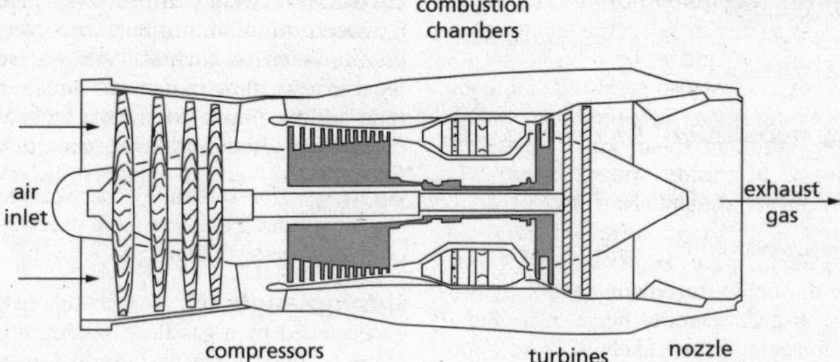

two-spool turbojet engine

Diagram courtesy of Rolls-Royce plc

two-dimensional flow A fluid flow in which the velocity at any time depends upon two spatial coordinates, such as the radial and axial locations in developing pipe flow.

two-phase flow A flow in which two phases are present, for example gas bubbles in a liquid, liquid bubbles in another liquid with which it is immiscible, solid particles in a liquid or gas. *See also* MULTI-PHASE FLOW.

two-port system *See* SINGLE INPUT SINGLE OUTPUT.

two-row turbine stage An impulse steam-turbine stage in which there are two rotors separated by a stator. *See also* CURTIS STAGE.

two-spool turbofan A turbofan engine with two concentric shafts; for example, one to carry the fan and low-pressure turbine, the other the high-pressure compressor and turbine.

two-spool turbojet (twin-shaft turbojet) A gas-turbine engine with two concentric shafts; for example, one to carry the low-pressure compressor and turbine, the other the high-pressure compressor and turbine.

two-stage compressor A machine in which gas is compressed from low pressure to an intermediate pressure in a low-pressure cylinder, and then to final pressure in a high-pressure cylinder. Efficiency is improved if the two cylinders are separated by an intercooler.

two-start thread *See* SCREW.

two-stroke engine (two-cycle engine) A petrol or diesel engine in which an air/fuel charge is introduced through an induction port, compressed and burned, expanded and then exhausted through an exhaust port. The two ports in the cylinder wall are opened and closed by the piston. There are two strokes in each revolution of the crankshaft. In total-loss lubrication petrol engines, the lubricating oil is mixed with the fuel.

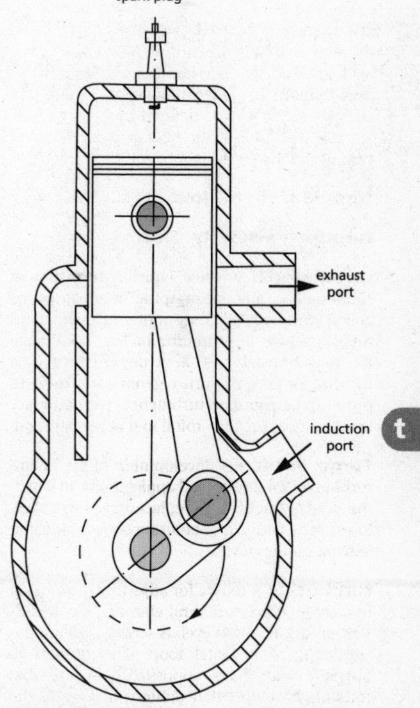

two-stroke engine

SEE WEB LINKS

- Animation and explanation of the principle of a two-stroke engine

two-way generation *See* TIDAL ENERGY.

two-way spool valve *See* PISTON VALVE.

type *See* SYSTEM TYPE.

tyre A flexible ring-shaped cover mounted on a wheel rim that supports the weight of a vehicle, such as a motor vehicle, aircraft, or bicycle, and transmits power and torque to a road surface. Some pneumatic tyres are inflated directly, others have an inner tube that is inflated. For certain applications, solid tyres are used.

t

U-bend A 180° bend in a pipe.

Ubblehode viscometer (Ubblehode suspended-level viscometer) A type of U-tube capillary viscometer similar to the Ostwald viscometer but in which the lower end of the vertical capillary is maintained at atmospheric pressure.

U-bolt A rod threaded at both ends and bent into a U-shape. It is used for clamping.

ullage The empty volume at the top of a container partially filled by liquid.

ultimate analysis *See* PLASTIC DESIGN.

ultimate-cycle method *See* ZIEGLER AND NICHOLS METHOD.

ultimate load (ultimate-load design) *See* PLASTIC DESIGN.

ultimate strength (ultimate tensile strength, UTS) *See* TENSILE STRENGTH.

ultrafiltration A filtration technique in which molecules larger than the pore size of a membrane are retained on its surface and may form a concentrated gel layer.

ultrasonic A term referring to any process or device in which ultrasound plays a key role.

ultrasonic atomizer An atomizer in which liquid is ejected through small orifices in a spray nozzle to form droplets. The droplets then pass through an intense high-frequency sound field, where they form a fine spray.

ultrasonic cleaning A method of cleaning in which items are immersed in a tank of water or solvent that is vibrated at ultrasonic frequencies.

ultrasonic concentration analyser A device that determines the concentration of substances diluted or mixed into a fluid by measuring the speed of sound in the fluid at ultrasonic frequencies.

ultrasonic cutting (ultrasonic machining) Cutting of materials in which a blade is vibrated at ultrasonic frequencies, usually 20 or 35 kHz. In **ultrasonic drilling**, a drill is percussed at ultrasonic frequencies. For brittle materials, abrasive particles (typically in slurry form) are placed between the workpiece and the ultrasonic tool to do the actual cutting.

ultrasonic depth finder (ultrasonic thickness gauge) A device that estimates the depth of a liquid by measuring the time for ultrasonic waves to propagate to the bottom and back. Used similarly to determine the thickness of solid objects.

ultrasonic flaw detector (ultrasonic testing) The use of ultrasonic waves in the non-destructive determination of defects in bodies and their location by means of reflected ultrasonic waves, and the determination of layer thicknesses.

ultrasonic flow meter (time-of-flight flow meter) A device that estimates the average velocity in a flow between two ultrasonic transducers by measuring the time difference for ultrasonic waves propagating against and with the flow.

ultrasonic leak detector A device used to detect leaks by analysing high-frequency sound compared to low-frequency background noise.

ultrasonic machining *See* ULTRASONIC CUTTING.

ultrasonics The science and application of ultrasonic waves.

ultrasonic sealing *See* ULTRASONIC WELDING.

ultrasonic separation A technique using an ultrasonic standing wave field for the separation of small particles from a fluid.

ultrasonic testing *See* ULTRASONIC FLAW DETECTOR.

ultrasonic thickness gauge *See* ULTRASONIC DEPTH FINDER.

ultrasonic transducer A device that converts an electrical input to ultrasound or vice versa, typically using a piezoelectric or magnetostrictive material.

ultrasonic welding (ultrasonic sealing) Welding of thermoplastic materials in which the required heat is generated by application of ultrasonic waves to the joint.

ultrasound (ultrasonic waves) Inaudible sound waves, with frequencies greater than about 20 kHz up to 10^{13} Hz.

ultraviolet radiation Electromagnetic radiation with wavelengths in the range of about 10 nm to 400 nm, so shorter than those of visible light but longer than those of X-rays.

unavailable energy In thermodynamics, energy converted by an irreversible process into a form that cannot do work. It is the difference between the total energy of a system and the exergy of that energy.

uncertainty analysis (error analysis) A method of estimating the uncertainty in a measurement, taking into account the uncertainties of all contributing factors including instrument errors due to calibration, random fluctuations, and resolution. If the measured quantity F depends upon individual measurements $A, B, C\ldots$ etc. according to the relation $F = A^aB^bC^c\ldots$ where a, b, c etc. are known constants, then the overall uncertainty in F can be calculated from

$$\frac{\delta F}{F} = a\left|\frac{\delta A}{A}\right| + b\left|\frac{\delta B}{B}\right| + c\left|\frac{\delta C}{C}\right| + \ldots\ldots$$

where $\delta A/A, \delta B/B, \delta C/C$, etc. represent the individual uncertainties in the measurements of A, B, C, etc. As can be seen, if $a = 2$ the error contribution is doubled, if $a = 0.5$, it is halved, etc. An alternative formulation is

$$\frac{\delta F}{F} = \sqrt{a\left(\frac{\delta A}{A}\right)^2 + b\left(\frac{\delta B}{B}\right)^2 + c\left(\frac{\delta C}{C}\right)^2 + \ldots\ldots}$$

SEE WEB LINKS

- Essentials of expressing measurement uncertainty

uncouple To separate two or more objects or processes. *See also* DECOUPLE.

UNC thread *See* SCREW.

undamped forced response The vibrational behaviour of an undamped system when subjected to a forcing frequency.

undamped natural frequency *See* NATURAL FREQUENCY.

undercarriage *See* LANDING GEAR.

undercut A recess or groove machined, moulded, or etched into the inside or outside of a component.

undercutting *See* TOOTHED GEARING.

underdamped system A system that oscillates at a frequency lower than the undamped case, with amplitudes gradually decreasing to zero.

under-expanded jet A supersonic gas jet exhausting from a nozzle into an environment where the pressure is lower than the pressure in the exit plane of the nozzle. Expansion waves form at the exit, followed by a series of oblique shocks and expansion waves confined within the boundaries of a sausage-like jet. *See also* OVER-EXPANDED JET.

underflow *See* HYDROCYCLONE.

undershoot A term used where the output of a system subjected to a change in input briefly drops below the eventual steady-state value or, for a control system, is below the set point.

undersquare engine A piston engine in which the cylinder bore is less than the stroke.

understeer The undesirable situation that arises when the set up of the steering gear, suspension, tyres, etc., of a vehicle results in a turn of larger radius than intended. *See also* OVERSTEER.

unducted fan engine (propfan engine) A type of turbofan engine with the fan situated outside the engine nacelle.

UNF thread *See* SCREW.

uniaxial stress A state of stress in one direction only, i.e. where the other two principal stresses are zero.

unidirectional ply A layer in a reinforced-composite material where all filaments are aligned.

unified atomic mass unit A unit defined as $1/12^{th}$ the mass of one atom of the isotope carbon-12. *See also* MOLECULAR MASS.

u

unified screw thread *See* SCREW.

unifilar suspension The suspension of a body by a single filament.

uniform circular motion Motion around a circle at constant angular velocity.

uniform load (uniform stress) The state where the load (stress) on a component or structure is distributed evenly over the affected region.

unilateral tolerance *See* TOLERANCE.

union (union joint) A threaded pipe fitting that allows two pipes to be connected and detached without the need for either to be rotated, and without damaging the pipe ends.

unipolar drive A method of switching the current in the coils of a stepper motor such that only one coil is energized at any given time. *See also* BIPOLAR DRIVE.

uniqueness of solution In elasticity theory, there is only one solution corresponding to given surface and body forces when a loaded body is initially free of residual stresses.

unison ring In a turbomachine with adjustable guide or stator vanes, the ring to which a set of vanes is linked such that when the ring is rotated, all vane angles are adjusted.

unit A numerical indication of the magnitude of a physical quantity. The basic units of mass, length, time, etc. are relative to agreed standards. *See also* DIMENSIONS; INTERNATIONAL SYSTEM OF UNITS.

unitary air conditioner An air-conditioning system in which all the components needed to heat, cool, dehumidify, filter, and transport air are incorporated in one or more factory-made assemblies. In a split system, the fan, heating, and cooling coils are located inside the building while the condenser and compressor are outside.

unit heater A heat exchanger in which air is blown over electrically-heated bars or tubes through which there is a flow of hot fluid. Used in the heating, ventilating, and air-conditioning industry.

unitized body *See* MONOCOQUE STRUCTURE.

unitized tooling Tooling (usually for sheet-metal operations) that is compatible with adjacent stages of forming, thus enabling easy transfer between operations.

unit step function A function having value unity for times t given by $0 \leq t \leq \infty$ and zero, otherwise used as a common test signal for control systems.

units used in dimensioning The dimensions of an object (lengths, radii, etc.) on an engineering drawing are normally shown in mm or m.

unity feedback A control system where the feedback signal applied to the summing junction is equal to the system output, no mathematical operations being applied to the signal.

universal dividing head A device that rotates a workpiece on a machine tool through known angles in between cuts.

universal gas constant (molar gas constant, $\mathcal{R}$) The gas constant for one mole of any gas with the value 8.31451 J/mol.K or 8.31451 kJ/kmol.K. *See also* PERFECT GAS.

universal grinding machine A grinding machine having a rotatable base.

universal gripper A robot end effector designed to be capable of holding most types of components.

universal joint (Hooke's joint) A double-pivoted connection that allows power and torque to be transmitted between two shafts at an angle to each other. For constant driving angular velocity of the input yoke, the angular velocity of the ouput yoke fluctuates by amounts depending on the angle of intersection of the shafts. Speeds of the driving and driven shafts may be made identical (giving a constant-velocity joint) when an intermediate shaft, at each end of which there is a universal joint, is interposed. The driving and driven shafts must be equally inclined to the intermediate shaft, the two forks of which must lie in the same plane.

universal vice A vice mounted on swivels, enabling a workpiece to be orientated at any angle.

unrelaxed modulus (Unit Pa) In a visco-elastic material, the modulus observed from the load-displacement diagram when loading

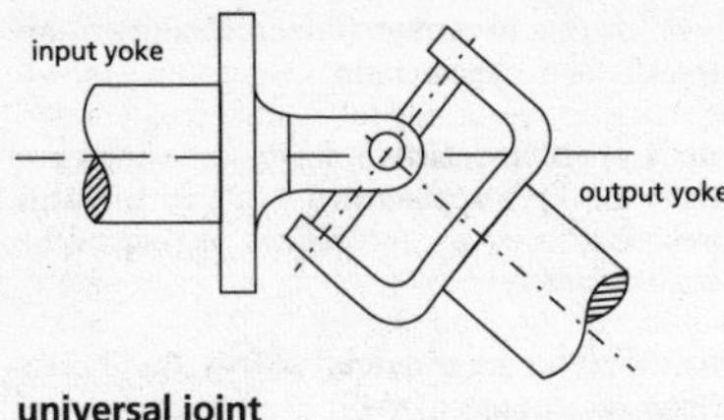

universal joint

is so rapid that time-dependent processes have insufficient time to act. *See also* DYNAMIC MODULUS.

unresisted expansion *See* FREE EXPANSION.

unsprung axle An axle of a motor vehicle which is not suspended by the vehicle springs and therefore is sprung only through the tyres.

unsprung mass (unsprung weight) (Unit kg (N)) Those parts of a vehicle suspension, such as axles, brakes, and wheels, not held by the springing. *See also* SPRUNG MASS.

unstable equilibrium The condition where a body, slightly displaced from its initial position, continues to move further from it. *See also* NEUTRAL EQUILIBRIUM; STABLE EQUILIBRIUM.

unstable flow A laminar flow in which infinitesimal disturbances are not damped but grow, leading to transition to turbulence or the development of a well-defined flow pattern such as Taylor vortices in Couette flow or counter-rotating vortices in Bénard convection. Unstable flow occurs when the governing flow parameter (e.g. the Grashof, Rayleigh, Reynolds, or Taylor number) exceeds a critical value.

unstalled flow The flow in an adverse pressure gradient, such as in a diffuser or over the suction surface of a lifting aerofoil, where the flow does not separate.

unsteady flow A flow in which the flow velocity, pressure, etc. at any point in the flow varies with time.

updraught carburettor *See* CARBURETTOR.

updraught furnace A furnace in which cold combustion air is introduced at the bottom.

upmilling (conventional milling) Cutting where the peripheral speed of the cutters of a rotating cylindrical tool in contact with the workpiece and the linear feed of a workpiece have opposite directions. It results in chips that are thin initially, but become thicker. *See also* DOWN MILLING.

upper-bound theorem A kinematically-admissible displacement field for a loaded body or structure is one that satisfies displacement compatibility but does not satisfy equilibrium, and for which the predicted stiffness is smaller than the exact stiffness. The theorem states that loads which produce kinematically-admissible displacement fields in a body or structure are equal to, or greater than, the true load that will produce plastic flow. Thus an overestimate of working loads in plasticity (an **upper bound**) is obtained by employing such a field. The lowest upper bound of a series of displacement (velocity) fields will be closest to the true load (work rate). *See also* LOWER BOUND.

upper consolute temperature (upper critical solution temperature) *See* CONSOLUTE TEMPERATURE.

upset forging Compression of parts of a workpiece to form, for example, bolt heads.

upstream Flow approaching an object or location is said to be upstream of it. *See also* DOWNSTREAM.

upthrust *See* AEROFOIL.

useful energy That part of final energy that is available after final conversion for the intended use. For example, in final conversion, electricity becomes light, mechanical energy, or heat.

utilization factor *See* CAPACITY FACTOR.

UTS *See* TENSILE STRENGTH.

U-tube A device consisting of two vertical tubes connected at either the top or bottom by a length of tubing such that a U-shape is formed.

U-tube heat exchanger A shell-and-tube heat exchanger in which tubes in the form of a U are connected to inlet and outlet plenums and typically supported by baffles that also determine the shell-side flowpath. *See also* SHELL-AND-TUBE CONDENSER.

valve gear Any of various mechanisms by which valves in a piston engine are opened and closed at the correct time in a cycle.

valve guide The cylindrical hole, usually containing a bushing, in the cylinder head or block of a piston engine, which aligns the stem of a poppet valve.

valve head The mushroom-shaped disc of a poppet valve containing the sealing face.

valve insert A hard-wearing alloy-steel valve seat pressed into the head of a high-performance piston engine.

valveless engine *See* PULSE-JET ENGINE.

valve lifter 1. A device for compressing the valve springs of piston engines in order to permit valve removal. **2.** A similar device employed on motor-cycle engines when stopped on top-dead centre to enable the crankshaft to be rotated, and hence the engine started. **3.** *See* TAPPET.

valve overlap In a reciprocating engine, the crankshaft rotation between the opening of the inlet valve and the closing of the exhaust valve.

valve plug A conical or cylindrical plug, in which there are transverse holes, in a valve, such as a plug valve. Flow through the holes occurs when the plug is rotated.

valve port Any opening in a valve through which fluid can flow.

valve positioner A pneumatically powered mechanical amplifier which increases the force available to actuate a valve.

valve rocker *See* ROCKER ARM.

valve seat The surface of a valve against which the moveable valve head contacts in order to provide sealing.

valve spring The spring that restores a valve to its closed position after having been opened, and is also intended to prevent valve bounce.

valve stem (valve rod, valve spindle) The rod connected to the valve head through which motion is given to a valve. It is usually integral with the valve head, but may be connected separately.

valve-timing diagram For a piston engine, a diagram showing the crank angles relative to top- (TDC) and bottom-dead centre (BDC) at which the valves open and close. In the diagram, I refers to the inlet valve and E to the exhaust valve, O to opening and C to closing. Lead refers to a valve opening before either BDC or TDC, and lag to closing after BDC or TDC. Overlap is the condition where the inlet and exhaust valves are open at the same time. *See also* TIMING.

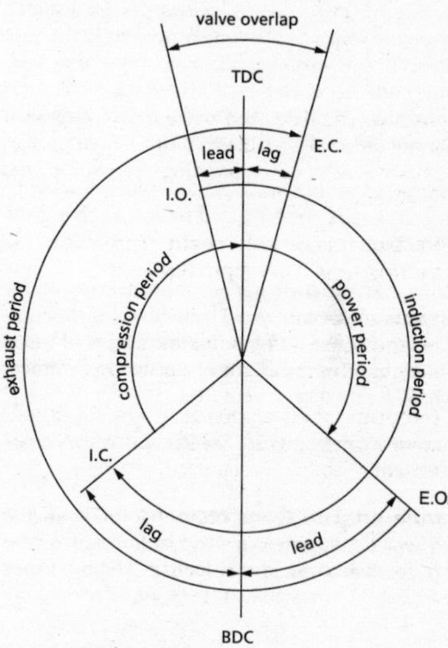

valve-timing diagram

valve train The operating mechanism for valves, including the valves themselves.

van der Waals gas (van der Waals equation) A model equation of state for a real gas given by

$$p = \frac{\rho RT}{1 - b\rho} - a\rho^2$$

where p is the gas pressure, ρ is its density, T is its absolute temperature, and R is the specific gas constant. For a given gas, the constants a and b are obtained in terms of the critical temperature and pressure by fitting the equation to the isotherm passing through the critical point.

van der Waals surface-tension formula An equation for the dependence of the surface tension of a liquid σ on its absolute temperature T, given by

$$\sigma = \sigma_0 \left(1 - \frac{T}{T_C}\right)^n$$

where T_C is the critical temperature, σ_0 is a constant for each liquid dependent on T_C and the critical pressure, and n is an empirical factor.

van Driest damping function A near-wall modification of the formula $l = \kappa y$ for the mixing length l in turbulent flow, given by $l = \kappa y[1 - exp(-y^+/A)]$ where y is the distance from the wall, κ is the von Kármán constant, y^+ is the non-dimensional distance from the wall, $u_\tau y/\nu$, u_τ being the friction velocity and ν the kinematic viscosity of the fluid. The empirical constant A is given the value 26.

vane 1. *See* BLADE. **2.** A flat or curved sheet of material used to deflect a flow and so change its direction.

vane anemometer A simple instrument for determining wind speed by measuring the rotation speed of a number (typically four) of vanes at the ends of radial arms attached to a vertical shaft. Calibration is required.

vane compressor *See* SLIDING-VANE COMPRESSOR.

vane engine (vane motor) A rotary engine in which high-pressure hydraulic fluid in the spaces between spring-loaded sliding vanes held in an offset rotor acts on the vanes, causing the rotor to revolve within a cylinder. The design is much like a sliding-vane compressor, as is that of a **vane pump**, used to pump liquids.

vane-type separator *See* BAFFLE-TYPE SEPARATOR.

vaporization *See* EVAPORATION.

vaporization cooling The cooling associated with the heat required to vaporize a liquid. *See also* LATENT HEAT OF VAPORIZATION.

vapour The gas-like phase of a substance at a temperature below its critical point. A vapour can be condensed to a liquid or a solid by increasing its pressure or reducing its temperature. *See also* SATURATED VAPOUR; SUPERCOOLED VAPOUR; SUPERHEATED VAPOUR; UNSATURATED VAPOUR; WET VAPOUR.

vapour-absorption refrigeration cycle (absorption cycle) A refrigeration cycle in which the circulating refrigerant is dissolved in a liquid at low temperature. The most widely used refrigerants in an absorption cycle are ammonia with water as the absorber, as shown in the diagram, or water as the refrigerant with lithium bromide as the absorber. The solution

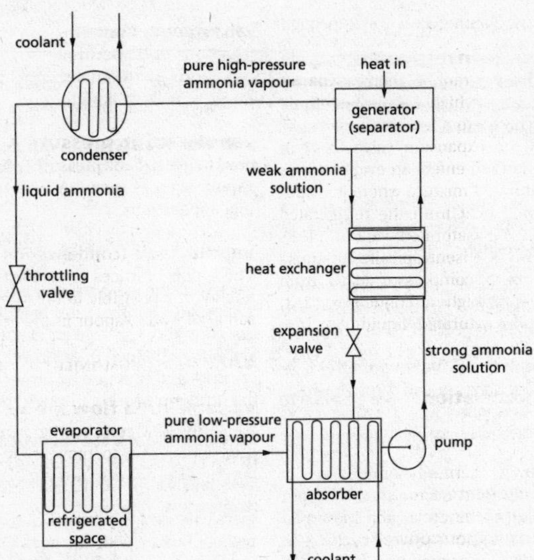

vapour-absorption refrigeration cycle

V

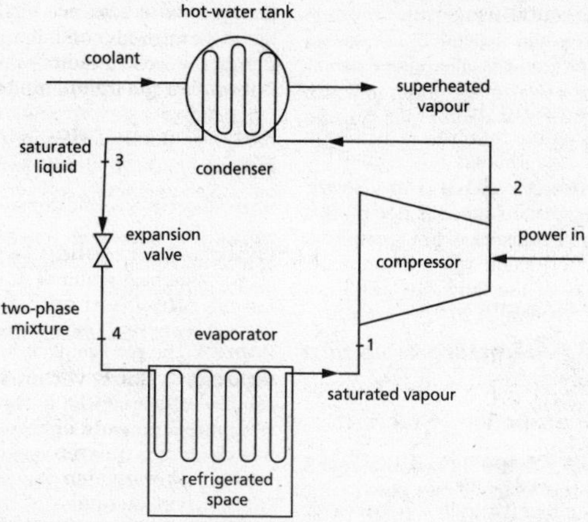

vapour-compression refrigeration cycle

is pumped to a higher pressure and heated to release the refrigerant vapour. The vapour is condensed, passed through a throttling valve and finally an evaporator. Condensation (2–3), throttling (3–4) and evaporation occur in conventional components, while the absorber, pump, generator, heat exchanger, and expansion valve replace the compressor in a refrigeration cycle.

vapour-compression refrigeration cycle As shown in the diagram, a thermodynamic refrigeration cycle in which a saturated-liquid refrigerant is throttled to a low temperature by passing through an expansion valve (3–4) or capillary tube and then enters an evaporator as a low-quality saturated mixture where it evaporates by absorbing heat from the refrigerated space and leaves as saturated vapour (4–1). It is then compressed isentropically to superheated vapour in a compressor (1–2) from where it passes through a condenser (2–3) which it leaves as saturated liquid. *See also* REFRIGERATION CYCLE.

vapour concentration *See* SPECIFIC HUMIDITY.

vapour cycle A thermodynamic cycle in which the working fluid occurs in the vapour phase during part of the cycle, and the liquid phase in the rest. A **vapour power cycle**, such as the Rankine cycle, is a vapour cycle in which power is produced.

vapour lock A condition where fuel evaporates in a fuel line, forming a vapour bubble, and so reducing or stopping the fuel flow to an engine.

vapour–pressure curve For a pure substance, the curve of saturation pressure plotted *vs* saturation temperature.

vapour-pressure thermometer (vapour-filled thermometer) A type of fluid-expansion thermometer in which the working fluid is a volatile liquid.

vapour static pressure A pressure analogous to hydrostatic pressure, where the fluid is a vapour such as steam. A term used in geothermal applications.

vapour trail (condensation trail, contrail) The trailing vortices or exhaust flow from a jet aircraft, made visible in the sky due to condensation of water vapour in the exhaust gases.

VARI *See* VACUUM-INJECTED MOULDING.

variable-area flow meter A flow meter in which the flow cross section increases with flow rate, either due to movement of a float or a sprung disc. *See also* ROTAMETER.

variable-area propelling nozzle A propelling nozzle with an adjustable outlet area required for a turbojet with an afterburner to cope with the wide range of mass flow rates.

variable displacement pump A pump designed in such a way that the flow rate can be varied at a fixed speed by altering the geometry; for example, a swashplate pump in which the piston stroke can be varied by changing the swashplate angle.

variable-frequency drive An electric motor where the rotation speed is determined by controlling the frequency of the applied AC voltage.

variable gear *See* SECTOR GEAR.

variable-head flow meter *See* BERNOULLI OBSTRUCTION FLOW METER.

variable-pitch propeller *See* PROPELLER.

variable-speed drive (variable-speed gear) A mechanism which allows the velocity ratio between driving and driven shafts to be continuously varied.

variable-stroke engine A piston engine in which the stroke can be adjusted; for example, by using a linkage between the big end and the crankshaft.

variance (s^2, σ^2) The variation in a data set from the mean value. For a set of N values of a quantity x_i with mean value:

$$\bar{x} = \frac{1}{N}\sum_{i=1}^{N} x_i,\ \sigma^2 = \frac{1}{N}\sum_{i=1}^{N} (x_i - \bar{x})^2$$

If $1/N$ is replaced by $1/N$-1 in this equation, the quantity σ^2 is sometimes referred to as the **bias-corrected variance.** *See also* STANDARD DEVIATION.

VAWT *See* VERTICAL-AXIS WIND TURBINE.

V-belt A drive belt having a trapezoidal cross section which runs in pulleys with V-shaped grooves. Higher torques can be transmitted than with a flat belt.

V-block A block having a 90° V-shaped recess; used in a workshop to hold round workpieces.

vector Any physical quantity, such as velocity, acceleration, force, or momentum, that is specified in terms of both its magnitude and its direction.

vectored thrust *See* THRUST VECTORING.

vee engine (V-engine) A piston engine in which there is a single crankshaft but two cylinder banks with an angle between them. Such engines are often designated by VN where N is the number of cylinders (e.g. V5, V6, V8, and V12), although for one cylinder the term single cylinder is used, and twin cylinder for two. In a **flat engine (pancake engine, boxer engine, opposed-cylinder engine)** the two cylinder banks are horizontally opposed. The arrangement is typically, but not exclusively, used for air-cooled engines.

vehicle A means of conveyance, often with wheels and often self-propelled, for transporting goods and people.

velocimeter *See* ANEMOMETER.

velocity (velocity vector, *V*) (Unit m/s) The velocity of a particle is the rate of change of its displacement $\boldsymbol{x}$ with respect to time t, i.e. $\boldsymbol{V} = d\boldsymbol{x}/dt$. It is a vector quantity with both direction and magnitude (the speed).

velocity analysis *See* APPLIED MECHANICS.

velocity boundary layer *See* BOUNDARY LAYER.

velocity coefficient *See* COEFFICIENT OF DISCHARGE.

velocity-compounded steam turbine An impulse steam turbine in which all the pressure drop occurs through a single ring of nozzles before the steam passes through a series of rotor/stator stages in which the velocity progressively reduces. *See also* COMPOUNDING; CURTIS STAGE.

velocity controller A controller used to control the time derivative of the plant output. The **velocity error (velocity-lag error)** is the difference between the time derivative of the set point and that of the plant output. This can be determined from the **velocity-error constant (velocity constant, K_v)** which is the product of the Laplace transform variable s and the open-loop transfer function of controller and plant evaluated as $s \to 0$. For a unity feedback type 1 system responding to a ramp input, the steady-state velocity error will be given by $\varepsilon_{ss} = 1/K_v$. For type 0 systems the steady-state velocity error is infinite, and for systems of type 2 and higher, it is zero. *See also* VELOCITY FEEDBACK; VELOCITY SERVOMECHANISM.

velocity defect In a viscous flow, such as a boundary layer or wake, the reduction in velocity, due to friction, compared with that of the free stream.

velocity diagram *See* HODOGRAPH.

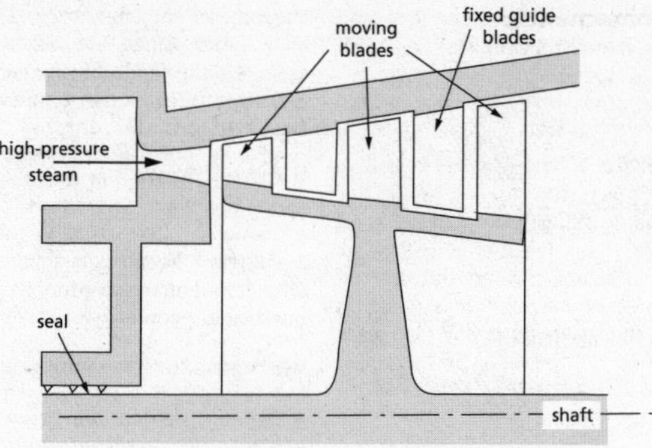

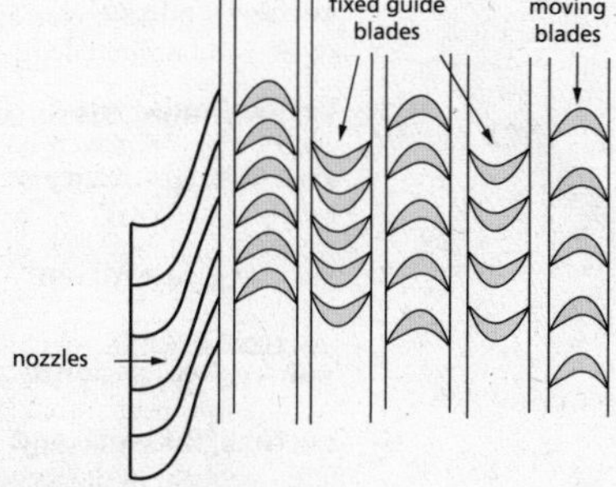

velocity-compounded steam turbine

velocity distribution (velocity profile) In a viscous flow, such as a boundary layer, pipe flow, jet, or wake, the variation of velocity with transverse or radial position at any downstream location.

velocity feedback Where one feedback signal in a closed loop controller is proportional to the time derivative of the plant output. *See also* VELOCITY CONTROLLER; VELOCITY SERVOMECHANISM.

velocity field A representation of the velocity vector at discrete points throughout a fluid flow.

velocity flow meter A flow meter in which flow rate is determined from the measured velocity distribution by weighted integration.

velocity lag error The velocity error in the output of a control system responding to a ramp input.

velocity-modified temperature (T_{mod}) (Unit K) For some metals, such as plain carbon steels, it is found that the combined effects of strain rate and temperature on the yield stress is a unique function of $T_{mod} = T[1 - A ln\ (\dot{\varepsilon}/\dot{\varepsilon}_0)]$ where T is the absolute temperature of interest, $\dot{\varepsilon}$ is the strain rate, and A and $\dot{\varepsilon}_0$ are constants.

velocity potential (ϕ) In an irrotational flow, a quantity defined by $\boldsymbol{V}=\nabla\phi$, where $\boldsymbol{V}$ is the vector velocity, which combined with the continuity equation, leads to Laplace's equation for ϕ, i.e. $\nabla^2\phi=0$. *See also* POTENTIAL FLOW.

velocity pressure *See* DYNAMIC PRESSURE.

velocity profile *See* VELOCITY DISTRIBUTION.

velocity ratio 1. The ratio between the input velocity to a machine, train of gears, etc. and the output velocity. **2.** The ratio between the displacement of an applied force at one part of a mechanism and the movement of the load at a different part.

velocity servomechanism A control system employing velocity feedback. *See also* VELOCITY CONTROLLER.

velocity triangle A triangle used in the analysis of turbomachinery to determine the velocity w of the working fluid relative to a rotor, usually at the inlet or outlet of a stage. In the diagram for a reaction turbine, the sides of the triangle represent vectors corresponding to w, the actual fluid velocity c, and the rotor peripheral speed U.

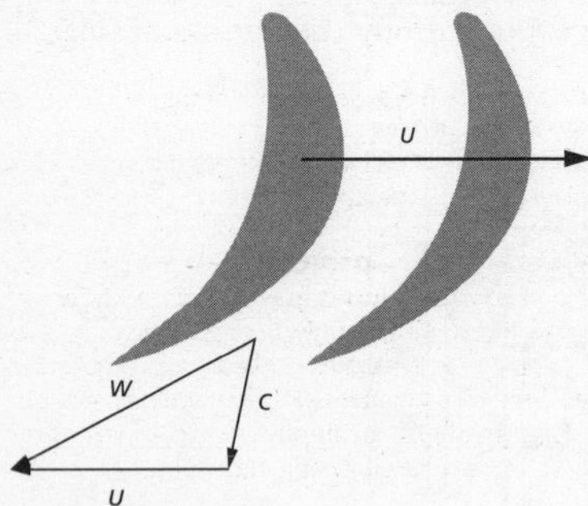

velocity triangle

velocity vector *See* VELOCITY.

V-engine *See* VEE ENGINE.

vent A small valve that allows the release of pressurized fluid from a pipe, pressure vessel, etc., often to the atmosphere.

ventilation A system for circulating fresh air in a room, building, passenger vehicle, or other enclosed space.

Venturi A convergent-divergent flow nozzle, usually circular in cross section, with a relatively short convergent section (the confuser) upstream of a throat followed by a gradually diverging section (the diffuser). Such nozzles usually have flanges at either end for installation in a pipeline. Applications include flow meters and ejectors. When a fluid flows through a convergent duct at subsonic speed, there is an increase in velocity accompanied by a decrease in pressure **(Venturi effect)**. A **Venturi flow meter (Venturi meter, Venturi tube, Herschel-type Venturi tube)** is a type of Bernoulli obstruction flow meter using a Venturi. If the pressure drop between the upstream section and the throat is Δp, the mass flow rate $\dot{m}$ for frictionless, incompressible flow is given by $\dot{m} = A_T\sqrt{2\rho\Delta p/(1-r^2)}$ where A_T is the throat area, ρ is the fluid density, and r is the ratio of the throat area to the upstream area. *See also* DALL TUBE.

Venturi flume A reduction in the width of an open channel, sometimes accompanied by a rise in the bed. Used to measure flow rate. A **Parshall flume** is a type of Venturi flume used for water-flow monitoring, the water flow rate being related to water depth at the throat.

vernier A short auxiliary graduated scale, sliding in contact with the principal graduated scale, by means of which fractions of the least division of the main scale may be estimated.

vernier calliper A calliper having a vernier scale.

vertical-axis wind turbine (VAWT) A wind turbine in which the rotor rotates about a vertical axis. Examples include the Banki, Darrieus, H-VAWT, Savonius–Darrieus and V-VAWT turbines. *See also* HORIZONTAL-AXIS WIND TURBINE.

vertical engine An engine having the cylinders above the crankshaft.

vertical take-off and landing (VTOL) A term applied to helicopters and certain jet aircraft equipped with vectored engines that can swivel to direct thrust vertically downwards for take-off, landing, and hover.

vertical traverse Controlled movement of a device in the vertical direction, typically using a stepper or DC servo motor.

vertical turbine pump A centrifugal pump designed to pump water from deep wells, the impeller(s) being at depth.

VHN *See* VICKERS HARDNESS NUMBER.

via point An intermediate location through which the end effector of a robot must pass en route to a programmed end point.

vibration 1. A periodic change with time of the displacements of elements making up a component or structure. **2.** The study of the oscillatory motion of bodies and systems and the frequencies, amplitudes, and forces associated with them. *See also* MODE OF VIBRATION.

vibration isolation The prevention of transmission of vibration from one component of a system to another part of the same system, such as a building or other structure. Isolation may be achieved using dampers (**vibration damping**) or by active feedback-control methods. Mechanical vibration is often attenuated by means of components immersed in oil such as in dashpots (**viscous damping**). **Vibration suppression** can be achieved (a) using dampers and absorbers tuned to a particular frequency to suppress vibratory forces in structures and other systems (**passive suppression**) or (b) by the measurement of vibration at key locations in a structure and the application of cancellation forces (**active suppression**).

vibration-testing machine (vibrator) Any machine that subjects components or systems to vibration at known amplitudes and frequencies in order to determine the response.

vibratory equipment Vibrating process equipment, such as shakers, used to separate small and large particles, or feeders to supply particulate or granular material that may otherwise clog.

vibrograph An instrument that records vibrations in a system over time.

vibrometer A device used to measure the motion of a vibrating surface, typically using a contactless laser-based technique.

vice A workshop tool used to hold a workpiece and consisting of two jaws, one fixed and the other moved by turning a screw.

Vickers hardness number (diamond hardness number, DHN, VHN, VPN) (Unit kg/mm^2 originally, sometimes now Pa) Indentation hardness given by load divided by the surface area of the permanent impression obtained when the indenter is in the form of a square pyramid whose opposite faces make an angle of 136° with one another. The mean length d of the diagonals of the indentation is determined, from which VPN = $0.927(2W/d^2)$ where W is the load, since the base of the pyramid has an area equal to 0.927 times the surface area.

view factor (angle factor, configuration factor, geometrical factor, shape factor) In radiant interchange among diffusely reflecting surfaces, the fraction of radiation from one surface that arrives at another. *See also* EXCHANGE FACTOR.

VIM *See* VACUUM-INJECTION MOULDING.

virial equation of state An equation of state for a substance in series form, i.e.

$$p = \frac{RT}{v} + \frac{a(T)}{v^2} + \frac{b(T)}{v^3} + \frac{c(T)}{v^4} + \dots$$

where p is the pressure, T the absolute temperature, R is the specific gas constant, v is the specific volume, and the **virial coefficients** a, b, c, etc. are functions of temperature.

virtual centre *See* INSTANTANEOUS CENTRE.

virtual displacement Any hypothetical displacement of a body, framework, or system of particles, not necessarily actually experienced. They can satisfy or violate constraints.

virtual entropy *See* PRACTICAL ENTROPY.

virtual inertia *See* ADDED MASS.

virtual work The work done in a virtual displacement by the applied forces.

virtual-work principle A system with workless constraints is in equilibrium under applied forces if, and only if, zero virtual work is done by the applied forces in an arbitrary infinitesimal displacement satisfying the constraints. In this context, an applied force is any force other than a reaction at a constraint.

viscoelasticity The behaviour of substances that exhibit both elastic and viscous characteristics.

viscoelastic theory The relationships between stress, strain, and strain rate in viscoelastic materials.

viscometer Any instrument for measuring fluid viscosity. *See also* OSTWALD VISCOMETER; RHEOMETER; UBBLEHODE VISCOMETER.

viscometric flow A laminar shear flow in which fluid particles on any streamline are always the same distance apart, i.e. there is no stretching or elongation of the fluid. Examples are Couette flow, fully-developed duct flow (including flow through a concentric annulus with relative rotation and an eccentric annulus without rotation), and flow in a cone-and-plate or parallel-plate geometry. In certain geometries, such as square ducts, viscoelastic fluids lead to situations in which there are secondary flows so that the overall flow is not viscometric.

viscoplasticity The behaviour of substances, such as a Bingham plastic, that exhibit both plastic and viscous characteristics.

viscosity *See* DYNAMIC VISCOSITY (1).

viscosity gauge *See* FRICTION VACUUM GAUGE.

viscous A fluid having viscosity (i.e. any real fluid) is said to be viscous.

viscous damping *See* VIBRATION ISOLATION.

viscous dissipation (viscous heating) **1.** In a laminar flow, the conversion of mechanical energy to heat. *See also* BRINKMAN NUMBER. **2.** In a turbulent flow, the reduction in turbulent kinetic energy due to viscous stresses and its conversion to heat.

viscous filter A type of filter, typically used in air-conditioning systems and engine intakes, in which the filter material is coated in a viscous substance, such as a gel, to collect fine particles in the air flow.

viscous fingering The appearance of finger-like patterns that arise when a low-viscosity fluid is injected into a high-viscosity fluid.

viscous slip coefficient (σ_P) In flow problems where the length scale is comparable with or smaller than the mean-free path of the fluid λ, a factor relating the slip velocity u_S to the near-wall velocity gradient $\partial u/\partial y|_S$ according to $u_S = \sigma_P \lambda(\partial u/\partial y|_S)$.

viscous sublayer The thin, quasi-laminar region in a turbulent wall flow in the immediate vicinity of the surface. The thickness is given by $y^+ \approx 5$ where $y^+ = u_\tau y/v$, y being the physical thickness, u_τ is the friction velocity, and v is the kinematic viscosity of the fluid.

visible radiation Electromagnetic radiation visible to the human eye, with wavelengths from about 380 nm (violet) to 740 nm (red).

vision system A computer-based system that uses a camera and processing software to provide a visual input to a robot or other manufacturing system.

visioplasticity An experimental method for determining the velocity vector field (strain rates) in plastic flow fields from which stresses may be calculated. It is accomplished by measuring the incremental movement of grids marked on the exterior of specimens, or on the interior of split specimens by hand or image-recognition techniques.

vitrification Conversion of a substance into a glassy form.

VM *See* VACUUM FORMING.

void An unwanted bubble, usually small, in a solid component. Typically filled with air or another gas, depending upon the manufacturing process.

void sheet In ductile fracture by shear, the initiation, growth, and coalescence of voids along shear bands, which appear as elongated dimples on fracture surfaces.

Voigt material *See* KELVIN–VOIGT MATERIAL.

volatility A qualitative measure of how readily a substance evaporates, particularly at room temperature and pressure.

volatilization *See* EVAPORATION.

volume ($\mathcal{V}$) (Unit m^3) The amount of space occupied by a specified mass of substance or by an object. *See also* MOLAR PROPERTY; PARTIAL VOLUME; SPECIFIC PROPERTY.

volume flow rate ($\dot{Q}$, $\dot{\mathcal{V}}$) (Unit m^3/s) The volume of a material, usually a fluid or powder, that flows across a surface or through a pipe or other duct per unit time.

volume fraction The proportion by volume of each constituent in a mixture of substances. *See also* MASS FRACTION.

volume thermal expansion coefficient (volume thermal expansivity) *See* THERMAL EXPANSION.

volumetric analysis In thermodynamics, the analysis of a mixture of gases by volume according to Amagat's law. *See also* GRAVIMETRIC ANALYSIS.

volumetric efficiency For a piston engine, the ratio of the volume of the induced charge per induction stroke, determined at a reference pressure and temperature, to the swept volume.

volumetric strain (volume strain, dilatation, θ) The fractional change in volume of a substance, i.e. $\theta = (\mathcal{V} - \mathcal{V}_0)/\mathcal{V}_0$ where $\mathcal{V}_0$ is the initial volume.

volume viscosity *See* BULK VISCOSITY.

volute A spiral casing surrounding the impeller of a centrifugal pump or fan that collects the flow from the impeller and acts as a diffuser. In some designs, the volute is split into two (a **double volute**).

volute spring A coil spring that is both spiral and helical, made of broad thin strip, adjacent coils of which overlap and slide over each other when loaded axially.

von Kármán constant (κ) An empirical constant in the logarithmic law of the wall with a value of about 0.41.

von Kármán momentum-integral equation *See* MOMENTUM-INTEGRAL EQUATION.

von Mises–Hencky theory *See* FLOW RULES.

von Mises stress *See* EFFECTIVE STRESS.

von Mises yield criterion The proposition that yielding of a solid, having yield stress Y in simple tension, will occur when the combination of normal stresses σ and shear stresses τ referred to Cartesian axes x, y, z satisfies:

$$Y = \frac{1}{\sqrt{2}}\sqrt{(\sigma_x - \sigma_y)^2 + (\sigma_y - \sigma_z)^2 + (\sigma_z - \sigma_x)^2 + 6\left(\tau_{xy}^2 + \tau_{yz}^2 + \tau_{zx}^2\right)}$$

or, in terms of principal stresses σ_1, σ_2, and σ_3:

$$Y = \frac{1}{\sqrt{2}}\sqrt{(\sigma_1 - \sigma_2)^2 + (\sigma_2 - \sigma_3)^2 + (\sigma_3 - \sigma_1)^2}.$$

vortex A flow structure in which fluid particles rotate around a centre. The fluid pressure is lowest at the centre and increases progressively with radial distance. *See also* RANKINE VORTEX.

vortex amplifier *See* FLUIDICS.

vortex breakdown A flow phenomenon in which an intense vortex undergoes a sudden increase in diameter and decrease in axial velocity, often accompanied by recirculation.

vortex burner *See* SWIRL BURNER.

vortex chamber A cylindrical chamber with an axial outlet and one or more tangential inlets such that there is an intense swirl flow within the chamber. *See also* FLUIDICS.

vortex core The central region of a vortex in which the tangential velocity increases with radius. *See also* RANKINE VORTEX.

vortex decay The reduction in swirl of a trailing vortex with downstream distance, due to viscous dissipation.

vortex diode *See* FLUIDICS.

vortex generator A small angled vane or protrusion mounted on a surface over which there is fluid flow to promote transition to turbulence or increased turbulence to delay separation. The most common application is to an aircraft wing, where a row of vortex generators is mounted close to the leading edge.

vortex pair Two parallel vortices, either with the same rotation sense, or counter-rotating as in the case of trailing vortices.

vortex-precession flow meter A flow meter in which the fluid enters the meter body through vanes that cause it to swirl and form a vortex, then flow through a convergent-divergent nozzle. In the divergence, the vortex core develops a helical structure that precesses at a frequency roughly proportional to the flow rate.

vortex ring A toroidal flow structure, the cross section of which has the form of a vortex, created when a mass of fluid is suddenly ejected from an orifice into a body of the same or different fluid. A smoke ring is an example.

vortex shedding The phenomenon whereby, in flow past a transverse cylinder, shear layers develop on either side, roll up and, over a range of Reynolds numbers, alternately break away and travel downstream. The Strouhal number based upon the frequency of shedding, the cylinder diameter and the flow velocity is approximately equal to 0.2. If the frequency is close to a natural frequency of a structure, such as a bridge, chimney, mast, high building, antenna, overhead cable, etc., flow-induced vibration may result.

vortex-shedding meter A flow meter in which the frequency of vortex shedding from a cylinder (usually not circular) installed across the diameter of a pipe is used to determine the flow rate. The device must be calibrated.

vortex sheet A tangential surface of discontinuity in an inviscid fluid across which there is a jump in velocity. *See also* KELVIN–HELMHOLTZ INSTABILITY.

vortex street *See* KÁRMÁN VORTEX STREET.

vortex strength The magnitude of the circulation over the cross section of a vortex.

vortex turbine A small-scale (up to about 150 kW) vertical-axis hydraulic turbine in which the rotor is situated at the centre of a swirling flow in the outflow of a large tank.

vortical flow *See* ROTATIONAL FLOW.

vorticity (ω) (Unit s^{-1}) A measure of rotation within a fluid flow, defined as the curl of the velocity vector **V** at any point, $\omega = \nabla \times \boldsymbol{V}$.

VPN *See* VICKERS HARDNESS NUMBER.

V-thread *See* SCREW.

VTOL *See* VERTICAL TAKE-OFF AND LANDING.

vulcanization A process of increasing the strength and durability of rubber by heating with sulphur so as to form cross-links.

V-VAWT A type of vertical-axis wind turbine with straight blades in the form of a V, attached at their lower ends. *See also* H-VAWT.

Wahl factor (K_W) An empirical factor employed in the design of coil springs to account for the effects of wire curvature and transverse shear induced owing to the different inner and outer radii of the coils. $K_W = (4C - 1)/4(C - 1) + 0.615/C$ where C is the spring index.

waist The joint in a robot corresponding to the human waist, i.e. providing rotation about a vertical axis. In an articulated robot, the first joint mounted at the base frame.

wait An instruction in a programmable-logic controller or other controller which delays execution for a specified period of time.

wake The narrow region of reduced velocity downstream of an object around which there is viscous-fluid flow.

walkthrough programming Programming a robot by making it move slowly through a series of steps, normally using a teach pendant. The steps are then replayed at greater speed when the robot is operating.

wall In fluid mechanics or heat transfer, any solid boundary in contact with a fluid. *See also* NO-SLIP CONDITION.

wall-attachment amplifier *See* FLUIDICS.

wall coordinates *See* LAW-OF-THE-WALL VARIABLES.

wall functions *See* LAW-OF-THE-WALL VARIABLES.

wall jet The flow created by injecting a high-velocity jet of fluid parallel to a surface. There

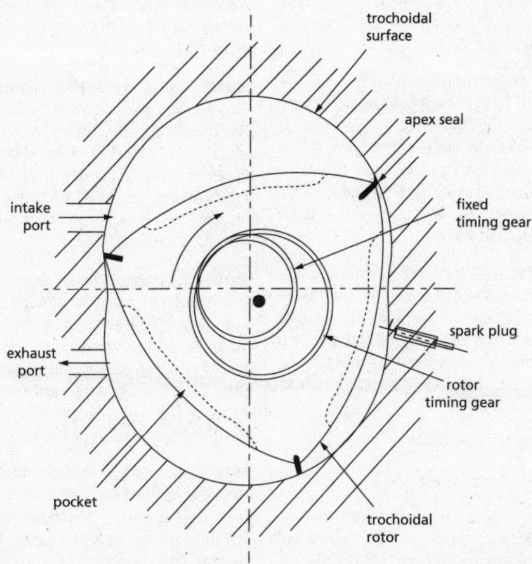

Wankel engine

may or may not be a free-stream flow. Wall jets are analysed in essentially the same way as boundary layers.

wall superheat *See* BOILING.

wall units *See* LAW-OF-THE-WALL VARIABLES.

wall variables *See* LAW-OF-THE-WALL VARIABLES.

Wankel engine A type of rotary internal-combustion engine in which a trochoidal rotor orbits around an eccentric on a shaft within a stationary casing with which the rotor tips are in sliding contact.

SEE WEB LINKS

- Animation and explanation of the principles of the Wankel engine

warm working (warm forging) A method of component manufacture by hammering of a metal in a temperature range between about $0.1T_M$ and $0.4T_M$ where T_M is the metal's melting point.

warp 1. To move a load attached to a cable, chain, or rope by winding on to a drum. **2.** The lengthwise threads in a fabric. *See also* WEFT. **3.** The mooring rope of a ship.

warping Distortion of an object either naturally (timber) or under load (such as of cross sections of prismatic sections in bending and torsion). *See also* SOAP-FILM ANALOGY.

Washburn's equation An equation for the interface speed V when one liquid in a capillary tube displaces another. In the limiting case of an empty capillary, it may be written $V = \sigma R\cos\theta/2\mu x$ where σ is the surface tension, R is the tube radius, θ is the meniscus contact angle at the capillary wall, μ is the viscosity of the fluid, and x is the penetration distance.

washer An annular disc of metal, rubber, plastic, ceramic, etc., placed between two surfaces in contact either to spread the load (for example, between a surface and a tightened nut or a bolt head), to provide a seal, or to separate or align components. *See also* GASKET; LOCKWASHER; TABWASHER.

washout thread *See* SCREW.

wastegate *See* TURBOCHARGING.

waste heat 1. Heat generated by internal-combustion engines, gas turbines, electrical generators, electrical equipment, and industrial processes that is not used directly but is expelled to the environment, often in hot flue or exhaust gases. **2.** Heat generated from waste matter, such as municipal (including landfill), industrial, hospital, or agricultural.

waste-heat recovery The recovery of thermal energy from flue and exhaust gases, or from liquids heated in industrial processes. Recovery devices include pre-heaters, recuperators, regenerators, and waste-heat boilers.

watchdog timer A system running in parallel with a controller which monitors the output and, if the output remains constant or zero for a specified period, restarts the controller or shuts down the plant. It thus prevents damage to the plant should the controller fail.

water brake A type of absorption dynamometer in which the input mechanical power is dissipated by turbulence generated by an impeller rotating in water contained in a housing.

water calorimeter *See* CALORIMETER.

water column Water in a tube, which may be vertical or inclined. If the tube is open to the atmosphere, the vertical height from a datum level to the water surface is a measure of the water pressure at the datum level.

water-cooled furnace A furnace in which the walls are cooled by water circulated through attached tubes.

water-flow pyrheliometer A pyrheliometer using a water-cooled cylindrical chamber blackened on the interior and designed to completely absorb the incident radiation, the intensity of which is determined from the electrical heating required to cause the same temperature increase in the water flowing through a second chamber.

water hammer The reflected pressure surge that occurs in a liquid flowing through a pipe, usually as a consequence of sudden closure of a valve. The surge may cause the pipe to vibrate and a hammering noise to be heard. Key factors affecting the surge amplitude are the compressibility of the liquid (especially if it contains undissolved gas) and the elasticity of the pipe wall.

water jacket A casing, typically of sheet metal, surrounding a machine, such as an engine, that requires cooling. Water is circulated through the jacket and a heat exchanger where the heat is removed.

water jet *See* PUMP JET.

water-jet cutting A method of cutting materials by means of a high-speed jet of water containing abrasive particles.

water power *See* HYDROPOWER.

water-sealed holder A gas holder utilizing a **water seal** wherein a cylindrical shell, closed at the top, sits in water to prevent gas within the shell from escaping.

water-tube boiler A boiler in which steam is produced from water flowing through tubes heated by hot combustion gases in a furnace. Heat transfer is by a combination of radiation and convection. In a **waterwall boiler**, close-packed tubes cover the furnace wall. The feedwater is usually preheated in an economizer in the exhaust-gas stream. Saturated steam leaving the steam drum may be superheated by a further pass through the furnace. *See also* FIRE-TUBE BOILER.

water tunnel An experimental facility similar to a recirculating wind tunnel but with water as the working fluid. Because the kinematic viscosity of water (10^{-6} m^2/s at 20°C) is less than that of air (1.51×10^{-5} m^2/s at STP), for the same flow speed and scale, the Reynolds number is about seven times higher. *See also* CAVITATION TUNNEL.

water turbine *See* HYDRAULIC TURBINE.

water vapour The phase of water that can be produced from a pool of liquid by raising its temperature or reducing the pressure, causing water molecules to leave the free surface.

waterwall boiler *See* WATER-TUBE BOILER.

watt (W) The SI unit of power equivalent to 1 J/s or 1 N.m/s.

watt-peak *See* PEAK WATTS.

Watt's mechanism (Chebyshev's mechanism) A linkage consisting of two bars OA and CD pinned at their fixed outer ends and joined by a shorter bar AD such that when OA and CD are parallel, the angle between them and AD is a right angle. When the links OA and CD oscillate, the path of any point on AD is a figure-of-eight (a lemniscate) but for small movements, to a

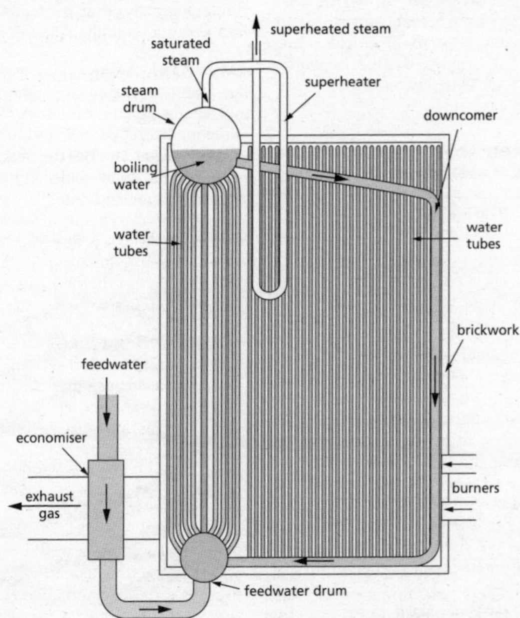

water-tube boiler

Watt's mechanism

good approximation the short bar moves in a straight line perpendicular to the line between the end points.

wave A periodic phenomenon in a medium or a vacuum. *See also* LONGITUDINAL WAVE; PROGRESSIVE WAVE; RAYLEIGH WAVE; SHEAR WAVE; STATIONARY WAVE; TRANSVERSE WAVE; TRAVELLING WAVE.

wave diagram *See* WAVEFRONT.

wave drag *See* DRAG.

wave-energy converter A device used to extract energy from ocean waves (**wave energy**) and convert it to electrical power (**wave power**). Examples include oscillating water column, TAPCHAN, waveplane, and various floating devices.

wave equation The equation representing the propagation of a wave, given by

$$\frac{\partial^2 u}{\partial t^2} = c^2 \nabla^2 u$$

where ∇^2 is the Laplace operator, u is either the particle displacement or the pressure, t is the time, and c is the wave speed.

waveform The shape taken by undulations in a steady or unsteady wave at any moment.

wavefront A continuous line or surface that is a locus of points in a waveform where the phase is the same at a given instant. At large distances from a small source in a uniform medium, the fronts are part of a sphere of very large radius, and may be considered as plane waves in practice. A **wave diagram** shows the position of a wave front at any instant of time.

wave gauge A device used to monitor the height of ocean waves.

wave gait A pattern of walking motion in a multi-legged robot having two sets of parallel legs.

wave height (Unit m) The difference between the heights of adjacent peaks and troughs of a surface wave. If the wave is sinusoidal, the wave height is twice the amplitude.

wavelength (λ) (Unit m) The distance from crest to crest in a waveform. If c is the wave speed and f is the frequency, $\lambda = c/f$. Its reciprocal $1/\lambda$ or, sometimes, $2\pi/\lambda$ is the **wave number (*k*)**.

wave maker A machine for making surface waves in water, such as in a wave tank.

waveplane A wave-energy converter consisting of a horizontal circular tube with a tangential inlet through which flow incoming waves, creating an intense vortex in the tube. Electrical power is generated by a turbine at the tube exit.

wave power *See* WAVE-ENERGY CONVERTER.

wave speed *See* PHASE SPEED.

wave spring A type of compression spring, similar to a coil spring, but made of strip shaped into waves around the circumference.

wave tank A water tank, typically rectangular, equipped with a wave maker at one end and a wave absorber at the other. Applications include the hydrodynamic testing of scale models of ships towed through the tank, and the study of the spread of oil slicks and of surface-wave phenomena generally.

wave velocity *See* PHASE SPEED.

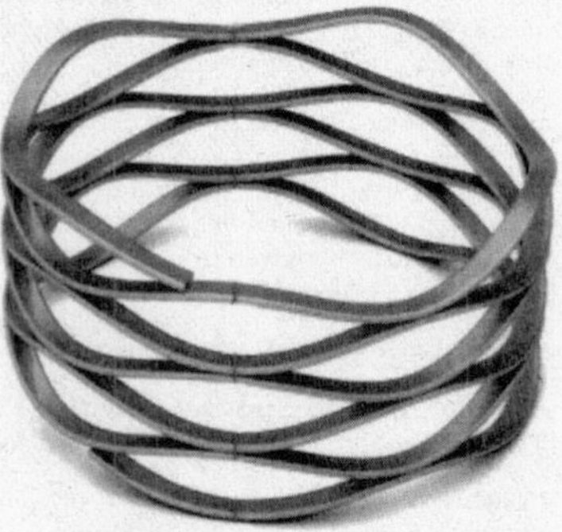

wave spring

waviness A surface defect in the form of undulations on a larger scale than surface roughness and usually the result of the manufacturing process. *See also* SURFACE TOPOGRAPHY.

weak mixture *See* STOICHIOMETRIC MIXTURE.

weak shock *See* SHOCK WAVE.

weak-shock solution *See* OBLIQUE SHOCK WAVE.

wear The deterioration of a component or structure with time and usage, often impairing the function for which it was designed, owing to abrasion, corrosion, fatigue, friction, etc. during relative motion of parts. The **wear factor (*K*)**, with unit mm^3/N.m, is an empirical dimensional factor that quantifies surface wear due to mechanical frictional contact, and defined by $K = \mathcal{V}/Fs$ where $\mathcal{V}$ is the worn volume, F is the contact load, and s is the sliding distance.

web 1. The vertical part of an I-beam joining the flanges. **2.** The projecting sides of a crankshaft which hold the crankpins.

Weber number (*We*) A non-dimensional parameter that arises in the study of free-surface flow, such as in droplets, capillary flows, ripple waves, and very small hydraulic models. It is defined by $We = \rho V^2 L/\sigma$ where ρ is the fluid density and σ is its surface tension, V is a characteristic flow velocity, and L is a characteristic length.

WECS *See* WIND ENERGY.

wedge A short triangular prism whose major surfaces subtend an acute angle, that can be driven between two objects or parts of an object to split, tighten, or secure them, or to widen an opening or raise a heavy object. The acute angle gives a high mechanical advantage.

wedge flow An inviscid fluid flow over a wedge for which the variation of free-stream velocity U_∞ with distance along the surface x is $U_\infty = Kx^m$ where K and m are constants. The wedge half angle θ is given by $\theta = m\pi/(m+1)$. *See also* FALKNER–SKAN SOLUTIONS.

wedge gate valve *See* GATE VALVE.

weft The threads in a woven fabric that are at right angles to the warp.

Weibull function In the absence of being able to predict an exact value for the fracture stress of a brittle material, the probability p of rupture, or the probability of survival p_s, at a given stress S_u is used. The fraction of the total number N of identical testpieces in a batch for which the strength exceeds S_u is represented by p_s; p is the fraction for which the strength falls below S_u. Thus $p + p_s = 1$. $p_s = exp[-(S_u/S_{uo})^m]$ where m is the **Weibull modulus** and S_{uo} is the stress level for which the cumulative probability of fracture of specimens in a batch at S_{uo}, or lower, is 63%. For $m \to \infty$, $p_s \to 0$ and $p \to 1$ and all specimens break at the same stress. For $m \to 0$, $p_s \to 1$ and $p \to 0$ and fracture is likely to occur at any stress level. Thus high m indicates little variability, low m high variability. Most materials exhibit 5 < m < 25. m may be determined from the slope of a plot of the double log of $[1/(1-p)]$ *vs* $log\,(S_u/S_{uo})$. A **Weibull chart** is a graph of the probability of rupture p *vs* normalized stress (S_u/S_{uo}), with experimental data superimposed.

weight (Unit N) The force exerted on the mass of a body when in the gravitational field of another body. On earth, the acceleration due to gravity is 9.81 m/s^2, so that the weight of a 1 kg mass is 9.81 N.

weightlessness *See* ZERO GRAVITY.

weir A device used for the measurement and control of liquid flow through an open channel. It consists typically of a thin metal plate placed transverse to the flow, with a rectangular or Vee notch. The volumetric flow rate is given by $\dot{Q} = Cg^{\frac{1}{2}}H^{\frac{5}{2}}$ where C is a constant dependent upon the notch geometry, g is the acceleration due to gravity, and H is the liquid height above the lowest part of the notch.

Weissenberg effect (rod-climbing effect) The phenomenon whereby a viscoelastic liquid moves vertically up a rotating rod partially immersed in the liquid due to tension in the circular streamlines associated with the normal-stress difference.

Weissenberg number (*Wi*) A non-dimensional group, closely related to the Deborah number, used to characterize the flow of a viscoelastic fluid, in particular the degree to which elasticity manifests itself in response to deformation. It can be defined as the product of a characteristic relaxation time of the fluid λ and a characteristic shear rate in the flow $\dot{\gamma}$. The inverse of $\dot{\gamma}$ can also be regarded as a characteristic time for the flow, and for many problems is the same as that used to define the Deborah number.

Weissenberg rheogoniometer *See* RHEOMETER.

welding A joining process in which the mating surfaces are at least softened, or more usually melted, unlike soldering and brazing. In solid-state (non-melting) welding, similar or dissimilar metals or thermoplastics may be joined by applying pressure to hot interfaces. In **forge welding**, joining is achieved by compression across the join between pre-heated workpieces. In **friction welding (spin welding)** one of the contacting surfaces is rotated rapidly before being loaded against the other. On contact, heat generated by friction softens the materials and permits severe plastic deformation that seals the joint. In **friction stir welding**, a wear-resistant tool is rotated rapidly, indents the surfaces and is traversed to form a join line. **Resistance welding** involves simultaneous pressure and electrically-generated heat, as in **spot welding** where a heavy electric current is passed for a short time through metal sheets which are pressed together between electrodes.

In **fusion welding**, joining is achieved by local melting along the contacting surfaces and filling the gap with materials metallurgically compatible with the parent metals. Fusion-welding processes are classified according to the heat source: gas flame; electric arc; plasma arc; laser; and electron beam. Oxidation of carbon and hydrogen in fuel gases such as butane, propane, and acetylene provide high-temperature gas flames: an oxyacetylene flame can just melt tungsten (3410°C). In **electric-arc welding**, the workpiece is usually the anode and the **filler rod** that is melted into, or along, a joint is the cathode (called straight polarity); for welding thin sheets, reverse polarity is employed to prevent the sheet melting before filler. The gap to be welded can, in practice, be quite large (about the radius of the welding rod). An arc of length between 2 and 10 mm is struck with a potential difference of 30–75 volts, giving an impinging power density of 10–100 MW/m^2. However, up to 10% of metal is lost because of evaporation, condensing as dust; and 10–20% of filler is lost by spatter.

In gas-flame and arc welding, protection against deleterious gas entrainment in joints is achieved either by (i) flooding with an inert gas (argon is most common, as nitrogen is not sufficiently inert in the plasma) that minimizes the amount of oxides and nitrides in the weld metal; or (ii) feeding a flux into the weld region. **Tig (tungsten inert gas) welding** is an arc-welding process that uses a non-consumable tungsten electrode shielded by argon; **mig (metal inert gas) welding** is a similar process, developed originally for aluminium and other non-ferrous metals, in which a consumable wire acting as the electrode is fed into the weld zone. Tig and mig are examples of **gas-shielded arc welding (GSAW)**. Fluxes perform as for soldering and brazing, and often are to be found as coatings on electrode or filler material in hand-welding operations. In older automatic filler-wire feed machinery, flux is in powder form and covers the moving arc, hence the name **submerged arc welding**. Newer designs employ hollow tubes of filler filled with flux. The compounds that result from the use of fluxes in arc welding are called slag.

Three other heat sources are used in fusion welding: **plasma arc**, where an ionized stream of a gas mixture such as 75% argon and 25% hydrogen passes along a plasma arc that is directed at any cool surface, where the ionized gas evolves heat on returning to molecular form, giving a power density in the range 1–10 GW/m^2; **laser**, where a coherent CO_2 laser produces infrared radiation with wavelengths of about 10 μm that is absorbed by metals, resulting in power densities of 10–100 GW/m^2, but the process is very inefficient in use of energy; and **electron beam**, where a focussed beam of electrons emitted from a thorium-tungsten source (giving non-coherent radiation with wavelengths in the range 0.01 to 10 nm) bombards the target in a vacuum of 10^{-3} to 10^{-4} Torr (0.1–0.01 Pa) under several thousand volts with power density of 100–1000 GW/m^2. **Induction welding** uses electrical induction heating to melt a conductive material. In **explosive welding**, metal plates are joined using a controlled explosive charge that impacts obliquely one plate (the 'stand-off' or 'flyer' plate) against the other at a moving collision front. A plasma jet running just ahead of the front cleans the two surfaces, which then bond under high pressure along a characteristically wavy interface. The flyer plate is typically one-third the thickness of the second plate. The process is useful for welding metallurgically-incompatible plates.

Weld penetration is the depth below the surfaces, as revealed in microstructural sections of joined metals, which is melted during welding.

Joint strength of a simple butt weld is determined mainly by penetration (and to a lesser extent by **bead height**) as it determines the unwelded area between plates. Full penetration

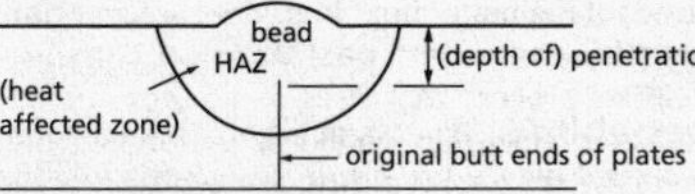

weld nomenclature

results from melting across the whole joint leaving no air gaps. However, undesirable metallurgical changes may occur in the **heat-affected zone (HAZ)** which reduce the strength. Temperatures in the HAZ can overage and soften hard aluminium alloys, and the rapid cooling that follows heating may be fast enough to produce brittle martensite in steels. Furthermore, differential heating and cooling rates within a joint give rise to weld **residual stresses** in as-welded components (**weldments**). **Preheating** of high plain-carbon and alloy-steel parts reduces the level of residual stress that results from subsequent welding, and post-weld heat treatment reduces the higher level of residual stress formed in simple as-welded joints. Slag inclusions can act as starter cracks in fatigue of weldments. The locations of welds required in assembly are indicated on drawings by an arrowhead that touches the surfaces where they are to be joined. Connected to the arrow are two lines (the reference line that relates to welding the near side of the plate, and a dashed line below concerning the far side). The type of weld required is indicated by a symbol, some of which are shown in the diagram. *See also* STRESS RELIEVING.

SEE WEB LINKS
- Welding terminology

welding torch *See* OXYACETYLENE TORCH.

well-regulated system A system employing a regulator controller which is able to maintain the plant output within a specified tolerance despite external influences.

Wells turbine An axial flow turbine that rotates in one direction irrespective of the flow direction. Used in oscillating water column wave-energy systems.

well-type manometer A U-tube manometer in which the unknown pressure is applied to the surface of the manometer liquid in a large-area reservoir (the well) incorporated into one leg of the U. Changes in the liquid height in the well are negligible and the pressure can be estimated from the height change *h* in the other leg.

wet-back boiler A shell boiler in which the hot-gas direction is reversed in a chamber con-

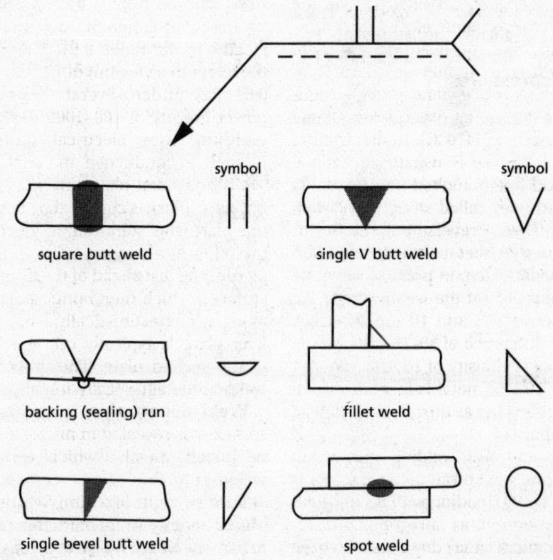

weld symbols

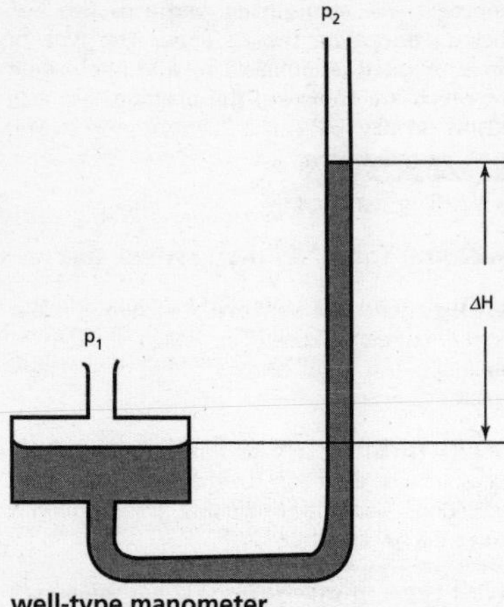

well-type manometer

tained within the boiler. *See also* DRY-BACK BOILER.

wet-bulb temperature The temperature indicated by a thermometer (**wet-bulb thermometer**) the bulb of which is covered with a cotton wick saturated with water when air is blown over the wick. The **dry-bulb temperature** is the ordinary temperature of atmospheric air indicated by a **dry-bulb thermometer**. The difference between these two temperatures is the **wet-bulb depression.**

wet clutch A friction clutch that runs immersed in oil.

wet cooling tower A cooling tower in which the temperature of the water which has been used to cool process equipment is reduced by spraying into the air flowing through the tower, which leads to evaporation of some of the water. *See also* DRY COOLING TOWER.

wet liner *See* CYLINDER LINER.

wetness fraction *See* DRYNESS FRACTION.

W

wet scrubber A device for removing pollutants from flue, exhaust, or process gas by bringing the gas into contact with a liquid, for example in the form of a spray or a pool through which the gas is bubbled. *See also* SCRUBBER.

wet-sump engine A piston engine where the lubricating oil is stored in an oil pan (the sump) at the bottom of the engine. *See also* DRY-SUMP ENGINE.

wettability The ease or otherwise of a liquid to spread over a surface. It is quantified by the contact angle ϕ, measured through the liquid, between the interface of a liquid droplet and the surface: wetting corresponds to $0° < \phi < 90°$, non-wetting to $90° < \phi < 180°$. A **wetting agent** is a substance, such as a surfactant, that when added to a liquid, improves the latter's wettability by reducing its surface tension. **Wetting** is the ability of a liquid to maintain contact with a solid surface. *See also* YOUNG'S EQUATION.

wetted perimeter For flow through a duct, at any cross section, any part of the surface in contact with the fluid. The term applies to all fluids, not just liquids. *See also* HYDRAULIC DIAMETER.

wet-test meter A positive displacement gas-flow meter in which a drum is caused to rotate within a low-viscosity liquid, usually water or an oil. The volumetric flow rate is proportional to the number of rotations in a given time.

wetting *See* WETTABILITY.

wetting agent *See* WETTABILITY.

wet vapour A mixture of saturated vapour and saturated liquid. *See also* DRYNESS FRACTION.

wheel A solid disc, or a circular ring with spokes radiating from a central hub, either attached to an axle around which it revolves or which revolves with a rotating axle.

wheel base The distance between the leading and trailing axles in a vehicle.

whirling Rapid rotation about an axis or centre, applied particularly to the motion of out-of-balance shafts when the rotation speed equals the natural frequency of transverse vibrations, termed the **whirling speed (critical speed)**. The amplitude of the shaft deflexion is limited by damping and shaft stiffness.

whirl velocity The tangential component of fluid passing through a turbomachine rotor.

whiskers Short fibres of some metals, alumina, graphite, silicon carbide, etc., having diameters of about 1 μm whose strength approaches the defect-free theoretical value of $E/10$, where E is Young's modulus. Used to reinforce epoxies.

white body A hypothetical body with zero absorptivity that absorbs no radiation at any wavelength. The opposite of a blackbody.

white iron *See* CAST IRON.

Whitworth quick-return linkage (turning-block linkage) An inversion of the slider-crank chain in which the crank is fixed and the connecting rod becomes the driver that results in unequal speeds on forward and backward strokes.

Whitworth screw thread *See* SCREW.

whole depth *See* TOOTHED GEARING.

wicket gate A ring of pivoted guide vanes used to regulate the flow of water into a hydraulic turbine.

Wien's displacement law An equation derived from Planck's law of radiation for the wavelength λ_{max} at which the spectral energy distribution is a maximum for a given absolute temperature T. It can be written as $\lambda_{max}.T = 2898$ μm.K.

Willans line **1.** For a steam turbine, a plot of steam consumption *vs* output power. **2.** For piston engines and gas turbines, a plot of fuel consumption *vs* brake power. The power loss due to internal friction can be obtained by extrapolating the curve to zero fuel consumption.

Wilson line In the expansion of steam flowing through a nozzle, the locus of points where condensation occurs independent of the temperature and pressure at the nozzle inlet.

winch A hand- or power-driven lifting machine based on a rotatable drum on which is coiled a cable, chain, or rope and often incorporating gears and a ratchet.

windage loss **1.** The air resistance of a moving object, especially a rotating machine part. Also the force of wind on a stationary object. **2.** The allowance made for the effect of wind in deflecting a missile.

wind energy (Unit kJ) The kinetic energy associated with wind that can be converted by a **wind energy conversion system (WECS)** into electrical or mechanical power by a rotor, such as a multi-bladed propeller, exposed to the wind. A **wind turbine (wind generator)** generates electrical power, a **windmill** generates mechanical power. A **wind farm** is an array of wind turbines, typically ten to several hundred, at a single location, either onshore or offshore. If the air density is ρ and the wind speed is V, then the kinetic energy flux is $\rho V^3/2$. The actual **wind power** that can be extracted by a wind turbine intercepting a cross section of wind A is $C_P \rho A V^3/2$ where C_P is an empirical efficiency factor termed the power coefficient. The **wind-energy distribution** is a histogram of the calculated wind power that can be generated annually from the wind-speed frequency distribution at a given location. *See also* HORIZONTAL-AXIS WIND TURBINE; VERTICAL-AXIS WIND TURBINE.

winding angle The angle at which fibres are laid in pressure vessels made of fibre-reinforced composites. In a pressurized thin cylindrical vessel, filaments aligned at angle θ to the axis when loaded under tensile stress σ give a hoop stress of $\sigma\cos^2\theta$ and an axial stress of $\sigma\sin^2\theta$. To obtain the same 2:1 hoop:axial stress ratio found in an unreinforced vessel, it follows that $\cot^2\theta = 2$, i.e. $\theta \approx 55°$ to give 'equal strength'.

winding gear The machinery employed for hoists, lifts, mineshafts, etc.

wind generator **1.** *See* WIND ENERGY. **2.** A propeller used to provide an airflow, for example in motor-vehicle testing and film production.

windmill *See* WIND ENERGY.

windmill anemometer *See* PROPELLER ANEMOMETER.

windmilling Rotation of aircraft propellers, compressor blading, etc. caused by air movement when the engine is not operating.

wind power *See* WIND ENERGY.

wind pressure The dynamic pressure associated with the wind. The static pressure of the wind is the atmospheric pressure.

wind pump A pump driven directly by a windmill on a tower. Widely used in remote locations.

wind shear *See* WIND VELOCITY.

wind speed *See* WIND VELOCITY.

wind-speed frequency distribution For a given location, a histogram of the number of hours in a given period, typically one year, for which the wind speed lies within a given band, such as 0–1 m/s, 1–2 m/s, 2–3 m/s, etc. *See also* WIND ENERGY.

wind tunnel A duct in which a controlled flow of air is used for testing and research. Types include blow down, closed circuit, open return, and open section. Closed-circuit tunnels may be pressurized or evacuated. The working section is usually situated immediately downstream of the flow-conditioning elements, which may include a plenum chamber, honeycomb, screens, turbulence grids, and a contraction. Immediately downstream there is usually a diffuser. For basic research the usual requirement is for uniform steady flow in the working section with low swirl and turbulence intensity. For some applications the flow may be density stratified, sheared, unsteady, or of high turbulence intensity.

wind turbine *See* WIND ENERGY.

windup The twist of a shaft (a) due to inadequate torsional stiffness for the design torque, and/or (b) due to overloading. It may lead to vibration should the shaft alternately wind up and unwind.

wind vane A vertical plate, which may be flat or have a symmetrical aerofoil-shaped cross section, at the end of a horizontal arm or body pivoted to allow rotation about a vertical axis. It is used either to show wind direction or to ensure the pivoted body is aligned with the wind direction.

wind velocity (Unit m/s) **1.** In the earth's atmosphere, a vector quantity that quantifies both the magnitude of the **wind speed** and its direction at a given altitude and location. For convenience, the unit kph is often used. It could in principle also include information about large- and small-scale unsteadiness. Spatial variation in wind velocity, either with altitude or in a horizontal plane, is termed **wind shear**. **2.** The airspeed in the working section of a wind tunnel.

wing A relatively long and thin body with a cross section designed to produce lift when there is motion relative to a fluid. The principal applications are to aircraft and high-performance motor vehicles. *See also* AEROFOIL.

winglet A vertical or inclined endplate attached to a wing tip to block some of the leakage of air from the high-pressure to the low-pressure side, thereby reducing the strength of wing-tip vortices and induced drag.

wing nut A nut having two opposite protruding wings to permit hand tightening.

wire drawing Production of wire by drawing through convergent dies of decreasing diameter.

wire gauge Standardized numbered scales to which the diameter of wire and the thickness of sheet material may be referred. The American Wire Gauge (AWG), Birmingham Wire Gauge (BWG), Brown and Sharpe Gauge (B & S), and Standard Wire Gauge (Imperial Standard Wire Gauge, British Standard Wire Gauge, SWG) are all based on the inch; the Paris Wire gauge is metric. In all cases, the higher the number, the smaller the size. The scales are not linear; the difference between SWG gauges 20 and 21 is 0.101 mm, but between gauges 30 and 31 it is 0.020 mm.

wire rope A type of cable formed from individual strands of wire with a helical twist.

Wobbe index (Wobbe number, I_W) (Unit kJ/m^3) A dimensional parameter used to compare the thermal-energy content of different fuel gases, defined by $I_W = HCV/relative\ density$ where *HCV* is the higher calorific value of a fuel.

wobble-plate *See* SWASH PLATE.

Wöhler test *See* FATIGUE.

Wollaston wire Very fine platinum wire, typically 1–5 μm diameter, that can be used as one leg of a thermocouple or the sensor wire of a hot-wire anemometer.

Womersley number *See* SHEAR NUMBER.

wood gas (producer gas) Gas produced from biomass by heating to reduce volatiles followed by reaction with steam and oxygen. It consists mainly of carbon monoxide and hydrogen with some methane, higher hydrocarbons, and tars, together with carbon dioxide and water. The heat of combustion is about 5 MJ/m^3 at STP.

Woodruff key A key in the form of a segment of a disc, the curved part of which fits into a slot of the same radius cut into a shaft and the straight part into a normal keyway in a hub.

work (W) (Unit J) **1.** In thermodynamics, the form of energy transfer associated with the

movement of a force at the boundary of a system. Like heat, work is energy in transition and is never contained in or possessed by a body. **2.** The incremental work δW done by a force F acting on a particle is given by $\boldsymbol{F}.\delta\boldsymbol{s} = X\delta x + Y\delta y + Z\delta z$ where $\delta\boldsymbol{s}$ is the distance moved by the particle in the direction of $\boldsymbol{F}$, X, Y, and Z being the components of $\boldsymbol{F}$ in Cartesian coordinates and δx, δy, and δz the corresponding components of $\delta\boldsymbol{s}$. The work done on a rigid body is $\delta W = \boldsymbol{F}.\delta\boldsymbol{s} + \boldsymbol{M}.\delta\boldsymbol{\omega}$ where $\boldsymbol{M}$ is the applied moment and $\boldsymbol{\omega}$ is the corresponding rotation (angular displacement). The area under a stress (σ)–strain (ε) curve is given by $\int\sigma d\varepsilon$ and represents work done per unit volume. Work done may be stored as reversible elastic strain energy, or irreversibly dissipated as plastic flow.

work-done factor (Y) For an axial-compressor stage, an empirical factor introduced to account for non-uniformity of the velocity profile in the blade passages. It can be defined by $Y = P/\dot{m}c_B\Delta c_w$ where P is the actual power input, $\dot{m}$ is the mass flow rate, c_B is the rotor-blade velocity, and Δc_w is the change in axial velocity.

workhardening *See* STRAIN HARDENING.

working depth *See* TOOTHED GEARING.

working fluid In thermodynamics, the liquid, vapour, or gas contained within the system boundaries. It is the fluid that experiences changes in its properties during any fluid dynamic or thermodynamic process.

working load The load under which a component or structure is designed to operate under normal conditions.

working pressure The pressure at which a pressure vessel, such as a boiler, is designed to operate under normal conditions.

working section That part of a wind tunnel in which models are placed for testing, or where detailed measurements are performed.

working-space volume *See* WORKSPACE.

working stress The stress under which a component or structure is designed to operate under normal conditions.

work–kinetic-energy theorem (principle of work and energy) The work W done by a force acting on a particle is equal to the change in the kinetic energy T of the particle; alternatively, the rate of increase of kinetic energy $\dot{T}$ of a particle is equal to the rate of working of the force $\dot{W}$ acting on it. In a conservative (path-independent) field of force with potential energy V, $T + V =$ constant, which is the principle of the conservation of energy, that may be extended to elastically-deformable bodies. In a non-conservative field (path-dependent with irreversibilities), $\dot{T} \neq \dot{W}$.

work of fracture *See* FRACTURE TOUGHNESS.

workpiece programming *See* PART PROGRAMMIMG.

work ratio The ratio of the net work out to the actual work out for a thermodynamic cycle.

workspace (working-space volume) The volume around the base frame of a robot, defined by the reach of the robot. The robot ca thus only perform tasks within this volume.

workstation The location at which a rol operates, including, for example, convey moving parts to the robot, or a machine loaded by the robot. A robot has **worksta independence** if it is so programmed tha same program may be successfully used v the robot is moved to another workstation

world coordinates The fixed x, y, z c nate system from which robot motion is sured, i.e. the coordinates of the base fra

world modelling The process t which an adaptive robot gains knowled environment so that it can respond to in that environment.

worm *See* TOOTHED GEARING.

worm conveyor *See* SCREW CONV

worm gearing *See* TOOTHED GEAR

wrench *See* SPANNER.

wringing The process of joining th slip gauges together by a sliding an motion in order to remove all air be mating faces. A similar wiping mot ployed to separate slip gauges.

wrist The final three joints on a r simulate rotations provided by the h and thus allow orientation of the en which it is attached by the **wrist sc**

wrist pin A stud projecting from attachment for a connecting rod.

wrought iron A highly ductile (but anisotropic) type of iron containing elongated slag fibres that resulted from the method of manufacture in which excess carbon in pig iron was burnt and worked out. The yield strength is some 200 MPa, tensile strength 320 MPa, and reduction of area on a 50 mm gauge length up to 35%. Now replaced by steel.

wye-fitting *See* PIPE FITTING.

X-ray thickness gauge A device comprising an X-ray source and a detector used to determine the thickness of material in sheet or plate form, including metals, paper, plastics, rubber, and ceramics.

***X–Y* recorder** A device that plots a graph of a variable *Y* against a variable *X*, both in the form of voltages.

yaw angle The angle, for an aircraft or missile, between its longitudinal axis and the direction of flight.

Y-fitting (wye fitting, Y-branch) *See* PIPE FITTING.

yield moment *See* PLASTIC HINGE.

yield point (yield strength, yield stress) The stress at which the onset of permanent (plastic) deformation (**yielding**) occurs in a body under increasing loading. The **yield criterion** is the combination of normal and shear stresses which produces yielding under multiaxial loading. *See also* ELASTIC LIMIT; FAILURE; MOHR-COULOMB YIELD CRITERI;ON; OFFSET YIELD STRENGTH; PERMANENT SET; TRESCA CRITERION; VON MISES CRITERION.

yield-stress fluid A material which flows once the applied shear stress τ exceeds a yield stress τ_Y but acts like a solid for $\tau < \tau_Y$. A **Bingham plastic** is a yield-stress fluid satisfying the linear relation (for $\tau > \tau_Y$) $\tau = \tau_Y + \mu_P\dot{\gamma}$ where μ_P is the plastic viscosity and $\dot{\gamma}$ is the shear rate. Certain paints, greases, pastes, and clay behave approximately as a Bingham plastic. The **Bingham number (B_N)** is a non-dimensional parameter which arises in the analysis of the flow of a Bingham plastic. It is defined by $B_N = \tau_Y D/\mu_P V$ where D is a characteristic dimension and V is a characteristic velocity.

yoke A component having a forked end, as in a clevis or Hooke's joint.

Young–Laplace equation *See* SURFACE TENSION.

Young's equation An equation for the contact angle ϕ at a gas–liquid–solid interface. For a perfectly flat, smooth, clean surface, $\sigma_{SG} - \sigma_{SL} - \sigma_{LG}\cos\phi = 0$ where σ represents the surface tension and the subscripts SG, SL, and LG refer to gas-solid, solid–liquid and liquid–gas interfaces, respectively.

Young's modulus (elastic modulus, modulus of elasticity, tensile modulus, *E*) (Unit Pa) The ratio of stress to strain in the initial linear, reversible part of a uniaxial (tensile or compressive) stress–strain diagram.

zero *See* POLE.

zero bevel gear *See* TOOTHED GEARING.

zero-crossing period In a periodic signal, the time between successive crossings of the zero line where the slope of the signal has the same sign, i.e. from positive to negative or negative to positive. The zero-crossing period is a less useful measure in non-sinusoidal signals.

zero drift The tendency of the zero point of an instrument to become non-zero with time, thus requiring a compensating adjustment (**zero setting**) or recalibration before taking measurements.

zero-gravity The situation in which there is no force of gravity (i.e. **weightlessness**).

zero initial conditions The assumption normally made in control engineering that up to a start time defined as $t = 0$, all variables have zero value.

zero-lift line An imaginary line through the cross section of an aerofoil showing the flow direction when the section gives no lift.

zero-order hold *See* SAMPLE AND HOLD.

zeroth law of thermodynamics If two bodies are in thermal contact with a third body, they are also in thermal equilibrium with each other. A consequence is that if two bodies have equality of temperature with a third body, they have equality of temperature with each other.

Z-factor *See* COMPRESSIBILITY FACTOR.

Ziegler and Nichols method Two techniques developed by Ziegler and Nichols to select the proportional gain, derivative-time, and integral-time constants in a PID controller, based on observation of the plant output. The **process-reaction-curve method** plots the response of the plant to a small input with no controller connected. Assuming the plant to be stable (neither method can be applied if it is not), two parameters which represent the maximum slope and speed of response are obtained from the plot and these are used to determine the gain and the time constants using a fixed table of conversions. The **ultimate-cycle method** uses initial testing with only the controller proportional term active to determine the gain at which continuing oscillation occurs and the period of that oscillation from which the gain and time constants are determined, again using a fixed table of conversions.

zirconia-toughened alumina composites (ZTA composites) Ceramic composites based on aluminium oxide to which zirconium oxide is added, resulting in improved fracture toughness.

zone melting (zone refining) A process of purification of materials in which a narrow molten zone is moved along the length of the material, resulting in impurities being segregated at one end.

zone of action The disturbed region within the Mach cone in supersonic flow of a sound source. The **zone of silence** is the undisturbed region outside the Mach cone.

z-transform The discrete time equivalent of the Laplace transform, derived from the generalization of the Fourier transform of a sampled signal.